Dictionary of Analytical Chemistry

edited by Grażyna Maludzińska

Dictionary of Analytical Chemistry

english·german·french·polish·russian

ELSEVIER
Amsterdam—Oxford—New York—Tokyo 1990

Editor in Chief: G. MALUDZIŃSKA M.Sc. (Chem.)

Editorial Staff: Z. AUGUSTOWSKA-BRIZE M.Sc. (Chem.),
W. BIERNACKI Ph.D.†, D. KRYT M.Sc. (Chem.),
E. SOZAŃSKA M.A., M. WOJCIECHOWSKA M.Sc. (Chem.),
M. WOLIŃSKA, W.M. WRZOŁEK M.Sc. (Chem.)

Graphic Design M. STAŃCZYK

Published in coedition with Wydawnictwa Naukowo-Techniczne,
Warsaw

Distribution of this book is handled by the following publishers:
    for the U.S.A. and Canada
ELSEVIER SCIENCE PUBLISHING COMPANY, INC.
655 Avenue of the Americas
New York, NY 10010
    for the East European Countries, China, Northern Korea,
    Cuba, Vietnam and Mongolia
WYDAWNICTWA NAUKOWO-TECHNICZNE
ul. Mazowiecka 2/4, 00-048 Warsaw, Poland
    for all remaining areas
ELSEVIER SCIENCE PUBLISHERS
25. Sara Burgerhartstraat 25
P. O. Box 211, 1000 AE Amsterdam, The Netherlands

Library of Congress Cataloging-in-Publication Data

Dictionary of analytical chemistry / edited by Grażyna Maludzińska.
    p. cm.
   English, French, German, Polish, and Russian.
   Includes index.
   ISBN 0-444-98729-0
   1. Chemistry, Analytic—Dictionaries—Polyglot.
   2. Dictionaries, Polyglot. I. Maludzińska, Grażyna.
QD71.5.D53 1990
543′.003 — dc20   90-42784
                                          CIP

PREFACE

The aim of this dictionary is to present the terminology used in analytical chemistry. The subject has become very much an interdisciplinary science in the last two decades. Chemistry, physics, biology, electronics, mathematics and computer science have supplied analytical chemistry with theoretical bases, modern methods and sophisticated instruments and apparatus. For its part, analytical chemistry has brought fresh information on various chemical systems to numerous scientific disciplines and technologies.

The terminology used in analytical chemistry is very rich and diversified. Hence the selection of scientific and technical terms for a dictionary is a difficult task. The precise definition and proper adjustment of meaning of the terms encountered in analytical chemistry present an additional problem. This difficulty is increased by the spontaneous and independent creation of new terminology in various research centres in different countries.

The present dictionary is based on the terminological recommendations of the *Analytical Section of the International Union of Pure and Applied Chemistry*, collected in the *Compendium of Analytical Nomenclature* (published by Pergamon Press, 1978), and the papers published currently in *Pure and Applied Chemistry*. The present literature on analytical chemistry, comprising current publications and papers, textbooks, monographs, and lexicons, also served as a source for preparation of the dictionary.

The International System of Units (SI) has been taken into account. As a result, new concepts have been introduced into the dictionary, such as the amount of a substance and its unit — the mole, the elementary entity, the molar mass, and the equivalence factor. Much care has been taken to ensure the proper application of the above terms, especially in the definitions of various physicochemical quantities, laws, equations, constants and coefficients. Slow adaptation by chemists to new terms justifies the need to include in the dictionary terms which are not recommended by the International System. Their definitions are provided with appropriate remarks or indications regarding their correctness.

The dictionary covers the following methods of analysis: qualitative, quantitative, elemental, optical, chromatographic, radiochemical, X-ray, electrochemical, spectrophotometric, and surface analysis. Terms covering laboratory equipment, statistics and error estimation, optimization of analytical methods, and design of experiments are also included.

The dictionary is intended for scientists, researchers, engineers, students and all those who are concerned with modern analytical chemistry and scientific literature. Any comments on the usefulness of the dictionary and on the gaps and errors noticed will be gratefully welcomed by the publisher. Such comments will serve to improve future editions.

THE EDITOR

EXPLANATORY NOTES

The dictionary contains approximately 2400 English terms arranged in
alphabetical order. Each term is followed by its synonym and the English
definition; next follow equivalent terms in German (D), French (F),
Polish (P) and Russian (R).
Synonymous terms are separated by commas.
The part of a term in parentheses denotes that it can be omitted, thus
two synonymous forms are provided. The term "infrared (radiation)" should
therefore be read as two synonyms: infrared radiation and infrared.
Similarly, the German term "a(nti)symmetrische Schwingungen" reads as:
antisymmetrische Schwingungen and asymmetrische Schwingungen.
Any German, French, Polish and Russian equivalent of an English term
can easily be found in the bilingual alphabetical indexes at the end of the dictionary.
The gender of the terms is also given. The following abbreviations are used:
m — masculine, *f* — feminine, *n* — neuter, *pl* — plural.

SYMBOLS USED IN THE DICTIONARY

A — activity
A — affinity
A — area
A_r — relative atomic mass
a — mean-ionic diameter
a — specific activity
a — specific decadic absorption coefficient
$a_\pm$ — mean activity of an electrolyte
a_B — activity of electrolyte B
a_B — relative activity of substance B
B — susceptance
$[B]$ — amount-of-substance concentration of substance B
b — path length
C — differential capacitance of double layer
c — velocity of light (in vacuum)
$c_\pm$ — mean concentration of electrolyte
c_B — amount-of-substance concentration of substance B
$\mathbf{D}$ — covariance matrix
D — diffusion coefficient
D — distribution coefficient
D_c — concentration distribution ratio
D_m — mass distribution ratio
d — diameter
d — interplanar spacing
E — electromotive force
E — expected value
$\%E$ — extraction percentage
E° — standard electrode potential
$E_{1/2}$ — half-wave potential
E_d — diffusion potential
E_e — irradiance
E_f° — conditional electrode potential
E_j — liquid-junction potential
E_m — membrane potential
E_{mix} — mixed polyelectrode potential
E_p — peak potential
$E_{p/2}$ — half-peak potential
E^z — potential at the point of zero charge
e — electron
e — elementary charge
F — Faraday constant
F — formality
F — nominal linear flow (of the mobile phase)
F_a — volumetric flow rate of mobile phase (specified at column outlet at ambient temperature)

F_c — volumetric flow rate of mobile phase (specified at column outlet, corrected to column temperature)
$f_\pm$ — mean activity coefficient of electrolyte (the concentration expressed in mole fractions)
f_B — activity coefficient of substance B (the composition expressed in mole fraction)
$f_{eq}(B)$ — equivalence factor of component B
G — conductance
G — thermodynamic potential
g — geometric mean
H — enthalpy
H — height equivalent to an effective plate
H — radiant exposure
h — height
h — height equivalent to a theoretical plate
h — Planck constant
h_r — reduced plate height
I — electric current
I — ionic strength (of a solution)
I_0 — exchange current
I_a — partial anodic current
I_c — partial cathodic current
I_d — diffusion current
I_f — faradaic current
I_f — photocurrent
I_l — retention index
I_l — limiting current
$\hat{I}_l$ — average limiting current
I_p — peak current
I_v — luminous intensity
i_d — dark current
j — current density
j — pressure-gradient correction factor
K — integral capacitance of double layer
K — partition coefficient
K_D — distribution constant
K_D° — partition constant
K_{ex} — extraction constant
K_l — limiting current constant
K_n — stepwise stability constant
K_{so} — solubility product
k — Boltzmann constant
k — cell constant
k — mass distribution ratio
k — reaction rate constant
k° — standard rate constant of an electrode reaction

k_f° — conditional rate constant of an electrode reaction

$k_{A/B}$ — selectivity coefficient

$k_{A/B}^{a}$ — corrected selectivity coefficient

$k_{A,B}^{pot}$ — potentiometric selectivity coefficient

k_e — electrochemical equivalent

L — conductance of an electrolyte

L_v — luminance

l — length

ln — natural logarithm

log — common logarithm

M — molar mass

M — metal

M_r — relative molecular mass

m — mass

m — sample median

m_B — molality

N — normality

N — number of effective plates

n — amount-of-substance

n — order of diffraction

n — neutron

n — number of theoretical plates

n — refractive index

n — spectral order

n_{12} — relative refractive index

P — power

P — probability

p — pressure

p — proton

Q — heat

Q — quantity of electricity

Q_0 — total specific capacity

Q_B — breakthrough capacity of an ion exchanger bed

Q_e — radiant energy

Q_v — volume capacity

R — correlation matrix

R — gas constant

R — recovery factor

R — resistance

R — Rydberg constant

R_n — peak resolution

$R_{\%}$ — extraction percentage

r — specific refraction

$r_{1,2}$ — relative retention

S — blackening

S — enrichment factor

S — entropy

S — Sandell's sensitivity index

s — sticking probability

$s_{\bar{x}}$ — standard deviation of the mean

T — period

T — thermodynamic temperature

T — titre

T — transmittance

t — empirical temperature

t — time

t_B — transport number of ionic species B

t_M — mobile phase hold-up time

t_N — net retention time

t_R — total retention time

t_R' — adjusted retention time

U — electric potential difference

U — internal energy

u — unified atomic mass unit

u — interstitial velocity (*of the mobile phase*)

$\bar{u}$ — mean interstitial velocity of carrier gas

u_B — electric mobility of ion B

V — volume

$\bar{V}$ — peak elution volume

V_c — column volume

V_d — extra-column volume

V_{ext} — extra-column volume

V_G — interstitial volume

V_g — specific retention volume

V_I — interstitial volume

V_i — stationary liquid volume

V_L — liquid phase volume

V_M — mobile phase hold-up volume

V_N — net retention volume

V_o — interstitial volume (*in permeation chromatography*)

V_R — retention volume

V_R' — adjusted retention volume

V_R^{o} — corrected retention volume

V_s — stationary phase volume

V_t — total liquid volume

ΔV — electric potential difference

v — velocity

$w_{1/2}$ — peak width at half-height

w_B — mass fraction (of substance B)

w_b — peak width at base

w_s — weight swelling in solvent

X — bed volume

X — reactance

$\bar{x}$ — arithmetic mean

x_B — mole fraction (of substance B)

Y — admittance

$y_{\pm}$ — mean activity coefficient of electrolyte (*the composition expressed in the amount-of-substance concentration*)

y_B — activity coefficient of substance B (*the composition expressed in the amount-of-substance concentration*)

Z — impedance

z — charge number

α — apparent degree of dissociation

α — side-reaction coefficient

$\alpha_{A/B}$ — separation factor

$[\alpha]_{\lambda}^{t}$ — specific rotation

β — blaze angle

β_n — cumulative stability constant

γ — gamma contrast factor

$\gamma_{\pm}$ — mean activity coefficient of electrolyte (*the composition expressed in molalities*)

γ_B — activity coefficient of substance B (*the composition expressed in molalities*)

δ — solubility parameter

ε — electric permittivity

ε^0 — solvent strength
ε_a — efficiency of atomization
ε_1 — interstitial fraction
ε_n — efficiency of nebulization
ε_r — relative electric permittivity
ε_s — stationary-phase fraction
ζ — electrokinetic potential
η — overpotential
η_e — activation overpotential
η_d — diffusion overpotential
θ — angle of reflection
θ — glancing angle
θ — surface coverage
$\varkappa$ — electrolytic conductivity
Λ — molar conductivity of an electrolyte
Λ^0 — limiting molar conductivity
Λ^* — equivalent conductivity of an electrolyte
Λ^{*0} — limiting equivalent conductivity of an electrolyte
λ — wavelength
λ_B — molar conductivity of ionic species B
λ_B^0 — limiting molar conductivity of ionic species B
λ_B^* — equivalent conductivity of ionic species B
λ_B^{*0} — limiting equivalent conductivity of ionic species B
λ_β — blaze wavelength
μ — attenuation coefficient

μ_B^α — chemical potential (of species B in phase α)
$\tilde{\mu}_B^\alpha$ — electrochemical potential (of ionic component B in phase α)
ν — frequency
v — reduced velocity of mobile phase
$\tilde{\nu}$ — wave number
$\Delta\nu$ — chemical shift
$\Delta\bar{\nu}$ — Raman frequency shift
$\Delta\nu_{1/2}$ — half-intensity width
ρ — coefficient of correlation
ρ — density
ρ — resistivity
Σ — macroscopic cross section
σ — microscopic cross section
σ — wave number
σ^2 — variance
τ — transition time
τ — transmission factor
Φ_e — radiant flux
Φ_v — luminous flux
ϕ_B — volume fraction (of substance B)
$\Delta_\alpha^\beta\phi$ — Galvani potential difference
$\Delta\phi_m$ — membrane potential
ϕ^β — inner electric potential of phase β
χ^β — surface electric potential
ψ^β — outer electric potential of phase β
$\Delta_\alpha^\beta\psi$ — Volta potential difference
ω — electrochemical reaction order
ω — X-ray fluorescence yield

GREEK ALPHABET

A α — alpha
B β — beta
Γ γ — gamma
Δ δ — delta
E ε — epsilon
Z ζ — zeta
H η — eta
Θ θ — theta
I ι — iota
K κ — kappa
Λ λ — lambda
M μ — mu

N ν — nu
Ξ ξ — xi
O o — omicron
Π π — pi
P ρ — rho
Σ σ — sigma
T τ — tau
Υ υ — upsilon
Φ φ — phi
X χ — chi
Ψ ψ — psi
Ω ω — omega

A

AAS → atomic absorption spectroscopy

aberration → optical aberration

absolute detector sensitivity
The change required in the physical parameter that will produce a full-scale deflection of the recorder at maximum detector sensitivity and for a defined background noise level.
D absolute Detektorempfindlichkeit
F sensibilité absolue d'un détecteur
P czułość detektora absolutna, czułość detektora bezwzględna
R абсолютная чувствительность детектора

absolute (electrode) potential
The difference between the inner potential of an electrode and that of the electrolyte; unmeasurable quantity.
D absolutes Elektrodenpotential
F potentiel absolu d'électrode
P potencjał (elektrody) absolutny, potencjał (elektrody) bezwzględny
R абсолютный (электродный) потенциал

absolute error
The difference between the observed and the true value of the measured quantity.
D Absolutfehler
F erreur absolue
P błąd bezwzględny
R абсолютная ошибка

absolute full energy peak efficiency
(*of a radiation spectrometer*)
The counting efficiency when considering only the events recorded in the full energy peak.
D ...
F efficacité absolue du pic d'absorption totale
P wydajność absolutna piku całkowitej absorpcji
R абсолютная эффективность пика полного поглощения

absolutely dry ion exchanger
An ion exchanger dried at elevated temperatu-

res over phosphorous pentoxide or Anhydrone to constant mass.
D getrockneter Ionenaustauscher
F échangeur d'ions sec
P jonit (bezwzględnie) suchy
R сухой ионит, сухой ионообменник

absolute permittivity → permittivity

absolute photopeak efficiency (*of a γ-ray spectrometer*)
The counting efficiency when considering only the events recorded in the photopeak.
D Photoansprechvermögen
F efficacité photoélectrique absolue
P wydajność fotopiku absolutna
R абсолютная фотоэффективность

absolute potential → absolute electrode potential

absolute refractive index → refractive index

absorbance, internal absorbance, *A*
The negative common logarithm of the internal transmittance *T*

$$A = \log \frac{I_0}{I} = -\log T$$

where I_0 − intensity of the radiation transmitted by a standard or blank solution and I − intensity of the radiation transmitted by a sample.
Remark: The terms absorbancy, extinction and optical density are not recommended.
D Absorbanz, Extinktion, optische Dichte
F absorbance, densité optique, extinction
P absorbancja, wartość absorpcji, gęstość optyczna, ekstynkcja
R внутренняя поглощательная способность, оптическая плотность, экстинкция, погашение

absorbancy → absorbance

absorbancy index → specific decadic absorption coefficient

absorbate, absorptive
The gaseous substance being absorbed.
Remark: In some cases the term absorbate refers to a substance which has already been absorbed whereas the term absorptive is restricted to the substance which is capable of being absorbed.
D Absorbat, Absorptiv
F absorbat, substance absorbée
P absorbat
R абсорбат, абсорбированное вещество, абсорбтив

absorbent
The liquid or solid capable of absorbing.
D Absorbens, Absorptionsmittel
F absorbant

P absorbent
R абсорбент, поглотитель

absorber
The apparatus in which the absorption is effected.
D Absorber
F absorbeur
P absorber
R абсорбер

absorptance, absorption factor, α
The ratio of the radiant or luminous flux absorbed by a substance to the flux incident upon that substance.
Remark: The term absorptivity is not recommended.
D Absorptionsgrad, Absorptionsfaktor
F facteur d'absorption
P współczynnik pochłaniania
R коэффициент поглощения

absorptiometric gas analysis
The determination of a constituent (or constituents) of gaseous mixtures based on the volume decrease caused by specific absorption, taking place in an appropriate reagent solution.
D absorptiometrische Gasanalyse, Gasabsorptiometrie
F analyse absorptiométrique des gaz
P analiza gazowa absorpcjometryczna
R абсорбционный газовый анализ

absorption
A process of mass transfer in which an absorbate is taken up by the whole volume of an absorbent.
D Absorption
F absorption
P absorpcja
R абсорбция

absorption (*of radiation*)
Transformation of radiant energy by interaction with matter.
D Absorption
F absorption
P absorpcja
R абсорбция

absorption cell
In radiation absorption tests, the vessel containing the sample being investigated. The distance between the two parallel windows, perpendicular to the direction of radiation, defines the thickness of the investigated sample.
D Absorptionsküvette
F cuve absorbante
P kuweta absorpcyjna
R абсорбционная кювета

absorption cross section
The cross section of an atomic nucleus for the absorption of bombarding particles.

D Absorptionsquerschnitt
F section efficace d'absorption
P przekrój czynny na absorpcję
R сечение поглощения

absorption curve
A diagram showing the dependence of the absorbance, transmittance, absorption coefficients or any functions of these quantities on the wavelength of an electromagnetic radiation, wave number or frequency of vibration.
D Absorptionskurve
F courbe d'absorption
P krzywa absorpcji, krzywa spektrofotometryczna
R абсорбционная кривая

absorption edge (*in X-ray spectroscopy*)
An abrupt change of the X-ray mass absorption coefficient characteristic for each chemical element which takes place when the energy of absorbed radiation corresponds to the binding energy of an electron in a given energy state (atomic shell K, L, M, etc.). The absorption edges K, L, M, etc. are distinguished.
D Absorptionskante
F discontinuité d'absorption
P krawędź absorpcji, próg absorpcji
R край поглощения

absorption effect (*in* X-*ray fluorescence analysis*)
The absorption of exciting and excited radiations of an analyte by all elements constituting the sample analysed, inclduing the analyte itself. It is proportional to their mass fractions and mass absorption coefficients, and together with the enhancement effect constitutes the matrix effect.
D Absorptionseffekt
F effet d'absorption
P efekt absorpcji
R абсорбционный эффект

absorption factor → absorptance

absorption filter
An optical filter which absorbs electromagnetic radiation in a certain region of the spectrum.
D Absorptionsfilter
F filtre d'absorption
P filtr absorpcyjny
R абсорбционный фильтр

absorption path length → path length

absorption spectrochemical analysis
Spectrochemical analysis based on the investigation of the absorption spectra.
D Absorptionsspektralanalyse
F analyse spectrochimique d'absorption, spectranalyse d'absorption
P analiza spektralna absorpcyjna
R абсорбционный сп트рохиекмический анализ

absorption spectrophotometry
A branch of spectrophotometry in which the
radiation absorbed upon passage through
a given medium in the ultraviolet, visible and
infrared ranges is measured as a function of
wavelength by means of a spectrophotometer.
D Absorptionsspektralphotometrie
F spectrophotométrie d'absorption
P spektrofotometria absorpcyjna
R абсорбционная спектрофотометрия

absorption spectroscopy, AS
A branch of spectroscopy concerned with the
production, measurement and interpretation
of the absorption spectra of atoms and
molecules.
D Absorptionsspektroskopie
F spectroscopie d'absorption
P spektroskopia absorpcyjna
R абсорбционная спектроскопия

absorption spectrum
A spectrum produced by the passage of
radiant or particle energy from a continuous
source through a selectively absorbing medium.
D Absorptionsspektrum
F spectre d'absorption
P widmo absorpcyjne
R абсорбционный спектр, спектр поглощения

absorption spectrum → excitation spectrum

absorption tube
A tube used for the absorption of moisture,
carbon dioxide or other gases, depending on
its filling.
D Absorptionsröhrchen
F tube absorbeur
P rurka absorpcyjna
R поглотительная трубка

absorptive → absorbate

absorptivity → absorptance

absorptivity → specific decadic absorption
coefficient

abundance, isotopic abundance
The number of atoms of a particular
isotope in a mixture of the isotopes of an
element expressed as a fraction of all the atoms
of this element.
D Isotopenhäufigkeit
F teneur isotopique, abondance isotopique,
richesse en isotopes
P rozpowszechnienie izotopu, abundancja
R распространённость изотопа

ac arc → alternating current arc

accumulator → electrical accumulator

accuracy
The discrepancy between a result (or mean)
and the true value.
D Richtigkeit
F exactitude, justesse
P dokładność
R правильность

acid-base indicator
An indicator which is itself an acid or base
and which exhibits a colour change on
neutralization by a base or acid at or near
the equivalence-point of a titration.
D pH-Indikator, Säure-Base-Indikator
F indicateur acide-base, indicateur de pH
P wskaźnik alkacymetryczny, wskaźnik pH
R кислотно-основный индикатор,
индикатор pH

acid-base reaction
The reaction of proton exchange between
an acid supplying it and a base being the
acceptor, taking place according to the scheme

$$HA^{n+1} + B^m \rightleftarrows A^n + BH^{m+1}$$

$(acid_1 + base_2 \rightleftarrows base_1 + acid_2)$

D Säure-Base-Reaktion
F réaction acide-base
P reakcja kwas-zasada
R реакция кислота-основание

acid-base titration
A titration involving the transfer of protons
(Brönsted-Lowry theory) or of electron-pairs
(Lewis theory) from one reacting species to
the other in solution.
D Säure-Base-Titration, Neutralisationstitration
F titrage acide-base
P miareczkowanie alkacymetryczne
R кислотно-основное титрование

acid-base titration curve → neutralization
titration curve

acid form of a cation exchanger
The ionic form of a cation exchanger in which
the counter-ions are hydrogen ions (H^+ −form)
or the ionogenic groups have added a proton
to yield forming an undissociated acid, e.g.
—COOH.
D saure Form des Kationenaustauschers
F forme acide d'un échangeur de cations
P forma kwasowa wymieniacza kationów
R кислотная форма катионообменника

acidic group (*of chelating agent*)
A group containing hydrogen which can be
replaced by a metal, e.g. hydroxyl group —OH,
carboxyl group —COOH, oxime group =NOH,
imine group =NH, thiol group —SH,
arsonium group —AsO(OH)$_2$.
D saure Gruppe
F groupement acide

P grupa kwasowa,
R солеобразующая группа

acidic solvent → protogenic solvent

acidimetric titration
An acid-base titration in which a base is titrated with a standard solution of an acid.
D acidimetrische Titration
F titrage acidimétrique
P miareczkowanie acydymetryczne
R ацидиметрическое титрование

acidimetry
The determination of a substance by titration with a standard solution of an acid.
D Acidimetrie
F acidimétrie
P acydymetria
R ацидиметрия

acidimetry and alkalimetry
The determination of a substance by titration with a standard solution of a base or of an acid.
D Neutralisationsanalyse
F acido-alcalimétrie
P alkacymetria
R ацидиметрия и алкалиметрия

ac polarography → conventional alternating-current polarography

activation
The process of inducing radioactivity by irradiation with neutrons, charged particles or γ-rays.
D Aktivierung
F activation
P aktywacja
R активация

activation analysis, nuclear activation analysis
A method of elemental analysis based on the measurement of nuclear radiation induced in the analysed sample by irradiation with neutrons, charged particles or γ-rays.
D Aktivierungsanalyse
F analyse par activation
P analiza aktywacyjna
R активационный анализ

activation cross section
The cross section for the formation of a radionuclide by a given reaction.
D Aktivierungsquerschnitt, Wirkungsquerschnitt
F section efficace d'activation
P przekrój czynny na aktywację
R сечение активации

activation energy
The energy that must be supplied to a system to allow a particular process to occur.
D Aktivierungsenergie
F énergie d'activation
P energia aktywacji
R энергия активации

activation of an adsorbent
A special procedure used to modify an adsorbent in order to increase its adsorptive properties. Usually this involves an appropriate drying procedure enabling the removal of adsorbed water.
D Adsorbensaktivierung
F activation d'un adsorbant
P aktywacja adsorbentu
R активация адсорбента

activation overpotential, charge-transfer overpotential, η_e
An overpotential of an electrode process caused by a slow, in comparison with other elementary steps, charge transfer step over the electrode-electrolyte interface.
D Durchtrittsüberspannung, Aktivierungsüberspannung
F surtension de transfert, surtension d'activation
P nadpotencjał elektroaktywacyjny, nadpotencjał reakcji
R перенапряжение стадии переноса заряда, перенапряжение перехода

active experiment
An experiment carried out according to the previously established design.
D aktiver Versuch
F expérience active
P eksperyment czynny, eksperyment planowany
R ...

active solid (*in chromatography*)
A porous solid with sorptive properties by means of which chromatographic separation may be achieved.
D aktiver Festkörper
F solide actif, solide activé
P sorbent stały
R активное твёрдое вещество

activity, disintegration rate, A
The number of disintegrations occurring in a given quantity of a radioactive substance per unit of time. SI unit: becquerel, Bq.
D Aktivität
F activité (nucléaire)
P aktywność (promieniotwórcza), szybkość rozpadu promieniotwórczego
R активность

activity coefficient of substance B, y_B, γ_B, f_B
The deviation of the thermodynamic properties of component B in a given solution from the properties of this component in the ideal solution, defined by the ratio

$$y_B = a_B/c_B$$

where a_B − activity of component B,

c_B — amount-of-substance concentration of this component; the activity coefficient is a function of pressure, temperature and concentration of other components present in the solution.

Remark: It is recommended to use the following symbols: f, γ, y, when the composition is given in mole fractions, molality and amount-of-substance concentration, respectively.

D Aktivitätskoeffizient der Substanz B
F coefficient d'activité du constituant B
P współczynnik aktywności składnika B
R коэффициент активности вещества B

activity of electrolyte B, a_B
The effective concentration of electrolyte $B = X_{\nu_+} Y_{\nu_-}$ defined as

$$a_B = \exp((\mu_B - \mu_B^\ominus)/RT) = a_+^{\nu_+} a_-^{\nu_-} = a_\pm^\nu$$

where μ_B — chemical potential of electrolyte B in a solution containing B and other species, $\mu_B^\ominus$ — chemical potential of electrolyte B in its standard state, R — gas constant, T — absolute temperature, a_+ and a_- — activities of the cation and anion, ν_+ and ν_- — numbers of cations and anions formed from one molecule of electrolyte B, $a_\pm$ — mean ionic activity, and $\nu = \nu_+ + \nu_-$.

D Aktivität des Elektrolytes B
F activité de l'électrolyte B
P aktywność elektrolitu B
R активность электролита B

activity of substance B → relative activity of substance B

actor (*in induced reactions*)
A substance which is the common substrate in both primary and secondary reactions in the system of induced reactions.

D Aktor
F ...
P aktor
R актор

additivity law (*of absorbance*)
Absorbance A of a multicomponent system is equal to the sum of absorbances A_i of the individual components forming that system.

$$A = \sum_{i=1}^n A_i$$

Remark: The law is fundamental in the spectrophotometric method of quantitative analysis of the multicomponent systems.

D Additivität
F additivité
P prawo addytywności
R закон аддитивности

adhesion
Sticking together of the surfaces of two different phases — solid/liquid or solid/solid — due to the intermolecular forces.

D Adhäsion
F adhésion, adhérence
P adhezja
R адгезия, прилипание

adjusted retention time, t_R'
The total retention time t_R less the mobile phase hold-up time t_M,

$$t_R' = t_R - t_M$$

D reduzierte Retentionszeit
F temps de rétention réduit
P czas retencji zredukowany
R приведённое время удерживания

adjusted retention volume, V_R'
The total retention volume V_R less the mobile phase hold-up volume, V_M,

$$V_R' = V_R - V_M = t_R' \cdot F_c$$

where t_R' — adjusted retention time and F_c — volumetric flow rate of the mobile phase.

D reduziertes Retentionsvolumen
F volume de rétention réduit
P objętość retencji zredukowana
R приведённый объём удерживания

admittance, complex admittance, Y
A measure of the ability of an alternating-current circuit to pass an electric current, equal to the reciprocal of its impedance. Admittance is a complex quantity algebraically given by the equation: $Y = G + iB$, where i is equal to $\sqrt{-1}$, G is the conductance, and B is the susceptance. SI unit: siemens, S.

D Admittanz, (komplexer) Scheinleitwert
F admittance électrique, admittance complexe
P admitancja, przewodność (elektryczna) pozorna
R полная проводимость, комплексная проводимость (*электрической цепи*)

adsorbate, adsorptive
The substance which is adsorbed onto the surface of the adsorbent.

Remark: In some cases the term adsorbate refers to a substance which has already been adsorbed, whereas the term adsorptive is restricted to the substance (present in gaseous or liquid bulk phases) which can be adsorbed under given conditions.

D Adsorbat, Adsorptiv
F adsorbat, adsorpt, adsorbendum
P adsorbat
R адсорбат, адсорбтив

adsorbent
The substance at the surface of which adsorption may occur.

D Adsorbens, Adsorptionsmittel, Adsorber
F adsorbant
P adsorbent
R адсорбент

adsorbent activity (*in chromatography*)
A measure of adsorptive properties of an
adsorbent as determined from the retention
time (volume) of a standard substance
for a given set of the operating conditions.
D Adsorbensaktivität
F activité d'un adsorbant
P aktywność adsorbentu
R активность адсорбента

adsorption
A process of accumulation of an adsorbate on
the surface of an adsorbent due to the
intermolecular attractive forces or chemical
interactions (*comp.* physisorption and
chemisorption).
D Adsorption
F adsorption
P adsorpcja
R адсорбция

adsorption chromatography
Chromatography in which the separation of the
sample components is based on the
differences in the adsorption affinities of the
components towards the surface of an active
solid (e.g. adsorbent).
D Adsorptionschromatographie
F chromatographie par adsorption,
chromatographie d'adsorption
P chromatografia adsorpcyjna
R адсорбционная хроматография

adsorption coefficient → sticking probability

adsorption current
In polarography, the current which is observed
when the substrate or the product of an
electrode reaction is adsorbed at the electrode
surface.
D Adsorptionsstrom
F courant d'adsorption
P prąd adsorpcyjny
R адсорбционный ток

adsorption indicator
An indicator which is adsorbed or desorbed,
with concomitant colour change, by
a precipitation system at or near the
equivalence-point of a titration.
D Adsorptionsindikator
F indicateur d'adsorption
P wskaźnik adsorpcyjny
R адсорбционный индикатор

adsorption isobar
Temperature dependence of the surface
concentration of an adsorbate, determined at
constant gas pressure in the adsorption
system and under equilibrium conditions.
D Adsorptionsisobare
F isobare d'adsorption
P izobara adsorpcji
R изобара адсорбции

adsorption isostere
Temperature dependence of the pressure of
a gas to be adsorbed, determined at constant
concentration of an adsorbate in the
adsorption system and under equilibrium
conditions.
D Adsorptionsisostere
F isostère d'adsorption
P izostera adsorpcji
R изостера адсорбции

adsorption isotherm
A curve presenting the relationship between
the amount of substance adsorbed by unit mass
of adsorbent and its equilibrium pressure
(orequilibrium concentration of adsorbate)
determined at a constant temperature.
D Adsorptionsisotherme
F isotherme d'adsorption
P izoterma adsorpcji
R изотерма адсорбции

adsorption probability → sticking probability

adsorption wave
The part of a polarographic curve, in the
shape of a wave, corresponding to the
adsorption of reagents at the surface of the
electrode; it may appear before the diffusion
polarographic wave (adsorption of substrates)
or after the diffusion polarographic wave
(adsorption of products of the electrode
process).
D Adsorptionsstufe
F ...
P fala adsorpcyjna
R адсорбционная волна

adsorptive → adsorbate

AEM → Auger-electron microscopy

AES → atomic emission spectroscopy

AES → Auger-electron spectroscopy

affinity (*of chemical reaction*), *A*
The quantity which determines the intensity
or force of a chemical reaction

$$A = -\sum_B \nu_B \, \mu_B$$

where ν_B — stoichiometric coefficient of
reagent B (positive for products, negative for
substrates), μ_B — chemical potential of
reagent B, and $\sum_B$ denotes summation over

all components.
D Affinität
F affinité chimique
P powinowactwo chemiczne
R химическое сродство

affinity chromatography
A form of chromatography, in which the
separation of sample macromolecules results

from specific intermolecular interactions between the immobilized, active ligand and biomolecules. Only molecules that are bonded by the specific ligand are retarded and retained on the biospecific sorbent which is prepared by coupling an active ligand (such as an enzyme, antigen, hormone, inhibitor) to the water-insoluble support (Sephadex, agarose).

D Affinitätschromatographie
F chromatographie d'affinité
P chromatografia powinowactwa
R аффинная хроматография

AFS → atomic fluorescence spectroscopy

agarose
The neutral component of agar-agar, a polysaccharide which forms with water a gel used for separating substances of high-molecular mass.

D Agarose
F agarose
P agaroza
R агароза

ageing (of a precipitate)
The time-dependent change of the properties of a precipitate, e.g. loss of water, growth of crystals, recrystallization, decrease of the specific surface, loss of coprecipitated substances or improvement of the filtering properties. The process of ageing is very often promoted by keeping the precipitate and precipitation medium together at elevated temperatures for a period of time.

Remark: The terms chemical, physical and thermal ageing may be used in cases when some of the (usually combined) effects named above are to be emphasized specifically.

D Alterung, Altern (eines Niederschlages)
F vieillissement (d'un précipité)
P starzenie (się osadu)
R старение (осадка)

ageing of sols
The change of physical and chemical properties of colloidal solutions with time.

D Alterung der Sole
F vieillissement des colloïdes
P starzenie (się) roztworów koloidalnych
R старение коллоидных растворов

agglomeration
The formation and growth of aggregates ultimately leading to phase separation by the formation of precipitates of larger than collodal size.

D Agglomeration
F agglomération
P aglomeracja
R агломерация

aggregate
A group of particles held together at random.

D Aggregat
F agrégat
P agregat, zespół cząstek
R агрегат

aggregation
Formation of an aggregate.

D Aggregation
F agrégation
P agregacja
R агрегирование

air-gap electrode
A gas sensing electrode in which an ion selective electrode, the reference electrode, and a thin-layer of internal electrolyte are separated from the sample solution by an air-gap.

D Luftspalt-Elektrode
F électrode à bulle d'air
P elektroda z przerwą powietrzną
R электрод с воздушным зазором, электрод с газовым зазором, электрод с воздушным промежутком

air peak (*in gas chromatography*)
The peak due to the presence of small amounts of air during the sample injection. *See*: differential chromatogram and integral chromatogram.

D Inertpeak, Luftpeak, Luftberg
F pic de l'air
P pik powietrza
R пик воздуха

alcoholometry
The determination of ethyl alcohol content (percentage by weight or volume) in water-ethanol solutions by means of an alcoholometer.

D Alkoholometrie
F alcoométrie
P alkoholometria
R алкоголиметрия

alkalimetric titration
An acid-base titration in which an acid is titrated with a standard solution of an alkali.

D alkalimetrische Titration
F titrage alcalimétrique
P miareczkowanie alkalimetryczne
R алкалиметрическое титрование

alkalimetry
The determination of a substance by titration with a standard solution of a base.

D Alkalimetrie
F alcalimétrie
P alkalimetria
R алкалиметрия

allowed transition
The transition between two energy levels whose probability, according to the selection rules, differs from zero.

D erlaubter Übergang
F transition permise
P przejście dozwolone
R разрешённый переход, допустимый переход

all-solid-state ion-selective electrode
An ion-selective electrode in which a solid-state
membrane is directly connected to a solid
metallic contact. (The electrode does not
contain an internal reference solution).
D ...
F ...
P elektroda jonoselektywna stała ze stałym
   kontaktem
R полностью твердофазный ионоселективный
   электрод

altered layer
The near surface region of a solid with
a composition different from that of the bulk,
due to the preferential sputtering of one
component in the bombarded system and/or
implantation of incoming ions.
D ...
F couche altérée
P warstwa zmodyfikowana
R деформированный слой, модифицированный
   слой

alternating current arc, ac arc
An electrical arc fed by an alternating current
supply, normally having the mains network
voltage and frequency, but without an energy
storing capacitor in the arc circuit. Sometimes
voltages of up to 5000 V are necessary to
achieve ignition of the arc, the operating level
is then maintained at about 50 V.
D Wechselstrombogen
F arc à courant alternatif
P łuk prądu zmiennego
R дуга переменного тока

alternating-current chronopotentiometry
A modification of chronopotentiometry, in
which, instead of a direct current, an
alternating current of constant amplitude
is imposed on the indicator electrode and
the potential E of this electrode is measured
as a function of the electrolysis time t.
Remark: In a narrow sense the term is often used to denote
the sinusoidal alternating-current chronopotentiometry.

 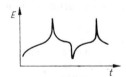

sinusoidal alternating-current
chronopotentiometry —
typical response curve

D Wechselstrom-Chronopotentiometrie
F chronopotentiométrie à courant alternatif
P chronopotencjometria zmiennoprądowa
R переменно-токовая хронопотенциометрия

alternating-current polarography, polarography
with superimposed periodic voltage
The variety of polarography in which a
potential linearly changing with time and
modulated by a small alternating voltage of
constant amplitude is applied to the indicator
electrode.
Remark: In a narrow sense, the term denotes the
conventional alternating-current polarography.

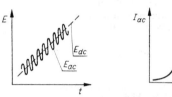 

D Wechselstrompolarographie, Polarographie
   mit überlagerter periodischer Spannung
F polarographie à tension alternative,
   polarographie à tension périodique
   surimposée
P polarografia zmiennoprądowa
R переменно-токовая полярография,
   полярография переменного тока,
   полярография с наложением периодически
   меняющегося напряжения

alternating-current polarography → conventional
alternating-current polarography

alternating-voltage chronopotentiometry
An electrochemical technique, in which the
constant current is imposed on the indicator
electrode whose surface is renewed, e.g. the
mercury dropping electrode, and a small
alternating voltage, such as a sine-wave,
is simultaneously applied to this electrode.
The alternating component of the current I_{ac}
is then measured as a function of the
electrolysis time t.
Remark: In a narrow sense, the term is often used to denote
sinusoidal alternating-voltage chronopotentiometry.

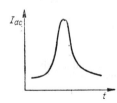

sinusoidal alternating-voltage chronopotentiometry —
typical response curve

D Wechselspannungschronopotentiometrie
F chronopotentiométrie à tension alternative
P chronopotencjometria zmiennonapięciowa
R хронопотенциометрия с переменным
   напряжением

alternating-voltage polarography, av
polarography

Polarography in which an alternating current (e.g. sine-wave) of small and constant amplitude is made to flow into the indicator electrode while the undirectional potential linearly changing with time is also applied to it. The alternating component of the electrode potential E_{ac} is investigated as a function of the direct component of the electrode potential E_{dc}.

Remark: In a narrow sense, the term is often used to denote the sinusoidal alternating-voltage polarography.

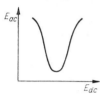

polarization curve recorded under conditions of the sinusoidal alternating-voltage polarography

D av-Polarographie, Wechselspannungspolarographie
F polarographie à courant alternatif imposé
P polarografia zmiennonapięciowa
R полярография с измерением переменного напряжения

alternative hypothesis, H_1
Any admissible statistical hypothesis alternative to the null hypothesis.
D Alternativhypothese
F hypothèse alternative
P hipoteza alternatywna
R альтернативная гипотеза

amalgam
The alloy of mercury with other metal(s) where the former is the main constituent; the amalgam may be solid or liquid, depending on its composition.
D Amalgam
F amalgame
P amalgamat
R амальгама

amalgam electrode
A version of a metal-metal ion electrode in which the metal is present as an amalgam rather than in the pure form; used as an indicator electrode.
D Amalgamelektrode
F électrode à amalgame
P elektroda amalgamatowa
R амальгамный электрод

amalgam polarography
Polarography in which a dropping amalgam electrode is used instead of a dropping mercury electrode.
D Amalgampolarographie
F ...

P polarografia amalgamatowa, polarografia amalgamatów
R амальгамная полярография

American degree of water hardness
The unit of water hardness corresponding to the content of 1 mg calcium carbonate per 1 l of water.
D amerikanischer Härtegrad, amerikanische Härte
F degré hydrotimétrique américain
P stopień twardości wody amerykański
R американский градус жёсткости воды

amorphous precipitate → colloid precipitate

amount of substance, n
SI basic physical quantity proportional to the number of specified elementary entities contained in a defined amount of a given substance (*see* elementary entity). The unit of the amount of substance is the mole, mol.
D Stoffmenge
F quantité de matière
P liczność materii, ilość materii
R количество вещества

amount-of-substance concentration of substance B, concentration of substance B, c_B, [B]
The amount of substance B divided by the volume of solution. The SI unit is mol m^{-3}, but the practical units are mol dm^{-3} or mol l^{-1}.
Remark: The term molarity should be abandoned as clearly redundant. A solution with an amount-of-substance concentration of 0.1 mol dm^{-3} is often called a 0.1 molar solution and written as a 0.1 M solution.
D Stoffmengenkonzentration, Konzentration eines Stoffes B, molare Konzentration
F concentration du constituant B, concentration en quantité de matière du constituant B, concentration molaire, molarité
P stężenie molowe (składnika B w roztworze)
R концентрация растворенного вещества B, молярная концентрация, молярность

amperometric analysis *See* amperometry
D amperometrische Analyse
F analyse ampérométrique
P analiza amperometryczna
R амперометрический анализ

amperometric titration → amperometric titration with one indicator electrode

amperometric titration with a dropping mercury electrode
Remark: This term is recommended in preference to polarometric titration or polarographic titration
D amperometrische Titration einer Quecksilbertropfelektrode

F titrage ampérométrique à une électrode
à gouttes de mercure
P miareczkowanie amperometryczne
z elektrodą rtęciową kroplową
R амперометрическое титрование с капающим
ртутным электродом

amperometric titration (with one indicator electrode)

Titration in which the end point is determined
from the abrupt change in the current I,
flowing between the indicator electrode and the
suitable reference electrode while a small
but constant potential difference is maintained
between them.

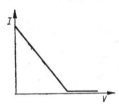

curve recorded in the titration of a substance forming
reversible redox system with the titrant
V — volume of a titrant

D amperometrische Titration, amperometrische
Titration mit einer polarisierbaren Elektrode,
amperometrische Indikation
F titrage ampérométrique (à une électrode
indicatrice), titrage ampérométrique à une
électrode polarisée
P miareczkowanie amperometryczne (z jedną
elektrodą spolaryzowaną)
R амперометрическое титрование (с одним
поляризуемым электродом),
амперометрическое титрование с одним
индикаторным электродом

amperometric titration with two indicator electrodes

Titration in which the end point is determined
from the abrupt change in the current I,
flowing between two indicator electrodes
immersed in the same solution while
a small and constant potential difference is
maintained between them.

Remark: The terms biamperometric titration and dead-stop
titration are no longer recommended.

curve recorded in the titration of a substance forming
reversible redox system with the titrant which yields no
reversible system
V — volume of a titrant

D amperometrische Titration mit zwei
polarisierbaren Elektroden, amperometrische

Titration mit zwei Indikator-Elektroden,
Polarisationsstromtitration,
biamperometrische Titration, dead-stop
Titration
F titrage ampérométrique à deux électrodes
indicatrices, titrage biampérométrique,
titrage à arrêt net
P miareczkowanie amperometryczne z dwiema
elektrodami spolaryzowanymi,
miareczkowanie biamperometryczne,
miareczkowanie do punktu martwego
R амперометрическое титрование с двумя
поляризуемыми электродами,
амперометрическое титрование с двумя
индикаторными электродами,
амперометрическое титрование
с поляризационным током,
биамперометрическое титрование,
титрование до мёртвой точки

amperometry

A number of electrochemical techniques in
which the potential applied to the indicator
electrode or the potential difference applied
between two indicator electrodes, is kept
constant and the current flowing through the cell
is measured as a function of the concentration
of an electroactive substance.

Remark: In chemical analysis the term amperometric
analysis is used.

D Amperometrie
F ampérométrie
P amperometria
R амперометрия

amperometry with one indicator electrode

The branch of amperometry in which the
potential of the indicator electrode (with
respect to the reference electrode) is set and
kept constant, and the current I flowing
through the cell is measured as a function of
the concentration c of the electroactive
substance.

Remark: Terms like stirred-mercury-pool amperometry and
rotating-platinum-wire electrode amperometry are
recommended to denote the indicator electrode employed.

D Amperometrie mit einer Indikator-Elektrode,
Amperometrie mit einer polarisierbaren
Elektrode
F ampérométrie à une électrode indicatrice,
ampérométrie à une électrode polarisée
P amperometria z jedną elektrodą
spolaryzowaną
R амперометрия с поляризуемым электродом,
амперометрия с индикаторным электродом

amperometry with two indicator electrodes

The branch of amperometry, in which a small

and constant potential difference is applied to
two similar indicator electrodes immersed in
the same solution and the current I passing
through the cell is measured as a function of
the concentration c of the electroactive
substance.

Remark: The term biamperometry is no longer
recommended.

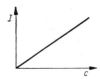

D Amperometrie mit zwei Indikator-Elektroden,
Amperometrie mit zwei polarisierbaren
Elektroden, Biamperometrie
F ampérométrie à deux électrodes indicatrices,
ampérométrie avec deux électrodes polarisées,
biampérométrie
P amperometria z dwiema elektrodami
spolaryzowanymi, biamperometria
R амперометрия с двумя индикаторными
электродами, амперометрия с двумя
поляризуемыми электродами,
биамперометрия

amperostat → galvanostat

amphion → ampholyte ion

amphiprotic solvent, amphoteric solvent
A solvent whose molecules may be both donors
and acceptors of protons, e.g. water, organic
acids, alcohols, ammonia.

D amphiprotisches Lösungsmittel,
ampholytisches Lösungsmittel
F ampholyte
P rozpuszczalnik amfiprotyczny,
rozpuszczalnik amfoteryczny
R амфипротный растворитель, амфотерный
растворитель

ampholyte, amphoteric electrolyte
An electrolyte which in a solution can behave
both as an acid and as a base.

D Ampholyt amphoterer Elektrolyt,
amphoterer Stoff
F ampholyte
P amfolit, elektrolit amfoteryczny
R амфолит, амфотерный электролит

ampholyte ion, amphion, zwitterion,
amphoteric ion, dipolar ion, hybrid ion
An ion which carries both positive and
negative charge, e.g. aminoacids can form the
ampholyte ion by ionization of the carboxyl
group and protonation of the amino group
$H_2NRCOOH \rightleftarrows {}^+H_3NRCOO^-$.

D Zwitterion, amphoteres Ion
F ion ampholyte, amphion, ion amphotère,
zwitterion, ion hermaphrodite

P jon obojnaczy, jon dwubiegunowy
R амфион, амфотерный ион, цвиттерион

amphoteric electrolyte → ampholyte

amphoteric ion → ampholyte ion

amphoteric ion exchanger
An ion exchanger containing both cation- and
anion-exchange groups.

D amphoterer Ionenaustauscher
F échangeur d'ions amphotère
P wymieniacz jonowy amfoteryczny
R амфотерный ионообменник

amphoteric solvent → amphiprotic solvent

amplification reaction
An amplification reaction is one which
replaces the conventional reaction used in
a particular determination so that a more
favourable measurement can be made.

Remark: Sometimes the term multiplication reaction is
used synonymously with amplification reaction, but this
usage is not recommended.

D Verstärkungsreaktion
F réaction d'amplification
P reakcja amplifikacji
R ...

analyser
A device carrying out analysis in a continuous
or semi-continuous way, usually provided with
an automatic recorder.

D Analysator
F analyseur
P analizator
R анализатор

analysing crystal (*in an* X-*ray spectrometer*)
The plate of a single crystal, cut or cleaved
parallelly to the required reflecting planes,
used for detection and analysis of the X-rays.

D Analysatorkristall
F cristal analyseur
P kryształ analizujący
R кристалл-анализатор

analysis by neutron slowing-down
A method of radiometric analysis which takes
advantage of the neutron slowing-down
processes caused by elastic collision with light
nuclei. The method is generally used for moisture
determination.

D Analyse durch Neutronenabbremsung
F analyse par ralentissement de neutrons
P metoda spowalniania neutronów
R анализ на замедляющих нейтронах

analysis of variance, ANOVA
A statistical method used in the case when
the empirical problem can be reduced to
comparison of mean values of specified sets of
observations. The procedure consists in

splitting the total variation into components which can be assigned to specified sources of variation according to which the observations are classified. This enables testing the significance of influence of these sources of variation and their interactions.

D Varianzanalyse, Dispersionsanalyse
F analyse de variance, analyse de dispersion
P analiza wariancji
R дисперсионный анализ

analysis via functional groups → functional group analysis

analyte
The constituent to be detected and/or determined.
D ...
F analyte
P analit
R ...

analyte signal → analytical signal

analytical balance
A balance with a weighing capacity from 50 to 200 g and with a sensitivity of 0.1 mg.
D Analysenwaage
F balance analytique
P waga analityczna
R аналитические весы

analytical chemistry
The applied science dealing with the discovery and formulation of laws, criteria and methods, enabling the determination with definite precision and accuracy of the qualitative and quantitative composition of substances.
D analytische Chemie
F chimie analytique
P chemia analityczna
R аналитическая химия

analytical concentration
The total concentration of all existing forms of a given constituent in solution, analysed by any suitable method.
D Totalkonzentration, stöchiometrische Konzentration
F concentration analytique
P stężenie analityczne
R аналитическая концентрация

analytical curve
A graph illustrating the measured quantity as a function of concentration of the constituent being determined.
D Eichkurve
F courbe analytique, courbe d'étalonnage
P krzywa wzorcowa, krzywa analityczna
R калибровочный график, калибровочная кривая

analytical factor
The ratio of relative atomic mass A_r of the constituent being determined to the relative

molecular mass M_r of the compound weighed; e.g.

$$\frac{A_{r_{Ba}}}{M_{r_{BaSO_4}}}, \qquad \frac{2A_{r_{Fe}}}{M_{r_{Fe_2O_3}}}$$

The relative molecular mass M_r constitutes the basis for calculation of the analytical results.

D analytischer Faktor, stöchiometrischer Faktor
F facteur d'analyse
P mnożnik analityczny, mnożnik przeliczeniowy
R аналитический фактор, аналитический множитель, фактор пересчёта

analytical gap, electrode gap
The gap between two spectral electrodes where the sample is excited during an arc discharge.
D Analysenstrecke, Elektrodenabstand
F espace inter-électrode, entrode
P przerwa analityczna, przerwa międzyelektrodowa
R аналитический промежуток, межэлектродное расстояние

analytical group
A group of cations (or anions) giving under definite conditions, identical reactions with a group reagent, or failing to react with group reagents.
D Gruppe (von Elementen)
F groupe (analytique)
P grupa analityczna
R (аналитическая) группа

analytical line (*in spectrochemical analysis*)
The spectral line of an element used for the determination and identification of that element.
D Analysenlinie
F raie d'analyse
P linia analityczna
R аналитическая линия

analytical method
The method of detection or determination of a component (or components) in a sample.
D Analysenverfahren
F méthode analytique
P metoda analityczna
R аналитический метод

analytical reaction
A chemical reaction which may be used for analytical purposes.
D analytische Reaktion
F réaction analytique
P reakcja analityczna
R аналитическая реакция

analytical reagent
A reagent bearing the producer's certificate specifying the maximum content of

the given admixtures, and in some instances,
the percentage content of the declared
substance. Analytical reagents are usually
used for preparation of standard solutions.
D Reagens zur Analyse, pro-analytisches
   Reagens
F réactif analytique
P odczynnik czysty do analizy
R реактив чистый для анализа

analytical sample
The portion of a product isolated (or taken)
from the examined or average samples, which
is designed for a single determination,
or investigation.
D Analysen-Probe
F prise d'essai
P próbka analityczna
R аналитическая проба

analytical signal, analyte signal
The recorded signal which can be separated from
the noise level and which is used for detection
and/or determination of a component (e.g. the
change of the electrode potential, the change
of the current flowing through the system,
the change of absorbance).
D analytisches Signal
F signal de l'analyte
P sygnał analityczny
R аналитический сигнал

analytical wavelength (*in absorptiometry*)
A selected wavelength at which the interferences
from other components are as small as possible
relative to the main band intensities being
measured. Thus, the wavelength at which
absorbances are read.
D analytische Wellenlänge
F longueur d'onde analytique
P długość fali analityczna
R аналитическая длина волны

analytical weights
The reference masses used for weighing objects.
D analytische Gewichte
F poids analytiques
P odważniki analityczne
R аналитические гири

angular dispersion (*of a prism or diffraction
grating*)
The angular measure of the resolution of
spectral lines given by the ratio $d\phi/d\lambda$,
where ϕ — deflection angle of a given spectral
line and λ — wavelength of a spectral line.
D Winkeldispersion
F dispersion angulaire
P dyspersja kątowa
R угловая дисперсия

angular dispersion of an analysing crystal
The ability of an analysing crystal to separate
two nearest X-ray wavelengths in a given

angular range of reflection. It is defined by
the equation
$$\frac{d\theta}{d\lambda} = \frac{n}{2\,d\cos\theta}$$
where θ — angle of reflection, λ — wavelength,
n — order of diffraction, d — interplanar spacing
of the crystal.
D Winkeldispersion des Analysatorkristalls
F dispersion angulaire d'un cristal analyseur
P dyspersja kątowa kryształu analizującego
R угловая дисперсия кристалл-анализатора

Anhydrone, Dehydrite
Trademark for granular anhydrous magnesium
perchlorate used for absorption of water.
D Anhydrone
F Anhydrone
P Anhydron
R ангидрон

anion → negative ion

anion exchange
The process of exchanging anions between
a solution and an anion exchanger.
D Anionenaustausch
F échange d'anions
P wymiana anionowa
R анионный обмен

anion exchanger
An ion exchanger containing anions which
may be exchanged for other anions present
in the solution.
Remark: The term anion-exchange resin may be used in the
case of solid organic polymers. In Polish and Russian
nomenclature the terms anionit and анионит are used to
denote the solid anion exchanger.
D Anionenaustauscher
F échangeur d'anions
P wymieniacz anionowy, anionit
R анионообменник, анионит

anion-exchange resin *See* anion exchanger
D Anionenaustauschharz
F résine échangeuse d'anions
P żywica anionowymienna
R анионообменная смола

anode
An electrode through which a net positive
current flows. The chemical reaction taking
place at the anode is an oxidation.
D Anode
F anode
P anoda
R анод

anodic stripping voltammetry *see* stripping
voltammetry
D Voltammetrie mit anodischer Auflösung,
   anodische Inversvoltammetrie

F voltampérométrie avec redissolution
anodique
P woltamperometria inwersyjna
z rozpuszczaniem anodowym
R анодная инверсионная вольтамперометрия,
вольтамперометрия с анодным
растворением

anolyte
The part of electrolyte surrounding the
anode, usually separated from the bulk of the
electrolyte by a semipermeable membrane.
D Anolyt
F anolyte
P anolit
R анолит

anomalous dispersion
Extraordinary deviation from the normal
refractive index versus wavelength curve which
occurs in the vicinity of the absorption lines
or bands in the absorption spectrum of a
substance (the refractive index increases with
increasing wavelength).
D anomale Dispersion
F dispersion anormale
P dyspersja anomalna
R аномальная дисперсия

ANOVA → analysis of variance

anticoincidence technique
A technique for measuring ionizing radiation
based on a principle reverse to that of
the coincidence method, i.e. a given type of
radiation is registered only then when no time
correlation exists between the emission of this
radiation and any radiation emitted by
the same nucleus, or produced in secondary
processes.
D Antikoinzidenzverfahren
F méthode d'anticoïncidences
P metoda antykoincydencji
R метод антисовпадений

antimony-antimony oxide electrode → antimony
electrode

antimony electrode, antimony-antimony oxide
electrode
An electrode of the second kind, made of pure
antimony dipped in an aqueous solution of an
electrolyte. A layer of antimony oxide forms
on the surface of the metal, and this is in
equilibrium with the hydroxide ions in solution.
The half-cell is represented as
$OH^-_{(aq)} | Sb_2O_{3(s)}, Sb_{(s)}$; used as an indicator
electrode to determine pH of a solution in the
range 4-12.
D Antimonelektrode
F électrode d'antimoine, électrode
d'antimoine-oxyde antimonieux
P elektroda antymonowa
R сурьмяный электрод

anti-Stokes band → anti-Stokes lines

anti-Stokes lines, anti-Stokes band
In the Raman spectrum, the Raman lines
occurring on the shorter wavelength
(high-frequency) side of the exciting line and
corresponding to vibrational energy transitions
in the scattering molecule in which the final
vibrational level, to which the excited molecule
drops back, is lower than the initial vibrational
level.
D anti-Stokessche Linien
F raies antistokes, bande antistokes
P linie antystokesowskie, pasmo
antystokesowskie
R антистоксовы линии

a(nti)symmetrical vibrations
Vibrations of atoms in atomic groups occurring
asymmetrically.
D a(nti)symmetrische Schwingungen
F vibrations antisymétriques
P drgania antysymetryczne
R а(нти)симметричные колебания

apparent degree of dissociation, degree of
ionization, dissociation fraction, α
The quantity expressed by the equation
$\alpha = \Lambda/\Lambda^0$ where Λ — molar conductivity of
the electrolyte, Λ^0 — limiting molar conductivity
of the electrolyte; this quantity is not identical
with the true degree of dissociation.
D Dissoziationsgrad nach Arrhenius
F degré de dissociation, fraction de dissociation,
facteur de dissociation, degré d'ionisation,
coefficient d'ionisation
P stopień dysocjacji pozorny, stopień
dysocjacji (wg Arrheniusa)
R степень электролитической диссоциации
(по Аррениусу)

apparent (half-) width
The experimentally observed spectral line width
larger than the true width because of
imperfection of the instrument (optical,
electrical and mechanical errors); it is usually
given as apparent half-width $\Delta v^a_{1/2}$.
D scheinbare Halbwertsbreite
F largeur apparente à mi-absorption
P szerokość linii spektralnej pozorna
R кажущаяся ширина спектральной линии

apparent width → apparent half-width

appearance-potential spectroscopy, APS, soft
X-ray appearance-potential spectroscopy,
SXAPS
Spectroscopy based on the measurement of
the threshold at which X-ray radiation appears
due to the bombardment of a solid with
an electron beam of continuously increasing
energy. This technique gives information on
the density of states at the Fermi level and on
the unoccupied part of the conduction band
in metals.

D Erscheinungspotential-Spektroskopie,
Appearancepotential-Spektroskopie
F ...
P spektroskopia potencjału wzbudzania,
spektroskopia potencjału pojawiania
(*miękkiego promieniowania rentgenowskiego*)
R спектроскопия внешнего потенциала
(*мягкого рентгеновского излучения*)

applicator (*in planar chromatography*)
A device used to apply samples to paper or a
plate, in the form of spots (for small samples)
or streaks (for large samples).
D Auftragegerät
F applicateur
P aplikator
R аппликатор

aprotic solvent, inert solvent, indifferent solvent
A solvent whose molecules are neither donors
nor acceptors of protons, e.g. aliphatic and
aromatic hydrocarbons, carbon tetrachloride.
D aprotisches Lösungsmittel, indifferentes
Lösungsmittel
F solvant aprotique, solvant inerte
P rozpuszczalnik aprotyczny, rozpuszczalnik
bierny
R апротонный растворитель

APS → appearance-potential spectroscopy

aquametry
The direct titrimetric determination of water
present in organic and inorganic substances,
carried out with Karl Fischer's reagent, or
other similar reagents; also the indirect
determination of constituents of a sample,
calculated from the amount of water liberated
or consumed in the course of a reaction.
D Aquametrie
F aquamétrie
P akwametria
R акваметрия

arc → electrical arc

arc spectrum
Emission spectrum of a wide variety of
materials obtained by exciting the sample in
an electrical arc; it comprises mainly the
emission lines of neutral atoms, but also the
spectral lines of singly-ionized species (in the
case of atoms with low ionization potentials),
molecular bands and also continuous spectra
due to radiant particles.
D Bogenspektrum
F spectre d'arc
P widmo łukowe
R дуговой спектр

argentimetric titration
The precipitation titration with a standard
solution of silver(I) salt.
D argentometrische Titration
F titrage argentimétrique

P miareczkowanie argentometryczne
R аргентометрическое титрование

argentimetry, argentometry
The determination of a substance by titration
with a standard silver solution.
D Argentometrie
F argentométrie
P argentometria
R аргентометрия

argentometry → argentimetry

argon detector → argon ionization detector

argon (ionization) detector (*in chromatography*)
An ionization detector in which solute
molecules entering the detector cell with the
argon carrier gas are ionized due to energy
transfer in their collisions with metastable
argon atoms. An increase in current across the
detector occurs which is then amplified and
recorded. The excitation of argon gass to
a metastable state is achieved by collision with
secondary electrons produced in ionization of
other argon atoms by a radioactive β-emitter
(e.g. ^{90}Sr, ^{63}Ni).
D Argon-Detektor, Argon-Ionisationsdetektor
F détecteur (à ionisation) d'argon, détecteur
à argon
P detektor (jonizacyjny) argonowy
R аргоновый (ионизационный) детектор,
аргон-детектор

arithmetic mean, mean, $\bar{x}$
Statistic defined by the formula

$$\bar{x} = \frac{1}{n} \sum_{i=1}^{n} x_i$$

where x_i — value of the *i*th element of the
sample, n — sample size. Arithmetic mean is
often applied as an estimator of the expected
value of a population.
D arithmetisches Mittel
F moyenne arithmétique
P średnia arytmetyczna
R среднее арифметическое

**Arrhenius' and Ostwald's theory of
electrolytic dissociation**
The theory based on the assumption that part of
electrolyte dissociates into free ions due to
the interactions with a solvent, while the
solution becomes diluted, and hence must
obey the classical law of mass action.
D Arrhenius Ostwald-Theorie der
elektrolytischen Dissoziation
F théorie (de la dissociation électrolytique)
d'Arrhenius-Ostwald
P teoria dysocjacji elektrolitycznej Arrheniusa
i Ostwalda
R теория электролитической диссоциации
Аррениуса (и Оствальда)

AS → absorption spectroscopy

AS → atomic spectroscopy

Ascarite
A trademark for sodium hydrate-asbestos used for rapid and quantitative absorption of carbon dioxide.
D Ascarit
F Ascarite, amiante sodé
P Askaryt
R аскарит

ascending chromatography → ascending development

ascending development, ascending chromatography, ascending technique
In planar chromatography, a technique of developing a chromatogram, in which the mobile phase is introduced to the lower edge of a sheet of paper or thin-layer plate and flows upwards, owing to capillary action, with gradually decreasing velocity due to gravitation.
D aufsteigende Entwicklung, aufsteigende Chromatographie, aufsteigende Methode
F développement ascendant, chromatographie ascendante, méthode ascendante
P rozwijanie wstępujące, chromatografia wstępująca, metoda wstępująca
R восходящее проявление, восходящая хроматография, восходящий метод

ascending technique → ascending development

ascorbimetry
A branch of reductometry in which standard solutions of ascorbic acid are used for determination of substances.
D ...
F ascorbinométrie
P askorbinometria
R аскорбинометрия

ashing technique
The heating and calcination of a sample in an open crucible in order to burn organic matter.
D Veraschung, Veraschen
F méthode de calcination, méthode d'incinération
P spopielanie próbek
R кальцинирование пробы; озоление пробы

association degree of ions
Fraction of ions in solution which associate into stable ion pairs or higher ionic associates. *See* Bjerrum's theory of ionic association.
D Assoziationsgrad der Ionen
F degré d'association des ions
P stopień asocjacji jonów
R степень ассоциации ионов

astigmatism
A deffect of lenses and mirrors with a spherical surface on account of which the image of a point off the optical axis is not a point, but forms two mutually perpendicular short lines located in different planes.
D Astigmatismus
F astigmatisme
P astygmatyzm
R астигматизм

asymmetrical vibrations → antisymmetrical vibrations

asymmetry effect → relaxation-time effect

asymmetry factor of a peak, peak asymmetry factor (*in chromatography*)
The measure of deformation of a peak usually expressed by the ratio of the lengths of segments ED and DF into which the straight line perpendicular to the peak base and drawn from the peak maximum, divides the peak width at the base. *See* peak.
D Peakasymetriekoeffizient, Peakasymetriefaktor
F facteur d'asymétrie du pic
P współczynnik asymetrii piku
R фактор асимметрии пика

atomic absorption spectroscopy, AAS
Absorption spectroscopy concerned with the investigation of the atomic spectra.
D Atom-Absorptions-Spektroskopie
F spectroscopie d'absorption atomique
P spektroskopia absorpcyjna atomowa
R атомно-абсорпционная спектроскопия

atomic attenuation coefficient *See* attenuation coefficient
D atomarer Schwächungskoeffizient
F coefficient d'atténuation atomique
P współczynnik osłabienia atomowy
R атомный коэффициент ослабления

atomic core
An atom deprived of the valence electrons.
D Atomrumpf
F tronc d'atome
P rdzeń atomowy, zrąb atomowy
R атомный остов

atomic cross section
The product of the isotopic cross section and the relative abundance of a given isotope in a natural element.
D atomarer Wirkungsquerschnitt
F section efficace atomique
P przekrój czynny atomowy
R атомное (эффективное) сечение

atomic emission spectroscopy, AES
Emission spectroscopy concerned with the investigation of atomic spectra.
D Atom-Emissions-Spektroskopie
F spectroscopie d'emission atomique
P spektroskopia emisyjna atomowa
R атомная эмиссионная спектроскопия

atomic fluorescence spectroscopy, AFS
Fluorescence spectroscopy concerned with the
investigation of atomic spectra.

D Atom-Fluoreszenz-Spektroskopie
F spectroscopie de fluorescence atomique
P spektroskopia fluorescencyjna atomowa
R атомно-флуоресцентная спектроскопия

atomic mass unit → unified atomic mass unit

atomic refraction
The product of the specific refraction of an
element and its atomic mass.

D Atomrefraktion
F réfraction atomique
P refrakcja atomowa
R атомная рефракция

atomic spectrochemical analysis
Spectrochemical analysis based on the
investigation of atomic spectra.

D Atomspektralanalyse
F analyse spectrochimique atomique
P analiza spektralna atomowa
R атомный спектрохимический анализ

atomic spectroscopy, AS
A branch of spectroscopy concerned with the
production, measurement and investigation of
atomic spectra, either emission or absorption
ones.

D Atomspektroskopie
F spectroscopie atomique
P spektroskopia atomowa
R атомная спектроскопия

atomic spectrum
The emission spectrum with a line structure
corresponding to the transitions between
different atomic energy states.

D Atomspektrum
F spectre atomique
P widmo atomowe
R атомный спектр

atomization
A process of conversion of the volatilized
analyte into free atoms.

D Atomisierung
F atomisation
P atomizacja
R атомизация

atomizer
A device for converting the analyte into free
atoms, usually preceded by volatilization
of a sample.

D Atomisator, Atomisierungseinrichtung
F atomiseur
P atomizer
R атомизатор

atomizer → nebulizer

ATR → attenuated total reflectance

ATR spectrum → attenuated total reflectance
spectrum

attenuated total reflectance, internal reflectance,
internal reflection technique, ATR
An infrared technique used particularly for
examining opaque solids in which the sample
is placed in optical contact with the inside
face of a prism. A beam of radiation passing
through the wall of the prism is directed so as
to cause the total reflection off the inside face.
If the sample absorbs any of the incident
wavelengths of radiation the reflected beam
will be attenuated in these regions. This
attenuated radiation when plotted as a function
of wavelength will give an absorption spectrum
characteristic of the sample (ATR spectrum).

D abgeschwächte Totalreflexion, ATR-Methode
F réflexion totale atténuée
P metoda osłabionego całkowitego odbicia
R метод ослабления полного отражения

attenuated total reflectance spectrum, ATR
spectrum *See* attenuated total reflectance

D abgeschwächte Totalreflexions-Spektrum
F spectre de réflexion totale atténuée
P widmo osłabionego całkowitego odbicia
R спектр ослабления полного отражения

attenuation → attenuation of radiation

attenuation coefficient, μ (*of substance for
a parallel beam of specified radiation*)
The coefficient μ in the expression $\mu\Delta x$ for the
fraction of radiation removed by attenuation in
passing through a thin layer of thickness Δx of
that substance. According as Δx is expressed in
terms of length, mass per unit area or atoms
per unit, μ is called respectively the linear,
mass, or atomic attenuation coefficient.

D Schwächungskoeffizient
F coefficient d'atténuation
P współczynnik osłabienia
R коэффициент ослабления

attenuation (of radiation)
The reduction of a radiation quantity upon
passage of radiation through matter resulting
from interactions of the radiation with the
matter it traverses.

D Strahlenschwächung
F atténuation de rayonnement
P osłabienie promieniowania
R ослабление излучения

Auger effect
The emission of an electron (Auger electron)
from an atom accompanying the filling of
a vacancy in an inner electron shell.

D Auger-Effekt, Auger-Übergang
F effet Auger
P efekt Augera
R эффект Оже

Auger-electron microscopy, AEM, scanning Auger-electron spectroscopy, SAES
A type of Auger-electron spectroscopy used for the observation of the topography of element distribution in the outermost layer of a solid, performed by scanning an electron beam over the solid surface.
D Auger-Elektronenmikroskopie
F microscopie d'electrons Auger
P mikroskopia (skaningowa) elektronów Augera
R (сканирующая) спектроскопия оже-электронов

Auger-electron spectroscopy, AES
A branch of spectroscopy concerned with the observation of Auger electron intensities versus their energies; useful for chemical analysis of very thin surface layers and for chemical analysis of light elements; used also in the study of catalysts, corrosion and impurity segregation at surfaces.
D Auger-Elektronenspektroskopie
F spectroscopie d'électrons Auger
P spektroskopia elektronów Augera
R спектроскопия электронов Ожé, электронная оже-спектроскопия

automatic balance
A balance effecting at least the following operations: seeking the equilibrium range, mass equilibration and balance indication without the intervention of the analyst.
D automatische Waage
F balance automatique
P waga automatyczna
R автоматические весы

automatic determination of carbon, hydrogen and nitrogen
Simultaneous determination of carbon, hydrogen and nitrogen mainly by a milligram method, in which the sample mixed with oxidizing agents is subjected to flash-combustion in a helium stream. The combustion products — carbondioxide, water and nitrogen — are separated chemically or in chromatographic columns and determined by means of a thermal conductivity detector.
D automatische Kohlenstoff-, Wasserstoff- und Stickstoffbestimmung
F dosage automatique du carbone, de l'hydrogène et de l'azote
P oznaczanie automatyczne węgla, wodoru i azotu
R автоматическое определение углерода, водорода и азота

automatic zero burette
D Bürette mit automatischer Nullpunkteinstellung
F burette à zéro automatique
P biureta z automatycznym nastawianiem zera
R бюретка с автоматической установкой нулевой метки

automatic zero pipette
D automatische Pipette
F pipette à zéro automatique
P pipeta z automatycznym nastawianiem zera
R автоматическая пипетка

autoprotolysis, self-dissociation
The spontaneous electrolytic dissociation in which some of the solvent molecules behave like an acid (supplying protons) while others behave like a base (accepting protons). This effect is observed (to a limited extent) in the amphiprotic solvents, e.g.:

$$H_2O + H_2O \rightleftarrows H_3O^+ + OH^-$$
$$NH_3 + NH_3 \rightleftarrows NH_4^+ + NH_2^-$$
$$CH_3COOH + CH_3COOH \rightleftarrows$$
$$\rightleftarrows CH_3COOH_2^+ + CH_3COO^-$$

D Autoprotolyse
F autoprotolyse
P autoprotoliza, autojonizacja, autodysocjacja
R автопротолиз

autoradiogram → autoradiograph

autoradiograph, autoradiogram
An image obtained by placing an object containing a radioactive substance in contact with a photographic plate or film for a suitable exposure time and developing. The image shows the distribution of the radioactive element in the object.
D Autoradiogramm
F autoradiogramme, autoradiographe
P autoradiogram
R авторадиограмма, авторадиограф

auxiliary electrode, counter electrode, secondary electrode (*in electrochemistry*)
An electrode which serves to carry the current that passes through the indicator or working electrode, usually separated from them by a porous membrane, e.g. a glass frit. The electrode processes taking place at the auxiliary electrode do not influence the electrochemical characteristics of the indicator or working electrode.
D Hilfselektrode, Gegenelektrode
F électrode auxiliaire, contre-électrode
P elektroda pomocnicza
R вспомогательный электрод, противоэлектрод

auxochrome
An atom or group of atoms having no absorption property in the visible or ultraviolet regions which, in the presence of a chromophore, cause a shift of the absorption spectrum towards longer wavelengths; e.g. $-CH_3 < -OH < -OCH_3 < -NH_2 < -NHCH_3 < -N(CH_3)_2$
(in the order of increasing shifting power).
D Auxochrom, auxochrome Gruppe
F auxochrome

P auksochrom, grupa auksochromowa
R ауксохром

average deviation → mean deviation

average interstitial velocity of the carrier gas →
mean interstitial velocity of the carrier gas

average mobile phase interstitial velocity →
mean interstitial velocity of the carrier gas

average sample
A sample taken from the gross sample
representing the properties of the bulk of the
given material and carefully stored to secure
preservation of its identity. The sample should
have the average composition and properties
of the examined material.
D Durchschnittsprobe
F échantillon pour laboratoire , échantillon
moyen
P próbka laboratoryjna średnia
R средняя лабораторная проба

av polarography → alternating voltage
polarography

axial eddy diffusion → eddy diffusion term

axial molecular diffusion → molecular diffusion
term

azotometer → nitrometer

back-extraction, retrograde extraction,
stripping
The process opposite to extraction consisting
in transferring a substance from an extract to
another liquid phase (usually into the aqueous
phase).
D Rückextraktion Rückschüttelung, Strippen
F extraction en retour
P reekstrakcja
R реэкстракция

back-flush → backflushing

backflushing, back-flush (*in gas chromatography*)
The technique of reversing the flow of carrier
gas in chromatographic columns for flushing
out slowly moving components from the
column (after elution and determination of
fast migrating components).
D Zurückspülung Rückspülung
F contre-balayage
P wymywanie zwrotne
R обратная продувка

background (*of a device*)
The value indicated by a radiation measuring
device in the absence of the source whose
radiation is to be measured when the device
is under the normal conditions of its operation.
D Nulleffekt, Nullwert, Untergrund
F mouvement propre, bruit de fond
P tło przyrządu
R фон детектора

background → background radiation

background corrector (*in atomic absorption
spectroscopy*)
An electrooptical device for compensation of
the background which is generated due to
non-specific absorption.
D Untergrundkompensator
F correcteur de fond, correcteur d'absorption
non spécifique
P korektor tła
R корректор фона

background (radiation)
The radiation from any source other than the one to be detected and/or measured.
D Strahlungsuntergrund
F fond de rayonnement
P tło promieniowania
R фон радиации

backscattering
Scattering of radiation in a generally backward direction. In the assay of radioactivity, it applies to the scattering of radiation into the radiation detector from any material except the sample and the detector.
D Rückstreuung
F rétrodiffusion, diffusion en retour
P rozpraszanie wsteczne, rozpraszanie zwrotne
R обратное рассеяние

backscattering factor, r (*in Auger-electron spectroscopy*)
The ratio of the total number of ionizations in the surface layer of a solid to the number of izonizations in this layer caused by primary electrons. This parameter determines the influence of scattered secondary and primary electrons upon the yield of Auger electrons and it depends on the atomic number of the irradiated material, the energy of the primary electrons and their incidence angle.
D Rückstreukoeffizient
F facteur de diffusion en retour
P współczynnik wstecznego rozpraszania
R коэффициент обратного рассеяния

backscatter peak
In γ-radiation spectrum the peak of radiation scattered at the angle of 180° (approx.).
D Rückstreumaximum, Rückstreu-Peak
F pic de rétrodiffusion
P pik rozpraszania wstecznego, pik rozpraszania zwrotnego
R пик обратного рассеяния

back-titration
A titration of an unreacted standard solution that has been added in excess to a sample.
D Rücktitration
F titrage en retour
P miareczkowanie odwrotne, odmiareczkowanie nadmiaru
R обратное титрование

balanced-density slurry packing (*in high performance liquid chromatography*)
The packing procedure performed by pumping a suspension of the sorbent prepared in a liquid medium of the same density into a column at a pressure much higher than operating pressure.
D balancierte Viskositätsmethode
F remplissage (de la colonne) par une suspension à densité compensée
P napełnianie kolumn zawiesiną o wyrównanej gęstości
R заполнение колонки методом сбалансированной плотности суспензии

balanced filters, Ross filters
A pair of X-ray filters made of elements with adjacent or close atomic numbers with the thicknesses so matched that the transmission of both filters is practically equal in the energy (wavelength) range beyond the absorption edges of the elements constituting the pair.
D Absorptions-Kantenfilter
F filtres balancés, filtres équilibrés
P filtry zrównoważone, filtry zbalansowane, filtry Rossa
R балансные фильтры

balance sensitivity, sensitivity of a balance (*for a given load*)
The response per unit mass in terms of the pointer scale per unit mass.
D Empfindlichkeit der Waage
F sensibilité d'une balance
P czułość wagi
R чувствительность весов

band → zone

band broadening See peak broadening
D Bandverbreiterung
F étalement de bande
P rozmycie pasma
R расширение полосы, размывание полосы

band edge → band head

band group, band system
The whole set of bands in the molecular spectrum corresponding to a given electronic transition.
D Bandengruppe, Bandensystem
F système de bandes
P układ pasm
R система полос

band head, band edge (*in a band spectrum*)
The limit to which groups of lines approach as they come closer and closer together.
D Bandenkopf, Bandenkante
F téte de bande
P głowica pasma
R край полосы, голова полосы

band spectrum
An emission or absorption molecular spectrum consisting of a series of closely spaced spectral lines, resulting from vibrational, rotational and electronic transitions.
D Bandenspektrum
F spectre de bande
P widmo pasmowe
R полосатый спектр

band system → band group

barn, b
Illegal unit of area used for expressing nuclear cross sections. In the SI system $1b = 10^{-28}$ m².
D Barn
F barn
P barn
R барн

barrier-layer cell → photovoltaic cell

barrier-layer photocell → photovoltaic cell

base form of an anion exchanger
The ionic form of an anion exchanger in which the counter-ions are hydroxide ions (OH^- — form) or the ionogenic groups form an uncharged base, e.g.—NH_2.
D basische Form des Anionenaustauschers
F forme basique d'un échangeur d'anions
P forma zasadowa wymieniacza anionów
R основная форма анионообменника

base-line
The section of a chromatogram corresponding to the outflow of the pure eluent from the column.
D Basislinie, Null-Linie, Grundlinie
F ligne de base
P linia podstawowa
R нулевая линия, основная линия

base-line drift (*in detector*)
A gradual displacement of the base-line signal from the detector perpendicular to the time axis over a considerable period of time.
D Basislinienabweichung, Basislinienänderung, Basisliniendrift
F dérive de la ligne de base, déplacement de la ligne de base
P dryf linii podstawowej
R дрейф нулевой линии, дрейф базисной линии

base-line method
A method for calculation of the absorbance of the given constituent in quantitative spectrophotometric analysis. An absorption band of the substance under analysis is chosen and the base-line is drawn as a straight line tangent to the absorption band. The transmittance value P is measured at the bottom of the band at the point of maximum absorption and the transmittance P_0 is measured at the same wavenumber at the chosen base-line and the absorbance is calculated as $A = \lg P_0/P$. The base-line method automatically provides for cell absorbance and reflection losses.
D Basislinie-Methode
F méthode à ligne de base
P metoda linii podstawowej
R метод основной линии

base peak (*in mass spectrometry*)
The most intense line in a mass spectrum,

corresponding to the most abundant ion, dependent on the nature of compound; it may be either a fragment ion peak or the parent peak.
D Basispeak
F pic de base
P pik główny, pik podstawowy
R основной пик

basic solvent → protophilic solvent

bathochrome
An atom or group of atoms which, when substituted into a molecule of organic compound, shift its absorption spectrum towards longer wavelengths; e.g.—NH_2, —OH,—OCH_3.
D bathochrome Gruppe
F groupement bathochrome
P batochrom, grupa batochromowa
R батохромная группа

bathochromic shift, red shift
The displacement of the characteristic maximum of the absorption band towards longer wavelengths (lower frequency, lower energy) due to the structural modification in a molecule.
D bathochrome Verschiebung, Farbtonverschiebung
F déplacement bathochrome, effet bathochrome, migration bathochrome
P przesunięcie batochromowe, efekt batochromowy, przesunięcie czerwone
R батохромное смещение (окраски), батохромный эффект

battery, secondary battery, storage battery
A source of direct current, consisting of two or more secondary cells connected together and used as one unit.
D Sekundarbatterie
F batterie
P bateria
R батарея

B band
In the electronic spectrum of benzene derivatives and homologues, the band which corresponds to the $\pi - \pi^*$ electronic transition, by analogy to the benzene band of 254 nm; B band shows low intensity and a well developed oscillation structure.
D B Band
F bande B
P pasmo B
R полоса B

becquerel, Bq
SI unit of activity equal to one disintegration per second.
D Becquerel
F becquerel

P bekerel
R бекерель

bed volume, X
The volume of that part of a chromatographic
column which is occupied by the column filling.
It is calculated from the column internal
diameter and the filling bed height.
D Bettvolumen
F volume du lit, volume du remplissage
P objętość złoża
R объём слоя

bed volume capacity, volume capacity of an
ion exchanger bed
The number of milliequivalents of ionogenic
groups per 1 cm^3 of the volume of ion
exchanger bed under specified conditions
(ionic form of the ion exchanger, medium, etc.).
D ...
F capacité du lit de résine
P zdolność wymienna objętościowa złoża
    jonitu, pojemność wymienna właściwa
    objętościowa złoża jonitu
R объёмная ёмкость слоя

Beer's law
The law according to which the intensity of a
beam of a parallel, monochromatic radiation
decreases exponentially with the increasing
concentration c of the absorbing species in a
homogeneous medium

$$\log \frac{I_0}{I_t} = Kc$$

where I_0 − intensity of the incident beam,
I_t − intensity of the transmitted beam, and
K − constant depending on the wavelength of
the radiation, the nature of the medium, and
the sample path length.
Remark: The combined law: Bouguer-Lambert-Beer law
is commonly called Beer's law.
D Beersches Gesetz
F loi de Beer
P prawo Beera
R закон Бэра

Beer's law → Bouguer-Lambert-Beer law

bending vibrations, deformation vibrations
Vibrations which imply displacement of atoms
out from the bonding axis and thus a change
in bond angles either between two atoms
bonded to a third atom or between a group of
atoms and the rest of the molecule.
D Deformationsschwingungen
F vibrations de déformation
P drgania deformacyjne, drgania zginające
R деформационные колебания

Bernoulli distribution → binomial distribution

beta-particle absorptiometry, beta-particle
absorption technique *See* nuclear radiation
absorptiometry

D Methode der β-Durchstrahlung
F analyse par absorption des particules β
P metoda absorpcji promieniowania β
R метод анализа по поглощению β-лучей

beta-particle absorption technique → beta-
particle absorptiometry

beta-scintillation spectrometer
An instrument with a liquid scintillator as
a detector for simultaneous activity
measurement of two or three beta-emitters
(generally ^{14}C and ^3H).
D Szintillations-β-Spektrometer
F spectromètre à scintillations β
P spektrometr scyntylacyjny β
R сцинтилляционный β-спектрометр

biamperometric titration → amperometric
titration with two indicator electrodes

biamperometry → amperometry with two
indicator electrodes

bias of estimator
The difference between the real value of
a parameter and the expected value of its
estimator.
D Verzerrung der Schätzfunktion
F biais de l'estimateur
P obciążenie estymatora
R смещение

bifunctional ion exchanger
An ion exchanger containing two types of
ionogenic groups.
D bifunktioneller Ionenaustauscher
F échangeur d'ions difonctionnel, échangeur
    d'ions bifonctionnel
P wymieniacz jonowy dwufunkcyjny
R бифункциональный ионообменник

binomial distribution, Bernoulli distribution
Probability distribution of a discrete random
variable X with the probability function defined
by the formula

$$P(X=k) = \binom{n}{k} p^k (1-p)^{n-k}$$

for $k = 0, 1, 2, ..., n$ $(0 < p < 1)$.

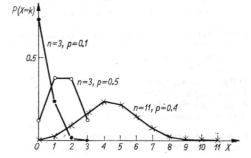

D Binomialverteilung, Bernoullische
  Verteilung
F distribution binomiale, distribution de
  Bernoulli
P rozkład dwumianowy, rozkład Bernoullego
R биномиальное распределение,
  распределение Бернулли

bioluminescence
Luminescence occurring during biochemical
reactions which produce atoms and molecules
in excited state and light emission results.
D Biolumineszenz
F bioluminescence
P bioluminescencja
R биолюминесценция

bipotentiometric titration → controlled-current
potentiometric titration with two indicator
electrodes

bipotentiometry → controlled-current
potentiometry with two indicator electrodes

BIS — bremsstrahlung isochrome spectroscopy

biuret reaction
A characteristic reaction for the detection of
proteins, peptides, and some aminoalcohols,
involving treatment of the above with copper
salts in alkaline medium, whereupon a violet
coloration develops caused by copper — biuret
complex formation.
D Biuretreaktion
F réaction du biuret
P reakcja biuretowa
R биуретовая реакция

Bjerrum's ion-association theory → Bjerrum's
theory of ionic association

Bjerrum's theory of ionic association,
Bjerrum's ion-association theory
The theory according to which the ions of
opposite signs being in a solution at a distance l,
where

$$a \leqslant l \leqslant \frac{|z_+ z_-| e^2}{8\pi \, \varepsilon_r \, \varepsilon_0 \, kT} \quad \text{(in SI system)}$$

form (under the influence of Coulomb
forces) an ion pair with life time long enough
to be treated as a species resistant to thermal
movements in a solution; a — mean ionic
diameter, for other symbols see Debye-Hückel
limiting equation.
D Bjerrumsche Theorie der Ionenassoziation,
  Assoziationstheorie der Ionen von Bjerrum
F théorie de l'association des ions de Bjerrum
P teoria Bjerruma asocjacji jonów
R теория ассоциации ионов Бьеррума, теория
  ионной ассоциации Бьеррума

blackening (*of photographic emulsion*), S
The common logarithm of the inverse of the
transmittance of the developed photographic
image, $S = -\log T$.

Remark: The use of the term optical density in
spectrochemical analysis is discouraged by the IUPAC
nomenclature recommendations.
D Schwärzung
F noircissement, densité optique
P zaczernienie, gęstość optyczna
R почернение, оптическая плотность

blank (solution)
A solution that intentionally does not contain
the analyte, but in other respects has the
same composition as the sample solution.
D Blindlösung
F blanc
P roztwór ślepej próby
R холостой раствор

blank test
A test carried out in the course of an analysis
to ascertain the purity correction factor for
the reagents used.
D Blindprobe, Blindreaktion, Leerversuch
F essai à blanc
P próba ślepa, próba zerowa
R холостой опыт, слепой опыт

blank titration
A titration carried out on a solution identical
to the sample solution (i.e. in volume,
acidity, amount of indicator, etc.) except for
the sample itself.
D Blindtitration
F titrage à blanc
P miareczkowanie ślepe
R холостое титрование

blaze angle (*of a diffraction grating*), β
The angle between the facet of the groove on
which the radiation is incident and the
grating plane.
D Glanzwinkel
F angle de miroitement
P kąt błysku
R угол блеска

blaze wavelength, λ_β
The wavelength at which the intensities
of the radiations split on and reflected
from the grating are maximum.
D Glanzwinkel-Wellenlänge
F longueur d'onde de blaze
P długość fali błysku
R длина волны блеска

bleeding (*in gas-liquid chromatography*)
A progressive loss of the liquid stationary
phase from the column in the course of
mobile phase flow.
D Abdampfen der Trennflüssigkeit
F fuite de la phase stationnaire
P wymywanie fazy nieruchomej
R вымывание неподвижной фазы, испарение
  неподвижной фазы

block
A part of experimental design obtained by
splitting the full design into smaller units in a
manner which allows to avoid the influence of
uncontrolled factors on the observed value.
D Block
F bloc
P blok
R блок

blocking of metal indicator
The formation of a complex by the action of
the indicator on a metal; this complex has a
similar or higher stability than that formed by
the same metal with the titrant (e.g. EDTA).
Consequently, the colour change at the
end-point of the complexometric titration is
not distinct, or may not be observed at all.
D Blockierung des Metallindikators
F blocage d'un métal indicateur
P blokowanie metalowskaźnika
R блокирование металлоиндикатора

blow-pipe test
The qualitative test which consists in heating
the examined sample and a flux on a charcoal
block in a reducing blow-pipe flame, followed
by chemical analysis of the residue.
D Lötrohrprobe
F analyse pyrognostique (au chalumeau)
P próba dmuchawkowa
R проба паяльной трубкой

Blue Dextran
A high-molecular substance employed in
gel-permeation chromatography as the
coloured marker to determine the interstitial
volume.
D Dextran Blau
F ...
P dekstran błękitny
R синий декстран

blue shift → hypsochromic shift

BN-chamber, Brenner-Niederwieser chamber
A horizontal sandwich chamber in which the
eluent is introduced to the sorbent layer by
means of filter paper or a special distributor.
It enables continuous development of
chromatograms by evaporation of the eluent
from the opposite edge of the sorbent layer.
D BN-Kammer, Brenner-Niederwieser-Kammer
F chambre BN, chambre de Brenner et
  Niederwieser
P komora (chromatograficzna) Brennera
  i Niederwiesera, komora BN
R БН-камера, камера Бреннера-Нидервизера

boiling point apparatus
D Siedepunktbestimmer
F appareil de détermination du point
  d'ébullition
P aparat do oznaczania temperatury wrzenia
R прибор для определения температуры
  кипения

bolometer (*in IR spectroscopy*)
A miniature resistance (metal or semiconductor)
thermometer which on change of resistance of
the thermistor with temperature gives an
electrical signal.
D Bolometer
F bolomètre
P bolometr
R болометр

bomb calorimeter
A pressure vessel with hermetic cover designed
for measuring heat of combustion.
D Bombenkalorimeter, Kalorimeterbombe,
  Verbrennungsbombe
F bombe calorimétrique
P bomba kalometryczna
R калориметрическая бомба

bonded stationary phase → chemically bonded
stationary phase

borax bead
The transparent vitreous mass formed by
fusing borax on platinum wire. The borax
bead gives a characteristic coloration when
fused with compounds containing heavy
metals; hence its use for identification in
inorganic qualitative analysis.
D Boraxperle, Boraxsalzperle
F perle au borax
P perła boraksowa
R перл буры

Born's theory of solvation
The theory describing the solvation process as
purely electrostatic.
D Bornsche Theorie, Bornsche Solvatation-
  Theorie
F théorie de la solvatation de Born
P teoria solwatacji Borna
R теория сольватации Борна

Bouguer-Beer law → Bouguer-Lambert-Beer law

Bouguer-(Lambert-)Beer law, Lambert-Beer law,
Beer's law
The law saying that if a parallel beam of
monochromatic light enters an absorbing
medium at a right angle to the plane, parallel to
the surfaces of the medium, the intensity of the
beam decreases exponentially with the sample
path length b (cuvette interior) and with the
concentration c (in grams per litre) of absorbing
material

$$T = 10^{-A} = 10^{-abc}$$

where T — internal transmittance,
A — absorbance, and a — specific absorption
coefficient. When concentration c is in mole per
litre the law is $A = \varepsilon bc$, where ε — molar
absorption coefficient.
D (Bouguer-)Lambert-Beersches Gesetz
F loi de Bouguer-Beer, loi de Lambert-Beer

P prawo (Lamberta i) Beera, prawo Bouguera
i Beera
R закон Бугера-Ламберта-Бэра

Bouguer-Lambert law → Lambert's law

Bouguer's law → Lambert's law

Bragg's law
The law describing the diffraction of X-rays
in a crystal
$$n\lambda = 2d \sin \theta$$
where n — order of diffraction, λ — wavelength,
d — interplanar spacing, θ — glancing angle.
D Braggsche Gleichung, Braggsches
Reflexionsgesetz
F formule de Bragg, loi de Bragg
P prawo Bragga
R закон Вульфа-Брэгга

branching decay
Radioactive decay which can proceed in two
or more different ways.
D Verästelung, Aufspaltung
F ramification
P rozpad rozgałęziony
R разветвлённый распад, разветвление

breadth of spectral line → line width

breakthrough (*of a chromatographic column*)
Appearance of a given component in the
column effluent.
D Durchbruch
F percée
P przebicie
R проскок

**breakthrough capacity of an ion exchanger
bed,** Q_B
The total amount of ions, expressed in
milliequivalents or millimoles, taken up by 1 g of
dry ion exchanger or 1 cm^3 of bed volume. The
practical capacity of an ion exchanger bed in
a dynamic system is determined experimentally
under flow-through conditions and refers to
the amount of ions which have been taken up
before the species is first detected in the effluent
or before its concentration in the effluent
reaches some arbitrarily established value.
D ...
F capacité de fixation d'un lit d'échangeur
d'ions
P zdolność wymienna do chwili przebicia
złoża jonitu, zdolność wymienna robocza
złoża jonitu, pojemność wymienna robocza
złoża jonitu
R ёмкость слоя ионообменника до проскока

breakthrough curve
In frontal chromatography, the curve obtained
by plotting the concentration of a substance
in the effluent (or the concentrations ratio in
the effluent and influent) as a function of
time or volume.

D Durchbruchskurve
F courbe de percée
P krzywa przebicia
R кривая проскакивания

bremsstrahlung
The electromagnetic radiation with a continuous
spectrum arising from the momentum change
of a charged particle interacting with the
electrostatic field of the atomic nucleus and
electrons.
D Bremsstrahlung
F rayonnement de freinage
P promieniowanie hamowania
R тормозное излучение

bremsstrahlung isochrome spectroscopy, BIS
A method which utilizes electron bombardment
for generation of X-rays. The emitted
bremsstrahlung is recorded with an
selective energy detector as a function of the
accelerating voltage across the tube. This
technique gives information on the intrinsic
empty electronic surface states in (or near)
the fundamental gap of semiconductors.
D ...
F spectroscopie du rayonnement
monochromatique de freinage
P metoda izochromat rentgenowskiego
promieniowania hamowania
R изохромная спектроскопия тормозного
излучения

Brenner Niederwieser chamber → BN-chamber

brightening agent
Organic compounds added in trace amounts to
salt solutions to modify, mainly by
adsorption processes, the crystal growth
of the deposited metal.
D Glanzbildner
F (agent) brillanteur
P substancja wybłyszczająca
R блескообразователь

brightness → luminance

Brockmann scale of activity (*in adsorption
chromatography*)
The five-step scale of adsorption activity of
alumina. Six azo dyes of different polarities are
used as test solutes and the activity of adsorbents
is classified according to which azo dye is
adsorbed at the top of the column or
determined from the R_F values obtained for
these dyes.
D Aktivitätsskale nach Brockmann
F échelle d'activité de Brockmann
P skala aktywności Brockmanna
R шкала активности по Брокману

bromatometry
The determination of a substance by titration
with a standard bromate solution
(e.g. potassium bromate).

D Bromatometrie
F bromatométrie
P bromianometria
R броматометрия

bubble counter → bubble gauge

bubble gauge, bubble counter, bubbler
A device for controlling the gas flow rate.
D Blasenzähler
F compte-bulles
P licznik pęcherzyków
R счётчик пузырьков

bubbler → bubble gauge

Büchner funnel
A funnel with a flat perforated bottom
on which a circular piece of filter paper is
placed, used for suction filtering of liquids.
D Büchner-Trichter, Filternutsche nach
  Büchner, Büchner-Nutsche
F entonnoir de Büchner
P lejek Büchnera, lejek sitowy
R воронка Бюхнера

buffer (solution), pH buffer
Solution of a weak acid (base) and its salt
with a strong base (acid), exhibiting a constant
concentration of hydrogen ions. The pH of
such a buffer mixture remains practically
constant on dilution, or after addition of
a small quantity of a strong acid (base).
D Puffer(lösung), Puffergemisch
F (solution) tampon, mélange tampon
P bufor pH, roztwór buforowy pH
R буферный раствор

bulk of a solution
The part of an electrolyte solution in the
electrochemical cell whose composition
does not change when the current flows.
D Inneres einer Lösung
F sein d'une solution
P głębia roztworu, wnętrze roztworu
R глубина раствора, внутренняя часть
  раствора

bulk of material, lot of material
The total amount of a given product, packed
or unpacked, supplied by the manufacturer
to the customer in one batch. The lot of material
that should be uniform in composition,
is estimated on the ground of analysis of its
average laboratory sample.
D Partie
F partie de matériau
P partia produktu
R партия материала

Bunte gas burette
A burette designed for collecting gases and
measuring their volume.
D Gasbürette nach Bunte, Bunte-Bürette
F burette de Bunte

P biureta Buntego
R газовая бюретка Бунте

burette
A long graduated tube usually with a tap at
one end, used to deliver known volume of
liquid in titration.
D Bürette
F burette
P biureta
R бюретка

burner (*in flame spectroscopy*)
A device producing a flame, to which fuel and
oxidant (usually in the form of gases) are
supplied.
D Brenner
F brûleur
P palnik
R горелка

burning voltage
The potential difference across the analytical
gap during an arc discharge.
D Brennspannung, Arbeitsspannung
F tension de fonctionnement
P napięcie robocze
R рабочее напряжение

burn-up (*in radiochemistry*)
Destruction or transformation of specified
nuclei as a result of neutron capture.
D Abbrand, Ausbrand
F consommation
P wypalanie
R выгорание

Butler-Volmer equation
The equation describing (for given
concentrations of reagents c_P and c_S)
the dependence of the faradaic current I_f on the
potential E of the electrode; for the electrode
reaction

$$\nu_S \, S^{z_S} + ne \rightleftarrows \nu_P \, P^{z_P}$$

$$I_f = \frac{n}{\nu} \, FAk^\circ \left\{ c_P^{\nu_P/\nu} \exp\left[\frac{\nu\alpha_a F}{nRT} (E-E^\circ) \right] + \right.$$
$$\left. - c_S^{\nu_S/\nu} \exp\left[- \frac{\nu\alpha_c F}{nRT} (E-E^\circ) \right] \right\}$$

where ν_S and ν_P — stoichiometric coefficients
of substrates S and products P, z_S and
z_P — charge numbers of S and P, respectively,
n — number of electrons; ν — stoichiometric
number of the transfer reaction (number of
identical activated complexes formed and
destroyed in the course of the overall
reaction), A — area of the electrode, k° — standard
(or formal) rate constant of the electrode
reaction, α_a and α_c — transfer coefficients
anodic and cathodic, respectively,
E° — standard (or formal) potential of the
electrode reaction.

D Butler-Volmer-Gleichung
F relation de Butler-Volmer
P równanie Butlera i Volmera
R уравнение Бутлера-Вольмера

by-pass injector, injection valve
A sample injector used in chromatography by
means of which the mobile phase may be
temporarily diverted through a sample
chamber, shut off from the main stream by
valves, so that the sample is carried to the
column.
D Bypass-Probengeber, Bypass-Injektor,
   Teilstromprobengeber, Dosierschleife
F injecteur à dérivation, injecteur avec by-pass,
   vanne à boucle d'iniection
P dozownik bocznikowy, dozownik
   dwustrumieniowy
R байпасный дозатор, обходный дозатор

cadmium ratio
The ratio of the activity of an irradiated
material (in a nuclear reactor) to that
of the material protected by cadmium foil
0.75 mm thick.
D Kadmium-Verhältnis
F rapport cadmique, relation de cadmium
P stosunek kadmowy
R кадмиевое отношение

calcium hardness
The hardness of water due to the presence
of calcium ions therein.
D Kalkhärte
F dureté calcicité
P twardość wapniowa
R кальциевая жёсткость

calibration error → scale error

calibration of weights
Comparison of weights to be tested with
reference ones, or at least with one reference
weight.
D Prüfung der Gewichte
F vérification des poids
P sprawdzanie odważników
R проверка гирь

calomel electrode, mercury-mercurous chloride
electrode
An electrode of the second kind, consisting of
mercury covered with a paste of mercury(I)
chloride and mercury, in contact with
a chloride solution of specified concentration
and saturated with mercurous chloride; used
as a reference electrode. The half-cell may be
represented as $Cl^-_{(aq)}|Hg_2Cl_{2_{(s)}}, Hg_{(1)}$.
D Kalomelelektrode, Quecksilber-Quecksilber(I)-
   -chlorid-Elektrode
F électrode au calomel
P elektroda kalomelowa
R каломельный электрод

canonical analysis
A generalization of multiple regression
methods used in the investigation of

relations between a set of explained
(dependent) variables and a set of explanatory
variables.

D ...
F analyse canonique
P analiza kanoniczna
R канонический анализ

capacity (*of a balance*)
The maximum safe load claimed by the
manufacturer.

D Tragfähigkeit
F capacité
P obciążenie maksymalne, udźwig wagi,
   nośność wagi
R предельная нагрузка

capacity current, charging current
(*in electrochemistry*)
The current corresponding to the charging
of a condenser consisting of the electric double
layer generated by the imposed change of the
electrode potential, or by change of the
indicator electrode surface.

D Kapazitätsstrom, Ladestrom, Ladungsstrom,
   Kondensatorstrom
F courant capacitif, courant de charge
P prąd pojemnościowy
R ёмкостный ток, ток заряжения,
   конденсаторный ток

capacity factor → mass distribution ratio

capacity ratio → mass distribution ratio

capillary column
A chromatographic column of capillary
dimensions with internal diameter usually
smaller than 1 mm.

D Kapillarsäule, Trennkapillare
F colonne capillaire
P kolumna kapilarna
R капиллярная колонка

capillary tubes (*in elemental analysis*)
Quartz or hard glass capillaries, at one
end sealed and drawn into a thin rod, used for
weighing liquid samples and subsequent
transferring them to a combustion tube.

D Kapillaren
F tubes capillaires
P mikrokapilarki
R капилляры

carbonate hardness
The hardness of water due to the presence
of calcium and magnesium bicarbonates,
carbonates and hydroxides dissolved therein.

D Carbonathärte
F dureté temporaire
P twardość węglanowa
R карбонатная жёсткость

carbonization of filter paper → charring of
filter paper

carbon-paste electrode
An electrode made from a carbon paste
(a mixture of a spectrally pure graphite with,
e.g. paraffin oil) placed in a receptacle made of
Teflon or other insulating material; used as the
indicator or working electrode.

D Kohlepasteelektrode
F ...
P elektroda z pasty węglowej
R электрод из угольной пасты

carbon-tube furnace → graphite-tube furnace

Carbowax
The trade name for polyethylene glycols of
molecular mass ranging from 200 to 20 000,
used as a liquid stationary phase in gas
chromatography.
Remark: The number included in the trade name of
Carbowax, e.g. Carbowax 1500, relates to the average
molecular mass.

D Carbowax
F Carbowax
P Carbowax
R карбовакс

Carius method
A method of mineralization of organic
compounds in which a sample is oxidized with
concentrated nitric acid in a thick-walled
sealed tube at a 250-300°C.

D Carius-Methode, Carius-Aufschluß
F méthode de Carius
P metoda Cariusa
R метод Кариуса

carrier
A substance that carries trace amounts
of a specified substance through a chemical or
physical process.

D Trägersubstanz, Träger
F entraîneur, porteur
P nośnik
R носитель

carrier → solid support

carrier distillation (*in spectrochemical analysis*)
Fractional distillation of a sample in the
presence of a spectrochemical carrier.

D Trägerdestillation
F distillation avec un entraîneur
P destylacja nośnikowa
R фракционная дистилляция с носителем

carrier gas, eluent gas
A gas used to elute a sample from the
chromatographic bed in gas chromatography.

D Trägergas
F gaz vecteur, gaz porteur

P gaz nośny, gaz wymywający
R газ-носитель, газ-элюент

catalytic combustion
A method of quantitative elemental analysis involving the combustion of a sample in oxygen atmosphere at red heat temperature in the presence of platinum; the method is used for determination of sulfur and halogens in organic compounds.

D katalytische Verbrennung
F combustion catalitique
P spalanie katalityczne
R каталитическое сожжение

catalytic current (*in electrochemistry*)
The current corresponding to the catalytic reaction taking place in the vicinity of the electrode surface, resulting in the regeneration of the substrate of an electrode reaction (from the product of the reaction).

D katalytischer Strom
F courant catalytique
P prąd katalityczny
R каталитический ток

catalytic reaction
A chemical reaction whose rate is modified by a catalyst.

D katalytische Reaktion
F réaction catalytique
P reakcja katalityczna
R каталитическая реакция

cathode
An electrode through which a net negative current flows. The chemical reaction taking place at the cathode is reduction.

D Kathode
F cathode
P katoda
R катод

cathode layer (*in arc spectroscopy*)
A region of high vapour concentration near the cathode in the direct current arc.

D Katodenschicht
F couche cathodique
P warstwa katodowa
R прикатодный слой

cathode-ray polarography → single-sweep polarography

cathodoluminescence, CL
An electromagnetic radiation in the visible, infrared or ultraviolet spectral region emitted when an electron beam strikes a solid surface. This interaction creates electron-hole pairs; the energy of their recombination is emitted in the form of light.

D Katodolumineszenz
F cathodoluminescence
P katodoluminescencja
R катодолюминесценция

catholyte
The electrolyte surrounding the cathode, usually separated from the bulk of the electrolyte by a semi-permeable partition.

D Kat(h)olyt
F catholyte
P katolit
R католит

cation → positive ion

cation exchange
The process of exchanging cations between a solution and a cation exchanger.

D Kationenaustausch
F échange de cations
P wymiana kationowa
R катионный обмен

cation exchanger
An ion exchanger containing cations that can be exchanged for other cations present in the solution.

Remark: The term cation-exchange resin may be used in the case of solid organic polymers. In Polish and Russian nomenclature the terms kationit and катионит are used to denote the solid cation exchanger.

D Kationenaustauscher
F échangeur de cations
P wymieniacz kationowy, kationit
R катионообменник, катионит

cation-exchange resin *See* cation exchanger

D Kationenaustauscherharz
F résine échangeuse de cations
P żywica kationowymienna
R катионообменная смола

cavity resonator, resonant cavity (*in EPR spectroscopy*)
A spectroscopic absorption cell in the form of a cylindrical or rectangular cavity in which the quartz microtube containing the sample under examination is placed.

D Resonator, Hohlraumresonator
F cavité résonante
P wnęka rezonansowa, rezonator
R резонансная полость, резонатор

CC → column chromatography

CD → circular dichroism

Celite
The trade name for diatomaceous earth fused with a small amount of sodium carbonate. In gas chromatography Celite is used as a solid support or an adsorbent having a large specific surface area, high porosity, but only weak adsorptive properties.

D Celite
F Célite
P Celit
R целит

cell → electrochemical cell

cell constant (of a conductivity cell), k
Ratio of the distance between two parallel cell electrodes of equal area to the electrode area. The constant is specific for a particular cell and is used for comparative measurements of electrolytic conductance. The constant is determined from the equation $k = \varkappa R$, where R — resistance of the cell filled with an electrolyte of known electrolytic conductance $\varkappa$. SI unit: reciprocal of metre, m^{-1}, or cm^{-1}.
D Zellkonstante, Widerstandskonstante, Widerstandskapazität
F constante de cellule
P stała naczynka konduktometrycznego, pojemność oporowa naczynka
R постоянная (кондуктометрической) ячейки

cell without transference, cell without transport
An electrochemical cell in which two reversible galvanic cells whose electrodes are reversible with respect to each of the ions constituting the electrolyte are combined in opposition; e.g.

$Pt_{(s)}, H_{2(g)}|HCL_{(aq)}|AgCl_{(s)}, Ag_{(s)}|Pt'_{(s)}$

D überführungsfreie Zelle, Zelle ohne Überführung, Kette ohne Überführung
F pile sans transport
P ogniwo bez przenoszenia
R цепь без переноса

cell without transport → cell without transference

cell with transference, cell with transport
The galvanic cell in which two different electrolytic solutions A and B are in contact, e.g. by means of a porous partition through which the electric charge is transferred; e.g.
$Pt_{(s)}|H_{2(g)}|$ solution A; solution $B|H_{2(g)}|Pt_{(s)}$.
Remark: Solutions A and B may differ in respect of type of a solvent, solute and concentration.
D Zelle mit Überführung, Kette mit Überführung
F pile avec transport
P ogniwo z przenoszeniem
R (электрохимическая) цепь с переносом

cell with transport → cell with transference

CELS → transmission energy loss spectroscopy

centigram method *See* mesoanalysis
D Centigramm-Methode
F méthode centigrammique
P metoda centygramowa
R полумикрометод

centrifuging
The separation of two phases (solid phase —
— liquid phase, two liquid phases) effected by centrifugal force.

D Zentrifugieren
F centrifugation
P odwirowanie
R центрифугирование

Cerenkov radiation
Polarized electromagnetic radiation (in the visible and ultraviolet range) generated when a charged particle crosses a medium with a velocity greater than that of light in that medium.
D Cerenkov-Strahlung, Tscherenkow-Strahlung
F rayonnement de Cerenkov
P promieniowanie Czerenkowa
R излучение Вавилова-Черенкова, черенковское излучение

cerimetric titration
Oxidimetric titration in which the reducing agent is treated with a standard solution of cerium(IV) salt.
D cerimetrische Titration
F titrage cérimétrique
P miareczkowanie cerometryczne
R цериметрическое титрование

cerimetry
The titrimetric determination of a substance by means of a standard ceric salt solution (e.g. ceric sulfate).
D Cerimetrie
F cérimétrie
P cerometria
R церометрия

certified burette
A burette officially calibrated, bearing a legalization mark or provided with an appropriate certificate.
D Bürette amtlich geprüft
F burette vérifiée avec certificat d'étalonnage
P biureta legalizowana
R ...

chamber saturation (*in planar chromatography*)
The uniform distribution of the eluent vapour throughout the chamber until reaching equilibrium prior to the start of chromatographic separation.
D Kammersättigung, Sättigung der Kammer
F saturation de la chambre (de développement)
P nasycenie komory
R насыщение камеры

characteristic curve → emulsion characteristic curve

characteristic energy loss spectroscopy → transmission energy loss spectroscopy

characteristic frequency (*in spectroscopy*)
The frequency of oscillation taking place in a part of the molecule, relatively independent of the vibrations occurring at other sites of same molecule.

D charakteristische Frequenz
F fréquence caractéristique
P częstość charakterystyczna, częstość grupowa
R характеристическая частота

characteristic X-radiation, characteristic X-rays
X-radiation consisting of discrete wavelengths
which are characteristic for the emitting
element.
D charakteristische Röntgenstrahlung
F rayonnement X caractéristique
P promieniowanie rentgenowskie
  charakterystyczne
R характеристическое рентгеновское излучение

characteristic X-rays → characteristic
X-radiation

charged-particle activation
Activation effected by a stream of protons,
deuterons or α-particles.
D Aktivierung mit Ladungsträger
F activation par porteurs électrisés
P aktywacja za pomocą cząstek naładowanych
R активация заряженными частицами

charged-particle activation analysis
Activation analysis in which a stream of
charged particles, e.g. protons or α-particles, is
used for activation.
D Aktivierungsanalyse mit geladenen
  Kernteilchen, Aktivierungsanalyse mit
  schnellen Ionen
F analyse par activation aux particules
  chargées
P analiza aktywacyjna za pomocą cząstek
  naładowanych
R активационный анализ с применением
  заряженных частиц

charge-step polarography → incremental-charge
polarography

charge-transfer band, CT-band (*in an electronic
spectrum*)
The absorption band which corresponds to
the transfer of an electron from one atom
(or group) to another.
D Ladungsaustauschband, CT-Band
F bande de transfert de charge
P pasmo przeniesienia ładunku, pasmo CT
R полоса переноса заряда

charge-transfer coefficient, (electrochemical)
transfer coefficient (*anodic or cathodic*)
A coefficient equivalent to the slope of
the curve presenting the variation of the
logarithm of the partial current (anodic I_a
or cathodic I_c) with the potential E of the
electrode

$$\alpha_c = -\frac{vRT}{nF}\left(\frac{\partial \ln I_c}{\partial E}\right)_{T,p,c_i\ldots}$$

or

$$\alpha_a = \frac{vRT}{nF}\left(\frac{\partial \ln I_a}{\partial E}\right)_{T,p,c_i\ldots}$$

where n — number of electrons exchanged in
the overall reaction, T — thermodynamic
temperature, v — the stoichiometric number
of the transfer reaction.
D (elektrochemischer) Durchtrittsfaktor
F coefficient de transfert
P współczynnik przeniesienia ładunku,
  współczynnik przejścia
R коэффициент переноса заряда

charge-transfer complex → donor-acceptor
complex

charge-transfer overpotential → activation
overpotential

charge-transfer step
An elementary step in which transfer of an
electric charge (electron or ion) across the
electrode — electrolyte interface takes place.
D Ladungs-Durchtrittsreaktion,
  Ladungs-Transferreaktion
F processus de transfert de charge
P etap przeniesienia ładunku, reakcja wymiany
  ładunku
R стадия процесса разряда, стадия разряда-
  ионизации, реакция переноса заряда

charging current → capacity current

charging effect
Change in the surface potential due to the
charging of the sample, caused by the escape
of electrons or ions under bombardment
of a solid with a primary beam of photons,
electrons or ions.
D Aufladungs-Effekt
F effet de charge
P efekt ładowania elektrycznego
R эффект заряжения

charring of filter paper, carbonization of
filter paper
The gentle ignition of a filter paper, followed
by its incineration.
D Verkohlung des Filters
F carbonisation d'un filtre
P zwęglanie sączka
R обугливание фильтра

check test
The preliminary checking of the apparatus,
or the test of applicability of a method,
carried out with reference substances.
D Kontrollprobe
F essai de contrôle, examen de contrôle
P próba kontrolna
R контрольная проба

chelate (compound)
A complex in which the metal is incorporated
into a ring formed by coordination of the
polydentate ligand.

D Chelat, Chelatkomplex, Chelatverbindung, Metallchelat
F chélate, molécule chélatée
P kompleks chelatowy, związek chelatowy, związek kleszczowy
R циклический комплекс, хелат

chelate ring
Heterocyclic ring, usually five- or sixmembered, formed by coordination of the central ion (atom) of a metal with a polydentate ligand.
D Chelatring
F cycle chélaté
P pierścień chelatowy
R хелатное кольцо

chelating agent
An organic compound forming chelate complexes with cations.
D Chelatligand
F agent chélatant
P (od)czynnik chelatujący
R хелатообразующий лиганд

chelating ion exchanger
A selective ion exchanger containing chelating functional groups, e.g. $-CH_2,-N(CH_2COOH)_2$, showing high affinity towards certain metal ions.
Remark: Practically available chelating ion exchangers are of organic type and are also known as chelating resins.
D Chelat-Ionenaustauscher
F échangeur d'ions chélateur
P wymieniacz jonowy chelatujący
R хелатный ионообменник

chelating resin *See* chelating ion exchanger
D Chelatharz
F résine chélatante
P żywica chelatująca
R хелатная (ионообменная) смола

chemical adsorption → chemisorption

chemical analysis
A sequence of actions carried out in order to ascertain the qualitative and/or quantitative composition of the sample examined.
D chemische Analyse
F analyse chimique
P analiza chemiczna
R химический анализ

chemical balance
A laboratory balance of weighing capacity 250 to 5000 g and sensitivity within 2×10^{-3} and 2×10^{-2} g.
D technische Waage
F trébuchet
P waga techniczna
R технические весы

chemical cell
A galvanic cell in which chemical energy is converted into electrical energy.

D chemische Zelle
F pile chimique
P ogniwo chemiczne
R химическая цепь

chemical coulometer, coulombmeter
An electrolytic cell used for measuring the quantity of electrical charge (number of coulombs) which has passed trough an electrical circuit by the amount of material deposited electrochemically. The electrochemical reaction must occur with 100% current efficiency and the total quantity of electricity passed can be calculated from the increase in mass of the cathode and the electrochemical equivalent of the metal.
D chemisches Coulometer, Coulombmeter
F coulomètre chimique
P kulometr chemiczny
R химический кулонометр

chemical effect (*in X-ray fluorescence analysis*)
Change of the characteristic X-ray wavelength of an element due to the oxidation state of its atom, or depending on its particular environment in the chemical compound (the kind and number of atoms bonded to it chemically). The effect may be detected only by instruments of very high resolution power.
D chemischer Effekt
F effet chimique
P efekt chemiczny
R химический эффект

chemical electroanalysis → electroanalysis

chemical equivalent
The number of mass units of a reagent required by the stoichiometry of a given chemical reaction.
Remark: Neither the concept, nor the term have been accepted for use by the SI system.
D chemisches Äquivalent
F équivalent chimique
P równoważnik chemiczny
R химический эквивалент

chemical ionization, CI
Ionization resulting from the collision of gaseous molecules with positively or negatively charged ions.
Remark: When a positive ion is produced in the chemical ionization, the term may be used without qualification. When a negative ion is obtained, the term negative ion chemical ionization should be used
D chemische Ionisation
F ionisation chimique
P jonizacja chemiczna
R химическая ионизация

chemically bonded phase → chemically bonded stationary phase

chemically bonded (stationary) phase, bonded stationary phase (*in chromatography*)

A variety of column packing materials in which the active stationary phase is an organic substance chemically bonded to the surface of an inactive solid support. Usually it consists of a monomolecular layer of bonded material with polar groups (brush type) or of a thin layer of bonded polymer (polymeric type).

D chemisch gebundene stationäre Phase
F phase stationnaire greffée
P faza (nieruchoma) związana chemicznie
R химически связанная фаза

chemically pure reagent, CP reagent
A reagent of purity higher than analytical. Chemically pure reagents are provided with certificates specifying the kind and maximum possible contents of contaminants.

D chemisch reines Reagens
F réactif chimiquement pur
P odczynnik chemicznie czysty
R химически чистый реактив

chemical potential (of species B in phase α**),** μ_B^α
The intensive thermodynamic quantity

$$\mu_B^\alpha = \left(\frac{\partial U}{\partial n_B^\alpha}\right)_{S,V,n_c} = \left(\frac{\partial H}{\partial n_B^\alpha}\right)_{S,p,n_c} =$$
$$= \left(\frac{\partial F}{\partial n_B^\alpha}\right)_{T,V,n_c} = \left(\frac{\partial G}{\partial n_B^\alpha}\right)_{T,p,n_c}$$

where n_B^α — number of moles of species B in phase α, n_c — number of moles of the remaining components of phase α, U — internal energy of phase α, H — enthalpy of phase α, F — free energy of phase α, S — entropy of phase α, V — volume of phase α, T — thermodynamic temperature, p — pressure. SI unit: joule per mole, J mol^{-1}.

D chemisches Potential (der Substanz B)
F potentiel chimique (du constituant B dans la phase α)
P potencjał chemiczny (składnika B w fazie α)
R химический потенциал (вещества B в фазе α)

chemical shift (*in NMR spectroscopy*), $\Delta\nu$
The difference between the positions of the absorption signal of a measured nucleus and the absorption signal of the substance taken as a reference. This difference may be measured in hertz, or in δ-scale. Chemical shift values for different nuclei depend on their chemical and magnetic environments.

D chemische Verschiebung
F déplacement chimique
P przesunięcie chemiczne
R химический сдвиг

chemical sputtering → reactive sputtering

chemiionization
Ionization resulting from the collisions of gaseous molecules with other internally excited gaseous molecules or molecular species, e.g. metastable helium atoms.

D Chemiionisation
F chimiionisation
P chemijonizacja
R хемиионизация

chemiluminescence
Luminescence effected by a chemical reaction. Part of the energy of the chemical reaction is absorbed producing excited atoms and molecules which decay with emission of light.

D Chemilumineszenz
F chimi(o)luminescence
P chemiluminescencja
R хемилюминесценция

chemiluminescent indicator
An indicator (acid-base or other) exhibiting chemiluminescence or a quenching of chemiluminescence at or near the equivalence-point.

D Chemilumineszenzindikator
F indicateur par chimiluminescence, indicateur chimiluminescent
P wskaźnik chemiluminescencyjny
R хемилюминесцентный индикатор

chemisorption, chemical adsorption
The adsorption in which the forces involved are valence forces of the same kind as those operating in the formation of chemical compounds.

D Chemisorption, chemische Adsorption
F chimisorption, adsorption chimique
P chemisorpcja, adsorpcja chemiczna
R хемосорбция, химическая адсорбция

chi-squared distribution, χ^2-distribution
Probability distribution of a continuous random variable X with probability density function given by the formula

$$f(x) = \begin{cases} \dfrac{1}{2^{n/2}\,\Gamma\left(\dfrac{n}{2}\right)}\,e^{-x/2}\,x^{n/2-1} & \text{for } x > 0 \\ 0 & \text{for } x \leqslant 0 \end{cases}$$

where n — number of degrees of freedom. Chi-squared distribution finds application, e.g. in tests for goodness of fit.

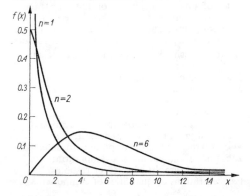

D χ^2-Verteilung
F distribution de χ^2
P rozkład χ^2
R распределение χ^2

chi-squared test for goodness of fit
The test for goodness of fit used for verification of hypothesis about the consistency of a sample with a given distribution of the population; elements of the sample are classified on mutually exclusive categories and the observed numbers of elements in each category are compared with theoretical numbers in categories (calculated on the base of the given distribution) by use of the following statistic having the chi-squared distribution

$$\chi^2 = \sum_{i=1}^{r} (n_i - np_i)^2 / np_i$$

where n_i — observed number of elements in ith category, $n = \sum n_i$, p_i — probability that the random variable will assume values obtaining values belonging to ith category $(i = 1, 2, ..., r)$ of the random variable X; the calculated value of χ^2 is compared with tabulated values.

D χ^2-Anpassungstest
F test χ^2 de validité de l'ajustement
P test zgodności χ^2
R критерий согласия χ^2

chromathermography, temperature gradient chromatography
A chromatographic technique in which a temperature gradient is produced, e.g. by slowly passing an external source of heat down a column in the direction of flow of the mobile phase.

D Thermochromatographie, nichtstationäre Temperaturgradientenchromatographie
F chromathermographie, thermochromatographie
P chromatermografia
R хроматермография

chromatic aberration
An aberration of an optical system containing refractive elements due to that the refractive index of a material depends on the wavelength of light, so different colours are focused at different points. The images formed in that way are surrounded by the fringes of spectral colours.

D chromatische Aberration, Farbenabweichung
F aberration chromatique
P aberracja chromatyczna
R хроматическая аберрация

chromatogram
The results of chromatographic separations presented in the form of: (a) a sheet, strip, disc of paper or thin-layer plate with detected zones of substances under examination; (b) a record (graph) of detector response (concentrations or quantities of components

either emerging from the chromatographic column or separated on the chromatographic plate or paper) against time or volume of the effluent; (c) a packed column after separation and detection (visualization) of the separated zones.

D Chromatogramm
F chromatogramme
P chromatogram
R хроматограмма

chromatograph (*noun*)
An apparatus for carrying out chromatographic separations; usually the assembly for column chromatography, equipped with a device for automatic recording of the results, e.g. gas chromatograph, liquid chromatograph.

D Chromatograph
F chromatographe
P chromatograf
R хроматограф

chromatograph (*verb*)
To separate by chromatography.

D chromatographieren
F chromatographier
P chromatografować
R хроматографировать

chromatographic analysis
Qualitative or quantitative analysis by means of chromatography.

D chromatographische Analyse
F analyse chromatographique
P analiza chromatograficzna
R хроматографический анализ

chromatographic bed
Any of the different forms in which the stationary phase may be used (e.g. packed in a column, spread as a layer on a support plate or distributed as a film on a solid support).

D chromatografisches Bett
F lit chromatographique
P złoże chromatograficzne
R хроматографический слой

chromatographic chamber, developing chamber, chromatographic tank, development tank (*in thin-layer and paper chromatography*)
Tightly closed container of any form or size in which the chromatograms are developed on a paper sheet or thin-layer plate. Chambers of large volumes saturated or non-saturated with eluent vapours and flat sandwich chambers are most commonly used.

D Trennkammer, chromatographische Kammer, Trogkammer, Entwicklungskammer
F cuve de développement, cuve à chromatographie, chambre de développement
P komora chromatograficzna
R хроматографическая камера, камера для проявления хроматограмм

chromatographic column, column
A tube (e.g. of glass, stainless steel or plastic)
generally cylindrical, containing the
stationary phase and provided with an
inlet and outlet for the mobile phase.
D chromatographische Säule,
Chromatographiesäule, Trennsäule
F colonne chromatographique
P kolumna chromatograficzna
R хроматографическая колонка

chromatographic support → solid support

chromatographic tank → chromatographic
chamber

chromatography
A separation method in which the different
affinities of the sample components towards
the two phases, one of which is immobile
(stationary phase) and the other moves
(mobile phase) are taken advantage of.
D Chromatographie
F chromatographie
P chromatografia
R хроматография

chromatometry
The determination of a substance by titration
with a standard solution of a dichromate
(e.g. potassium dichromate).
D Chromatometrie, Dichromatometrie
F chromométrie
P chromianometria
R хроматометрия

chromatopolarography
An electroanalytical method which involves
the chromatographic separation of components
of a mixture and polarographic determination
of the components in the effluent.
D Chromatopolarographie
F chromatopolarographie
P chromatopolarografia
R хроматополярография

chromophore, chromophoric group
An atom, a group of atoms, or a conjugated
system of multiple bonds in an organic
compound exhibiting characteristic absorption
in the visible and near ultraviolet regions
which is practically independent of the
remaining components of the molecule.
In inorganic compounds, the chromophoric
groups may include several atoms, e.g.
permanganate or dichromate groups, or may
be a single atom with electron shell,
vacancies (rare earths or transition elements).
D Chromophor, chromophore Gruppe
F chromophore, groupe chromophique
P chromofor, grupa chromoforowa
R хромофор, хромофорная группа

chromophoric group → chromophore

Chromosorb
The trade name for series of chromatographic
solid supports and sorbents. The series of
Chromosorbs numbered from 101 to 108
consists of a range of porous, synthetic
polymers based on either a styrene-
divinylbenzene copolymer or other aromatic
or aliphatic resins; Chromosorbs A, G, P
and W are different modifications of
diatomaceous earth, Chromosorb T is
a porous Teflon with an inert surface.
D Chromosorb
F Chromosorb
P Chromosorb
R хромосорб

chronoamperometry
A number of eletrochemical techniques in
which a given constant potential E is applied
upon a stationary indicator electrode immersed
in an unstirred solution and the current I flowing
through this electrode is measured as a function
of the electrolysis time t.

Remark: The term polarographic chronoamperometry is
recommended to denote the technique in which
measurements are made during the lifetime of a single
drop on a dropping electrode.

 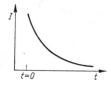

E_0 − open circuit
electrode potential

D Chronoamperometrie
F chronoampérométrie
P chronoamperometria
R хроноамперометрия

chronoamperometry with linear potential sweep,
stationary-electrode voltammetry, linear-sweep
voltammetry
The voltammetry in the conditions of
a diffusion mass transport to/from an indicator
electrode which has a non-renewable surface,
while the potential E of the electrode is changed
linearly with time t.

 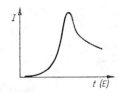

D Chronoamperometrie mit linearem
Potentialanstieg, Voltammetrie an stationärer
Elektrode, Voltammetrie mit linearem
Spannungsanstieg

F chronoampérométrie avec balayage linéaire
du potentiel, voltampérométrie à une
électrode stationnaire, voltampérométrie
à balayage linéaire
P woltamperometria z liniową zmianą
potecjału, chronowoltamperometria
R хроноамперометрия с линейной развёрткой
потенциала, вольтамперометрия с линейной
развёрткой потенциала, вольтамперометрия
со стационарным электродом

chronocoulometry
A number of electrochemical techniques in
which a known constant potential E is imposed
upon a stationary indicator electrode immersed
in an unstirred solution, and the electric charge
Q which passed through this electrode is
measured as a function of the electrolysis time t.

Remark: The commonly used term potential-step
chronocoulometry is not recommended.

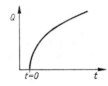

E_0 — open circuit
electrode potential

D Chronocoulometrie,
Potentialstufenchronocoulometrie
F chronocoulométrie (à échelon de potentiel)
P chronokulometria, chronokulometria ze
skokiem potencjału
R хронокулонометрия, хронокулонометрия
со ступенчатым изменением потенциала

chronopotentiometry
A number of electrochemical techniques in
which a constant current I is imposed on the
stationary indicator electrode submerged in
an unstirred solution, and the potential E of
this electrode is measured as a function of the
electrolysis time t.

Remark: The chronopotentiometric technique which
employs the rotating disk electrode is also known.

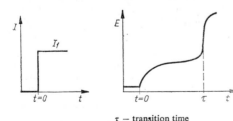

τ — transition time

D Chronopotentiometrie
F chronopotentiométrie, transitométrie
P chronopotencjometria
R хронопотенциометрия

chronopotentiometry with linearly increasing
current → chronopotentiometry with linear
current sweep

chronopotentiometry with linear current sweep,
chronopotentiometry with linearly increasing
current
Chronopotentiometry in which a current I
varying linearly with time, is imposed on the
indicator electrode and the potential E of
this electrode is measured as a function of the
electrolysis time t.

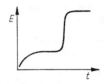

D Chronopotentiometrie mit linearem
Stromanstieg, Chronopotentiometrie mit
linearem Strom-Zeit-Verlauf
F chronopotentiométrie avec balayage linéaire
du courant
P chronopotencjometria z liniowo
narastającym prądem
R хронопотенциометрия с линейной
развёрткой тока

chronopotentiometry with superimposed
alternating current
Chronopotentiometry in which the current
(being the sum of the alternating current of
a small amplitude, such as a sine-wave one,
and the direct current I_{dc}) is imposed on the
indicator electrode and the potential E of
this electrode is measured as a function of
the electrolysis time t.

Remark: In a narrow sense, the term is often used to
denote the chronopotentiometry with superimposed
sinusoidal alternating current.

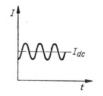

chronopotentiometry with
superimposed alternating
current — typical response
curve

D Chronopotentiometrie mit überlagertem
Wechselstrom
F chronopotentiométrie à courant alternatif
surimposé
P chronopotencjometria z nałożonym prądem
zmiennym
R хронопотенциометрия с наложением
переменного тока

Chugaev reagent
One per cent alcoholic solution of
dimethylglyoxime used for detection and
determination of nickél (pink precipitate), or
palladium (yellow precipitate).
D Tschugajews Reagens
F réactif de Tschugaeff
P odczynnik Czugajewa
R реактив Чугаева

CI → chemical ionization

circular birefringence
The difference in indices of refraction for
right- and left-circularly polarized light.
D zirkulare Doppelbrechung
F biréfringence circulaire
P dwójłomność kołowa
R двойное лучепреломление по кругу

circular chromatography, radial development,
ring chromatography
A technique in paper and thin-layer
chromatography, in which the spot of the
sample to be separated is applied at the centre
of a paper circle or square plate and the
chromatogram is developed from the centre
to the edges.
D Zirkularchromatographie,
   Ringchromatographie, Circular-Entwicklung
F chromatographie circulaire, développement
   radial
P chromatografia krążkowa
R круговая хроматография, радиальная
   хроматография

circular dichroism, CD
A phenomenon in which circularly polarized
light, after traversing an optically active
medium, emerges elliptically polarized. This is
due to differences in the molar absorption
coefficients for left- and right-handed
circularly polarized light.
D Circulardichroismus
F dichroïsme circulaire
P dichroizm kołowy
R круговой дихроизм

circularly polarized light *See* circular
polarization
D zirkular polarisiertes Licht
F lumière polarisée circulairement, lumière
   à polarisation circulaire
P światło spolaryzowane kołowo
R циркулярно поляризованный свет

d **circularly polarized light** → right-handed
circularly polarized light

l-**circularly polarized light** → left-handed
circularly polarized light

circular polarization
Polarization of light that occurs when the
electric field vector of a ray of light describes,

either clockwise or counterclockwise, at any
point in space a cirle about the direction of
propagation of the light; the light is then said
to be circularly polarized.
D zirkulare Polarisation, Zirkularpolarisation
F polarisation circulaire
P polaryzacja kołowa
R круговая поляризация

CL → cathodoluminescence

Clark oxygen electrode → Clark oxygen-
sensing probe

Clark oxygen-sensing probe, oxygen sensor,
Clark oxygen electrode, oxygen polarographic
probe
A gas sensing membrane electrode in which
gaseous oxygen diffuses through a gas-
permeable membrane of Teflon or a silicone.
The membrane isolates the platinum (or gold)
cathode, silver anode, and the reference
solution (saturated solution of potassium
chloride) from the environment. When an
external potential is applied across the cell,
oxygen is reduced at the cathode surface with
the formation of the oxidation product
(silver oxide) at the anode, and the resultant
current is proportional to the amount of
oxygen reduced.
D ...
F ...
P elektroda tlenowa Clarka, czujnik tlenowy
R кислородный зонд

Clark's method
A method of determination of water hardness
involving titration of the water sample with
a standard solution of potassium soap to the
appearance of permanent lather on stirring.
D Clarksche Methode
F méthode de Clark
P metoda Clarka
R (*определение жёсткости воды мыльным*
   раствором)

classical packed column → packed column

clock time
The time of measurement as indicated by a clock,
without taking account of the intervals of
inactivity of the measuring device, i.e. without
excluding the dead time of the instrument.
D Uhrzeit
F temps réel
P czas zegarowy
R часовое время

CMA → cylindrical mirror analyser

coagulation, flocculation
The formation and growth of aggregates
ultimately leading to phase separation on
a macroscopic scale.

D Koagulation, Flockung
F coagulation, floculation
P koagulacja, flokulacja
R коагуляция, флокуляция

coagulation concentration
The smallest concentration of an electrolyte,
usually in millimoles per litre, that causes a
quick visible coagulation of a sol.

D Flockungskonzentration, Flockungs
   (schwellen) wert
F valeur de coagulation
P wartość koagulacyjna, wartość progowa
   koagulacji, stężenie koagulujące
R порог коагуляции, число коагуляции

coated wire ion-selective electrode
A type of an ion-selective electrode in which
a metallic conductor, usually platinum wire
or graphite rod, is coated with a thin polymeric
layer (PVC, epoxy resin) in which an
electroactive substance is incorporated.

D Drahtüberzug-Elektrode
F ...
P elektroda jonoselektywna z powleczonym
   drutem
R проволочный электрод с чувствительным
   покрытием, проволочный электрод
   с электродноактивным покрытием

coefficient of correlation, ρ
Correlation between two random variables as
given by the formula

$$\rho = \frac{\text{cov}(X, Y)}{\sigma_X \, \sigma_Y}$$

where cov (X, Y) − covariance of the random
variables X and Y, σ_X, σ_Y − standard
deviations. The correlation coefficient assumes
values from the interval $[-1, 1]$.

D Korrelationskoeffizient
F coefficient de corrélation
P współczynnik korelacji
R коэффициент корреляции

coefficient of multiple correlation, R
A measure of the ccorelation between the observed
(measured) values and their estimates found on
the basis of the multifactor model; the coefficient
R is given by the formula

$$R = 1 - \frac{\sum\limits_{i=1}^{n} (\hat{y}_i - y_i)^2}{\sum\limits_{i=1}^{n} (y_i - \bar{y})^2}$$

where n − number of observations, y_i − ith
observed value, $\hat{y}_i$ − value estimated from the
model, $\bar{y}$ − mean (arithmetic) of observed values.

D multipler Korrelationskoeffizient
F coefficient de corrélation multiple
P współczynnik korelacji wielokrotnej,
   współczynnik korelacji wielowymiarowej
R множественный коэффициент корреляции

coefficient of variation, CV, relative standard deviation, RSD
A quantity defined by the formula

$$v = \frac{\sigma}{\text{E}(X)}$$

where σ − standard deviation,
E (X) − expected value of the random
variable X. Coefficient of variation is a relative
measure of dispersion; used mainly for
comparing the dispersion of distributions
having different expected values.

D Variationskoeffizient, relative
   Standardabweichung
F coefficient de variabilité, écart-type relatif
P współczynnik zmienności, odchylenie
   standardowe względne
R коэффициент вариации

coherent scattering (*of photons*)
The scattering of X- or γ-ray photons by
electrons of atoms where the scattered
beam has the same energy (and hence the same
wavelength) as the incident beam. There is
a relationship between the phase of
the scattered beam and that of the incident
beam.

D kohärente Streuung
F diffusion cohérente
P rozpraszanie spójne
R когерентное рассеяние

coil of Nichrome wire, Nichrome helix, Nichrome source
A nickel and chromium alloy wire coil
(unsupported or wound on a ceramic support)
raised to incandescence by resistive heating;
employed as a source of infrared radiation in
the range of 2.5 to 50 µm. Its operating
temperature is about 350 K.

D Nichrom-Spirale
F filament de nichrome
P spirala z nichromu
R нихромовая спираль

coincidence technique
A method for measuring ionizing radiation in
which a given type of radiation is registered
only then, when its emission occurs
simultaneously with another radiation
produced by the same nucleus (or by secondary
processes).

D Koinzidenzverfahren
F méthode de comptage de coïncidences
P metoda koincydencji
R метод совпадений

co-ions (*in ion exchange*)
The mobile ions (able to diffuse freely) with
a charge of the same sign as the fixed ions.

D Coionen
F co-ions
P kojony, współjony
R коионы

cold neutrons
Neutrons of a very low energy.
D kalte Neutronen, unterthermische
  Neutronen
F neutrons froids
P neutrony zimne
R холодные нейтроны, подтепловые
  нейтроны

collection
The removal of a micro- or macrocomponent
from solution by coprecipitation with an
intentionally added substance or by adding a
solid that adsorbs or traps the component to
be removed.
D Sammeln
F entraînement, collection
P strącanie z nośnikiem, zbieranie
R собирание

collimation
Transformation of a beam of electromagnetic
or corpuscular radiation into a set of parallel
rays.
D Bündelung, Kollimation
F collimation
P kolimacja
R коллимация

collimator
The optical device producing a parallel beam
of electromagnetic or corpuscular radiation.
D Kollimator
F collimateur
P kolimator
R коллиматор

colloid precipitate, amorphous precipitate
A precipitate consisting of particles with no
lattice ordering which generally yields a
colloidal solution, (sol) upon dissolving.
D kolloider Niederschlag
F précipité colloïdal
P osad koloidalny, osad bezpostaciowy
R коллоидный осадок

colorimeter
An instrument for the measurement of the
selective absorption of radiation in a given
medium in the visible spectral region.
D Kolorimeter, Tintometer
F colorimètre
P kolorymetr
R колориметр

colorimetric analysis, colorimetry
A method of analysis in which the quantity
of a substance is determined from the intensity
of colour produced by a reagent, by comparing
it with a reference solution.
D kolorimetrische Analyse, Kolorimetrie
F analyse colorimétrique, colorimétrie
P analiza kolorymetryczna, kolorymetria
R колориметрический анализ, колориметрия

colorimetric titration
A method of titration in which a coloured
reaction in the sample solution is in one
absorption cell or flat-bottomed cylinder while
to a second cell or cylinder all the reagents are
added in the same proportions as to the first.
Then a standard solution of the analyte is
added from a burette to the second cell until
the colours in both cells or cylinders are judged
to match.
D kolorimetrische Titration
F titrage colorimétrique
P miareczkowanie kolorymetryczne
R колориметрическое титрование

colorimetry → colorimetric analysis

colour change interval
The concentration range of protons, metal
ions, or the like, generally expressed in the
exponential form (pH, pM), in which temporary
coloration of the indicator takes place.
D Umschlagsintervall des Farbindikators
F intervalle de virage
P zakres zmiany barwy wskaźnika
R интервал перехода окраски

column → chromatographic column

column chromatography, CC
Chromatography in which the stationary phase
is uniformly packed into a chromatographic
column.
D Säulenchromatographie
F chromatographie sur colonne
P chromatografia kolumnowa
R колоночная хроматография, хроматография
  на колонке

column efficiency → column performance

column filling → column packing

column packing, packing, column filling
The active sorbent, liquid deposited on a solid
support, or swollen gel used in a column
chromatographic procedure.
Remark: The term column packing refers to the material
being introduced into the column before the chromatographio
process is started, whereas the term stationary phase refers
to the packing in the course of the process.
D Säulenpackung, Säulenfüllung
F remplissage de la colonne, garnissage de
  colonne
P wypełnienie kolumny
R материал для набивки колонки, набивка
  колонки

column performance, column efficiency
(*in chromatography*)
The measure of band broadening expressed
by the number of theoretical plates.
D Trennleistung der Säule, Säulenwirksamkeit
F performance d'une colonne, efficacité d'une
  colonne

P sprawność kolumny
R эффективность колонки

column temperature *See* separation temperature
D Säulentemperatur
F température de la colonne
P temperatura kolumny
R температура колонки

column volume, geometric volume of the column, volume of the column tube, V_c
The total volume of that part of the empty column that contains the packing:

$$V_c = d_c^2 \, \pi L/4$$

where d_c — inside diameter, L — height or length of the column filled with the packing.
D (geometrisches) Säulenvolumen
F volume (géométrique) de la colonne
P objętość kolumny (geometryczna)
R объём колонки

coma
An aberration of lenses in which a point object off the axis of the system gives an asymmetric image (which appears as a pear-shaped spot).
D Komma, Asymmetriefehler
F coma
P koma
R кома

combination band
In the infrared spectrum of a molecule, a band of a frequency corresponding to the sum or the difference of the frequencies of two or more fundamental vibrations.
D Kombinationsbande
F bande de combinaison
P pasmo kombinacyjne
R комбинационная полоса

combination electrode
An electrochemical apparatus which incorporates an indicator electrode, e.g. ion-selective electrode, and a reference electrode. The outer electrolyte solution containing the compound being determined serves as a salt bridge between the reference and indicator electrodes.
D kombinierte Elektrode
F électrode combinée
P elektroda kombinowana
R комбинированный электрод

combined sample → gross sample

combined techniques
The investigation of a substance by multiple techniques in each of which is a separate sample used.
D kombinierte Untersuchungsmethoden
F techniques combinées
P techniki kombinowane
R комбинированные методы

combustion
A rapid exothermal reaction usually accompanied by emission of light; most often a reaction with oxygen.
D Verbrennung
F combustion
P spalanie
R сжигание, сгорание

combustion analysis
The combustion of samples or their constituents under specified conditions, followed by determination of the gaseous combustion product on the ground of the generated gas volume gain.
D Verbrennungsanalyse
F analyse par combustion
P analiza przez spalanie
R анализ сжиганием

combustion boat, sampling boat
A small vessel for weighing samples and transferring them into a combustion tube. It can be made of porcelain, quartz, platinum or aluminium.
D Schiffchen, Glühschiffchen, Verbrennungsschiffchen
F nacelle à combustion
P łódeczka do spalań
R лодочка для сожжения, лодочка для сжигания

combustion bomb
A small pressure vessel made of nickel or stainless steel, with hermetic cover designed for mineralization of organic compounds by heating with sodium peroxide.
D Universalbombe
F bombe au peroxyde de sodium
P bomba do mineralizacji
R бомба для сжигания

combustion furnace
A stationary or movable electric furnace used for heating combustion tubes.
D Verbrennungsofen
F four à tubes
P piec do spalań
R печь сжигания

combustion tube
A tube used for the pyrolysis of samples of organic or inorganic compounds in a stream of oxygen or inert gas. The combustion tubes are usually made of quartz, occasionally of nickel or porcelain.
D Verbrennungsrohr
F tube à combustion
P rura do spalań
R трубка для сожжения, сжигательная трубка

common ion effect
The decrease of solubility of a precipitate due to the presence of a common (for precipitate and solution) ion in the solution.

D ...
F effet d'un ion commun
P efekt wspólnego jonu
R ...

compact layer
A part of the electrical double layer at the metal-electrolyte interface which adheres to the electrode surface.
D Helmholtzsche Doppelschicht, Helmholtz-Schicht
F couche rigide d'Helmholtz
P warstwa sztywna, warstwa Helmholtza
R конденсированный двойной слой, гельмгольцевский двойной слой

comparison solution
A solution with a colour identical to that which a titrated solution should have at the end-point of titration.
D Vergleichslösung
F solution de comparaison
P świadek miareczkowania
R раствор сравнения, свидетель

complete analysis, total analysis
The determination of all constituents in the sample examined.
D Gesamtanalyse
F analyse totale
P analiza całkowita
R полный анализ

complete factorial design, factorial design, design of the n^p type
Experimental design consisting of n^p experimental points whose coordinates are defined by all possible combinations of n levels of each p factors (variables).
D vollständiger faktorieller Versuchsplan
F plan factoriel plein
P plan czynnikowy pełny, plan całkowitego eksperymentu czynnikowego
R полный факторный план

completely polarizable electrode → ideal polarizable electrode

complex admittance → admittance

complex electrode
An electrode (half-cell), at which two or more (potential determining) electrode reactions occur simultaneously.
D Mehrfachelektrode
F électrode multiple
P elektroda złożona, elektroda mieszana
R смешанный электрод, полиэлектрод

compleximetric titration, complexometric titration
A titration involving the formation of a soluble complex between the metal ion and the complexing agent.

D komplexometrische Titration, Komplexbildungstitration
F titrage complexométrique
P miareczkowanie kompleksometryczne
R комплексиметрическое титрование, комплексометрическое титрование

compleximetric titration curve
A diagram presenting the concentration of a metal (expressed as the exponent $pM = -\log [M^{n+}]$) as a function of volume of the standard solution of the complexing agent added.
D komplexometrische Titrationskurve
F courbe de titrage complexométrique
P krzywa miareczkowania kompleksometrycznego
R комплексиметрическая кривая титрования, комплексометрическая кривая титрования

compleximetry, complexometry
A titration with, or of, a substance capable of forming a soluble weakly dissociated complex.
D Komplexometrie, Kompleximetrie
F complexométrie
P kompleksometria
R комплексометрия

complex impedance → impedance

complexometric titration → compleximetric titration

complexometry → compleximetry

complexones
The complexforming reagents of aminopolycarboxylic acid type, applied for the titration and masking of metals with which they form stable chelate complexes. The characteristic group in complexones is the nitrogen atom bonded with two methylenecarboxylic groups $-N(CH_2COOH)_2$.
D Komplexonen
F complexones
P kompleksony
R комплексоны

compressibility correction factor → pressure-gradient correction factor

Compton continuum, Compton spectrum
The continuous energy spectrum of γ-rays scattered in the Compton effect.
D Comptonverteilung
F fond Compton, distribution Compton
P kontinuum komptonowskie, widmo komptonowskie
R комптоновское распределение

Compton edge
The maximum energy in a Compton continuum.
D Compton-Kante

F front Compton
P krawędź komptonowska
R комптоновский край

Compton effect, Compton scattering,
incoherent scattering (*of photons*)
The scattering of X- or γ-ray photons
resulting from the inelastic collisions with
loosely bound electrons (orbital electrons)
which can be treated as immobile free electrons.
In the collision, a part of the photon energy
and momentum is transferred to the electron,
which gains kinetic energy (recoil energy),
whereas the photon looses that part of energy
and changes momentum, concurrently
changing its direction (the scattering angle
changes from 0 to 180°).
D Compton-Effekt, Compton-Streuung,
inkohärente Streuung von Photonen
F effet de Compton, diffusion de Compton,
diffusion incohérente
P efekt Comptona, rozpraszanie
komptonowskie, rozpraszanie niespójne
R эффект Комптона, комптоновское
рассеяние, некогерентное рассеяние

Compton scattering → Compton effect

Compton spectrum → Compton continuum

concave grating
A diffraction reflection grating in which the
grooves are made on the concave surface of a
spherical mirror.
D Konkavgitter, Hohlgitter
F réseau concave
P siatka wklęsła
R вогнутая решётка

concentration cell
The galvanic cell in which the electrical
energy is generated from the free energy change
accompanying the transfer of a substance from
a system of high concentration to one of low
concentration.
D Konzentrationszelle, Konzentrationskette
F pile de concentration
P ogniwo stężeniowe
R концентрационная цепь, концентрационный
элемент

concentration distribution ratio, distribution
ratio, distribution coefficient, D_c
(*in chromatography*)
The ratio of the analytical concentration of
a component in the stationary phase to its
analytical concentration in the mobile phase.
D Konzentrationsverteilungsverhältnis,
Verteilungsverhältnis, Verteilungskoeffizient
F rapport de distribution (de concentration),
coefficient de distribution (de concentration)
P stosunek podziału (stężeniowy), iloraz
podziału (stężeniowy), współczynnik podziału
R (концентрационное) отношение
распределения, коэффициент распределения

concentration distribution ratio, distribution
ratio, distribution coefficient, extraction
coefficient, D_e (*in extraction*)
The ratio of the analytical concentration of
a substance in the organic phase to its
analytical concentration in the aqueous phase
usually measured at equilibrium.
Remark: The word concentration can be omitted when
there is no ambiguity with the mass distribution ratio, D_m.
D Konzentrationsverteilungsverhältnis,
Verteilungsverhältnis, Extraktionskoeffizient,
Verteilungskoeffizient
F rapport de distribution (de concentration),
coefficient d'extraction, coefficient de
distribution
P stosunek podziału (stężeniowy),
współczynnik ekstrakcji, współczynnik
podziału
R (концентрационное) отношение
распределения, коэффициент экстракции,
коэффициент распределения

concentration limit
The smallest mass concentration of a substance
in a sample which may be detected by a given
method.
D Grenzkonzentration
F concentration limite
P stężenie graniczne
R предельная концентрация

concentration method
In general, the increasing of concentration
of a substance. In trace analysis, the
increasing of concentration of the trace
components to obtain a concentration higher
than the limit of determination of the method
applied. The methods of concentration are:
precipitational, extractional, chromatographic,
electrochemical, and those taking advantage of
the differences in volatility.
D Anreicherungsmethode
F méthode de concentration, méthode
d'enrichissement
P metoda zagęszczania, metoda wzbogacania
R метод концентрирования, метод
обогащения

concentration of substance B → amount-of-
substance concentration of substance B

concentration overpotential, mass transfer
overpotential
An overpotential which is due to changes in the
concentration of reagents in the vicinity of the
electrode surface. The concentration
overpotential is the sum of the diffusion,
reaction, and crystallization overpotentials.
D Konzentrationsüberspannung
F surtension de concentration
P nadpotencjał stężeniowy
R концентрационное перенапряжение

concentration profile → depth profile

conditional (electrode) potential, formal (electrode) potential, $E_f^{\circ'}$
The electrode (half-cell) potential in the case when the concentrations of the oxidized and reduced forms are equal. It is characteristic for a given electrode reaction taking place in a given medium.
D formales Potential
F potentiel formel (d'électrode), potentiel apparent (d'électrode)
P potencjał (elektrody) warunkowy, potencjał (elektrody) formalny
R условный (электродный) потенциал, формальный (электродный) потенциал, реальный потенциял

conditional potential → conditional electrode potential

conditional rate constant of an electrode reaction, k_f°, $k^{\circ'}$
The electrode reaction rate constant at the conditional (formal) potential of the electrode reaction.
D konventionelle Geschwindigkeitskonstante der Elektrodenreaktion
F ...
P stała szybkości (reakcji elektrodowej) warunkowa, stała szybkości (reakcji elektrodowej) formalna
R условная константа скорости электродной реакции

conductance, electric(al) conductance, G
The ability of a direct-current circuit to conduct a steady current, equal to the reciprocal of the resistance.
For an alternating-current circuit, the real part of the admittance. SI unit: siemens, S.
D Wirkleitwert, elektrischer Leitwert, Konduktanz, (*früher auch* elektrische Leitfähigkeit)
F conductance électrique
P konduktancja, przewodność elektryczna (czynna), przewodnictwo elektryczne
R электрическая проводимость, активная проводимость (электрической цепи)

conductance of electrolyte, electrolytic conductance, L
The quantity L for a segment of solution placed in an electric field is directly proportional to the cross-sectional area, A, perpendicular to the field vector and it is inversly proportional to the length l of the segment along the field

$$L = \varkappa \frac{A}{l}$$

where $\varkappa$ is the electrolytic conductivity.
D elektrolytischer Leitwert (*früher auch* elektrolytische Leitfähigkeit)
F conductance (électrique) d'un électrolyte
P konduktancja elektrolityczna, przewodnictwo elektrolityczne
R электрическая проводимость электролита

conductimetry → conductometry

conduction → electrical conduction

conductivity
The ability of a substance to conduct heat or electricity.
D Leitfähigkeit
F conductibilité, conductivité
P przewodność
R проводимость

conductivity bridge → conductometer

conductivity meter → conductometer

conductometer, conductivity bridge, conductivity meter
An instrument used for measuring the conductivity of an electrolyte.
D Leitfähigkeitsmesser
F conductivimètre
P konduktometr
R мостик для измерения электропроводности

conductometric analysis *See* conductometry
D konduktometrische Analyse
F analyse conductométrique
P analiza konduktometryczna
R кондуктометрический анализ

conductometric titration
Titration in which the end-point is determined from the abrupt change in the conductance G of a substance, or its solution. Measurements are carried out at the frequency of the alternating voltag not exceeding ca 10^5 Hz.

V — volume of a titrant

D konduktometrische Titration, konduktometrische Indikation
F titrage conductométrique
P miareczkowanie konduktometryczne
R кондуктометрическое титрование

conductometry
An electrochemical technique in which an alternating voltage, of frequency not exceeding ca 10^5 Hz, is applied to the conductance cell and the conductance G of a substance, or of a solution containing this substance, is measured as a function of the concentration c of this substance.

Remark: The spelling conductimetry is not recommended. In chemical analysis the term conductometric analysis is also used.

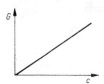

D Konduktometrie, Konduktimetrie
F conductométrie, conductimétrie
P konduktometria
R кондуктометрия, кондуктиметрия

conductor → electrical conductor

confidence interval
A random interval (t_1, t_2) obtained on the base of a sample in such a manner that the probability of covering the unknown value of a parameter θ by this interval is $1-\alpha$, that is

$$P\{t_1 \leqslant \theta \leqslant t_2\} = 1-\alpha$$

where t_1, t_2 — limits of the interval,
$1-\alpha$ — confidence level.
D Konfidenzintervall, Vertrauensintervall
F intervalle de confiance
P przedział ufności
R доверительный интервал

confidence level
The probability $1-\alpha$ of covering the estimated parameter by the confidence interval (or statistical tolerance limits).
D Konfidenzniveau
F niveau de confiance
P poziom ufności
R доверительный уровень

confidence limits
Numbers t_1 and t_2 determining the left and right limits of the confidence interval.
D Konfidenzgrenzen, Vertrauensgrenzen
F limites de confiance
P granice ufności
R доверительные пределы, доверительные
границы

conformation
One of the non-identical spatial arrangements of atoms in a molecule, formed by rotation about the single bond (or bonds) without its (their) rupture. From the among infinite number of possible conformations, the most stable is the one in which all interactions between the mutually non-bonded substituents are the weakest.
D Konformation, Konstellation
F conformation
P konformacja
R конформация

conformational analysis
Ascertaining the spatial arrangement of atoms or groups of atoms in a molecule in which free rotation about a single bond is possible.

D Konformationsanalyse
F analyse conformationnelle
P analiza konformacyjna
R конформационный анализ

conjugate reaction
The reaction which may be considered as consisting of two simultaneous processes, e.g. the acid-base reaction

$$HA^{n+1} \rightleftarrows A^n + H^+$$
$$acid_1 \rightleftarrows base_1 + proton$$
$$B^m + H^+ \rightleftarrows BH^{m+1}$$
$$base_2 + proton \rightleftarrows acid_2$$
_____
$$HA^{n+1} + B^m \rightleftarrows A^n + BH^{m+1}$$
$$acid_1 + base_2 \rightleftarrows base_1 + acid_2$$

D konjugierte Reaktion
F réaction conjuguée
P reakcja sprzężona
R сопряженная реакция

consecutive titration → successive titration

constant current coulometry → controlled current coulometry

constant-current electrogravimetry
The electrogravimetry in which the electrodeposition of a substance at the working electrode is conducted at constant current.
D Elektrogravimetrie mit konstantem Strom, Elektrogravimetrie bei konstanter Stromstärke
F électrogravimétrie à tension d'électrolyse constante
P elektrograwimetria przy stałym prądzie, elektrograwimetria stałoprądowa
R электрогравиметрия при постоянной плотности тока

constant-current electrolysis
The eletrolysis carried out with the current kept constant in time.
D Elektrolyse mit konstantem Strom, Elektrolyse bei konstanter Stromstärke
F électrolyse à courant constant, électrolyse à tension d'électrolyse constante
P elektroliza przy stałym prądzie, elektroliza stałoprądowa
R электролиз при постоянной силе тока

constant mass (*in gravimetric analysis*)
The mass of a precipitate which does not change by more than $\pm 0.000\,02$ g in two successive weighings before and after drying or calcination.
D konstante Masse
F masse constante
P masa stała
R постоянная масса

contamination of a precipitate
The presence of minor amounts of at least one
chemically different species in a precipitate.
D Verunreinigung des Niederschlages
F contamination d'un précipité, impureté d'un
   précipité
P zanieczyszczenie osadu
R загрязнение осадка

content
The total amount of a constituent in the
sample analysed.
D Gehalt
F teneur
P zawartość
R содержание

continuous extraction
Extraction in which the substance to be
extracted and the extractant are continuously
fed, the extract being simultaneously removed
from the system.
D kontinuierliche Extraktion
F extraction continue
P ekstrakcja ciągła
R непрерывная экстракция

continuous random variable
A random variable which can take any value
from the given real numbers interval.
D stetige Zufallsgröße, stetige Variable
F variable aléatoire continue
P zmienna losowa ciągła
R непрерывная случайная величина

continuous simultaneous techniques
The analysis of a sample by combined
techniques in which the sampling for the
consecutive techniques is effected by a
continuous process; e.g. the thermal
differential analysis combined with the
thermomanometric (or thermovolumetric)
determination of the evolved gases.
D simultane kontinuierliche
   Untersuchungsmethoden
F techniques simultanées continues
P techniki jednoczesne współdziałające ciągle
R ...

continuous spectrum
A spectrum in which the quantity being
studied no longer has discrete values, but may
take any value over a continuous range. Such
a spectrum is characteristic of an unquantized
process.
D kontinuierliches Spektrum
F spectre continu
P widmo ciągłe
R сплошной спектр

contrast factor → gamma

control chart
Graphic way of testing (in time) the variation
of the controlled parameter. Usually the

control chart consists of a central line (target
level of the parameter) and two parallel lines —
upper and lower control limits; on such a
diagram experimental data are plotted in time
order. Control chart is used, e.g, for quality
control of product output, analysis of processing
conditions, control of accuracy and precision
of analytical method.
D Kontrollkarte
F carte de contrôle
P karta kontrolna
R контрольная карта

controlled-current coulometry, constant-current
coulometry, coulometric titration
The coulometric method for the determination
of a substance, conducted under the
conditions of a convective mass transport
to/from the working electrode and a constant
(controlled) current throughout the reaction
period. The titrant is generated in the solution
as a result of an electrode reaction, and
various physico-chemical methods are used to
locate the end-point of the titration.
Remark: Terms like potentiometric coulometric titration
or controlled-current coulometry with potentiometric end
point detection are recommended when the technique of
end-point location is to be specified.
D stromkontrollierte Coulometrie,
   galvanostatische Coulometrie, coulometrische
   Titration, Coulometrie bei konstantem
   Strom
F coulométrie à courant imposé, coulométrie
   à courant contrôlé, coulométrie à intensité de
   courant constante, titrage coulométrique
P kulometria amperostatyczna, kulometria
   galwanostatyczna, kulometria z
   kontrolowanym natężeniem prądu,
   miareczkowanie kulometryczne
R кулонометрия с контролируемым током,
   гальваностатическая кулонометрия,
   амперостатическая кулонометрия,
   кулонометрия при постоянном токе,
   кулонометрическое титрование

**controlled-current coulometry with
potentiometric end-point detection** *See*
controlled-current coulometry
D stromkontrollierte Coulometrie mit
   potentiometrischer Endpunktsbestimmung
F coulométrie à courant imposé avec détection
   potentiométrique du point de fin de titrage
P kulometria z kontrolowanym prądem
   i potencjometryczną detekcją punktu
   końcowego
R кулонометрия с контролируемым
   потенциалом и потенциометрической
   индикацией конечной точки

controlled-current potentiometric titration
A modification of the potentiometric titration
in which the end-point is determined from
changes in the potential E of the indicator
electrode to which a controlled, usually

constant, current is imposed from an external source.

V — volume of a titrant

D stromkontrollierte potentiometrische Titration, potentiometrische Titration mit gesteuertem Strom, voltametrische Indikation
F titrage potentiométrique à courant imposé, titrage potentiométrique à intensité constante
P miareczkowanie potencjometryczne z kontrolowanym prądem
R потенциометрическое титрование с контролируемым током

controlled-current potentiometric titration with two indicator electrodes
A modification of potentiometric titration in which the end-point is determined from changes in the potential difference E between two indicator electrodes immersed in the same solution while a controlled, usually constant, current flows between them.

Remark: The term bipotentiometric titration is no longer recommended.

curve recorded in the titration of a substance forming reversible redox system with the titrant which also yields reversible system
V — volume of a titrant

D stromkontrollierte potentiometrische Titration mit zwei Indikator-Elektroden, potentiometrische Titration bei konstantem Strom mit zwei polarisierbaren Elektroden, potentiometrische Titration mit gesteuertem Strom und zwei Indikatorelektroden, voltametrische Indikation mit zwei Indikatorelektroden, bipotentiometrische Titration
F titrage potentiométrique à courant imposé à deux électrodes indicatrices, titrage bipotentiométrique.
P miareczkowanie potencjometryczne z kontrolowanym prądem z dwiema elektrodami spolaryzowanymi, miareczkowanie bipotencjometryczne
R потенциометрическое титрование с контролируемым током и двумя индикаторными электродами, бипотенциометрическое титрование

controlled-current potentiometry →
controlled-current potentiometry with one indicator electrode

controlled-current potentiometry (with one indicator electrode)
A modification of potentiometry with the application of a controlled and (usually) constant current to the indicator electrode.

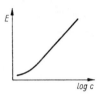

D stromkontrollierte Potentiometrie, Potentiometrie mit gesteuertem Strom
F potentiométrie à courant imposé, potentiométrie à intensité constante
P potencjometria z kontrolowanym prądem
R потенциометрия с контролируемым током

controlled-current potentiometry with two indicator electrodes
A modification of potentiometry in which the quantity measured is the difference between the potentials of two indicator electrodes immersed in the same solution, with controlled current applied to these electrodes.

Remark: The term bipotentiometry is no longer recommended.

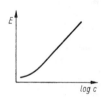

D stromkontrollierte Potentiometrie mit zwei Indikator-Elektroden, Potentiometrie mit gesteuertem Strom und zwei Indikatorelektroden, Bipotentiometrie
F potentiométrie à courant imposé à deux électrodes indicatrices, bipotentiométrie
P potencjometria z kontrolowanym prądem z dwiema elektrodami spolaryzowanymi, bipotencjometria
R потенциометрия с контролируемым током с двумя индикаторными электродами, бипотенциометрия

controlled-porosity glass, CPG
Porous granules of glass in which the pore size is carefully controlled during the manufacture. In permeation chromatography, they are used as a rigid column packing.
D Gläser mit kontrollierter Porosität
F verre de porosité contrôlée

P szkło o kontrolowanej porowatości
R стекло с контролируемой пористостью

controlled-potential coulometric titration →
controlled-potential coulometry

controlled-potential coulometry
The coulometric method carried out under the
conditions of a convective mass transport
to/from the working electrode and at constant
(controlled) potential of the working electrode
throughout the reaction period.
Remark: The term controlled-potential coulometric
titration is inappropriate.
D potentialkontrollierte Coulometrie,
 potentiostatische Coulometrie, Coulometrie
 mit gesteuertem Potential, Coulometrie bei
 konstantem Potential,
 potentialkontrollierte coulometrische
 Titration
F coulométrie potentiostatique, coulométrie
 à potentiel contrôlé, coulométrie à potentiel
 constant, titrage coulométrique à potentiel
 contrôlé
P kulometria potencjostatyczna, kulometria
 z kontrolowanym potencjałem,
 miareczkowanie kulometryczne
 z kontrolowanym potencjałem
R кулонометрия с контролируемым
 потенциалом, кулонометрия при постоянном
 потенциале, потенциостатическая
 кулонометрия

controlled-potential electrogravimetry
The electrogravimetry in which
electrodeposition of a substance at the working
electrode is carried out with the potential of
this electrode (with respect to the reference
electrode) kept constant in time, or varried
according to an imposed program.
D potentialkontrollierte Elektrogravimetrie,
 Elektrogravimetrie mit gesteuertem Potential,
 Elektrogravimetrie mit kontrolliertem
 Potential
F électrogravimétrie à potentiel contrôlé,
 électrogravimétrie à tension d'électrode
 contrôlée
P elektrograwimetria z kontrolowanym
 potencjałem
R электрогравиметрия с контролируемым
 потенциалом, электрогравиметрия при
 регулируемом потенциале

controlled-potential electrolysis
The electrolysis carried out under the
conditions in which the potential of the
working electrode (with respect to the
reference electrode) is kept constant or varried
according to an imposed program.
D Elektrolyse bei kontrolliertem Potential
F électrolyse à potentiel contrôlé, électrolyse
 potentiostatique, électrolyse à tension
 d'électrode constante
P elektroliza z kontrolowanym potencjałem
R электролиз при контролируемом потенциале

controlled surface porosity support →
superficially porous support

control line pair *(in emission spectrochemical
analysis)* → fixation pair

control titration
A titration of a known amount of substance
with a titrant, made to determine the effect of
variable factors and foreign substances on the
accuracy of the titration.
D Kontrolltitration
F titrage de contrôle
P miareczkowanie kontrolne
R контрольное титрование

convection
Motion of molecules or ions of a given
component in a given phase under the
influence of displacements of some parts of
this phase with respect to other parts.
D Konvektion
F convection
P konwekcja
R конвекция

convective chronoamperometry
A modification of chronoamperometry
conducted under the conditions of a convective
mass transfer to/from the indicator electrode.
The current I flowing through this electrode
is measured as a function of the electrolysis
time t.

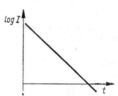

D konvektive Chronoamperometrie
F chronoampérométrie convective
P chronoamperometria konwekcyjna
R конвективная хроноамперометрия

convective chronocoulometry
A modification of chronocoulometry conducted
under the conditions of convective mass transfer
to/from the indicator electrode. The electric
charge Q which has passed through the
indicator electrode is measured as a function of
the electrolysis time t.
D konvektive Chronocoulometrie
F chronocoulométrie convective
P chronokulometria konwekcyjna
R конвективная хронокулонометрия

conventional alternating-current polarography,
alternating-current polarography, ac
polarography
The alternating-current polarography in which
a sinusoidal voltage E_{ac} of low frequency

(usually 50 Hz) and small amplitude (5-50 mV) is used. The alternating current component of the total current I_{ac} is measured as a function of the direct potential E_{dc} of the indicator electrode.

D klassische Wechselstrompolarographie
F polarographie à tension sinusoïdale surimposée, polarographie à tension alternative
P polarografia zmiennoprądowa sinusoidalna
R классическая полярография переменного тока, переменно-токовая полярография

conventional analysis
The analysis consisting in determination of the content of a constituent, or a property of an investigated sample under conventional, but strictly specified conditions. The results obtained from conventional analysis are not usually in agreement with the real content of the constituent in the examined sample, being merely comparable with the results obtained under the given conditions.

D konventionelle Analyse
F analyse d'usage, analyse ordinaire
P analiza umowna, analiza konwencjonalna
R конвенциональный анализ

conventional packed column → packed column

conventional polarography → direct current polarography

coordinating group (*of chelating agent*)
A group containing a donor atom, e.g. amine group —NH$_2$, —NHR or —NR$_2$, azo group —N=N—, carbonyl group =CO, ether group —O—.

D koordinierende Gruppe
F groupe coordonnant
P grupa koordynująca
R координирующая группа

copper coulometer
A chemical coulometer consisting of a copper anode and a copper cathode immersed in a solution of cupric sulphate containing sulphuric acid, ethanol and water. The cathode is weighed before and after electrolysis, and the electric charge which has passed is calculated from the increased weight of the cathode (3.295×10^{-4} g of copper is deposited by 1 C).

D Kupfercoulometer
F coulomètre à cuivre
P kulometr miedziowy
R медный кулонометр

coprecipitation
The simultaneous precipitation from a solution of a component, normally soluble under the given conditions, with a macro-component owing to the formation of mixed crystals, adsorption, occlusion or mechanical trapping.

D Mitfällung
F coprécipitation
P współstrącanie
R соосаждение

core level electron spectroscopy
The spectroscopy of electrons displaced from the inner atomic levels.

D Elektronenspektroskopie der innersten atomaren Niveaus, Rumpfniveau-Elektronenspektroskopie
F spectroscopie d'électrons déplacés des niveaux de coeur d'atome
P spektroskopia elektronów rdzenia atomowego
R ...

corpuscular radiation
The radiation consisting of particles whose rest mass is greater than zero, e.g. α, β, neutron, proton, fission fragment radiation.

D korpuskulare Strahlung
F rayonnement corpusculaire
P promieniowanie korpuskularne
R корпускулярное излучение

corrected R_f value → R'_f value

corrected R_M value → R'_M value

corrected retention volume, V_R^0 (*in gas chromatography*)
The total retention volume V_R multiplied by the pressure-gradient correction factor j,
$$V_R^0 = jV_R.$$

D korrigiertes Retentionsvolumen
F volume de rétention limite, volume de rétention corrigé
P objętość retencji poprawiona, objętość retencji skorygowana
R исправленный удерживаемый объём, исправленный объём удерживания

corrected selectivity coefficient, $k^a_{A/B}$, $k'_{A/B}$, k'^A_B
The selectivity coefficient of an ion exchange reaction, corrected to provide for the ratio of the mean activity coefficients of ions in the solution.

D korrigierter Selektivitätskoeffizient
F coefficient de sélectivité corrigé
P współczynnik selektywności poprawiony
R исправленный коэффициент селективности

correlation
Mutual relationship of two or more random variables such that one variable reacts to changes of the other variables by changes of its expected value.

D Korrelation
F corrélation
P korelacja
R корреляция

correlation matrix, R
A square matrix with elements r_{ij} equal to coefficients of correlation between the ith and jth random variable from the set $X_1, X_2, ..., X_n$ of random variables ($i, j = 1, 2, ..., n$).

D Korrelationsmatrix
F matrice de corrélation
P macierz korelacji
R корреляционная матрица

corrosion potential
The mixed potential of an electrode undergoing corrosion.

D Korrosionspotential
F potentiel de corrosion
P potencjał korozyjny
R потенциал коррозии

Cottrell equation
An equation describing the dependence of the diffusion current I_d on the bulk concentration c_i^0 of the electroactive substance and the time of electrolysis t under potentiostatic conditions and linear semi-infinite diffusion

$$I_d = nFAD_i^{1/2} c_i^0 (\pi t)^{-1/2}$$

where n — number of electrons exchanged in the electrode reaction, A — surface area of the electrode, D_i — diffusion coefficient of the electroactive substance.

D Cottrell-Gleichung
F équation de Cottrell
P równanie Cottrella
R уравнение Котрелла

coulogravimetry
An electrochemical method for quantitative determination of a substance, or a mixture of substances, subjected electrodeposition at an electrode, in which the total mass of metal deposit and the electric charge passed are measured.

D Coulogravimetrie
F coulogravimétrie
P kulograwimetria
R кулоногравиметрия

coulombmeter → chemical coulometer

coulometer
An instrument for measuring the electric charge which has passed through an electric circuit.

D Coulometer
F coulomètre
P kulometr
R куло(но)метр

coulometric analysis *See* coulometry

D coulometrische Analyse
F analyse coulométrique
P analiza kulometryczna
R кулонометрический анализ

coulometric titration → controlled-current coulometry

coulometry
A range of electrochemical techniques based on the measurement of the electric charge.
Remark: In chemical analysis the term coulometric analysis is also used.

D Coulometrie
F coulométrie
P kulometria
R кулонометрия

coulostatic method
An electroanalytical method in which a given electric charge is passed through an indicator electrode and the potential of the electrode is measured as a function of time.

D coulostatische Methode
F méthode coulostatique
P metoda kulostatyczna
R кулоностатический метод

2π-counter
A nuclear radiation counter recording the emitted particles or photons at a 2π solid angle.

D 2π-Zähler
F compteur 2π
P licznik 2π
R счётчик с геометрией 2π, 2π-счётчик

4π-counter
A nuclear radiation counter recording the emitted particles or photons at a 4π solid angle.

D 4π-Zähler
F compteur 4π
P licznik 4π
R счётчик с геометрией 4π, 4π-счётчик

countercurrent distribution, Craig distribution (*in extraction*)
A method usually applied for the separation of binary mixtures whose components have significantly different distribution coefficients in two immiscible liquid phases. In successive units of the apparatus, after every step displacement of a portion of the mobile phase, containing some of the solutes to be separated, a new equilibrium between the stationary and mobile phases is established; a discontinuous model of liquid-liquid partition chromatography.

D Gegenstromverteilung, Craig-Verteilung
F distribution à contre-courant
P rozdział przeciwprądowy, metoda Craiga
R противоточное распределение, метод Крейга

countercurrent extraction
Continuous extraction in which the directions of flow of the extractant and the substance to be extracted are opposite.

D Gegenstromextraktion
F extraction en contre-courant
P ekstrakcja przeciwprądowa
R противоточная экстракция

counter electrode (*in arc and spark spectroscopy*)
The non-sample carrying electrode, which is
used opposite to the self-electrode or to the
supporting electrode.
D Gegenelektrode
F contre-électrode
P przeciwelektroda
R противоэлектрод

counter electrode → auxiliary electrode

counter-ions (*in ion exchange*)
The mobile exchangeable ions able to diffuse
freely which have a charge opposite to that of
the fixed ions.
Remark: In ion-pair chromatography the specified ions
which form hydrophobic ion pairs with the ions to be
analysed.
D Gegenionen
F contre-ions
P przeciwjony
R противоионы

counter tube, gas-filled radiation-detection tube
Radiation detector consisting of a gas-filled
tube or valve whose gas amplification is much
greater than unity, and in which the individual
ionization events give rise to discrete electrical
pulses. The three main types of gas filled
detectors are the ionization chamber, the
proportional counter and the Geiger counter.
The three types of tube differ as regards their
operating conditions.
D Zählröhre
F tube compteur
P licznik promieniowania
R счётная трубка

counting efficiency
The ratio of the number of particles or photons
counted by a radiation counter and the number
of similar particles or photons emitted by the
radiation source.
D Zählausbeute, Quantenausbeute
F efficacité de comptage, rendement de
  comptage
P wydajność zliczania, wydajność liczenia
R эффективность счёта, эффективность
  счётчика

counting geometry, source-detector geometry
The arrangement in space of the various
components in an experiment, specyfically
the source and the detector in radiation
measurements.
D Zählgeometrie
F géométrie de comptage
P geometria liczenia, geometria pomiaru
R геометрия счёта, геометрия измерения

counting rate (*in radiometry*)
The number of counts recorded in unit time.
D Zählrate
F taux de comptage
P szybkość zliczania, szybkość liczenia
R скорость счёта

coupled columns (*in chromatography*)
A chromatographic technique in which
mixtures of components difficult to separate,
e.g. with different polarities, are analysed on
a series of combined columns of different
length and/or filled with different packings.
The columns are connected to the system
successively or simultaneously depending on
the predetermined program.
D gekoppelte Säulen, kombinierte Säulen
F couplage de colonnes
P technika kolumn łączonych, technika kolumn
  sprzężonych, kolumny łączone
R сборные колонки, дополнительные колонки,
  соединённые колонки

coupled (simultaneous) techniques
The investigation of the same sample by two
or more different techniques, in which the
analytical recording units are coupled, e.g.
the coupling of differential thermal analysis
and mass spectrometry.
D (simultane) gekuppelte
  Untersuchungsmethoden
F techniques (simultanées) associées
P techniki (jednoczesne) sprzężone
R сопряжённые (одновременные) методы

coupled techniques → coupled simultaneous
techniques

covariance
Quantity characterizing the stochastic relation
between two random variables; for random
variables X_1 and X_2 the covariance is given by
the equation

$$\text{cov}\,(X_1, X_2) = \text{E}\,\{[X_1 - \text{E}\,(X_1)]\,[X_2 - \text{E}\,(X_2)]\}$$

where E — operator of the expected value.
D Kovarianz
F covariance
P kowariancja
R коварианция

covariance matrix D (*in regression analysis*)
A matrix whose elements are variances
and covariances of regression coefficients

$$D = \sigma^2 [X^T X]^{-1}$$

where X — design matrix, σ^2 — variance of
output (independent) variable, T — denotes
transposition of the matrix.
D Kovarianzmatrix, Streuungsmatrix
F matrice de covariance, matrice de dispersion
P macierz kowariancji, macierz dyspersji
R ковариантная матрица

Covell's method
A method of calculating the photopeak area in
a γ-ray spectrum used in activation analysis

$$N = a_0 + \sum_{i=1}^{n} a_i + \sum_{i=1}^{n} b_i - \left(n + \frac{1}{2}\right)(a_n + b_n)$$

where N — photopeak area in counts/s,
a_0 — the heighest count rate measured in one
channel within the photopeak, a and b — count
rates measured in the neighbouring channels on
the left and right side of the photopeak,
n — number of channels on one side of the
photopeak.

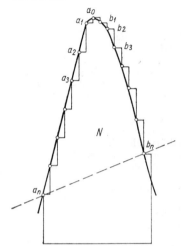

D Methode von Covell
F méthode de Covell
P metoda Covella
R метод Ковеля

CPG→ controlled-porosity glass

CP reagent → chemically pure reagent

cracking pattern → fragmentation pattern

Craig distribution → countercurrent distribution

crater (*in surface analysis*)
A pit formed by the erosion of a solid surface
under ion bombardment.
D Kratergrube
F cratère
P krater
R кратер распыления

criterion of optimality of experimental design
A requirement to be fulfilled by the experimental
design to impose some desired features on the
data obtained in the experiment; usually
criterions of optimality relate to the way the
data obtained in the experiment are processed
or to some properties that the regression
equations determined on the basis of the data
obtained in the experiment should exhibit.
D Optimalitätskriterium eines Versuchsplanes
F ...

P kryterium optymalności planu eksperymentu
R критерий оптимальности плана
эксперимента

critical determination
The determination allowing to establish the
difference in elementary composition of
two organic compounds, when the contents
of other determined elements are so similar
that the differences fall within the limits of
the permissible error, e.g. for compounds
containing:
C = 70,03% and C = 70,31%
H = 5,84% and H = 6,25%
N = 5,45% and N = 10,39%
the determination of nitrogen would be
critical.
D kritische Bestimmung
F dosage critique
P oznaczenie krytyczne
R критическое определение

critical region → rejection region

critical supersaturation ratio
The supersaturation ratio at which spontaneous
formation of the crystalline phase takes place.
D kritisches Übersättigungsverhältnis
F rapport de sursaturation critique
P stosunek przesycenia krytyczny
R критическое отношение пересыщения

cross-linking
A process of forming many intermolecular
covalent bonds (formation of skeleton) in,
e.g., polymer.
D Vernetzung
F réticulation
P sieciowanie
R сшивание полимера, образование сетчатых
молекул

cross section
A measure of the probability of a given
process (e.g. nuclear reaction).
D Wirkungsquerschnitt
F section efficace
P przekrój czynny
R эффективное сечение

crucible holder
A glass device for mounting a glass filter
crucible in a filter flask.
D Vorstoß für Filtertiegel, Tulpe
F entonnoir allonge pour creuset filtrant
P tulipan
R (*форштос для укрепления стеклянного тигля
в вакуумной колбе*)

crucible
A conical vessel made of porcelain, quartz or
metal for igniting precipitates and fusing
chemical substances.

D Tiegel
F creuset
P tygiel
R тигель

crucible with sintered disk, glass filter crucible
A glass or quartz crucible designed for
filtering by suction and drying the collected
precipitate. A symbol etched on the crucible
wall gives the porosity grade of the filtering
disk.
D Glasfiltertiegel
F creuset filtrant
P tygiel Schotta, tygiel z filtrem ze spiekanego
szkła
R шоттовский тигель

cryoscopic determination of the molar mass
Determination of the molar mass of a chemical
compound from the difference in the freezing
temperatures of a pure solvent and the solution
of the tested compound in that solvent.
D kryoskopische Bestimmung der Molmasse
F détermination cryoscopique de la masse
molaire
P oznaczanie masy molowej kriometryczne
R криоскопическое определение молярной
массы

crystalline electrode → crystalline membrane
ion-selective electrode

crystal(line) membrane
A solid-state membrane made either from a
single- or polycrystalline compound, e.g.
Ag_2S, or from a homogeneous mixture of such
compounds, e.g. AgI/Ag_2S. An active electrode
material can be incorporated in the inert
support.
D Kristallmembran
F membrane cristalline
P membrana krystaliczna
R кристаллическая мембрана

crystalline membrane electrode → crystalline
membrane ion-selective electrode

**crystal(line) membrane (ion-selective)
electrode,** crystalline electrode
An ion-selective electrode with a crystalline
membrane.
D Kristallmembran-Elektrode
F électrode à membrane (sélective) cristalline,
électrode cristalline
P elektroda (jonoselektywna) z membraną
krystaliczną, elektroda membranowa
krystaliczna
R кристаллический (мембранный) электрод,
ионоселективный электрод
с кристаллической мембраной

crystalline precipitate
A precipitate consisting of particles with an
ordered lattice structure generally yielding a
real solution on dissolving.

D kristalliner Niederschlag
F précipité cristallin
P osad krystaliczny
R кристаллический осадок

crystallization
The process of crystal formation and growth
in solutions or molten substances.
D Kristallisation
F cristallisation
P krystalizacja
R кристаллизация

crystallization overpotential
An overpotential of an electrode process
caused by a slow, as compared with to other
elementary steps, formation of crystallization
nuclei at the electrode surface and building in
or detachment of an ion, molecule or atom
from the crystalline lattice.
D Kristallisationsüberspannung
F surtension de cristallisation
P nadpotencjał krystalizacji
R перенапряжение кристаллизации,
кристаллизационное перенапряжение

crystallizer → crystallizing dish

crystallizing dish, crystallizer
D Kristallisierschale
F cristallisoir
P krystalizator
R кристаллизатор

crystal membrane → crystalline membrane

crystal membrane electrode → crystalline
membrane ion-selective electrode

crystal membrane ion-selective electrode →
crystalline membrane ion-selective electrode

crystal resolving power
The ability of an analysing crystal to resolve
two characteristic X-ray lines of very similar
wavelength; the resolving power of the crystal
is the better the greater its angular dispersion,
the smaller the width of the diffraction lines
and the smaller the number of lattice deffects.
D Auflösungsvermögen des Kristalls
F pouvoir de résolution de cristal
P zdolność rozdzielcza kryształu
R разрешающая способность
кристалл-анализатора

CT-band → charge-transfer band

cumulative formation constant → cumulative
stability constant

cumulative stability constant, cumulative
formation constant, β_n
The equilibrium constant of complex formation
reaction $M + nL \rightleftarrows ML_n$. Under conditions of
stability of the activity coefficients

$$\beta_n = \frac{[ML_n]}{[M][L]^n}$$

where $[ML_n]$, $[M]$, $[L]$ — concentrations of the complex, metal and ligand at equilibrium.

D ...
F constante globale de formation
P stała trwałości ogólna, stała trwałości całkowita, stała tworzenia kompleksu całkowita
R общая константа образования

cupellation
The isolation and determination of small quantities of noble metals from their ores, concentrates, and alloys. The method involves melting the sample with a metal (usually lead), followed by heating the separated alloy formed in air in a porous crucible.
The molten lead oxide penetrates into the crucible pores, leaving the noble metal as the residue.

D Kupellation
F coupellation
P kupelacja
R купеляция, купелирование

curie, Ci
A unit of activity of a radioactive substance, temporarily recognized as legal, equivalent to 3.7×10^{10} disintegrations per second. In the SI system 1 Ci $= 3.7 \times 10^{10}$ Bq.

D Curie
F curie
P kiur
R кюри

current-cessation chronopotentiometry
The current-step chronopotentiometry in which $I_2 = 0$.

D Stromunterbrechungschronopotentiometrie, Chronopotentiometrie mit unterbrochenem Strom
F chronopotentiométrie avec arrêt de courant
P chronopotencjometria z wyłączeniem prądu
R хронопотенциометрия с прерыванием тока

current density, electric current density, j
A vector quantity that is equal to the ratio of the electric current flowing through a current carrying medium to the cross-sectional area of the medium. SI unit: ampere per square metre, Am^{-2}.

D elektrische Stromdichte
F densité de courant
P gęstość prądu
R плотность электрического тока

current efficiency
The quotient of the electric charge related to a given partial reaction at an electrode and the overall charge that has passed through this electrode during electrolysis. In the case of

a single electrode reaction, the current efficiency is usually equal to 100%.

D Stromausbeute
F rendement de courant
P wydajność prądowa
R выход по току

current-reversal chronopotentiometry
The current-step chronopotentiometry in which $I_2 = -I_1$.

D Chronopotentiometrie mit umgekehrtem Strom, Stromumkehr-Chronopotentiometrie
F chronopotentiométrie à inversion de courant
P chronopotencjometria z odwróceniem kierunku prądu
R хронопотенциометрия с реверсом тока

current-scanning polarography, polarography with linear current sweep
Polarography in which the current, linearly increasing with time, is applied to the indicator electrode and the potential E of the latter is measured as a function of the electrolysis time t or current I.

 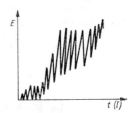

D Strom-Abtast-Polarographie, Polarographie mit linearer Stromabtastung, Polarographie mit linearem Strom-Zeit-Verlauf
F polarographie à balayage linéaire de courant
P polarografia z prądem narastającym liniowo
R полярография с развёрткой тока

current-step chronopotentiometry
The chronopotentiometry, in which a constant current I_1 is imposed on the indicator electrode and then, after a sudden change, the current $I_2 \neq -I_1$. The potential E of the electrode, is measured as a function of the electrolysis time t.

 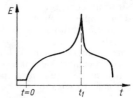

D Chronopotentiometrie mit Stromstufen, Stromstufen-Chronopotentiometrie
F chronopotentiométrie à échelon de courant
P chronopotencjometria ze skokową zmianą prądu
R хронопотенциометрия со ступенчатым изменением тока

current-voltage curve → polarization curve

curvilinear regression
The regression whose the functional form is
linear with respect to parameters $b_1, b_2, ..., b_k$
and non-linear with respect to the independent
variables $X_1, X_2, ..., X_n$

$$E\,(Y|X_1 = x_1, ..., X_n = x_n) = \sum_{i=1}^{k} b_i\,f_i(x_1, ..., x_n)$$

where at least one of the functions f_i
$(i = 1, ..., k)$ is non-linear.

D curvilineare Regression
F régression curvilinéaire, régression
curviligne
P regresja krzywoliniowa
R криволинейная регрессия

CV → coefficient of variation

cyclic chronopotentiometry
The chronopotentiometry in which periodically
reversed, symmetrical or asymmetrical, current I
is imposed on the indicator electrode, and the
potential E of this electrode is measured as
a function of the electrolysis time t.

Remark: The term cyclic current-reversal
chronopotentiometry may be used to signify that $I_2 = -I_1$,
and the term cyclic current-step chronopotentiometry to
signify that $I_2 \neq -I_1$. These two terms should not be
used except to emphasize the difference between them.

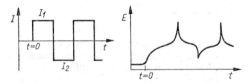

D cyclische Chronopotentiometrie
F chronopotentiométrie cyclique
P chronopotencjometria cykliczna,
chronopotencjometria wielocykliczna
R циклическая хронопотенциометрия

cyclic current-reversal chronopotentiometry
See cyclic chronopotentiometry

D cyclische Chronopotentiometrie mit
Stromumkehr, cyclische Stromumkehr-
-Chronopotentiometrie
F chronopotentiométrie à inversion cyclique
de courant
P chronopotencjometria cykliczna
z odwróceniem kierunku prądu
R циклическая хронопотенциометрия
с реверсом тока

cyclic current-step chronopotentiometry *See*
cyclic chronopotentiometry

D cyclische Stromstufen-Chronopotentiometrie
F chronopotentiométrie cyclique à échelon
de courant
P chronopotencjometria cykliczna ze skokową
zmianą prądu
R циклическая хронопотенциометрия со
ступенчатым изменением тока

cyclic triangular-wave polarography
Polarography in which during the life time
of one drop, a cycle of triangular-wave
potential pulses is applied to the indicator
electrode. The slopes of the linear sections of
the potential-time curve (a and b) may be equal
or unequal.

 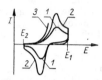

D cyclische Dreieckwellen-Polarographie,
cyclische Dreiecksspannungspolarographie
F polarographie à balayage triangulaire
cyclique
P polarografia cykliczna z falą trójkątną
R циклическая полярография с треугольной
развёрткой потенциала

cyclic triangular-wave voltammetry
The voltammetry involving diffusion mass
transfer to/from the electrode with a renewable
surface when the cyclic triangular potential is
applied to this electrode. The slopes of the two
linear segments of the potential-time curve a
and b may be different.

 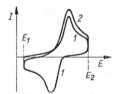

D cyclische Dreieckwellen-Voltammetrie,
cyclische Dreiecksspannungs-Voltammetrie
F voltampérométrie à balayage triangulaire
cyclique
P woltamperometria cykliczna (z falą
trójkątną), chronowoltamperometria
cykliczna
R циклическая вольтамперометрия
с треугольной развёрткой потенциала

cylindrical mirror analyser, CMA
A device for rapid analysis of the energy
of electrons based on their deflection in
a cylindrical electrostatic field. The deflection
is caused by the potential difference between
two coaxial cylinders: the inner at a ground
potential, and the outer one at a linearly
increasing negative potential. The device is
used in atomic emission and photoelectron
spectroscopies.

D Zylinder-Spiegelanalysator
F analyseur à miroir cylindrique
P cylindryczny analizator zwierciadlany
R анализатор с цилиндрическим зеркалом

D

Daffert automatic zero burette
D Bürette nach Daffert
F burette à zéro automatique de Daffert
P biureta z automatycznym nastawianiem zera Dafferta
R бюретка с автоматической установкой по Дафферту

damped balance
A balance in which the swings of the beam are damped pneumatically (by coaxial cylinders moving one inside another), or magnetically (a magnetic plate moving between the poles of a permanent magnet), thus securing rapid establishment of equilibrium.
D aperiodische Waage, Dämpfungswaage, gedämpfte Waage
F balance à amortisseurs
P waga aperiodyczna, waga tłumikowa, waga uchylna
R апериодичные весы, весы с успокоителями

DAPS → disappearance potential spectroscopy

dark current, i_d
A component of the output current of a photodetector that flows when no radiant energy is incident upon the detector.
D Dunkelstrom
F courant d'obscurité
P prąd ciemny
R темновой ток

daughter ion, product ion
An electrically charged product of reaction of a particular parent ion. The reaction need not necessarily involve fragmentation. It could, e.g. involve a change in the number of charges carried.
Remark: All fragment ions are daughter ions but not all daughter ions are necessarily fragment ions.
D Tochterion
F ion fils
P jon pochodny
R дочерний ион

daughter nuclide
A nuclide formed by radioactive decay of another nuclide called the parent one.
D Tochternuklid
F produit de filiation, substance fille
P nuklid pochodny
R дочерный нуклид

dc arc → direct current arc

dc polarography → direct current polarography

deactivation, desactivation, inactivation (*of an adsorbent*)
A process in which the activity of an adsorbent is decreased (the R_f value increases and the retention volume V_R decreases) due to adsorption of water vapour, washing with acid, etc.
D Entaktivierung, De(s)aktivierung
F désactivation, inactivation
P dezaktywacja
R дезактивирование, дезактивация, инактивация

dead-stop titration → amperometric titration with two indicator electrodes

dead time (*in chromatography*)
The time needed by the mobile phase molecules to pass through the extra-column volumes.
D Totzeit
F temps mort
P czas martwy
R мёртвое время

dead time (*of a device*)
Time immediately after a stimulus has been received during which an electrical device is insensitive to another pulse or stimulus.
D Totzeit
F temps mort
P czas martwy
R мёртвое время

dead volume → extra-column volume

Debye-Falkenhagen's effect, dispersion of conductance
An enhancement of the molar conductivity of an electrolyte when measurements are made at high frequencies of alternating electric field (of the order of 1 MHz) as a result of the disappearance of the relaxation-time effect.
D Debye-Falkenhagen-Effekt, Dispersionseffekt (der Leitfähigkeit)
F effet Debye-Falkenhagen, dispersion de conductance
P efekt Debye'a i Falkenhagena
R эффект Дебая-Фалькенгагена, дисперсия электропроводности

Debye-Hückel-Brönsted equation
A semiempirical equation describing the relation between the mean activity coefficient $y_\pm$ of a strong electrolyte and the ionic strength I of a solution of this electrolyte

$$\log y_\pm = -\frac{A|z_+ \, z_-|\sqrt{I}}{1 + Ba\sqrt{I}} + CI$$

where C — empirical constant. For the remaning notation see Debye-Hückel equation and Debye-Hückel limiting law.

D Debye-Hückel-Brönsted-Gleichung
F équation de Debye-Hückel-Brönsted
P równanie Debye'a, Hückela i Brönsteda
R третье приближение теории Дебая-Гюккеля

Debye-Hückel equation
Semiempirical equation describing the relation between the mean activity coefficient $y_\pm$, of a strong electrolyte and the ionic strength I of its solution

$$\log y_\pm = \frac{A|z_+ \, z_-|\sqrt{I}}{1 + Ba\sqrt{I}} \qquad B = e\left(\frac{2N_A}{\varepsilon_r \, \varepsilon_0 \, kT}\right)^{1/2}$$

where a — effective diameter of the ion. For the remaining notation see under Debye-Hückel limiting law.

D Debye-Hückelsche Gleichung
F formule de Debye-Hückel, équation de Debye-Hückel
P równanie Debye'a i Hückela
R (расширенное) уравнение Дебая и Хюккеля, второе приближение теории Дебая-Гюккеля

Debye-Hückel limiting law
The equation describing the relation between the mean activity coefficient $y_\pm$ of a strong electrolyte and the ionic strength I of its dilute solution

$$\log y_\pm = -A|z_+ \, z_-|\sqrt{I}$$

$$A = \frac{e^3 N_A^{1/2}}{2.303 \times 4\pi \times 2^{1/2}(\varepsilon_r \varepsilon_0 kT)^{3/2}} \quad \text{(in SI system)}$$

where z_+ and z_- — charge number of cation and anion of an electrolyte, respectively, e — elementary charge, N_A — Avogadro's number, ε_r — relative permittivity, ε_0 — permittivity of vacuum, k — Boltzmann constant; agreement between the experimental results and those predicted by the equation is obtained if the ionic strength does not exceed (in aqueous solutions) $0.02-0.03$ mol dm^{-3}.

D Debye-Hückelsches Grenzgesetz, Debye-Hückelsche Grenzbeziehung
F loi limite de Debye et Hückel
P równanie Debye'a i Hückela graniczne, prawo Debye'a i Hückela graniczne
R предельный закон Дебая-Хюккеля, предельное уравнение Дебая-Хюккеля, предельный закон Дебая-Гюккеля

Debye-Hückel-Onsager's limiting law for conductivity, Onsager's equation
The equation describing the molar conductance Λ (earlier known as the equivalent conductivity) of dilute solution of a strong electrolyte $A_{\nu_+}B_{\nu_-}$

$$\frac{\Lambda}{z_i \nu_i} = \frac{\Lambda^0}{z_i \nu_i} - \frac{|z_+ \, z_-| \, F^2 q \varkappa}{z_i \nu_i 12\pi N_A \varepsilon_0 \varepsilon_r \, RT \, (1 + \sqrt{q})} +$$

$$- \frac{(z_+ + |z_-|) \, F^2 \varkappa}{6\pi N_A \, \eta z_i \nu_i} \quad \text{(in SI system)}$$

where $z_i \nu_i = z_+ \, \nu_+ = |z_0| \, \nu_-$

$$q = \frac{z_+ \, z_- \, (\lambda_+^0 + \lambda_-^0)}{(z_+ + |z_-|) \, (z_- \, \lambda_+ + z_+ \, \lambda_-^0)}$$

$$\varkappa = e \left(\frac{2N_A \, I}{\varepsilon_0 \, \varepsilon_r \, kT}\right)^{1/2}$$

Λ_0 — limiting molar conductance, z_+ and z_- — charge number of cation and anion of the electrolyte, respectively, ν_+ and ν_- — number of cations and anions, respectively, formed from one molecule of the electrolyte, F — Faraday constant, N_A — Avogadro's number, ε_r — relative permittivity, ε_0 — permittivity of vacuum, k — Boltzmann's constant, T — thermodynamic temperature, η — dynamic viscosity, λ_+^0 and λ_-^0 — limiting conductivity of ions (cations and anions, respectively), e — elementary charge, I — ionic strength.

D (Debye-Hückel-) Onsagersches Grenzgesetz
F loi limite de Debye, Hückel et Onsager, équation d'Onsager
P równanie (Debye'a, Hückela i Onsagera) graniczne
R уравнение (Дебая-Гюккеля-)Онзагера, предельный закон Онзагера

Debye-Hückel's theory of strong electrolytes
The theory based on the simplified model of ionic cloud, which describes the physico-chemical properties of dilute solutions of strong electrolytes.

D Debye-Hückelschen Theorie der starken Elektrolyte, Debye-Hückel-Theorie, Theorie von Debye und Hückel
F théorie des électrolytes forts de Debye et Hückel
P teoria elektrolitów mocnych Debye'a i Hückela
R теория (сильных электролитов) Дебая и Хюккеля, теория Дебая и Гюккеля

decantation
The removal of the supernatant liquid from over the precipitate by pouring off the liquid without disturbing the precipitate.

D Dekantation
F décantation
P dekantacja
R декантация

decay chain, radioactive chain, radioactive series
A series of nuclides in which each member transforms into the next through radioactive decay until a stable nuclide formed.
D Zerfallsreihe, Umwandlungsfolge
F série de désintégrations
P szereg promieniotwórczy
R цепочка распадов, радиоактивный ряд

decay curve
A plot of the activity of a radioactive sample versus time, generally in a semilogarithmic coordinate system.
D Zerfallskurve
F courbe de désintégration, courbe de décroissance
P krzywa rozpadu promieniotwórczego
R кривая радиоактивного распада

decay scheme
A graphic presentation of the energy levels of the members of a decay chain showing the way along which radioactive decay proceeds.
D Zerfallsschema
F schéma de désintégration
P schemat rozpadu
R схема распада

decomposition potential
The electrode potential at which the electrolysis current begins to increase appreciably.
D Zersetzungspotential
F potentiel de décomposition
P potencjał rozkładu
R потенциал разложения

decomposition voltage
The extrapolated to $I = 0$ value of voltage corresponding to characteristic bend of the electrolysis current I versus electrolysis voltage U curve
Remark: The term is used mainly in technological electrochemistry; it does not have a clear physical interpretation

D Zersetzungsspannung
F tension de décomposition
P napięcie rozkładu
R напряжение разложения

decontamination (*in nuclear science*)
Removal of radioactive matter from skin, textiles, instruments and wrappings, i.e. from places where radioactivity threatens health and safety.
D Dekontamination, Entseuchung
F décontamination, désactivation
P dekontaminacja, odkażanie
R дезактивация

decontamination factor
The ratio of the amount of radioactive impurities present in a material before and after purification.
D Dekontaminationsfaktor, Entseuchungsgrad
F facteur de décontamination
P współczynnik dekontaminacji
R коэффициент очистки

deformation vibrations → bending vibrations

degree of extraction → recovery factor

degree of ionization → apparent degree of dissociation

degree of surface coverage → surface coverage

dehomogenisation of a mobile phase → solvent demixing

Dehydrite → Anhydrone

deionized water, demineralized water
Water from which foreign ions have been removed usually by multistep distillation or use of ion-exchangers.
D deionisiertes Wasser, entionisiertes Wasser
F eau déionisée, eau déminéralisée
P woda dejonizowana, woda demineralizowana
R обессоленная вода, деминерализованная вода

delayed neutrons
Neutrons emitted by some fission fragments of heavy nuclei not directly after fission (contrary to prompt neutrons) but only after the β-decay of these fragments, i.e. several dozen seconds after fission.
D verzögerte Neutronen
F neutrons retardés
P neutrony opóźnione
R запаздывающие нейтроны

delta scale, δ-scale (*in NMR spectroscopy*)
A commonly used dimensionless scale to express the chemical shifts with respect to a standard, independently of the magnetic field strength of the instrument. The resonance signal values are expressed in parts per million, ppm.
D δ-Skala
F échelle δ
P skala δ
R шкала δ

demasking agent
A reagent which converts masked compounds into reactive forms; e.g. formaldehyde decomposes cyanide complexes, previously formed for masking.

D Demaskierungsmittel
F réactif démasquant
P odczynnik demaskujący
R демаскировочный реактив

demineralized water → deionized water

demodulation polarography
The alternating current polarography in which a sinusoidal alternating voltage of angular frequency of ω_0 and small amplitude modulated with another sinusoidal voltage of angular frequency ω_m and of (usually) the same amplitude is used. The faradaic demodulation current I_{FD} is measured as a function of the direct potential E_{dc} of the indicator electrode. The faradaic demodulation current of the angular frequency ω_m is due to non-linearity of the faradaic admittance of the indicator electrode.

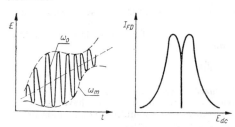

D Demodulationspolarographie
F polarographie par démodulation faradaïque, polarographie DF
P polarografia demodulacyjna
R демодуляционная полярография

densitometer → microphotometer

density bottle → pycnometer

density curve → emulsion characteristic curve

depolarizer
A substance which undergoes reduction or oxidation reaction sat an electrode and minimizes the polarization effects in electrochemical cell.
D Depolarisator
F dépolarisant, dépolariseur
P depolaryzator
R деполяризатор

deposition potential
The smallest potential producing electrolytic deposition when applied to an electrolytic cell.
D Abscheidungspotential
F potentiel de précipitation
P potencjał wydzielania
R потенциал выделения

depth profile, concentration profile
The depth dependence of the concentration of elements or compounds in a solid in the direction normal to the surface determined by a layer-by-layer sputtering of the solid surface and analysis of the sputtered species or composition of the newly exposed surface.
D Tiefenprofil, Konzentrationsprofil
F profil de profondeur
P profil stężenia, profil koncentracji
R профиль распределения по глубине

derivative chronopotentiometry
The chronopotentiometry in which a plot of the derivative dE/dt against the electrolysis time t is used for the determination of the transition time τ.

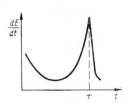

D derivative Chronopotentiometrie
F chronopotentiométrie dérivée
P chronopotencjometria różniczkowa, chronopotencjometria pierwszej pochodnej
R производная хронопотенциометрия

derivative polarography
The polarography in which the relationship between the derivative of the current with respect to time (dI/dt) — or alternatively (dI/dE) — and the potential E of the indicator electrode, is investigated.

Remark: The term is often used in a narrow sense to denote derivative dc polarography.

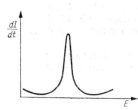

polarization curve recorded under conditions of the dc polarography

D Derivativ-Polarographie, derivative Polarographie
F polarographie dérivée
P polarografia różniczkowa, polarografia pierwszej pochodnej
R производная полярография

derivative potentiometric titration
The potentiometric titration in which the first derivative dE/dV, where E is the potential of the indicator electrode and V is the volume of the added titrant, is measured.

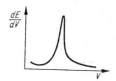

V — volume of a titrant

D derivative potentiometrische Titration
F titrage potentiométrique dérivé
P miareczkowanie potencjometryczne
   różniczkowe (pierwszej pochodnej)
R производное потенциометрическое
   титрование

derivative pulse polarography
Polarography in which square potential pulses
of a small and constant amplitude ΔE are
superimposed on the linearly varying potential
E_0 of the indicator electrode. The duration
of the pulses is small, and it is much less than
the drop life time. The measured response is
the difference ΔI between the direct currents
that flow through the indicator electrode at the
end of two successive pulses, for two successive
drops. One drop corresponds to one pulse.

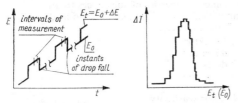

D derivative Puls-Polarographie
F polarographie dérivée à impulsions
P polarografia pulsowa różniczkowa,
   polarografia pulsowa pochodna
R производная импульсная полярография

derivative thermodilatometry
The thermodilatometry in which the first
derivative of the dimension of a substance
under investigation is measured versus time,
e.g. $dl/dt = f(T)$, where l — length of the
examined substance, T — temperature.

D Derivativ-Dilatometrie
F dilatométrie dérivée
P termodylatometria różniczkowa
R производная дилатометрия

derivative thermogravimetry, DTG
A technique yielding the first derivative of the
gravimetric curve with respect to either time or
temperature.

D Differentialthermogravimetrie
F thermogravimétrie dérivée
P termograwimetria różniczkowa
R производная термогравиметрия

derivative voltammetry
The voltammetry in which the relationship is
measured between the rate of current variation

in time dI/dt (or dI/dE) and the potential E
of the indicator electrode with a non-renewable
surface are measured.

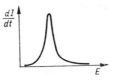

polarization curve recorded under conditions of the
hydrodynamic voltammetry

D derivative Voltammetrie
F voltampérométrie dérivée
P woltamperometria różniczkowa
R производная вольтамперометрия

derivatization (*in chromatography*)
The preparation of a derivative of the
component under examination, usually to
obtain a more volatile, coloured, or electroactive
substance in order to enhance selective response
of the detector.

D Derivatisierung
F dérivatisation
P derywatyzacja
R образование производных

derivatograph
The apparatus used in thermal analysis for
simultaneous investigation of a substance by
the thermogravimetric method and by
differential thermal analysis (simultaneous
recording of curves DTA, DTG and TG; *cf.*
simultaneous techniques).

Remark: The terms derivatography, derivatographic
analysis, differential and thermogravimetric analyses are
not recommended for methods where the derivatograph is
applied.

D Derivatograph
F dérivatograph
P derywatograf
R дериватограф

desactivation → deactivation

descending chromatography → descending
development

descending development, descending
chromatography, descending technique
(*in planar chromatography*)
A technique of developing a chromatogram
in which the mobile phase is introduced to the
upper edge of the sheet of paper or thin-layer
plate and flows downwards due to both
capillary action and gravitation.

D absteigende Entwicklung, absteigende
   Chromatographie, absteigende Methode
F développement vertical descendant,
   chromatographie descendante, méthode
   descendante

P rozwijanie spływowe, rozwijanie zstępujące, chromatografia zstępująca, metoda zstępująca
R нисходящее проявление, нисходящая хроматография, нисходящий метод

descending technique → descending development

desiccant
A substance capable of combining water and its vapour.
D Trockenmittel
F desséchant, agent de séchage
P środek suszący
R сушильный агент, осушитель

desiccator
A glass vessel for drying or for keeping materials over a suitable desiccant such as calcium chloride, phosphorus pentoxide, sulfuric acid, etc.
D Exsikkator
F dessiccateur, exciccateur
P eksykator
R эксикатор

designated volume
The volume at the temperature at which the volumetric glassware was calibrated (usually 20°C, but possibly 25° or 27°C in tropical countries).
D markiertes Volumenmeßgefäß
F volume nominal
P objętość deklarowana, objętość wyznaczona
R обозначенный объём

design matrix
A matrix whose elements x_{ij} denote values of the jth factor (independent variable) in the ith experimental run; one row of the design matrix presents the conditions of a single experimental run.
D Versuchsmatrix
F ...
P macierz planu, macierz planowania, macierz doświadczenia, macierz wejść
R матрица плана, матрица планирования

design of the n^p-type → complete factorial design

desorption
The process reverse to sorption, consisting in the liberation of the adsorbate from the adsorbent surface, or the absorbate from the total volume of the absorbent.
D Desorption
F désorption
P desorpcja
R десорбция

destructive activation analysis
An activation analysis in which the activity of the element (or a group of elements) to be determined is measured after its (their) separation from an irradiated sample by chemical methods.
D chemische Aktivierungsanalyse
F analyse destructive par activation
P analiza aktywacyjna destrukcyjna
R ...

detection
The process by which the presence of substances is recognized by means of chemical reactions or physical processes.
D Nachweis
F détection
P wykrywanie
R обнаружение

detection of radiation
Detection and characterization of the nature and energy of a radiation.
D Strahlennachweis
F détection de la rayonnement
P detekcja promieniowania
R обнаружение излучения

detector
A device applied to convert physical or chemical properties of substances to an analytical signal (digit, curve, colour, deflection of the recorder).
D Detektor
F détecteur
P detektor
R детектор

detector sensitivity
The increase of the detector signal per unit increase of concentration or mass of the test substance.
D Detektorempfindlichkeit
F sensibilité d'un détecteur
P czułość detektora
R чувствительность детектора

determinand
The substance being determined.
D bestimmte Substanz
F substance dosée, substance déterminée
P substancja oznaczana
R определяемое вещество

determination
The ascertainment of the quantity or concentration of a specific substance in a sample.
D Bestimmung
F dosage
P oznaczanie
R определение

deuterium discharge lamp, deuterium discharge tube
A discharge lamp filled with deuterium, usually employed as a continuous source of

high-intensity radiation of ultraviolet frequency
in the range of 165 to 360 nm.

D Deuteriumlampe
F lampe à deutérium
P lampa deuterowa
R дейтериевая лампа

deuterium discharge tube → deuterium
discharge lamp

developing chamber → chromatographic
chamber

development (*in photography*)
The chemical process producing a visible
image on the exposed light-sensitive material.

D Entwicklung
F développement
P wywoływanie
R проявление

development *See* elution

D Entwicklung
F développement
P rozwijanie
R проявление

development of a chromatogram
(*in planar chromatography*)
The process of passing an appropriate mobile
phase over the thin layer of the stationary
phase or paper to obtain separation as a result
of different migration rates of bands or spots
of components being separated.

D Chromatogrammsentwicklung,
   Entwicklung des Chromatogramms
F développement d'un chromatogramme
P rozwijanie chromatogramu
R проявление хроматограммы

development of a chromatogram → running
a chromatogram

development tank → chromatographic chamber

dextrarotatory substance, *d*-substance,
(+)-substance
An optically active substance that rotates the
plane of polarization of polarized light to the
right, or in the clockwise sense, as viewed
looking toward the light source.

D rechtsdrehende Substanz, *d*-Substanz,
   (+)-Substanz
F substance dextrogyre, *d*-substance,
   (+)-substance
P substancja prawoskrętna, *d*-substancja,
   (+)-substancja
R правовращающее вещество, *d*-вещество,
   (+)-вещество

diatomaceous earth, diatomite, kieselguhr,
infusorial earth
A mineral consisting mainly of amorphous,
hydrated silica contaminated with metal oxides;
its origin are frustules of diatoms. After
chemical treatment the diatomaceous earth

may be used in chromatography as a solid
support of weak adsorptive properties (Celite,
Chromosorb).

D Diatomenerde, Kieselguhr, Infusorienerde
F terre de diatomées, kieselguhr, terre
   d'infusoires
P ziemia okrzemkowa
R диатомитовая земля, диатомит, кизельгур,
   инфузорная земля

diatomite → diatomaceous earth

dielcometric titration → dielectrometric
titration

dielcometry → dielectrometry

dielectric constant → relative
permittivity

dielectrometric analysis *See* dielectrometry

D dielektrometrische Analyse
F analyse diélectrométrique
P analiza dielektrometryczna
R диэлектрометрический анализ

dielectrometric titration
Titration in which the end point is determined
from the abrupt change in the relative
permittivity ε_r of solution.

Remark: The term dielcometric titration is not
recommended.

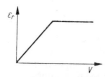

V — volume of a titrant

D dielektrometrische Titration, dekametrische
   Titration, dielkometrische Titration
F titrage diélectrométrique, titrage
   dielcométrique
P miareczkowanie dielektrometryczne
R диэлектрическое титрование,
   декаметрическое титрование,
   диэлькометрическое титрование

dielectrometry
An electrochemical technique in which the
relative permittivity ε_r of a solution is
measured as a function of the concentration
c of a substance.

Remark: The term dielcometry is not recommended. In the
field of chemical analysis the term dielectrometric analysis
is also used.

D Dielektrometrie, Dekametrie, DK-Metrie,
   Dielkometrie
F diélectrométrie, diélcométrie
P dielektrometria
R диэлектрометрия, декаметрия,
   диэлькометрия

differential amperometry

The amperometry in which the difference ΔI of
currents flowing between two indicator
electrodes immersed in two solutions separated
from each other by a salt bridge is measured
as a function of concentration c of the
electroactive substance.

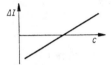

D Differentialamperometrie,
   Differenz-Amperometrie
F ampérométrie différentielle
P amperometria różnicowa
R разностная амперометрия

differential capacitance of double layer, C

The nonlinear capacitance of the capacitor
formed by the electric double layer at the
electrode-electrolyte interface given by the
equation

$$C = (\partial q / \partial E)_E$$

where q — electric charge of electrode,
E — potential of the electrode.

D differentielle Kapazität der Doppelschicht
F capacité différentielle de la couche double
P pojemność warstwy podwójnej różniczkowa
R дифференциальная ёмкость двойного слоя

differential chromatogram

A record of the analytical signal from the
differential detector, consisting of a series of
successive peaks.

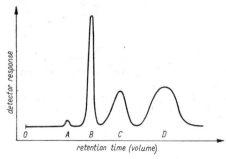

0 — start point, A — air peak, B, C, D — peaks of the
resolved components

D Differentialchromatogramm
F chromatogramme différentiel

P chromatogram różnicowy
R дифференциальная хроматограмма

differential detector (*in chromatography*)

A detector which measures instantaneous
concentration of the solute in the effluent.
The chart record from such a detector consists
of a series of peaks, the surface area of each
peak corresponding to the quantity of each
component of a mixture being separated.

D Differentialdetektor
F détecteur différentiel
P detektor różnicowy
R дифференциальный детектор

differential polarography

The polarography in which the difference ΔI
of currents flowing through two indicator
electrodes placed in two separate solutions
is measured.

Remark: The term is often used in a narrow sense to
denote the differential dc polarography.

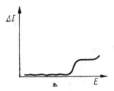

polarization curve recorded under conditions of the dc
polarography

D Differenz-Polarographie,
   Differentialpolarographie
F polarographie différentielle
P polarografia różnicowa
R разностная полярография,
   дифференциальная полярография

differential potentiometric titration

The potentiometric titration in which the
potential difference between two indicator
electrodes immersed in two solutions connected
by a salt bridge is measured; one of the
indicator electrodes is often placed in the
titrant.

V — volume of a titrant

D differenzpotentiometrische Titration,
   differentialpotentiometrische Titration,
   differentialpotentiometrische Indikation
F titrage potentiométrique différentiel
P miareczkowanie potencjometryczne różnicowe

R разностное потенциометрическое
титрование, дифференциальное
потенциометрическое титрование

differential potentiometry
The potentiometry in which the quantity
measured is the difference of potentials between
two indicator electrodes immersed in two
separate solutions connected by a salt bridge.

Remark: The term precision null-point potentiometry is not
recommended.

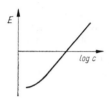

D Differenz-Potentiometrie,
   Differentialpotentiometrie,
   Präzisions-Nullstrom-Potentiometrie
F potentiométrie différentielle, potentiométrie
   de précision par compensation
P potencjometria różnicowa
R разностная потенциометрия,
   дифференциальная потенциометрия,
   прецизионная потенциометрия с нулевой
   точкой

differential pulse polarography
The polarography in which square potential
pulses of a small but constant amplitude
(usually of 30 mV) are superimposed on the
linearly varying potential of the indicator
electrode. The measured response is the
difference between the direct current which
flows at the end of a pulse and the one flowing
immediately before the pulse. One drop
corresponds to one pulse.

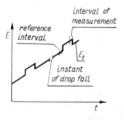
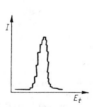

D Differenz-Puls-Polarographie, differentielle
   Pulspolarographie,
   Differentialpulspolarographie
F polarographie différentielle à impulsions
P polarografia pulsowa różnicowa
R разностная импульсная полярография,
   дифференциальная импульсная
   полярография

differential scanning calorimetry, DSC
A technique in which the difference in energy
inputs into a substance and a reference

material is measured as a function of
temperature whilst the substance and reference
material are subjected to a controlled
temperature programme.

D Differential-Scanning-Kalorimetrie
F analyse calorimétrique différentielle, analyse
   enthalpique différentielle
P kalorymetria skaningowa różnicowa
R дифференциальная сканирующая
   калориметрия

differential spectrophotometry
A spectrophotometric method for the
determination of concentration of a substance
in which the intensity of radiation transmitted
by a sample solution of unknown concentration
is measured and compared with that of a sample
solution of accurately known concentration.

D Differential-Spektrophotometrie
F spectrophotométrie différentielle
P spektrofotometria różnicowa
R дифференциальная спектрофотометрия

differential thermal analysis, DTA
A technique in which the temperature
difference between a substance and a reference
material is measured as a function of
temperature or time whilst the substance and
reference material are subjected to a controlled
temperature programme.

D Differentialthermoanalyse, differentiale
   thermische Analyse
F analyse thermique différentielle
P analiza termiczna różnicowa
R дифференциальный термический анализ

differential thermodilatometry
Thermodilatometry in which the difference of
dimension(s) between the examined and
reference substances are measured as a function
of temperature; e.g. $l_X - l_R = f(T)$ where
l_X — length of the examined substance,
l_R — length of the reference substance,
T — temperature.

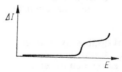

polarization curve recorded under conditions of the
hydrodynamic voltammetry

D Differentialdilatometrie
F dilatométrie différentielle
P termodylatometria różnicowa
R дифференциальная дилатометрия

differential voltammetry
The voltammetry in which the difference ΔI
of currents flowing through two indicator
electrodes with non-renewable surfaces and
placed in two separate solutions is measured.

D Differenz-Voltammetrie,
  Differentialvoltammetrie
F voltampérométrie différentielle
P woltamperometria różnicowa
R разностная вольтамперометрия

differentiating effect (*in acidimetry and alkalimetry*)
The differentiation in the protogenic ability of acids caused by a solvent.
D Differenzierung, differenzierender Effekt
F effet sélectif
P efekt różnicujący
R дифференцирующий эффект

differentiating solvent
A solvent exhibiting a differentiating effect on the protonodonating or protonoaccepting properties of definite groups of acids or bases.
D differenzierendes Lösungsmittel
F solvant différenciant, solvant capable de différencier des acides ou des bases
P rozpuszczalnik różnicujący
R дифференцирующий растворитель

diffraction grating, grating
An optical device with a planar or spherical polished surface on which equidistant parallel slits (or grooves) heve been ruled. The grating reflects or transmits incident radiation with interference, thus producing spectra.
D Beugungsgitter, Diffraktionsgitter
F réseau de diffraction
P siatka dyfrakcyjna
R дифракционная решётка

diffraction of light → light diffraction

diffractometry
Method of determining crystal or atomic structure of solids, liquids or gases by diffraction of X-rays or other beams.
D Diffraktometrie, Beugungsanalyse
F diffractométrie, analyse par diffraction
P dyfraktometria, analiza dyfrakcyjna
R дифрактометрия, дифракционный анализ

diffuse double layer
The part of the electrical double layer which occurs in a solution of an electrolyte separated from the electrode surface by a compact layer.
D diffuse Doppelschicht, diffuse Schicht, Gouy-Chapman-Schicht
F couche diffuse, couche de Gouy
P warstwa (podwójna) rozmyta, warstwa Gouya i Chapmana
R диффузионная часть двойного слоя, диффузионный (двойной) слой, слой Гуи-Чэпмена

diffusion (*in electrolyte solutions*)
Motion of ions and molecules in electrolyte solutions under the influence of the chemical potential gradient.

D Diffusion
F diffusion
P dyfuzja
R диффузия

diffusion coefficient, D
The quantity defined by Fick's law of diffusion, characteristic for a given substance (molecules or ions) and medium; it is equal to the quantity of the substance diffusing through a unit of surface area in unit time when the concentration changes by a unit along a unit path.
Remark: The term diffusion constant is not recommended.
D Diffusionskoeffizient, Diffusionskonstante
F coefficient de diffusion, diffusivité, constante de diffusion
P współczynnik dyfuzji, stała dyfuzji
R коэффициент диффузии, диффузионная постоянная

diffusion constant → diffusion coefficient

diffusion control → diffusion controlled process

diffusion control(led process)
A process, whose rate is determined by a slow (as compared with other steps) rate of the diffusion transfer of reagents to/from the interface (in the case of the electrode process to/from the surface of an electrode).
D Diffusions-Kontrolle
F processus contrôlé par diffusion
P proces kontrolowany dyfuzyjnie, kontrola dyfuzyjna
R электродный процесс лимитируемый стадией диффузии

diffusion current, I_d (*in elektrochemistry*)
The current which under given conditions is determined by the rate of diffusion transfer of reagents through the diffusion layer to the surface of an electrode.
D Diffusionsstrom
F courant de diffusion
P prąd dyfuzyjny
R диффузионный ток

diffusion flame
The flame of a burner fed solely with a fuel gas, which mixes subsequently with the oxidant gas (e.g. the surrounding air) at the burner outlet by diffusion.
D Diffusionsflamme
F flamme de diffusion
P płomień dyfuzyjny
R диффузионное пламя

diffusion layer
Region of phase in which the chemical potential of a component changes linearly with distance; in solution of an electrolyte the diffusion layer is located at the surface of the electrode when the rate of the electrode reaction is much higher than the rate of diffusion.

D Diffusionsschicht
F couche de diffusion
P warstwa dyfuzyjna
R диффузионный слой

diffusion overpotential, η_d
An overpotential of an electrode process
caused by a slow, as compared with the charge
transfer step, mass transfer of reagents to or
from the electrode surface.
D Diffusionsüberspannung
F surtension de diffusion
P nadpotencjał dyfuzyjny
R перенапряжение диффузии, диффузионное
    перенапряжение

diffusion potential, E_d
The difference of electric potentials of two
electrolytes in the same liquid phase, which
arises at the interface of two electrolytes
as a result of differences in mobilities and
activities of ions.
D Diffusionspotential
F potentiel de diffusion
P potencjał dyfuzyjny
R диффузионный потенциал

digestion
The passing of a substance into solution as
a result of a chemical reaction.
D Aufschluß
F digestion
P roztwarzanie
R растворение

dilatometer
An apparatus for measuring the coefficient of
expansion, or the change of density of a solid
or liquid due to variation of temperature. It
consists of a small stoppered glass bulb with
a measuring capillary tube attached to it.
D Dilatometer
F dilatomètre
P dylatometr
R дилатометр

diluent (in extraction)
An inert (organic) solvent used to improve
the physical properties (e.g. density, viscosity)
or the extractive properties (e.g. selectivity)
of the extractant.
D Verdünnungsmittel
F diluant
P rozcieńczalnik
R разбавитель

dilution correction factor
The factor by which the result of titration must
be multiplied to obtain a result independent
of the volume change of the sample taking
place in instrumental methods of titration.
$$P = (V + V_1)/V$$
where P – correction factor, V – initial volume
of titrated solution, V_1 – added volume of the
titrant.

D Korrektion, Verdünnungskorrektion
F facteur de dilution
P poprawka na rozcieńczenie
R поправка на разбавление

dilution effect (in titrimetric analysis)
The change of the measured quantity caused
by dilution of the sample due to addition of
the titrant.
D Verdünnungseffekt
F effet de dilution
P efekt rozcieńczania
R эффект разбавления

dilution limit
The reciprocal of the concentration limit.
The ratio of the volume of solution to the mass
of the substance to be detected. Its numerical
value is the number of millilitres of solution
containing 1 g of the said substance in
which the latter may still be detected by a given
method.
D Verdünnungsgrenze
F limite de dilution
P rozcieńczanie graniczne
R предельное разбавление

dimethyldichlorosilane, DMDCS
A silylation reagent used in gas chromatography
either for deactivation of supports and
adsorbents or for derivatization of the sample
components.
D Dimethyldichlorsilan
F dimethyldichlorosilane
P dimetylodichlorosilan
R диметилдихлорсилан

dipolar ion → ampholyte ion

direct current arc, dc arc
The most common type of electrical arc fed by
a dc source having a total available potential
difference of 100 to 300 V and a power of
several kW. In most instruments the arc is
ignited between graphite electrodes.
D Gleichstrombogen
F arc à courant continu
P łuk prądu stałego
R дуга постоянного тока

direct current polarography, dc polarography, conventional polarography
The branch of polarography in which the
potential E applied to the indicator electrode
is varied linearly with time.

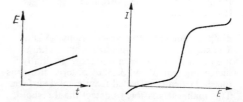

D Gleichstrompolarographie, dc-Polarographie
F polarographie en courant continu,
  cc polarographie
P polarografia stałoprądowa, polarografia
  klasyczna
R постоянно-токовая полярография,
  полярография постоянного тока,
  классическая полярография

direct-injection burner
A device combining the functions of
a nebulizer and burner. The fuel and oxidant
are fed to the burner separately and mixed
above the burner orifices thanks to their
turbulent motion. Most commonly, the oxidant
is also used for aspirating and nebulizing
the sample.
Remark: The term total consumption burner is not
recommended.

D direktzerstäubender Brenner
F brûleur à injection directe, brûleur
  à consommation totale
P palnik z bezpośrednim wtryskiem
R прямоточная горелка, горелка с полным
  потреблением

direct isotope dilution analysis
A method of isotope dilution analysis in
which a substance with a natural isotopic
composition is determined by adding a known
amount of the same substance labelled with
a radioactive tracer.

D einfache Isotopenverdünnungsanalyse,
  direkte Isotopenverdünnungsanalyse
F analyse par (la méthode de) dilution
  isotopique simple
P analiza metodą rozcieńczenia izotopowego
  prostego
R анализ методом прямого изотопного
  разбавления

direct method
An analytical method in which the signal
characteristic of the component to be detected
or determined is the basis of measurement.

D direktes Verfahren, Direktverfahren
F méthode directe
P metoda bezpośrednia
R прямой метод

direct titration
Titration in which the analysed substance
reacts directly with the added titrant.

D direkte Titration
F titrage direct
P miareczkowanie bezpośrednie
R прямое титрование

disappearance potential spectroscopy, DAPS
The surface sensitive spectroscopy, making
use of an electron beam of energy gradually
increasing from 0 to 2 keV to observe electrons
quasi-elastically reflected from a solid surface.
This makes it possible to detect the threshold,

at which the primary electrons produce
excitation of the core electrons and disappear
from the reflected beam.
D ...
F ...
P spektroskopia potencjału zanikania
  elektronów
R спектроскопия затухающего потенциала

discharge lamp, electric-discharge lamp
A lamp in which light is produced by an
electrical discharge between electrodes in a gas
(or vapour) at low or high pressure.
D Entladungslampe
F lampe à décharge
P lampa wyładowcza
R разрядная лампа

discharge polarography → incremental-charge
polarography

discontinuous simultaneous techniques
Coupled techniques in which collection of the
sample for testing by the second technique
is periodic, e.g. gas-chromatographic analysis
of volatile products evolved from a sample
investigated in the apparatus for differential
thermal analysis.
D simultane diskontinuierliche
  Untersuchungsmethoden
F ...
P techniki jednoczesne współdziałające
  nieciągle
R ...

discontinuous spectrum, discrete spectrum
A spectrum in which the quantity being studied
takes on discrete values.
Remark: In spectroscopic terminology, the term is used for
the line and band spectra.
D diskontinuierliches Spektrum
F spectre discontinu
P widmo nieciągłe
R дискретный спектр

discrete random variable, discrete variate
Random variable which can take a finite or
countable number of values from the given
interval.
D diskrete Zufallsgröße, diskrete Variable
F variable aléatoire discrète
P zmienna losowa nieciągła, zmienna losowa
  skokowa, zmienna losowa dyskretna
R дискретная случайная величина

discrete spectrum → discontinuous spectrum

discrete variate → discrete random variable

disintegration rate → activity

dispersion
A system of particles dispersed and suspended
in a solid, liquid or gas.

D Dispersion
F dispersion
P dyspersja
R дисперсия

dispersion (*in statistics*)
Variation of the elements of a population
(e.g. measurements or observations) caused by
random reasons.
D Dispersion, Streuung
F dispersion (statistique)
P rozrzut, dyspersja statystyczna
R дисперсия

dispersion of light → light dispersion

dispersion of conductance →
Debye-Falkenhagen's effect

displacement (*in chromatography*)
A process in which the components of the
mixture are displaced from the bed by the
mobile phase whose affinity to the stationary
phase is greater than that of any of the
components being separated.
D Verdrängung
F déplacement
P rugowanie, wypieranie
R вытеснение

displacement chromatography
A chromatographic technique in which
a sample after being introduced into the
chromatographic bed, is then displaced from
it by the mobile phase containing a displacing
agent (a substance whose affinity to
the stationary phase is greater than that of
the components under examination). The
particular sample components emerge from
the bed partially separated from each other,
in the order of increasing affinity to the
stationary phase.
D Verdrängungschromatographie
F chromatographie par déplacement
P chromatografia przez rugowanie,
   chromatografia rugująca, metoda rugująca
R вытеснительная хроматография,
   вытеснительный метод

displacement titration → replacement titration

dissociation
Splitting of a molecule into smaller fragments.
D Dissoziation
F dissociation
P dysocjacja
R диссоциация

dissociation field effect, second Wien effect
An increase of the molar conductivity of weak
electrolytes due to the enhancement of the
degree of dissociation of the electrolyte when
measurements are made at a high electric field
strength ($10^4 - 10^5$ V/cm).

D Dissoziationsfeldeffekt, zweiter Wien-Effekt,
   Felddissoziation
F effet dissociant du champ électrique
P efekt dysocjacyjny pola elektrycznego,
   dysocjacja pod wpływem pola elektrycznego,
   drugi efekt Wiena
R эффект диссоциации в поле, второй эффект
   Вина

dissociation fraction → apparent degree of
dissociation

dissolution
The mixing of two phases with the formation of
one new homogenous phase (i.e. solution).
D Auflösen, Lösen
F dissolution
P rozpuszczanie
R растворение

distilled water
Water that has been purified by distillation.
D destilliertes Wasser
F eau distillée
P woda destylowana
R дистиллированная вода

χ^2-distribution → chi-squared distribution

distribution coefficient → concentration
distribution ratio

distribution coefficient D_g
The ratio of the amount of a component
per gram of the dry stationary phase
to its analytical concentration in the mobile
phase. This coefficient is applicable in ion-
exchange and gel-permeation chromatography,
where swelling occurs, and in adsorption
chromatography with adsorbents of unknown
surface area.
D Verteilungskoeffizient D_g
F coefficient de distribution D_g
P współczynnik podziału D_g
R коэффициент распределения D_g

distribution coefficient D_s
The ratio of the amount of a component
adsorbed by 1 m^2 of surface area of the
stationary phase to its analytical concentration
in the mobile phase. This coefficient is
applicable in adsorption chromatography
with a well characterized adsorbent of known
surface area.
D Verteilungskoeffizient D_s
F coefficient de distribution D_s
P współczynnik podziału D_s
R коэффициент распределения D_s

distribution coefficient D_v, volume distribution
coefficient
The ratio of the amount of a component in
the stationary phase per cm^3 of the bed
volume to its analytical concentration in the

mobile phase. This coefficient is applicable in chromatography when it is not practicable to determine the mass of the solid stationary phase.

D Verteilungskoeffizient D_v
F coefficient de distribution D_v, coefficient de distribution en volume
P współczynnik podziału D_v, współczynnik podziału objętościowy
R коэффициент распределения D_v, объёмный коэффициент распределения

distribution constant, partition coefficient, K_D
(*in chromatography*)
The ratio of the concentration of component A in a single definite form in the stationary phase (s) to its concentration in the same form in the mobile phase (m) at equilibrium. The concentrations are calculated per unit volume of the phase

$$(K_D)_A = \frac{[A]_s}{[A]_m}$$

D Verteilungskonstante
F constante de distribution, coefficient de partage
P stała podziału
R константа распределения

distribution constant, partition coefficient, K_D
(*in extraction*)
The ratio of the concentration of a substance A to be extracted in a single definite form in the organic solvent phase (o) to its concentration in the same form in the aqueous phase (w) at equilibrium

$$(K_D)_A = \frac{[A]_o}{[A]_w}$$

Remark: The use of the inverse concentration ratio (aqueous/organic) or the ratio of the concentration of the less dense phase to the concentration of the denser phase is not recommended.

D Verteilungskonstante, (Nernstscher) Verteilungskoeffizient
F constante de distribution, coefficient de partage
P stała podziału
R константа распределения

distribution function
A function $F(x)$ determining the probability that a random variable X will assume a value smaller than the fixed real value x. Usually the distribution function is defined by the equation

$$F(x) = P(X < x)$$

D Verteilungsfunktion
F fonction de distribution
P dystrybuanta
R функция распределения

distribution isotherm (*in chromatography*)
A relationship between the concentration of a given substance in the stationary phase and

its concentration in the mobile phase at equilibrium under specified conditions, and at constant temperature.

D Verteilungsisotherme
F isotherme de distribution
P izoterma podziału
R изотерма распределения

distribution law → Nernst distribution law

distribution ratio → concentration distribution ratio

DMDCS → dimethyldichlorosilane

DME → dropping-mercury electrode

Donnan potential → membrane potential

donor-acceptor complex, charge-transfer complex
A complex of molecules which are electron donors with molecules being electron acceptors. Such combined molecules are able to exist independently.

D Elektronen-Donator-Akzeptor-Komplex, Landungsübertragungskomplex
F complexe de transfert de charge
P kompleks donorowo-akceptorowy
R донорно-акцепторный комплекс, комплекс с переносом заряда

Dorn effect → sedimentation potential

double-beam spectrophotometer
A spectrophotometer in which the monochromatic beam is split into two components. One beam passes through the reference material, the other through the investigated sample. These two beams may be separated in time (double beam-in-time spectrophotometer) or in space (double beam-in-space spectrophotometer).

D Zweistrahlspektralphotometer
F spectrophotomètre bifaisceau, spectrophotomètre à double faisceau
P spektrofotometer dwuwiązkowy
R двухлучевой спектрофотометр

double escape peak *See* pair escape peak

D zwei-Quanten-Escape-Peak, zwei-Quanten-Escape-Linie
F pic de deuxième échappement
P pik ucieczki podwójnej
R пик двойного вылета, пик вылета двух квантов

double-indicator titration
The titration in which two end points are observed, corresponding to the colour changes of two indicators added to the sample; e.g. phenolphthalein and methyl orange in the determination of sodium hydroxide and sodium carbonate.

D Titration mit zwei Indikatoren
F titrage avec deux indicateurs

P miareczkowanie wobec dwóch wskaźników
R титрование с применением двух индикаторов

double isotope dilution analysis
A method of isotope dilution analysis in which
a substance labelled with a radioactive tracer
is determined by adding known but different
quantities of the inactive carrier to two
aliquots of the sample examined.

D doppelte Isotopenverdünnungsanalyse
F analyse par (la méthode de) dilution
isotopique double
P analiza metodą rozcieńczenia izotopowego
podwójnego
R анализ методом двойного изотопного
разбавления

double-potential-step chronoamperometry
Chronoamperometry in which the potential
of the indicator electrode is suddenly changed
to a constant value E_1, and next to a constant
value E_2. The current I flowing through the
electrode is measured as a function of the
electrolysis time t.

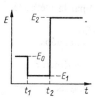

E_0 — open circuit
electrode potential

D Doppelpotentialstufenchronoamperometrie,
Chronoamperometrie mit doppelter
Potentialstufe
F chronoampérométrie à double échelon de
potentiel
P chronoamperometria z podwójnym skokiem
potencjału, chronoamperometria z podwójną
zmianą potencjału
R хроноамперометрия с двойным ступенчатым
изменением потенциала

double-potential-step chronocoulometry
Chronocoulometry in which the potential
of the indicator electrode is suddenly changed
to a constant value E_1, and next to a constant
value E_2. The electric charge Q which has
passed through this electrode is measured
as a function of the electrolysis time, t.

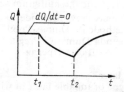

E_0 — open circuit
electrode potential

D Doppelpotentialstufenchronocoulometrie,
Chronocoulometrie mit doppelter
Potentialstufe
F chronocoulométrie à double échelon de
potentiel
P chronokulometria z podwójnym skokiem
potencjału, chronokulometria z podwójną
zmianą potencjału
R хронокулонометрия с двойным ступенчатым
изменением потенциала

double-tone polarography
The alternating current polarography in which
two different sinusoidal alternating voltages
of small and equal amplitudes, typically
$\leqslant$ 25 mV, and of slightly unequal and low
frequencies, $f_i = \omega_i/2\pi <$ 100 Hz, are used
simultaneously. The alternating current I_{ac} is
measured as a function of the direct component
of the indicator electrode potential E_{dc}.
Usually, signals due to the frequency
differences $\omega_2-\omega_1$, 2 $(\omega_2-\omega_1)$, and $2\omega_2-\omega_1$
(where ω_1, and ω_2 — the angular frequencies of
the alternating sinusoidal voltages) are
recorded.

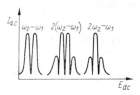

D Doppeltonpolarographie
F polarographie à double tonalité
P polarografia dudnieniowa
R двухчастотная полярография

drain (*in a field-effect transistor*)
One of the two current electrodes of
a field-effect transistor, being that through which
charge carriers leave the interelectrode space.
D Drain
F drain
P dren
R сток

Drechsel gas washing bottle
A washing bottle for removing the soluble
components from gases.
D Gaswaschflasche nach Drechsel,
Drechsel-Waschflasche
F flacon laveur de Drechsel
P płuczka Drechsela
R склянка Дрекселя

drift, time drift, trend
A slow non-random change of value of the
observed quantity (analytical signal) with time.
D Trend
F tendance
P dryf (czasowy), trend
R тренд

drift of a potential (*in electrochemical analysis*)
The slow, non-random variation with time of
the potential of an indicator electrode caused
by uncontrolled side processes taking place
at the electrode surface.
D Drift eines Potentials
F dérive d'un potentiel
P dryf potencjału
R дрейф потенциала

drop lifetime → drop time

dropper, dropping pipette
D Tropfpipette, Tropfer, Tropfenzähler
F compte-gouttes, flacon compte-gouttes
P pipetka wkraplająca, kroplomierz
R капельница

**dropping electrode chronoamperometry with
linear potential-voltage sweep** *See* single sweep
polarography
D Chronoamperometrie mit linearem
    Potential/Spannungsanstieg an der
    Quecksilbertropfelektrode
F chronoampérométrie avec balayage linéaire
    du potentiel à une électrode à gouttes
P ...
R хроноамперометрия с капающим
    электродом с линейной развёрткой
    потенциала/напряжения

dropping electrode coulometry *See* polarographic
coulometry
D Coulometrie mit tropfender Elektrode
F coulométrie à une électrode à gouttes
P kulometria z elektrodą kapiącą
R кулонометрия с капающим электродом

dropping-mercury electrode, DME
An electrode produced by passing a stream of
mercury through a very fine glass capillary
(usually with an internal diameter of 0.03
to 0.05 mm). The mercury emerges from the
tip of the capillary and forms spherical drops
which fall into the solution being investigated.
D Quecksilbertropfelektrode, tropfende
    Quecksilberelektrode
F électrode à gouttes de mercure
P elektroda rtęciowa (kroplowa) kapiąca
R ртутный капающий электрод, ртутный
    капельный электрод, капельный ртутный
    электрод

dropping pipette → dropper

drop time, drop lifetime, lifetime of a drop
(*in polarography*)
The time interval in seconds between the
falling of two successive drops of mercury
emerging from the dropping mercury electrode.
D Tropfzeit, Tropfenabstand
F durée de vie d'une goutte, temps de goutte
P czas trwania kropli
R время жизни капли, период жизни капли,
    время образования одной капли

drying of precipitate
The removal of liquid from precipitate by
evaporation in the air, by heating, or by adding
drying agents often under lowered pressure.
D Trocknen des Niederschlages
F séchage d'un précipité
P suszenie osadu
R высушивание осадка

dry method
A method of examination of a solid sample
without passing it into solution.
D trockenes Verfahren
F procédé à sec, procédé par voie sèche
P metoda sucha
R сухой путь (анализа)

dry-packing
Dry methods of packing chromatographic
columns in which the packing material (dry
sorbent) is added to the column in portions,
with agitation (e.g. vibration, tapping, rotating
the column) between additions, so as to achieve
the required packing density.
D Trockenpackung
F remplissage (de la colonne) par la méthode
    séche
P napełnianie kolumn na sucho
R заполнение колонок сухим способом

DSC → differential scanning calorimetry

DTA → differential thermal analysis

DTG → derivative thermogravimetry

Dumas method
A method of nitrogen determination in
chemical compounds in which the sample is
mixed with copper(II) oxide combusted in an
atmosphere of carbon dioxide. The nitrogen
oxides thus produced are subsequently reduced
to elementary nitrogen by passing them over
metallic copper. Nitrogen is collected in
a nitrometer over a potassium hydroxide
solution.
D Dumas Methode
F méthode de Dumas
P metoda Dumasa
R метод Дюма

Dumas-Pregl method
A classical Dumas method of determination
of nitrogen in chemical compounds modified
by Pregl for the milligram scale analysis.
D Dumas-Pregl-Methode
F méthode de Dumas-Pregl
P metoda Dumasa i Pregla
R метод Дюма-Прегля

dye laser
A type of tunable laser in which the active
material is a dye such as acridine red or
esculin, with very large molecules.

D Farbstofflaser
F laser à colorant
P laser barwnikowy
R лазер на красителе

dynode *See* electron multiplier
D Dynode
F dynode
P dynoda
R динод

ebullioscopic determination of the molar mass
Determination of the molar mass of a chemical
compound based on the measurement of the
difference in the boiling temperatures of the
pure solvent and the solution of the compound
in that solvent.
D ebullioskopische Bestimmung der Molmasse
F détermination ébullioscopique de la masse
   molaire
P oznaczanie masy molowej ebuliometryczne
R эбулиоскопическое определение молярной
   массы

ECD → electron capture detector

eddy diffusion term, axial eddy diffusion
A factor contributing to the total band
broadening represented by the first term (A)
of the van Deemter equation; its contribution
is the greatest at approximately optimum
interstitial velocities of the mobile phase.
This term accounts for the unequal paths
travelled by solute molecules in a packed
column.
D (axiale) Wirbeldiffusion, (axiale)
   Streudiffusion, Eddy-Diffusionsterm
F diffusion turbulente
P dyfuzja wirowa (osiowa)
R вихревая диффузия

EDL → electrodeless-discharge lamp

EDXRF → energy dispersive X-ray fluorescence
analysis

EE → electron emission

EELS → electron-energy loss spectroscopy

effective depth (*of characteristic* X-*ray
production*)
The distance from the specimen surface
along the incident beam direction to the
depth at which the energy of electrons of the
incident beam is equal to the critical excitation
energy for the given characteristic X-rays.

D effektive Reichweite
F profondeur effective
P głębokość czynna
R эффективная глубина

effective layer (*in* X-*ray fluorescence analysis*)
The effective specimen surface layer
contributing to the measured analyte-line
X-rays. The thickness of the effective layer
depends on the values of the absorption
coefficient for the exciting and excited radiations
of the specimen analysed.
D effektives Probevolumen
F couche active
P warstwa czynna
R эффективный слой

effective plate height → height equivalent to an
effective plate

effective plate number → number of effective
plates

effective theoretical plate number → number
of effective plates

effect of electrostatic charges
An apparent change in the mass of glassware
(e.g. absorption tubes) after wiping it with a dry
cloth or chamois leather, caused by
electrostatic charge developing on the glass
surface, especially at low humidity of air.
D Effekt elektrostatischer Aufladungen
F effet des charges électrostatiques
P efekt elektrostatyczny
R электростатический эффект

efficiency of atomization, ε_a
The ratio of the amount of analyte that passes
through the flame cross section in the form of
free neutral or ionized atoms at the observation
height to the amount of analyte aspirated.
D Atomisierungsgrad
F rendement d'atomisation
P wydajność atomizacji
R эффективность атомизации

efficiency of nebulization, ε_n
The ratio of the amount of analyte entering
the flame and the amount of analyte aspirated.
D Zerstäubungsgrad
F rendement de la nébulisation
P wydajność nebulizacji
R эффективность распыления

effluent
The liquid or gas issuing from the column.
D Effluent, Ablauf
F effluent
P wyciek
R эффлюент

EGA → evolved gas analysis

EGD → evolved gas detection

EID → electron stimulated desorption

EIID → electron stimulated desorption of ions

EIND → electron stimulated desorption of
neutrals

elastic collision
A collision in which the total kinetic energy of
the colliding species remains unchanged.
D elastischer Stoß
F choc élastique, collision élastique
P zderzenie elastyczne
R упругое столкновение

elastic low-energy electron diffraction →
low-energy electron diffraction

elastic scattering
A scattering due to an elastic collision or
interaction between species. No change of
wavelength relative to the incident radiation
occurs.
D elastische Streuung
F diffusion élastique
P rozpraszanie elastyczne, rozpraszanie
sprężyste
R упругое рассеяние

electrical accumulator, accumulator, secondary
cell, storage cell
An electrochemical cell which acts as a source
of direct current until its materials are
exhausted, and can then be recharged by
passing a current through it from an external
source.
D Akkumulator, Sekundärelement, Sammler
F accumulateur, pile secondaire, générateur
secondaire, élément secondaire
P akumulator, ogniwo wtórne
R аккумулятор, вторичный элемент

electrical arc, arc
A self-sustaining luminous electrical discharge
between at least two electrodes, characterized
by a comparatively low burning voltage and
relatively high current density. The electrodes
are heated by the discharge and their
evaporation helps to maintain it.
D Lichtbogen
F arc électrique
P łuk elektryczny
R электрическая дуга

electrical conductance → conductance

electrical conduction, conduction
The transmission of electric charge through
a medium under the influence of an electric
field. Conduction can occur by various
processes, e.g. by migration of electrons, ions.
D elektrische Leitung, Elektrizitätsleitung
F conduction électrique
P przewodnictwo elektryczne
R электрическая проводимость,
электропроводимость

electrical conductivity, $\varkappa$ (*formerly called specific conductance*)
A quantity which characterizes materials and media from the point of view of their ability to conduct electric current, $j = \varkappa E$, where j — current density, E — intensity of the electric field; electric conductivity is the reciprocal of resistivity.
SI unit: siemens per metre, $S\ m^{-1}$.
D elektrische Leitfähigkeit, Konduktivität (früher auch spezifische Leitfähigkeit)
F conductivité électrique, conductance spécifique
P konduktywność, przewodność elektryczna właściwa, przewodnictwo elektryczne właściwe
R удельная электрическая проводимость

electrical conductor, conductor
A material which when placed between terminals of different electrical potentials allows the passage of an electrical current.
D elektrischer Leiter, Konduktor
F conducteur électrique
P przewodnik elektryczny
R электрический проводник

electrical double layer → electrochemical double layer

electrical spark, spark (*in emission spectrochemical analysis*)
A series of electrical discharges of comparatively high instantaneous current, resulting from a sudden breakdown of the analytical gap between at least two electrodes of opposite high electric potentials, accompanied by a momentary flash of light.
Remark: The term is used also in mass spectroscopy to denote the ionic source.
D Funken
F étincelle
P iskra
R искра

electrical tension → electric potential difference

electric conductance → conductance

electric current, I, i
The rate of flow of electric charge.
SI unit: ampere, A.
D elektrische Stromstärke, elektrischer Strom
F (intensité de) courant électrique
P natężenie prądu elektrycznego, prąd elektryczny
R сила тока, электрический ток

electric current density → current density

electric-discharge lamp → discharge lamp

electric mobility of ion B, ionic mobility, u_B
The ratio of the migration velocity of an ion in electric field to the intensity of that field.
SI unit: square metre per volt per second, $m^2\ V^{-1}s^{-1}$.
D elektrische Beweglichkeit der Ionen B, Ionenbeweglichkeit
F mobilité (électrique) de l'ion B, mobilité ionique
P ruchliwość jonu B
R (электрическая) подвижность иона B, ионная подвижность

electric potential difference, electric(al) tension, ΔV, U, E
The potential difference between two points; equal to the linear integral of the electric field intensity along the way from the first to the second point in the electric field.
Remark: The term voltage is not recommended.
D Differenz der elektrischen Potentiale, elektrische Potentialdifferenz, elektrische Spannung
F différence de potentiel (entre deux points), tension électrique
P różnica potencjałów elektrycznych, napięcie elektryczne
R разность электрических потенциалов, электрическое напряжение, падение напряжения

electric potential difference of a galvanic cell, U, E
The difference in electric potential between a metallic terminal attached to the right-hand electrode (in the cell diagram) and an identical metallic terminal attached to the left-hand electrode measured under conditions of current flow through the cell, or in the absence of current.
Remark: This definition should not suggest that the absolute electrode potential is measurable.
D elektrische Potentialdifferenz einer galvanischen Zelle, Differenz der elektrischen Potentiale einer galvanischen Zelle, Zellspannung, Klemmenspannung
F différence de potentiel d'un cellule galvanique, tension électrique d'un cellule galvanique
P różnica potencjałów elektrycznych ogniwa galwanicznego, napięcie ogniwa galwanicznego
R разность электрических потенциалов для гальванического элемента, электрическое напряжение электрохимической цепи

electric tension → electric potential difference

electroactivation control
A process, whose rate is determined by the slow (as compared with other steps of this process) rate of the charge transfer through the electrode-electrolyte interface.

D ...
F processus contrôlé par le transfert de charge
P proces kontrolowany szybkością
przeniesienia ładunku, kontrola
elektroaktywacyjna
R электродный процесс лимитируемый
стадией разряда-ионизации

electroactive substance
A molecule, ion or radical which takes part in
an electrode reaction; in potentiometry — the
substance which determines the electrode
potential; in current measurements — the
substrate of an elementary electrochemical
reaction or an ion transferring charge through
the electrode-electrolyte interface.
D ...
F corps électro-actif, substance électro-active,
espèce électro-active
P substancja elektroaktywana
R электроактивное вещество

electroanalysis, electrochemical analysis,
chemical electroanalysis
Method of chemical analysis based on
the phenomena occurring during an electrode
reaction or when a current is passed through
a solution of an electrolyte.
D Elektroanalyse, elektrochemische Analyse
F électro-analyse, analyse électrochimique
P elektroanaliza (chemiczna), analiza
elektrochemiczna
R электроанализ, электрохимический анализ

electrocapillarity → electrocapillary phenomena

electrocapillary phenomena, electrocapillarity
The variation with electrode potential of the
surface tension of an electrode at the
electrode-electrolyte interface.
D Elektrokapillareffekte, Elektrokapillarität,
elektrokapillare Erscheinungen
F électrocapillarité, phénomènes
électrocapillaires
P efekty elektrokapilarne, elektrokapilarność
R электрокапиллярность, электрокапиллярные
явления

electrochemical analysis → electroanalysis

electrochemical cell, cell
A device consisting of at least two electrodes
immersed in one or more solutions of
electrolytes.
D elektrochemische Zelle
F pile électrochimique
P ogniwo elektrochemiczne
R электрохимическая цепь

electr(ochem)ical double layer
A layer which extends on both sides of
the boundary between two condensed phases
and in which the charge distribution is different
from that in the bulk of either phases.

D elektrochemische Doppelschicht,
(elektrische) Doppelschicht
F couche double électrochimique, couche
double (électrique)
P warstwa podwójna (elektryczna), warstwa
podwójna elektrochemiczna
R двойной (электрический) слой,
электрохимический двойной слой

electrochemical equivalent, k_e
The ratio of the molar mass M of a given
substance to the product of the Faraday
constant F and the number of electrons n,
taking part in the reduction or oxidation of one
molecule, ion or atom of this substance at the
electrode
$$k_e = M/nF$$
Numerically it is equal to the mass of
substance formed as a result of the electrode
reaction during the passage of an electric charge
equal to one coulomb.
D elektrochemische Wertigkeit
F équivalent électrochimique
P równoważnik elektrochemiczny
R электрохимический эквивалент

electrochemical generation of a titrant, titrant
generation (*in coulometry*)
Generation of a titrant by electrolytic
reduction or oxidation, often in situ or
externally immediately before use. It is
important that the generating proces is either
100 percent efficient or that the efficiency is
known.
D elektrochemische Reagenzerzeugung,
elektrochemische Erzeugung des Reagenzes
F génération électrochimique d'un réactif
P wytwarzanie elektrochemiczne titrantu,
generowanie elektrolityczne titrantu
R электрогенерирование титранта

electrochemical potential (of ionic component B
in phase α**),** $\tilde{\mu}_B^\alpha$
The work done to transfer an of ion B from
infinity into the bulk of phase α, expressed in
potential units. The intensive thermodynamic
magnitude
$$\tilde{\mu}_B^\alpha = \mu_B^\alpha + z_B\,F\varphi^\alpha$$
where μ_B^α — chemical potential of the ion B in
phase α, z_B — electric charge of the ion (with
the sign), F — Faraday's constant, φ^α — internal
potential of phase α.
D elektrochemisches Potential (einer ionischen
Komponente B)
F potentiel électrochimique (de l'espèce ionique
B dans la phase α)
P potencjał elektrochemiczny (jonu B w fazie α)
R электрохимический потенциал (компонента
B в фазе α)

electrochemical sensor
An electrode or a system of electrodes
that serves as a source of electric signals which

are used to determine the activity
(concentration) of an electroactive substance
(e.g. ion-selective electrode) or some
electrochemical parameters of a medium being
investigated (e.g. electrolytic conductivity).

D elektrochemischer Sensor
F capteur électrochimique
P czujnik elektrochemiczny
R электрохимический сенсор,
электрохимический датчик, чувствительный
элемент

electrochemical series, electromotive series
A series of chemical elements (mostly
metals) arranged in the order of magnitudes of
their electrode potentials.

D elektrochemische Spannungsreihe
F série électrochimique, série des potentiels
P szereg napięciowy, szereg potencjałów
normalnych
R электрохимический ряд активности
металлов, (электрохимический) ряд
напряжений, (электрохимический) ряд
потенциалов

electrochemical stripping analysis, stripping
analysis
Any electroanalytical technique in which the
determination of trace amounts of a substance
are determined by a procedure which involves
two steps: preconcentration in which the
substance is deposited on the electrode, and
redissolution (stripping) — which may be
observed by different voltammetric techniques.
Remark: In a narrow sense the term is used to denote
anodic stripping voltammetry.

D Auflösungsanalyse
F analyse par redissolution
P analiza inwersyjna, analiza stripingowa
R инверсионный анализ

electrochemical transfer coefficient →
charge-transfer coefficient

electrochemistry
The branch of chemistry concerned with the
study of systems and processes in which
electrolytes take part.

D Elektrochemie
F électrochimie
P elektrochemia
R электрохимия

electrode
An electroconducting material used to emit,
collect, or control the flow of, charged particles
in a liquid, gas, semiconductor or vacuum.

D Elektrode
F électrode
P elektroda
R электрод

electrode, half-cell (*in electrochemistry*)
A system composed of two or more
electrically conducting phases (at least one of
which is an electrolyte); the phases are in
contact with one another in such way that the
flow of electrons or ions through the boundary
is possible.

D Elektrode, Halbzelle
F électrode, demi-pile, demi-cellule
P elektroda, półogniwo
R электрод, полуэлемент

electrode gap → analytical gap

electrodeless-discharge tube →
electrodeless-discharge lamp

electrodeless discharge lamp, high-frequency
excited lamp, electrodeless discharge tube, EDL
A lamp containing a noble gas at low pressure
and some volatile metal or metal salt (e.g.
iodide or chloride). A discharge produced in
the noble gas by high frequency fields generates
electrons which by collisions excite the analyte
atoms.

D elektrodenlose Entladungslampe
F lampe à décharge sans électrode, lampe
à excitation haute fréquence sans électrode
P lampa z wyładowaniem bezelektrodowym
R безэлектродная разрядная лампа,
высокочастотная безэлектродная лампа

electrode memory → hysteresis

electrode of the first kind, electrode of the first
order
An electrode (half-cell) consisting of a piece of
metal in equilibrium with a solution containing
ions which determine the electrode potential,
e.g. metal electrode, amalgam electrode,
hydrogen electrode.

D Elektrode erster Art
F électrode de premier ordre, électrode de
première espèce
P elektroda pierwszego rodzaju
R электрод первого рода, электрод первого
типа

electrode of the first order → electrode of the
first kind

electrode of the second kind, electrode of the
second order
An electrode (half-cell) consisting of a piece
of metal coated with a layer of its sparingly
soluble salt in equilibrium with a solution of
a salt which has a common anion. The
electrode is reversible with respect to the
common anion, e.g. calomel electrode.

D Elektrode zweiter Art
F électrode de second ordre, électrode de
deuxième espèce
P elektroda drugiego rodzaju
R электрод второго рода, электрод второго
типа

electrode of the second order → electrode of
the second kind

electrode of the third kind, electrode of the third order
An electrode (half-cell), consisting of a piece of metal (M) in equilibrium with a layer of its sparingly soluble salt (MA) and a layer of an insoluble salt with a common anion but different cation (M′A), immersed in a solution containing the common cation (M′), e.g. calcium electrode

$$Ca^{2+}_{(aq)}|Ca\ (COO)_{2_{(s)}}|Pb\ (COO)_{2_{(s)}}|Pb_{(s)}.$$

The electrode is reversible with respect to the common cation.
D Elektrode dritter Art
F électrode de troisième ordre, électrode de troisième espèce
P elektroda trzeciego rodzaju
R электрод третьего рода, электрод третьего типа

electrode of the third order → electrode of the third kind

electrodeposition, electrolytic deposition
A process in which a substance is deposited at either the anode or the cathode as a result of the passage of an electric current through the solution, or suspension of the substance.
D elektrolytische Abscheidung, elektrolytische Fällung
F électrodéposition
P elektrowydzielanie
R электролитическое осаждение, электроосаждение

electrodeposition of metals, electroreduction of metals, electroplating
An electrode process in which metal ions are reduced to the metal at the electrode.
D katodische Metallabscheidung
F électrodéposition de métaux
P wydzielanie metali elektrolityczne, elektrowydzielanie metali
R электроосаждение (металлов)

electrode potential, relative electrode potential
The electric potential difference of a galvanic cell when the left-hand electrode acts as a reference electrode, e.g.

$$Pt\ |H_2|\ H^+\ |\ Zn^{2+}\ |Zn|\ Pt'$$

left-hand   right-hand
electrode   electrode
D Elektrodenpotential, relatives Elektrodenpotential
F potentiel (relatif) d'électrode, tension électrique relative d'électrode
P potencjał elektrody (względny)
R (относительный) электродный потенциал, электродное напряжение

electrode process
A process including the electrode reaction and the transport of reagents to/from the electrode-electrolyte interface.

D Elektrodenprozeß
F processus d'électrode
P proces elektrodowy
R электродный процесс

electrode reaction, potential determining reaction
Oxidation or reduction of a substance taking place at the surface of an electrode.
D Elektrodenreaktion, potentialbestimmende Reaktion
F réaction d'électrode
P reakcja elektrodowa
R электродная реакция

electrode response
The change of the equilibrium or rest potentials of an indicator electrode, e.g. ion-selective electrode, with the change of concentration (activity) of the electroactive substance under given experimental conditions.
D ...
F réponse d'une électrode
P odpowiedź (potencjałowa) elektrody
R электродная функция, отклик потенциала электрода

electrodialysis
A form of dialysis in which a potential difference, applied across platinum electrodes, aids the selective diffusion of electrolyte ions through a semipermeable membrane.
D Elektrodialyse
F électrodialyse
P elektrodializa
R электродиализ

electro-(endo)osmosis
Flow of a liquid through a stationary solid, e.g. through a porous diaphragm or capillary tube, due to an electric field gradient.
D Elektro(endo)osmose
F électro-osmose, électroendoosmose
P elektroosmoza, (elektro)endoosmoza
R электроосмос

electrography
A method for accurate identification and determination of substances which consists in anodic or cathodic stripping of the test substance from a solid electrode onto a piece of filter paper or gelatin-coated paper saturated with a suitable electrolyte.
D Elektrographie
F électrographie
P elektrografia
R электрография

electrogravimetric analysis → electrogravimetry

electrogravimetry, electrogravimetric analysis
A method of chemical analysis in which the mass of a substance deposited on the working electrode is determined. Electrodeposition is carried out under conditions of convective

mass transfer to the working electrode and the deposit may be a metal or an insoluble material.

D Elektrogravimetrie
F électrogravimétrie
P elektrograwimetria
R электрогравиметрия, электровесовой анализ

electrokinetic effects → electrokinetic phenomena

electrokinetic phenomena, electrokinetic effects
A group of phenomena occurring when the electrical double layer is present at the interface. They manifest themselves by the displacement of one phase with respect to the other due to the electric field applied across the system (*see* electrophoresis and electroosmosis), or by a measurable potential difference generated by induced relative mowement of the two phases (*see* sedimentation potential and streamingpotential).

D elektrokinetische Erscheinungen, elektrokinetische Effekte
F phénomènes électrocinétiques
P efekty elektrokinetyczne, zjawiska elektrokinetyczne
R электрокинетические явления

electrokinetic potential, zeta-potential, ζ
The electrical potential existing across the solid|liquid interface. It may be calculated (in millivolts) from the formula

$$\zeta = 4\pi\eta/v\varepsilon_r\, E$$

where η — viscosity of the liquid medium, v — liquid flow velocity, ε_r — relative permittivity, and E — electric field.

D elektrokinetisches Potential, Zetapotential
F potentiel électrocinétique, potentiel zêta
P potencjał elektrokinetyczny, potencjał zeta
R электрокинетический потенциал, дзета-потенциал

electroluminescence
Luminescence produced by the application of an electric field to a material, usually solid.

D Elektrolumineszenz
F électroluminescence
P elektroluminescencja
R электролюминесценция

electrolysis
The process consisting in the flow of an electric current (supplied from an external source) through an ionic conductor which is accompanied by electrochemical reactions at the electronic conductor-ionic conductor interface.

D Elektrolyse
F électrolyse
P elektroliza
R электролиз

electrolyte
A substance which when molten or dissolved in a suitable solvent will conduct electric current by means of free and stable ions.

D Elektrolyt
F électrolyte
P elektrolit
R электролит

electrolytic cell
An electrochemical cell in which electrical energy supplied from an external source produces the desired electrochemical reaction.

D Elektrolysezelle, Elektrolysierzelle, elektrolytische Zelle
F cellule d'électrolyse, cellule électrolytique, cuve à éléctrolyse, électrolyseur
P naczynko elektrolityczne, elektrolizer
R электролитическая ячейка, электролизная ячейка, электролизёр

electrolytic conductance → conductance of electrolyte

electrolytic conduction → ionic conduction

electrolytic conductivity (*formerly called specific conductance*), $\varkappa$, σ
The ability of an electrolyte solution to carry an electric current, defined as the conductance of a 1 m³ of electrolyte at a given temperature. The conductivity depends on the concentration of the electrolyte. SI unit: siemens per metre, S m⁻¹

D elektrolytische Leitfähigkeit (früher auch spezifische Leitfähigkeit)
F conductivité d'un électrolyte, conductivité électrolytique, conductance spécifique d'un électrolyte
P konduktywność elektrolityczna, przewodność elektrolityczna właściwa, przewodnictwo elektrolityczne właściwe
R удельная электрическая проводимость электролита, удельная электропроводность электролита

electrolytic conductor → ionic conductor

electrolytic deposition → electrodeposition

electrolytic dissociation
Decomposition of a substance into thermodynamically stable free ions due to the interaction between the dissolved substance and the solvent.

D elektrolytische Dissoziation
F dissociation électrolytique
P dysocjacja elektrolityczna
R электрическая диссоциация

electrolytic thermocell → galvanic thermocell

electromagnetic radiation
Waves of energy consisting of electric and magnetic fields vibrating at a right angle to the

direction of propagation of the waves. They have a velocity c, equal to $2.997\,92458 \times 10^8$ metres per second and their properties depend on their frequency (or wavelength).
D elektromagnetische Strahlung
F radiation électromagnétique, rayonnement électromagnétique
P promieniowanie elektromagnetyczne
R электромагнитное излучение

electromagnetic spectrum
The arrangement of all known electromagnetic radiations according to their frequency, wavelength or wave number, extending from the longest radio waves to the shortest known cosmic rays.
D elektromagnetisches Spektrum
F spectre électromagnétique
P widmo elektromagnetyczne
R электромагнетический спектр

electromechanical balance
A balance which determines the mass of an object as a function of the electrical signal generated in the electromechanical transducer by the applied load.
D elektrische Waage, Elektrowaage
F balance électromécanique
P waga elektromechaniczna
R электромеханические весы

electromotive force (of a galvanic cell), EMF, E
The limiting value of the electric potential difference of a galvanic cell when no current is flowing through the cell.
D elektromotorische Kraft (einer galvanischen Zelle), EMK der Zelle
F force électromotrice d'une cellule galvanique, F.e.m. d'une pile, tension chimique d'une cellule galvanique
P siła elektromotoryczna (ogniwa galwanicznego), SEM ogniwa
R электродвижущая сила гальванического элемента, э.д.с. гальванического элемента

electromotive series → electrochemical series

electron bombardment ion source → electron-impact ion source

electron capture detector, ECD
An ionization detector in which free electrons and positive ions are produced in the carrier gas (nitrogen or argon) under the influence of radiation from either a ^{3}H or ^{63}Ni source. When the electron-capturing solute enters the detector ionization chamber a current decrease occurs.
D Elektroneneinfangdetektor, Electron-Capture-Detektor
F détecteur à capture d'électrons
P detektor wychwytu elektronów
R электронно-захватный детектор, детектор электронного захвата

electron conduction → electronic conduction

electron conductor → electronic conductor

electron emission, EE
The liberation of electrons from a solid (metal or semiconductor) or liquid into the surrounding space (vacuum or gas), usually under the influence of heat, light, or a high electric field.
D Elektronenemission
F émission d'electrons
P emisja elektronów
R электронная эмиссия

electron-energy loss spectroscopy, EELS, low-energy electron loss spectroscopy, LEELS
The surface sensitive vibrational spectroscopy based on energy analysis of electrons inelastically backscattered from the surface; these electrons suffer energy losses attributed to the excitation of surface interband and intraband electronic transitions and surface vibrations in the adsorbed molecules. On this basis information is obtained on the unfilled electronic states at the surface, on the degree of dissociation, binding energies, binding states and lateral interactions of the adsorbate.
D Elektronenenergieverlustspektroskopie
F spectroscopie de pertes d'energie d'électrons
P spektroskopia charakterystycznych strat energii elektronów
R спектроскопия низкоэнергетических характеристических потерь

electron exchanger *See* redox polymer
D Elektronenaustauscher
F échangeur d'électrons
P wymieniacz elektronów
R электронообменник

electronic band
In the electronic molecular spectrum the band corresponding to the transition among different rotational-vibrational-electronic levels. The band is observed in the visible, ultraviolet and near infrared regions.
D Elektronenband
F bande électronique
P pasmo elektronowe
R электронная полоса

electron(ic) conduction
Electrical conduction resulting from the migration of electrons in metals, semiconductors or gases.
D Elektronenleitung
F conduction électronique, conduction par des électrons
P przewodnictwo elektronowe
R электронная проводимость

electron(ic) conductor, metallic conductor, first-class conductor
The electrical conductor in which the current

is carried by the migrating electrons, as in metals.

D Elektronenleiter, metallischer Leiter, Leiter 1. Ordnung, Leiter erster Klasse
F conducteur électronique, conducteur métallique, conducteur de première espèce
P przewodnik elektronowy, przewodnik metaliczny
R металлический проводник, проводник первого рода, проводник с электронной проводимостью

electronic spectroscopy
The branch of spectroscopy concerned with the production, measurement, and interpretation of electronic spectra of atoms and molecules.

D Elektronenspektroskopie
F spectroscopie électronique
P spektroskopia elektronowa
R электронная спектроскопия

electronic spectrum
A spectrum resulting from emission or absorption of electromagnetic radiation during changes in the electron configuration of atoms, ions or molecules

D Elektronenspektrum
F spectre électronique
P widmo elektronowe
R электронный спектр

π-π^* electronic transition
According to the one-electron excitation model, the transition of an electron from the bonding π-orbital to the antibonding π^*-orbital.

D π-π^*-Elektronenübergang
F transition électronique π-π^*
P przejście elektronowe typu π-π^*
R электронный переход π-π^*

σ-π^* electronic transition
According to the one-electron excitation model, the transition of an electron from the bonding σ-orbital to the antibonding π^*-orbital.

D σ-π^*-Elektronenübergang
F transition électronique σ-π^*
P przejście elektronowe typu σ-π^*
R электронный переход σ-π^*

n-π^* electronic transition
According to the one-electron excitation model, the transition of an electron from the nonbonding n-orbital of the free electron pair to the antibonding π^*-orbital.

D n-π^*-Elektronenübergang
F transition électronique n-π^*
P przejście elektronowe typu n-π^*
R электронный переход n-π^*

n-σ^* electronic transition
According to the one-electron excitation model, the transition of an electron from the nonbonding n-orbital of the free electron pair to the antibonding σ^*-orbital.

D n-σ^*-Elektronenübergang
F transition électronique n-σ^*
P przejście elektronowe typu n-σ^*
R электронный переход n-σ^*

electronic-vibration-rotation spectrum
A molecular spectrum corresponding to the transitions between electronic energy levels which occur simultaneously with the transitions between vibrational and rotational energy levels; the electronic spectrum consists of a series of bands representing different vibrational transitions, the initial state being at one electronic level and the final state at another level, and each band is made up of closely spaced lines due to changes in the rotational energy. The electronic spectrum appears in the visible and ultraviolet regions.

D Elektronen-Rotationsschwingungsspektrum
F spectre électronique de rotation-vibration
P widmo elektronowo-oscylacyjno-rotacyjne
R вращательно-колебательно-электронный спектр

electron impact desorption → electron stimulated desorption

electron impact ion desorption → electron stimulated desorption of ions

electron-impact ionization
Ionization resulting from the interaction of an energetic electron with any particle, e.g. a molecule or atom.

D Elektronenstoßionisation
F ionisation par impact électronique
P jonizacja elektronowa
R ионизация электронным ударом

electron-impact ion source, electron bombardment ion source (*in mass spectrometry*)
An ion source in which the sample molecules in the gas phase are ionized by bombarding the molecules at right angles to their line of propagation with a stream of rapidly moving electrons emitted from the heated cathode filament and accelerated by an electric field maintained between positively charged slits through which the electrons pass into the ionization chamber. Tungsten, rhenium, or thoriated iridium are generally used as filament material.

D Elektronenbeschuß-Ionenquelle, Elektronenstoß-Ionenquelle
F source (d'ions) à impact électronique, source à impact d'électrons
P źródło jonów uzyskanych przez bombardowanie elektronami
R источник для ионизации электронным ударом, ионный источник с электронной бомбардировкой

electron induced desorption → electron stimulated desorption

electron induced ion desorption → electron stimulated desorption of ions

electron induced neutrals desorption → electron stimulated desorption of neutrals

electron microprobe X-ray analyser,
electron-probe microanalyser, EPMA
An instrument used for qualitative and quantitative analysis of the chemical composition of the sample on a microscopic scale, based on the measurement of the wavelengths and intensities of characteristic X-ray spectra excited by an electron beam having a diameter of the order of 1 μm.
D Castaingsche Mikrosonde,
  Elektronenstrahl-Mikroanalysator
F microsonde (de Castaing) électronique
P mikroanalizator rentgenowski, mikrosonda elektronowa
R рентгеновский микроанализатор

electron multiplier
An electronic device for detecting electrons from secondary emission. It consists of an evacuated tube containing a number of electrodes called dynodes. An electron hitting the first electrode ejects two or more secondary electrons from the dynode plate. These are further amplified by impact on successive dynode plates arranged in series.
D Elektronenvervielfacher
F multiplicateur électronique
P powielacz elektronowy
R фотоэлектронный умножитель

electron-multiplier phototube → photomultiplier tube

electron nuclear double resonance, ENDOR
A technique for improving the EPR spectra involving irradiation of a sample with nuclear resonance frequencies, thus the broadening of the EPR signal is obtained with hyperfine details enhanced.
D Elektron-Kern-Doppelresonanz
F résonance paramagnétique électronique double
P rezonans elektronowo-jądrowy podwójny
R двойной электронный парамагнитный резонанс

electron paramagnetic resonance, EPR, paramagnetic resonance, electron-spin resonance, ESR
A phenomenon exhibited by atoms, ions, free radicals and molecules possessing an odd number of electrons or unpaired electrons. When the substance is placed in a static magnetic field and simultaneously subjected to a microwave alternating field at right angles to the former, absorption of radiation

(resonance) occurs at a particular frequency with the effect that electrons pass from lower to higher energy level.
D paramagnetische Elektronenresonanz, Elektronenspin-Resonanz
F résonance paramagnétique électronique, résonance de spin électronique
P rezonans magnetyczny elektronowy, rezonans paramagnetyczny elektronowy, rezonans spinowy elektronowy
R электронный парамагнитный резонанс, электронный спиновый резонанс

electron paramagnetic resonance spectroscopy,
EPR-spectroscopy, electron spin resonance spectroscopy, ESR-spectroscopy
A technique for the study of atoms, ions, molecules and free radicals possessing an odd number of electrons or unpaired electrons by taking advantage of the electron paramagnetic resonance.
D Elektronenresonanzspektroskopie, Elektronen-Spin-Resonanzspektroskopie
F spectroscopie de résonance (para)magnétique électronique, RPE-spectroscopie, spectroscopie de résonance de spin électronique
P spektroskopia elektronowego rezonansu paramagnetycznego, spektroskopia elektronowego rezonansu spinowego
R спектроскопия электронного парамагнитного резонанса

electron-probe microanalyser → electron microprobe X-ray analyser

electron spectroscopy
A branch of spectroscopy in which the energy of electrons ejected from atoms and molecules is investigated.
D Elektronenspektroskopie
F spectroscopie électronique
P spektroskopia elektronów
R спектроскопия электронов

electron spectroscopy for chemical analysis → X-ray photoelectron spectroscopy

electron-spin resonance → electron paramagnetic resonance

electron stimulated desorption, ESD, electron induced desorption, electron impact desorption, EID
Desorption of neutral molecules or ions by bombardment of the surface of an adsorbent with an electron beam of energy from 75 to 100 eV. The composition of the desorbing species is determined by mass spectroscopy.
D elektronenstrahl-induzierte Desorption
F désorption par impact d'électrons, désorption par choc d'électrons
P desorpcja wywołana bombardowaniem elektronowym
R десорбция при электронном облучении

electron stimulated desorption of ions, ESDI,
electron impact ion desorption, electron
induced ion desorption, EIID
The electron stimulated desorption with mass
analysis of the species that desorb in an ionic
form.

D elektronenstrahl-induzierte Desorption von
 Ionen
F désorption d'ions par impact d'électrons,
 désorption d'ions par choc d'électrons
P desorpcja jonów wywołana bombardowaniem
 elektronowym
R десорбция ионов при электронном
 облучении, электронное стимулирование
 десорбции ионов

electron stimulated desorption of neutrals,
ESDN, electron induced neutrals desorption,
EIND
The electron stimulated desorption which
gives information on the neutral atoms and
molecules removed from a solid surface under
electron bombardment. The neutrals are
subsequently ionized in an auxiliary ionization
chamber in order to make the mass analysis
possible.

D elektronenstrahl-induzierte Desorption von
 Neutralteilchen
F désorption de particules neutres par impact
 d'électrons, désorption de particules neutres
 par choc d'électrons
P desorpcja cząstek obojętnych wywołana
 bombardowaniem elektronowym
R десорбция нейтральных частиц при
 электронном облучении, электронное
 стимулирование десорбции нейтральных
 частиц

electronvolt, eV
A unit of energy to the energy acquired by an
electron, when it is accelerated ober a potential
difference of one volt in a vacuum;

$$1\text{eV} = 1.602\ 1892 \times 10^{-19}\ \text{J}.$$

Remark: This unit does not belong to the International
System of Units (SI) and its use is to be progressively
discouraged.

D Elektronenvolt
F électron-volt
P elektronowolt
R электрон-вольт

electro-osmosis → electro-endoosmosis

electropherogram → electrophorogram

electrophoresis
The movement of charged particles of the
dispersed phase in a medium under
the influence of an electric field.

D Elektrophorese
F électrophorèse
P elektroforeza
R электрофорез

electrophoresis diagram → electrophorogram

electrophoretic effect
A decrease of the net velocity of an ion in an
electric field due to the movement of the
ionic atmosphere in the direction opposite to
that ion.

D electrophoretischer Effekt
F effet électrophorétique
P efekt elektroforetyczny
R электрофоретический эффект,
 электрофоретическое торможение,
 катафоретический эффект

electrophoretic potential → sedimentation
potential

electrophorogram, electropherogram,
electrophoresis diagram
Term employed in electrophoresis, which
is equivalent to the term chromatogram in
chromatography.

D Elektropherogramm
F électrophorogramme, diagramme
 d'électrophorèse
P elektroforogram
R электрофорограмма, электрофорезная
 диаграмма

electroplating → electrodeposition of metals

electroreduction of metals → electrodeposition
of metals

electrothermal atomizer, non-flame atomizing
device
A flameless atomizer, e.g. a carbon rod,
graphite tube or other device, which is
electrically heated to a high temperature to
cause volatilization and atomization of
samples in atomic spectroscopy.

D thermoelektrischer Atomisator,
 flammenlose Atomisierungseinrichtung
F atomiseur électrothermique
P atomizer elektrotermiczny, atomizer
 bezpłomieniowy
R электротермический атомизатор,
 безпламенный атомизатор

ELEED → low-energy electron diffraction

elemental analysis
Detection and/or determination of elements
present in the sample analysed.

D Elementaranalyse
F analyse élémentaire
P analiza pierwiastkowa
R элементный анализ

elemental analysis, elementary analysis
(*of organic compounds*)
Methods of chemical analysis used for
detection and/or determination of individual
elements in organic compounds, mainly of
carbon, hydrogen, nitrogen, sulfur, phosphor
and halogens.

D (organische) Elementaranalyse
F analyse élémentaire organique
P analiza elementarna (związków organicznych)
R ¦элементный анализ, элементарный анализ
   (органических соединений)

elementary analysis → elemental analysis
(*of organic compounds*)

elementary entity
A molecule, atom, ion, electron, proton, or
some other species, also a defined group of
species whose number may be expressed
in units of the amount of substance.
D Elementarindividuum, Einzelteilchen
F entité élémentaire
P cząstka
R элементарная единица

elementary event
D Elementarereignis
F événement élémentaire
P zdarzenie elementarne
R элементарное событие

elementary reaction
A one-step and direct, i.e. occurring without
any intermediate steps, conversion of
substrates into products.
D Elementarreaktion
F réaction élémentaire.
P reakcja elementarna, reakcja prosta
R элементарная реакция, простая реакция

elementary step (*in electrochemistry*)
The part of an overall process (e.g. an electrode
process) being a one step and direct transition
of a component of a system from one state to
another. The elementary step may be an
elementary reaction, the transfer of a unit of
charge across the electrode/electrolyte interface,
or a transfer of a reagent to or from the
electrode surface (e.g. diffusion).
D Elementarstufe
F étape du processus
P etap elementarny
R стадия процесса

ellipsometry
An optical technique for the characterization
of phenomena at an interface, based on the
change of the state of polarization of light
after reflection from or transmission through
the interface.
D Ellipsometrie
F ellipsométrie
P elipsometria
R эллипсометрия

elliptically polarized light *See* elliptical
polarization
D elliptisch polarisiertes Licht
F lumière polarisée elliptiquement, lumière
   à polarisation elliptique
P światło spolaryzowane eliptycznie
R эллиптически поляризованный свет

elliptical polarization
Polarization of light that occurs when the
electric field vector of a ray of light describes,
either clockwise or counterclockwise, at any
point in space an ellipse about the direction of
propagation of the light; the light is then said
to be elliptically polarized.
D elliptische Polarisation
F polarisation elliptique
P polaryzacja eliptyczna
R эллиптическая поляризация

eluant → eluent

eluate
The effluent emerging from a chromatographic
bed when elution is carried out.
Remark: Now, this term is also used in a wider sense to
refer to the effluent in all forms of chromatography.
D Eluat
F éluat
P eluat
R элюат

eluent, eluant
A liquid or a gas entering the chromatographic
bed and used to carry out the separation by
elution. *See* mobile phase.
Remark: Sometimes this term is also used in a wider
sense to describe the mobile phase in all forms of
chromatography.
D Eluent, Eluant, Elutionsmittel
F éluant
P eluent, eluant
R элюент

eluent gas → carrier gas

eluotropic series, mixotropic series
The arrangement of liquid eluents in the order
of their increasing elution strength.
Note that this arrangement is specific for a
given separation system.
D eluotrope Reihe
F série éluotropique
P szereg eluotropowy, seria eluotropowa
R элюотропный ряд, элюотропная серия

elution
The process of passing the mobile phase
through the stationary phase and of
transporting at the same time the sample
components along the chromatographic bed
until the components have left the
chromatographic bed.
Remark: The term elution is preferred to the term
development, but in planar chromatography where the
separated sample components remain on the paper or
plate, the latter is still in use.
D Elution
F élution
P elucja, wymywanie
R элюирование, элюция

elution band → peak

elution chromatography
A chromatographic technique in which
a sample earlier introduced into the
chromatographic bed is eluted with an eluent.
The particular sample components migrate with
different velocities along the stationary phase
and are separated thanks to their different
affinities to the stationary phase. They emerge
in the successive fractions of the eluate.
D Elutionschromatographie, Elutionsanalyse
F chromatographie par élution,
  chromatographie d'élution, analyse d'élution
P chromatografia elucyjna
R элюентная хроматография, элюционная
  хроматография, проявительная
  хроматография

elution curve
A curve obtained by plotting the values of the
solute concentration in the eluate against the
elution time/volume.
D Elutionskurve
F courbe d'élution
P krzywa elucji
R кривая элюирования, элюционная кривая,
  кривая проявления

elution volume → peak elution volume

elutriation (*in chromatography*)
The method of separating adsorbent or ion
exchanger particles into fraction of uniform
particle size, by taking advantage of the
different rates of sedimentation of the particles
in an upward stream of liquid.
D Elutriation
F élutriation
P elutriacja
R жидкостная сортировка, отмывание от пыли

emanation method → emanometric method

emanometric method, emanation method
A method of determining radon, or indirectly
other members of the natural radioactive
series, based on the measurement of
radioactivity of the gas released from the sample.
D Emaniermethode
F méthode d'émanation
P metoda emanacyjna
R эманационный метод

EMF → electromotive force of a galvanic cell

emission monochromator (in *a spectrofluorimeter*)
A monochromator for selecting the required
wavelength of luminescent radiation emitted
from the sample.
D Emissionsmonochromator
F monochromateur d'émission
P monochromator promieniowania
  wzbudzonego
R эмиссионный монохроматор

emission spectrochemical analysis
Spectrochemical analysis based on the
investigation of emission spectra.
D Emissionsspektralanalyse
F analyse spectrochimique d'émission
P analiza spektralna emisyjna
R эмиссионный спектрохимический
  анализ

emission spectroscopy, ES
A branch of spectroscopy concerned with the
production, measurement and interpretation
of the emission spectra of atoms and molecules.
D Emissionsspektroskopie
F spectroscopie d'émission
P spektroskopia emisyjna
R эмиссионная спектроскопия

emission spectrum
A spectrum produced by emission of
electromagnetic radiation resulting from the
decay of excited atoms or molecules to some
lower energy level.
D Emissionsspektrum
F spectre d'emission
P widmo emisyjne
R эмиссионный спектр, спектр испускания

emitter
A substance emitting a specified type of
radiation.
D Strahler
F émetteur
P emiter
R излучатель, радиатор

empirical coefficient method (*in* X-*ray*
fluorescence analysis)
A mathematical method for the elimination
of matrix effects. It accounts for the influence
of a given matrix element on the intensity of
a characteristic X-ray line of the analyte in
the form of a numerical coefficient. The values
of the coefficients are calculated by solving
a set of linear equations combining the
measured line intensities of the elements
with their concentrations in a suitable
set of standards. The calculated coefficients
are substituted in the set of equations and the
concentrations of elements in the analysed
samples are determined.
D Einflußkoeffizientenverfahren
F méthode des coefficients d'influence
P metoda współczynników empirycznych
R метод коэффициентов влияния

empirical distribution
Probability distribution obtained empirically
from a given sample.
D empirische Verteilung
F distribution empirique
P rozkład empiryczny
R эмпирическое распределение

empirical variation coefficient, relative standard deviation of the sample
Standard deviation of the sample divided by the arithmetic mean of the sample.

D empirischer Variationskoeffizient, relative empirische Standardabweichung
F coefficient de variation empirique
P współczynnik zmienności w próbie, odchylenie standardowe względne z próby
R выборочный коэффициент вариантности

empty tube combustion
Combustion of a sample in a rapid stream of oxygen, usually effected at a temperature of $900-950°C$ in a quartz tube with no catalyst or oxidizing filling.

D Leerrohrverbrennung, Verbrennung im leeren Rohr
F combustion en tube vide
P spalanie w pustej rurze
R сожжение в пустой трубке

emulsion calibration curve
The plot presenting the quantity which characterizes the photographic emulsion, e.g. blackening, transmittance, as a function of the common logarithm of exposure.

D Gradationskurve
F courbe d'étalonnage d'emulsion
P krzywa wzorcowania emulsji
R калибровочная кривая фотоэмульсии

emulsion characteristic curve, characteristic curve, Hurter and Driffield curve, H and D-curve
The emulsion calibration curve in which the blackening S is plotted as a function of the logarithm of exposure H.

Remark: The term density curve is not recommended.

D Schwärzungskurve
F courbe caractéristique (d'une émulsion), courbe de noircissement (d'une émulsion), courbe de Hurter et Driffield
P krzywa zaczernienia emulsji, krzywa charakterystyczna emulsji
R характеристическая кривая эмульсии, кривая Хартера и Дриффилда

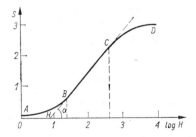

ENDOR → electron nuclear double resonance

end-point of titration
The point in titration at which some property of the solution shows a pronounced change corresponding more or less closely to the equivalence-point, e.g. the colour change exhibited by the indicator. In the graphical method of end-point determination the end-point may be represented by the intersection of two lines or curves.

D Titrationsendpunkt
F point de fin de titrage
P punkt końcowy miareczkowania
R точка конца титрования, конечная точка титрования

end window counter → window counter

energy dispersive X-ray fluorescence → energy dispersive X-ray fluorescence analysis

energy dispersive X-ray fluorescence (analysis), EDXRF
A method of X-ray fluorescence analysis based on the energy measurement in an emission spectrum of characteristic X-rays. The analysing unit consists of a detector which transforms single radiation photons into electrical pulses whose amplitudes are proportional to their energies.

D energiedispersive Röntgenfluoreszenzanalyse
F analyse par fluorescence X avec dispersion de l'énergie
P analiza fluorescencyjna rentgenowska z dyspersją energii
R рентгенофлуоресцентный анализ с энергетической дисперсией

energy resolution of a gamma spectrometer
(1) In the case of a scintillation detector: the ratio of the half-width of the total absorption peak (in energy units) to the nominal energy of that peak; expressed in per cent is usually given for the ^{137}Cs peak of 0,662 MeV nominal energy.
(2) In the case of a semiconducting detector: the half-width of the total absorption peak, (in energy units, keV or eV) usually given for the ^{60}Co peak of 1332 keV nominal energy.

D Energieauflösungsvermögen, Auflösungsvermögen
F résolution de détection de l'énergie du photopic
P zdolność rozdzielcza spektrometru gamma
R энергетическое разрешение гамма-спектрометра

English degree of water hardness
The unit of water hardness corresponding to the content of 100 mg of calcium carbonate per 1 gallon (4.55 l) of water, i.e. 14.3 mg of calcium carbonate per litre.

D englischer Härtegrad, englische Härte
F degré hydrotimétrique anglais
P stopień twardości wody angielski
R английский градус жёсткости воды

enhancement effect (*in X-ray fluorescence analysis*)
The increase in the intensity of the characteristic X-ray radiation of an analyte

excited by primary radiation. The increase is due to additional excitation by the characteristic radiation of matrix elements of energy higher than the absorption edge of the analyte, excited in the analysed sample simultaneously with the analyte. Together with the absorption effect it constitutes the matrix effect.

D Interelementanregung, Sekundäranregung
F effet d'exaltation
P efekt wzmocnienia
R эффект избирательного возбуждения

enrichment factor, S
The factor by which the original ratio of the concentration of two substances to be separated must be multiplied to give the ratio after separation (usually in the organic phase).

$$\frac{Q_B}{Q_A} = S_{B/A} \frac{(Q_B)'}{(Q_A)'}$$

Hence

$$S_{BA} = \frac{Q_B(Q_A)'}{Q_A(Q_B)'} = \frac{R_B}{R_A}$$

where Q_A and Q_B — amount of substance A and B in the extracts, $(Q_A)'$ and $(Q_B)'$ — initial quantities of substances A and B, R_A and R_B — recovery factors for substances A and B.

Remark: In the liquid-liquid extraction the enrichment factor should not be given as $(D_c)_B/(D_c)_A$, where $(D_c)_A$ and $(D_c)_B$ — concentration distribution ratios of substances A and B, respectively.

D Anreicherungsfaktor
F facteur d'enrichissement
P współczynnik wzbogacenia
R коэффициент обогащения

enthalpimetric titration → thermometric titration

enthalpimetry
A method of examining a substance by measuring the temperature difference corresponding to enthalpy changes.

Remark: In chemical analysis, the term calorymetric analysis is used.

D Enthalpiemetrie
F enthalpimétrie
P entalpimetria
R энтальпиметрия

enzyme electrode → enzyme-substrate electrode

enzyme(-substrate) electrode
A sensitized ion-selective electrode in which the ion-selective electrode is coated with a layer containing an enzyme that initiates the reaction of an organic or inorganic substance (substrate) to yield a species to which the electrode responds, or with a layer of substrate which reacts with the enzyme to be determined.

D Enzym-Elektrode
F électrode à substrat enzymatique
P elektroda enzymatyczna
R ферментный электрод, энзимный электрод, электрод на основе субстрата энзима

epicadmium neutrons
Slow neutrons, not absorbed by a cadmium foil of thickness of ca. 1 mm (i.e. neutrons of kinetic energy greater than 0.4 eV).

D Epikadmiumneutronen
F neutrons épicadmiques, neutrons épicadmium
P neutrony epikadmowe, neutrony nadkadmowe
R закадмиевые нейтроны

epithermal neutrons
Neutrons of kinetic energy from 0.1 to 100 eV.

D epithermische Neutronen
F neutrons épithermiques
P neutrony epitermiczne
R эпитепловые нейтроны, надтепловые нейтроны

EPMA → electron microprobe X-ray analyser

EPR → electron paramagnetic resonance

equalizing solvent → levelling solvent

equieluotropic series
A graphical arrangement of liquid binary eluents in the order of their increasing elution strength in the form of horizontal scales of composition. A line perpendicular to the scales intersects at points of compositions of the binary mixtures of similar elution strengths but different selectivities.

D äquieluotrope Reihe
F série équiéluotropique
P szereg ekwieluotropowy, seria ekwieluotropowa
R эквиэлюотропный ряд

equilibrium (electrode) potential
The potential of an electrode (with respect to the reference electrode) at which only one reaction is significant and the net current is equal to zero.

D Gleichgewichtspotential
F potentiel d'électrode à l'équilibre, potentiel d'équilibre, tension d'électrode à l'équilibre
P potencjał (elektrody) równowagowy
R равновесный (электродный) потенциал

equilibrium potential → equilibrium electrode potential

equivalence factor in acid-base reactions
The ratio of the amount of substance n_B of component B to the amount of substance n_{H^+} of protons (or n_{OH^-} hydroxylic ions) taking part in a specified reaction

$$f_{eq}(B) = \frac{n_B}{n_{H^+}} \quad \text{or} \quad f_{eq}(B) = \frac{n_B}{n_{OH^-}}$$

D Äquivalenz-Faktor in den
Neutralisationsreaktionen
F facteur d'équivalence dans les réactions
acide-base
P współczynnik równoważności w reakcjach
zobojętniania
R фактор эквивалентности для
кислотно-основных реакций

**equivalence factor in precipitation and
complex formation reactions**
For the reaction:

$$\nu_A\, A + \nu_B\, B \rightarrow \text{reaction products}$$

the equivalence factor of constituent B
is defined as

$$f_{eq}(B) = \frac{\nu_B}{\nu_A} = \frac{n_B}{n_A}$$

where ν_A, ν_B — stoichiometric coefficients of
constituents A and B, respectively, n_A, n_B — the
amounts of substance of constituents A and B.
It is assumed that $\nu_A \geqslant \nu_B$, so $f_{eq}(B) \leqslant 1$, and
the equivalence coefficient of constituent A
is always $f_{eq}(A) = 1$.
D ...
F facteur d'équivalence dans les réactions de
précipitation et de complexation
P współczynnik równoważności w reakcjach
strącania i kompleksowania
R ...

equivalence factor in redox reactions
The ratio of the amount of substance n_B of
constituent B to the amount of substance n_e
of electrons taking part in a specified reaction

$$f_{eq}(B) = \frac{n_B}{n_e}$$

D Äquivalenz-Faktor in den Redoxreaktionen
F facteur d'équivalence dans les réactions
d'oxydo-réduction
P współczynnik równoważności w reakcjach
redoks
R фактор эквивалентности для
окислительно-восстановительных реакций

equivalence factor of component B, $f_{eq}(B)$
The coefficient characteristic for a reacting
constituent, resulting from the stoichiometry
of the reaction, equal to the ratio of the amount
of substance of one reagent to the amount of
substance of the other. It is calculated for
a given type of reaction.
D Äquivalenz-Faktor der Komponente B
F facteur d'équivalence du réactif B
P współczynnik równoważności składnika B
R фактор эквивалентности компонента B

**equivalence-point, stoichiometric end-point,
theoretical end-point (*of titration*)**
The point in a titration in which the amount
of titrant added is chemically equivalent to
the amount of substance titrated.

D Äquivalenzpunkt, stöchiometrischer Punkt,
theoretischer Endpunkt (der Titration)
F point d'équivalence, point stoichiométrique,
point théorique de fin de titrage
P punkt równoważnikowy, punkt
stechiometryczny, punkt końcowy
teoretyczny (miareczkowania)
R точка эквивалентности, стехиометрическая
точка, теоретическая конечная точка
(титрования)

equivalent conductivity at infinite dilution →
limiting equivalent conductivity of electrolyte

equivalent conductivity of electrolyte, Λ^*
It is recommended that the use of the terms
"equivalent" and the "equivalent
conductivity" be discontinued.
The equivalent conductivity is related to the
recommended quantity called "molar
conductivity of an electrolyte" (Λ) by the
equation

$$\Lambda^* = \frac{\Lambda}{z_+\nu_+} = \frac{\Lambda}{|z_-|\nu_-}$$

where z_+ and z_- — charge number of cation
and anion, respectively, ν_+, ν_- — numbers of
cations and anions formed from one molecule of
electrolyte $B = X_{\nu_+}Y_{\nu_-}$.
D Äquivalentleitfähigkeit eines Elektrolyten
F conductivité équivalente d'un électrolyte
P przewodnictwo równoważnikowe elektrolitu
R эквивалентная электропроводность
электролита, эквивалентная электрическая
проводимость электролита

**equivalent conductivity of electrolyte at
infinite dilution** → limiting equivalent
conductivity of electrolyte

equivalent conductivity of ionic species B;
equivalent ionic conductivity, λ_B^*
The part of the total equivalent conductivity
of an electrolyte which corresponds to a given
ionic species B.
Remark: It is recommended that the use of these terms be
discontinued.
D Äquivalentleitfähigkeit eines Ions,
Ionenäquivalentleitfähigkeit
F conductivité ionique équivalente
P przewodnictwo równoważnikowe jonowe
R эквивалентная электропроводность иона,
эквивалентная ионная электропроводность

equivalent ionic conductivity → equivalent
conductivity of ionic species B

**equivalent ionic conductivity at infinite
dilution** → limiting equivalent conductivity of
ionic species B

error of first kind, type I error (*in statistics*)
An error consisting in rejecting the null
hypothesis when in fact it is true and should
have been accepted.

D Fehler erster Art
F erreur de première espèce
P błąd pierwszego rodzaju
R ошибка первого рода

error of method
Systematic error connected with the
physico-chemical properties of the analytical
system and the measurement procedure
(analytical determination) applied.
D metodischer Fehler, Verfahrensfehler
F erreur de méthode
P błąd metody
R ошибка метода, методическая ошибка

error of second kind, type II error (*in statistics*)
An error consisting in accepting the null
hypothesis when in fact it is false and should
have been rejected.
D Fehler zweiter Art
F erreur de seconde espèce
P błąd drugiego rodzaju
R ошибка второго рода

ES → emission spectroscopy

ESCA → X-ray photoelectron spectroscopy

ESD → electron stimulated desorption

ESDI → electron stimulated desorption of ions

ESDN → electron stimulated desorption of
neutrals

ESR → electron paramagnetic resonance

estimator
Any statistic used for the estimation of the
unknown value of the general population
parameter.
D Schätzfunktion
F estimateur
P estymator, oszacowanie
R оценка

evaporation
Conversion of a liquid to the vapour state.
D Evaporieren, Abdampfen
F évaporation
P odparowanie
R выпаривание

Everhart-Thornley detector
The scintillator-photomultiplier detector of the
energy of secondary and backscattered
electrons applied especially in scanning
electron microscopy equipment.
D Everhart-Thornley-Detektor
F detecteur de Everhart et Thornley
P detektor Everharta i Thornleya
R детектор Эверхарта-Торнли

evolved gas analysis, EGA
A method of thermal analysis in which the
nature and/or amount of volatile product(s)

released by a substance is/are measured as
a function of temperature whilst the substance
is subjected to a controlled temperature
programme.
D Untersuchung der flüchtigen
Zersetzungsprodukte
F analyse des gaz émis
P analiza wydzielanego gazu
R анализ выделенных газов

evolved gas detection, EGD
In thermal analysis, any method enabling the
detection of the gaseous product formed.
D Gasentwicklungsanalyse
F détection des gaz émis
P wykrywanie wydzielanego gazu
R обнаружение выделенных газов

exact design, experimental design
A program, according to which the values of
the particular factors are varied in the
consecutive experimental runs; the exact design
can be characterized by a design matrix.
D (konkreter) Versuchsplan
F plan d'expérience
P plan eksperymentu
R (точный) план эксперимента

EXAFS → extended X-ray absorption fine
structure

exchange adsorption
The adsorption taking place as a result of
exchange of ions from the adsorbent surface
and these present in solution and carrying
the same charge.
D Austauschadsorption
F adsorption d'échange
P adsorpcja wymienna
R обменная адсорбция

exchange current, I_0
The faradaic current whose intensity is
determined by the rate of the electrode process
at the equilibrium potential.
D Austauschstrom
F courant d'échange
P prąd wymiany
R ток обмена

excitation
The supply of energy to an atom (molecule,
nucleus, etc.) in the ground state to obtain
a state of higher energy called the excited state.
D Anregung
F excitation
P wzbudzenie
R возбуждение

excitation frequency → exciting frequency

excitation function
The relation between the cross section for
a given nuclear reaction and the energy of the
bombarding particles.

D Anregungsfunktion
F fonction d'excitation
P funkcja wzbudzenia
R функция возбуждения

excitation monochromator
(*in a spectrofluorimeter*)
A monochromator which selects the required
wavelength of radiation emitted from the
excitation source for exciting the analyte.
D Anregungsmonochromator
F monochromateur d'excitation
P monochromator promieniowania
wzbudzającego
R монохроматор возбуждения

excitation source, spectrochemical source
(*in emission spectroscopy*)
A device used to excite the analyte, so that
emission of electromagnetic radiation follows.
D Anregungsquelle, Erregungs-Quelle
F source d'excitation
P źródło wzbudzenia, wzbudzalnik
R источник возбуждения

excitation spectrum, absorption spectrum
(*of a luminescent compound*)
The variation of the luminescence flux from a
sample at the wavelength of the luminescence
maximum with the wavelength of the
exciting radiation.
D Erregungs-Spektrum
F spectre d'excitation, spectre d'absorption
P widmo wzbudzenia
R спектр возбуждения

exciting frequency, excitation frequency, v_0
(*in luminescence and Raman spectroscopy*)
The frequency of the incident radiation.
D Erregerfrequenz
F fréquence excitatrice
P częstotliwość wzbudzająca
R частота возбуждения

exciting line
One of the intense spectral lines emitted from
a non-continuous radiation source for
exciting the luminescence or Raman scattering.
D Erregerlinie
F raie excitatrice
P linia wzbudzająca
R возбуждённая линия

exciting line → Rayleigh line

exclusion (*in chromatography*)
Inhibition of penetration of molecules or ions
into the internal pore structure of the sorbent
due primarily to such physical factors as the size
and/or shape of these species.
D Ausschluß
F exclusion
P ekskluzja, wykluczanie
R исключение, эксклюзия

exclusion limit
In permeation chromatography the
molecular-mass limit above which the molecules
of the sample under investigation are unable
to penetrate into the molecular sieve or gel
pores and migrate with a velocity equal to that
of the mobile phase.
D Ausschlußgrenze
F limite d'exclusion
P granica ekskluzji
R предел исключения

exoelectron spectroscopy
The investigation of the energy of electrons
emitted from a solid surface freshly formed
by phase transition, fracture or abrasion.
D Exoelektronenspektroskopie
F spectroscopie d'exoélectrons
P spektroskopia egzoelektronów
R экзоэлектронная спектроскопия

expected value (*of a random variable*), E
Mean value of a random variable X defined as

$$E(X) = \sum_i [x_i \, P(X = x_i)]$$

for the discrete random variable,

$$E(X) = \int_{-\infty}^{+\infty} xf(x) \, dx$$

for the continuous random variable,
where $P(X = x_i)$ − probability of assuming
by the random variable X the value x_i
$f(x)$ − probability density function of the
random variable X.
D Erwartungswert
F espérance mathématique
P wartość oczekiwana, wartość przeciętna
R математическое ожидание

experiment
A process of taking controlled
(by an experimenter) observations in order
to estimate the true value and/or to verify
a hypothesis.
D Versuch
F expérience
P eksperyment
R эксперимент, опыт, испытание

experimental design (*in statistics*)
The term used to describe the stages:
1. identifying the factors which may affect
the result of an experiment,
2. designing the experiment so that the effects
of uncontrolled factors are minimized,
3. using statistical analysis to separate the
effects of the various factors involved.
D (statistische) Versuchsplanung
F planification d'expérience
P planowanie eksperymentalne
R планирование эксперимента

experimental design → exact design

experimental point (*in experimental design*)
A point in an experimental region, the
coordinates of which determine the
conditions of execution of one experimental
run (the coordinates are values of independent
variables in this run).
D Versuchspunkt
F point d'expérience
P punkt doświadczalny
R точка эксперимента

experimental region, factor space
(*in an experimental design*)
A multidimensional space whose points are
the experiments.
D Versuchsbereich
F espace factoriel
P przestrzeń czynnikowa
R область измерений

exponential distribution
Probability distribution of the continuous
random variable X whose probability density
function defined is by the formula
$$f(x) = \begin{cases} ae^{-ax} & \text{for } x > 0 \\ 0 & \text{for } x \leqslant 0 \end{cases}$$
where a — parameter of the population $(a > 0)$
D Exponentialverteilung
F distribution exponentielle
P rozkład wykładniczy
R экспоненциальное распределение

exposure → radiant exposure

exposure time
The time for which a material is illuminated or
irradiated.
D Exponierungszeit, Expositionszeit
F temps de pose, pose
P czas naświetlania
R время экспозиции

external indicator
An indicator used externally; it reacts with
a drop of the titrated solution taken from
the vessel in which the titration is carried out.
D äußerer Indikator, Außen-Indikator
F indicateur externe
P wskaźnik zewnętrzny
R внешний индикатор

external standard (*in NMR spectroscopy*)
A sealed capillary tube containing a reference
substance, e.g. tetramethylsilane, located
coaxially in the microtube filled with the sample
being analysed.
D äußerer Standard
F étalon externe
P wzorzec zewnętrzny
R внешний эталон

extinction → absorbance

extinction coefficient → specific decadic
absorption coefficient

extra-column volume, dead volume, V_{ext}, V_d
(*in chromatography*)
The volume occupied by the mobile phase
between the effective injection point and the
detection point less the bed volume: i.e. any
section of the chromatographic flow system in
which diffusion of solutes can occur without
separation.
Remark: The term dead volume often defines the volume
occupied by the mobile phase between the lower level of
the bed and either the outlet of the column or the detection
point.
D Säulenvorvolumen, Totvolumen
F volume hors de la colonne, volume mort
P objętość pozakolumnowa, objętość martwa
R предколоночный объём, мёртвый объём

extract
A separated phase (usually organic)
containing a substance extracted from another
phase.
D Extrakt
F extrait
P ekstrakt, wyciąg
R экстракт

extractant, extraction solvent
The liquid phase (usually an organic solvent,
mixture of solvents, or a solution of the
extracting agent in an organic solvent) used
for extraction of a substance from another
liquid phase (usually aqueous) or from a solid
phase.
D Extraktionsmittel
F solvant d'extraction, extractant
P ekstrahent
R экстрагент, экстрагирующее средство

extracting agent
A reagent forming complex salts or adducts
with the extracted substance; such products
may be transferred through the interface of
the extraction system.
D Extraktionsreagent
F extractant, agent extractant
P odczynnik ekstrahujący
R экстракционный реагент

extraction
The process of mass transfer in a
multicomponent system of limited solubility,
involving the dissolution of the constituent(s)
of a liquid or solid phase in a liquid extractant.
D Extraktion
F extraction
P ekstrakcja
R экстракция, экстрагирование

extraction chromatography, reversed-phase
partition chromatography
Partition chromatography in which the

stationary phase is an organic extractant
deposited on a porous, hydrophobic support,
and the mobile phase is a suitable aqueous
solution of an acid, base, or salt.

D Extraktionschromatographie,
Umkehrphasen-Verteilungschromatographie
F chromatographie par extraction,
chromatographie de partage en phases
inversés
P chromatografia ekstrakcyjna,
chromatografia podziałowa z odwróconymi
fazami
R экстракционная хроматография,
распределительная хроматография
с обращёнными фазами

extraction coefficient → concentration
distribution ratio

extraction constant, K_{ex}
The equilibrium constant of the distribution
reaction, e.g. for most reactions

$$M_w^{n+} + nHL_o \rightleftarrows ML_{no} + nH_w^+$$

where w — aqueous phase, o — organic phase
in which the reagent HL initially dissolved in
the organic phase reacts with a metal ion M^{n+}
in the aqueous phase to yield the product ML_n
much better soluble in the organic phase than
in the water one and the extraction constant
may be written as

$$K_{ex} = \frac{[ML_n]_o \, [H^+]_w^n}{[M^{n+}]_w \, [HL]_o^n}$$

In solutions in which the ionic strength
equals zero, the extraction constant is denoted
as K_{ex}^o.

D Extraktionskonstante
F constante d'extraction
P stała ekstrakcji
R константа экстракции

exctraction curve
A curve presenting the correlation between the
function describing the extraction effectiveness
and the parameters which specify the extraction
conditions, e.g. in the system: logarithm of the
extraction coefficient versus pH, or extraction
percentage versus pH.

D Extraktionskurve
F courbe d'extraction
P krzywa ekstrakcji
R кривая экстракции

extraction efficiency → recovery factor

extraction indicator
An indicator (acid-base or other type) which is
abruptly extracted from one liquid phase into
another at or near the equivalence-point of
a titration. The indicator should not change
its colour in the course of titration. In some
instances the titrant may serve as the indicator.

D Extraktionsindikator
F indicateur par extraction
P wskaźnik ekstrakcyjny
R экстракционный индикатор

extraction percentage, $\%E$, $R\%$
Recovery factor expressed in per cent.

D Extraktionsausbeute in %,
Extraktionsgrad in %
F pourcentage d'extraction
P procent ekstrakcji
R процент экстракции

extraction solvent → extractant

extractive titration
Titration in which the titrant makes one phase
and the analysed solution the other; e.g.
titration of the aqueous zinc salt solution with
dithizone solution in carbon tetrachloride.
After each addition of the titrant, sufficient
time must be allowed for the phases to separate.

D extraktive Titration
F titrage extracteur
P miareczkowanie ekstrakcyjne
R экстракционное титрование

extractor
The apparatus in which the process of
extraction is carried out.

D Extrakteur
F extracteur
P ekstraktor
R экстрактор

F

FAAS → flame atomic absorption spectroscopy

factor (*in experimental design*)
Any independent variable controlled in the
course of experiment.

D Faktor
F facteur
P czynnik
R фактор

factor analysis
Analysis of and testing the significance of factors
and their interactions.

D Faktor(en)-Analyse
F analyse factorielle
P analiza czynnikowa
R факторный анализ

factorial design → complete factorial design

factor space → experimental region

FAES → flame atomic emission spectroscopy

FAFS → flame atomic fluorescence spectroscopy

faradaic current, I_f
The current which corresponds to the reaction
of charge transfer through the electrode-
electrolyte interface.

D Faradayscher Strom
F courant faradaïque
P prąd faradajowski
R фарадеевский ток

Faraday constant, F
The constant numerically equal to the charge
of one mole of electrons, $F = eN_A$
where e — elementary charge, N_A — Avogadro
constant, the Faraday constant equals
96 487 C mol^{-1}.

D Faraday-Konstante, Faradaysche Zahl
F constante de Faraday
P stała Faradaya
R число Фарадея, постоянная Фарадея

Faraday's laws (*of electrolysis*)
(1) The mass of any substance liberated in an
electrochemical reaction is proportional to the

quantity of electric charge which has passed
through the electrode.
(2) The masses of different substances liberated
in electrochemical reactions by a given
quantity of electric charge which
has passed through the electrode in the same
time are proportional to their molar masses
corresponding to one mole of electrons taking
part in the electrochemical reaction (earlier, to
their chemical equivalents).

D Faradaysche Gesetze
F lois de Faraday
P prawa Faradaya
R законы Фарадея

far infrared, FAR-IR
The range of the infrared radiation of
wavelengths from 50 µm to 1 mm.

D fernes Infrarot, fernes Ultrarot
F infrarouge lointain
P podczerwień daleka
R дальнее инфракрасное излучение

FAR-IR → far infrared

far ultraviolet, FAR-UV, FUV
The range of ultraviolet radiation of
wavelengths from 10 to 200 nm.

D fernes Ultraviolett
F ultraviolet lointain
P nadfiolet daleki
R далёкий ультрафиолет

FAR-UV → far ultraviolet

FAS → flame absorption spectroscopy

fast neutrons
Neutrons of kinetic energy greater than some
specified value dependent upon the given
application, e.g. in activation analysis from
0.5 to 20 MeV.

D schnelle Neutronen
F neutrons rapides
P neutrony prędkie
R быстрые нейтроны

FD → flash desorption

F-distribution, Fisher's distribution
Probability distribution of the continuous
random variable X whose probability density
function is defined by the formula

$$f(x) = \frac{\Gamma\left[(k_1+k_2)/2\right]}{\Gamma(k_1/2)\,\Gamma(k_2/2)}\left(\frac{k_1}{k_2}\right)^{k_1/2} \times$$

$$\times \frac{x^{(k_1-2)/2}}{[1+(k_1/k_2)\,x]^{(k_1+k_2)/2}} \quad \text{for } x > 0$$

$$f(x) = 0 \quad \text{for } x \leqslant 0$$

where k_1 and k_2 — numbers of degrees of
freedom. The F-distribution is applied very
frequently in analysis of variance for testing
hypotheses concerning variances.

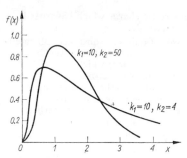

D Fishersche Verteilung, F-Verteilung
F distribution de Fisher
P rozkład F(-Snedecora)
R распределение Снедекора

Fehling's reagent
A mixture of two aqueous solutions: the first
containing copper (II) sulfate, the second —
sodium-potassium tartrate and potassium
hydroxide; the reagent is used, inter alia, for
sugar determination (red precipitate of
copper(I) oxide).
D Fehlingsche Lösung
F réactif de Fehling
P odczynnik Fehlinga
R реактив Фелинга

Fermi resonance (*in IR-spectroscopy*)
Abnormality observed in polyatomic molecules
in which two different vibrational states
possess the same energy and mutually interact.
This phenomenon is often observed as a new
band between the fundamental band and an
overtone.
D Fermi-Resonanz
F résonance Fermi
P rezonans Fermiego
R резонанс Ферми

FES → flame emission spectroscopy

FET → field effect transistor

FFS → flame fluorescence spectroscopy

FI → field ionization

Fick's laws *(of diffusion)*
Fick's first law states that the amount ∂m of
substance crossing a unit cross-sectional area A
in a given direction during the time interval ∂t
is proportional to the concentration gradient
$\partial c/\partial t$ of that substance;

$$\frac{\partial m}{\partial t} = - DA \left(\frac{\partial c}{\partial x}\right)_t$$

where D — diffusion coefficient of the
substance, x — the distance in the direction of
diffusion. Fick's second law states that the
amount of the substance in a horizontal layer
beetween x and $x+\partial x$ increases in time

$$\left(\frac{\partial c}{\partial t}\right) = - D \left(\frac{\partial c}{\partial x^2}\right)_t$$

D Ficksche Gesetze
F lois de Fick
P prawa Ficka
R законы Фика

FID → flame ionization detector

field desorption
The desorption process induced by a strong
external electric field (10^8 V cm^{-1}) reducing
the width of the surface energy barrier and
enabling electron tunneling from a metal to
the adsorbate.
D Felddesorption
F désorption de champ
P desorpcja polowa
R десорбция полем

field effect transistor, FET
A unipolar multielectrode semiconductor
device in which current flows through a narrow
conducting channel, between two electrodes
(source and drain) and is modulated by an
electric field produced in the channel by the
potential applied to a third electrode (gate).
D Feldeffekttransistor
F transistor à effet de champ
P tranzystor polowy
R полевой транзистор

field emission ion source → field ionization ion
source

field ionization, FI
Ionization resulting from the effect of a very
strong electrical field on any particle. The
strong electrical field may produce ionization
in a space or in a region very close to a metal
or other surface.
D Feldionisation
F ionisation par champ
P jonizacja polem
R ионизация полем

field ionization ion source, field emission ion
source (*in mass spectrometry*)
An ion source in which the sample is subjected
to a high voltage field of about 10 kV cm^{-1} with
a sharp edge, a sharp point, or fine wire anode
placed about 1 mm away from the cathode.
The high voltage leads to a loss of electrons by
the molecules, resulting in a high yield of
molecular ions.
D Feldionenquelle
F source à émission de champ
P polowe źródło jonów
R источник ионизаций полем

field strength effect → Wien's effect

field-sweep method (*in NMR spectrometry*)
A method of scanning the NMR spectrum by
sweeping the external magnetic field while
keeping the frequency of the radio-frequency
radiation constant.

D Feldsweep-Methode
F méthode de balayage du champ magnétique
P metoda przemiatania polem
R метод развёртки магнитного поля

filter fluorimeter, filter fluorometer
A fluorimeter which consists of a source of
excitation energy, a sample cuvette, a detector
to measure the intensity of the fluorescent
radiation, and a pair of filters. The primary
filter is selected so as to transmit only specific
excitation wavelengths and the secondary filter
is selected so as to transmit only wavelengths
corresponding to maximum fluorescent
emission and to arrest any scattered excitation
light.
D Filterfluorimeter
F fluorimètre à filtre
P fluorymetr z filtrem
R фильтровый флуориметр

filter fluorometer → filter fluorimeter

filter funnel with sintered disk, fritted disk
funnel
A glass funnel with a flat sintered glass
bottom used for filtering by suction. A symbol
etched on the funnel wall gives the porosity
grade of the filtering disk.
D Glasfiltertrichter
F entonnoir à plaque de verre fritté
P lejek z filtrem ze spiekanego szkła, lejek
  Schotta
R фильтр со стеклянной пористой пластинкой

filter paper
Circular piece of porous paper used for
filtration.
D Papierfilter
F papier-filtre
P sączek
R бумажный фильтр

filter tube
A type of filter with sintered glass disc, used
for separating solid particles from gases.
D Gasfilter
F filtre à gaz
P rurka filtracyjna, filtr gazowy
R трубка для фильтрования

filtrate
The solution left after filtering
a precipitate.
D Filtrat
F filtrat
P przesącz
R фильтрат

filtration
Mechanical separation of a solution-precipitate
mixture on a porous layer (filter paper, or
sintered glass plate)
D Filtrieren, Filtration
F filtration

P sączenie
R фильтрование

fine structure (*in atomic spectroscopy*)
The splitting of the spectral lines or terms
caused by the spin angular momentum of
electrons and the coupling of the spin to the
orbital angular momentum.
D Feinstruktur
F structure fine
P struktura subtelna
R тонкая структура

fingerprint region
The range of the infrared spectrum
(from 1300 to 650 cm^{-1}) of a substance
in which a large number of absorption
peaks characteristic of the given substance
occur.
D Fingerabdruck
F empreinte digitale
P przedział daktyloskopowy
R область отпечатка пальцев

first-class conductor → electronic conductor

first order design
An experimental design which allows to
approximate the response function by a first
order polynomial.
D Versuchsplan 1. Ordnung
F plan de premier degré
P plan pierwszego rzędu
R план первого порядка

first overtone (band) *See* overtone band
D erste Oberschwingung
F première harmonique
P nadton pierwszy, pasmo nadtonowe pierwsze
R первый обертон

first Wien effect → Wien's effect

Fisher's distribution → F-distribution

fission, nuclear fission
The splitting of a nucleus into two or more
fragments with masses of the same order of
magnitude, usually accompanied by the emission
of neutrons, γ-radiation and, rarely, small
charged nuclear fragments.
D Kernspaltung, Spaltung
F fission (nucléaire)
P rozszczepienie (jądrowe)
R (ядерное) деление, деление ядра

fission fragments (*of atomic nucleus*)
Nuclei with high kinetic energy, produced in
a nuclear fission process.
D Kernbruchstücke, Spaltprodukte
F fragments de fission
P fragmenty rozszczepienia
R осколки деления

fission neutrons
Neutrons originating in the fission process
which have retained their original energy.

D Spaltneutronen
F neutrons de fission
P neutrony rozszczepieniowe
R нейтроны деления

fission yield
The fraction of fissions yielding fission
products of a given mass number.

D Spaltausbeute
F rendement de fission
P wydajność produktu rozszczepienia
R выход продуктов деления, выход осколков
деления

fixation pair, control line pair (*in emission
spectrochemical analysis*)
A pair of lines of similar intensities that change
their intensity in a different way as the
conditions of excitation vary; these lines are
used to check whether the excitation conditions
remain constant during each exposure.

D Fixierungspaar
F paire de raies de fixation
P para linii kontrolna
R фикспара линий

fixed effects model, type I model
A model of analysis of a variance in which the
effects of fixed levels of factors are
investigated and the changes of output values are
treated as depending on changes of the factor
levels (e.g. determination of differences of
analytical results obtained by particular
analysts).

D Modell mit festen Effekten, Modell I
F modèle à effets fixes, modèle I
P model stały, model I
R модель с фиксированными эффектами

fixed ionic groups → ionogenic groups

fixed ions
In an ion exchanger, the non-exchangeable
ions which have a charge opposite to that of
the counter-ions.

D Festionen, gebundene Ionen
F ions fixes, ions retenus
P jony związane
R фиксированные ионы

flame (*in flame spectroscopy*)
A continuously flowing gas mixture at
atmospheric pressure which emerges from
a burner and is heated by combustion.

D Flamme
F flamme
P płomień
R пламя

flame absorption spectroscopy, FAS
A branch of absorption spectroscopy in which
a flame is used to vaporize and atomize
the sample.

D Flammen-Absorptions-Spektroskopie
F spectroscopie d'absorption de flamme,
spectroscopie de flamme par absorption
P spektroskopia absorpcyjna płomieniowa
R пламенная абсорпционная спектроскопия,
абсорпционная спектроскопия пламени

flame atomic absorption spectroscopy, FAAS
Flame absorption spectroscopy in which both
a flame is used and the atomic lines are
observed.

D Flammen-Atom-Absorptions-Spektroskopie
F spectroscopie d'absorption atomique de
flamme
P spektroskopia absorpcyjna atomowa
płomieniowa
R пламенная атомно-абсорпционная
спектроскопия, атомно-абсорпционная
спектроскопия пламени

flame atomic emission spectroscopy, FAES
Flame emission spectroscopy in which
both a flame is used and atomic lines are
observed.

D Flammen-Atom-Emissions-Spektroskopie
F spectroscopie d'émission atomique de flamme
P spektroskopia emisyjna atomowa
płomieniowa
R пламенная атомно-эмиссионная
спектроскопия, атомно-эмиссионная
спектроскопия пламени

flame atomic fluorescence spectroscopy, FAFS
Flame fluorescence spectroscopy in which
both a flame is used and atomic lines are
observed.

D Flammen-Atom-Fluoreszenz-Spektroskopie
F spectroscopie de fluorescence atomique de
flamme
P spektroskopia fluorescencyjna atomowa
płomieniowa
R атомно-флуоресцентная спектроскопия
пламени, пламенная
атомно-флуоресцентная спектроскопия

flame coloration (*in qualitative analysis*)
The coloration of the laboratory burner
flame caused by the emission of
electromagnetic radiation of a specific
wavelength by the atoms of certain elements
when they are heated on a platinum wire
in the flame.

D Flammenfärbung
F coloration de flamme
P barwienie płomienia
R окраска пламени

flame emission spectroscopy, FES
A branch of emission spectroscopy in which
a flame is used to vaporize, atomize and
excitate of the sample.

Remark: The use of the term flame photometry
has been abandoned by the IUPAC nomenclature
recommendations.

D Flammen-Emissions-Spektroskopie
F spectroscopie d'émission de flamme, photométrie de flamme
P spektroskopia emisyjna płomieniowa, fotometria płomieniowa
R эмиссионная спектроскопия пламени, пламенная фотометрия

flame fluorescence spectroscopy, FFS
A branch of fluorescence spectroscopy in which a flame is used to vaporize and atomize of the sample.
D Flammen-Fluoreszenz-Spektroskopie
F spectroscopie de fluorescence de flamme
P spektroskopia fluorescencyjna płomieniowa
R пламенная флуоресцентная спектроскопия, флуоресцентная спектроскопия пламени

flame ionization detector, FID
(*in chromatography*)
An ionization detector consisting of a hydrogen burner whose flame is burning between two electrodes with potential difference between them. When the column effluent gas is burned with air or oxygen in the hydrogen flame, an increased ionization current is obtained as a result of ionization of the solute molecules.
D Flammenionisationsdetektor
F détecteur à ionisation de flamme
P detektor płomieniowo-jonizacyjny, detektor jonizacyjny płomieniowy
R пламенно-ионизационный детектор

flame photometer
An instrument used to measure the intensities of radiation in the flame-excited spectra of samples.
Remark: The use of this term has been abandoned by the IUPAC nomenclature recommendations.
D Flammenphotometer
F photomètre de flamme
P fotometr płomieniowy
R пламенный фотометр

flame photometric detector, photometric detector (*in chromatography*)
The detector measuring the intensity of the light (at fixed wavelength) which accompanies the combustion of the eluate in a hydrogen-rich flame.
D (flammen) photometrischer Detektor
F détecteur photométrique, détecteur à photométrie de flamme
P detektor fotometryczny płomieniowy, detektor płomieniowo-fotometryczny
R пламенно-фотометрический детектор

flame photometry → flame emission spectroscopy

flame spectrophotometer
Flame photometer which measures the intensity of radiation of a given wavelength after its separation by monochromator.
D Flammenspektrophotometer, Flammenspektralphotometer

F spectrophotomètre à flamme
P spektrofotometr płomieniowy
R пламенный спектрофотометр

flash combustion
Rapid combustion of a sample in a capsule made of aluminium, silver or tin foil, introduced directly into the hot zone of a combustion tube. The combustion of the metal provides additional heat, thus causing considerable increase of the temperature, which promotes complete combustion of the sample.
D Blitzverbrennung
F combustion éclair
P spalanie błyskawiczne, mineralizacja zapłonowa
R сожжение со вспышкой

flash desorption, FD, flash filament technique
The sudden desorption caused by a rapid rise of temperature of the adsorbent, the latter being in the form of a thin ribbon or wire.
D Flash-filament-Methode, Impulsdesorption, Heizdraht-Methode
F technique du filament chauffé
P desorpcja błyskawiczna, desorpcja natychmiastowa
R флеш-десорбция, импульсная десорбция

flash filament technique → flash desorption

flash pyrolysis
The pyrolysis technique in which the temperature of the sample is raised in a fraction of a second to $800-1000°C$.
D Flash-Pyrolyse, Blitzpyrolyse
F pyrolyse par flash
P piroliza błyskowa
R импульсный пиролиз

flocculation → coagulation

flow-programmed chromatography, flow programming
A chromatographic technique in which the rate of flow of the mobile phase is varied systematically during a part of (or the whole) separation process. The flow programming may be linear, with a constant increase of the flow rate, or stepwise with periodic increases of the flow rate.
D flußprogrammierte Chromatographie, durchflußprogrammierte Chromatographie, strömungsprogrammierte Chromatographie
F chromatographie avec programmation du débit, chromatographie à écoulement programmé, programmation de débit
P chromatografia z programowaniem prędkości przepływu
R хроматография с программированием потока

flow programming → flow-programmed chromatography

flow rate of mercury, rate of flow of mercury from capillary (*in polarography*)
The mass of mercury flowing out from a capillary in unit time.
D Ausströmgeschwindigkeit des Quecksilbers an der Kapillare
F vitesse d'écoulement du mercure, vitesse de sortie du mercure, débit du mercure par le capillaire
P wydajność kapilary
R скорость вытекания ртути из капилляра

fluorescence
Luminescence that ceases as soon as the source of excitation is removed, i.e. the immediate emission of light from an atom or molecule after it has absorbed some form of energy. Fluorescence results from the spin - allowed transition.
D Fluoreszenz
F fluorescence
P fluorescencja
R флуоресценция

fluorescence analysis → fluorimetry

fluorescence emission spectrum
The variation of the fluorescence flux from a sample at the excitation wavelength for the analyte with the wavelength of the emitted fluorescent radiation.
D Fluoreszenz-Emissionsspektrum
F spectre d'émission de fluorescence
P widmo emisyjne fluorescencyjne
R флуоресцентный эмиссионный спектр

fluorescence excitation spectrum
The variation of the fluorescence flux from a sample at the wavelength of the fluorescence maximum with the wavelength of the exciting radiation.
D Fluoreszenz-Erregungsspektrum
F spectre d'excitation de fluorescence
P widmo wzbudzenia fluorescencyjne
R флуоресцентный спектр возбуждения

fluorescence spectrochemical analysis
The analysis based on the investigation of fluorescence spectra.
D Fluoreszenz-Spektralanalyse
F analyse spectrochimique de fluorescence
P analiza spektralna fluorescencyjna
R флуоресцентный спектрохимический анализ

fluorescence spectroscopy, FS
A branch of spectroscopy concerned with the production, measurement and interpretation of the fluorescence spectra of atoms and molecules.
D Fluoreszenz-Spektroskopie
F spectroscopie de fluorescence
P spektroskopia fluorescencyjna
R флуоресцентная спектроскопия

fluorescence spectrum
A spectrum of emitted electromagnetic radiation due to de-excitation of an atom or a molecule excited by absorption of radiation of higher frequency (energy).
D Fluoreszenzspektrum
F spectre de fluorescence
P widmo fluorescencyjne, widmo fluorescencji
R флуоресцентный спектр

fluorescent indicator
An indicator (acid-base or other type) which when activated by radiation of a suitable] wavelength exhibits a change of the intensity of fluorescent radiation at or near to the equivalence-point of a titration.
D Fluoreszenzindikator
F indicateur par fluorescence
P wskaźnik fluorescencyjny
R флуоресцентный индикатор

fluorescent X-rays → X-ray fluorescence radiation

fluorimeter, fluorometer
An instrument designed for measuring the total fluorescent radiation emitted by a sample exposed to monochromatic radiation.
D Fluorimeter
F fluorimètre
P fluorymetr
R флуориметр, флуорометр

fluorimetric analysis → fluorimetry

fluorimetric titration
Titration in which the end point is determined visually or photometrically from the change of the intensity of fluorescent radiation of the titrated solution.
D fluorimetrische Titration, Fluoreszenztitration
F titrage fluorimétrique
P miareczkowanie fluorymetryczne
R флуориметрическое титрование

fluorimetry, fluorometry, fluorimetric analysis, fluorescence analysis
A method based on the measurement of the intensity of fluorescent radiation emitted by an appropriately excited sample; in chemical analysis a method taking advantage of the fluorescent properties of the substance to be determined
D Fluorimetrie, Fluoreszenzanalyse
F fluorimétrie
P fluorymetria, analiza fluorescencyjna
R флуориметрия, флуорометрия, флуоресцентный анализ

fluorometer → fluorimeter

fluorometry → fluorimetry

97

fractional precipitation

fluted filter
A filter prepared by repeated folding a circular
piece of filter paper along its radii, used for
quick separation of the precipitate from a
solution.
D Faltenfilter
F filtre plissé, filtre à plis
P sączek fałdowany, sączek karbowany
R плоёный фильтр

flux, fusing agent
A substance which when added to another
substance facilitates its fusion or prevents the
formation of oxides.
D Flußmittel
F fondant, flux
P topnik
R плавень, флюс

flux density (of particles or photons)
The number of particles or photons incident
in a time interval on a suitably small sphere
centered at a given point in space, divided
by the cross-sectional area of that sphere and
by that time interval. The particle flux density
is identical with the product of the particle
density and the average velocity of the particles.
D Flußdichte (der Teilchen oder Photonen)
F densité de flux (de particules ou de photons)
P gęstość strumienia (cząstek lub fotonów)
R плотность потока (частиц или фотонов)

flux gradient (of particles or photons)
The variations of the irradiating flux with the
position of the sample (in a reactor, neutron
generator, etc.).
D Gradient des Flusses (der Teilchen oder
   Photonen)
F gradient de flux (de particules ou de photons)
P gradient strumienia (cząstek lub fotonów)
R градиент потока (частиц или фотонов)

focusing circle (*in an* X-*ray spectrometer*)
The circle of diameter R on whose
circumference, in the focusing geometry of the
spectrometer, the source of analysed
radiation (slit), the crystal and detector are
located in such a way that the distances L
from the crystal centre to the source and
detector are equal, hence

$L = n\lambda R/2d = R \sin \theta$

where n — order of the diffraction,
λ — wavelength, d — spacing between
neighbouring crystallographic planes, θ — angle
of diffraction.
D Fokussierkreis,
F cercle de focalisation
P koło ogniskujące
R окружность фокусировки, окружность
   изображения

fog (*of a photographic emulsion*)
Unintentional blackening of a photographic
emulsion caused by improper storage,
aging, or processing.

D Schleier, Schleierschwärzung
F voile
P zadymienie
R вуаль

forbidden transition
The transition between two energy levels of a
system whose probability, according to
the selection rules, is equal to zero.
D verbotener Übergang
F transition interdite
P przejście wzbronione
R запрещённый переход

force constant, f (*in IR spectroscopy*)
The restoring force per unit displacement
from the equilibrium position for a system of
two point masses.
D Kraftkonstante
F constante de force
P stała siłowa (wiązania)
R силовая константа

formal electrode potential → conditional
electrode potential

formality, F
The number of gram formula weights of the
reacting substance in one litre of solution, the
formula being specified whenever there is
a risk of ambiguity.
D Formalität
F formalité
P formalność (roztworu)
R формальность

formal potential → conditional electrode
potential

fraction
A part of a whole having strictly specified
properties, e.g. the fraction with a given boiling
temperature range.
D Fraktion
F fraction
P frakcja
R фракция

fractional analysis
The separation of a multicomponent system
into fractions containing components of
similar physical properties, e.g. boiling,
freezing, melting points or grain size.
D ...
F analyse fractionnée
P analiza frakcyjna
R фракционный анализ

fractional free volume → interstitial fraction

fractional precipitation
The precipitation by gradual addition of
the precipitation reagent which forms
with ions present in solution compounds of

different solubilities; e.g. the quantitative
separation of iodide from chloride ions,
effected by a addition of a silver nitrate
solution.
D fraktionierte Fällung
F précipitation fractionnée
P strącanie frakcjonowane, strącanie różnicowe
R фракционное осаждение

fraction collector (*in chromatography*)
A device used for collecting of fractions of
eluate emerging from the chromatographic
column, according to a predetermined time or
volume program.
D Fraktionssammler
F collecteur de fractions
P kolektor frakcji
R коллектор фракций, сборник фракций

fragmentation (*in mass spectrometry*)
Cleavage of bonds in parent ions which
leads to the formation of ions of smaller
molecular mass and one or more neutral
fragments.
D Fragmentierung, Bruchstückbildung
F fragmentation
P fragmentacja
R фрагментация

fragmentation pattern, cracking pattern
(*in mass spectrometry*)
The series of values of the ion beam intensities
calculated for a given compound, as a
percentage of the most intense peak which is
assignated the value of hundred.
Fragmentation pattern is characteristic of
compound under specific conditions of
excitation and serves for its identification.
D Fragmentierungsmuster
F mécanisme de fragmentation
P schemat fragmentacji
R схема фрагментации

fragment ion (*in mass spectrometry*)
An electrically charged dissociation product of
the fragmentation process. The fragment ion
may dissociate further to form other
electrically charged molecular or atomic
moieties of sucessively decreasing formula
mass.
D Fragmention, Bruchstückion
F ion fragment
P jon fragmentacyjny
R фрагментный ион

free volume → interstitial volume

French degree of water hardness
The unit of water hardness equivalent to
the content of 10 mg of calcium carbonate
in 1 l of water.
D französischer Härtegrad, französische Härte
F degré hydrotimétrique français
P stopień twardości wody francuski
R французский градус жёсткости воды

frequency, ν
In periodical processes, the number of
complete cycles per unit time. For
electromagnetic radiation $\nu = c/\lambda = \bar{\nu}c$,
where c — velocity of light in vacuum,
λ — wavelength, and $\bar{\nu}$ — wave number.
SI unit: hertz, Hz.
D Frequenz, Schwingungsfrequenz,
Schwingungszahl
F fréquence
P często(tliwo)ść
R частота

frequency-sweep method (*in NMR spectroscopy*)
A method of scanning the NMR spectrum by
changing the frequency of the radio-frequency
transmitter while keeping the external field
constant.
D Frequenzsweep-Methode
F méthode de balayage de fréquence
P metoda przemiatania częstotliwości
R метод частотной развёртки

Friedrichs gas washing bottle
D Gaswaschflasche nach Friedrichs
F flacon laveur de Friedrichs
P płuczka Friedrichsa
R склянка Фридрихса

fritted disk funnel → filter funnel with sintered
disk

front → mobile phase front

frontal analysis → frontal chromatography

frontal chromatography, frontal analysis
A chromatographic technique in which the
sample is continuously fed to the
chromatographic bed. Only the component
least retained can be obtained
in the pure state.
D Frontalchromatographie, Frontalanalyse
F analyse frontale
P chromatografia czołowa, analiza czołowa,
metoda czołowa
R фронтальная хроматография, фронтальный
анализ, фронтальный метод

fronting
Asymmetry of a chromatographic peak such
that the ascending part of the peak, the front,
is less steep than the descending part of the
peak, the rear. In planar chromatography, the
appearance of a diffused zone in front of
the spot
D Fronting
F front diffus (situé sur le pic avant le sommet)
P rozmycie przedniej części piku/pasma/plamy,
tworzenie odwróconej komety
R размытие фронта

FS → fluorescence spectroscopy

F-test
The test of significance allowing to decide
(on a given significance level) if the variances
estimated on the basis of two samples taken
from two normally distributed populations
differ significantly.
D F-Test
F test F (de Snedecor)
P test F, test Snedecora
R критерий Снедекора

full energy peak, total absorption peak
The peak in a γ-radiation spectrum
corresponding to the total absorption of γ-rays
in the detector by various interaction
mechanisms, such as the photoelectric effect,
multiple Compton scattering, pair production
without escape of the annihilation photons.
D Gesamtenergiepeak
F pic d'absorption totale
P pik całkowitej absorpcji
R пик общей абсорбции, пик полного
    поглощения

functional group
An atom or a group of atoms which, on
substitution of the hydrogen atom at the
carbon atom in an organic molecule, confers
new properties to the latter, different from those
of the original molecule; e.g. —Cl, —OH,
—SH, —NO$_2$, —CN, —CHO, —CO—, —NO,
—COOH, —SO$_2$OH, —N=N—, —CO—O—.
D funktionelle Gruppe
F groupe fonctionnel, groupemant fonctionnel
P grupa funkcyjna
R функциональная группа

functional group analysis, analysis via functional
groups
The analysis of organic compounds based on
the identification and determination of the
functional groups present in the molecule.
D (funktionelle) Gruppenanalyse, funktionelle
    Analyse
F analyse fonctionnelle, analyse par
    groupements fonctionnels
P analiza według grup funkcyjnych
R функциональный анализ, анализ по
    функциональным группам

fundamental band → fundamental vibration
band

fundamental infrared → middle infrared

fundamental parameter method (*in X-ray
fluorescence analysis*)
A mathematical method of calculation of the
sample composition in which account is taken
of the spectral distribution of the primary
radiation mass absorption coefficients for all
sample elements and of fluorescence yields for
these elements as well as of the measured
intensities of characteristic X-rays of sample
elements in the sample and in pure element
standards.
D Fundamentalparameter-Verfahren
F méthode des paramètres fondamentaux
P metoda parametrów podstawowych
R метод основных параметров

fundamental (vibration) band
A band in the infrared spectrum of a molecule
corresponding to the transitions between the
ground state in which all quantum numbers
are zero to an excited state in which the
vibrational quantum number is 1.
D Grundschwingungsbande, Grundbande
F bande fondamentale
P pasmo podstawowe, pasmo fundamentalne
R основная полоса, полоса основных
    колебаний

funnel heater, funnel heating mantle
D Heizmantel für Trichter, Heißwassertrichter
F chauffe-entonnoir, entonnoir chauffant
P lejek do sączenia na gorąco, podgrzewacz do
    lejków
R воронка для горячего фильтрования

funnel heating mantle → funnel heater

fusing agent → flux

fusion → melting

fusion method (*in X-ray fluorescence analysis*)
A method of elimination of the grain size
effect, mineralogical and heterogeneity effects,
and substantial decrease of the matrix effect
consisting in the fusion of the analysed sample
with a flux (e.g. sodium tetraborate) and in
investigation of the obtained solid solution.
D Aufschließenverfahren, Aufschlußverfahren
F méthode de fusion
P metoda stapiania
R метод плавления

FUV → far ultraviolet

G

galvanic cell, voltaic cell
A system composed of two half-cells so
constructed that when the electrodes are
connected externally by a conductor, a flow
of electrons occurs.

D galvanische Zelle, galvanische Kette,
galvanisches Element
F pile électrique, pile galvanique, cellule
galvanique
P ogniwo galwaniczne
R гальванический элемент, гальваническая
цепь

galvanic thermocell, thermogalvanic cell,
electrolytic thermocell
The galvanic cell consisting of two identical
electrodes (half-cells) with a temperature
gradient between them.

D thermogalvanische Zelle, Thermoelement,
elektrolytische Thermozelle
F pile thermoélectrique, cellule
thermoélectrique
P termoogniwo galwaniczne, ogniwo
termoelektryczne
R термогальванический элемент,
электрохимическая термоцепь,
гальванический термоэлемент

Galvani potential → inner electric potential
of phase β

Galvani potential difference, $\Delta_\alpha^\beta \phi$
The difference in the internal electrical
potentials of two phases

$$\Delta_\alpha^\beta \phi = \phi^\beta - \phi^\alpha$$

D Differenz der Galvani-Potentiale,
Galvani-Potentialdifferenz, Galvani-Spannung
F différence des potentiel de Galvani, tension
de Galvani
P różnica potencjałów Galvaniego, napięcie
Galvaniego
R разность Гальвани-потенциалов,
напряжение Гальвани

galvanometer
An instrument used for detecting or
measuring small electric currents.

D Galvanometer
F galvanomètre
P galwanometr
R гальванометр

galvanostat, amperostat
An instrument which induces a flow of current
of a programmed and time-constant value
through an electrochemical cell.

D Galvanostat
F galvanostat
P galwanostat, amperostat
R гальваностат, амперостат

galvanostatic method
An electroanalytical method consisting in
applying a constant current to the indicator
electrode and in measuring the potential of this
electrode as a function of time.

D galvanostatische Methode, amperostatische
Methode
F méthode galvanostatique
P metoda galwanostatyczna, metoda
amperostatyczna
R гальваностатический метод,
амперостатический метод

gamma, contrast factor (*in photography*)
Tangent of the angle between the straight-line
section of the emulsion characteristic curve
and the abscissa, on which the values of the
common logarithm of the exposure values
(in conventional units) are plotted.

D Gamma-Wert, Kontrastfaktor, Gradation γ
F gamma, facteur de contraste
P współczynnik gamma, współczynnik
kontrastowości
R гамма коэффициент, коэффициент
контрастности

gamma quantum, γ-quantum
High energy photon.

D γ-Quant
F quantum γ
P kwant γ
R γ-квант

gamma radiation
The electromagnetic radiation emitted in the
process of a nuclear transition or annihilation
of particles.

D γ-Strahlung
F rayonnement γ
P promieniowanie γ
R γ-лучи

gamma-ray absorptiometry, gamma-ray
absorption technique *See* nuclear radiation
absorptiometry

D Durchstrahlungsverfahren mit
Quantenstrahlung
F analyse par absorption des rayons γ
P metoda absorpcji promieniowania γ
R методы анализа по поглощению γ-излучения

gamma-ray absorption technique → gamma-ray absorptiometry

gamma-ray spectrometer
An instrument designed for measuring the energy spectrum of gamma radiation.
D γ(-Strahl)-Spektrometer
F spectromètre (des rayons) γ!
P spektrometr (promieniowania) γ
R спектрометр-γ излучения

gas amplification, gas multiplication
Multiplication of the charge value collected at the electrodes of a gas-filled counter tube as a result of the secondary ionization of the gas.
D Gasverstärkung
F amplification gazeuse
P wzmocnienie gazowe
R газовое усиление

gas analysis
The qualitative and/or quantitative ascertainment of the composition of the gas sample examined.
D Gasanalyse
F analyse des gaz
P analiza gazowa
R газовый анализ

gas chromatography, GC
Chromatography in which a gas is the mobile phase.
D Gaschromatographie
F chromatographie en phase gazeuse
P chromatografia gazowa
R газовая хроматография

gas electrode
A redox electrode, consisting of a chemically inert conductor (usually platinized platinum) which is immersed in a solution saturated with an appriopiate gas and containing ions or molecules arising from the oxidation or reduction of the gas at the electrode surface, e.g. the hydrogen gas electrode.
D Gaselektrode
F électrode à gaz
P elektroda gazowa
R газовый электрод

GASFET → gas-sensitive field effect transistor

gas-filled radiation-detection tube → counter tube

gas-liquid chromatography, GLC
A form of gas chromatography in which the liquid deposited on a solid support is the stationary phase.
D Gas-flüssig-Chromatographie, Gas-Flüssig(keits)-Chromatographie
F chromatographie gaz-liquide
P chromatografia w układzie gaz-ciecz
R газо-жидкостная хроматография

gas multiplication → gas amplification

gasometric analysis
The determination of the constituents of a sample by measuring the volume under constant pressure, or the pressure at constant volume of a gas liberated from the examined sample in the course of a chemical reaction.
D Gasvolumetrie und Gasmanometrie
F gazométrie
P analiza gazometryczna
R газометрический анализ

gasometric titration
Determination of a gas constituent by adding to the sample a definite volume of another gas. The end point of titration is usually determined thermometrically.
D ...
F titrage gazométrique
P miareczkowanie gazometryczne
R газометрическое титрование

gas-permeable membrane See gas sensing electrode
D gaspermeable Membran, gasdurchlässige Membran
F membrane perméable aux gaz
P membrana przepuszczalna dla gazów
R газопроницаемая мембрана

gas sampling pipette, gas sampling tube
A glass device for sampling and keeping a fixed volume of gas.
D Gaspipette, Gassammelröhre
F tube échantillonneur de gaz, pipette pour les gaz
P pipeta do gazu
R газовая пипетка

gas sampling tube → gas sampling pipette

gas sensing electrode, gas-sensing (membrane) probe
A sensitized ion-selective electrode in which the ion-selective electrode, the reference electrode and the thin-layer of internal electrolyte are separated from the sample solution by a gas-permeable membrane or an air gap. The gas diffuses through the membrane and the internal electrolyte interacts with the gaseous species in such a way as to produce a change of the measured value, e.g. pH, of the internal electrolyte which is then sensed by the ion-selective electrode. The e.m.f. of the cell is a function of the activity of the gas to be determined.
D gas-sensitive Elektrode, Gas-Sensor
F électrode indicatrice de gaz
P elektroda gazowa selektywna, czujnik gazowy selektywny
R электрод чувствительный к газам, газочувствительный электрод

gas-sensing membrane probe → gas sensing electrode

gas-sensing probe → gas sensing electrode

gas-sensitive field effect transistor, GASFET
An electrochemical sensor consisting of a field
effect transistor in which the gate is replaced
by (or contacted with) a gas-sensitive material;
the drain current is proportional to the activity
of the gaseous substance.
D ...
F transistor à effet de champ sensible au gaz
P tranzystor polowy czuły na gazy
R полевой транзистор селективный к газам

gas-solid chromatography, GSC
A form of gas chromatography in which the
stationary phase is an active solid.
D Gas-fest-Chromatographie,
  Gas-Fest(körper)-Chromatographie
F chromatographie gaz-solide
P chromatografia w układzie gaz-ciało stałe
R газо-твёрдая хроматография,
  газо-твердофазная хроматография

gas titration analysis
The titrimetric determination of a gas previously
absorbed in a titrant; the excess of the latter
is determined by back titration.
D Gastitrimetrie
F analyse titrimétrique des gaz
P analiza miareczkowa gazów
R титриметрический газовый анализ

gas washing bottle with sintered head
D Gaswaschflasche mit Glasfritte
F flacon laveur à plaque frittée
P płuczka z bełkotką ze spiekanego szkła
R промывная склянка со стеклянной пористой
  пластинкой

gate (*in a field-effect transistor*)
The electrode that affects the flow of current
through the channel between a source and
a drain.
D Gate
F grille
P bramka
R затвор

Gaussian distribution → normal distribution

GC → gas chromatography

GDMS → glow discharge mass spectroscopy

GDOES → glow-discharge optical emission
spectroscopy

GDOS → glow-discharge optical emission
spectroscopy

Geiger counter → Geiger-Müller counter tube

Geiger(-Müller) counter (tube)
A counter tube that is used to detect ionizing
radiation (especially α-particles) and to count
particles.

D Geiger(-Müller)-Zähler, Geiger-Müller-
  Zählrohr, GM-Zählrohr
F tube compteur de Geiger(-Müller),
  compteur GM
P licznik Geigera (i Müllera), licznik G-M
R счётчик Гейгера(-Мюллера)

gel
A dispersed system in which crosslinked,
porous, spatial structures filled with
dispersion medium are formed.
D Gel
F gel
P żel
R гель

gel chromatography *See* gel-permeation
chromatography
D Gelchromatographie
F chromatographie sur gel
P chromatografia żelowa
R гелевая хроматография,
  гель-хроматография

gel filtration *See* gel-permeation chromatography
D Gelfiltration
F chromatographie par filtration de gel,
  filtration sur gel
P filtracja żelowa
R гель-фильтрация

Ge(Li) detector → lithium-drifted germanium
detector

gel-permeation chromatography, GPC
Permeation chromatography performed in
non-aqueous media in which hydrophobic gels
are used as the stationary phase; the most
popular gels are those based on polystyrene.
Remark: The term gel filtration, is not recommended for
describing the separation of the sample molecules in aqueous
media with hydrophilic gels (agarose, cross-linked dextran)
as the stationary phase. The term gel chromatography
is proposed as a joint name for gel filtration and
gel-permeation chromatography).
D Gel-Permeationschromatographie
F chromatographie par perméation sur gel
P chromatografia żelowo-permeacyjna
R гель-проникающая хроматография

geometric mean, g
Statistic defined by the formula

$$g = (x_1 x_2 ... x_n)^{1/n}$$

where $x_1, x_2 ..., x_n$ — values of the elements of
the sample, n — sample size.
D geometrisches Mittel
F moyenne géométrique
P średnia geometryczna
R среднее геометрическое

geometric volume of the column → column
volume

geometry factor
The average solid angle in sterradians at the source subtended by the aperture or sensitive volume of the detector, divided by 4π.
D Geometriefaktor
F facteur de géométrie
P wydajność zliczania geometryczna, wydajność liczenia geometryczna
R геометрический коэффициент, геометрический фактор

German degree of water hardness
The unit of water hardness equivalent to the content of 10 mg of calcium oxide in 1 l of water.
D deutscher Härtegrad, deutsche Härte
F degré hydrotimétrique allemand
P stopień twardości wody niemiecki
R немецкий градус жёсткости воды

ghosts (*in a spectrum*)
False images of a spectral line due to the imperfections of the optical system.
D Geister
F fantômes, ghosts
P duchy
R духи

glancing angle, θ
The angle between the direction of the analysed beam of radiation and the crystal plane.
D Glanzwinkel
F angle de réflexion
P kąt połysku, kąt odbłysku
R угол скольжения

glass electrode → glass membrane electrode

glass filter crucible → crucible with sintered disk

glass funnel
D Glastrichter
F entonnoir en verre
P lejek szklany
R стеклянная воронка

glass (membrane) electrode, ion-selective glass electrode
An ion-selective electrode which comprises a thin-walled bulb of special glass (membrane) at the end of a glass tube and containing the solution of a salt of constant composition and an internal reference electrode. Depending on the composition of the glass the electrode response may be to the hydrogen ion (glass pH electrode) or to other cations such as alkali cations, alkaline earth cations, or even certain organic cations.
Remark: In a narrow sense the term glass electrode is used to denote the glass pH electrode.
D (ionenselektive) Glasmembranelektrode, Glaselektrode
F électrode (à membrane sélective) de verre

P elektroda (membranowa) szklana, elektroda jonoselektywna szklana
R стеклянный (мембранный) электрод

glass pH electrode, hydrogen(-sensitive) glass electrode, pH-sensitive glass electrode *See* glass membrane electrode
D pH-Glaselektrode
F électrode de verre indicatrice des ions hydrogène
P elektroda szklana wodorowa
R водород-мувствительный стеклянный электрод, водородселективный стеклянный электрод

glassy carbon electrode
An electrode made of glassy carbon, i.e. of an organic polymer carbonized under high pressure and at high temperature. The glassy carbon electrode is usually used in the from of a cylinder with the walls shrouded by an insulator; one end of the cylinder is placed in a solution, the other contacts with a metallic conductor (e.g. mercury). It is used as the indicator or working electrode.
D Glaskohlen(stoff)elektrode
F électrode de graphite vitrifié
P elektroda z węgla szklistego
R стекло-углеродный электрод

GLC → gas-liquid chromatography

globar
One of the most common sources of infrared radiation, made of sintered silicon carbide in the form of a rod or cylinder that glows in the range 750-1500°C when heated electrically.
D Globar, Silitstift, Silitheizstab
F globar
P globar
R глобар, штифт Глобара, силитовый стержень

glow discharge
A type of electrical discharge occurring between two electrodes in gases at low pressures. The discharge consists of a number of dark and bright bands in the region between the electrodes and is characterized by the non-uniform decrease of voltage with distance down the tube, the high current, and the potential drop in the vicinity of the cathode which is much greater than the ionization potential of the gas.
D Glimmentladung
F décharge luminescente
P wyładowanie jarzeniowe
R тлеющий разряд

glow discharge mass spectroscopy, GDMS, sputteved neutrals mass spectroscopy, SNMS
The spectroscopy in which the glow discharge in noble or reactive gases causes the sputtering of a solid surface. This method makes it possible to detect the volatile products of plasma etching of a solid surface.

D Massenspektroskopie nachionisierter
Neutralteilchen an Oberflächen
F spectroscopie de masse sous plasma
P spektroskopia mas z jonizacją jarzeniową
R ...

glow-discharge optical (emission) spectroscopy,
GDO(E)S
A method involving a gas discharge for the
sputtering of the target particles in an excited
state, and subsequent measurement of the light
emitted due to their de-excitation.
D optische Spektroskopie von Glimmentladung
an Oberflächen
F spectroscopie d'émission optique sous plasma
P spektroskopia optyczna (emisyjna) cząstek
wzbudzonych wyładowaniem jarzeniowym
R ...

Golay cell → Golay pneumatic detector

Golay detector → Golay pneumatic detector

Golay(pneumatic) detector, Golay cell
(*in IR spectroscopy*)
A thermal detector which consists of a small
metal cylinder closed by a rigid blackened metal
plate at one end and a flexible silver-plated
diaphragm at the other. The chamber is filled
with xenon. The gas either absorbs the radiation
or is heated indirectly by contact with the
blackened plate and expands against the
diaphragm whose motion is detected and
measured.
D Golay-Zelle, Golay-Detektor
F cellule pneumatique de Golay
P detektor Golaya
R ячейка Голея, элемент Голея

Gooch crucible
A porcelain crucible with perforated bottom
used for filtering by suction.
D Gooch-Tiegel
F creuset de Gooch
P tygiel Goocha
R фильтровальный тигель Гуча

GPC → gel-permeation chromatography

gradient elution, solvent programming
An elution procedure in which the composition
of the eluent (concentration, pH) is changed
continuously during a chromatographic run.
D Gradient(en)elution
F élution par gradient de composition,
gradient d'élution, programmation de
solvant
P elucja gradientowa, programowanie
rozpuszczalnika
R градиентное элюирование

gradient layer, gradient packing
(*in chromatography*)
A layer of active solid with a continuous change

of the property affecting the separation of
a mixture, e.g. of the pH, composition of
a mixed adsorbent, etc.
D Gradientschicht, Gradientpackung
F gradient de composition de la couche mince,
gradient de composition du remplissage
P warstwa gradientowa
R градиентный слой, градиентное заполнение

gradient packing → gradient layer

graduated pipette, measuring pipette
A graduated glass tube for measuring out the
required quantities of a liquid.
D Meßpipette
F pipette graduée
P pipeta z podziałką, pipeta wielomiarowa
R градуированная пипетка

grain size effect (*in X-ray fluorescence analysis*)
The variation of the intensity of the scattered,
diffracted or excited radiation with the grain size
of the powdered or polycrystalline specimen.
D Korngrößeeffekt
F effet granulométrique
P efekt uziarnienia, efekt ziarnistości, efekt
granulacji
R эффект размера зерна

gram equivalent, val
The number of grams of a substance equal to
its chemical equivalent.
Remark: The International System of Units (SI) does not
recognize the quantity gram equivalent. The product of
the molar mass and the equivalence factor should be used
instead: $val = Mf_{eq}$.
D Grammäquivalent
F équivalent-gramme, valence-gramme
P gramorównoważnik, wal, val
R грамм-эквивалент

gram mole → gram molecule

gram-molecular weight → gram molecule

gram mole(cule), mole, gram-molecular weight
Number of grams of a substance equal
numerically to its relative molecular mass.
Remark: Since the mole is the SI unit of the amount of
substance, the term molar mass should be used in preference
the to mole = gram molecule.
D Gramm-Molekül, Mol
F molécule-gramme, mole
P gramocząsteczka, mol
R грамм-молекула, грамм-моль, моль

Gran's method → Gran's plot

Gran's plot, Gran's method
A method of treating potentiometric titration
data in which the end point cannot be located
far from the inflection point of the titration
curve. The data are transformed into a function
(F) which, when plotted as ordinate against

the volume of titrant added as abscissa, give straight lines intersecting the volume (V) axis at the end point, e.g. in the case of a potentiometric alkalimetric titration the dependence of $F = (V_0 + V) \, 10^{-pH}$ on V is analysed (V_0 is the volume of the solution titrated).

D Gran-Methode
F méthode de Gran
P metoda Grana
R метод Грана

graphite cuvette
Graphite-tube furnace with closed ends.

D Graphit-Küvette, Graphitrohrküvette
F cellule de graphite, cuvette de graphite
P kuweta grafitowa
R графитовая кювета

graphite-tube furnace, carbon tube furnace
A device containing an electrically heated element made of graphite or carbon, usually of the shape of a tube, used for atomization of a sample.

D Graphitrohrofen
F four à tube de graphite, four à tube de carbone, tube de graphite, four graphite
P piec grafitowy
R графитовая трубчатая печь

grating → diffraction grating

gravimetric analysis, gravimetry
The determination of the constituents of a sample from the mass of the precipitate formed during analysis.

D Gravimetrie, Gewichtsanalyse
F gravimétrie, analyse pondérale
P analiza wagowa, analiza grawimetryczna, grawimetria
R гравиметрический анализ, гравиметрия, весовой анализ

gravimetric gas analysis
The determination of a constituent (or constituents) of a gaseous mixture based on the absorbent mass gain.

D Gasgravimetrie
F analyse gravimétrique des gaz, analyse pondérale des gaz
P analiza gazowa wagowa
R гравиметрический газовый анализ

gravimetry → gravimetric analysis

grey filter → neutral-density filter

gross error
Error whose value differs so much from the values of other errors in the sample that it may be ascribed to a factor acting temporarily, e.g. improper measurement, use of a dameged instrument, error in computations.

D grober Fehler
F erreur aberrante
P błąd gruby
R промах, грубая ошибка

gross sample, combined sample
The part of a batch of material composed of all the primary samples taken from the same batch.

D Sammelprobe
F échantillon global
P próbka ogólna
R общая проба, генеральная проба

group reagent
A reagent capable of reacting under specified conditions with an analytical group of cations/anions, or with a group of organic compounds.

D Gruppenreagens
F rèactif de groupe
P odczynnik grupowy
R групповый реагент

growth factor → saturation factor

GSC → gas-solid chromatography

H

half-cell → electrode

half-intensity width, half-width, $\Delta\nu_{1/2}$
The total width at the half-peak height of
a spectral line profile that appears in
a spectrum.

D Halbwertsbreite
F largeur à mi-intensité, (de)mi-largeur
P szerokość linii spektralnej połówkowa
R полуширина линии

half-life (period), $T_{1/2}$
The time required for the activity of
a radioactive nuclide to decrease to half of its
original value.

D Halbwertszeit
F période radioactive
P okres połowicznego zaniku, okres półrozpadu,
okres półtrwania
R период полураспада

half-peak potential, $E_{p/2}$
The potential of an indicator electrode at
which the current is equal to one-half of the
peak current. *See* peak current.

D Halbpeakpotential
F ...
P potencjał półpiku
R ...

half-thickness (*in radiometry*)
The thickness of a substance which, when
placed in the path of a given beam of radiation,
reduces the value of a specified radiation
quantity by one half.

D Halbwertsdicke, Halbwertschicht
F couche de demi-absorption, épaisseur-moitié
P warstwa połówkowa, warstwa połowicznego
osłabienia
R слой половичного поглощения

half-wave potential, $E_{1/2}$
The potential of an indicator electrode with
respect to a suitable reference electrode at
which the current is equal to one-half of the
limiting current. *See* polarographic wave.

D Halbstufenpotential, Halbwellenpotential
F potentiel de demi-vague, potentiel de
demi-onde, potentiel de demi-palier
P potencjał półfali
R потенциал полуволны

half-width → half-intensity width

H and D-curve → emulsion characteristic curve

hanging-drop method (*in electrochemistry*)
A modification of inverse voltammetry, in
which the deposition and dissolution processes
of components to be investigated are carried
out on the hanging mercury drop electrode.

Remark: The term hanging-drop method may be used as
a general term to denote any electrochemical technique
using the hanging electrode, usually mercury, as the
indicator electrode.

D Methode des hängenden Tropfen
F ...
P metoda wiszącej kropli
R метод висячей капли

hanging mercury drop electrode, HMDE
An electrode which consists of a drop of
mercury (of diameter of 0.4−1.5 mm) hanging
at the tip of a mercury-filled capillary.

D hängende Quecksilbertropfelektrode,
stationäre Quecksilbertropfelektrode
F électrode à goutte de mercure pendante
P elektroda rtęciowa kroplowa wisząca
R ртутный висячий электрод, стационарный
ртутный капельный электрод

harmonic → overtone band

harmonic band → overtone band

Hartmann diaphragm
A metal plate with a fishtail shaped aperture at
one end and with several staggered holes in the
centre. It is placed facing the slit of
a spectrometer and is used to expose the selected
areas of the slit.

D Hartmannblende
F diaphragme de Hartmann
P przesłona Hartmanna, diafragma Hartmanna
R дифрагма Гартмана

HCL → hollow-cathode lamp

headspace analysis
The determination of the amount of volatile
components in a liquid solution by the
ascertainment of the composition of the
vapour being in equilibrium with that
solution, under specified conditions
(temperature, phase ratio).

D Headspace-Technik, Dampfraumanalyse
F analyse de tête espace, analyse de head space
P analiza fazy gazowej nad roztworem
R анализ равновесного пара, анализ паровой
фазы над жидкостью

heating curve
A record of the temperature T of a substance plotted againt time t in an environment heated at a controlled rate. T should be plotted on the ordinate increasing upwards and t on the abscissa increasing from left to right.
D Erhitzungskurve
F courbe d'échauffement
P krzywa ogrzewania
R кривая нагревания

heat radiation → infrared radiation

HEED → high energy electron diffraction

HEEP → height equivalent to an effective plate

HEETP → height equivalent to an effective plate

Hehner measuring cylinder
A clear glass measuring cylinder with a drain stopcock and polished bottom. It is used for visual matching of the colour intensity of the solution to be analysed with that of the standard. The measurement is carried out by adjusting the layer thickness of one of the solutions.
D Kolorimeterzylinder nach Hehner
F tube d'Hehner
P cylinder kolorymetryczny Hehnera
R цилиндр Генера

height equivalent to an effective plate, HEEP, effective plate height, height equivalent to an effective theoretical plate, HEETP, H (*in chromatography*)
The column length L divided by the number of effective plates N, $H = L/N$.
D effektive Trennstufenhöhe, Höhenäquivalent eines effektiven theoretischen Bodens
F hauteur équivalente à un plateau (théorique) effectif
P wysokość równoważna półce (teoretycznej) efektywnej
R высота эквивалентная эффективной теоретической тарелке

height equivalent to an effective theoretical plate → height equivalent to an effective plate

height equivalent to a theoretical plate, HETP, theoretical plate height, h (*in chromatography*)
The column length L divided by the number of theoretical plates n, $h = L/n$.
D Trennstufenhöhe, Höhenäquivalent eines theoretischen Bodens
F hauteur équivalente à un plateau théorique
P wysokość równoważna półce teoretycznej
R высота эквивалентная теоретической тарелке

height of mercury column (*in polarography*)
Difference between the level of the meniscus of the mercury column and the outlet of the

capillary, which defines the pressure at the end of the capillary.
D Hg-Niveaudifferenz
F hauteur de la colonne de mercure
P wysokość słupa rtęci
R высота столба ртути

HEISS → high-energy ion scattering spectroscopy

HELC → high-performance liquid chromatography

Henderson's equation
The equation describing the diffusion potential E_d, derived under the assumption that the concentrations of all ions vary linearly in the boundary layer between points p and q in the direction of flow;

$$E_d = - \frac{RT}{F} \frac{\sum_i |z_i| u_i (c_{iq} - c_{ip})}{\sum_i z_i^2 u_i (c_{iq} - c_{ip})} \times \ln \frac{\sum_i z_i^2 u_i c_{iq}}{\sum_i z_i^2 u_i c_{ip}}$$

where z_i — charge number of ion i, u_i — mobility of ion i, c_{iq} and c_{ip} — concentrations of ion i at points q and p, respectively, F — Faraday constant.
D Hendersonsche Gleichung
F équation d'Henderson, formule d'Henderson
P równanie Hendersona
R уравнение Гендерсона, формула Гендерсона

heterogeneous membrane
A solid-state membrane in which the active substance, e.g. a crystal or a sparingly soluble salt, or a mixture of active substances, is incorporated into an inert matrix (e.g. silicone rubber, PVC) or deposited on hydrophobized graphite.
D heterogene Membran
F membrane hétérogène
P membrana heterogeniczna
R гетерогенная мембрана

heterogenous membrane electrode → heterogenous membrane ion-selective electrode

heterogeneous membrane (ion-selective) electrode
An ion-selective electrode with a heterogeneous membrane.
D heterogene Membran-Elektrode
F électrode à membrane (sélective) hétérogène
P elektroda (jonoselektywna) z membraną heterogeniczną, elektroda membranowa heterogeniczna
R гетерогенный мембранный электрод, (ионселективный) электрод с гетерогенной мембраной

heterogeneous nucleation
The nucleation of a compound on a foreign substance.

D heterogene Keimbildung
F nucléation hétérogène
P zarodkowanie niejednorodne, zarodkowanie
  heterogeniczne
R гетерогенное образование зародышей

heterometric titration → turbidimetric titration

heteropolyacid
A polyacid containing more than one
acid-forming element, e.g. phosphomolybdic
acid, $H_3(PMo_{12}O_{40})$.
D Heteropolysäure
F hétéropolyacide
P heteropolikwas
R гетерополикислота

HETP → height equivalent to a theoretical
plate

hexamethyldisilazane, HMDS
A silylation reagent used in gas chromatography
either for deactivation of supports and
adsorbents or for derivatization of the
compounds under examination.
D Hexamethyldisilazan
F hexaméthyldisilazane
P heksametylodisilazan
R гексаметилдисилазан

hf polarography → radio-frequency polarography

high-efficiency liquid chromatography →
high-performance liquid chromatography

high-energy electron diffraction, HEED
The diffraction method in which electrons,
usually of energy of about 10^2 keV, are used for
investigation of the bulk structure of thin films
or surface layers from 10 to 100 nm thick.
D Beugung schneller Elektronen
F diffraction d'électrons à grande vitesse
P dyfrakcja elektronów szybkich, dyfrakcja
  elektronów o dużych energiach
R дифракция быстрых электронов

high-energy ion scattering spectroscopy, HEISS,
Rutherford backscattering spectroscopy, RBS
Probing of the subsurface layers of a solid
with a beam of helium ions (having an energy
of 2 MeV) and the analysis of the energy and
spatial distribution of the backscattered ions.
D Spektroskopie hochenergetischer
  rückgestreuter Ionen
F spectroscopie de dispersion d'ions de haute
  énergie
P spektroskopia rozpraszania jonów o dużych
  energiach
R спектроскопия обратного рассеяния
  быстрых ионов

higher-harmonic ac polarography
The conventional alternating-current
polarography in which one or more components

of the sinusoidal alternating current (higher
harmonics), usually the second and third
harmonics, are filtered out and measured.

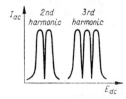

D Wechselstrompolarographie höherer
  Harmonischer, ac-Polarographie höherer
  Harmonischer, Obertonpolarographie
F polarographie à tension alternative
  surimposée par détection d'harmonique
  supérieur
P polarografia zmiennoprądowa sinusoidalna
  wyższych harmonicznych
R переменно-токовая полярография высших
  гармоник

**higher-harmonic ac polarography with phase
sensitive rectification**
The higher-harmonic ac polarography in
which rectification with phase-sensitive detection
provides measurement which is independent of
the capacity current.

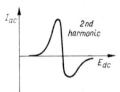

D Wechselstrompolarographie höherer
  Harmonischer mit phasenempfindlicher
  Gleichrichtung, Obertonpolarographie mit
  Phasengleichrichtung
F polarographie à tension alternative
  surimposée par détection d'harmonique
  supérieur et redressement de phase
P polarografia zmiennoprądowa sinusoidalna
  wyższych harmonicznych z fazoczułym
  prostowaniem
R переменно-токовая полярография высших
  гармоник с фазовой селекцией

high-frequency conductometric titration,
oscillometric titration
Titration in which the end-point is determined
from changes in the conductance, admittance
or susceptance of a substance, or its solution.
Measurements are carried out at frequencies of
the alternating voltage greater than ca 10^5 Hz.
D hochfrequenzkonduktometrische Titration,
  Hochfrequenztitration, oszillometrische
  Titration, oszillometrische Indikation
F titrage conductométrique haute fréquence,
  titrage oscillométrique
P miareczkowanie konduktometryczne wielkiej
  częstości, miareczkowanie oscylometryczne

R высокочастотное кондуктометрическое
титрование, осциллометрическое
титрование

high-frequency conductometry, oscillometry
A modification of conductometry, in which
an alternating voltage, of a frequency greater
than ca 10^5 Hz, is applied to the conductance
cell and the conductance G, admittance Y,
or susceptance B, is measured as a function of
the substance concentration c.

Remark: The recommended term is inexact when susceptance
or admittance is measured, but names like „susceptometry"
cannot be encouraged.

D Hochfrequenzkonduktometrie, Oszillometrie
F conductométrie haute fréquence, oscillométrie
P konduktometria wielkiej częstości,
   oscylometria
R высокочастотная кондуктометрия,
   осциллометрия

high-frequency excited lamp →
electrodeless-discharge lamp

high-frequency polarography → radio-frequency
polarography

high-level faradaic rectification
A modification of alternating current
polarography, in which direct potential E_{dc},
with a superimposed train of asymmetrical
square pulses is applied to the indicator

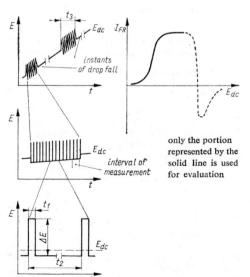

only the portion
represented by the
solid line is used
for evaluation

electrode in the second part of the lifetime of
the drop t_3, and the faradaic rectification
current I_{IR} is measured versus the direct
component of the potential of the indicator
electrode E_{dc}. The pulse amplitude ΔE is
typically $0,3 - 1$ mV, the pulse duration
$t_1 = 1 - 30$ μs, and the interval between
successive pulses $t_2 = 1$ ms.

D Faradaysche Gleichrichtung
F redressement faradaïque de haut niveau
P prostowanie faradajowskie
   wysokopoziomowe
R фарадеевское выпрямление высокого уровня

high-performance liquid chromatography,
HPLC, high-pressure liquid chromatography,
high-efficiency liquid chromatography, HELC,
high-speed liquid chromatography, HSLC
Liquid chromatography in which high-efficiency
columns filled with fine particle packings
of uniform particle size are employed and the
liquid mobile phase is usually fed under high
pressure.

Remark: The generally accepted name is now high-
performance liquid chromatography.

D Hochleistungs-Flüssig(keits)chromatographie,
   Hochdruckflüssig(keits)chromatographie,
   schnelle Flüssig(keits)chromatographie
F chromatographie en phase liquide à haute
   performance, chromatographie en phase
   liquide rapide, chromatographie en phase
   liquide sous haute pression
P chromatografia cieczowa wysokosprawna,
   chromatografia cieczowa wysokociśnieniowa,
   chromatografia cieczowa szybka
R высокопроизводительная жидкостная
   хроматография, высокоэффективная
   жидкостная хроматография, жидкостная
   хроматография при высоких давлениях,
   высокоскоростная жидкостная
   хроматография

high-pressure liquid chromatography →
high-performance liquid chromatography

**high-resolution nuclear magnetic resonance
spectrum**
A spectrum whose signals (lines) are not wider
than a few hertz.
D hochauflösendes kernmagnetisches
   Resonanzspektrum
F spectre de résonance magnétique nucléaire
   à haute résolution
P widmo magnetycznego rezonansu jądrowego
   wysokiej rozdzielczości
R спектр ядерного магнитного резонанса
   высокого разрешения

high-speed liquid chromatography →
high-performance liquid chromatography

histogram
Simplified, graphic way of presenting the
empirical distribution of the investigated
random variable. The histogram is obtained by

dividing the horizontal axis into intervals and constructing rectangles over these intervals with areas proportional to the number (frequency) of observations in that interval.

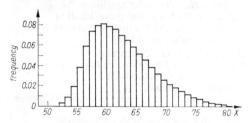

D Histogramm
F histogramme
P histogram
R гистограмма

HMDE → hanging mercury drop electrode

HMDS → hexamethyldisilazane

^{1}**H-NMR** → proton magnetic resonance

hold-back carrier
A carrier added to a solution in order to arrest a microcomponent, e.g. to prevent its coprecipitation or sorption on the main precipitate.
D Rückhalteträger
F agent de rétention, anti-entraîneur
P nośnik zwrotny, nośnik zatrzymujący
R удерживающий носитель, удерживающий агент

hold-up time → mobile phase hold-up time

hold-up volume → mobile phase hold-up volume

hole conduction
Electrical conduction occurring in a semiconductor, in which electrons move and fill the vacancies (holes), leaving a corresponding vacancies (holes) behind. The effective movement of the hole is equivalent to the movement of a positive charge in the same direction.
D Löcherleitung, Lückenleitung
F conduction par des trous, conduction par des lacunes
P przewodnictwo dziurowe
R дырочная проводимость

hollow-cathode discharge
A glow discharge in a carrier gas (usually noble) between an anode and a cathode, the latter in the shape of a hollow cylinder. Atoms of the cathode material, released by cathodic sputtering, are excited in the discharge to produce a characteristic radiation.
D Hohlkatodenentladung
F décharge à cathode creuse
P wyładowanie w katodzie wnękowej
R разряд в полом катоде

hollow-cathode lamp, hollow-cathode tube, HCL
A discharge lamp with a cathode in the shape of a hollow cylinder, filled with noble gas under low pressure, and supplied with direct current. The luminescent part of the discharge is localized inside the cathode and the radiation emitted is characteristic of the cathode filling.
D Hohlkatodenlampe
F lampe à cathode creuse
P lampa z katodą wnękową
R лампа с полым катодом

hollow-cathode tube → hollow-cathode lamp

homogeneous distribution law
The law determining the distribution of micro-component B between the solution and the precipitate in the process of formation of mixed crystals from a solution containing components A and B

$$K_{A,B} = b\,(a_0 - a)/a\,(b_0 - b)$$

where $K_{A,B}$ — separation coefficient, a_0, b_0 — concentration of components A and B in solution before crystallization, a, b — concentration of components A and B after crystallization.
D homogenes Verteilungsgesetz
F loi de distribution homogène
P prawo podziału homogeniczne
R гомогенный закон распределения

homogeneous membrane
The solid-state membrane formed either from a single compound, e.g. a crystal or compacted disc of an insoluble metal salt, or from a homogeneous mixture of two or more compounds; also liquid membranes which are homogeneous in nature.
D homogene Membran
F membrane homogène
P membrana homogeniczna
R гомогенная мембрана

homogeneous membrane electrode → homogeneous membrane ion-selective electrode

homogeneous membrane (ion-selective) electrode
An ion-selective electrode with a homogeneous membrane.
D homogene Membran-Elektrode
F électrode à membrane (sélective) homogène
P elektroda (jonoselektywna) z membraną homogeniczną, elektroda membranowa homogeniczna
R гомогенный мембранный электрод, (ионселективный) электрод с гомогенной мембраной

homogeneous nucleation
The nucleation of a single chemical compound.
D homogene Keimbildung
F nucléation homogène
P zarodkowanie jednorodne, zarodkowanie homogeniczne
R гомогенное образование зародышей

homologous lines
Spectral lines selected so that the variation of
the excitation conditions affects minimally the
ratio of their relative intensities. The pair:
the analytical line and the internal reference
line should be homologous lines.

D homologe Linien
F raies homologues
P linie homologiczne
R гомологические линии

Hooke's law (*in IR spectroscopy*)
The molecular vibrations are not random events
but can occur only at specific frequencies
controlled by the atomic masses and strengths
of the chemical bonds. In mathematical terms:

$$\bar{v} = \frac{1}{2\pi \, c}(f/\mu)^{1/2}$$

where $\bar{v}$ — frequency of the vibration,
c — velocity of light, f — force constant,
and μ — reduced mass of the atoms involved.

D Hookesches Gesetz
F loi de Hooke
P prawo Hooke'a
R закон Гука

Hopcalite
A trademark for a mixture of oxides of
copper(II), cobalt(III), manganese(IV) and
silver(I) used as an oxidizing catalyst in the
combustion tube.

D Hopkalit
F Hopcalite
P Hopkalit
R гопкалит, марганцевомедный катализатор

hot cell
A heavily shielded container for highly
radioactive materials. It may be used for their
handling, processing or storage.

D heiße Zelle
F cellule à haute activité, enceinte étanche
P komora gorąca
R горячая камера

hot laboratory
Laboratory designed and equipped for
handling of highly radioactive materials.

D heißes Laboratorium
F laboratoire chaud, laboratoire de haute
    activité
P laboratorium gorące
R горячая лаборатория

HPLC → high-performance liquid
chromatography

HSLC → high-performance liquid
chromatography

Hurter and Driffield curve → emulsion
characteristic curve

byhrid ion → ampholyte ion

hydrodynamic voltammetry
Voltammetry under conditions of convective
mass transfer to/from the surface of an
indicator electrode with a non-renewable
surface, when the potential E applied to the
electrode is varied linearly (in time).

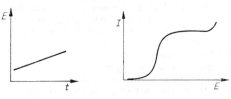

D hydrodynamische Voltammetrie
F voltampérométrie hydrodynamique
P woltamperometria hydrodynamiczna
R гидродинамическая вольтамперометрия

hydrogel
A gel in which water is the dispersion medium.

D Hydrogel
F hydrogel
P hydrożel
R гидрогель

hydrogen discharge lamp, hydrogen discharge
tube
A discharge lamp filled with hydrogen,
operated under low pressure and dc conditions,
usually employed as a source of continuous
radiation in the ultraviolet range . The lower
range limit being usually determined by the
type of glass used for the envelope, e.g. if the
lamp walls are made of fused silica the limit
is 180 nm.

D Wasserstoff-Entladungs-Lampe
F lampe à hydrogène
P lampa wodorowa
R водородная лампа

hydrogen discharge tube → hydrogen discharge
lamp

hydrogen electrode → hydrogen-gas electrode

hydrogen(-gas) electrode
An electrode (half-cell), made of platinum (or
palladium) coated with platinum black, which
is immersed in a solution containing hydrogen
ions whilst hydrogen is bubbled through the
solution and over the electrode surface. The
electrode acts as a metal electrode and produces
a pontetial which is dependent upon the
hydrogen ion activity in the solution. The
half-cell may be denoted

as $H^{+}_{(aq)}|H_{2(g)}, Pt_{(s)}$.

D Wasserstoffelektrode
F électrode à hydrogène
P elektroda wodorowa
R водородный (газовый) электрод

hydrogen glass electrode → glass pH electrode

hydrogen ion exponent → pH

hydrogen ion-indicating electrode → pH electrode

hydrogen overpotential
The overpotential of the electroreduction reaction of solvated hydrogen ions dependent to a large extent on the experimental conditions (mostly on the material of which the electrode is made).
D Überspannung des Wasserstoffs
F surtension de l'hydrogène
P nadpotencjał wydzielania wodoru
R перенапряжение водорода, водородное перенапряжение

hydrogen-sensitive glass electrode → glass pH electrode

hydrometer
D Aräometer
F aréomètre, hydromètre
P areometr, gęstościomierz
R ареометр

hydrosol
A sol in which water is the dispersion medium.
D Hydrosol
F hydrosol
P hydrozol
R гидрозоль

hydrostatic balance
A balance for determining the buoyancy of a body immersed in a fluid.
D hydrostatische Waage
F balance hydrostatique
P waga hydrostatyczna
R гидростатические весы

hydroxylic solvent
An amphiprotic solvent possessing acidic and basic properties similar to water, e.g. the aliphatic alcohols.
D wasserähnliches Lösungsmittel
F solvant hydroxylé
P rozpuszczalnik wodopodobny
R гидроксильный растворитель

hygroscopic water
The water adsorbed by the precipitate due to its hygroscopic property.
D hygroskopisches Wasser
F eau non combinée
P woda higroskopijna
R гигроскопическая вода

hyperchromic effect
An increase of the intensity of the absorption band for a particular chromophoric group, due to the structural modification of a molecule.
D hyperchromer Effekt
F effet hyperchromique
P efekt hyperchromowy
R гиперхромный эффект

hyperfine structure (*in atomic spectroscopy*)
Structure of the spectral lines or terms due to the interaction of the electronic and nuclear spins.
D Hyperfeinstruktur
F structure hyperfine
P struktura nadsubtelna
R сверхтонкая структура

hypersorption
A continuous process of adsorption in which the particles of an adsorbent migrate towards the lower section of adsorption column with simultaneous upward flow of the liquid mixture subjected to adsorption.
D Hypersorption
F hypersorption
P hipersorpcja
R гиперсорбция

hypochromic effect
A decrease of the intensity of the absorption band for a particular chromophoric group due to the structural modification of a molecule.
D hypochromer Effekt
F effet hypochromique
P efekt hipochromowy
R гипохромный эффект

hypsochrome
An atom or group of atoms which when introduced into an organic molecule shift its absorption spectrum towards shorter wavelengths.
D hypsochrome Gruppe
F groupement hypsochrome
P hipsochrom, grupa hipsochromowa
R гипсохромная группа

hypsochromic shift, blue shift
The displacement of the characteristic maximum of the absorption band towards a shorter wavelength (higher frequency, higher energy), due to the structural modification of a molecule.
D hypsochrome Verschiebung
F déplacement hypsochrome, migration hypsochrome
P przesunięcie hipsochromowe, efekt hipsochromowy, przesunięcie niebieskie
R гипсохромное смещение, гипсохромный эффект

hysteresis, electrode memory
Difference between the character of the relationship between the electrode potential and a specific variable (e.g. temperature, concentration) observed when these relationships are recorded for the increasing and decreasing values of the variable.
D Hysterese, Hysteresis
F hystérésis, mémoire de l'électrode
P histereza
R гистерезис, память электрода

I

ICP → inductively coupled plasma

ideal depolarized electrode → ideal
non-polarisable electrode

ideal non-polarizable electrode, ideal
depolarized electrode
An electrode through which large currents can
pass in either direction without shifting the
electrode potential from its equilibrium value.
D ideale nicht polarisierbare Elektrode
F électrode idéalement non-polarisable,
électrode impolarisable
P elektroda doskonale niepolaryzowalna
R неполяризуемый электрод,
неполяризующийся электрод

ideal polarizable electrode, IPE, completely
polarizable electrode
An electrode at which, in a given solution
and in a given range of imposed potentials,
no charged species are transferred from/to the
solution.
D ideale polarisierbare Elektrode
F électrode idéalement polarisable, électrode
totalement polarisable
P elektroda doskonale polaryzowalna
R идеально поляризуемый электрод,
идеально проляризующийся электрод

identification
The determination of the identity of a
substance by qualitative analysis.
D Identifizierung
F identification
P identyfikacja
R идентификация

identification reaction
A chemical reaction that enables the
detection of a given ion or compound.
D Identitätsreaktion, Nachweisreaktion
F réaction d'identification
P reakcja charakterystyczna
R характерная реакция

IEC → ion-exchange chromatography

ignition of precipitate
Heating a filter paper with a precipitate to
a high temperature (above 400°C).
D Glühen des Niederschlages
F grillage d'un précipité
P prażenie osadu
R прокаливание осадка

IIAES → ion-induced Auger electron
spectroscopy

IID → ion impact desorption

IIXE → ion induced X-ray spectroscopy

IIXS → ion induced X-ray spectroscopy

ILEED → inelastic low-energy electron
diffraction

Ilkovič equation (*in dc polarography*)
The equation describing the average limiting
current $\bar{I}_1$

$$\bar{I}_1 = 607 n m^{2/3} t^{1/6} D_i^{1/2} c_t^0$$

where n — number of electrons exchanged in
the electrode reaction, m — mercury flow rate
from the capillary, t — drop time, D_i —
diffusion coefficient of an electroactive
substance, c_i^0 — concentration of an
electroactive substance in the bulk of solution.
D Ilkovič-Gleichung
F équation de Ilkovič
P równanie Ilkoviča
R уравнение Ильковича

illuminance, illumination, luminous flux density
The amount of luminous flux that flows
through or is incident on a surface, per
unit surface area. SI unit: lux, lx.
D Beleuchtungsstärke
F éclairement lumineux
P natężenie oświetlenia
R освещённость

illumination → illuminance

immunoelectrophoresis
The method involving a combining of
electrophoresis and specific immunochemical
reactions by which proteins can be detected
and identified. The electrophorogram of
initially separated serum proteins is placed on
a glass plate, and a strip of filter paper soaked
with antibody is placed alongside touching the
edge of the electrophorogram. The mutual
diffusion of proteins and the antibodies
leads to the formation of crescent-shaped
precipitates at the sites where the two species
meet.
D Immunelektrophorese
F immunoélectrophorèse
P immunoelektroforeza
R иммуноэлектрофорез

impedance, complex impedance, Z
A measure of the ability of an electric circuit
to resist the flow of alternating current, equal
to the ratio of the complex value of the
potential difference applied across the circuit
to that of the current flowing through the circuit.
The impedance is a complex quantity,
algebraically given by the equation: $Z = R + iX$
where $i = \sqrt{-1}$, R — the resistance, and
X — the reactance. SI unit: ohm, Ω.

D Impedanz, Scheinwiderstand, komplexer
  Scheinwiderstand
F impédance électrique, impédance complexe
P impedancja, opór (elektryczny) pozorny
R импеданс, полное сопротивление,
  комплексное сопротивление (электрической
  цепи)

IMXA → ion induced X-ray spectroscopy

inactivation → deactivation

incineration of filter paper
Gradual conversion of the filter paper into ash,
effected with the precipitate (or without it)
with (or without) preliminary drying.

D Glühen des Filters
F combustion du filtre
P spalanie sączka
R сжигание фильтра

incoherent scattering → Compton effect

incremental-charge polarography
A modification of polarography in which
the electric charge applied to the dropping
indicator electrode is increased stepwise from
drop to drop. The potential E of the indicator
electrode is investigated as a function of the
square root of the time difference $\sqrt{t - t'}$. The
charge is injected at a given time of the drop
lifetime t'. The relationships of the first
derivatives of the quantities are also investigated.

Remark: The terms charge-step polarography and
discharge polarography are not recommended.

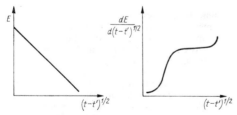

D Ladungsinkrement-Polarographie,
  Ladungsstufenpolarographie,
  Entladungspolarographie
F polarographie à incrément de charge,
  polarographie à sauts de charge,
  polarographie par décharge
P polarografia kulostatyczna, polarografia
  rozładowania

R полярография с нарастающим зарядом,
  полярография со ступенчатым изменением
  заряда, разрядная полярография

indicator, visual indicator
A substance used to detect the end point of
titration by observing the change of colour,
appearance or disappearance of fluorescence,
turbidity, etc.

D visueller Indikator
F indicateur visuel
P wskaźnik wizualny
R визуальный индикатор

indicator blank, indicator correction
A test carried out to determine the amount of
titrant (usually its volume) that causes the
same change of colour (or of another property)
of an indicator as was observed at the end-point
of titration of the analysed sample under the same
analytical conditions.

D Blindwert, Blindversuch
F essai à blanc de l'indicateur, correction
  d'indicateur
P próba ślepa wskaźnika, poprawka wskaźnika
R индикаторная поправка

indicator correction → indicator blank

indicator electrode, test electrode
An electrode (half-cell) used for the
determination of the concentration (activity),
or detection of changes in the concentration
(activity), of an electroactive substance by
measuring the electrode potential or the current
flowing through this electrode. In principle
processes occurring at the indicator electrode
do not affect the change of the bulk
composition of the solution being investigated.

D Indikator-Elektrode, Test-Elektrode
F électrode indicatrice, électrode de mesure
P elektroda wskaźnikowa
R индикаторный электрод

indicator error
A systematic error of titration caused by the
consumption of the titrant in its reaction
with the indicator.

D Indikatorfehler
F erreur d'indicateur
P błąd wskaźnika
R индикаторная ошибка

indicator exponent
The negative common logarithm of the
dissociation equilibrium constant of an
indicator, e.g. for the acid-base indicator
$HInd \rightleftharpoons H^+ + Ind^-$ the equilibrium constant is

$K_{HInd} = [H^+][Ind^-]/[HInd]$

and the indicator exponent

$pK_{HInd} = -\log K_{HInd}$

D Indikatorexponent
F exposant de l'indicateur

P wykładnik wskaźnika
R показатель индикатора

indicator paper, test paper
A strip of filter paper saturated with a
suitable indicator, e.g. pH indicator.
D Reagenzpapier, Indikatorpapier
F papier réactif
P papierek wskaźnikowy
R реактивная бумага

indifferent electrolyte → supporting electrolyte

indifferent solvent → aprotic solvent

indirect method
A method of analysis based on the analytical
signal characteristic for the constituent
reacting in a specific way with the constituent
being determined.
D indirekte Methode, indirektes Verfahren
F méthode indirecte
P metoda pośrednia
R косвенный метод

indirect titration
A titration (acid-base or other type) in which
the entity being determined does not react
directly with the titrant, but indirectly via an
intermediary stoichiometric reaction
with a titratable entity.
D indirekte Titration
F titrage indirect
P miareczkowanie pośrednie
R косвенное титрование

induced reaction
A reaction whose rate is increased by another
reaction (inducing reaction) taking place in
the same medium. Both reactions have
a common reagent (actor), e.g.

$2NaN_3 + I_2 \rightarrow 3N_2 + 2NaI$ (induced reaction)
$2Na_2S_2O_3 + I_2 \rightarrow Na_2S_4O_6 + 2NaI$
(inducing reaction)

In the above reactions, iodine is the actor and
sodium thiosulfate is the inductor.
D Sekundärreaktion, induzierte Reaktion
F réaction secondaire
P reakcja indukowana
R вторичная реакция, индуцированная
  реакция

inductively coupled plasma, ICP
A plasma generated by an electric field of
radio frequency (usually 27 MHz) coupled
inductively with that field.
D induktivgekoppeltes Plasma
F plasma induit par haute fréquence, plasma
  à couplage inductif
P plazma sprzężona indukcyjnie
R индуктивно-связанная плазма

inductor (*in induced reactions*)
The substance which causes or accelerates the
induced reaction.

D Induktor
F ...
P induktor
R индуктор

inelastic collision
A collision in which a part of kinetic energy
of the colliding species is converted into another
form of energy, e.g. into internal energies of
the species, and hence the total kinetic energy of
the colliding species is changed.
D unelastischer Stoß
F choc inélastique, collision inélastique
P zderzenie nieelastyczne
R неупругое столкновение

inelastic low-energy electron diffraction, ILEED
The diffraction method based on the spatial
analysis of inelastically backscattered electrons
(of energy from 20 to 200 eV).
D unelastische Diffraktion der energiearmen
  Elektronen
F diffraction inélastique des électrons de faible
  énergie
P dyfrakcja elektronów o małych energiach
  rozproszonych nieelastycznie
R неупругое рассеяние медленных электронов

inelastic scattering
Scattering due to inelastic collision or
interaction of species. The scattering process
involves a change of wavelength.
D unelastische Streuung
F diffusion inélastique
P rozpraszanie niesprężyste, rozpraszanie
  nieelastyczne
R неупругое рассеяние

inert solvent → aprotic solvent

infinately thick layer (*in radiometry*)
A layer in which total (or nearly total)
absorption of a given radiation takes place.
D unendlich dicke Schicht
F couche infiniment épaisse
P warstwa nieskończenie gruba
R бесконечно толстый слой

infinitely thin layer (*in radiometry*)
A layer in which practically no absorption of
a given radiation takes place.
D unendlich dünne Schicht
F couche infiniment mince
P warstwa nieskończenie cienka
R бесконечно тонкий слой

infinite thickness specimen → thick specimen

inflection points (of a peak) *See* peak
D Inflexionspunkte, Wendepunkte (des Peak)
F points d'inflexion (du pic)
P punkty przegięcia (piku)
R точки перегиба (пика)

infrared (radiation), IR
Electromagnetic radiation of wavelengths in
the range from ca 780 nm to 1 mm.
Remark: The term heat radiation is not recommended.

D Infrarot, Ultrarot
F (rayonnement) infrarouge, chaleur
  rayonnante, rayonnement calorifique
P podczerwień, promieniowanie podczerwone
R инфракрасное излучение, ИК излучение

**infrared reflectance-absorption spectroscopy,
IRAS**
Vibrational spectroscopy involving the
irradiation of a surface with infrared radiation of
varying wavelength in order to excite the
vibrational states in the adsorbed monolayer.
Measurement of the reflected signal makes it
possible to determine the adsorption bands and
the structure of the molecules adsorbed.

D Reflektions-Absorptions-
  Infrarotspektroskopie
F spectroscopie infrarouge d'absorption-
  réflexion
P spektroskopia odbiciowo-absorpcyjna
  w podczerwieni
R инфракрасная отражательно-абсорбционная
  спектроскопия

infrared region, IR region
The region of the electromagnetic spectrum with
wavelengths of 780 nm to 1 mm, which
lies between the visible radiation and radio
waves.

D Infrarotgebiet, Ultrarotgebiet
F région infrarouge
P zakres podczerwieni
R инфракрасная область (спектра)

infrared spectrum, IR spectrum
An emission or absorption spectrum of
a molecule in the infrared wavelength region,
due to transitions between different vibrational
and rotational levels within the same
electronic level.

D Infrarotspektrum, IR-Spektrum,
  Ultrarotspektrum
F spectre infrarouge
P widmo w podczerwieni
R инфракрасный спектр, ИК спектр

infusorial earth → diatomaceous earth

injection port → sample injector

injection valve → by-pass injector

inlet splitter → sample splitter

inner electric potential of phase β**, Galvani
potential,** ϕ^β
The electric potential between any point within
phase β and a point at an infinite distance
$\phi^\beta = \psi^\beta + \chi^\beta$, where ψ^β — the outer electric
potential, and χ^β — the surface electric potential
of phase β.

D inneres elektrisches Potential der Phase β,
  Galvani-Potential
F potentiel électrique interne de la phase β,
  potentiel de Galvani, potentiel électrique
  intérieur
P potencjał elektryczny wewnętrzny fazy β
R внутренний электрический потенциал
  фазы β, потенциал Гальвани

inner zone (*of a flame*) → primary combustion
zone

in-plane vibrations
A type of bending vibrations in the symmetry
plane of a molecule.

D Schwingungen in der Ebene
F vibrations dans le plan
P drgania płaskie
R плоскостные колебания, плоские колебания

INS → ion neutralization spectroscopy

insensitive volume of a detector
The volume of a radiation detector from which
the output signal does not originate.

D Totvolumen des Detektors
F volume insensible d'un détecteur
P objętość martwa detektora
R мёртвый объём датчика

instrumental activation analysis →
non-destructive activation analysis

instrumental analysis
Analysis carried out with the use of instruments
which measure physical or physico-chemical
quantities related to the content (concentration)
of the component of the sample being
determined.

D Instrumentalanalyse, Instrumentenanalyse
F analyse instrumentale
P analiza instrumentalna
R инструментальный анализ

instrumental error
Any error due to the improper use or faulty
operation of the instrument used.

D Instrumentenfehler, Gerätefehler
F erreur d'instrument
P błąd przyrządu, błąd instrumentalny
R инструментальная ошибка

integral capacitance of double layer, K
The nonlinear capacitance of the capacitor
formed by the electric double layer at an
electrode-electrolyte interface, defined by the
equation:

$$K = \frac{1}{E-E^z} \int_{E^z}^{E} C\mathrm{d}E = \frac{q}{E-E^z}$$

where E — electrode potential, E^z — potential
at the point of zero charge, q — electrode
charge, C — the differential capacitance of the
double layer.

D integrale Kapazität der Doppelschicht
F capacité intégrale de la couche double
P pojemność warstwy podwójnej całkowa
R интегральная ёмкость двойного слоя

integral chromatogram
A record of the analytical signal from the
integral detector, consisting of a series of
successive steps (step curve).

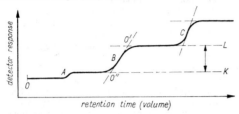

O — start point, *A* — air peak, *B, C* — steps, *KL* — step
height, *O'K, O'L* — base lines

D Integralchromatogramm
F chromatogramme intégral
P chromatogram całkowy
R интегральная хроматограмма

integral detector (*in chromatography*)
A detector which continuously measures the
total quantity of a sample accumulated from
the beginning of the analysis. The record
from such detector consists of a series of
ascending steps, the height of each step
corresponding to the quantity of each solute.
D Integraldetektor
F détecteur intégral
P detektor całkowy
R интегральный детектор

interface
The boundary between two phases.
D Grenzfläche
F interface
P granica faz, powierzchnia rozdziału faz
R граница фаз

interference filter
An optical filter whose operation is based on
the principle of light interferenca in thin plates
or membranes.
D Interferenz-Filter
F filtre d'interférence, filtre interférentiel
P filtr interferencyjny
R интерференционный фильтр

interfering ions
Ions which when present in the solution
interfere with the detection and/or
determination of the constituent analysed by
the given method.
D störende Ionen
F ions interférants
P jony przeszkadzające
R мешающие ионы

interfering nuclear reaction (*in activation
analisis*)
A nuclear reaction of a sample constituent
other than that to be determined (analyte),
which yields the same product as the analyte
(the result of analysis is then too high) or a
reaction in which the activation product is the
substrate (the result is then too low).
D störende Kernreaktion
F réaction nucléaire d'interférence
P reakcja jądrowa przeszkadzająca
R конкурирующая ядерная реакция

interfering substance
Every substance, except the one to be detected
and/or determined, which gives the analytical
signal.
D störende Substanz
F substance interférente
P substancja przeszkadzająca
R мешающее вещество

interferometer → optical interferometer

interferometry
A method of investigation of substances based
on the measurement of the difference of
refractive indices of the investigated and
standard samples by using the interferometer.
Used mostly in gas analysis, sometimes also
for liquids.
Remark: When applied in chemical analysis the term
interferometric analysis is used.
D Interferometrie
F interférométrie
P interferometria
R интерферометрия

interlaboratory research
The measurement of the same quantity in
different laboratories according to a
pre-established programme.
Interlaboratory research is conducted mainly
on:
— estimation of the true (real) value (analysis
   of standards);
— estimation of the repeatability and
   reproducitivity of any analytical method;
— analysis of the suitability of a method
   for the given analytical task.
D ...
F ...
P badania międzylaboratoryjne
R межлабораторные исследования

internal absorbance → absorbance

internal electrogravimetry, spontaneous
electrogravimetry
The electrogravimetry in which the
electrodeposition of a substance at the working
electrode proceeds spontaneously in
a short-circuited cell (without application
of an external voltage).
D interne Elektrogravimetrie, spontane
   Elektrogravimetrie

F électrogravimétrie interne, électrogravimétrie
  spontanée
P elektrograwimetria wewnętrzna,
  elektrograwimetria samorzutna
R внутренняя электрогравиметрия,
  самопроизвольная электрогравиметрия

internal electrolysis
A spontaneous electrochemical reaction which
proceeds at the electrode – electrolyte
interface in a short-circuited cell (without
application of an external voltage).
D innere Elektrolyse
F électrolyse interne
P elektroliza wewnętrzna
R внутренний электролиз

internal indicator
An indicator added to the titrated solution.
D innerer Indikator, Innen-Indikator
F indicateur interne
P wskaźnik wewnętrzny
R внутренний индикатор

internal reference electrode
A reference electrode which is inside of
an ion-selective electrode.
D innere Bezugselektrode
F électrode de référence interne
P elektroda porównawcza wewnętrzna,
  elektroda odniesienia wewnętrzna
R внутренний электрод сравнения

internal reference line, internal standard line
(*in spectrochemical analysis*)
A spectral line of a suitably selected reference
element with which the intensity of the
analytical line of the element to be determined is
compared.
D Bezugslinie, Innenstandardlinie
F raie de l'élément de référence interne, raie
  de l'étalon interne
P linia wzorca wewnętrznego
R линия элемента сравнения

internal reflectance → attenuated total
reflectance

internal reflection technique → attenuated total
reflectance

internal standard
A compound added in known concentration
to the sample, for example, to eliminate the
need to measure the size of the sample in
quantitative analysis.
D innerer Standard, Innenstandard
F étalon interne
P wzorzec wewnętrzny
R внутренний стандарт, внутренний эталон

internal standard (*in spectrochemical analysis*)
A substance present in, or added to a sample
whose spectral line or lines serve(s) as internal
reference line(s) for analytical purposes.

D innerer Standard, Innenstandard
F étalon interne
P wzorzec wewnętrzny
R внутренний стандарт, внутренний эталон

internal standard line → internal reference line
(*in spectrochemical analysis*)

internal-standard method
A method of instrumental analysis, based on
the addition to the analysed sample of a known
quantity of a foreign substance (internal
standard) for which the analytical curve in the
given conditions of determination is
identical to that of the determined
constituent. Making use of the linear
relationship between the quantity measured for
the standard and for the analysed substances,
and knowing the concentration of the standard
we can calculate the amount of the
analysed substance.
D Methode des inneren Standards
F méthode de l'étalon interne
P metoda wzorca wewnętrznego, metoda
  standardu wewnętrznego
R метод внутреннего стандарта

internal transmission density → absorbance

internal transmission factor → internal
transmittance

internal transmittance, internal transmission
factor, T
The ratio of the radiant flux transmitted by
a sample to the radiant flux transmitted by
a blank in an equivalent cell.
Remark: The terms transmittancy or transmission are not
recommended.
D innere Durchlässigkeit, inneres
  Transmissionsvermögen
F transmittance interne, facteur de
  transmission interne
P transmitancja
R внутреннее пропускание, внутренний
  коэффициент пропускания

interparticle porosity → interstitial fraction

interpolation
A procedure leading to determining the value
of a function $f(x)$ for any argument x from the
interval (x_0, x_n), when the values of that function
for arguments $x_0 \leqslant x_1 \leqslant x_2 \leqslant ... \leqslant x_n$ are
known.
D Interpolation
F interpolation
P interpolacja
R интерполяция

interstitial fraction, interparticle porosity,
fractional free volume, ε_I
The interstitial volume V_I per unit volume of
the column packing X, $\varepsilon_I = V_I/X$.

D Hohlraumanteil, Zwischenraumanteil
F fraction interstitielle, porosité
  interparticulaire
P objętość międzyziarnowa ułamkowa
R доля свободного объёма

interstitial velocity, *u* (*of the mobile phase*)
The linear flow velocity of the mobile phase
inside a packed column, calculated as the
average over the entire cross section

$$u = F/\varepsilon_I = F_C/\varepsilon_I A_C \ [\text{cm min}^{-1}]$$

where F — nominal linear flow rate of the
mobile phase, ε_I — interstitial fraction, F_C —
volumetric flow rate of the mobile phase, A_C —
cross-sectional area surface of the column.

Remark: In gas chromatography, the interstitial velocity
of the mobile phase is calculated at the outlet pressure
and the symbol u_o is used.

D Geschwindigkeit im Hohlraumbereich,
  Geschwindigkeit im Zwischenraum
F vitesse interstitielle
P prędkość przepływu międzyziarnowa
R истинная скорость

interstitial volume, free volume, void volume,
outer volume, V_I, V_G (*in chromatography*)
The volume occupied by the mobile phase in
the column (between the grains of the packing)
under static conditions,

Remark: In gas chromatography the gas occupying the
interstitial volume, V_G, expands to the volume V_G/j at the
outlet pressure where measurements are normally made;
j is the pressure-gradient correction factor. In permeation
chromatography the symbol V_o is used.

D Hohlraumvolumen, Zwischenraumvolumen
F volume interstitiel
P objętość międzyziarnowa, objętość wolna,
  objętość swobodna
R свободный объём

interval estimation
A method of estimating the unknown
parameter θ of a population consisting in
determining, on the basis of a random sample,
an interval which covers with the given
probability the true value of parameter θ.

D Bereichschätzung
F estimation (par) intervalle
P estymacja przedziałowa
R оценивание с помощью доверительного
  интервала

interzonal region (*of a flame*)
A region of a flame contained between the
primary- and secondary-combustion zones,
where in many instances the conditions for
flame analysis are optimal.

Remark: The interzonal region of a flame where the
combustion zones have the form of cones, is called the
interconal zone.

D Zwischenzone
F région intermédiaire
P obszar przejściowy
R межзональная область

intrastitial volume → stationary liquid volume

intrinsic detector efficiency
The ratio of the number of particles or
photons detected to the number of similar
particles or photons which have entered the
sensitive volume of a radiation detector.

D Ansprechwahrscheinlichkeit des Detektors
F efficacité du détecteur, rendement du
  détecteur
P wydajność detektora (wewnętrzna)
R эффективность детектора

intrinsic full energy peak efficiency
The detector efficiency when considering only
events where the total energy of the radiation
is absorbed in the sensitive volume of the
detector.

D ...
F ...
P wydajność piku całkowitej absorpcji
  wewnętrzna
R ...

intrinsic photopeak efficiency
The ratio of the number of photons detected
in the total absorption peak to the number of
photos of the same kind which have entered the
sensitive volume of the detector.

D inneres Photoansprechvermögen
F efficacité photoélectrique intrinsèque
P wydajność fotopiku wewnętrzna
R внутренняя фотоэффективность

inverse derivative potentiometric titration
The potentiometric titration, in which the first
derivative dV/dE, where V — volume of added
titrant and E — potential of the indicator
electrode, is measured.

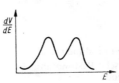

dV/dE — first derivative of the titrant volume versus the
electrode potential

D inverse derivative potentiometrische
  Titration
F titrage potentiométrique dérivé inverse
P miareczkowanie potencjometryczne
  różniczkowe odwrotne
R потенциометрическое титрование
  с регистрацией обратной производной

inverse titration
The process in which a known quantity of the
titrant is titrated with the solution of the
substance to be determined.

D inverse Titration
F titrage inverse
P miareczkowanie inwersyjne
R инверсионное титрование

iodate titration
The oxidimetric titration in which the a standard potassium iodate solution is the oxidant.
D ...
F titrage d'iodate
P miareczkowanie jodanometryczne
R ...

iodatometry
The determination of a substance by titration with a standard iodate solution (e.g. potassium iodate).
D Iodatometrie
F ...
P jodanometria
R ...

iodimetric titration
The titration of the reductant with a standard solution of iodine (usually I_3^-), or the titration of iodine (usually I_3^-) with a standard solution of the reductant.
D iodimetrische Titration
F titrage iodométrique
P miareczkowanie jodometryczne
R иодометрическое титрование

iodimetry, iodometry
The determination of a substance by titration with or of a standard solution of iodine (usually I_3^-).

Remark: Some authors restrict iodimetry to titration with a standard solution of iodine, and iodometry to titration of iodine; such restriction is not recommended.

D Iodometrie
F iodimétrie, iodométrie
P jodometria
R иодометрия

iodine flask
A flask designed for use in iodimetry.
D Kolben für jodometrische Bestimmung
F flacon d'iode, fiole à iode
P kolba jodowa
R колба для иодометрического титрования

iodine number, iodine value
The measure of iodine absorbed in a given time by a chemically unsaturated material used to establish the degree of unsaturation.
D Iodzahl
F indice d'iode
P liczba jodowa
R иодное число

iodometry → iodimetry

ion analyser → mass analyser

ion-association system
The system in which neutral molecules are formed by mutual association of inversely charged ions or groupings.

D Ionenassoziationssystem
F ...
P układ jonowo-asocjacyjny
R ...

ion exchange
The process of exchanging ions between a solution and an ion exchanger.
D Ionenaustausch
F échange d'ions
P wymiana jonowa
R ионный обмен

ion-exchange chromatography, IEC
Chromatography in which the separation of the components of a sample is based on the differences in the ion-exchange affinities of these components towards an ion exchanger which is used as the stationary phase.
D Ionenaustauschchromatographie
F chromatographie par échange d'ions, chromatographie d'échange d'ions
P chromatografia jonowymienna, chromatografia jonitowa
R ионообменная хроматография

ion-exchange isotherm
A relationship between the concentration of a counter-ion in the ion exchanger and its concentration in the external solution at equilibrium, under given conditions, and at constant temperature.

Remark: Concentrations both in the ion exchanger and in solution are usually given as equivalent fractions.

D Ionenaustauschisotherme
F isotherme d'échange d'ions
P izoterma wymiany jonowej
R изотерма ионного обмена

ion-echange membrane
An ion-exchange material of any geometrical form (usually sheet or film) which may be used as a partition between two solutions and which gives preference to the allows the preferential transport of either cations or anions.
D Ionenaustauschmembrane, Ionenaustauscher-Membrane
F membrane échangeuse d'ions
P membrana jonowymienna
R ионообменная мембрана

ion exchanger
A solid or liquid (inorganic or organic) containing ions exchangeable (in stoichiometric ratio) for others with a charge of the same sign present in the surrounding solution in which the ion exchanger is insoluble.
D Ionenaustauscher
F échangeur d'ions
P wymieniacz jonowy, wymieniacz jonów
R ионообменник

ion-exchange resin, organic ion-exchanger
An ion exchanger consisting of organic polymer

or polycondensate type matrix containing ionogenic groups.

D Ionenaustauscherharz, organischer Austauscher
F résine échangeuse d'ions
P żywica jonowymienna, jonit organiczny
R ионообменная смола, (синтетический) органический ионит

ion exchanger matrix
The molecular spatial network of an ion exchanger to which the ionogenic groups are bonded.

D Grundkörper des Ionenaustauschers, Matrix des Ionenaustauschers
F matrice d'échangeur d'ions
P szkielet jonitu
R матрица ионита, каркас ионита

ion exclusion
The separation of strong electrolytes from weak electrolytes and non-electrolytes on a synthetic ion-exchange resin. The non-ionic solutes, such as glycerine or sugar are appreciably absorbed by the ion-exchange resin, whereas the strong electrolites are excluded from the resin phase due to the Donnan effect, and emerges from the column, before the weak electrolytes and non-ionic compounds.

D Ionenausschluß
F exclusion d'ions
P ekskluzja jonów, wykluczanie jonów
R исключение ионов

ion-exclusion chromatography
A form of permeation chromatography with the ion exchanger acting as the stationary phase, used for separating strong electrolytes from weak electrolytes and non-electrolytes. *See* ion exclusion.

D Ionenausschlußchromatographie
F chromatographie par exclusion d'ions
P chromatografia jonowo-ekskluzyjna
R ион-эксклюзионная хроматография, эксклюзионная хроматография ионов

ionic atmosphere, ionic cloud
According to the Debye and Hückel theory of strong electrolytes, the vicinity close to an ion (central ion) in a solution in which the time average distribution of ions of the opposite sign is greater than the time average distribution of ions of the same sign as the charge of the central ion.

D Ionenatmosphäre, Ionenwolke
F atmosphère ionique, nuage ionique
P atmosfera jonowa, chmura jonowa
R ионная атмосфера, ионное облако

ionic cloud → ionic atmosphere

ionic conduction, electrolytic conduction
Electrical conduction consisting in the migration of ions in electrolytes or gases.

D Ionenleitung, elektrolytische Leitung
F conduction ionique, conduction électrolytique, conduction par des ions
P przewodnictwo jonowe, przewodnictwo elektrolityczne
R ионная проводимость, электролитическая проводимость

ionic conductor, electrolytic conductor, second-class conductor
An electrical conductor in which the current is carried by ions, as in solutions of acids, bases, and salts and in many fused compounds.

D Ionenleiter, Leiter zweiter Klasse, Leiter 2. Ordnung, elektrolytischer Leiter
F conducteur ionique, conducteur électrolytique, conducteur de seconde classe, conducteur de deuxième espèce
P przewodnik jonowy, przewodnik elektrolityczny
R ионный проводник, электролитический проводник, проводник с ионной проводимостью, проводник второго рода

ionic mobility → electric mobility of ion B

ionic refraction
The part of the experimental value of molar refraction per ion in an ionic compound computed in accordance with the additivity principle.

D Ionenrefraktion
F réfraction ionique
P refrakcja jonowa
R ионная рефракция

ion(ic) spectrum
A spectrum of a singly- or multiply-ionized atom, consisting of a series of separate lines (ionic lines).

D Ionenspektrum
F spectre ionique
P widmo jonowe
R ионный спектр

ionic strength (of a solution), *I*
The quantity which characterizes the electrostatic interactions in a strong electrolyte solution given by the formula:

$$I = \frac{1}{2} \sum_i c_i z_i^2$$

where c_i — concentration of ion i (in mol per litre); z_i — charge number of the ion i.

D Ionenstärke (der Lösung)
F force ionique (d'une solution)
P siła jonowa, moc jonowa (roztworu)
R ионная сила (раствора)

ion impact desorption, ion induced desorption, IID, ion stimulated desorption, ISD
The desorption of adsorbed molecules induced by the bombardment of a solid surface with a primary ion beam having an energy of about 500 eV.

D Ionenstoßdesorption
F désorption par impact d'ions, désorption
par choc d'ions
P desorpcja wywołana bombardowaniem
jonowym, desorpcja stymulowana jonowo
R десорбция при ионной бомбардировке,
стимулирование десорбции ионной
бомбардировкой

ion implantation
The modification of the surface region of
a solid due to the introduction of alien
components by ion bombardment.

D Ionenimplantation
F implantation d'ions
P implantacja jonów
R имплантация ионов

ion-induced Auger electron spectroscopy, IIAES
A type of Auger electron spectroscopy
involving ion bombardment of a solid surface
in order to induce Auger electron emission.

D ionenstrahl-induzierte
Auger-Elektronenspektroskopie
F spectroscopie d'électrons Auger excités par
bombardement ionique
P spektroskopia elektronów Augera wybjianych
bombardowaniem jonowym
R спектроскопия оже-электронов
индуцированных ионной бомбардировкой

ion induced desorption → ion impact
desorption

ion induced X-ray emission → ion induced
X-ray spectroscopy

ion induced X-ray spectroscopy, IIXS, ion
microprobe X-ray analysis, IMXA, ion
induced X-ray emission, IIXE
Analysis of X-rays emitted due to bombardment
of the surface with a primary ion beam in
order to determine the composition of a solid
surface layer.

D ...
F émission des rayons X par bombardement
ionique
P spektroskopia promieniowania
rentgenowskiego wzbudzanego
bombardowaniem jonowym
R рентгеновская спектроскопия
индуцированная ионной
бомбардировкой

ionization chamber
A gas filled chamber containing two oppositely
charged electrodes so arranged that when the
gas in the chamber is ionized, the ions formed
are drawn to the electrodes, creating an
ionization current which is measured.

D Ionisationskammer
F chambre d'ionisation
P komora jonizacyjna
R ионизационная камера

ionization detector
A detector which measures the changes of
electrical conductivity of gases, flowing
through a measuring chamber, prouced by
ionization of their molecules.

D Ionisationsdetektor
F détecteur à ionisation
P detektor jonizacyjny
R ионизационный детектор

ion(ization) source (in mass spectrometry)
The section of the mass spectrometer in which
the molecules of the sample under examination
become ionized and from which they are
accelerated towards the separating device,
the spread of kinetic energies prior to
acceleration being small.

D Ionenquelle
F source d'ions
P źródło jonów
R источник ионов

ionizing radiation
The electromagnetic or corpuscular radiation
of energy sufficient for the ejection of an
electron from an atom or molecule.

D ionisierende Strahlung
F rayonnement ionisant
P promieniowanie jonizujące
R ионизирующее излучение

ion microprobe X-ray analysis → ion induced
X-ray spectroscopy

ion neutralization spectroscopy, INS
Spectroscopy involving measurement of the
energy distribution of the electrons which are
emitted when ions of energy 4 to 10 eV are
neutralized on the adsorbent surface. This
method gives information on the surface
potential.

D Ionenneutralisierungs-Spektroskopie
F spectroscopie par neutralisation des ions
P spektroskopia (elektronów towarzyszących)
neutralizacji jonów
R спектроскопия нейтрализации ионов

ionogenic groups, fixed ionic groups
The groups fixed to an ion exchanger matrix
which are either ionized or capable of
dissociation into the fixed ions and mobile
counter-ions.

D ionogene Gruppen, (austauschaktive)
Ankergruppen
F groupements ionogènes
P grupy (funkcyjne) jonogenne
R ионогенные группы

ion-pair chromatography → ion-pair partition
chromatography

ion-pair (partition) chromatography, paired-ion
chromatography
Chromatography in liquid-solid or liquid-liquid
systems containing hydrophobic counter-ions
capable of forming with the ions to be analysed

ion pairs or more complex associated molecules
which are sorbed on hydrophobic sorbents
or extracted by the liquid stationary phase.
D Ionenpaarchromatographie
F chromatographie (par formation) de paires
  d'ions
P chromatografia jonowo-asocjacyjna
R ион-парная хроматография

ion scattering spectroscopy → low-energy ion
scattering spectroscopy

ion-selective electrode
An electrochemical sensor, whose potential
in a particular range is usually linearly
dependent on the logarithm of the activity of
a given ion in the solution. Such a device is
distinct from systems which involve redox
reactions.
Remark: The term ion-specific electrode is not
recommended.
D ionenselektive Elektrode, ionensensitive
  Elektrode, ionenspezifische Elektrode
F électrode à membrane sélective, électrode
  sélective indicatrice d'ion, électrode
  spécifique
P elektroda jonoselektywna, elektroda
  jonospecyficzna
R ион(о)селективный электрод, селективный
  ионочувствительный электрод,
  ионспецифический электрод

ion-selective glass electrode → glass membrane
electrode

ion-selective membrane
A homogeneous or heterogeneous layer of
electroactive substance separating two
electrolytic solutions or covering a surface of
a metallic conductor. It allows only the
transfer of the ion molecule of interest through
the membrane or between the membrane and
the solution of the electrolyte.
D ionenselektive Membran
F membrane sélective
P membrana jonoselektywna, membrana
  elektrodowa
R ионселективная мембрана

ion-selective membrane electrode → membrane
ion-selective electrode

ion-sensitive field effect transistor, ISFET
An electrochemical sensor consisting of a field
effect transistor with a layer of an active
material having an ion-exchanging property
with respect to the ion being determined in the
solution. The intensity of the current flowing
through the source-drain circuit is proportional
to the activity of the ion being determined.
D ionenselektive Transistorelektrode
F électrode indicatrice d'ions à transistor
  à effet de champ
P tranzystor polowy czuły na jony
R ионоселективный полевой транзистор

ion separator → mass analyser

ion source → ionization source

ion-specific electrode → ion-selective electrode

ion spectrum → ionic spectrum

ion stimulated desorption → ion impact
desorption

ion triplet, triple ions
According to Bjerrum's theory, an associate
formed from three ions as a result of Coulomb
forces; the lifetime of such an associate is
long enough, to allow to treating it as a species
resistant to the thermal movements in the
solution.
D Ionendrilling, Ionentriplett
F triplet ionique
P trójka jonowa
R ионный тройник

IPE → ideal polarizable electrode

IR → infrared radiation

IRAS → infrared reflectance-absorption
spectroscopy

irradiance, radiant flux density, E_e
The amount of radiant flux Φ_e, that flows
through or incident on a surface, per unit
surface area S
$$E_e = \Phi_e/S$$
SI unit: watt per square metre, W m^{-2}.
D Bestrahlungsstärke
F irradiance, éclairement énergétique
P natężenie napromienienia
R энергетическая освещённость

IR region → infrared region

irreversible coagulation
Coagulation in which the product
(a precipitate or a gel) cannot be reversibly
converte into the sol, e.g. coagulation of
lyophobic colloids by the action of electrolytes,
denaturation of proteins.
D irreversible Koagulation
F coagulation irréversible
P koagulacja nieodwracalna
R необратимая коагуляция

irreversible electrode
An electrode at the surface of which, for a net
zero current, irreversible electrode processes
take place that cause the deviation of the
electrode potential from its equilibrium value.
D nicht umkehrbare Elektrode, irreversible
  Elektrode
F électrode irréversible
P elektroda nieodwracalna, elektroda
  nierównowagowa
R необратимый электрод

irreversible electrode reaction
The electrode reaction in which the charge
transfer step is the slowest step (irreversibility
is dependent on the rate of transport of an
electroactive substance to the electrode surface).
In thermodynamics, the irreversible electrode
reaction is defined as the electrode reaction
which is accompanied by an increase of the
entropy of the system, or at the boundary of
the system.

Remark: In electrochemistry the irreversible electrode
reaction is considered, for the practical reasons, as the
electrode reaction which occurs in one direction only.

D ...
F réaction d'électrode irréversible
P reakcja elektrodowa nieodwracalna
R необратимая электродная реакция

irreversible redox indicator
The indicator which is irreversibly oxidized with
a change of colour by a given redox potential.

D irreversibler Redoxindikator
F indicateur d'oxydo-réduction irréversible
P wskaźnik redoks nieodwracalny
R необратимый окислительно-
    восстановительный индикатор,
    необратимый редокс-индикатор

IR spectrum → infrared spectrum

ISD → ion impact desorption

ISFET → ion-sensitive field effect transistor

isobaric mass-change determination
A technique of thermal analysis for obtaining
a record of the equilibrium mass of a substance
as a function of temperature T at a constant
partial pressure of the volatile product
(or products).

D ...
F thermogravimétrie isobare
P oznaczanie izobarycznych zmian masy
R изобарное определение изменения массы

isochrome
The plot which shows the variation
of radiation intensity as a function of the
accelerating voltage of the anode in
an X-ray tube, as recorded for a constant
wavelength.

D Isochrome
F isochrome
P izochromata
R изохрома

isocratic elution (*in chromatography*)
An elution procedure in which the composition
of the eluent is constant during a single
run.

D isokratische Elution
F élution isochratique
P elucja izokratyczna
R изократное элюирование

isoelectric point
The state of a colloidal system when the
electrokinetic potential of the colloidal particles
is equal to zero; the isoelectric point is
obtained by discharging the colloidal particles
(by addition of an electrolyte) or
(in the case of proteins) by changing the pH of
the solution.

D isoelektrischer Punkt
F point iso-électrique
P punkt izoelektryczny
R изоэлектрическая точка

isopolyacid
A polyacid containing only one acid-forming
element, e.g. heptamolybdic acid $H_6[Mo_7O_{24}]_6$

D Isopolysäure
F isopolyacide
P izopolikwas
R изополикислота

isopotential point
(1) The point on the graph showing
the dependence of the EMF of a galvanic cell
(composed of a ion-selective electrode and
a reference electrode) on the activity (or the
logarithm of activity) of an electroactive
substance, at which the EMF (corresponding
to the given activity) is not temperature
dependent.
(2) The point at which the surface
(e.g. of a colloidal particle) is completely
discharged, the electrical double layer has
collapsed, and the zeta-potential and the
surface charge are equal to zero.

D ...
F point isopotentiel
P punkt izopotencjałowy
R изопотенциальная точка

isosbestic point
The point on wavelength scale of equal
absorption cofficients of two or more
compounds existing in chemical
equilibrium

D isosbesticher Punkt
F point isobestique
P punkt izozbestyczny
R изобестическая точка

isotachophoresis
Electrophoresis in which an electrolyte
whose ions move faster than those of the
same charge in the sample is employed.
Zones of the sample ions are formed
according to the decreasing mobility and
migrate with a velocity equal to that of the
leading zone. In one run it is possible to
separate ions of different dimensions but of
the same sign only.

D Isotachophorese
F isotachophorèse
P izotachoforeza
R изотахофорез

isotherm
A set of states of a system of equal temperature.

D Isotherme
F isotherme
P izoterma
R изотерма

isothermal mass-change determination
A method of thermal analysis in which the
dependence of the mass of a substance on time
t is measured at constant temperature.

D ...
F thermogravimétrie isotherme
P oznaczanie izotermicznych zmian masy
R изотермическое определение изменения
   массы

isotope dilution (analysis)
A method of determining a substance by tracing
the change in its isotopic composition caused
by the addition of a known amount of this
substance of a different isotopic composition.

D Isotopen-Verdünnungsanalyse, Methode der
   Isotopenverdünnung
F analyse par dilution isotopique, méthode de
   dilution isotopique
P analiza metodą rozcieńczenia izotopowego,
   metoda rozcieńczenia izotopowego
R анализ методом изотопного разбавления,
   метод изотопного разведения

isotope effect
The differences in the physical and chemical
properties of istopes and their compounds due
to the differences in the atomic masses of these
isotopes.

D Isotopeneffekt
F effet isotopique
P efekt izotopowy
R изотопный эффект

isotope exchange
The exchange of the isotopes of a given element
between different phases, compounds or inside
a molecule.

D Isotopenaustausch
F échange isotopique
P wymiana izotopowa
R изотопный обмен

isotopic abundance → abundance

isotopic carrier
A carrier for a radioactive substance
chemically identical with that substance,
e.g. the inactive SO_4^- ions for the $^{35}SO_4^-$ ions.

D isotop(isch)er Träger
F entraîneur isotopique, porteur isotopique
P nośnik izotopowy
R изотопный носитель

isotopic cross section
The cross section of an isotope of
a given element for a definite nuclear reaction.

D Isotopenwirkungsquerschnitt
F section efficace isotopique
P przekrój czynny izotopowy
R изотопное (эффективное) сечение

isotopic neutron source
A source emitting neutrons (usually
accompanied by of radiation of some other type)
as a result of a nuclear reaction, or some
spontaneous fission of a heavy nuclei
(e.g. ^{252}Cf) occurring in the material of the
source.

D isotopische Neutronenquelle
F source isotopique de neutrons
P źródło neutronów izotopowe
R изотопический источник нейтронов

isotopic tracer
A tracer which differs only in isotopic
composition from the substance te be traced.

D Leitisotop, isotoper Indikator
F indicateur isotopique, traceur isotopique
P wskaźnik izotopowy, znacznik izotopowy
R изотопный индикатор

ISS → low-energy ion scattering spectroscopy

IUPAC convention → sign convention

J

K

Johann crystal
The analysing crystal, in an X-ray spectrometer with wavelength disperison, curved to a radius equal to the diameter of the focusing circle.

D Johann-Kristall
F cristal de Johann
P kryształ Johanna
R ...

Johannson crystal
The analysing crystal, in an X-ray spectrometer with wavelength dispersion, curved to a radius equal to the diameter of the focusing circle with the surface ground to the radius of the focusing circle; it reveals a very good focusing power.

D Johannson-Kristall
F cristal de Johannson
P kryształ Johannsona
R ...

Kalousek polarography
The polarography in which the pulses of potential E_t of an amplitude linearly varying with time from the initial potential E_0 are used. The current I flowing through the indicator electrode is measured during the pulse time. Usually 5—50 pulses are applied during the life of one drop.

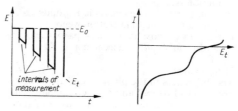

D Kalousek-Polarographie
F polarographie de Kalousek
P polarografia Kalouska
R полярография Калоусека

Karl Fischer reagent
A solution of iodine and sulfur dioxide in a pyridine-methanol mixture, used for water determination by titration. The following reactions take place in the process:

$$SO_2 + I_2 + H_2O + 3C_5H_5N \rightarrow$$
$$\rightarrow 2C_5H_5N \cdot HI + C_5H_5N \cdot SO_3$$
$$C_5H_5N \cdot SO_3 + CH_3OH \rightarrow$$
$$\rightarrow C_5H_5NH \cdot OSO_2OCH_3$$

D Karl-Fischer-Lösung
F réactif de Fischer
P odczynnik Karla Fischera
R реактив Фишера

katharometer → thermal conductivity detector

kayser, K
Obsolete unit of wavenumber, $1\ K = 1\ cm^{-1}$.

D Kayser
F kayser
P kajzer
R кайзер

K band
Intensive absorption band (characteristic of
conjugated systems) corresponding to the
allowed π-π^* electronic transition.
D K-Band
F bande K
P pasmo K
R полоса К, диапазон К

Kelvin method
The surface sensitive method of determination
of the surface potentials of metals (clean or
covered with an adsorbate) based on
the measurement of the contact potential
difference between two electrodes of
a capacitor − the reference electrode of
stable potential inert to the adsorbate and the
investigated one whose potential is dependent
on the gas adsorbed on the metal.
D Kelvin-Methode
F méthode de Kelvin
P metoda Kelvina
R метод Кельвина

kieselguhr → diatomaceous earth

kinetic analysis
Chemical analysis involving the measurement
of the rate of reactions that are accelerated by
homogeneous catalysis when the determined
constituent acts catalytically.
D kinetische Analyse, katalytische Analyse
F méthode d'analyse cinétique
P analiza kinetyczna
R кинетический метод анализа

kinetic control(led process)
A process whose rate is determined by the slow
(as compared with other steps of this process)
rate of a chemical reaction following or
preceding the charge transfer step.
D kinetische Kontrolle
F processus contrôlé par une réaction chimique
P proces kontrolowany kinetycznie, kontrola
  kinetyczna
R электродный процесс в условиях медленной
  химической реакции, электродный процесс
  лимитируемый стадией химической реакции

kinetic current (*in electrochemistry*)
The current corresponding to the slow
(as compared with the charge transfer step)
chemical reaction taking place in the vicinity
of the electrode surface. The reaction may
consist in the formation of an
electroactive substrate from an electroinactive
substance, or the decomposition of the
product of the electrode reaction.
D kinetischer Strom
F courant cinétique
P prąd kinetyczny
R кинетический ток

Kipp gas generator, Kipp's apparatus
Laboratory apparatus for generating certain
gases, usually hydrogen, hydrogen sulfide and
carbon dioxide, by chemical reaction of
acids with a metal or a suitable salt.
D Kippscher Apparat, Kippscher Gasapparat,
  Gasentwicklungsapparat nach Kipp
F appareil de Kipp
P aparat Kippa
R аппарат Киппа

Kipp's apparatus → Kipp gas generator

Kjeldahl method
A method of determination of nitrogen in
organic compounds.
D Kjeldahl-Methode
F méthode de Kjeldahl
P metoda Kjeldahla
R метод Кьельдаля

known addition method → standard addition
method

Kofler micro melting point apparatus, Kofler
micro hot stage
An apparatus for microdetermination of the
melting point consisting of an electrically
heated hot stage connected to an electric
device for controlling temperature. The heated
sample is viewed through an objective lens and
eye-piece.
D Heiztisch nach Kofler
F banc (chauffant) Kofler
P mikroskop Koflera, stolik Koflera
R предметный столик Кофлера для
  микроскопа

Kohlrausch's (additive) law, law of independent
migration of ions
The limiting molar conductivity of an
electrolyte Λ^0 is the sum of the limiting ionic
conductivities λ^0_+ and λ^0_-

$$\Lambda^0 = \lambda^0_+ + \lambda^0_-$$

D Kohlrausches Gesetz von der unabhängigen
  Ionenwanderung, Kohlrauschsche Regel der
  unabhängigen Ionenwanderung
F loi de Kohlrausch, loi de l'indépendance de
  la migration des ions
P prawo (niezależnego ruchu jonów)
  Kohlrauscha
R закон Кольрауша, закон независимости
  движения ионов

Körbl catalyst
A product of thermal decomposition of silver
permanganate, used in combustion tubes as a
catalyst and an oxidizer, which combines
with halogens and oxides of sulfur present in
the combustion products.
D Körbl-Katalysator
F catalyseur de Körbl
P katalizator Körbla
R катализатор Кэрбля

Kovats index → retention index

L

labelled compound
A chemical compound with some molecules containing an isotopic tracer (radioactive or stable).

D markierte Verbindung
F composé marqué, combinaison marquée
P związek znaczony
R меченое соединение

laboratory balance
A device used in the laboratory for measuring mass of the analysed samples.

D gleicharmige Hebelwaage
F balance de laboratoire
P waga laboratoryjna
R лабораторные весы

l(a)evorotatory substance, *l*-substance, (−)-substance
An optically active substance that rotates the plane of polarization of polarized light to the left, or in the counter-clockwise sense, as viewed looking toward the light source.

D linksdrehende Substanz, *l*-Substanz, (−)-Substanz
F substance lévogyre, *l*-substance, (−)-substance
P substancja lewoskrętna, *l*-substancja, (−)-substancja
R левовращающее вещество, *l*-вещество, (−)-вещество

Lambert-Beer law → Bouguer-Lambert-Beer law

Lambert's law, Bouguer's law, Bouguer-Lambert law
When a beam of parallel monochromatic radiation enters a homogeneous absorbing medium at right angles to the flat parallel surfaces of the medium, each infinitesimally small layer of the medium decreases the intensity of the beam I entering that layer by a constant fraction

$$\frac{-\mathrm{d}I}{I} = k\,\mathrm{d}b$$

where b − thickness of the absorbing layer and k − constant depending on the wavelength of radiation for a given absorbing medium and on the nature of the solvent, and on concentration (if the medium is a solution).

D (Bouguer-)Lambertsches Gesetz
F loi de Lambert, loi de Bouguer(-Lambert)
P prawo (Bouguera i) Lamberta
R закон (Бугера-) Ламберта

laminar flame
A flame obtained when the fuel and oxidant gases leave the burner ports with a laminar flow.

D laminare Flamme
F flamme laminaire
P płomień laminarny
R ламинарное пламя

laser
A device for producing a narrow, parallel beam of high intensity of monochromatic and coherent electromagnetic radiation in the visible, ultraviolet, or infrared regions by the stimulated emission.

Remark: The term laser is an acronym constructed from the name: light amplification by stimulated emission of radiation.

D Laser
F laser
P laser
R оптический квантовый генератор, лазер

laser fluorescence spectroscopy, LFS
The spectroscopy involving laser excitation of atoms in the gas phase in order to investigate the fluorescence due to the de-excitation of atoms.

D Laser-Fluoreszenzspektroskopie
F spectroscopie par fluorescence à laser
P spektroskopia fluorescencji wzbudzanej laserowo
R лазерная флуоресцентная спектроскопия

laser Raman spectroscopy
The branch of Raman spectroscopy in which a laser is used as the source of excitation. The advantages of the laser are that it produces a powerful, parallel beam of monochromatic, coherent radiation and thus enhances sensitivity of the method and makes possible the determination of Raman spectra for much smaller amounts of samples in various states.

D Laser-Raman-Spektroskopie
F spectroscopie Raman à laser, spectroscopie Raman à excitation laser
P spektroskopia ramanowska laserowa
R лазерная спектроскопия комбинационного рассеяния, лазерная рамановская спектроскопия

latin square
An experimental design used for estimation of the influence of three factors on the observed output variable. Each factor varies on n levels.

Instead of n^3 experimental runs determined by
all possible combinations of the factor levels in
the latin square, n^2 experimental runs are chosen
so that each level of one factor appears with
every level of other factors only once.

D lateinisches Quadrat
F carré latin
P kwadrat łaciński
R латинский квадрат

latitude (*of a photographic emulsion*)
The range of exposure corresponding to the
linear portion of the emulsion characteristic
curve, lying between the under-exposure and
over-exposure regions of the latter.

D Empfindlichkeitsbereich
F latitude d'exposition
P tolerancja naświetlenia
R широта эмульсии, фотографическая широта

law of independent migration of ions →
Kohlrausch's additive law

law of radioactive decay
The number of nuclei that undergo radioactive
disintegration in unit time (dN/dt) is
proportional to the number N_t of unchanged
nuclei present at moment t

$$-\frac{dN}{dt} = \lambda N_t$$

where λ — radioactive decay constant.

D Zerfallsgesetz
F loi de désintégration radioactive
P prawo rozpadu promieniotwórczego
R закон радиоактивного распада

layer equilibration (*in planar chromatography*)
Saturation of the stationary phase with the
eluent vapour.

D Schichtäquilibrierung
F équilibrage de la couche
P kondycjonowanie warstwy
R приведение слоя в равновесие,
уравновешивание слоя

LC → liquid chromatography

leaching, liquid-solid extraction
Extraction in the system: liquid
phase — solid phase resulting in the
dissolution of a given component of the
solid phase.

D Laugung
F lessivage
P ekstrakcja (w układzie) ciecz-ciało stałe,
ługowanie
R выщелачивание, экстракция из твёрдой
фазы

least-squares method
A method of estimating the unknown
parameters $b_{i=1,2,...,k}$ of a function
$y = f(X_1, X_2, ..., X_n, b_1, b_2, ..., b_k)$ of the

independent variables $X_1, X_2, ..., X_n$ in the
manner satisfying the requirement

$$\sum_{j=1}^{m}[y_j - f(x_{1j} x_{2j} ..., x_{nj}, b_1, b_2, ..., b_k)]^2 = \min$$

where y_j — observed value obtained in the
jth experimental run carried out under the
conditions: $X_1 = x_{1j}, X_2 = x_{2j}, ..., X_n = x_{nj}$,
m — the number of experimental runs.

D Methode der kleinsten Quadrate
F methode des moindres carrés
P metoda najmniejszych kwadratów, metoda
najmniejszej sumy kwadratów
R метод наименьших квадратов

LEC → ligand-exchange chromatography

Leclanché cell
A primary galvanic cell consisting of a carbon
cathode surrounded by a mixture of
manganese dioxide and powdered carbon with
a zinc rod as the anode, and of a solution of
amonium chloride as the electrolyte;
schematically represented as

$C_{(s)}|MnO_{2(s)}|NH_4Cl_{(aq)}(20\%)|Zn_{(s)}|C'_{(s)}$

D Leclanché-Element, Leclanché-Zelle
F pile de Leclanché, élément de Leclanché
P ogniwo Leclanchégo
R элемент Лекланше

LEED → low-energy electron diffraction

LEELS → electron-energy loss spectroscopy

left-handed circularly polarized light, l-circularly
polarized light
Circularly polarized light in which the electrical
vector describes a circle counterclockwise
around the direction of the light beam.

D links-zirkular polarisiertes Licht
F lumière à polarisation circulaire gauche
P światło spolaryzowane kołowo w lewo
R свет поляризованный по кругу влево,
левополяризованный по кругу свет

LEID → low-energy ion diffraction

LEISS → low-energy ion scattering
spectroscopy

levelling effect (*in acidimetry and alkalimetry*)
The effect of a solvent on the equalization
of the protogenic ability of acids, or
protophilic ability of bases.

D Nivellierung, nivellierender Effekt
F effet égalisant
P efekt wyrównujący
R нивелирующий эффект

levelling solvent, equalizing solvent
A solvent exhibiting a levelling effect on the
protogenic or protophilic properties of definite
groups of acids or bases.

D nivellierendes Lösungsmittel
F solvant-niveleur
P rozpuszczalnik wyrównujący
R нивелирующий растворитель

Levich equation
The equation describing the limiting current I_l flowing through the rotating disk electrode

$$I_l = 0.62nFAD_i^{2/3}v^{1/6}\omega^{1/2}c_i^0$$

where n − number of electrons exchanged in the electrode reaction, A − surface area of the disk electrode, D_i − diffusion coefficient of the electroactive substance, v − kinematic viscosity of the solution, ω − angular velocity of disk rotation, c_i^0 − bulk concentration of the electroactive substance.

D Lewitsch-Gleichung
F équation de Levich
P równanie Lewicza
R уравнение Левича

levorotatory substance → laevorotatory substance

Lewis-Randall's ionic strength law
In dilute solutions of strong electrolytes of the same ionic strength $I < 0.02$ mol dm⁻³, the value of the average activity coefficient of a given electrolyte is the same irrespective of the type of solution.

D Ionenstärke-Gesetz von Lewis und Randall
F règle (de la force ionique) de Lewis et Randall
P reguła siły jonowej Lewisa i Randalla
R закон ионной силы Льюиса и Рендалла

Lewis-Sargent's relation, Lewis-Sargent's equation
The equation expressing the diffusion potential E_d generated at the boundary between two electrolytes having the same concentration and a common anion (−) or cation (+)

$$E_d = \pm \frac{RT}{F} \ln \frac{\Lambda^{(1)}}{\Lambda^{(2)}}$$

where $\Lambda^{(1)}$ and $\Lambda^{(2)}$ − molar conductivities of solutions (1) and (2), respectively.

D Lewis-Sargentsche Gleichung
F équation de Lewis et Sargent
P wzór Lewisa i Sargenta, równanie Lewisa i Sargenta
R формула Льюиса-Саржента

LFS → laser fluorescence spectroscopy

LGC → liquid-gel chromatography

Liebig's method (*for determination of cyanides*)
Determination of cyanides by titration with silver nitrate solution. The following reactions take place successively

$$Ag^+ + 2\,CN^- \rightarrow Ag\,(CN)_2^- \tag{1}$$

$$Ag^+ + Ag\,(CN)_2^- \rightarrow 2\,AgCN\downarrow \tag{2}$$

An excess of silver nitrate makes the solution cloudy due to the formation of the poorly soluble silver cyanide (reaction 2).

D Liebig-Methode
F méthode de Liebig
P metoda Liebiga
R метод Либиха

lifetime of a drop → drop time

ligand-exchange chromatography, LEC
A chromatographic technique in which an ion exchanger loaded with metal ions is used as the stationary phase for separation of a mixture of ligands in solution.

D Ligandenaustauschchromatographie
F chromatographie par échange de ligands
P chromatografia z wymianą ligandów
R лигандообменная хроматография, хроматография обмена лигандов

light, light radiation, visible radiation
Electromagnetic radiation inducing a visual sensation through the eye, with wavelengths ranging approximately from 380 to 780 nm.

D Licht, sichtbare Strahlung, sichtbares Licht
F lumière, rayonnement visible, radiation visible
P światło, promieniowanie widzialne
R свет, видимое излучение, видимая радиация

light diffraction
The formation of light and dark bands around the boundary of a shadow cast by an object or aperture. The effect results from interference of the light diffracted from the edges of the object.

D Lichtbeugung
F diffraction de la lumière
P dyfrakcja światła, ugięcie światła
R дифракция света

light dispersion, dispersion of light
Splitting of a beam of white light in a transparent medium into separate colours, arising from the dependence of the refractive index of a given medium on the wavelength.

D Dispersion des Lichtes
F dispersion de la lumière
P dyspersja światła
R дисперсия света

light filter
An optical filter used to alter the frequency distribution of a beam of light.
Remark: The term is often used as a synonym for the optical filter.

D Lichtfilter
F filtre de lumière
P filtr świetlny
R светофильтр

light intensity → luminous intensity

light radiation → light

light refraction
The change of direction of propagation of
a light wave on passing from one medium
into another due to the difference in velocity of
light in each of these media.
D Lichtbrechung
F réfraction de la lumière
P załamanie światła
R преломление света

light scattering, LS
The deflection of light by fine particles of
solid, liquid or gaseous matter from the main
direction of a beam without appreciable
change of wavelength.
D Lichtstreuung, Lichtzerstreuung
F diffusion de la lumière
P rozpraszanie światła
R рассеяние света, светорассеяние

light source, primary source (*in atomic
absorption and atomic fluorescence spectroscopy*)
A device producing the radiation which is to
be absorbed (and partly re-emitted as
fluorescence) by the analyte in the atomizer.
D Lichtquelle
F source de lumière
P źródło światła
R источник света

limiting current, I_1 *(in electrochemistry)*
The current which, under specified conditions,
is controlled by the transport rate of the
reactant to the electrode.
D Grenzstrom
F courant limite
P prąd graniczny
R предельный ток, ток насыщения,
   граничный ток

limiting current constant, polarographic
diffusion current constant, K_1
The coefficient of proportionality between the
limiting current I_1 and the concentration c_t^0
of the electroactive substance in the bulk of
solution. This constant is characteristic for
a given electroanalytical method and depends
on the diffusion coefficient of the electroactive
substance and the experimental conditions

$$I_1 = K_1 \, c_t^0$$

Remark: In dc polarography, the limiting current is often
incorrectly called the diffusion current and the
constant K_1 the diffusion current constant.
D ...
F constante du courant de diffusion
P stała prądu granicznego
R константа диффузионного тока

limiting equivalent conductivity of electrolyte,
equivalent conductivity (of electrolyte)
at infinite dilution, Λ^{*0}

The value of the equivalent conductivity of an
electrolyte when its concentration in the
solution approaches zero.
Remark: It is recommended that the use of these terms be
discontinued.

D Äquivalentleitfähigkeit (eines
   Elektrolyten) bei unendlicher Verdünnung,
   Äquivalentleitfähigkeit (eines Elektrolyten)
   bei der Grenzverdünnung,
   Äquivalentgrenzleitfähigkeit eines
   Elektrolyten
F conductivité équivalente (d'un électrolyte)
   à une dilution infinie, conductivité équivalente
   limite d'un électrolyte
P przewodnictwo równoważnikowe graniczne
   elektrolitu
R эквивалентная электропроводность
   (электролита) при бесконечном
   разбавлении, предельная эквивалентная
   электропроводность электролита

limiting equivalent conductivity of ionic species B,
limiting equivalent ionic conductivity,
equivalent ionic conductivity at infinite
dilution, λ_B^{*0}
The value of the equivalent conductivity of
ionic species B when the concentration of its
ions in the solution approaches zero.
Remark: It is recommended that the use of these terms be
discontinued.

D Ionenäquivalentleitfähigkeit bei unendlicher
   Verdünnung, Äquivalentgrenzleitfähigkeit
   eines Ions
F conductivité ionique équivalente à une
   dilution infinie, conductivité ionique
   équivalente limite
P przewodnictwo równoważnikowe jonowe
   graniczne
R эквивалентная ионная электропроводность
   при бесконечном разбавлении, предельная
   эквивалентная электропроводность иона

limiting equivalent ionic conductivity →
limiting equivalent conductivity of ionic
species B

limiting molar conductivity (of electrolyte),
molar conductivity (of electrolyte) at
infinite dilution, Λ^0
The value of the molar conductivity of an
electrolyte when the concentration of the
electrolyte in the solution approaches zero.
D molare Grenzleitfähigkeit eines Elektrolyten,
   molare Leitfähigkeit (eines Elektrolyten)
   bei unendlicher Verdünnung
F conductivité molaire limite d'un électrolyte,
   conductivité molaire (d'un électrolyte)
   à une dilution infinie
P konduktywność molowa graniczna elektrolitu,
   przewodnictwo molowe graniczne elektrolitu
R предельная молярная электропроводность
   электролита, молярная электропроводность
   (электролита) при бесконечном разбавлении

limiting molar conductivity of ionic species B,
limiting molar ionic conductivity, molar ionic
conductivity at infinite dilution, λ_B^0
The value of the molar conductivity of ionic
species B when the concentration of ions B in
the solution approaches zero.
D molare Ionenleitfähigkeit bei unendlicher
    Verdünnung, molare Grenzleitfähigkeit
    eines Ions, molare Ionengrenzleitfähigkeit
F conductivité ionique molaire limite,
    conductivité ionique molaire à une dilution
    infinie
P konduktywność molowa jonowa graniczna,
    przewodnictwo molowe jonowe graniczne
R предельная молярная электропроводность
    иона, молярная ионная электропроводность
    при бесконечном разбавлении

limiting molar ionic conductivity → limiting
molar conductivity of ionic species B

limit of detection
The smallest concentration, or the smallest
quantity of a constituent in the sample being
analysed that can be detected with reasonable
certainty by a given analytical procedure.
D Nachweisgrenze, Erfassungsgrenze
F limite de détection, limite de décèlement
P granica wykrywalności, wykrywalność
R открываемый минимум, открываемый
    предел, минимально открываемое
    количество

limit of determination
The smallest concentration, or the smallest
quantity of a constituent in the sample being
analysed that can be determined in a given
analytical procedure.
D Bestimmungsgrenze
F limite de dosage, limite de détermination
P granica oznaczalności, oznaczalność
R предел определения, определяемый
    минимум

linear absorption coefficient → linear decadic
absorption coefficient

linear attenuation coefficient *See* attenuation
coefficient
D linearer Schwächungskoeffizient
F coefficient d'atténuation linéaire
P współczynnik osłabienia liniowy
R линейный коэффициент ослабления

linear (decadic) absorption coefficient, k
The ratio of the absorbance A of the
molecules to the path length b
in cm traversed by the radiation in the sample:
$k = A/b$.
D linearer (dekadischer) Absorptionskoeffizient
F coefficient d'absorption linéaire
P współczynnik absorpcji liniowy
R линейный (десятичный) показатель
    поглощения

linear dispersion (*of a spectrograph*)
The linear measure of the resolution of the
spectral lines given by the relation $dx/d\lambda$, where
x — the coordinate in the direction of
recording of the spectral lines, λ — wavelength
of the recorded spectral line.
D lineare Dispersion
F dispersion linéaire
P dyspersja liniowa
R линейная дисперсия

linearization
A procedure by which the system of
coordinates is converted in such a manner that
a function non-linear in the initial system will
be linear in the transformed system, e.g.
conversion of the $x - y$ system of coordinates to
a $\log x - \log y$ system, in which the originally
non-linear function $y = ax^b$ assumes the linear
form $Y = A + bX$, where $Y = \log y$, $X = \log x$,
$A = \log a$.
D Linearisierung
F linéarisation
P linearyzacja
R линеаризация

linear polarization → plane polarization

linear polarized light → plane-polarized light

linear regression
The regression in which a linear relation
between the explanatory (independent)
variables and the expected value of the
investigated random variable (experimental
outcome) is assumed.
D lineare Regression
F régression linéaire
P regresja liniowa
R линейная регрессия

linear-sweep voltammetry → chronoamperometry
with linear potential sweep

line coincidence (*in spectrochemical analysis*)
Partial or complete overlapping of two or
more spectral lines of close wavelength.
D Linienkoinzidenz
F coincidence de raies
P nakładanie się linii
R наложение линий, перекрытие линий,
    перекрывание линий

line pair (*in emission spectrochemical analysis*)
A pair of spectral lines consisting of the
analytical line and the internal reference line
whose intensities are compared for analytical
purposes.
D Linienpaar
F couple de raies
P para linii (analityczna)
R (аналитическая) пара линий

line spectrum
The electromagnetic radiation spectrum consisting of separate lines corresponding to the specific values of the radiation energy characteristic of ions and atoms.

D Linienspektrum
F spectre de raies
P widmo liniowe
R линейный спектр

line width, breadth of spectral line
A quantity which defines the degree of monochromaticity of a given line in the spectrum, usually given as the half-width of the spectral line in the units of frequency or wavelength.

D Linienbreite
F largeur d'une raie
P szerokość linii spektralnej, szerokość linii widmowej
R ширина спектральной линии

liquid chromatography, LC
The chromatography in which the mobile phase is a liquid.

D Flüssig(keits)chromatographie
F chromatographie en phase liquide
P chromatografia cieczowa
R жидкостная хроматография

liquid-gel chromatography, LGC
The liquid chromatography in which the stationary phase is a swollen gel. It includes both ion-exchange and gel-permeation chromatography.

D Flüssig-Gel-Chromatographie
F chromatographie liquide-gel
P chromatografia w układzie ciecz-żel
R жидкость-гелевая хроматография, жидкостно-гелевая хроматография

liquid ion-exchange electrode → liquid ion-exchange membrane electrode

liquid ion-exchange membrane → liquid ion exchanger membrane

liquid ion-exchange (membrane) electrode
An ion-selective electrode with a liquid ion—exchange membrane.

D Elektrode mit flüssiger Ionenaustauschermembran, ionenselektive Elektrode mit flüssigem Ionenaustauscher, Flüssigaustauscher-Elektrode
F électrode (à membrane sélective) à échangeur d'ions liquide
P elektroda (jonoselektywna) z membraną z wymieniaczem jonowym ciekłym, elektroda (membranowa) z wymieniaczem jonowym ciekłym
R (ионоселективный) электрод с мембраной из жидкого ионита

liquid ion exchanger
An ion exchanger in the form of water insoluble organic compounds containing ionogenic groups, and soluble in water-immiscible organic solvents. Long-chain aliphatic amines are commonly used as anion exchangers and long-chain organic acids (e.g. alkyl phosphoric and alkyl sulphonic acids) as cation exchangers.

D flüssiger Ionenaustauscher
F échangeur d'ions liquide
P wymieniacz jonowy ciekły
R жидкий ионообменник

liquid ion exchanger membrane, liquid ion-exchange membrane
A liquid membrane based on solutions of liquid ion exchangers with low water solubility, e.g. long-chain alkyl ammonium salts or salts of transition metal complexes such as derivatives of 1,10 phenanthroline as anion exchangers, or salts of long-chain alkyl phosphoric acid as cation exchangers.

D flüssige Ionenaustauschermembran
F membrane à échangeur d'ions liquide
P membrana z wymieniaczem jonowym ciekłym
R мембрана из жидкого ионита

liquid-junction potential, E_J
The difference of electric potentials between two liquid phases which arises at their interface as a result of differences of mobilities and activities of the cations and anions and differences of the standard chemical potentials of ions in both phases.

Remark: The liquid-junction potential is often called the diffusion potential.

D Flüssigkeits(grenzflächen)potential
F potentiel de jonction liquide
P potencjał cieczowy
R потенциал жидкостного соединения, жидкостный потенциал

liquid-liquid chromatography, LLC
A form of liquid chromatography in which the stationary phase is a liquid held on a solid support.

D Flüssig-flüssig-Chromatographie, Flüssig(keits)-Flüssig(keits)-Chromatographie
F chromatographie liquide-liquide
P chromatografia w układzie ciecz-ciecz
R жидкость-жидкостная хроматография, жидкостно-жидкостная хроматография

liquid-liquid distribution, partition between two liquids
The transfer of a substance from one liquid phase to another non-miscible with the first one. The liquid-liquid distribution finds application in separation and condensation processes.

D Flüssig-Flüssig-Verteilung
F distribution liquide-liquide
P rozdzielanie w układzie ciecz-ciecz
R жидкость-жидкостное распределение

liquid-liquid extraction
The extraction in a system composed of
two immiscible liquid phases; a special case of
separation in the liquid-liquid system.
D Flüssig-Flüssig-Extraktion
F extraction (liquide-liquide)
P ekstrakcja (w układzie) ciecz-ciecz
R жидкость-жидкостная экстракция

liquid membrane
An ion-selective membrane in which a liquid
ion exchanger (either anionic or cationic), or an
uncharged molecular carrier of ions, is
dissolved in an appropriate water-immiscible
solvent and held on an inert porous support
(e.g. Millipore filter) or incorporated in
a non-porous support (e.g. PVC).
D Flüssigmembran
F membrane liquide
P membrana ciekła
R жидкая мембрана

liquid membrane electrode → liquid membrane
ion-selective electrode

liquid membrane (ion-selective) electrode
An ion-selective electrode with a liquid
membrane.
D Flüssigmembran-Elektrode, ionenselektive
 Elektrode mit flüssiger Membran
F électrode à membrane (sélective) liquide
P elektroda (jonoselektywna) z membraną
 ciekłą, elektroda membranowa ciekła
R жидкостный-мембранный электрод,
 (ионоселективный) электрод с жидкой
 мембраной

liquid phase → stationary liquid phase

liquid(phase)volume, V_L (*in gas chromatography*)
The volume occupied in the column by the
stationary liquid phase
$$V_L = m_L \, \rho_L$$
where m_L — mass of the liquid in the column
and ρ_L — density of this liquid at the column
temperature.
D Volumen der flüssigen Phase,
 Flüssigkeitsvolumen
F volume du liquide, volume de (la) phase
 liquide
P objętość fazy (nieruchomej) ciekłej
R объём жидкой (неподвижной) фазы

liquid-solid chromatography, LSC
A liquid chromatography in which the
stationary phase is an active solid.
D Flüssig-Fest-Chromatographie,
 Flüssig(keits)-Fest(körper)-Chromatographie
F chromatographie liquide-solide

P chromatografia w układzie ciecz-ciało stałe
R жидкость-твёрдая хроматография,
 жидкостно-твёрдая хроматография

liquid-solid extraction → leaching

liquid stationary phase → stationary liquid
phase

liquid volume → liquid phase volume

lithium-drifted germanium detector, Ge(Li)
detector
A semiconductor detector of X- and/or
γ-radiation made of a *p*-type germanium
monocrystal, in which the effective layer has
been produced by the lithium drift method.
D lithiumgedrifteter Germaniumdetektor
F détecteur semi-conducteur Ge(Li)
P detektor germanowo-litowy
R литий-германиевый детектор,
 (полупроводниковый) Ge-(Li) детектор

lithium-drifted silicon detector, Si(Li) detector
A semiconductor detector of X- and low
energy γ-radiation, made of a *p*-type silicon
monocrystal, in which the effective layer has
been produced by the lithium drift method.
D lithiumgedrifteter Siliciumdetektor
F détecteur semi-conducteur Si(Li)
P detektor krzemowo-litowy
R литий-кремнёвый детектор,
 (полупроводниковый) Si(Li) детектор

live time (*in radiometry*)
The time during which a radiation measuring
assembly is capable of recording the events
occurring in the radiation detector. It equals
the clock time minus the resolving or dead time.
D wahre Zeit
F temps actif
P czas żywy
R живое время

LLC → liquid-liquid chromatography

load (*of a balance*)
The mass of the object to be weighed placed on
the balance pan.
D Belastung
F charge
P obciążenie
R нагрузка

local analysis
Analysis of a given site of the sample
usually carried out in order to ascertain the
heterogeneity of the investigated sample,
e.g. by means of an electron microprobe X-ray
analyser.
D Lokalanalyse
F analyse locale
P analiza lokalna
R локальный анализ

logarithmic distribution law
The law determining the distribution of micro-
component B between the solution and the
precipitate in the formation of mixed crystals
from a solution containing components A and B:
$\ln a_0/a = \lambda \ln b_0/b$
where λ — logarithmic distribution
coefficient, a_0, b_0—concentrations of
components A and B in the solution
before crystallization, a, b — concentrations of
components A and B in the solution after
crystallization.

D logarithmisches Verteilungsgesetz
F loi de distribution logarithmique
P logarytmiczne prawo podziału, prawo
   Doernera i Hoskinsa
R закон Дернера-Госкинса

logarithmic-normal distribution, log-normal
distribution
Probability distribution of a continuous random
variable X, which takes on only positive
values, such that log X has the normal
distribution.

D logarithmische Normalverteilung
F distribution logarithmico-normale
P rozkład logarytmiczno-normalny
R логарифмически нормальное распределение

log-normal distribution → logarithmic-normal
distribution

longitudinal diffusion → molecular diffusion
term

long-lived radioisotope
A radioisotope with a long half-life (generally
longer than several dozen hours).

D langlebiges Isotop
F radioisotope de période longue, radioisotope
   à longue période
P izotop długożyciowy
R долгоживущий изотоп

lot of material → bulk of material

low-energy electron diffraction, LEED, elastic
low-energy electron diffraction, ELEED
The surface sensitive diffraction technique
utilizing electrons of energy from 20 to 200 eV
elastically backscattered from the outermost
layer of a solid. This method gives information
on the periodic structure of monocrystalline
surfaces, clean or reconstructed by an adsorbate.

D Beugung langsamer Elektronen
F diffraction d'électrons lents, diffraction
   d'électrons à basse énergie, diffraction
   d'électrons de faible énergie
P dyfrakcja elektronów o małych energiach,
   dyfrakcja elektronów powolnych
R дифракция медленных электронов,
   упругое рассеяние медленных электронов

low-energy electron loss spectroscopy →
electron-energy loss spectroscopy

low-energy ion diffraction, LEID
The diffraction technique in which an ion beam
of energy in the range from 10 to 30 keV, is
used to determine the structure of a solid
surface.

D Beugung langsamer Ionen
F diffraction d'ions lents
P dyfrakcja jonów o małych energiach
R дифракция медленных ионов

low-energy ion scattering spectroscopy, LEISS,
ion scattering spectroscopy, ISS
Probing of a solid surface with a primary beam
of noble gas ions (usually of helium or neon)
of an energy from 1 to 10 keV followed by
analysis of the energy and spatial distribution
of the backscattered ions.

D Ionenstreuungsspektroskopie
F spectroscopie de diffusion des ions
P spektroskopia rozpraszania jonów
   o małych energiach
R спектроскопия обратного рассеяния
   (медленных) ионов

LS → light scattering

LSC → liquid-solid chromatography

Luggin(-Haber) capillary
The fine capillary tip of the reference electrode
or the salt bridge, which is used in
measurements of the electrode potential in the
three-electrode system. The tip is brought very
close to the surface of the working electrode
in order to minimize the resistance of the
electrolyte between the working and the
reference electrodes.

D Haber-Luggin-Kapillare, Luggin-Kapillare
F capillaire de Luggin, siphon de Haber et
   Luggin
P kapilara Ługgina
R капилляр Луггина

luminance, L_v
The brightness in a particular direction of
a surface that is emitting or reflecting light.
It is given at a particular point, by the
luminous intensity I_v per unit of area S,
projected onto an area at right angles to the
direction. $L_v = I_v/S \cos \theta$, where θ is the angle
the direction makes with the surface.
SI unit: candela per square metre, cd m^{-2}.
Remark: The term brightness is not recommended.

D Leuchtdichte
F luminance, densité lumineuse
P luminancja
R яркость

luminescence
The emission of light by any mechanism that
cannot be attributed merely to the high
temperature of the emitting body, but is
enhanced by chemical reactions at normal
temperature, electron bombardment,
electromagnetic radiation, and electric fields.

In luminescence atoms and molecules are produced in an excited state from which they decay, either directly back to the ground state or via an intermediate excited state, with the emission of light or other radiation.

D Lumineszenz
F luminescence
P luminescencja
R люминесценция

luminous energy, quantity of light, Q_v
The total radiant energy emitted by a light source, as measured by its capacity to produce visual sensation; the product of luminous flux and its duration.
SI unit: lumen-hour, lm h, or lumen-second, lm s.

D Lichtmenge, Lichtenergie
F quantité de lumière
P ilość światła
R световая энергия

luminous flux, Φ_v
The rate of flow of radiant energy from a light source, as measured by its capacity to produce visual sensations.
SI unit: lumen, lm.

D Lichtstrom
F flux lumineux
P strumień świetlny
R световой поток

luminous flux density → illuminance

luminous intensity, light intensity, I_v
Luminous energy emitted per second into a unit solid angle by a point source, in a given direction.
SI unit: candela, cd.

D Lichtstärke, Lichtintensität
F intensité lumineuse
P światłość
R сила света

Lunge-Rey weighing pipette
A device designed for weighing and proportioning volatile liquid samples.

D Wägepipette nach Lunge-Rey
F pipette gravimétrique de Lunge-Rey
P pipeta wagowa Lungego i Reya
R пипетка Лунге-Рея

M

macroanalysis
Analysis of a sample whose mass is greater than 0.1 g.

Remark: Macroanalysis embraces gram methods (1 – 10 g) and decigram methods (0.1 – 1 g).

D Makroanalyse
F macro-analyse
P makroanaliza
R макроанализ

macro-component (*in gravimetric analysis*)
A substance dissolved in a solution in such concentration that it may be precipitated with suitable reagents.

D Makrokomponente
F macroconstituant
P makroskładnik
R макрокомпонент

macroporous ion exchanger
An ion exchanger whose pores are large compared to atomic dimensions.

D makroporöser Ionenaustauscher
F échangeur d'ions macroporeux, échangeur d'ions à macropores
P jonit makroporowaty
R макропористый ионит, макропористый ионообменник

macroscopic cross section, Σ
The cross section per unit volume of a given material for a specified process. For a pure nuclide, it is the product of the microscopic cross section and the number of target nuclei per unit volume, for a mixture of nuclides, it is the sum of such products.

D makroskopischer Wirkungsquerschnitt
F section efficace macroscopique
P przekrój czynny makroskopowy
R макроскопическое (эффективное) сечение

magnesium hardness
The hardness of water due to the presence of magnesium ions in it.

D Magnesiahärte
F dureté magnésienne

P twardość magnezowa
R магниевая жёсткость

magnetic deflection (*in mass spectrometry*)
The deflection of an ion beam due to
the motion of the ions in a magnetic field.
Generally, the direction of motion of the ions
is at right angles to the direction of the
magnetic field, and the motion is uniform.
D magnetische Ablenkung
F déviation magnétique
P odchylenie magnetyczne
R отклонение в магнитном поле

magnetic-deflection analyser →
magnetic-deflection mass spectrometer

magnetic-deflection mass spectrometer,
magnetic-deflection analyser
A mass spectrometer in which the ions
produced are passed through a combination of
electrical and magnetic fields arranged so that
the beam of ions is deflected in the magnetic
field according to mass-to-charge ratios of the
ions and the succesive ion beams are then
focused onto a detector.
D Sektorfeldmassenspektrometer
F analyseur de champ magnétique
P spektrometr mas z sektorem magnetycznym,
    analizator z sektorem magnetycznym
R масс-анализатор с магнитным полем

magnetic resonance
A phenomenon exhibited by a substance
containing unpaired nuclear or electronic spin
or orbital magnetic moments which when
placed in a strong constant magnetic field and
subjected simultaneously to a radio-frequency
magnetic field at right angles to the former
absorbs energy from the oscillating magnetic
field at a certain characteristic frequency.
D magnetische Resonanz
F résonance magnétique
P rezonans magnetyczny
R магнитный резонанс

magnetic nuclear resonance → nuclear magnetic
ersonance

major constituent
A constituent whose content in the sample
varies from 1 to 100%.
D Hauptbestandteil
F constituant majeur
P składnik główny
R главный компонент, основной компонент

manganometry
The determination of a substance by titration
with a standard solution of permanganate
(e.g. potassium permanganate).
D Manganometrie, Permanganometrie
F manganométrie
P manganometria
R перманганатометрия

manometric gas analysis
The determination of a constituent (or
constituents) of gaseous mixtures based on the
pressure changes at constant volume and
temperature.
D manometrische Gasanalyse, Gasmanometrie
F analyse manométrique des gaz
P analiza gazowa manometryczna
R манометрический газовый анализ

manometric methods
Methods of quantitative elemental analysis
consisting in combustion of a substance in a
special apparatus and in measuring the pressure
of the combustion products. These methods
are used for the determination of carbon or
nitrogen and indirectly also of sulfur and
phosphorus in organic compounds.
D manometrische Verfahren
F dosages manométriques
P metody manometryczne
R манометрические методы

manual spectrophotometer → non-recording
spectrophotometer

marker (*in chromatography*)
A reference substance (usually added to the
sample) to assist in identifying the components
or to determine the flow rate of the mobile
phase.
D Bezugssubstanz
F marqueur
P znacznik, marker
R метка

masking agent
A reagent converting interfering ions into
forms inactive under the given conditions
(e.g. by forming stable soluble complexes, or
changing the oxidation number).
D Maskierungsmittel, Maskierungsreagens
F agent masquant
P odczynnik maskujący
R маскирующий агент

masking of ions
Transformations of the interfering ions into
forms inert under the given conditions.
Generally, the formation of stable, readily
soluble complexes whose ions are incapable of
engaging in reactions interfering in the
detection or determination.
D Maskierung von Ionen
F masquage d'ions
P maskowanie jonów
R маскирование ионов

mass analyser, ion analyser, ion separator
The section of the mass spectrometer in which
the beam of ions emerging from the ion source
is separated into a series of beams, according
to the mass-to-charge ratios.

D Massenanalysator
F analyseur de masse
P analizator mas
R масс-анализатор

mass attenuation coefficient *See* attenuation coefficient
D Massenschwächungskoeffizient
F coefficient d'atténuation massique
P współczynnik osłabienia masowy
R массовый коэффициент ослабления

mass/charge ratio → mass-to-charge ratio

mass concentration of substance B, ρ_B
The ratio of the mass m_B of constituent B to the volume V of the solution containing that mass $\rho_B = m_B/V$, where B — chemical formula or conventional symbol of constituent B.
SI unit: kilogram per cubic metre, kg m^{-3}; practical unit: gram per litre, g l^{-1}.
D Massenkonzentration eines Stoffes B
F concentration en masse du constituant B
P stężenie masowe składnika B w roztworze
R массовая концентрация

mass distribution ratio, D_m, k (*in chromatography*)
The ratio of the fraction of a component in the stationary phase to the fraction of this component in the mobile phase at any point in the column at equilibrium.
Remark: The terms capacity ratio, capacity factor, partition ratio are no longer recommended.
D Massenverteilungsverhältnis, Kapazitätsfaktor
F coefficient de distribution massique, facteur de capacité, rapport de partage
P stosunek podziału masowy, iloraz podziału masowy
R массовое отношение распределения, коэффициент ёмкости

mass exchange
The transfer of a substance (or several substances) from one phase to another.
D Stoffaustausch
F échange de masse
P wymiana masy
R массообмен

mass fraction (of substance B), w_B
The ratio of the mass m_B of constituent B to the mass m_G of the system G containing that component.
$$w_B = \frac{m_B}{m_G}$$
Unit: kilogram per kilogram, kg kg^{-1}, % (*m/m*).
D Massenanteil, Massengehalt
F fraction en masse
P ułamek masowy
R весовая доля

mass number → mass-to-charge ratio

mass spectrograph
An instrument in which beams of ions, produced from a sample under investigation, are separated according to the mass-to-charge ratio of the ions with concurrent recording on a photographic plate.
D Massenspektrograph
F spectrograph de masse
P spektrograf mas
R масс-спектрограф

mass spectrometer
An instrument in which molecules of a substance are broken down into fragments and/or ionized usually by means of an ionizing beam of electrons. The ions produced are separated according to their mass-to-charge ratios. The mass-separated ion beams are then focused onto a detector that measures the beam intensity electrically.
Remark: This term should also be used when a scintillation detector is employed.
D Massenspektrometer
F spectromètre de masse
P spektrometr mas
R масс-спектрометр

mass spectrometry, mass spectroscopy, MS
A branch of spectroscopy concerned with the production, measurement and interpretation of the mass spectra of a sample exposed to ionizing beam of electrons.
Remark: The term mass spectroscopy seems preferable, but mass spectrometry is widely used.
D Massenspektrometrie, Massenspektroskopie
F spectrométrie de masse, spectroscopie de masse
P spektrometria mas, spektroskopia mas
R масс-спектрометрия, масс-спектроскопия

mass spectroscope
A term which may refer to either a mass spectrometer or a mass spectrograph.
D Massenspektroskop
F spectroscope de masse
P spektroskop mas
R масс-спектроскоп

mass spectroscopy → mass spectrometry

mass spectrum
The record of ion beam intensities, usually expressed as the relative intensities, against the mass-to-charge ratios of the ions. It may be either in the form of a photographic record, as produced by a mass spectrograph, or as a chart record of the type obtained from a mass spectrometer.
Remark: Usually the term mass spectrum refers to a spectrum of positive ions.
D Massenspektrum
F spectre de masse
P widmo mas
R масс-спектр

mass-to-charge ratio, mass/charge ratio, mass number, m/e
D Quotient Masse/Ladung, Masse-Ladung-Verhältnis
F rapport masse sur charge, rapport masse/charge
P stosunek masa/ładunek
R отношение масса/заряд

mass transfer overpotential → concentration overpotential

mass transfer term, resistance to mass transfer
The third term (C) of the van Deemter equation accounting for the contribution to the total band broadening due to local lack of equilibrium in the process of distribution of the solute between the stationary and mobile phase. The term becomes significant in determining the theoretical plate height at high mobile phase flow rates.
D Massenübergangswiderstand, Verzögerung des Stoffaustausches
F résistance au transfert de masse
P opór przenoszenia masy
R сопротивление массопереносу, сопротивление массообмену

matrix (*in chemical analysis*)
The main constituents of a sample other than the detected and/or determined constituent.
D Matrix
F matrice
P matryca
R основа, матрица

matrix effect
The disturbing influence of other components of the analysed sample on the detection and/or determination of an analyte.
D Matrixeffekt
F effet de matrice
P efekt matrycy, efekt składników towarzyszących
R матричный эффект

matrix effect (*in X-ray fluorescence analysis*)
Change in the X-ray characteristic radiation intensity of an analyte due to the absorption (absorption effect) and enhancement (enhancement effect) of this radiation by other components of the sample.
D Matrixeffekt
F effet de matrice
P efekt matrycy
R матричный эффект

maximum likelihood method
A method of estimating the unknown parameters b_i of a function
$y = f(X_1, X_2, ..., X_n, b_1, b_2, ..., b_k)$ in a manner satisfying the requirement of maximization of the likelihood function L given by the formula

$$L = \prod_{j=1}^{p} f(x_{1j}, x_{2j}, ..., x_{nj}, b_1, b_2, ..., b_k)$$

where $x_{1j}, x_{2j}, ..., x_{nj}$ — realizations of random variables $X_1, X_2, ..., X_n$.
D Maximum-Likelihood-Methode
F méthode du maximum de vraisemblance
P metoda największej wiarygodności
R метод максимального правдоподобия

MBRS → molecular beam reactive scattering

MBSS → molecular beam surface scattering

MD → mean deviation

mean → arithmetic mean

mean activity coefficient of electrolyte, mean ionic activity coefficient, $y_\pm, \gamma_\pm, f_\pm$
For electrolyte $B = X_{v+} Y_{v-}$ the quantity defined by the formula

$$y_\pm = \frac{a_\pm}{c_\pm} = \sqrt[v]{y_+^{v+} y_-^{v-}}$$

where $a_\pm$ — mean activity of electrolyte, $c_\pm$ — mean concentration of electrolyte, y_+ and y_- — activity coefficients of cations and anions, respectively, v_+ and v_- — numbers of cations and anions, respectively, which are formed from one molecule of electrolyte, $v = v_+ + v_-$. For the symbols $\gamma_\pm$ and $f_\pm$ *see* activity coefficient of substance B.
D mittlerer Aktivitätskoeffizient der Elektrolyte, mittlerer Ionenaktivitätskoeffizient
F coefficient moyen d'activité ionique
P współczynnik aktywności średni elektrolitu, współczynnik aktywności jonowej średni
R средний коэффициент активности электролита, средний ионный коэффициент активности, среднеионный коэффициент активности

mean activity of electrolyte, mean ionic activity, $a_\pm$
For electrolyte $B = X_{v+} Y_{v-}$, the quantity defined as

$$a_\pm = \sqrt[v]{a_+^{v+} a_-^{v-}} = \sqrt[v]{a_B} = c_B y_\pm \sqrt[v]{v_+^{v+} v_-^{v-}}$$

where a_+ and a_- — activities of the cation and anion, respectively, v_+ and v_- — numbers of cations and anions, respectively, formed from one molecule of electrolyte B, $v = v_+ + v_-$, a_B — acti- vity of electrolyte B, c_B — molar concentration of electrolyte B, $y_\pm$ — mean ionic activity coefficient; mean activity of electrolyte may be also expressed in mole fractions — $x_\pm$ or molalities $m_\pm$.
D mittlere Aktivität eines Elektrolyten, mittlere Ionenaktivität
F activité moyenne d'un électrolyte
P aktywność średnia elektrolitu, aktywność jonowa średnia
R средняя ионная активность, средняя активность электролита

mean concentration of electrolyte, mean ionic concentration, $c_\pm$
For electrolyte $B = X_{v_+} Y_{v_-}$ the quantity given by the equation

$$c_\pm = \sqrt[v]{c_+^{v_+} c_-^{v_-}} = c_B \sqrt[v]{v_+^{v_+} v_-^{v_-}}$$

where c_+ and c_- — molar concentrations of cations and anions, respectively, v_+ and v_- — number of cations and anions, respectively, which are formed from one molecule of electrolyte B, $v = v_+ + v_-$, c_B — the molar concentration of electrolyte B. The composition of the solution may be also expressed in mole fractions — $x_\pm$ and molalities — $m_\pm$.

D mittlere Ionenkonzentration
F concentration ionique moyenne
P stężenie średnie elektrolitu, stężenie jonowe średnie
R средняя концентрация электролита

mean deviation, average deviation, MD
The statistic describing the dispersion of observations around a certain central value defined by the formula

$$MD = \frac{1}{n} \sum_{i=1}^{n} |x_i - C|$$

where x_i — value of the ith element of the sample, n — sample size, C — central value of the sample (usually arithmetic mean or median).

D mittlere Abweichung
F écart moyen
P odchylenie średnie
R среднее абсолютное отклонение

mean interstitial velocity of the carrier gas, average interstitial velocity of the carrier gas, average mobile phase interstitial velocity, $\bar{u}$
(*in gas chromatography*)
The interstitial velocity of the mobile phase u_0 multiplied by the pressure-gradient correction factor j, $\bar{u} = ju_0$. In liquid chromatography, $j = 1$ can be assumed, thus the interstitial velocity and the mean interstitial velocity of the mobile phase are identical.

D mittlere Trägergasgeschwindigkeit im Hohlraumbereich
F vitesse moyenne interstitielle du gaz-vecteur
P prędkość przepływu międzyziarnowa średnia gazu nośnego
R истинная средняя скорость газа-носителя

mean ionic activity → mean activity of an electrolyte

mean ionic activity coefficient → mean activity coefficient of electrolyte

mean ionic concentration → mean concentration of electrolyte

mean ionic diameter, a
In the theory of strong electrolytes, a parameter in the Debye-Hückel equation which corresponds approximately to the minimal distance between the ionic centres to which any ion of the ion cloud may approach the central ion.

D mittlerer Ionendurchmesser
F sphère d'activité, rayon ionique
P średnica jonu efektywna, średnica jonu średnia
R средний ионный диаметр

measured current → observed current

measurement of non-faradaic admittance
An electrochemical technique in which an alternating sinusoidal voltage E_{as}, usually with an amplitude of $1-5$ mV, is superimposed on the potential of the indicator electrode, and the alternating component of current I_{ac}, flowing through the indicator electrode, is measured as a function of the constant component of the electrode potential E_{dc}; used mainly for the investigation of adsorption and desorption processes.

Remark: Tensammetry, the name most widely used, is not recommended.

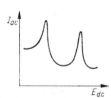

D nichtfaradaysche Admittanzmessung, Messung des nichtfaradayschen Leitwerts, Tensammetrie
F mesure d'admittance non faradaïque, tensamétrie
P metoda admitancji niefaradajowskiej, tensammetria
R измерение нефарадеевского адмиттанса, тензаметрия

measuring cylinder
A graduated cylindrical vessel designed for measuring the volume of liquids with a low accuracy.

D Meßzylinder
F éprouvette graduée
P cylinder pomiarowy
R измерительный цилиндр

measuring pipette → graduated pipette

mechanical entrapment
(1) The process of random incorporation of comparatively small quantities of other phases (e.g. water, dust, particles, etc.) in the bulk of a precipitate during its formation.

(2) The intentional capture of small quantities of such phases by intentional addition of solids to a liquid phase.

D ...
F piégeage mécanique
P zatrzymywanie mechaniczne
R механический захват

median → sample median

melting, fusion
Transition from the solid to the liquid state.
D Schmelzen
F fusion
P topnienie
R плавление

melting-point capillaries, melting-point tubes
Thin wall capillary glass tubes, sealed at one end, having 70−90 mm in length and 1 mm of inner diameter used in the melting point determination.
D Schmelzpunktröhrchen
F tubes capillaires pour mesurer le point de fusion
P kapilarki do oznaczania temperatury topnienia
R капилляры для определения точки плавления

melting-point tubes → melting-point capillaries

membrane electrode → membrane ion-selective electrode

membrane equilibrium
The thermodynamic state of a system composed of two phases separated by a semipermeable membrane in which the electrochemical potential of any component penetrating through the interface is the same in both phases.
D Membrangleichgewicht, Donnan-Gleichgewicht
F équilibre de membrane, équilibre de Donnan
P równowaga membranowa, równowaga przeponowa, równowaga Donnana
R мембранное равновесие, доннановое равновесие

membrane (ion-selective) electrode,
ion-selective membrane electrode
The term is used to denote the ion-selective electrode which comprises an ion-selective membrane.
D Membran-Elektrode
F électrode à membrane sélective
P elektroda z membraną jonoselektywną, elektroda (jonoselektywna) membranowa
R (ионселективный) мембранный электрод, ионселективный электрод с ионообменной мембраной

membrane potential, Donnan potential, $\Delta\phi_m$, E_m
The difference of the inner potentials of solutions (I and II) separated by a semi-permeable membrane; if the membrane is permeable to ions of B, then

$$\Delta\phi_m = -\frac{RT}{z_B F}\ln\frac{a_B^I}{a_B^{II}}$$

where a–activity of tons, z–their charge.
D Donnanpotential, Membranpotential
F potentiel de membrane
P potencjał membranowy, potencjał przeponowy, potencjał Donnana
R мембранный потенциал, потенциал Доннана, мембранная разность потенциалов

mercurimetric titration
The complexometric titration with a standard mercury(II) salt solution; e.g. the determination of chlorides, bromides, cyanides, and thiocyanates.
D merkurimetrische Titration
F titrage mercurimétrique
P miareczkowanie merkurymetryczne
R меркуриметрическое титрование

mercurimetry
The determination of a substance by titration with a standard solution of a mercury(II) salt.
D Merkurimetrie
F mercurimétrie
P merkurymetria
R меркуриметрия

mercurometric titration
The precipitation titration with a standard mercury(I) salt solution, e.g. the determination of chlorides in water.
D merkurometrische Titration
F titrage mercurométrique
P miareczkowanie merkurometryczne
R меркурометрическое титрование

mercurometry
The determination of a substance by titration with a standard solution of a mercury(I) salt.
D Merkurometrie
F mercurométrie
P merkurometria
R меркурометрия

mercury cell → mercury dry cell

mercury (dry) cell
The galvanic cell schematically represented as: $Zn_{(s)}|ZnO_{(s)}|KOH_{(aq)}|HgO_{(s)}|Hg_{(l)}|C_{(s)}|Zn'_{(s)}$ used as a source of current.
D Quecksilberzelle
F pile (sèche) à mercure
P ogniwo rtęciowe
R (сухой) ртутный элемент

mercury-mercuric oxide electrode
An electrode of the second kind, consisting of
mercury covered with a paste of mercuric
oxide and mercury, in contact with an aqueous
solution of an electrolyte; used as a reference
or indicator electrode to determine the pH of
a solution. The half-cell is represented by the
scheme: $OH^-_{(aq)}|HgO_{(s)}$, $Hg_{(l)}$.
D Quecksilber(II)-oxid-Elektrode
F électrode mercure-oxyde mercurique
P elektroda tlenkowortęciowa
R ртутно-окисный электрод, окиснортутный
электрод

mercury-mercurous chloride electrode →
calomel electrode

mercury-mercurous sulfate electrode
An electrode of the second kind, consisting of
mercury covered with a paste of mercurous
sulfate and mercury, in contact with
potassium sulfate solution of specified
concentration and saturated with mercurous
sulfate; used as a reference electrode. The
half cell is given by the scheme:
$SO_4{}^{2-}{}_{(aq)}|Hg_2SO_{4(s)}$, $Hg_{(l)}$.
D Quecksilber(I)-sulfat-Elektrode
F électrode mercure-sulfate mercureux,
électrode à sulfate mercureux
P elektroda siarczano(wo) rtęciowa
R ртутно-сульфатный электрод,
сульфатно-ртутный електрод

MES → Mössbauer spectroscopy

mesoanalysis
Analysis of a sample whose mass lies in the
range of 0.01 to 0.1 g.
Remark: Mesoanalysis comprises centigram methods; the
term semimicroanalysis is not recommended.
D Halbmikroanalyse
F mésoanalyse, analyse centigrammique
P mezoanaliza
R полумикроанализ

metal buffer, pM buffer
A solution, usually containing metal ions and
an excess of a ligand, exhibiting constant
concentration of free ions of a metal; the
concentration of such a solution remains
practically constant on dilution, or addition
of a small quantity of metal ions.
D Metallpuffer
F ...
P bufor pM
R ...

metal electrode → metal-metal ion electrode

metal indicators
Indicators forming complexes or precipitates
with ions of the metals being determined.
The most important group of metal indicators
are metallochromes.

D Metallindikatoren
F métal-indicateurs
P metalowskaźniki
R металлиндикаторы

metallic conductor → electronic conductor

metallic electrode → metal-metal ion electrode

metallochromic indicator
An indicator which is itself a complexing
agent and which exhibits a colour change when
it reacts with metal ions or has them removed
from its complex with them at or near the
equivalence-point of a complexometric or
precipitation titration.
D Metallindikator
F indicateur métallochrome
P wskaźnik metalochromowy
R металлохромный индикатор

metallofluorescent indicator
A metallochromic indicator which is itself
a complexing agent and which, when
excited by a suitable irradiation, exhibits
a change in its fluorescence emission at or near
the equivalence point of titration.
D Metallfluoreszenzindikator
F indicateur par métallofluorescence
P wskaźnik metalofluorescencyjny
R металлофлуоресцентный индикатор

metal-metal ion electrode, metal(lic) electrode
An electrode (half-cell), consisting of a piece
of metal in contact with a solution of its ions.
The electrode is reversible with respect to the
ions of the metal phase, e.g. silver electrode.
D Metall-Metallion-Elektrode,
Metall(ionen)elektrode
F électrode métal-ion métallique, électrode
métallique
P elektroda metalowa
R металлический электрод

metal-vapour lamp
A discharge lamp filled with a noble gas
and operating at low pressure of the metal
vapour which is generated and excited by
the thermal effect of the electric discharges.
D Dampf-Entladungslampe
F lampe à vapeur de métal
P lampa wyładowcza z parami metalu
R лампа с разрядом в парах металлов,
разрядная трубка с парами металлов

metastable decomposition, metastable ion
transition (*in mass spectrometry*)
The process of decomposition of metastable
ions occurring during their flight through the
field-free region of the mass spectrometer.
The metastable ion transition has the general
form: original ion → daughter ion + neutral
fragment. The process of decomposition is on
the time-scale longer than that generating
a normal fragment ion.

D metastabiler Abbau
F décomposition métastable
P rozpad metastabilny
R метастабильный распад

metastable equilibrium of an electrode
Thermodynamic state of an electrode (half-cell)
in which slow and irreversible processes
occurring at the electrode have no influence on
the potential (measured under specified
conditions), the latter being determined by
a fast and reversible electrode reaction.

D metastabiles Gleichgewicht einer Elektrode
F équilibre métastable d'une électrode
P równowaga metastabilna elektrody
R метастабильное равновесие электрода

metastable ion (*in mass spectrometry*)
An ion formed in a metastable state in the
ionization chamber of a mass spectrometer
and which decomposes in the field-free region
after it has been accelerated out of the ion
source but before entering the analyser; its
lifetime is about 10^{-6} s.

D metastabiles Ion
F ion métastable
P jon metastabilny
R метастабильный ион

metastable (ion) peak (*in the mass spectrum*)
A weak, diffuse peak resulting from
the metastable decomposition: original ion of
mass $m_1 \rightarrow$ ion of mass m_2 + neutral fragment.
The apparent mass m^* observed on the
spectrum is related to the masses m_1, m_2
by the relationship $m^* = \dfrac{m_2^2}{m_1}$.

D Peak eines metastablen Ions, metastabiler
  Peak
F pic (d'ion) métastable
P pik jonu metastabilnego
R пик метастабильного иона,
  метастабильный пик

metastable ion transition → metastable
decomposition

metastable peak → metastable ion peak

method of short swings
A generally used method of weighing on an
undamped balance in which the zero
and equilibrium points are determined
as the arithmetic mean of two swings of the
balance pointer by one to three
scale divisions to the right and left.

D Methode der kleinen Schwingungen
F ...
P metoda małych wahnień
R метод коротких колебаний

method of swings
A method of weighing on an undamped balance
in which the zero and the equilibrium points
are determined as the arithmetic mean of at

least three swings of the balance pointer to one
side and two swings to the other side.

D Schwingungsmethode
F méthode d'oscillations
P metoda wahnień
R метод колебаний

microanalysis
Analysis of a sample whose mass lies in
the range of 0.001 to 0.01 g.

Remark: Microanalysis comprises milligram methods.

D Mikroanalyse
F micro-analyse
P mikroanaliza
R микроанализ

microanalytical reagent
A reagent of a special purity grade used for
microanalytical determination, mainly in
elemental analysis.

D mikroanalytisches Reagens
F réactif micro-analytique
P odczynnik do mikroanalizy, odczynnik
  mikroanalityczny
R реагент для микроанализа

microburette
A burette of capacity up to 5 cm³ with a
0.01 cm³ graduation.

D Mikrobürette
F microburette
P mikrobiureta
R микробюретка

microchemical balance
A laboratory balance of a weighing capacity
of 5 to 20 g and a sensitivity of 10^{-6} g.

D mikrochemische Waage
F balance microchimique
P waga mikroanalityczna
R микровесы

microcircular chromatography
The method for preliminary selection of
a mobile phase, in which a series of spots of
samples are placed on a thin layer of adsorbent
and after applying the mobile phase to the
centres of the spots a series of circular
chromatograms are obtained having a diameter
of about 1 – 2 cm.

D Mikrozirkularchromatographie
F chromatographie microcirculaire
P chromatografia mikrokrążkowa
R микрокруговая хроматография

micro-component
In gravimetric analysis a substance present in
solution which is not normally precipitated
because of its low concentration.

D Mikrokomponente, Mikrobestandteil
F microconstituant, microcomposant
P mikroskładnik
R микрокомпонент

microcosmic salt bead
The transparent vitreous mass formed by
fusing sodium ammonium hydrogen phosphate
on platinum wire. The microcosmic bead gives
a characteristic coloration when fused with
compounds containing some heavy metals;
hence its use for identification in inorganic
qualitative analysis.
D Phosphorsalzperle
F perle de métaphosphate de sodium
P perła fosforanowa
R перл фосфорной соли

microcoulometry → polarographic coulometry

microcrystalloscopic analysis
A method of qualitative analysis of compounds
or ions by microscopic inspection of the
characteristic crystals formed as a result of
application of a suitable reagent.
D kristalloskopische Mikroanalyse,
  Mikrokristalloskopie
F analyse microcristalloscopique,
  microcristalloscopie
P analiza mikrokrystaloskopowa, analiza
  mikrokrystaliczna
R микрокристаллоскоповый анализ,
  микрокристаллоскопия

microgram method *See* submicroanalysis and
ultramicroanalysis
D Ultramikromethode
F méthode microgrammique,
  ultramicrométhode
P metoda mikrogramowa
R ультрамикрометод

micronitrometer
A type of gas burette with a measuring tube
of 1.5 cm³ capacity, used for collecting
nitrogen over aqueous potassium hydroxide
and measuring its volume.
D Mikroazotometer
F microazotimètre, micronitromètre
P mikroazotometr
R микроазотометр, микронитрометр

microphotometer
An optical instrument for measuring the
photographic transmittance at a given
site on a photographic emulsion.
Remark: The term densitometer is not recommended.
D Mikrophotometer
F microphotomètre, densitomètre
P mikrofotometr
R микрофотометр, денситометр

micropipette
D Mikropipette
F micropipette
P mikropipetka
R микропипетка

microscopic cross section, σ
The cross section as related to a single
atomic or nuclear process (e.g. collision of an
atomic nucleus with the bombarding particle).
D mikroskopischer Wirkungsquerschnitt
F section efficace microscopique
P przekrój czynny mikroskopowy
R микроскопическое (эффективное) сечение

microtraces
Trace constituents whose content in a sample
is 10^{-7} to 10^{-4} ppm, or 10^{-11} to 10^{-8} %.
D Mikrospuren
F microtraces
P mikroślady
R микроследы

microwave plasma
Plasma generated by an electromagnetic field
of microwave frequency (above 300 MHz),
the frequency of 2450 MHz being most
common.
D Mikrowellen-Plasma
F plasma micro-onde
P plazma mikrofalowa
R микроволновая плазма

microwave radiation → microwaves

microwave region
The range of the electromagnetic spectrum with
wavelengths from 1 mm to 30 cm, which
lies between infrared radiation and radiowaves.
D Mikrowellengebiet
F région des micro-ondes
P zakres mikrofalowy
R микроволновая область

microwaves, microwave radiation
Electromagnetic radiation with wavelengths
ranging from 1 mm to 30 cm.
D Mikrowellen
F micro-ondes
P mikrofale
R микроволны

microwave spectroscopy, MWS
The measurement and interpretation of the
selective absorption or emission of microwaves
of various frequencies by solids, liquids and
gases.
D Mikrowellen-Spektroskopie
F spectroscopie de micro-ondes
P spektroskopia mikrofalowa
R микроволновая спектроскопия

microwave spectrum
An emission or absorption spectrum of
a molecule in the microwave wavelength region
due to the transitions between different
rotational energy levels within the same
vibrational and electronic levels.
D Mikrowellenspektrum
F spectre de micro-ondes
P widmo mikrofalowe
R микроволновый спектр

middle infrared, mid-infrared, fundamental infrared
The range of infrared radiation of wavelengths 2.5 to 50 μm.
D mittleres Infrarot, mittleres Ultrarot
F infrarouge moyen
P podczerwień średnia
R среднее инфракрасное излучение

mid-infrared → middle infrared

mid-range
Arithmetic mean of the greatest and the smallest values in a sample.
D Mittelwert der Spannweite
F centre de l'intervalle de variation
P środek rozstępu
R середина размаха

migration current (*in electrochemistry*)
The current whose intensity is controlled by the rate of migration of ions (to or from the electrode surface), which undergo an electrode reaction, or arise from it.
D Migrationsstrom
F courant de migration
P prąd migracyjny
R миграционный ток

migration of ions
The motion of ions under the influence of an electric field.
D Migration von Ionen, Ionenwanderung
F migration des ions, électromigration
P migracja jonów
R миграция ионов

Miller indices (*in crystallography*)
Three numbers giving the position of any lattice plane in a chosen system of crystallographic axes a, b, c; e.g. the plane which intersects axis a at point $x = a$ and parallel to b, c has indices (100); the plane parallel to a, intersecting axis b at point $y = b/2$, and axis c at point $z = c/3$ has indices (023).
D Millersche Indizes
F indices de Miller
P wskaźniki Millera, wskaźniki płaszczyzny
R миллеровские индексы

millicoulometry → polarographic coulometry

milligram method *See* microanalysis
D Mikromethode, mikroanalytische Methode
F méthode milligrammique, microméthode
P metoda miligramowa, mikrometoda
R микрометод

mineralization
The process of total decomposition of organic constituents of the analysed sample by oxidizing or reducing methods. The mineralization may be effected by wet or dry procedures, in dynamic or static systems, under conditions close to adiabatic or isothermal.
D Mineralisation
F minéralisation
P mineralizacja
R минерализация

mineralogical effect (*in X-ray fluorescence analysis*)
Change of the X-ray characteristic radiation intensity of an element, with the kind of mineral in which the element occurs.
D mineralogischer Effekt
F effet minéralogique
P efekt mineralogiczny
R минералогический эффект

minor constituent
A constituent whose content in the sample ranges from 0.01 to 1%.
D Nebenbestandteil
F constituant mineur
P składnik uboczny
R побочный компонент, примесь

mixed bed
A mixture of different packings, e.g. cation and anion exchanger, placed in one column.
D Mischbett
F lit mixte
P złoże mieszane
R смешанный слой

mixed crystal, solid solution
A crystal which contains a second constituent which fits into and is distributed in the lattice of the host crystal.
Remark: The use of the term solid solution for amorphous materials is not recommended.
D feste Lösung, Mischkristall
F cristal mixte, solution solide
P kryształ mieszany, roztwór stały
R смешанный кристалл, твёрдый раствор

mixed indicator
A mixture of indicators of the same function chosen so that their transition regions are very close and of such a composition that the resultant overall colour change of the mixture is better distinguishable than that of either indicator used separately.
D Mischindikator
F indicateur mixte
P wskaźnik mieszany
R смешанный индикатор

mixed (polyelectrode) potential, E_{mix}
The resultant of potentials of electrode reactions proceeding at the complex electrode.
D Mischpotential
F potentiel mixte
P potencjał mieszany
R смешанный потенциал (полиэлектрода), компромиссный потенциал

mixed potential → mixed polyelectrode potential

mixed solvent
A liquid phase consisting of at least two
solvents, e.g. alcohol and water.
D Lösungsmittelgemisch
F solvant mixte
P rozpuszczalnik mieszany
R смешанный растворитель

mixotropic series → eluotropic series

mixture design
An experimental design used when
investigating the properties of mixtures as a
function of their components only. Usually
it is a simplex design what results from the
fact that the pure components (treated as
experimental points) define a simplex whose
points represent all physically possible
mixtures.
D Mixtur-Versuchsplan
F composition du projet
P plan dla mieszanin
R план для смесей

mobile phase (*in chromatography*)
A gaseous or liquid phase which moves
through the chromatographic bed during the
separation process. It includes the fraction of
the sample present in this phase.
D mobile Phase
F phase mobile
P faza ruchoma
R подвижная фаза

mobile phase distance, solvent migration –
distance (*in planar chromatography*)
The distance travelled by the mobile phase
front, as measured from the starting line or point.
D Laufstrecke der mobilen Phase,
   Fließmittelwanderungsstrecke
F distance de la phase mobile, distance de
   migration du solvant
P dystans rozwijania, droga rozwijania
R пройденное расстояние подвижной фазы,
   расстояние миграции растворителя,
   расстояние пройденное фронтом
   растворителя

mobile phase front, (solvent) front
(*in chromatography*)
The front line of the mobile phase.
D Front der mobilen Phase, Fließmittelfront,
   Laufmittelfront, Solvensfront,
   Lösungsmittelfront
F front de la phase mobile, front de solvant
P czoło rozpuszczalnika, linia czołowa fazy
   ruchomej, front fazy ruchomej
R фронт подвижной фазы, фронт растворителя

mobile phase hold-up time, hold-up time,
retention time of the mobile phase, t_M
(*in chromatography*)
The time required for the mobile phase
molecules to pass through the chromatographic
system. In practice, the mobile phase hold-up

time is the total retention time of non-retained
solute and includes contributions due to the time
needed to pass the interstitial volume of the
column and any extra-column volumes present
in the system.
Remark: While in general liquid chromatography both the
mobile phase hold-up time and the retention time of the
non-retained solute are the same, this is not so in
permeation chromatography.
D Mobilzeit
F temps de rétention d'un composé non-retenu
   temps de rétention de l'air
P czas retencji składnika nie zatrzymywanego
   w kolumnie
R время удерживания несорбирующегося
   вещества

mobile phase hold-up volume, hold-up
volume, V_M (*in chromatography*)
The total volume of a mobile phase in the
chromatographic system. The hold-up volume
is the total retention volume of a non-retained
solute and includes contributions due to the
interstitial volume of the column and the
effective volume of the sample injector and
detector.

$$V_M = t_M F_c = V_I + V_{ext}$$

where V_I – interstitial volume,
V_{ext} – extra-column volume, t_M – mobile phase
hold-up time, and F_c – corrected volumetric
flow rate of the mobile phase.
Remark: While in general liquid chromatography both the
mobile phase hold-up volume and the retention volume of
non-retained solute are the same, this is not so in
permeation chromatography.
D Mobilvolumen
F volume de rétention d'un composé non
   retenu, volume de rétention de l'air
P objętość retencji składnika nie
   zatrzymywanego w kolumnie
R объём задержки подвижной фазы, объём
   удерживания несорбирующегося вещества

mobile phase volumetric flow rate F_a →
volumetric flow rate F_a

mobile phase volumetric flow rate F_c →
volumetric flow rate F_c

mode
The value of a random variable for which the
probability (in the case of a discrete random
variable) or probability density function
(in the case of a continuous random variable)
reaches its maximum.
D Modalwert, Dichtemittel, Mode
F mode, dominante
P wartość modalna, moda, dominanta
R мода

modified active solid (*in chromatography*)
An active solid, whose adsorptive properties
have been changed by treatment with
a gas, liquid or another solid.

D modifizierter aktiver Festkörper
F solide active modifié
P sorbent stały modyfikowany
R модифицированное активное твёрдое
   вещество

modified Nernst equation → Nikolsky equation

modulation polarography
The alternating current polarography in which
two alternating voltages of different amplitudes
and of widely differing frequencies are used
simultaneously. The variable component of
the current I_{ac} is measured as a function of
the constant component of the indicator
electrode potential E_{dc}. The response curve,
obtained with the use of phase-sensitive
rectification, is shown.

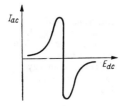

D Modulationspolarographie
F polarographie avec modulation
P polarografia modulacyjna
R модуляционная полярография

Mohr's method
The determination of chlorides by direct titration
of the neutral solution of a chloride with
a standard silver nitrate solution in the presence
of chromate ions.
D Mohrsche Methode
F méthode de Mohr
P metoda Mohra
R метод Мора

molality (*of a solution*), m_B
The amount of substance B divided by the
mass of solvent.
Unit: mole per kilogram, mol kg^{-1}.
D Molalität
F molalité
P molalność
R моляльность

molar absorbancy → molar absorptivity

molar absorbancy index → molar absorptivity

molar absorption coefficient → molar
absorptivity

molar absorptivity, ε, ε_{mol}
A measure of the efficiency with which
molecules absorb radiation, defined from the
Bouguer-Lambert-Beer law as the
absorbance A divided by the path length b

(in centimetres) tranversed in the sample by the
radiation and the concentration c (in moles per
litre) of the absorbing substance B: $\varepsilon = A/bc_B$.
Remark: The terms molar absorbancy, molar extinction
coefficient, molar (decadic) absorption coefficient and molar
absorbancy index are not recommended.
D molarer (dekadischer) Absorptionskoeffizient,
   molarer Extinktions-koeffizient
F coefficient d'absorption molaire
P współczynnik absorpcji molowy,
   współczynnik ekstynkcji molowy
R молярный (десятичный) показатель
   поглощения, молярная абсорбируемость,
   мольный (десятичный) коэффициент
   погашения

molar concentration → amount-of-substance
concentration of substance B

**molar conductivity of electrolyte at infinite
dilution** → limiting molar conductivity of
electrolyte

molar conductivity of electrolyte, Λ
The electrolytic conductivity, $\varkappa$, at
a concentration c, of 1 mol m^{-3}, $\Lambda = \varkappa/c$.
Numerically, it is equal to the electrolytic
conductance of a cell having electrodes 1 m
apart and with crosssectional area such that
between the electrodes there is a volume of
the electrolyte solution containing 1 mole of
the conducting substance.
SI unit: siemens square metre per mole,
S m^2 mol^{-1}.
Remark: The formula unit of the conducting substance
whose concentration is c must be specified and should be
given in brackets, e.g. Λ (MgCl$_2$), Λ (1/2 HgCl$_2$), Λ (KCl).
D molare Leitfähigkeit eines Elektrolyten
F conductivité molaire d'un électrolyte
P konduktywność molowa elektrolitu,
   przewodnictwo molowe elektrolitu
R молярная электропроводность электролита,
   молярная электрическая проводимость
   электролита, мольная электропроводность
   электролита

molar conductivity of ionic species B, molar
ionic conductivity, λ_B
The contribution of ionic species B to the total
molar conductivity of an electrolyte, defined as

$$\lambda_B = |z_B| \, F \, u_B$$

where z_B – charge number of ionic
species B, F – Faraday constant, and
u_B – electric mobility of ionic species B.
SI unit: siemens square metre per mole,
S m^2 mol^{-1}.
Remark: In most current practice the charge number of the
ionic species is taken as unity, i.e. the ionic conductivity is
taken as that of species such as Na$^+$, 1/2 Ca^{2+}, 1/3 La^{3+}
etc. To avoid ambiguity the species considered should be
clearly stated, e.g. as λ (¹/₂Ca^{2+}), λ (Mg^{2+}) = 2λ (¹/₂Mg^{2+}).
D molare Leitfähigkeit eines Ions, molare
   Ionenleitfähigkeit
F conductivité ionique-molaire

P konduktywność jonowa molowa,
przewodnictwo jonowe molowe
R молярная электропроводность иона,
молярная ионная электропроводность

molar decadic absorption coefficient → molar
absorptivity

molar extinction coefficient → molar
absorptivity

molar ionic conductivity → molar conductivity
of ionic species B

molar ionic conductivity at infinite dilution →
limiting molar conductivity of ionic species B

molarity → amount-of-substance
concentration of substance B

molar mass, M
The mass of one mole of a given type of
elementary entities (*cf.* elementary entity)
numerically equal to the mass m of the
substance divided by the amount n of substance
of the defined elementary entities contained in
that mass, $M = m/n$.
The basic SI unit: kg mol^{-1}; practical
unit: g mol^{-1}.
D molare Masse, Molmasse
F masse molaire
P masa molowa
R молярная масса

molar refraction, R_m
A characteristic, temperature independent
property of a substance defined as

$$R_m = \frac{n^2-1}{n^2+2} V_m$$

where n — refractive index and V_m — molar
volume of the substance.
D Molrefraktion
F réfraction molaire, réfractivité molaire
P refrakcja molowa
R молярная рефракция, мольная рефракция

molar rotation
The quantity equal to the product of the
specific rotation and 1/100th of the molar mass
of the given compound.
D molare Drehung
F rotation moléculaire
P skręcalność molowa
R молекулярное вращение

mole, mol
The unit of the amount of substance in
a system which contains as many elementary
entities as there are carbon atoms in 0.012 kg
of the pure nuclide ^{12}C; the elementary entities
must be specified.
D Mol
F mole
P mol
R моль

mole → grammolecule

molecular absorption spectroscopy
Absorption spectroscopy concerned with the
study of molecular spectra.
D Absorptions-Molekülspektroskopie
F spectroscopie d'absorption moléculaire
P spektroskopia absorpcyjna cząsteczkowa
R абсорбционная молекулярная
спектроскопия

molecular beam
A beam of gas molecules with a well-defined
Maxwellian energy distribution.
D Molekularstrahl
F faisceau moléculaire
P wiązka molekularna
R молекулярный пучок

molecular beam reactive scattering, MBRS
The surface sensitive technique involving the
use of a molecular beam of reactive gases. The
method gives information on the surface
reactions (especially chemisorption), energy
and momentum transfer from the gaseous
molecules to the solid surface.
D ...
F dispersion réactive par faisceau moléculaire
P rozpraszanie reaktywne wiązki molekularnej
R реактивное рассеяние молекулярного пучка

molecular beam surface scattering, MBSS
A method of determination of the energy
accomodation coefficients based on the
analysis of the density and velocity of the
scattered molecular beam as a function of the
scattering angle.
D ...
F dispersion de surface par faisceau
moléculaire
P rozpraszanie powierzchniowe wiązki
molekularnej
R поверхностное рассеяние молекулярного
пучка

molecular diffusion term, axial molecular
diffusion, longitudinal diffusion
A factor contributing to the total band
broadening represented by the second term (B)
of the van Deemter equation and being of
great importance especially at slow interstitial
velocities of the mobile phase. This term
accounts for the diffusion in the direction of
flow in the mobile phase.
In liquid chromatography, the longitudinal
diffusion in the stationary phase may be also of
importance.
D Molekulardiffusionsterm, (axiale)
Molekulardiffusion
F diffusion moléculaire (axiale), diffusion
longitudinale
P dyfuzja molekularna (osiowa), dyfuzja
podłużna
R молекулярная (продольная) диффузия

molecular emission spectroscopy
Emission spectroscopy concerned with the
study of molecular spectra.

D Emissions-Molekülspektroskopie
F spectroscopie d'émission moléculaire
P spektroskopia emisyjna cząsteczkowa
R эмиссионная молекулярная
   спектроскопия

molecular fluorescence spectroscopy
Fluorescence spectroscopy concerned with the
study of molecular spectra.

D Fluoreszenz-Molekülspektroskopie
F spectroscopie de fluorescence moléculaire
P spektroskopia fluorescencyjna cząsteczkowa
R флуоресцентная молекулярная
   спектроскопия

molecular ion (*in mass spectrometry*)
An ion formed by the removal from(positive
ions) or addition to (negative ions) a molecule
of one or more electrons without
fragmentation of the molecular structure.
The mass of this ion is the sum
of the masses of the most abundant naturally
occurring isotopes of the various atoms that
make up the molecule (with a correction for
the masses of the electrons lost or gained).

D Molekülion
F ion moléculaire
P jon cząsteczkowy
R молекулярный ион

molecular occlusion → occlusion

molecular peak, parent molecular peak
A peak in the mass spectrum at a mass-to-
charge ratio corresponding to the molecular
mass of the molecule.

D Molekülpeak
F pic moléculaire
P pik molekularny
R молекулярный пик

molecular radiation
A radiation that results from the rotational,
vibrational and electronic energy transitions in
a molecule.

D Molekülradiation
F radiation moléculaire
P promieniowanie cząsteczkowe
R молекулярное излучение

molecular-sieve chromatography
Permeation chromatography in which the
stationary phase is a molecular sieve.

Remark: This term is sometimes used as a synonym of
gel-permeation chromatography.

D Molekularsiebchromatographie
F chromatographie sur tamis moléculaires
P chromatografia sitowo-molekularna
R хроматография на молекулярных ситах

molecular sieves
Sorbents of defined pore structure, capable of
sorbing only those molecules whose
dimensions are smaller than the pores of the
sorbent structure, e.g. zeolites, controlled pore
glasses, porous polymers.

D Molekularsiebe
F tamis moléculaires
P sita molekularne, sita cząsteczkowe
R молекулярные сита

molecular spectrochemical analysis
Spectrochemical analysis concerned with the
investigation of molecular spectra.

D molekulare spektrochemische Analyse
F analyse spectrochimique moléculaire
P analiza spektralna cząsteczkowa
R молекулярный спектрохимический анализ

molecular spectroscopy, MS
The branch of spectroscopy concerned with
the measurement and interpretation of
molecular spectra.

D Molekülspektroskopie
F spectroscopie moléculaire
P spektroskopia cząsteczkowa, spektroskopia
   molekularna
R молекулярная спектроскопия

molecular spectrum
A general term referring to electromagnetic
spectra, either emitted or absorbed,
resulting from transitions between energy
levels in a molecule. It includes electronic,
vibrational and rotational spectra.

D Molekülspektrum
F spectre moléculaire
P widmo cząsteczkowe
R молекулярный спектр

molecular vibrations
All atoms or atomic groups in molecules are
assumed, in continuous motion with respect to
each other, vibrating at definite frequencies
which depend on the molecular structure.

D Molekülschwingungen
F vibrations moléculaires
P drgania cząsteczkowe
R молекулярные колебания, колебания
   молекул

mole fraction (of substance B), x_B
The ratio of the amount of substance n_B of
constituent B of a mixture to the total amount
of substance of all constituents of that
mixture $\sum_i n_i$

$$x_B = \frac{n_B}{\sum_i n_i}$$

SI unit: mole per mole, $mol\ mol^{-1}$.

D Stoffmengenanteil
F ...
P ułamek molowy
R молярная доля

monochromatic radiation
Any electromagnetic radiation of only one discrete wavelength, or a beam of particles of only one energy (or containing only one type of particles).
D monochromatische Strahlung
F radiation monochromatique
P promieniowanie monochromatyczne
R монохроматическое излучение

monochromator
A device for isolating one narrow wavelength region from a wide wavelength range of radiation.
D Monochromator
F monochromateur
P monochromator
R монохроматор

monofunctional ion exchanger
An ion exchanger containing only one type of ionogenic groups.
D monofunktioneller Ionenaustauscher
F échangeur d'ions monofonctionnel
P wymieniacz jonowy jednofunkcyjny
R монофункциональный ионообменник

monolayer
A monoatomic or unimolecular surface layer of an adsorbent or adsorbate.
D Monoschicht
F monocouche
P monowarstwa
R монослой

monostandard method, single-comparator method
A method used in multielement activation analysis allowing to determine on the basis of irradiation and measurement of the activation of a single element (monocomparator, monostandard) the contents of all other components of the sample.
D Komparatortechnik, Komparatormethode
F ...
P metoda monostandardów, metoda monokomparatorów
R метод применения мониторов

Moseley law
The square root of the frequency of the characteristic X-ray line belonging to the given series is linearly dependent on the atomic number Z of elements.
D Moseleysches Gesetz
F loi de Moseley
P prawo Moseleya
R закон Мозли

MOSS → Mössbauer spectroscopy

Mössbauer effect
A nuclear phenomenon defined as the elastic (recoil-free) emission of gamma rays by a nucleus which is built into a solid matrix followed by subsequent absorption (resonance absorption) of the gamma-rays by another atomic nucleus.
D Mößbauer-Effekt
F effet Mössbauer
P efekt Mössbauera
R эффект Мёссбауэра

Mössbauer effect spectroscopy → Mössbauer spectroscopy

Mössbauer spectroscopy, MOSS, Mössbauer effect spectroscopy, MES, nuclear gamma-ray resonance spectroscopy
The branch of nuclear spectroscopy based on the Mössbauer effect.
D Mößbauer-Spektroskopie
F spectroscopie Mössbauer
P spektroskopia Mössbauera
R спектроскопия Мёссбауэра

most efficient estimator
An unbiased estimator which has the smallest variance among all possible unbiased estimators of parameter Θ.
D wirksamste Schätzfunktion
F estimateur le plus efficient
P estymator najefektywniejszy
R наиболее эффективная оценка

MS → mass spectrometry

muffle furnace
An electric furnace with a muffle for heating and calcinating crucibles with precipitates or for mineralizing and fusing samples.
D Muffelofen, gemuffelter Ofen
F four à moufle
P piec muflowy
R муфельная печь

mulling → **mull technique**

mull technique, mulling (*in IR spectroscopy*)
A technique of sampling of solids which involves grinding the powdered sample with a greasy liquid medium, e.g. paraffin oil (Nujol), chlorofluorocarbon greases. A thick paste or mull is obtained which is then squeezed into a thin film between the IR transmitting windows of the sample cell.
D Mull-Verfahren, Nujol-Technik
F broyage dans le nujol
P metoda badania próbki w postaci zawiesiny
R суспендирование твёрдого вещества в нуйоле

multicomponent analysis → multiple analysis

multiple analysis, multicomponent analysis
(*in spectrophotometry*)
The simultaneous determination of two or
more absorbing substances in the same
solution. To analyse for n constituents,
n wavelengths must be chosen to set up
n simultaneous equations of the type
$A = \sum_n A_n = b \sum_n a_n c_n$, where A — absorbance
of the mixture, A_n — individual absorbance,
a_n — absorption coefficient, b — sample path
length, and c_n — concentration.
D Mehrkomponentenanalyse
F analyse spectrophotométrique simultanée
P analiza wieloskładnikowa
R многокомпонентный анализ

multiple range indicator → universal indicator

multiple regression
The regression in which the influence of more
than one independent variable on the expected
value of the investigated random variable
(experimental outcome) is taken into
consideration.
D multiple Regression, mehrfache Regression
F régression multiple
P regresja wielokrotna, regresja wieloraka
R множественная регрессия

multiple regression method (*in* X-*ray
fluorescence analysis*)
A statistical method of eliminating of the
matrix effects consisting in the determination of
the functional relationship between the
intensity of the characteristic radiation of an
analyte and its concentration, based on the
assumption that the concentration of
a particular element is also a function of the
radiation intensity of some (or all) elements
constituting the matrix, and/or the intensity
of radiation scattered from the analysed
sample.
D multiple Regressionsmethode
F méthode de régression multiple
P metoda regresji wielokrotnej
R ...

multiplet
(1) Spectral line formed by two or more
closely spaced lines.
(2) A group of electronic states in an atom of
close energies belonging to the same atomic
term.
D Multiplett
F multiplet
P multiplet
R мультиплет

multiple techniques
The investigation of a substance by two or
more techniques.
D kombinierte Untersuchungsmethoden
F techniques multiples
P techniki połączone
R исследование рядом методов

multiplication reaction → amplification reaction

multiplier phototube → photomultiplier tube

multisweep polarography
A modification of polarography in which,
during the life-time of one drop, a potential
linearly increasing with time is applied to the
indicator electrode and a series of the
current-potential curves is recorded.

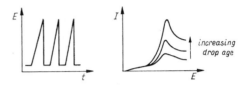

D Multisweep-Polarographie
F polarographie à balayages multiples
P polarografia ze zmianą potencjału
  wielokrotną
R полярография с многократной развёрткой
  потенциала

murexide reaction
A reaction used for the detection of uric acid,
consisting in converting the latter into
purple-red murexide (purpuric acid ammonium
salt) under the successive action of nitric acid
and aqueous ammonia. Many purine compounds
give a positive murexide reaction.
D Murexidreaktion
F réaction de la murexide
P reakcja mureksydowa
R мурексидная реакция

MWS → microwave spectroscopy

N

nanogram method *See* ultramicroanalysis
D Nanogrammethode
F méthode nanogrammique
P metoda nanogramowa
R ультрамикрометод

nanotraces
Trace constituents whose content in a sample ranges from 10^{-10} to 10^{-7} ppm, i.e. from 10^{-14} to $10^{-11}\%$.
D Nanospuren
F nanotraces
P nanoślady
R ...

natural convection
Diplacement of molecules in a liquid due to the density gradient in the solution.
D natürliche Konvektion
F convection naturelle
P konwekcja naturalna
R естественная конвекция

natural isotopic abundance (*of a specified isotope of an element*)
The isotopic abundance in the element as found in nature.
D natürliche Isotopenhäufigkeit
F abondance isotopique naturelle
P zawartość danego izotopu w naturalnym pierwiastku
R природная распространенность изотопов

natural line-width
An irreducible width of a spectral line defined from the Heisenberg uncertainty principle.
D natürliche Linienbreite
F largeur naturelle d'une raie spéctrale
P szerokość linii spektralnej naturalna
R естественная ширина спектральной линии

near infrared
The part of the infrared radiation with wavelengths ranging from ca 0.8 to 2.5 μm.

D nahes Infrarot, nahes Ultrarot
F infrarouge proche
P podczerwień bliska
R ближнее инфракрасное излучение

near ultraviolet
The part of ultraviolet radiation with wavelengths ranging from 200 to 380 nm.
D nahes Ultraviolett
F ultraviolet proche
P nadfiolet bliski
R ближний ультрафиолет

nebulization
Conversion of the sample solution into a mist or aerosol.
D Vernebeln, Vernebelung
F nébulisation
P nebulizacja
R распыление

nebulizer
A device for nebulization.
Remark: The term atomizer should not be used in this sense.
D Nebelblaser
F nébuliseur
P nebulizer
R распылитель

negative branch → P-branch

negative ion (*in mass spectrometry*)
An atom, radical, molecule or molecular moiety which has gained one or more electrons thereby acquring a negative electric charge.
Remark: The use of the term anion as an alternative is not recommended.
D negatives Ion
F ion négatif
P jon ujemny
R отрицательный ион

negative-ion mass spectrum
A mass spectrum of negative ions.
D Massenspektrum negativer Ionen
F spectre de masse d'ions négatifs
P widmo mas jonów ujemnych
R масс-спектр отрицательных ионов

nephelometer
An instrument that measures the intensity of light scattered by a suspended medium.
D Nephelometer
F néphélomètre
P nefelometr
R нефелометр

nephelometric titration
The titration in which the end point of titration is determined from the change in the ratio of the radiation intensity diffused by the titrated solution at a right angle to the incident radiation intensity.

D nephelometrische Titration,
   tyndallometrische Titration
F titrage néphélométrique
P miareczkowanie nefelometryczne
R нефелометрическое титрование

nephelometry
A method of quantitative analysis which
involves the measurement of the intensity of
a light beam scattered by suspended matter.
Measurements are usually made at right angles
to the direction of propagation of the light
beam from the source. The intensity of light
scattered at any particular angle is a function
of the concentration of the scattering particles,
their size, their shape, of the wavelength
of light, of the difference in refractive indices
of the particle and the medium.

D Nephelometrie, Tyndallometrie
F néphélométrie
P nefelometria
R нефелометрия, тиндалиметрия

Nernst distribution law, distribution law
In the equilibrium state, the dissolved substance
is distributed between two non-miscible
solvents in such a way that the ratio of
concentrations of that substance in the two
phases is constant at a given temperature,
provided the molecular mass of the dissolved
substance is the same in each phase

$$K_D = A_o/A_w$$

where K_D — distribution constant; A_o, A_w —
concentration of the substance in the organic
and aqueous phases, respectively. If different
kinds of molecules occur in both phases,
the particular types of molecules distribute so
as if no other molecules were present in the
system.

D Nernstscher Verteilungssatz
F loi de distribution, loi (de partage) de Nernst
P prawo podziału (Nernsta)
R закон распределения

Nernst equation
The equation describing the relation between
the equilibrium potential E of an electrode and
the activities of substances which take part in
the electrode reaction

$$Ox + ne \rightleftarrows Red$$

$$E = E° + \frac{RT}{nF} \ln \frac{a_{Ox}}{a_{Red}}$$

where $E°$ — standard potential of the electrode,
n — number of electrons exchanged in the
electrode reaction, a_{ox} — activity of the
oxidized form, and a_{Red} — activity of the
reduced form.

D Nernst-Gleichung
F équation de Nernst, formule de Nernst
P równanie Nernsta
R уравнение Нернста

Nernst glower
One of the most common sources of infrared
radiation. A hollow rod about 3 cm long and 2
to 3 mm in diameter, obtained by sintering
a mixture of cerium, zirconium, thorium and
yttrium oxides. It glows when maintained
at a high temperature (about 1800 K)
by electrical heating.

D Nernst-Lampe, Nernst-Stift
F lampe Nernst, filament de Nernst
P włókno Nernsta
R штифт Нернста, лампа Нернста

Nernstian electrode response
The behaviour of the electrode potential
being in agreement with the Nernst equation.

D ...
F réponse nernstienne
P odpowiedź elektrody nernstowska
R нернстовская электродная функция

Nernstian slope
A slope of the curve representing the relationship
between the potential of an electrode and the
logarithm of the activity of the potential-
determining ion which is in agreement with the
Nernst equation.

D Nernstsche Neigung
F pente nernstienne
P nachylenie nernstowskie
R нернстовский наклон

Nessler comparison tube → Nessler measuring
cylinder

Nessler measuring cylinder, Nessler comparison
tube
A clear glass cylinder with polished bottom
used for colorimetric titration, and also for
visual matching of the colouring or turbidity
of the solution to be analysed with the
colouring or turbidity of a standard. The
measurement is carried out at equal thickness
of the layers.

D Kolorimeterzylinder nach Neßler
F tube de Nessler (pour colorimétrie)
P cylinder kolorymetryczny Nesslera
R пробирка Несслера

Nessler reagent
Aqueous solution of mercury(II) iodide and
potassium iodide (the HgI_4^{2-} complex
is formed), to which potassium (or sodium)
hydroxide is added. The reagent finds
application in qualitative detection and
colorimetric determination of ammonia
and in the nephelometric analysis of higher
alcohols.

D Neßlers Reagens, Neßlers Reagenz
F réactif de Nessler
P odczynnik Nesslera
R реактив Несслера

net current
The algebraic sum of the partial currents. *See*
partial anodic/cathodic current.
D ...
F courant total, courant global
P prąd wypadkowy
R общий ток

net retention time, t_N
In gas chromatography, the adjusted retention
time t_R' multiplied by the pressure-gradient
correction factor j,
$$t_N = jt_R'$$
In liquid chromatography, the mobile phase
compressibility is negligible and thus the
pressure-gradient correction factor does not
apply, so the adjusted and net retention times
are identical.
D Nettoretentionszeit
F temps de rétention net
P czas retencji absolutny
R чистое время удерживания

net retention volume, V_N
In gas chromatography, the adjusted retention
volume V_R' multiplied by the pressure-gradient
correction factor j,
$$V_N = jV_R'$$
In liquid chromatography, the mobile phase
compressibility is negligible and thus the
adjusted and net retention volumes are
identical.
D Nettoretentionsvolumen
F volume de rétention absolu, volume de
rétention net
P objętość retencji absolutna
R чистый удерживаемый объём

neutral-carrier membrane
A liquid membrane based on solutions of
molecular carriers of ions, e.g. uncharged
macrocyclic molecules, antibiotics or other
sequestering agents, which form stoichiometric
complexes with ions, usually cations.
D Membran mit elektroneutralem
Ladungsüberträger
F ...
P membrana z nośnikiem obojętnym, membrana
z nośnikiem nienaładowanym
R мембрана на основе нейтральных
переносчиков, мембрана с электрически
нейтральным лигандом

neutral-carrier membrane electrode → neutral-
carrier membrane ion-selective electrode

**neutral-carrier membrane (ion-selective)
electrode**
An ion-selective electrode with a neutral-carrier
membrane.
D Elektrode mit elektroneutralem
Ladungsüberträger, Elektrode mit neutralem
Ligand
F ...

P elektroda (jonoselektywna) z membraną
z nośnikiem obojętnym, elektroda
membranowa z nośnikiem obojętnym
R (ионоселективный) электрод с мембраной
на основе нейтральных переносчиков

neutral chelate
A chelate in which the positive charge of the
central ion is compensated for by the negative
charge of the ligand(s).
D inners Komplexsalz
F complexe interne
P sól wewnątrzkompleksowa, chelat
wewnętrzny, kompleks wewnętrzny
R внутрикомплексное соединение

neutral-density filter, grey filter
A light filter which decreases the intensity of
light without appreciably changing its colour.
D Neutralfilter, Graufilter
F filtre neutre
P filtr neutralny, filtr szary
R нейтральный фильтр

neutral filter → neutral-density filter

neutralization
A reaction of acid and base which produces
a neutral solution, i.e. a solution for which the
activities of both the acidic and basic forms of
the solvent are equal.
D Neutralisation
F neutralisation
P zobojętnianie
R нейтрализация

neutralization titration curve, acid-base
titration curve
A plot of the pH of a solution as a function
of the volume of the added standard solution
of an acid or a base.
D Säure-Base-Titrationskurve,
Neutralisationskurve
F courbe de titrage acido-basique
P krzywa miareczkowania alkacymetrycznego
R кислотно-основная кривая титрования

neutron absorptiometry, neutron absorption
technique *See* nuclear radiation absorptiometry
D Neutronenabsorptionsmethode,
Neutronenabsorptionsverfahren
F analyse par absorption de neutrons
P metoda absorpcji neutronów
R нейтронно-абсорпционный метод анализа

neutron absorption technique → neutron
absorptiometry

neutron activation
Activation effected with a stream of thermal,
fast or resonance neutrons.
D Neutronenaktivierung
F activation par neutrons
P aktywacja neutronowa
R активация нейтронами

neutron activation analysis
Activation analysis in which a flux of neutrons is applied for activation.
D Neutronenaktivierungsanalyse
F analyse par activation de neutrons
P analiza aktywacyjna neutronowa
R нейтронный активационный анализ, нейтроноактивационный анализ

neutron detector
A device for detection of neutrons based on the neutron-induced nuclear processes in which an ionizing radiation is produced, e.g. slow neutrons are detected by recording α-particles from the $^{10}B(n, \alpha)^7Li$ reaction.
D Neutronendetektor
F détecteur de neutrons
P detektor neutronów
R детектор нейтронов

neutron generator
A device producing neutrons in an appropriate nuclear reaction, e.g.
$^2H(d,n)^3He$, $^3H(d,n)^4He$.
D Neutronengenerator
F générateur de neutrons
P generator neutronów
R нейтронный генератор

neutron output
Number of neutrons emitted from a source in unit time at a solid angle of 4π.
D Neutronenausbeute
F rendement de neutrons
P wydajność emisji neutronów
R выход нейтронов

neutron spectrum
Curve presenting the energy dependence of the neutron flux density.
D Neutronenspektrum
F spectre de neutrons
P widmo neutronów
R спектр нейтронов

neutron thermalization
The slowing down of neutrons to reduce their energy to a value corresponding to the thermal energy of atoms or molecules of the medium.
D Neutronenthermalisierung
F thermalisation des neutrons
P termalizacja neutronów
R термализация нейтронов

NHE → standard hydrogen electrode

Nichrome helix → coil of Nichrome wire

Nichrome source → coil of Nichrome wire

Nikolsky equation, modified Nernst equation
The semiempirical equation which describes the relationship between the potential of an

ion-selective electrode and the activities of the main substance M and the interfering substances N_i

$$E = E_{ISE}^{\circ} + S_M \log \left([M] + \sum_i k_{M,N_i}^{pot} [N_i]^{z_{M/z_{N_i}}} \right)$$

where k_{M,N_i}^{pot} — potentiometric selectivity coefficient, z — charge number of the ion, S_M — slope of the potential versus logarithm of activity of ion M curve, E_{ISE}° — potential of electrode extrapolated to unit activity, $[M] = 1$ mol dm^{-3} at $[N_i] = 0$.
D erweiterte Nernst-Gleichung
F équation de Nernst modifiée, formule de Nikolski
P równanie Nikolskiego, równanie Nernsta zmodyfikowane
R модифицированное уравнение Нернста

nitrometer, azotometer
Glass apparatus used to collect and measure nitrogen and other gases evolving in a chemical reaction (e.g. Lunge nitrometer).
D Nitrometer
F nitromètre
P azotometr, nitrometr
R нитрометр

NMR → nuclear magnetic resonance

NMR spectrometer → nuclear magnetic resonance spectrometer

NMR spectroscopy → nuclear magnetic resonance spectroscopy

noise (level), N
All random variations of the detector signals obtained for a given analytical system (recorded in the absence of the component to be analysed).
D Rauschen, Rauschpegel
F bruit, niveau de bruit
P szum, poziom szumu
R шум, уровень шумов

nominal linear flow, F (of the mobile phase)
The linear velocity of the mobile phase in a part of the column not containing any packing, as given by the equation

$$F = \frac{F_c}{A_c} \text{ [cm min}^{-1}]$$

where F_c — volumetric flow rate of the mobile phase and A_c — cross-sectional surface area of the column.
D nominelle lineare Flußgeschwindigkeit
F vitesse d'écoulement linéaire nominalle
P prędkość przepływu liniowa nominalna
R номинальный линейный поток

non-aqueous titration
A titration (acid-base or other type) in which
the solvent is other than water and in which
the concentration of the latter is minimal
(say less than 0.5 per cent).
D Titration in nichtwässriger Losung
F titrage en milieu non aqueux
P miareczkowanie w roztworze niewodnym
R неводное титрование

non-carbonate hardness
The hardness of water due to the presence
of calcium and magnesium compounds other
than hydrogenecarbonates, carbonates and
hydroxides, i.e. chlorides, sulfates and nitrates.
D Nichtcarbonathärte
F dureté permanente
P twardość niewęglanowa
R некарбонатная жёсткость

non-destructive activation analysis, instrumental
activation analysis
The activation method of analysis in which
the selective measurement of activity of the
element to be determined is carried out by
physical methods without chemical separation
of the irradiated sample.
D zerstörungsfreie Aktivierungsanalyse
F analyse non destructive par activation,
  analyse par activation instrumentale
P analiza aktywacyjna niedestrukcyjna, analiza
  aktywacyjna instrumentalna
R инструментальный активационный анализ

non-destructive analysis
A method of analysis in which the sample
examined is not destroyed.
D zerstörungsfreie Analyse
F analyse non destructive
P analiza niedestrukcyjna, analiza nieniszcząca
R неразрушающий анализ, недеструктивный
  анализ

non-flame atomizing device → electrothermal
atomizer

non-isotopic carrier
A carrier of a radioactive substance chemically
non-identical (usually isomorphic or
isodimorphic) with this substance, e.g. barium
ions Ba^{2+} for Ra^{2+} ions which coprecipitate
with barium salts.
D nichtisotoper Träger
F entraîneur non isotopique
P nośnik nieizotopowy
R неизотопный носитель

non-linear regression
The regression whose functional form is
non-linear with respect to both parameters and
independent variables.
D nichtlineare Regression
F régression non-linéaire
P regresja nieliniowa
R нелинейная регрессия

non-parametric test
A test for verification of statistical hypotheses
which can be applied without any
assumptions as to the type of probability
distribution.
D nichtparametrischer Test
F test non paramétrique
P test nieparametryczny
R непараметрический критерий

non-recording spectrophotometer, manual
spectrophotometer
A manually operated spectrophotometer, most
frequently single-beam, by means of which
a spectrophotometric curve is plotted point by
point at some chosen wavelength.
D handbetätigtes Spektralphotometer
F spectromètre non-enregistreur
P spektrofotometr punktowy
R нерегистрирующий спектрофотометр

normal conditions → standard conditions

normal dispersion
Dispersion of a material when the refractive
index decreases monotonically and
continuously with the increasing wavelength.
D normale Dispersion
F dispersion normale
P dyspersja normalna
R нормальная дисперсия

normal distribution, Gaussian distribution
Probability distribution of the continuous
random variable X, probability density
function being defined by the formula

$$f(x) = \frac{1}{\sigma\sqrt{2\pi}} \exp\left[-(x-\mu)^2/2\sigma^2\right]$$

where $-\infty < x < +\infty$, μ — expected value,
σ — standard deviation of X.

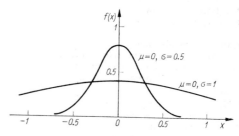

D Normalverteilung, Gaußsche Verteilung
F distribution normale, distribution de
  Laplace-Gauss
P rozkład normalny, rozkład Gaussa
  i Laplace'a
R нормальное распределение, распределение
  Гаусса

normal electrode potential → standard electrode
potential

normal electromotive force → standard electromotive force

normal equations
A system of linear equations resulting from the least squares method whose solution gives the coefficients of the multiple linear regression

$$X^T X b = X^T y$$

where X — design matrix, y — vector of values of the observed quantity corresponding to the design matrix, b — vector of regression coefficients, T — matrix transposition operation.
D Normalgleichungen
F équations normales, système d'équations normales
P równania normalne, układ równań normalnych
R нормальные уравнения

normal hydrogen electrode → standard hydrogen electrode

normality, N
The concentration expressed as the amount of gram-equivalents of a substance dissolved in one litre of a solution.
Remark: Neither the concept, nor the term have been accepted for use by the SI system.
D Normalität
F normalité
P normalność
R нормальность

normal potential → standard electrode potential

normal pulse polarography → pulse polarography

normal vibrations
In a system of material points the harmonic oscillations in which all points move with the same frequency and in a constant, time-independent phase. The normal vibrations in a molecule result from assuming that it is a classic set of material points.
D Normalschwingungen
F vibrations normales
P drgania normalne
R нормальные колебания

nuclear activation analysis → activation analysis

nuclear fission → fission

nuclear gamma-ray resonance spectroscopy → Mössbauer spectroscopy

nuclear magnetic resonance, NMR, magnetic nuclear resonance
A phenomenon exhibited by a large number of substances built of atoms whose nuclei possess a spin. When the substance is placed

in powerful uniform magnetic field and simultaneously subjected to much weaker radio-frequency alternating field at right angles to the former, absorption of radio-frequency (resonance) occurs at particular combination of the frequency and magnetic field strength with the result that nuclei are raised from the lower to the higher energy level.
D kernmagnetische Resonanz
F résonance magnétique nucléaire
P rezonans magnetyczny jądrowy
R ядерный магнитный резонанс

nuclear magnetic resonance spectrometer, NMR spectrometer
An instrument used for the excitation and registration of the nuclear magnetic resonance spectra.
D kernmagnetisches Resonanz-Spektrometer
F spectromètre de résonance magnétique nucléaire
P spektrometr jądrowego rezonansu magnetycznego, spektrometr NMR
R спектрометр ядерного магнитного резонанса, спектрометр ЯМР

nuclear magnetic resonance spectroscopy, NMR spectroscopy
A branch of spectroscopy that utilizes nuclear magnetic resonance in the study of molecular structure, autodiffusion processes, hindered rotation, and for qualitative and quantitative chemical analysis.
D Kernresonanzspektroskopie, magnetische Kernresonanzspektroskopie, NMR-Spektroskopie
F spectroscopie de résonance magnétique nucléaire
P spektroskopia jądrowego rezonansu magnetycznego, spektroskopia NMR
R спектроскопия ядерного магнитного резонанса, ЯМР-спектроскопия

nuclear purity
The absence in a material of elements of high absorption cross section for neutrons sufficient to allow its use in reactor technique.
D Kernreinheit, nukleare Reinheit
F pureté nucléaire
P czystość jądrowa
R ядерная чистота

nuclear radiation absorptiometry
A method of radiometric analysis based on measurements of the transmittance of γ-rays, X-rays, β-particles, or neutron radiation, mainly used for the analysis of two-component systems.
D Analyse durch Kernstrahlenabsorption
F analyse par absorption de rayonnement nucléaire
P metoda absorpcji promieniowania jądrowego
R метод абсорбции ядерного излучения

nuclear radiation spectrum
A diagram presenting the number of particles or photons against energy.
D Kernstrahlungsspektrum
F spectre de radiation nucléaire
P widmo promieniowania jądrowego
R спектр ядерного излучения

nucleation
The process of formation of crystallization nuclei in a solution.
D Keimbildung
F nucléation
P zarodkowanie, tworzenie się zarodków
R образование зародышей

nucleus
The smallest solid phase aggregate of atoms, molecules, or ions which is formed during a precipitation and which is capable of spontaneous growth.
D Keim
F germe
P zarodek
R зародыш

Nujol (*in IR spectroscopy*)
A trade name for a liquid parrafin used as a mulling agent.
D Nujol
F nujol
P olej parafinowy, nujol
R нийол

null-current potentiometric titration →
potentiometric titration

null-current potentiometry → potentiometry

null hypothesis, H_0
Basic statistical hypothesis verified in a given test.
D Nullhypothese
F hypothèse nulle
P hipoteza zerowa
R нулевая гипотеза

number of degrees of freedom
The number of independent random variables decreased by the number of relationships existing between these variables; e.g. the sampling variance has $n-1$ degrees of freedom because in the formula for computing the sampling variance one relationship given by the equation $\bar{x} = (x_1 + x_2 + ... + x_n)/n$ is involved.
D Anzahl der Freiheitsgrade, Zahl der Freiheitsgrade
F nombre des degrés de liberté
P liczba stopni swobody
R число степеней свободы

number of effective plates, effective (theoretical) plate number, N
(*in chromatography*)
The parameter characterizing the real column efficiency, as it corrects for the dead volume

$$N = 16 \left(\frac{t'_R}{w_b}\right)^2 = 5.545 \left(\frac{t'_R}{w_{1/2}}\right)^2 = n \left(\frac{k}{1+k}\right)^2$$

where t'_R — adjusted retention time, w_b — peak width at base, $w_{1/2}$ — peak width at half-height, n — number of theoretical plates, and k — mass distribution ratio.
D effektive Trennstufenzahl, effektive theoretische Bodenzahl
F nombre de plateaux (théoriques) effectifs
P liczba półek efektywnych, liczba półek teoretycznych efektywna
R число эффективных теоретических тарелок

number of theoretical plates, theoretical plate number, n (*in chromatography*)
The parameter characterizing column efficiency:

$$n = 16 \left(\frac{t_R}{w_b}\right)^2 = 5.545 \left(\frac{t_R}{w_{1/2}}\right)^2$$

where t_R — uncorrected retention time, w_b — peak width at base, and $w_{1/2}$ — peak width at half-height.
D theoretische Trennstufenzahl, Zahl der theoretischen Stufen, theoretische Bodenzahl, Zahl der theoretischen Boden
F nombre de plateaux théoriques
P liczba półek teoretycznych
R число теоретических тарелок

numerical titration
A tiration in which the quantity of the titrant is determined from the number of added capsules containing a known amount of the titrant. This method of titration finds application in the control of technological processes.
D numerometrische Titration
F titrage numérique
P miareczkowanie numerometryczne
R ...

O

observed current, measured current (*in electrochemistry*)
The sum of the net current and the residual current.
D ...
F courant observé, courant mesuré
P prąd obserwowany, prąd mierzony
R экспериментально измеренный ток

occlusion, molecular occlusion
The process of mechanical incorporation of foreign substances as molecular species within the precipitates as they are formed.
D Okklusion
F occlusion
P okluzja (molekularna)
R окклюзия

OC-curve → operating characteristic function

odorimetry (*in sensory analysis*)
A method of analysis of substances on the basis of their odour intensity.
D Odorimetrie
F odorimétrie
P odorymetria
R одориметрия

ohmic overpotential → pseudo resistance overpotential

ohmic polarization → pseudo resistance overpotential

oleophilic ion exchanger
A synthetic ion exchanger which swells in non-aqueous media (e.g. non-polar solvents) and whose exchange reaction rates are comparable to those observed for conventional ion exchangers in aqueous solutions.
D oleophiler Ionenaustauscher
F échangeur d'ions oléophile
P jonit oleofilowy
R олеофильный ионит, олеофильный ионообменник

one-colour indicator
An indicator which exhibits a colour on only one side of its transition interval and is colourless on the other, or which exhibits a deeper or a less intense shade of the same colour on one side of the interval.
D einfarbiger Indikator
F indicateur monocolore
P wskaźnik jednobarwny
R одноцветный индикатор

one mark pipette → transfer pipette

one-sided test
A statistical test in which the rejection region lies on one end of the distribution of the test statistic; i.e. if t is the test statistic, then this region comprises values for which either $t > t_1$ or $t < t_0$ (for given t_0 and t_1).
D einseitiger Test
F test unilatéral
P test jednostronny
R односторонний критерий

Onsager's equation → Debye-Hückel-Onsager's limiting law for conductivity

open-circuit (electrode) potential
The potential of an electrode (half-cell) in an open circuit.
D Ruhepotential
F potentiel de repos, potentiel en circuit ouvert
P potencjał (elektrody) spoczynkowy
R потенциал покоя

open-circuit potential → open circuit electrode potential

open-tube chromatography, open-tubular chromatography
Chromatography in which the inner open-tube column wall itself, a liquid or an active solid, respectively, supported on the column wall, acts as the stationary phase.
D Kapillarchromatographie
F chromatographie sur colonne capillaire
P chromatografia na kolumnach o otwartym przekroju, chromatografia na kolumnach otwartych, chromatografia kapilarna
R хроматография в открытой трубке, капиллярная хроматография

open-tube column, open tubular column
A chromatographic column, having an unobstructed axial gasflow channel and usually of capillary dimensions, in which the column wall, a liquid or an active solid supported on the column wall, acts as the stationary phase.
D offene röhrenartige Säule
F colonne tubulaire ouverte, colonne à tube ouvert
P kolumna o otwartym przekroju, kolumna o otwartym świetle, kolumna otwarta, kolumna OT
R открытая (трубчатая) колонка

open-tubular chromatography → open-tube chromatography

open tubular column → open-tube column

operating characteristic (function), OC-curve
Function $\beta(\theta_1)$ where β — probability of the error of second kind, θ_1 — value of the parameter θ postulated in the alternative hypothesis. This function expresses the probability that the null hypothesis, concerning the parameter θ in relation to the true value of the parameter, holds.
D Operationscharakteristik, OC-Funktion
F courbe caractéristique, courbe OC
P funkcja operacyjno-charakterystyczna, funkcja OC
R оперативная характеристика

operational amplifier
The constant voltage amplifier with a high amplification (e.g. 10^4) used for linear and non-linear transformation of the input signals, due to the application of a strong feedback.
D Funktionsverstärker, Operationsverstärker, Rechenverstärker
F amplificateur opérationnel
P wzmacniacz operacyjny
R операционный усилитель

operative error
An error, mostly physical in nature, associated with the manipulations of an analysis, independent of the instruments and devices employed, not related to the chemical properties of the system and its magnitude, depends more upon the analyst himself than on any other factor.
D ...
F erreur commise par l'operateur
P błąd (popełniany przez) analityka
R оперативная погрешность

optical aberration, aberration
The failure of an optical system to form an image of a point as a point, of a straight line as a straight line, and of an angle as an equal angle.
D Aberration
F aberration
P aberracja
R аберрация

optical activity
The property of certain substance to rotate the plane of polarization of plane-polarized light, or the major axis of the polarization ellipse of the elliptically polarized light, when it is passed through this substance. The property is exhibited by certain solids and liquids, and may also be observed in solutions of such substances, and in the vapour phase.

D optische Aktivität
F activité optique
P aktywność optyczna, czynność optyczna
R оптическая активность

optical axis
A line passing through a radially symmetrical optical system such that rotation of the system about this line does not alter it in any detectable way.
D optische Asche
F axe optique
P oś optyczna
R оптическая ось

optical bench
A metal bar, parallel to the optical axis of an intrument, used for mounting and moving optical elements.
D optische Bank
F banc d'optique
P ława optyczna
R оптическая скамья, рельс

optical density → absorbance

optical density → blackening

optical filter
A semi-transparent material, usually in the form of a thin plate, which eliminates (or reduces) certain waves (or frequencies) of electromagnetic radiation in the visible, ultraviolet, or infrared ranges and transmits a narrow band of the desired wavelengths.
D optisches Filter
F filtre optique
P filtr optyczny
R оптический фильтр

optical interferometer, interferometer
Any instrument whereby a beam of light from a source is split into two or more parts by partial reflections, the parts being subsequently reunited after traversing different optical paths and the two components then produce interference; used for precise measurements of wavelength, of very small distances and thicknesses, for the study of the hyperfine structure of spectral lines, and for precise determination of refractive indices.
D (optisches) Interferometer
F interféromètre optique
P interferometr optyczny
R (оптический) интерферометр

optically active substance
A substance which due to the lack of symmetry in its molecular or crystalline structure has the property of rotating the plane of polarization of polarized light.
D optisch aktive Substanz
F substance optiquement active
P substancja optycznie czynna
R оптически активное соединение

optical prism, prism
A piece of a transparent material, such as
glass, quartz or rock salt, cut in the shape of
a prism; it is used for dispersing or changing
the direction of light.
D Prisma
F prisme
P pryzmat
R призма

optical purity
The degree of contamination of one enantiomer
with the other enantiomer, expressed in
percentage of the excess of one of them with
respect to the total mixture. If the
enantiomeric mixture contains 95% of
($-$)enantiomer nad 5% of ($+$)enantiomer,
the former is said to be 90% optically pure.
D optische Reinheit
F pureté optique
P czystość optyczna
R оптическая чистота

optical rotation
Rotation of the plane of polarization of
plane-polarized light, or the major axis of the
polarization ellipse of elliptically polarized
light, when it is passed through an optically
active substance or medium. The magnitude of
the optical rotation depends on the wavelength
of the light used, the path in the medium
through which the light passes, the
temperature, the nature of the solvent, and
the concentration if the substance is dissolved
in a solution.
D optische Drehung, optisches Drehvermögen
F pouvoir rotatoire naturel
P skręcalność optyczna
R оптическое вращение

optical rotatory dispersion, rotatory dispersion,
ORD
The variation of the angle of optical rotation
of the plane of polarization of plane-polarized
light by an optically active molecule with the
wavelength of light.
D optische Rotationsdispersion
F dispersion rotatoire
P dyspersja skręcalności optycznej
R дисперсия оптического вращения

optical spectrometer, spectrometer
An instrument used for measuring the radiant
flux of electromagnetic radiation absorbed by
or emitted from a substance at one selected
wavelength within the spectral range or by
scanning the range, provided with an entrance
slit, dispersing device, one or more exit slits,
detector and display devices. The quantity
detected is a function of the radiant flux.
D Spektrometer
F spectromètre
P spektrometr
R спектрометр

optical spectroscopy, OS
A branch of spectroscopy concerned with the
measurement and interpretation of ultraviolet,
visible and infrared spectra associated with
excitations of valence electrons of atoms and
molecules, and vibrations and rotations of
molecules.
D optische Spektroskopie
F spectroscopie optique
P spektroskopia optyczna
R оптическая спектроскопия

optical spectrum
An emission or absorption spectrum of
wavelengths in the ultraviolet, visible and
infrared regions, ranging from about 10 nm
to 1mm, associated with excitations of the
valence electrons of atoms and molecules,
and vibrations and rotations of molecules.
D optisches Spektrum
F spectre optique
P widmo optyczne
R оптический спектр

optimization
The procedure with the objective to establish
conditions in which a criterion function (called
also objective function) achieves its maximum
or minimum.
D Optimierung
F optimisation
P optymalizacja
R оптимизация

optoacoustic spectroscopy → photoacoustic
spectroscopy

ORD → optical rotatory dispersion

order of reaction
The sum of the exponents to which the
concentrations or activities are raised in the
kinetic equation of a given chemical reaction.
D (kinetische) Reaktionsordnung,
   Gesamtordnung der Reaktion
F ordre d'une réaction, ordre global d'une
   réaction
P rząd reakcji
R (суммарный) порядок реакции, общий
   порядок реакции

**order of reaction with respect to a given
substance**
The exponent to which the concentration or
activity of a given substance is raised in the
kinetic equation of a given chemical reaction.
D Einzelordnung
F ordre partiel d'une réaction
P rząd reakcji względem danej substancji
R порядок реакции по данному веществу

organic ion-exchanger → ion-exchanger resin

organoleptic assessment (*in sensory analysis*)
The quality assessment effected by the senses,
with no other conditions specified.

D organoleptischer Test
F essai organoleptique
P ocena organoleptyczna
R органолептическая оценка

Orsat analyser → Orsat gas analyser

Orsat (gas) analyser
Gas analysis apparatus in which various gases
are absorbed selectively by passing them
through a series of preselected absorbents.
D Orsat-Apparat, Gasanalysenapparat nach
  Orsat
F appareil d'Orsat
P aparat Orsata
R абсорбционный газоанализатор

orthogonal design
An experimental design with an orthogonal
design matrix X, i.e. $X^T X = I$.
D orthogonaler Versuchsplan
F plan orthogonal
P plan ortogonalny
R ортогональный план

OS → optical spectroscopy

oscillographic polarograph → oscillopolarograph

oscillographic polarography →
oscillopolarography

oscillometric titration → high-frequency
conductometric titration

oscillometry → high-frequency conductometry

oscillopolarograph, oscillographic polarograph
A device for producing electrical signals and
recording on a screen of an oscilloscope
experimental curves which are usually
dE/dt versus E plots, where E − potential
of the indicator or working electrode,
t − time.
D oszillographischer Polarograph
F oscillopolarographe
P oscylopolarograf
R осциллографический полярограф

oscillopolarography, oscillographic polarography
A voltammetric method in which a sinusoidal
alternating current is applied to the indicator
electrode and the dependence of the rate of
change of the electrode potential with time
(dE/dt) on the electrode potential E is
investigated.

 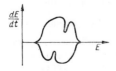

D Oszillopolarographie, oszillographische
  Polarographie
F polarographie oscillographique
P oscylopolarografia
R осциллополярография, осциллографическая
  полярография

Ostwald dilution law
The relation between the electrolytic
(concentration) dissociation constant K_c and
the molar conductivity Λ of an electrolyte of
molar concentration c; in the case of a $1:1$
electrolyte

$$K_c = \frac{\Lambda^2 c}{(\Lambda^0 - \Lambda)\,\Lambda^0} = \frac{\alpha^2 c}{1 - \alpha}$$

where Λ° − limiting molar conductivity,
α − degree of dissociation. Ostwald's dilution
law applies approximatively to diluted weak
electrolytes.
D Ostwaldsches Verdünnungsgesetz
F loi de dilution d'Ostwald
P prawo rozcieńczeń Ostwalda
R закон разведения Оствальда

Ostwald ripening
The growth of larger crystals from smaller
ones which have higher solubility than the
larger ones.
D Ostwald-Reifung
F mûrissement d'Ostwald
P dojrzewanie osadu
R оствальдовское созревание

Ostwald viscometer
A device for measuring the viscosity of liquids
based on the rate of flow through a calibrated
capillary tube.
D Ostwald-Viskosimeter
F viscosimètre d'Ostwald
P wiskozymetr Ostwalda
R вискозиметр Оствальда

outer electric potential of phase β, Volta
potential, ψ
The electric potential of phase β determined by
the total electric charge of the phase.
D äußeres elektrisches Potential der Phase β,
  Volta-Potential
F potentiel électrique externe de la phase β,
  potentiel Volta, potentiel électrique extérieur
P potencjał elektryczny zewnętrzny fazy β
R внешний электрический потенциал фазы β,
  Вольта-потенциал

outer volume → interstitial volume

outer zone (*of a flame*) → secondary combustion
zone

outlier
Result of an experiment which is probably
biased by a gross error.

D Ausreißer
F résultat aberrant
P wynik odbiegający, wynik nietypowy
R резко выделяющееся наблюдение

out-of-plane vibrations
A type of bending vibrations out of the symmetry plane of a molecule.
D Schwingungen aus der Ebene
F vibrations hors de plan
P drgania niepłaskie
R внеплоскостные колебания, неплоские колебания

over-exposure region (*of a photographic emulsion*)
The upper bended section of the emulsion characteristic curve correspo nding to the region of excessive exposure.
D Überexpositionsgebiet
F région de surexposition, zone de solarisation
P obszar prześwietleń
R область передержек

overpotential, overvoltage, η
The difference of potential of an electrode through which a current flows and that of the same electrode at equilibriu m that causes the flow of a given current.

$$\eta = E(I) - E(0) - I\Delta R$$

where $E(I)$ is the measured potential of the indicator electrode with respect to the reference electrode when the current I flows between the indicator electrode and the auxiliary electrode, $E(0)$ is the meas ured potential of the indicator electrode when the current is zero and the reaction at this electrode is at equilibrium, and ΔR is the uncompensated resistance of the cell between the indicator and the reference electrodes.
D Überspannung
F surtension, survoltage
P nadpotencjał, nadnapięcie
R перенапряжение

overtone (band), harmonic (band)
A band in the infrared spectrum of a molecule corresponding to the transitions between the ground state in which all the quantum numbers are zero to an excited state in which the vibrational quantum number changes by more than 1, e.g. by two (first overtone band), three (second overtone band).
D Oberschwingungsbande, Obserschwingung
F (bande) harmonique
P nadton, pasmo nadtonowe
R обертон

overvoltage → overpotential

oxidant, oxidizing agent, oxidizer
A substance that accepts electrons from another (oxidized) substance called the reductant; the oxidant is reduced in the electron transfer process.

D Oxidationsmittel, Oxidans
F (agent) oxydant
P utleniacz, środek utleniający
R окислитель

oxidation
The chemical reactio n involving the loss of an electron (or electrons) by a molecule (an atom or ion).
D Oxidation
F oxydation
P utlenianie
R окисление

oxidation-reduction buffer
A solution which contains a redox couple in such a concentration that its redox potential changes insignificantly on addition of a small quantity of the oxidant (reductant).
D Redoxpuffer
F solution tampon redox
P bufor redoks
R редоксибуфер

oxidation-reduction electrode → redox electrode

oxidation-reduction indicator, redox indicator
An indicator which is capable of being oxidized or reduced and which undergoes a colour change in the process at or near the equivalence-point.
D Redoxindikator
F indicateur d'oxydo-réduction, indicateur redox
P wskaźnik redoks
R окислительно-восстановительный индикатор, редокс-индикатор

oxidation-reduction potential, redox potential
The potential of an electrode (half-cell), made of platinum (or any other noble metal), in a solution containing a redox system. The redox potential of such a system is described by the Nernst equation

$$E = E° + \frac{RT}{nF} \ln \frac{a_{Ox}}{a_{Red}}$$

where $E°$ — standard potential of the given redox system ($E = E°$ when $a_{Ox} = a_{Red}$), R — gas constant, T — absolute temperature, n — number of electrons taking part in the reaction, F — Faraday's constant, a_{Ox} and a_{Red} — activities of the oxidizing and reducing agents, respectively.
D Oxidations-Reduktionspotential, Redoxpotential
F potentiel redox, potentiel d'oxydo-réduction, tension d'oxydo-réduction, tension redox
P potencjał redoks
R окислительно-восстановительный потенциал, редокс-потенциал

oxidation-reduction titration, redox titration
A titration involving the transfer of one or more electrons from a donor ion or molecule

(the reductant) to an acceptor (the oxidant)
and the related change of the degree of
oxidation.

D Oxidations-Reduktions-Titration,
   Redoxtitration
F titrage par oxydo-réduction
P miareczkowanie redoks(ymetryczne)
R окислительно-восстановительное
   титрование

oxide electrode
An electrode of the second kind which can be
prepared by atmospheric or anodic oxidation,
or by mixing the powdered metal (M) with the
powdered oxide (MO). The half-cell is
represented as $OH^-_{(aq)}|MO_{(s)}$, M. The
antimony electrode is the most common, but
silver, mercury, tungsten, molybdenum and
tellurium ions are also used.

D Metalloxidelektrode
F électrode métal-oxyde métallique
P elektroda tlenkowa
R металлооксидный электрод,
   металлоокисный электрод

oxidimetric titration
Oxidation-reduction titration in which the
reductant is titrated with a standard solution
of an oxidant. During the titration oxidation
occurs of the substance being determined.

D oxidimetrische Titration
F titrage par oxydation
P miareczkowanie oksydymetryczne
R оксидиметрическое титрование

oxidimetry
The determination of substances by titration
with standard solutions of oxidants.

D Oxidimetrie
F oxydimétrie
P oksydometria
R оксидиметрия

oxidizer → oxidant

oxidizing agent → oxidant

oxygen electrode → oxygen gas electrode

oxygen (gas) electrode
A gas electrode which consists of a noble
metal immersed in a solution of hydroxide ions
saturated with oxygen.

D Sauerstoffelektrode
F électrode à oxygène
P elektroda tlenowa
R кислородный электрод

oxygen polarographic probe → Clark
oxygen-sensing probe

oxygen sensor → Clark oxygen-sensing probe

oxy-hydrogen flame method (*in quantitative
elemental analysis*)
An effective mineralization method based on
gasification of a substance at elevated
temperature under hydrogen atmosphere with
subsequent combustion of the resulting
hydrogen-organic vapours mixture in
a counter-current of oxygen.

D Knallgasverbrennung
F méthode de combustion en chalumeau
   oxhydrique
P spalanie w płomieniu tlenowodorowym
R сожжение в водородно-кислородном
   пламени

P

packed column, classical packed column, regular packed column, conventional packed column
A chromatographic column filled with a packing, for which the ratio of the internal diameter of the column d_c to the average diameter of the solid particles d_p is greater than 10 ($d_c/d_p > 10$).
D gepackte Säule, gefüllte Säule
F colonne (régulièrmeent) remplie, colonne garnie classique
P kolumna z wypełnieniem zwykła, kolumna z wypełnieniem klasyczna
R заполненная колонка, насадочная колонка, набивная колонка

packing → column packing

paired-ion chromatography → ion-pair partition chromatography

pair escape peak
A peak in a γ-radiation spectrum caused by the formation of an electron-positron pair in the active volume of the detector, the positron annihilation and escape from the detector of one annihilation quantum (single escape peak), or two quanta (double escape peak).
D Escape-Peak
F pic de fuite, pic d'échappement
P pik ucieczki pary (elektronowej)
R пик утечки пары (электронов), пик вылета

paper chromatography, PC
Chromatography in which a paper strip or sheet acts both as the solid support and as the stationary phase.
D Papierchromatographie
F chromatographie sur papier
P chromatografia bibułowa, chromatografia na bibule
R бумажная хроматография, хроматография на бумаге

paramagnetic resonance → electron paramagnetic resonance

parameter (*in statistics*)
The term which occurs at least in the following meanings:
—a quality calculated on the basis of the results of sample investigation, e.g. median, standard deviation, slope; usually called sample parameter, sample statistic or briefly statistic;
—a quantity which is an estimator of the proper quantity in a population, called population parameter;
—a constant in a probability distribution of a random variable called distribution parameter; e.g. the normal distribution has two parameters: the expected value and the standard deviation.
D Parameter
F paramètre
P parametr
R параметр

parent ion, precursor ion (*in mass spectrometry*)
An electrically charged molecular moiety which may dissociate to form fragments, one or more of which may be electrically charged, and one or more neutral species. A parent ion may be a molecular ion or an electrically charged fragment of a molecular ion.
D Mutterion
F ion-parent
P jon macierzysty
R исходный ион

parent molecular peak → molecular peak

parent nuclide
A radionuclide which decays to produce another nuclide called daughter.
D Mutternuklid, Ausgangsnuklid
F nuclide père, parent nucléaire
P nuklid macierzysty
R материнский нуклид, исходный нуклид

Parnas-Wagner apparatus
An apparatus for steam distillation of ammonia, used for nitrogen determination. It consists of a steam generating flask, a tube for steam drying, a distillation flask, dephlegmator and condenser.
D Parnas-Wagner-Apparatur
F appareil de Parnas et Wagner
P aparat Parnasa i Wagnera
R прибор Парнаса и Вагнера

Parr bomb fusion
A rapid and effective method of decomposing organic compounds for the determination of elements such as phosphorus, boron, sulfur and halogens; the sample to be analysed is heated with sodium peroxide in a Parr bomb.
D Aufschluß in der Parr-Bombe
F fusion en bombe de Parr
P stapianie w bombie Parra
R сплавление в бомбе Парра

partial anodic/cathodic current, I_a or I_c
The faradaic current corresponding to
a specified anodic/cathodic reaction on the
electrode.

D partieller anodisch/katodischer Strom
F courant partiel anodique/cathodique
P prąd cząstkowy anodowy/katodowy
R частичный анодно-катодный ток

partial electrode reaction
Every electrode reaction of all the reactions
taking place at a given electrode which proceeds
in one direction. Such a reaction may consist
of a series of elementary steps.

D partielle Elektrodenreaktion
F réaction partielle
P reakcja elektrodowa cząstkowa
R частичная электродная реакция

particle induced X-ray emission, PIXE
Spectral emission X-ray analysis in which the
charged particles, mainly protons and α-particles,
are used for excitation of the characteristic
X-rays. This method finds application in the
analysis of thin samples, small grains, dust
particles and inclusions.

D partikel-induzierte
    Röntgenemissionsspektroskopie
F spectroscopie d'émission des rayons X
    excités par particules
P analiza spektralna rentgenowska ze
    wzbudzaniem cząstkami naładowanymi
R рентгенофлуоресцентный анализ
    с возбуждением ускоренными ионами

partition between two liquids → liquid-liquid
distribution

partition chromatography
Chromatography in which the separation of
the sample components is based on the
differences in their solubilities in the mobile
and stationary phases. The stationary phase
is either a liquid deposited on a solid support
or a sorbent in the form of a gel.

D Verteilungschromatographie
F chromatographie de partage
P chromatografia podziałowa
R распределительная хроматография

partition coefficient, K (*in gas chromatography*)
The ratio of the amount of a component
per unit volume of the stationary phase
(g cm^{-3}) to its amount per unit volume of
the mobile phase at constant temperature and
at equilibrium. The partition coefficient is
assumed to be independent of concentration
at the concentration prevailing in gas
chromatography, and is related to the net
retention volume V_N by

$$K = \frac{V_N}{V_L} = \frac{V_N \rho_L}{m_L} = \frac{V_g T \rho_L}{273}$$

where V_L — volume of the liquid stationary
phase, ρ_L — density of the liquid, m_L — mass
of the liquid, and V_g — specific retention volume.

D Verteilungskoeffizient
F coefficient de partage
P stała podziału
R коэффициент распределения

partition coefficient → distribution constant

partition constant, K_D°
The distribution constant K_D for infinite
dilution of the given compound.

D thermodynamische Verteilungskonstante
F constante de partage (thermodynamique)
P stała podziału termodynamiczna
R константа распределения

partition isotherm
The isotherm describing the dependence of
the concentration of a substance dissolved in
an organic phase on the equilibrium
concentration of that substance in an aqueous
(or other) phase, under given conditions.

D Verteilungsisotherme
F isotherme de partage
P izoterma podziału
R изотерма распределения

partition ratio → mass distribution ratio

parts per billion, ppb
The content of a constitutent in trace analysis
expressed in parts per billion.

Remark: According to the IUPAC recommendation, the
unit ppb is ambiguous and its use is discouraged.

D Milliardstel
F ...
P ...
R ...

parts per million, ppm
The content of a constituent in very small
concentrations expressed in parts per million
(by mass).

D Millionstel
F parties par million
P ...
R миллионная доля

PAS → photoacoustic spectroscopy

passivation of an electrode
Formation of layers on the surface of an
electrode which inhibit the electrode processes.

D Passivierung einer Elektrode
F passivation d'une électrode
P pasywacja elektrody
R пассивация электрода

passive experiment
An experiment in which the investigator has no
influence on the values of factors acting on the
output variable.

D passiver Versuch
F expérience passive
P eksperyment bierny
R ...

path length, sample path length, absorption path length, *b* (*in spectrophotometry*)
The distance between internal surfaces of windows of the absorption cell, or the thickness of a homogeneous, isotropic nonmetallic sample with smooth, plane, parallel surfaces. The path length is measured through the cell/sample in the direction of the beam of light which is being absorbed and is expressed in centimetres.

D Schichtdicke
F longueur du trajet d'absorption, épaisseur de l'échantillon
P grubość warstwy absorbującej, grubość warstwy próbki
R длина поглощающего слоя, толщина поглощающего слоя

P-branch, negative branch
In the rotational-vibrational band, the lower-frequency lines that result from transitions in which the rotational quantum number changes by −1 as the vibrational quantum number increases.

D P-Zweig, negativer Zweig
F branche P
P gałąź P, gałąź ujemna
R P-ветвь, отрицательная ветвь

PC → paper chromatography

peak, elution band (*in chromatography*)
A part of the differential chromatogram recording the detector response or eluate concentration while a single component emerges from the column, often similar in shape to the Gaussian curve. If separation is incomplete, two or more components may appear as one unresolved peak.

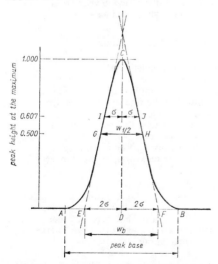

peak height at the maximum

Gaussian curve: *AB* — peak base, *EF* — peak width at base, *GH* — peak width at half-height, *CD* — peak height (at maximum), *I* and *J* — inflection points, *C* — peak maximum, and σ — standard deviation

D Peak, Pik, Berg, Elutionsbande
F pic, bande d'élution
P pik
R пик, полоса элюирования

peak area
The area lying between the peak outline and the peak base. *See* peak.

D Peakfläche, Bergfläche
F aire du pic, surface du pic
P powierzchnia piku
R площадь пика

peak asymmetry factor → asymmetry factor of a peak

peak base
The segment of the baseline between the extremities of the peak. *See* peak.

D Peakbasis, Bergbasis
F base du pic
P podstawa piku
R основание пика

peak broadening (*in chromatography*)
The broadening of the base of a peak with the decrease of its height (also the broadening of a band or spot) of a given component, due to the processes described by the van Deemter equation.

D Peakverbreiterung
F élargissement du pic
P rozmycie piku
R расширение пика

peak capacity
The number of completely resolved peaks recorded in a predetermined period of time (the term refers to the chromatographic column of defined efficiency). The peak of a non-retained compound is accepted as the first peak and the peak recorded in the predetermined period of time as the last one.

D Peakkapazität
F capacité (de séparation exprimée en nombre) de pics
P pojemność względem pików, pojemność pikowa
R ёмкость пика, пиковая ёмкость

peak current, I_p (*in electrochemistry*)
The current whose intensity is equal to the height of the peak (at maximum) recorded on the polarization curve.

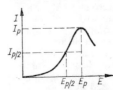

polarization curve recorded in the linear potential sweep voltammetry
$I_{p/2}$ — half-peak current, $E_{p/2}$ — half-peak potential, E_p — peak potential

D Peakstrom, Spitzenstrom
F courant de crête, courant maximum
P prąd piku
R пиковый ток

peak elution volume, elution volume, $\bar{V}$ *See* retention volume
D Peakelutionsvolumen
F volume d'élution du pic
P objętość elucji piku
R объём пика элюирования

peak height (*at maximum*), *h*
The distance between the peak maximum and the peak base, measured parallel to the axis representing the detector response. *See* peak.
D Peakhöhe
F hauteur du pic
P wysokość piku
R высота пика

peak maximum
The point on a peak at which the distance to the peak base, measured in the direction parallel to the axis representing the detector response, is maximum. *See* peak.
D Peakmaximum
F maximum du pic
P maksimum piku
R максимум пика

peak potential, E_p (*in electrochemistry*)
The value of the potential of an indicator electrode, with respect to a suitable reference electrode, at which the current reaches the value of the peak current. *See* peak current.
D Peakpotential, Spitzenpotential
F ...
P potencjał piku
R ...

peak resolution, R_n, $R_{1,2}$ (*in chromatography*)
The measure of the separation of two adjacent (Gaussian) peaks, expressed by the equation

$$R_n = \frac{t_{R2} - t_{R1}}{n(\sigma_1 + \sigma_2)}$$

where t_{R1} and t_{R2} — retention times for components 1 and 2, respectively,
n — number, greater than zero chosen arbitrarily, σ_1 and σ_2 — standard deviation of the Gaussian peaks for components 1 and 2, respectively. The separation is arbitrarily assumed to be complete when the resolution exceeds unity, $R_n \geqslant 1$. In gas chromatography and in some forms of liquid chromatography, the peak resolution is expressed by the equation

$$R_s = \frac{t_{R2} - t_{R1}}{\bar{w}_b} = \frac{2(t_{R2} - t_{R1})}{w_{b1} + w_{b2}}$$

where $\bar{w}_b$ — average peak width at base, w_{b1} and w_{b2} — peak widths at base for components 1 and 2, respectively. R_s is equal to R_n for $n = 2$.
D Peakauflösung
F résolution de deux pics
P zdolność rozdzielcza pików
R разрешение пиков, разделение пиков

peak-to-Compton ratio (*in γ-spectrum*)
The ratio of the full energy peak height to the Compton height measured at approximately 100 keV below the Compton edge.
D Peak-to-total-Verhältnis
F rapport du pic photoélectrique au Compton
P stosunek fotopiku do tła komptonowskiego
R соотношение пик-фон

peak width at base, w_b
The segment of the baseline intercepted by the tangents drawn to the inflection points on either side of the peak. *See* peak.
D Peakbreite an der Basislinie
F largeur du pic à la base
P szerokość piku przy podstawie
R ширина пика у основания

peak width at half-height, $w_{1/2}$
The length of the line parallel to the peak base that bisects the peak height, and terminates at the intersection with the two limbs of the peak. *See* peak.
D Peakbreite in halber Höhe, Halbwertsbreite des Peak
F largeur du pic à (de)mi-hauteur
P szerokość piku w połowie wysokości
R ширина пика на половине высоты

pelletizing → pellet technique

pellet technique, pelletizing (*in IR-spectroscopy*)
A technique of sampling of solids, which involves mixing the fine solid sample with potassium bromide powder, and pressing the mixture in an evacuable die at sufficient pressure to produce a transparent disc. Other alkali halides may also be used, especially CsI or CsBr for measurements at longer wavelengths.
D KBr-Preß(ling)technik
F technique de pastillage, pastillage
P pastylkowanie, metoda pastylkowania, metoda badania próbki w postaci pastylki
R суспендирование (вещества) в таблетке бромистого калия

pellicular ion exchanger, pellicular resin
An ion exchanger whose particles consist of a solid, inert, spherical core coated with a thin layer or film of an organic ion exchanger.

D Pellicular-Ionenaustauscher
F échangeur d'ion pelliculaire, résine
  pelliculaire
P jonit błonkowaty, jonit pelikularny, żywica
  błonkowata, żywica pelikularna
R плёночный ионит, плёночный
  ионообменник, пелликулярный
  ионообменник

pellicular resin → pellicular ion exchanger

pellicular support → superficially porous
support

PEM → photoelectron microscopy

peptization
Conversion of a gel or freshly precipitated
colloidal precipitate into a sol, e.g. under the
influence of washing with a pure solvent.
D Peptisation
F peptisation
P peptyzacja
R пептизация

percentage error
A relative error expressed in per cent.
D prozentualer Fehler
F erreur en pourcentage
P błąd procentowy
R процентная ошибка

per cent titrated
The ratio of the quantity of a titrated
substance which reacted with the titrant to the
initial quantity of that substance (before
titration), expressed in per cent.
D Titrationsgrad
F pourcentage de titrage
P procent zmiareczkowania
R процент титрования

periodic extraction
The extraction in which the introduction of
the substance to be extracted and the
extractant, and also the removal of the
products, are effected discontinuously.
D periodische Extraktion
F extraction périodique
P ekstrakcja okresowa
R периодическая экстракция

permanent hardness
The hardness of water remaining after boiling.
D permanente Härte, bleibende Härte
F dureté permanente
P twardość stała
R постоянная жёсткость

permeation chromatography
A form of liquid chromatography in which the
separation of the sample components is based

on the differences in the size and/or
shape of the molecules (e.g. molecular-sieve
chromatography) or in the charge
(e.g. ion-exclusion chromatography).
D Permeationschromatographie
F chromatographie par perméation,
  chromatographie de perméation
P chromatografia permeacyjna
R проникающая хроматография

permittivity, absolute permittivity, ε
The ratio of the dielectric shift to the electric
field intensity in a dielectric, or the ratio of
the capacity of a condenser filled with
a dielectric to the capacity of the empty
condenser.
SI unit: farad per metre, $F\ m^{-1}$.
D Permittivität, (absolute)
  Dielektrizitätskonstante
F permittivité (absolue)
P przenikalność elektryczna
  (bezwzględna)
R (абсолютная) диэлектрическая
  проницаемость

permselectivity
The permeation of certain ions in preference
to other ions through the ion-exchange
membrane.
D Perm(eations)selektivität
F perméation sélective
P selektywność permeacji
R селективная проницаемость

persistent lines → raies ultimes

PES → photoelectron spectroscopy

PFHS → precipitation from homogeneous
solution

pH, hydrogen ion exponent
The negative logarithm of the hydrogen-ion
(more precisely hydroxonium ion) activity
$pH = -\log a_{H_3O^+}$. Since the individual
ion-activities cannot be measured,
a conventional pH scale is used. The scale is
based on standard buffer solutions and the
electromotive force is measured with an
electrode reversible with respect to the
hydrogen ions:

$$pH_x = pH_s + \frac{(E_x - E_s)\ F}{2.303\ RT}$$

where pH_s and pH_x — the pH values for the
standard buffer and the test solution,
respectively, E_s, E_x — electromotive forces
of the standard buffer and the test solution,
respectively.
D pH(-Wert), Wasserstoffionenexponent
F pH, cologarithme de l'activité des ions
  hydrogènes
P pH, wykładnik jonów wodorowych
R pH, водородный показатель

phase analysis
The ascertainment of the qualitative and/or quantitative phase composition of a sample.
D Phaseanalyse
F analyse de phase
P analiza fazowa
R фазовый анализ

phase ratio (*in chromatography*)
The ratio of the volume of the mobile phase to the volume of the stationary phase in a column.
D Phasenverhältnis
F rapport de phase
P stosunek fazowy, iloraz fazowy, stosunek objętości faz
R отношение фаз, фазовое отношение

phase titration
Titration in which the entity being titrated is present in a two-phase (liquid) system which is caused to become a single phase at or near the equivalence-point, or one in which a monophase containing two miscible components is caused to separate into a two-phase system by addition of a third component.
D Zweiphasentitration
F titrage en présence de plusieurs phases, titrage de phase
P miareczkowanie fazowe
R фазовое титрование

phase transformation detector → transport detector

pH buffer → buffer solution

pH electrode, hydrogen ion-indicating electrode
An indicator electrode for the measurement of the pH value of a solution; the hydrogen electrode or other electrodes which respond quantitatively to variations in hydrogen ion activity, e.g. the glass pH electrode, quinhydrone electrode, antimony electrode.
D pH-Elektrode
F électrode indicatrice de pH
P elektroda pH, elektroda pehametryczna
R pH(-метрический) электрод

pH-meter
A device for measuring the electrode potential or pH values of a solution, and which is characterized by a high input impedance and large amplification step of the electric signal.
D pH-Meter, pH-Messer, pH-Anzeigegerät
F pH-mètre
P pehametr
R pH-метр

pH-metry
A method for pH measurement.

D pH-Metrie, Pehametrie
F pH-métrie
P pehametria
R pH-метрия

phosphorescence
Luminescence that continues for some time, of the order of seconds or more, after excitation is discontinued, i.e. the long-lived emission of light from an atom or molecule after it has absorbed some form of energy. Phosphorescence results from the spin-forbidden transition, e.g. triplet-singlet transition.
D Phosphoreszenz
F phosphorescence
P fosforescencja
R фосфоресценция

phosphorescence emission spectrum
A plot of the phosphorescence flux from a sample at the excitation wavelength for the analyte as a function of the wavelength of the emitted phosphorescence radiation.
D Phosphoreszenz-Emissionsspektrum
F spectre d'emission de phosphorescence
P widmo emisyjne fosforescencyjne
R фосфоресцентный эмиссионный спектр

phosphorescence excitation spectrum
A plot of the phosphorescence flux from a sample, observed at the wavelength of a phosphorescence maximum, as a function of the wavelength of the exciting radiation.
D Phosphoreszenz-Erregungsspektrum
F spectre d'excitation de phosphorescence
P widmo wzbudzenia fosforescencyjne
R фосфоресцентный спектр возбуждения

phosphorimetry
Measurements of the intensity of phosphorescent radiation emitted by an appropriately excited sample, usually cooled to liquid nitrogen temperature, also a method of chemical analysis based on the nature and intensity of the phosphorescent radiation.
D Phosphorimetrie
F phosphorimétrie
P fosforymetria
R ...

photoacoustic spectroscopy, PAS, optoacoustic spectroscopy
A spectroscopic technique whereby a chopped beam of electromagnetic radiation incident on the sample is converted to heat when absorption occurs. The heating causes pressure changes in the gas in contact with the sample which can be measured by either a microphone or by a piezoelectric transducer.
D optisch-akustische Spektroskopie
F spectroscopie opto-acoustique
P spektroskopia optyczno-akustyczna
R акустооптическая спектроскопия

photoactivation, photon activation
Activation effected by a beam of γ-quanta.

D Photoaktivierung
F photoactivation
P fotoaktywacja
R фотоактивация

photoactivation analysis, photon activation
analysis
Activation analysis in which a beam of
γ-quanta is used for activation.

D Aktivierungsanalyse mit γ-Quanten
F analyse par photoactivation
P analiza fotoaktywacyjna
R фотоактивационный анализ

photocell, photoelectric cell
Any device that converts light or other
electromagnetic radiation directly into an
electric current which flows in the external
circuit, and whose current-voltage
characteristic is a function of the incident
radiation intensity.

D Photozelle
F cellule photoélectrique
P komórka fotoelektryczna
R фотоэлемент

photocolorimeter, photoelectric colorimeter
A colorimeter wherein any of several types of
photosensitive devices are used for
measuring the fraction of the incident
radiation that is absorbed by the coloured
solutions.

D Photokolorimeter
F photocolorimètre
P kolorymetr fotoelektryczny, fotokolorymetr
R фотоколориметр

photoconductor
Semiconductor exhibiting under illumination
an increase in its electric conductivity.

D Photoleiter
F semiconducteur photoélectrique,
  photoconducteur
P fotoprzewodnik
R фотопроводник, светочувствительный
  полупроводник

photocurrent, I_t
A component of the output current of
a photodetector induced by electromagnetic
radiation.

D Photostrom, photoelektrischer Strom
F courant photoélectrique
P prąd fotoelektryczny
R фототок

photodesorption
A desorption process induced by excitation of
the adsorbate with light of wavelength
shorter than 200 nm.

D Photodesorption
F photodésorption
P fotodesorpcja
R фотодесорбция

photodetector, photoelectric detector,
photosensor
Any electronic device that detects or responds
to radiant energy.

D photoelektrisches Strahlungsmeßgerät,
  Photodetektor
F détecteur photoélectrique, photodétecteur
P detektor fotoelektryczny, fotodetektor
R фотоэлектрический детектор, фотодетектор,
  фотоприёмник

photodiode
A semiconductor diode that produces
a significant photocurrent under the action of
electromagnetic radiation.

D Photodiode
F photodiode
P fotodioda
R фотодиод

photo effect → photoelectric effect

photoelectric cell → photocell

photoelectric colorimeter → photocolorimeter

photoelectric detector → photodetector

photo(electric) effect
The absorption of a photon by an atom with
the emission of an orbital electron.

D Photoeffekt
F effet photoélectrique
P efekt fotoelektryczny
R фотоэффект

photoelectric peak, photopeak
Full energy peak recorded as a result of the
photoelectric effect.

D Photopeak, Photolinie
F pic photoélectrique, photopic, photo-ligne
P fotopik
R фотопик

photoelectron
An electron liberated by electromagnetic
radiation incident on a substance.

D Fotoelektron, Photoelektron
F photoélectron
P fotoelektron
R фотоэлектрон

photoelectron line
An analytical spectral line recorded by
a photoelectron spectrometer.

D Photoelektronenlinie
F signal de photoélectrons
P linia fotoelektryczna
R фотоэлектронная линия

photoelectron microscopy, PEM
A technique involving the irradiation of
a surface with soft X-rays or ultraviolet light
in order to generate photoelectrons, analyse
their energy and obtain maps of the
distribution of the respective surface species.
D Photoelektronenmikroskopie
F microscopie des photoélectrons
P mikroskopia fotoelektronów
R микроскопия фотоэлектронов

photoelectron spectroscopy, photoemission
spectroscopy, PES
The branch of spectroscopy concerned with
the investigation of the energy distribution of
photoelectrons emitted from materials under
irradiation with a monochromatic photon
beam.
D Photoelektronenspektroskopie
F spectroscopie des photoélectrons
P spektroskopia fotoelektronów
R фотоэлектронная спектроскопия

photoemission spectroscopy → photoelectron
spectroscopy

photofraction
The ratio of the number of counts in
a photopeak to the total number of counts in
a γ-radiation spectrum.
D Photoanteil
F photofraction
P fotofrakcja
R фоточасть

photographic plate
Glass plate coated with a photo-sensitive
emulsion which is frequently sensitized to
specific wavelength regions (ultraviolet and
visible) of electromagnetic radiation.
D Photoplatte
F plaque photographique
P płyta fotograficzna
R фотографическая пластинка

photoionization
The ionization of atoms or molecules initiated
by the absorption of photons of energy equal
to or greater than that of ionization.
D Photoionization, Photoionisierung
F photoionisation
P fotojonizacja
R фотоионизация

photoionization ion source (*in mass spectrometry*)
An ion source in which the molecules are
ionized by irradiating the sample with
ultraviolet radiation. The ultraviolet
radiation source is separated from the
ionization chamber by a lithium fluoride
window. The absorptive properties of lithium
fluoride require the sample to be ionized to
have an ionization potential lower than 11.8 eV.
D Photoionisations-Ionenquelle
F source d'ions par photoionisation

P źródło jonów wytwarzanych przez
fotojonizację
R фотоионизационный ионный источник

photoluminescence
Luminescence occurring as a result of
irradiation by visible, infrared or ultraviolet
radiation. Incident radiation is absorbed and
produces atoms and molecules in excited
states and light emission results.
D Photolumineszenz
F photoluminescence
P fotoluminescencja
R фотолюминесценция

photometer
An instrument for measuring luminous
quantity, e.g., the luminous intensity, usually
by comparing a light source with a standard
source.
D Photometer
F photomètre
P fotometr
R фотометр

photometric detector → flame photometric
detector

photometric titration
Any titration in which the titrant, a reactant
or reaction product, absorbs radiation and has
a sufficiently large molar absorptivity. During
titration the absorbance of the solution is
measured after each portion of titrant is
added and the absorbance plotted versus the
volume of titrant added. The titration curve
consists, if the reaction is complete, of two
straight lines intersecting at the end point.
D photometrische Titration
F titrage photométrique
P miareczkowanie fotometryczne
R фотометрическое титрование

photometry
The technique of measurement of light, i.e. the
measurement of quantities which are based on
the response of the human eye, such as
luminous intensity, illuminance, colour.
D Photometrie
F photométrie
P fotometria
R фотометрия

photomultiplier → photomultiplier tube

photomultiplier(tube), multiplier phototube,
electron-multiplier phototube, PMT
A photodetector, consisting of an electron
multiplier with a photocathode on the front.
The electron streem from the photocathode,
produced when the cathode is exposed to
light or other electromagnetic radiation, is
amplified by the secondary emission of
electrons and detected by the electron
multiplier.

D Photovervielfacher,
  Sekundärelektronenvervielfacher
F photomultiplicateur, tube
  photomultiplicateur
P powielacz fotoelektronowy, fotopowielacz
R фотоэлектронный умножитель,
  фотоумножитель

photon
A quantum of electromagnetic radiation.
It has an energy $E = hv$ where h — Planck
constant and v — frequency of the radiation.
Photons are usually considered as elementary
particles with rest mass equal to zero and spin
quantum number $J = 1$.
D Photon
F photon
P foton
R фотон

photon activation → photoactivation

photon activation analysis → photoactivation
analysis

photoneutron
A neutron produced in a (γ, n) reaction which
takes place when the energy of the incident
γ-quantum exceeds the neutron binding energy
in the nucleus.
D Photoneutron
F photoneutron
P fotoneutron
R фотонейтрон

photopeak → photoelectric peak

photopolarography
Investigation of photochemical processes by
means of polarography.
D Photopolarographie, Fotopolarographie
F photopolarographie
P fotopolarografia
R фотополярография

photosensor → photodetector

phototransistor
A transistor which produces an amplified
photocurrent under the action of
electromagnetic radiation, used as
a photodetector.
D Phototransistor
F phototransistor
P fototranzystor
R фототриод, фототранзистор

photovoltaic cell, barrier-layer (photo)cell,
A semiconductor device producing an
electromotive force between two layers of
different materials, when the surface is
irradiated with light or some other
electromagnetic radiation.
D Sperrschichtphotozelle, Sperrschichtelement
F cellule photovoltaïque, photopile, (photo)
  cellule à couche d'arrêt

P ogniwo fotoelektryczne, fotoogniwo,
  komórka fotowoltaiczna
R вентильный фотоэлектрический элемент,
  вентильный фотоэлемент, фотоэлемент
  с запирающим слоем

pH-sensitive glass electrode → glass pH
electrode

pH-stat
An instrument for maintaining a constant pH
of a solution by automatic addition of portions
of an acid or a base.
D ...
F ...
P pehastat, pH-stat
R pH-стат

physical adsorption → physisorption

physisorption, physical adsorption
Adsorption in which the molecules of an
adsorbate are held on the surface of an
adsorbent by intermolecular attractive forces.
Remark: The term van der Waals adsorption is not
recommended.
D physikalische Adsorption, van der
  Waalssche Adsorption
F physisorption, adsorption physique,
  adsorption de van der Waals
P adsorpcja fizyczna, adsorpcja (siłami) van
  der Waalsa
R физическая адсорбция

picogram method *See* ultramicroanalysis
D Submikromethode
F méthode picogrammique
P metoda pikogramowa
R ...

picotraces
Trace constituents whose content in a sample
ranges from 10^{-13} to 10^{-10} ppm,
i.e. from 10^{-17} to $10^{-14}\%$.
D Picospuren
F picotraces
P pikoślady
R ...

piggie
D Wägegläschen für hygroskopische
  Substanzen
F tube à tare avec pieds, pèse-substance forme
  cochonnet
P naczynko wagowe leżące
R собачка

pile-up
The addition by a radiation spectrometer of
pulses due to the simultaneous absorption of
independent particles or photons in a radiation
detector. As a result, they are counted as one
single particle or photon with energy between
the individual energies and the sum of these
energies.

D Impulsanhäufung, Aufeinandertürmen, pile-up-Effekt
F empilement d'impulsions, superposition d'impulsions
P spiętrzanie impulsów, nakładanie się impulsów
R наложение импульсов, кумуляция импульсов

pipette
D Pipette
F pipette
P pipeta
R пипетка

piston automatic burette
A precise burette with digital reading of measurement. The titrant is delivered by a piston operated by a servo-motor.
D automatische Kolbenbürette
F burette à piston à lecture numérique
P biureta automatyczna tłokowa
R автоматическая поршневая бюретка

PIXE → particle induced X-ray emission

plane grating
Diffraction grating whose surface is flat.
D Plangitter
F réseau plan
P siatka płaska
R плоская решётка

plane of polarization
In plane-polarized light, the plane in which lie the magnetic field vector and the wave propagation vector.
D Polarisationsebene
F plan de polarisation
P płaszczyzna polaryzacji
R плоскость поляризации

plane of vibration
In plane-polarized light, the plane in which lie the electric field vector and wave propagation vector.
D Vibrationsebene, Oszillationsebene
P plan de vibration
R płaszczyzna drgań
F плоскость колебаний

plane polarization, linear polarization
Polarization of light that occurs when the transverse vibrations in light are restricted, under certain conditions, to one particular plane perpendicular to the direction of propagation; the light is then said to be plane-polarized.
D lineare Polarisation
F polarisation rectiligne
P polaryzacja liniowa
R линейная поляризация

plane-polarized light, linear polarized light
See plane polarization.

D linear polarisiertes Licht
F lumière rectilignement polarisée, lumière à polarisation rectiligne
P światło spolaryzowane liniowo
R плоскостнополяризованный свет

plasma (*in spectroscopy*)
A gas which is at least partly ionized and contains particles of various types, e.g. electrons, atoms, ions, and molecules. The plasma as a whole is electrically neutral.
D Plasma
F plasma
P plazma
R плазма

plasma anodization
Production of a non-volatile oxide, halide or nitride on a solid surface exposed to an oxidizing plasma.
D Plasma-Anodisierung
F anodisation par plasma
P anodowanie plazmowe
R плазменное анодирование

plasma ashing
The process occurring when materials containing organic species are exposed to an oxidizing plasma. A useful technique for the mineralization of samples.
D Plasma-Veraschung
F combustion totale par plasma
P spopielanie plazmowe
R плазменное озоление

plasma burner → plasma jet

plasma cleaning
Slight etching of a solid surface by means of a low-pressure noble gas plasma in order to remove the surface contaminants to the gas phase.
D Plasma-Reinigung
F nettoyage par plasma
P oczyszczanie plazmowe
R плазменная очистка

plasma etching
The formation of volatile products due to interaction of a solid material with radicals generated usually by a glow discharge. Reactions of this type are frequently enhanced by the bombardment of the solid surface with ions, electrons and photons.
D Plasma-Ätzung
F décapage par plasma
P trawienie plazmowe
R плазменное травление

plasma jet
A device in which flame-like plasma is obtained when a direct current arc operating between electrodes in an enclosure is blown through an orifice from its normal discharge passage by a stream of gas. The resulting

stream of hot gases (plasma plume) emerges
from the orifice at a very high velocity.

Remark: The use of term plasma burner is discouraged by
the IUPAC nomenclature recommendations.

D Plasmabrenner
F jet de plasma, brûleur à plasma
P plazmotron łukowy, palnik plazmowy
R плазменная струя, плазменная горелка

plasma torch
A device which consists of a tube assembly
and an induction coil. It has the following
functions: to confine the plasma gas axially in
the induction coil where the plasma occurs and
to ensure that it is self-sustained after initiation,
to isolate the plasma from the induction coil
and to feed the sample into the plasma by
means of a carrier gas.

D Plasmafackel
F torche à plasma
P źródło plazmowe
R плазменный факел

plateau of a counter
The range of the supply voltage of a counter,
within which the counting rate is not
significantly dependent on the value of the
voltage applied.

D Plateau des Zählrohres,
   Zählrohrcharakteristik
F plateau du compteur, caractéristique du
   compteur
P plateau licznika
R плато счётчика, счётная характеристика

platinization of electrodes
A process of coating the surface of platinum
electrodes with a layer of platinum black
which reduces the hydrogen overpotential
evolution; used for the preparation of
hydrogen electrodes and electrodes for
conductometric measurements.

D Platinieren der Elektroden
F platinisation des électrodes
P platynowanie elektrod
R платинирование электродов

platinized asbestos
Asbestos containing from 5 to 30% of
plantinum used, e.g., as an oxidizing contact
for filling combustion tubes.

D Platinasbest
F amiante platiné
P azbest platynowany
R платинированный асбест

platinized carbon
Pelletized gas black containing 50% of
platinum, used as the tube filling for pyrolysis
of organic compounds in the determination
of oxygen.

D Platin-Kohle, Platin-Gasruß
F charbon platiné
P węgiel platynowany
R платинированный уголь

platinum electrode
An electrode consisting of a platinum wire,
plate or grid, sometimes covered with platinum
black; used in electroanalysis as the indicator,
working, or reference electrode, or for making
electrical contact.

D Platinelektrode
F électrode de platine
P elektroda platynowa
R платиновый электрод

pM
The exponent of the metal ion concentration
equal to the negative logarithm of that
concentration $pM = -\log [M^{n+}]$.

D Metallionenexponent
F pM
P pM
R ...

pM buffer → metal buffer

PMR → proton magnetic resonance

PMT → photomultiplier tube

pneumatic nebulizer
A nebulizer in which the sample is drawn up
a capillary under the action of a compressed
gas stream.

D pneumatischer Nebelblaser
F nébuliseur pneumatique
P nebulizer pneumatyczny
R пневматический распылитель

Poggendorff's compensation method
A method of measurement of the EMF
consisting in compensation of the cell EMF by
a known potential difference applied in the
opposite direction.

D Poggendorffsche Kompensationsmethode,
   Kompensationsmethode von Poggendorff
F méthode de compensation de Poggendorff
P metoda kompensacyjna pomiaru SEM,
   metoda Poggendorffa
R (компенсационный) метод Поггендорфа

pOH
Negative logarithm of the hydroxyl ion
activity $pOH = -\log a_{OH^-}$.

D pOH
F pOH
P pOH, wykładnik jonów wodorotlenowych
R pOH

point estimation
A method of estimating the unknown
parameter θ of the population consisting in
that instead of the unknown value of
parameter θ the value of its estimator,
obtained in a random sample from the
population under consideration, is used.

D Punktschätzung
F estimation ponctuelle
P estymacja punktowa
R точечная оценка

Poisson distribution
Probability distribution of the discrete random variable X whose probability function is defined by the formula

$$P(X = k) = \lambda^k \exp(-\lambda)/k!$$

for $k = 0, 1, 2, 3, \ldots$ ($\lambda > 0$).

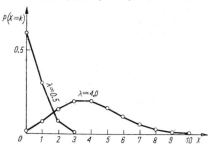

D Poisson-Verteilung
F distribution de Poisson
P rozkład Poissona
R распределение Пуассона

polarimeter
An instrument for determination of rotation of the plane of polarization of plane-polarized light when it passes through a layer of an optically active substance. It consists of a light source, polarizer, analyser, and sample tubes. The rotation is measured by rotating the analyser with respetc to the polarizer; it is also possible to measure the rotation by keeping the analyser permanently crossed with respect to the polarizer and compensating any rotation caused by the sample a piece of with quartz that rotates light in the direction opposite to that of the sample.

D Polarimeter
F polarimètre
P polarymetr
R поляриметр

polarimetry
The measurement of the change of the direction of vibration of polarized light when it interacts with an optically active substance by using a polarimeter; also a method of chemical analysis taking advantage of the properties of the optically active substance to be estimated.

Remark: The term polarimetry in its broadest sense comprises all investigations of optical phenomena in which polarized light is involved.

D Polarimetrie, Polarometrie
F polarimétrie
P polarymetria
R поляриметрия

polarization curve, current-voltage curve, voltammogram
A plot presenting the relationship between the current, or the mean current density, flowing throught the indicator electrode and the potential applied to this electrode.

D Polarisationskurve, Strom-Spannungskurve, voltammetrische Kurve
F courbe de polarisation, courbe intensité-potentiel, courbe courant-tension, courbe voltampérométrique
P krzywa polaryzacyjna, krzywa woltamperometryczna, krzywa prąd-potencjał
R поляризационная кривая, вольтамперометрическая кривая, кривая ток-напряжение, вольтамперограмм

polarization of electrode
The departure of the electrode potential from the rest potential upon passage of faradaic current. The overpotential is the measure of polarization.

D Polarisation der Elektrode
F polarisation d'électrode
P polaryzacja elektrody
R поляризация электрода

polarization of light
Restriction of the direction of vibrations in light or other electromagnetic radiation; the light is then said to be polarized.

D Polarisation des Lichtes
F polarisation de la lumière
P polaryzacja światła
R поляризация света

polarized light
A light in which the direction of the electric field vector is constant or varies in some definite way.

D polarisiertes Licht
F lumière polarisée
P światło spolaryzowane
R поляризованный свет

polarogram
The graph of the current-potential relationship obtained during a polarographic measurement.

D Polarogramm, polarographische Stromspannungskurve
F polarogramme
P polarogram, krzywa polarograficzna
R полярограмма, полярографическая кривая

polarograph
An instrument for programming the electrode potential and for recording the current-voltage relationship.

D Polarograph
F polarographe
P polarograf
R полярограф

polarographic analysis *See* polarography
D polarographische Analyse
F analyse polarographique
P analiza polarograficzna
R полярографический анализ

polarographic chronoamperometry *See*
chronoamperometry

D polarographische Chronoamperometrie
F chronoampérométrie polarographique
P chronoamperometria polarograficzna
R полярографическая хроноамперометрия

polarographic coulometry
The controlled-potential coulometry in which
the dropping mercury (or other liquid metal)
electrode is used as the working electrode.
The limiting current I_l flowing through this
electrode is measured against the time of
electrolysis t or the electric charge Q.

Remark: The terms microcoulometry and millicoulometry
are not recommended. The term dropping electrode
coulometry is more specific than the recommended one and
may be used when appropriate.

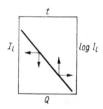

D polarographische Coulometrie,
  Mikrocoulometrie, Millicoulometrie
F coulométrie polarographique,
  microcoulométrie, millicoulométrie
P kulometria polarograficzna,
  mikrokulometria, milikulometria
R полярографическая кулонометрия,
  микрокулонометрия, милликулонометрия

polarographic diffusion current constant →
limiting current constant

polarographic maxima
The distortion of the polarographic wave by
more or less pronounced current maxima.
The maxima vary in shape from sharp peaks
to rounded humps and in all cases the current
increases abnormally until a critical value is
reached and then rapidly decreases to its
normal value.

D polarographische Maxima
F maxima polarographiques, maximums
  polarographiques
P maksima polarograficzne
R полярографические максимумы

polarographic titration *See* amperometric
titration with a dropping mercury electrode

D polarographische Titration
F titrage polarographique
P miareczkowanie polarograficzne
R полярографическое титрование

polarographic wave (*in* dc *polarography*)
The part of a polarographic curve
coresponding to a particular electrode

reaction; the height of the wave, equal to the
limiting current I_l of a substance under
investigation, is proportional to the
concentration of this substance.

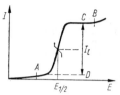

AB — polarographic wave, CD — step height,
$E_{1/2}$ — half-wave potential

D polarographische Stufe, polarographische
  Welle
F vague polarographique
P fala polarograficzna
R полярографическая волна,
  полярографический сдвиг

polarography
The voltammetry in which the liquid electrode
whose surface is periodically or continuously
renewed, e.g. dropping mercury electrode, is
used as the indicator electrode.

Remark: In a narrow sense, the term polarography is often
used to denote dc polarography; in chemical analysis
the term polarographic analysis is used.

D Polarographie
F polarographie
P polarografia
R полярография

polarography with linear current sweep →
current-scanning polarography

**polarography with superimposed periodic
voltage** → alternating-current polarography

polarometric titration *See* amperometric
titration with a dropping mercury electrode

D polarometrische Titration
F titrage polarométrique
P miareczkowanie polarometryczne
R полярометрическое титрование

polyacid
The condensation product of a number of
simple molecules of an oxy-acid which
contains more than one mole of the acid
anhydride per mole of water. Also an acid
whose molecule contains several hydrogen
atoms replaceable by bases.

D Polysäure
F polyacide
P polikwas
R поликислота

polychromatic radiation
Any electromagnetic radiation consisting of
a mixture of wavelengths, or a beam of
particles of a variety of energies (or containing
different type of particles).

D polychromatische Strahlung
F radiation polychromatique
P promieniowanie polichromatyczne
R полихроматическое излучение

polyelectrolyte
A macromolecular substance which dissociates
in an ionizing solvent with the formation
of multiple charge polyions (polyanions,
polycations, or zwitter-polyions) and an
equivalent number of mobile counterions
with a small opposite charge.
Remark: Dissociation of the polyelectrolyte entirely to
polycations and polyanions is also possible.

D Polyelektrolyt
F polyélectrolyte, électrolyte polimérique
P polielektrolit
R полиэлектролит

polyfunctional ion exchanger
An ion exchanger containing ionogenic groups
of more than one type.

D polyfunktioneller Ionenaustauscher
F échangeur d'ions polyfonctionnel
P wymieniacz jonowy wielofunkcyjny
R полифункциональный ионообменник

population
A set of elements (individuals, objects, events,
measurements) characterized by at least one
common property distinguishing these elements
from those which do not belong to this set and
at least one property that distinguishes the
elements within the set.

D Grundgesamtheit
F population (générale)
P populacja (generalna), zbiorowość generalna
R популяция, генеральная совокупность

Porapak
The trade name for porous synthetic polymers
obtained by copolymerization of styrene and
divinylbenzene. Porapaks have specific surface
areas of $50-300$ m^2 g^{-1} and they are
thermally stable up to $250-300°$C; they are
used in gas chromatography as column
packings.

D Porapak
F Porapak
P Porapak
R порапак

pore volume → stationary liquid volume

porode → porous cup electrode

porous cup electrode, porode (*in arc and spark
spectroscopy*)
A graphite supporting electrode shaped in the
form of a cup with a porous bottom through
which the sample solution penetrates slowly
into the interelectrode space.

D Sickerelektrode
F électrode poreuse, porode

P elektroda przesączalna, elektroda porowata
R пористый электрод

porous layer support → superficially porous
support

positive branch → R-branch

positive ion (*in mass spectrometry*)
An atom, radical, molecular, moiety which
having lost one or more electrons has thereby
obtained an electrically positive charge.
Remark: The use of the term cation as an alternative is not
recommended.

D positives Ion
F ion positif
P jon dodatni
R положительный ион

positive-ion mass spectrum
A mass spectrum of positive ions.

D Massenspektrum positiver Ionen
F spectre de masse d'ions positifs
P widmo mas jonów dodatnich
R масс-спектр положительных ионов

postprecipitation
The subsequent precipitation of a chemically
different species upon the surface of an initial
precipitate usually, but not necessarily,
including a common ion.

D Nachfällung
F postprécipitation
P strącanie następcze
R последующее осаждение

potassium bromide disk, potassium bromide
pelet, potassium bromide wafer
(*in IR-spectroscopy*)
A pelet obtained by pressing a solid sample
with potassium bromide in order to measure
its infrared spectrum; potassium bromide does
not adsorb in the mid-infrared region.

D Kaliumbromid-Tablette,
  Kaliumbromidscheibe, KBr-Preßling
F pastille de bromure de potassium,
  échantillon pastillé
P pastylka z bromku potasu
R таблетка бромистого калия

potassium bromide pelet → potassium bromide
disk

potassium bromide wafer → potassium bromide
disk

potential (at the point) of zero charge, zero-point
potential, E^z
The electrode (half-cell) potential at which the
external electrode potential is equal to zero.

D Nulladungspotential
F potentiel de charge au zéro
P potencjał ładunku zerowego
R потенциал нулевого заряда

potential-determining reaction
The reaction occurring at the electrode-electrolyte interface which determines the potential of the electrode material.
D potentialbestimmende Reaktion
F ...
P reakcja potencjałotwórcza
R потенциалопределяющая реакция

potential determining reaction → electrode reaction

potential of zero charge → potential at the point of zero charge

potential-step chronocoulometry → chronocoulometry

potentiometer (*in electrochemistry*)
An instrument for measurement of the e.m.f. of galvanic cells, e.g. the potentiometric bridge of the voltmeter with high input impedance (pH-meter, digital voltmeter).
D Potentiometer
F potentiomètre
P potencjometr
R потенциометр

potentiometric analysis *See* potentiometry
D potentiometrische Analyse
F analyse potentiométrique
P analiza potencjometryczna
R потенциометрический анализ

potentiometric coulometric titration *See* controlled-current coulometry
D potentiometrische coulometrische Titration
F titrage coulométrique potentiométrique
P miareczkowanie kulometryczne potencjometryczne
R потенциометрическое кулонометрическое титрование

potentiometric selectivity coefficient
(*for ion-selective electrodes*), $k_{A,B}^{pot}$
The coefficient in the Nikolsky equation determining the capacity of an ion-selective electrode to react to activity changes of a given electroactive substance in the presence of other (interfering) substances.
Remark: The term selectivity constant is not recommended.
D Selectivitätskonstante
F coefficient de sélectivité potentiométrique, constante de sélectivité, facteur de sélectivité
P współczynnik selektywności potencjometryczny
R потенциометрический коэффициент селективности, константа селективности, фактор селективности

potentiometric titration
Titration of which the end-point is determined from the abrupt change of the potential E of an indicator electrode measured against a suitable reference electrode under the conditions of no current ($I = 0$) flowing through the electrode.
Remark: The terms zero-current potentiometric titration and null-current potentiometric titration are not recommended.

V – volume of a titrant

D potentiometrische Titration, potentiometrische Indikation, Nullstrom-potentiometrische Titration
F titrage potentiométrique, titrage potentiométrique à courant nul
P miareczkowanie potencjometryczne
R потенциометрическое титрование

potentiometry
A group of electrochemical methods consisting in measurements of the potential of the indicator electrode E as a function of the logarithm of concentration $\log c$ of an electroactive substance under currentless conditions (when no current flows through the external circuit).
Remark: In application to chemical analysis the term potentiometric analysis is used;
the terms zero-current potentiometry and null-current potentiometry are not recommended.

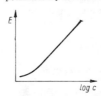

D (Nullstrom-)Potentiometrie
F potentiométrie (à courant nul), potentiométrie à intensité nulle
P potencjometria
R потенциометрия

potentiostat
An instrument which automatically maintains the potential of the working electrode constant at a chosen value.
D Potentiostat
F potentiostat
P potencjostat
R потенциостат

potentiostatic method
An electroanalytical method conducted under the conditions in which the indicator electrode is maintained at a suitable constant potential, and the current flowing through the electrode is measured as a function of time.

D potentiostatische Impulsmethode, potentiostatische Methode
F méthode potentiostatique
P metoda potencjostatyczna
R потенциостатический метод

powder funnel, solid addition funnel
A funnel with a short wide stem.
D Pulvertrichter
F entonnoir à poudre, entonnoir à solide
P lejek do materiałów sypkich
R воронка для порошков

power of a test
The probability which equals $(1 - \beta)$, where β — probability of the error of second kind.
D Macht eines Tests
F puissance d'un test
P moc testu
R мощность критерия

ppb → parts per billion

ppm → parts per million

practical specific capacity, Q_A (*of an ion exchanger*)
The total amount of ions, expressed in milliequivalents or millimoles, taken up per 1 g of dry ion exchanger under given conditions; (Q_A should always be given together with the specification of conditions).
D praktische spezifische Ionenaustauscherkapazität
F capacité spécifique pratique
P zdolność wymienna praktyczna, zdolność wymienna robocza, pojemność wymienna właściwa praktyczna (*względem określonego jonu*)
R практическая удельная ёмкость

pre-arc period (*in emission spectrochemical analysis*)
The time elapsed from the beginning of the arc discharge to the moment registration of the spectrum is started.
D Vorbrennungszeit
F temps de préflambage
P czas przedpalenia
R время предварительного обжига

precipitate
The solid phase formed inasolution.
D Niederschlag
F précipité
P osad
R осадок

precipitate-based ion-selective electrode,
precipitate-type membrane electrode, precipitate-impregnated electrode
An ion-selective electrode in which a solid-state membrane is formed from a finely divided sparingly soluble salt or from

a homogenous mixture of salts. An active electrode material can also be incorporated into the inert support.
D Niederschlagsmembran-Elektrode
F ...
P elektroda jonoselektywna z trudno rozpuszczalnym materiałem aktywnym
R осадочный мембранный электрод, ионоселективный электрод с осадочной мембраной

precipitate-impregnated electrode → precipitate-based ion-selective electrode

precipitate-type membrane electrode → precipitate-based ion-selective electrode

precipitating agent
A reagent which upon addition to a solution causes precipitation.
D Fällungsmittel
F réactif précipitant
P odczynnik strącający
R осаждающий реактив

precipitation
The formation of a precipitate.
D Fällung, Präzipitation
F précipitation
P strącanie, wytrącanie
R осаждение

precipitation from homogeneous solution, PFHS
The formation of a precipitate which is generated homogeneously and, usually, slowly by a precipitating agent within a solution.
D Fällung aus homogener Lösung
F précipitation (en milieu) homogène
P strącanie z roztworu jednorodnego, strącanie homogeniczne
R гомогенное осаждение, осаждение из гомогенного раствора

precipitation indicator
An indicator which precipitates from solution in a readily visible form at or near the equivalence-point.
D Fällungsindikator
F indicateur par précipitation
P wskaźnik strąceniowy
R осадительный индикатор

precipitation titration
A titration in which the entity being titrated is precipitated from solution by reaction with the titrant.
D Fällungstitration
F titrage par précipitation
P miareczkowanie strąceniowe
R осадительное титрование

precipitation titration curve
A diagram describing the concentration of the being determined ion X, (expressed as the

concentration exponent pX = −log [X]) as
a function of the volume of the standard
solution of a precipitating agent added.
D Fällungstitrationskurve
F courbe de titrage de précipitation
P krzywa miareczkowania strąceniowego
R кривая титрования с осаждением

precision
The agreement between the results of repeated
measurements.
D Präzision
F précision
P precyzja
R точность

precision null-point potentiometry →
differential potentiometry

precision of a balance
The standard deviation of the balance for
a given load. A statement regarding the
procedure, conditions, and experience of the
observer should be included.
D Präzision der Waage
F précision d'une balance
P precyzja wagi
R точность весов

precision of a weighing
The standard deviation of the balance for
a given load dependent on the method of
weighing and on the precision of the balance.
D Präzision der Wägung
F précision d'une pesée
P precyzja ważenia
R точность взвешивания

precursor ion → parent ion

preferential sputtering
The preferential removal of atoms of a certain
surface component due to differences in the
sputtering yields of the target components.
D bevorzugte Zerstäubung
F pulvérisation préférentielle
P rozpylanie preferencyjne, rozpylanie
   selektywne
R преимущественное распыление

Pregl procedure
Classical method of quantitative elementary
analysis of organic compounds for
determination of (inter alia): carbon, hydrogen,
halogens and sulfur, involving the combustion
of a sample in a slow stream of oxygen in
a horizontally placed tube, in the presence of
oxidizers or a platinum catalyst, and
ultimately ended by gravimetric determinations.
D Pregl-Verfahren der Elementarmikroanalyse
F procédé de Pregl de microanalyse
   élémentaire
P mikroanaliza elementarna według Pregla
R элементный микроанализ методом Прегля

premix(ed gas) burner (*in flame spectroscopy*)
A burner in which the fuel and oxidant are
thoroughly mixed before they leave the burner
ports and enter the primary-combustion zone
of the flame.
D Mischkammerbrenner
F brûleur à mélange préalable
P palnik ze wstępnym mieszaniem
R горелка с предварительным смешением

pre-spark period (*in emission spectrochemical
analysis*)
The time elapsed from the beginning of spark
discharge to the moment registration of the
spectrum is started.
D Vorfunkzeit
F temps de pré-étincelage
P czas przediskrzenia
R время предварительного обыскривания

pressure-gradient correction factor,
compressibility correction factor, *j* (*in gas
chromatography*)
A correction factor allowing for the
compressibility of the mobile phase. For
a homogeneously filled column of uniform
diameter pressure-gradient correction factor
is given by

$$j = \frac{3}{2} \frac{(p_i/p_o)^2 - 1}{(p_i/p_o)^3 - 1}$$

where p_i − pressure of the carrier gas at the
column inlet and p_o − pressure of the carrier
gas at the column outlet.
D Druckgradient-Korrekturfaktor,
   Korrekturfaktor für den Druckabfall
F facteur de correction du gradient de pression,
   facteur (de correction) de compressibilité,
   facteur de James et Martin
P współczynnik korekcyjny gradientu ciśnienia
R поправочный фактор на градиент давления

primary cell
An electrochemical cell that delivers electric
current as a result of a spontaneous
electrochemical reaction. The cell acts as
a source of electricity until its materials are
exhausted and cannot be recharged, e.g. the
Leclanché cell.
D Primärelement, Primärzelle
F pile primaire, élément primaire
P ogniwo pierwotne
R первичный элемент

primary combustion zone, primary reaction
zone, inner zone (*of a flame*)
The zone of the flame, situated directly above
the burner top, in which the combustion
process takes place.
D Innenkegel, innerer Kegel, primäre
   Reaktionszone
F zone de combustion primaire, zone interne,
   cône interne

P obszar wstępnego spalania, stożek
wewnętrzny
R зона первичного сгорания, первичная
реакционная зона, внутренняя зона

primary reaction zone → primary combustion
zone

primary sample
A part of material taken only once from
a single site of the bulk of a loose (unpacked)
product, or from a single site of a unit package.
D Stichprobe
F prélèvement élémentaire, échantillon brut
P próbka pierwotna
R первичная проба, частичная проба

primary source → light source (*in atomic
absorption and atomic fluorescence
spectroscopy*)

primary standard (*in titrimetric analysis*)
A substance of high purity used to prepare
a titrant.
D primäre Urtitersubstanz
F substance étalon primaire
P substancja podstawowa pierwotna
R первичный эталон

primary X-radiation, primary X-rays
The X-radiation emitted by a suitable source
(e.g. an X-ray tube or radioisotopic source)
used for excitation of fluorescent X-rays.
D primäre Röntgenstrahlen,
Röntgenprimärstrahlung
F rayons X primaires
P promieniowanie rentgenowskie pierwotne,
promieniowanie rentgenowskie wzbudzające
R первичное рентгеновское излучение

primary X-rays → primary X-radiation

prism → optical prism

probability
In simple terms probability may be defined
as the measure of the possibility of a given
random event occurring. There are different
definitions of probability, e.g. axiomatic,
classical, geometrical, statistical.
Axiomatic definition:
(1) To each random event A corresponds
a value $P(A)$ called the probability of the
a value $P(A)$ called the probability of
that event which satisfies the inequality
$0 \leqslant P(A) \leqslant 1$
(2) The probability of a sure event equals 1.
(3) The probability of finite or countable sum
of pairwise excluding random events is
equal to the sum of the probabilities of
these events.
Classical definition:
The probability $P(A)$ of a random event A
is given by the formula $P(A) = m/n$, where
m — number of elementary events in which

the event A occurs, n — finite number of all
possible elementary events.
D Wahrscheinlichkeit
F probabilité
P prawdopodobieństwo
R вероятность

probability density function (*of a continuous
random variable*)
First derivative of a distribution function of
a continuous random variable.
D Wahrscheinlichkeitsdichte, Verteilungsdichte
F densité de probabilité
P gęstość prawdopodobieństwa
R плотность вероятности, плотность
распределения

probability distribution (*of a discrete random
variable*)
A function determining the probability that
a discrete random variable X assumes
a particular value x_i, i.e.

$$P(X = x_i) = pl, \qquad i = 1, 2, ..., n$$

where $p_i > 0$ and $\sum_i p_i = 1$.

D Wahrscheinlichkeitsverteilung
F distribution de probabilité
P rozkład prawdopodobieństwa
R вероятностное распределение

probability theory
A branch of mathematics concerned with the
investigation of random events, especially with
the calculation of the probability distributions
of random variables.
D Wahrscheinlichkeitsrechnung,
Wahrscheinlichkeitstheorie
F calcul des probabilités, théorie des
probabilités
P rachunek prawdopodobieństwa, teoria
prawdopodobieństwa
R исчисление вероятностей, теория
вероятностей

process analysis
The analysis carried out for controlling
a technological process by a very rapid method.
D Prozeßanalyse
F analyse de plate-forme
P analiza ruchowa, analiza międzyoperacyjna
R производственный анализ

product ion → daughter ion

programmed-current chronopotentiometry
A modification of chronopotentiometry in
which a nonlinear but monotonic current I that
varies as a known function of time is imposed
on the indicator electrode and the potential E
of this electrode is measured as a function of
the electrolysis time t.

Remark: The nature of the current-time dependence must
be specified separately.

 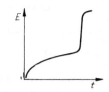

D stromprogrammierte Chronopotentiometrie, Chronopotentiometrie mit programmiertem Strom
F chronopotentiométrie à courant programmé
P chronopotencjometria z programowanym prądem
R хронопотенциометрия с заданной развёрткой тока

programmed temperature gas chromatography → temperature-programmed chromatography

prompt gamma radiation (in activation analysis)
The γ-radiation emitted in a nuclear reaction, other than the radiation emitted by the radioactive products of that reaction.
D prompte γ-Strahlung
F rayonnement γ instantané
P promieniowanie γ natychmiastowe
R мгновенное γ-излучение

prompt neutrons
Fission neutrons emitted in a time shorter than 10^{-13} s. They constitute about 99% of all neutrons emitted in the fission reactions.
D prompte Neutronen
F neutrons instantanés
P neutrony natychmiastowe
R мгновенные нейтроны

proportional counter (tube)
A counter tube operated under such conditions that the magnitude of each pulse is proportional to the number of ions produced in the initial ionizing effect.
D Proportionalzählrohr, Proportionalzähler
F compteur proportionnel
P licznik proporcjonalny
R пропорциональный счётчик

proportional gas-flow counter
A proportional counter with a forced continuous flow of gas through its active volume, used for detection of low energy radiation.
D Proportionalgaszähler, Durchflußzählrohr
F compteur proportionnel de courant gazeux
P licznik proporcjonalny przepływowy
R проточный пропорциональный газовый счётчик

protogenic solvent, proton donor solvent, acidic solvent
A solvent which supplies proton(s) to a dissolved substance, e.g. acetic acid.

D Donorlösungsmittel, saueres Lösungsmittel
F solvant acide, solvant proton-donneur
P rozpuszczalnik protogenny
R протоногенный растворитель, кислый растворитель

protolytic solvent
A solvent which adds and/or liberates protons.
D protolytisches Lösungsmittel
F solvant protolytique
P rozpuszczalnik, protolityczny, rozpuszczalnik czynny
R протолитический растворитель

proton acceptor solvent → protophilic solvent

proton donor solvent → protogenic solvent

proton magnetic resonance, PMR, ^{1}H-NMR
The most widely used technique of the nuclear magnetic resonance for investigation of organic compounds, involving the study of chemical shifts between the non-equivalent protons present in the molecule being examined.
D Protonenresonanz
F résonance magnétique nucléaire du proton
P rezonans magnetyczny protonowy
R протонный магнитный резонанс

protophilic solvent, proton acceptor solvent, basic solvent
A solvent with affinity to protons, e.g. liquid ammonia, water, ether.
D protonophiles Lösungsmittel, Akzeptorlösungsmittel, basisches Lösungsmittel
F solvant protophile, solvant proton-accepteur, solvant basique
P rozpuszczalnik protonofilowy, rozpuszczalnik protono-akceptorowy, rozpuszczalnik zasadowy
R протофильный растворитель, основный растворитель

proximate analysis → short analysis

pseudo-resistance overpotential, ohmic overpotential, ohmic polarization
The ohmic drop of potential in a layer of the electrolyte between the indicator electrode and the reference electrode (or a tip of the Luggin capillary) appearing during the flow of current through an electrolytic cell.
D ...
F surtension de résistance ohmique
P nadpotencjał pseudo-oporowy
R омическое перенапряжение, омическая поляризация

PTGC → temperature-programmed chromatography

pulse amplitude analyser, pulse height analyser
A device for determining the distribution function of a set of pulses in terms of their amplitude.

D Impulshöhenanalysator
F analyseur d'amplitudes
P analizator wysokości impulsów
R амплитудный анализатор импульсов

pulse amplitude analysis, pulse height analysis
The segregation of electrical pulses according
to their amplitudes and counting the pulses
with amplitudes within the intervals of
segregation (channels); used mainly for
determination of the amplitude distribution of
pulses from an energy sensitive detector.
D Impulshöhenanalyse
F analyse d'amplitude
P analiza amplitudy impulsów, analiza
wysokości impulsów
R анализ амплитуды импульсов

pulse height analyser → pulse amplitude
analyser

pulse height analysis → pulse amplitude
analysis

pulse polarography, normal pulse polarography
Polarography in which the pulses of potential E_t
of a small amplitude linearly increasing in
time from the base potential E_0, is used. The
current flowing through the indicator electrode
is measured at the end of each pulse. There is
one pulse to one drop.

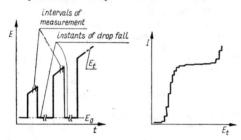

D (normale) Puls-Polarographie
F polarographie à impulsions
P polarografia (im)pulsowa (normalna)
R (нормальная) импульсная полярография

pure reagent
A reagent of the lowest purity grade whose use
in analysis requires a blank test.
D reines Reagens
F réactif pur
P odczynnik czysty
R чистый реактив

purity of a reagent
The purity expressed by the content
(in percentage by weight) of impurities present
in the reagent.
D Reinheit des Reagens
F pureté d'un réactif
P czystość odczynnika
R чистота реагента

pycnometer, density bottle, specific gravity
bottle
A small bottle of accurately determined
volume, with ground stopper or thermometer,
used for determination of the density of
liquids.
D Pyknometer, Wägefläschchen
F picnomètre, pycnomètre
P piknometr
R пикнометр

pyrochemical analysis
An identification method involving heating of
a solid substance or its mixture with an
appropriate reagent, and observing the
consecutive changes, e.g. the appearance of
a speck of mercury resulting from heating
mercuric nitrate with sodium carbonate.
D ...
F analyse pyrognostique
P analiza pirochemiczna
R пирохимический анализ

pyrolysis
The process of thermal decomposition of
a chemical compound resulting in the
formation of low molecular mass products.
D Pyrolyse
F pyrolyse
P piroliza
R пиролиз

pyrolyzer
A device used to carry out pyrolysis.
D Pyrolyser
F pyrolyseur
P pirolizer
R пиролизёр

Q

Q-branch, zero branch
The lines in rotational-vibration band that
result from transitions in which the rotational
quantum number does not change as the
vibrational quantum number increases.

D Q-Zweig, Nullzweig
F branche Q, raie zéro
P gałąź Q, gałąź zerowa
R Q-ветвь, нулевая ветвь

quadrupole (*electric or magnetic*)
The system of two electric dipoles (electric
quadrupole) or magnetic dipoles (magnetic
quadrupole) which does not display a resultant
dipole moment. It is the source of an electric
or magnetic field.

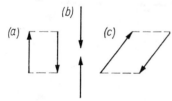

D Quadrupol
F quadripôle, quadrupôle
P kwadrupol
R квадруполь

quadrupole mass analyser, quadrupole mass
spectrometer
A mass spectrometer in which the ions
produced are passed along the centre axis of
four, parallel circular or hyperbolic quartz
rod. The rods are connected in pairs to
radio-frequency and direct-current supplies.
For any specified combination of rf and dc
voltages, ions with only a certain mass-to-charge
ratio traverse the path between rods without
striking an electrode.

D Quadrupolmassenspektrometer
F analyseur de masse quadripolaire
P analizator kwadrupolowy
R квадрупольный масс-анализатор

quadrupole mass spectrometer → quadrupole
mass analyser

quadrupole relaxation (*in nuclear magnetic
resonance*)
The relaxation caused by interaction of the
electric quadrupole moments of the nuclei
with the surrounding heterogeneous electric
field.

D Quadrupolrelaxation
F relaxation quadrupolaire
P relaksacja kwadrupolowa
R квадрупольная релаксация

qualitative analysis
The ascertainment of the qualitative
composition of a sample or the detection and
identification of the components of the sample
investigated.

D qualitative Analyse
F analyse qualitative
P analiza jakościowa
R качественный анализ

qualitative analysis of organic compounds
Methods of analysis involving: determination
of the elementary composition of the
investigated compound, establishing its molar
mass and empirical formula, identification of
the functional groups present in the molecule,
collecting information about the neighbourhood
of the functional groups and about the mutual
location of the molecule moieties, establishing
the structure of a compound by various
chemical and physical methods, finally, the
identification of the compound analysed.

D qualitative organische Analyse
F analyse organique qualitative
P analiza jakościowa związków organicznych
R качественный органический анализ

qualitative composition (*of a sample*)
The chemical constituents present in a sample.

D qualitative Zusammensetzung
F composition qualitative
P skład jakościowy
R качественный состав

quantile of a probability distribution
Value x of the random variable satisfying
the inequalities

$$P (X \leqslant x_p) \geqslant p, \qquad P (X \geqslant x_p) \geqslant 1-p$$

where X — random variable, P — probability,
p — order of the quantile $(0 < p < 1)$.

D Quantil (der Ordnung p)
F quantile (d'ordre p) de distribution de
probabilité
P kwantyl (rzędu p) rozkładu
prawdopodobieństwa
R квантиль (вероятностного распределения)

quantitative analysis
The determination of the quantitative
composition of a sample or the determination
of some constituents in it.

D quantitative Analyse
F analyse quantitative
P analiza ilościowa
R количественный анализ

quantitative analysis of organic compounds
Determination of the elemental composition
and molar mass of the compound analysed,
followed by identification of the functional
groups present, carried out by various chemical
and physical methods.

D quantitative organische Analyse
F analyse organique quantitative
P analiza ilościowa związków organicznych
R количественный анализ органических
соединений

quantitative composition (*of a sample*)
The proportion in which the chemical
individuals are present in a sample, expressed
usually in percentage.

D Mengenverhältnis
F composition quantitative
P skład ilościowy
R количественный состав

quantitative precipitation
The precipitation resulting in a (practically)
complete removal of a constituent from
solution.

D quantitative Fällungsreaktion
F précipitation quantitative
P strącanie ilościowe
R количественное осаждение

quantity of light → luminous energy

γ-**quantum** → gamma quantum

quenching
Any reduction in the intensity of fluorescence,
or phosphorescence, of a compound under
a given set of conditions. Some fluorescent
materials can be quenched by changes in the
chemical nature of the substance, e.g. pH of
solution, or by addition of another compound,
e.g. a transition metal with unfilled outer *d*
orbitals. Quenching can also be caused by
nonradiative loss of energy from the excited
molecules. The quenching agent may facilitate
conversion of the molecules from the excited
singlet to a triplet level from which emission
cannot occur.

D Löschung
F quenching
P wygaszanie
R тушение

quinhydrone electrode
A redox electrode consisting of an inert
electrode, e.g. platinum or gold, dipped into
a solution which is saturated with quinhydrone,
i.e. an equimolar mixture of quinone (Q) and
hydroquinone (H_2Q). The half-cell is
represented as H^+, H_2Q, $Q|Pt$; used as
a hydrogen ion indicating electrode in the pH
range $0-8.5$.

D Chinhydronelektrode
F électrode à quinhydrone
P elektroda chinhydronowa
R хингидронный электрод

R

radial development → circular chromatography

radiant energy, Q_e
The energy transmitted by waves through space or some medium without involving the transfer of matter; when unqualified, usually refers to electromagnetic radiation

$$Q_e = \int_0^t \Phi_e \, dt$$

where Φ_e — radiant flux and t — time.
SI unit: joule, J.

D Strahlungsenergie
F énergie rayonnante
P ilość energii promienistej
R энергия излучения

radiant energy flux → radiant flux

radiant energy source → radiation source

radiant exposure, exposure, H_e
A measure of the total radiant energy incident on a surface per unit area; equal to the integral over exposure time t of the irradiance E_e

$$H_e = \int_0^t E_e \, dt$$

SI unit: joule per square metre, J m^{-2}.

D Bestrahlung(sdauer), Exposition
F exposition (énergétique)
P naświetlenie
R экспозиция, количество освещения

radiant flux, radiant power, radiant energy flux, Φ_e
The time rate of flow of radiant energy, or the total radiant energy emitted by a source of electromagnetic radiation per unit time.
SI unit: watt, W.

D Strahlungsleistung, Strahlungsfluß
F flux énergétique, puissance rayonnante
P strumień promieniowania, moc promieniowania
R поток излучения, мощность излучения

radiant flux density → irradiance

radiant power → radiant flux

radiation
The emission of any rays, wave motion or particles (e.g. α-particles, β-particles, neutrons) from a source; it is usually applied to the emission of electromagnetic radiation.

D Strahlung, Radiation
F radiation
P promieniowanie
R излучение, лучеиспускание, радиация

radiation counter
Radiation measuring assembly comprising a radiation detector in which the individual ionizing events cause electrical pulses and the associated equipment for processing and counting the pulses.

D Strahlenzähler, Strahlenzählrohr
F compteur de rayonnement, tube de rayonnement
P licznik promieniowania
R счётчик радиоактивного излучения, счётчик излучений

radiation detector
A device for detection and registration of radiation in which the radiation energy is converted to some physical quantity easy to measure.

Remark: Human eye is considered sometimes as a radiation detector.

D Strahlendetektor, Strahlungsdetektor
F détecteur de rayonnement
P detektor promieniowania
R детектор излучения

radiation source, radiant energy source
A device or substance which produces radiation for various purposes.

D Strahlungsquelle
F source de radiation, source d'énergie rayonnante
P źródło promieniowania
R источник излучения

radioactive chain → decay chain

radioactive cooling
The decrease of activity of a radioactive material by radioactive decay.

D Aktivitätsverminderung
F refroidissement
P schładzanie
R остывание, охлаждение

radioactive decay, radioactive disintegration
The spontaneous transformation of an unstable nucleus into another nucleus accompanied by the emission of corpuscular and/or electromagnetic radiation.

D radioaktive Umwandlung, radioaktiver
Zerfall
F désintégration radioactive, décroissance
radioactive
P przemiana promieniotwórcza, rozpad
promieniotwórczy
R радиоактивное превращение,
радиоактивный распад

radioactive disintegration → radioactive decay

radioactive indicator → radioactive tracer

radioactive kryptonate
A solid into which radioactive krypton ^{85}Kr
has been introduced.
D radioaktives Kryptonat
F kryptonate radioactif
P kryptonat promieniotwórczy
R радиоактивный криптонат

radioactive reagent
A reagent containing a radioisotope.
D radioaktives Reagens, Radioreagens
F réactif radiométrique
P odczynnik promieniotwórczy
R радиоактивационный реагент

radioactive series → decay chain

radioactive tracer, radioactive indicator
Radioactive nuclide which has been introduced
into a system to be investigated.
D radioaktives Leitisotop, radioaktiver
Indikator
F indicateur radioactif, traceur radioactif
P wskaźnik promieniotwórczy, znacznik
promieniotwórczy
R радиоиндикатор, радиоактивный индикатор

radioactive tracer method
A method of tracing the course of chemical
and physical processes by means of radioactive
tracers.
D radioaktives Indikatorverfahren
F méthode de marquage radioactif, méthode
des radioéléments traceurs
P metoda wskaźników promieniotwórczych,
metoda znaczników promieniotwórczych
R метод изотопных радиоактивных
индикаторов

radioactivity
The property of certain nuclides of
undergoing radioactive decay.
D Radioaktivität
F radioactivité
P radioaktywność
R радиоактивность

radiochemical analysis, radiochemical methods
of analysis
Radiometric methods of analysis in which
chemical techniques play an important role.

D radiochemische Analyse
F analyse radiochimique
P analiza radiochemiczna
R радиохимический анализ

radiochemical methods of analysis →
radiochemical analysis

radiochemical purity (*of a radioactive
preparation*)
The absence in a material of impurities from
other radioactive substances.
D radiochemische Reinheit
F pureté radiochimique
P czystość radiochemiczna
R радиохимическая чистота

radiochemical separation
Chemical separation of radioactive substances.
D radiochemische Trennung
F séparation radiochimique
P rozdzielanie radiochemiczne
R радиохимическое разделение

radiochemical yield
The yield of a given isotope obtained in
a radiochemical separation, expressed as
a fraction of the activity originally present.
D radiochemische Ausbeute
F rendement radiochimique
P wydajność radiochemiczna
R радиохимический выход

radiochromatography
Chromatography of substances labelled with
radioactive tracers whose radioactivity is used
subsequently for qualitative and/or
quantitative evaluation of a chromatogram.
D Radiochromatographie
F radiochromatographie
P radiochromatografia
R радиохроматография

radioelectrophoresis
Electrophoresis of substances labelled with
radioactive tracers with subsequent qualitative
or quantitative evaluation (from
electrophoretograms).
D Radioelektrophorese
F radioélectrophorèse
P radioelektroforeza
R радиоэлектрофорез

radio-exchange method
A method of radiometric analysis based on the
measurement of the activity of a radioactive
tracer displaced by the (inactive) element to
be determined from a labelled compound, the
radioactive tracer passing to another phase.
D radioexchange Analyse
F méthode d'échange des réactifs radioactifs
P metoda wymiany wskaźnika
promieniotwórczego, metoda wymiany
promieniotwórczej
R ...

radiofrequency plasma, rf plasma
A plasma which is generated in a stream of
gas (or gases) by an externally applied
radiofrequency field.

Remark: It is recommended that the frequency at which the
source operates be given, e.g. 27.12 MHz ICP, and the
type of gas used be defined.

D Hochfrequenzplasma
F plasma à haute fréquence
P plazma o częstości radiowej
R высокочастотная плазма

radio-frequency polarography, rf polarography,
high-frequency polarography, hf polarography
The alternating current polarography in which
a sinusoidal alternating voltage (usually with a
frequency of $0.1 - 6.4$ MHz) and low
amplitude, modulated with an alternating
square-wave voltage with a frequency of
225 Hz, is used. The faradaic rectification
current I_{FR} is measured as a function of the
direct potential E_{dc} of the indicator electrode.
The faradaic rectification current is filtered out
by a low-pass filter and recorded during the
last stage of the drop life.

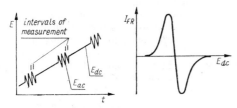

D Radiofrequenz-Polarographie,
rf-Polarographie,
Hochfrequenz-Polarographie,
hf-Polarographie
F polarographie aux radiofréquences,
polarographie RF
P polarografia częstości radiowych,
polarografia rf
R высокочастотная полярография,
радиочастотная полярография

radioisotope-excited X-ray fluorescence analysis
A method of energy dispersive X-ray
fluorescence analysis in which a suitable
radioisotope is used as the source of the
exciting X- or γ-radiation.

D radioisotopische Röntgenfluoreszenzanalyse
F analyse par fluorescence X
excitée par radio-isotope
P analiza fluorescencyjna rentgenowska
radioizotopowa
R радиоизотопный рентгенофлуоресцентный
анализ

radioluminescence
Luminescence caused by radioactive decay.
The incident beam (high energy electrons,
alpha particles or gamma rays) causes
excitation of atoms and light emission results.

D Radiolumineszenz
F radioluminescence
P radioluminescencja
R радиолюминесценция

radiometric analysis *See* radiometry
D radiometrische Analyse
F analyse radiométrique
P analiza radiometryczna
R радиометрический анализ

radiometric titration
Titration in which the substance to be titrated,
the titrant or the indicator, are labelled with
a radioactive tracer and the end point is
determined radiometrically.

D radiometrische Titration
F titrage radiométrique
P miareczkowanie radiometryczne
R радиометрическое титрование

radiometry
Methods of measuring radiation and
determining the activity of radioactive
preparations.

Remark: In chemical analysis, the term radiometric
analysis is used.

D Radiometrie, Strahlenmessung
F radiométrie
P radiometria
R радиометрия

radio-release method
A radiometric method of analysis involving
measurement of activity of the indicator
liberated from a radioactive reagent, due to its
reaction with the substance (inactive)
being determined; e.g. liberation of ^{85}Kr from
hydroquinone clathrate in its reaction with
ozone which is being determined.

D ...
F methode de dégagement de radioactivité
P metoda uwalniania wskaźnika
promieniotwórczego
R метод освобождения радиоиндикатора

radius of ionic atmosphere, $1/\varkappa$
The concept introduced to the theory of strong
electrolytes by Debye and Hückel. It is the
reciprocal of the quantity $\varkappa$ described by the
equation

$$\varkappa^2 = \frac{e^2 N_A}{\varepsilon_r \varepsilon_0 kT} \sum_i c_i z_i^2 = \frac{2e^2 N_A}{\varepsilon_r \varepsilon_0 kT} I$$

(in the SI system)

where e — elementary charge, N_A — Avogadro
constant, ε_r — relative permittivity,
ε_0 — permittivity of vacuum,
k — Boltzmann's constant, c_i — molar
concentration of the ion, z_i — charge number of
the ion, I — ionic strength.

D (effektiver) Radius der Ionenwolke,
(effektiver) Radius der Ionenatmosphäre,
Dicke der Ionenwolke, Debyescher Radius

F rayon de l'atmosphère ionique
P promień atmosfery jonowej, grubość atmosfery jonowej, promień Debye'a
R эффективный радиус ионной атмосферы, толщина ионной атмосферы, дебаевский радиус, дебаевская длина

raies ultimes, persistent lines, RU
The spectral lines of an element which are the last to disappear as the concentration of this element in the sample gradually decreases.
D letzte Linien
F raies ultimes, raies persistantes
P linie ostatnie
R последние линии

Raman effect, RE, Raman scattering, RS
An inelastic scattering of light or ultraviolet radiation when it has collided a with a molecule of a substance. The molecules may lose or gain energy so that the scattered radiation and the incident radiation differ in frequency by discrete amounts, corresponding to the vibrational or rotational changes in the energy level of the molecules. The Raman effect occurs when the polarizability of the molecule in the direction of the electromagnetic field of the light changes as a result of the vibration and rotation of the molecule.
D (normaler) Raman-Effekt, Raman-Strahlung
F effet Raman
P efekt Ramana, rozpraszanie ramanowskie
R комбинационное рассеяние, эффект Рамана

Raman (frequency), $\Delta \bar{v}$
The difference in frequency, more correctly in wavenumbers, between the Raman lines and the exciting line. This shift is proportional to the changes of vibrational or rotational energy in the molecule and independent of the exciting frequency.
D Raman-(Frequenz)-Verschiebung
F déplacement (en frequence) des raies Raman
P przesunięcie (częstotliwości) ramanowskie
R смещение комбинационного рассеяния света, рамановское смещение

Raman lines
Spectral lines in a Raman spectrum which appear at a frequency that is lower (Stokes lines) or higher (anti-Stokes lines) than the exciting frequency by an amount corresponding to the vibrational or rotational energy transitions in the molecules.
Conventionally, the positions of Raman lines are expressed as wavenumbers.
D Raman-Linien
F raies de Raman
P linia Ramana
R линии спектра комбинационного рассеяния света, линии спектра Рамана

Raman scattering → Raman effect

Raman shift → Raman frequency shift

Raman spectroscopy
The branch of molecular spectroscopy dealing with the measurement and interpretation of Raman spectra.
D Raman-Spektroskopie
F spectroscopie Raman
P spektroskopia ramanowska
R спектроскопия комбинационного рассеяния, рамановская спектроскопия

Raman spectrum
A spectrum of light or ultraviolet radiation scattered in inelastic collisions with molecules of a substance. It occurs as a series of lines of discrete frequencies, shifted symmetrically above and below the frequency of the exciting line and in a pattern characteristic of the molecule.
D Raman-Spektrum
F spectre Raman
P widmo Ramana
R спектр комбинационного рассеяния (света)

Randles-Ševčik equation
The equation describing the relationship between the peak current I_p and the bulk concentration c_i^0 of an electroactive substance which undergoes a reversible electrode reaction at constant in time scan rate of the potential (dE/dt) under conditions of linear semi-infinite diffusion

$$I_p = 2.72 \times 10^5 n^{3/2} D_i^{1/2} A (dE/dt)^{1/2} c_i^0$$

where n — number of electrons exchanged in the electrode reaction, D_i — diffusion coefficient of the electroactive substance, A — surface area of the electrode.
D Randles-Ševčik-Gleichung
F équation de Randles
P równanie Randlesa i Ševčika
R уравнение Шевчика

random balance method
A method of statistical investigation allowing to find which among all the investigated factors and their interactions have a significant influence on the observed value. Random balance method consists in carrying out an experiment according to the supersaturated design in which all factors are varying on only two levels but fully randomly; the significance of the influences of factors and interactions on the observed value may be estimated graphically or numerically.
D Zufallsbalancemethode, Methode der Zufallsbalance
F méthode de bilan de hasard
P metoda bilansu losowego
R ...

random digits tables → random number tables

random effects model, type II model
A model of analysis of variance in which the
levels of factors are treated as chosen at
random from the total available population of
levels and the influence of the factor variance
on the total variance is calculated
(e.g. determination of the influence of the
analysts on the variation of results).
D Modell mit zufälligen Effekten, Modell II
F modèle à effets aléatoires, modèle II
P model losowy, model II
R случайная модель, модель со случайными
  эффектами

random error
The difference between the observed and true
(expected) value of a given quantity caused by
accidental factors.
D Zufallsfehler
F erreur aléatoire
P błąd przypadkowy, błąd losowy
R случайная ошибка

random event
An event which cannot be predicted although
the conditions of its occurrence are precisely
defined. Random events can be regarded as
subsets of the space of elementary events.
D zufälliges Ereignis
F événement aléatoire
P zdarzenie losowe
R случайное событие

randomization
Random choice of some attributes of the
experiment, e.g. random fixing of the order of
experimental runs, random choice of
experimenters; the object of randomization is
to enssure as far as possible that the results
of the experiment are not burdened with
a systematic error.
D Randomisierung
F randomisation
P randomizacja
R рандомизация

random number tables, random digits tables
A set of digits which fulfil the specific
conditions of randomness (the digits are
randomly generated, e.g. by using computers).
Random number tables are used to select at
random the elements of a sample (from
a population).
D Zufallszahlentabellen
F tables de nombre aléatoires
P tablice liczb losowych
R таблицы случайных чисел

random sample
A set of elements which is obtained by random
selection from a population.
D mathematische Stichprobe, Zufallsstichprobe
F échantillon aléatoire
P próba losowa
R случайная выборка

random variable, stochastic variable
A function assigning any random event
(e.g. analytical determination) a definite
number.
D zufällige Variable, Zufallsvariable,
  zufällige Veränderliche
F variable aléatoire
P zmienna losowa
R случайная величина

range (*of the results*)
The difference between the greatest and the
smallest value measured for a given sample.
D Variationsbreite, Spannweite, Streubreite
F étendue
P rozstęp
R размах

range of applicability of a balance
The range between the minimum and maximum
loads that may be put in the balance pan.
D Anwendungsbereich der Waage
F domaine d'utilisation d'une balance
P zakres ważenia wagi, zakres pomiarowy wagi
R пределы измерения весов

Rast method
A cryoscopic method of determination of the
molar mass of a chemical compound, in which
camphor is used as a solvent. The high
value of the cryoscopic constant for camphor
enables the determination of the melting point
depression, even in an ordinary laboratory
device.
D Rast-Methode
F méthode de Rast
P metoda Rasta
R метод Раста

rate constant of electrode reaction
(oxidation or reduction), k_{Ox} or k_{Red}
The constant of the reaction

$$Ox + ne \underset{k_{Ox}}{\overset{k_{Red}}{\rightleftarrows}} Red$$

expressed by the formula

$$k_{Ox} = \frac{I_a}{nF \prod_i c_i^{\omega_{i,a}}} \quad \text{or} \quad k_{Red} = \frac{I_c}{nF \prod_i c_i^{\omega_{i,c}}}$$

The value of the rate constant of the electrode
reaction depends on the electrode potential E

$$k_{Ox} = k^{\circ} \exp\left[\frac{\alpha_a nF}{RT}(E - E^{\circ})\right],$$

$$k_{Red} = k^{\circ} \exp\left[\frac{-\alpha_c nF}{RT}(E - E^{\circ})\right]$$

where I_a and I_c — intensity of the partial
anode or cathode current, respectively,
n — number of electrones exchanged in the
electrode process, c_i — concentration of the
electroactive substance, $\omega_{i,a}$ and $\omega_{i,c}$ — order

of the electrochemical anodic or cathodic
reaction (with respect to the given substance),
$k°$ — standard rate constant of the electrode
reaction, α_c and α_c — anodic and cathodic
charge transfer coefficient, respectively.

Remark: The rate constant of an electrode reaction may be
determined by the conditional rate constant and the
conditional electrode potential. Instead of the symbols
k_{Ox} and k_{Red}, one may use k_{bh} and k_{fh}, k and $\overrightarrow{k}$, or k_a
and k_c.

D Geschwindigkeitskonstante der
   Elektrodenreaktion
F constante de vitesse de la réaction
   d'électrode
P stała szybkości reakcji elektrodowej
   (utleniania lub redukcji)
R константа скорости электродной реакции

rate of flow of mercury from capillary → flow
rate of mercury

rate of nucleation
The number of nuclei of crystallization formed
in unit time per unit volume of the liquid
phase.
D Keimbildungsgeschwindigkeit
F vitesse de nucléation
P szybkość tworzenia zarodków
R скорость образования центров
   кристаллизации

Rayleigh line, exciting line
A relatively intense center line which appears
in the Raman spectrum at a frequency equal
to the frequency of the incident radiation
(exciting frequency).
D Rayleigh-Linie, Erregerlinie
F ligne de Rayleigh, raie de Rayleigh, raie
   excitatrice
P linia rayleighowska, linia wzbudzająca
R линия рэлеевского рассеяния,
   возбуждённая линия

Rayleigh scattering
An elastic scattering of light in which radiation
is scattered in various directions by interaction
with the sample molecules in its path without
changing the vibrational or rotational energy
of the molecules. There is no difference in
energy and wavelength between the scattered
and incident light.
D Rayleigh-Streuung, Rayleighsche Streuung
F diffusion de Rayleigh
P rozpraszanie rayleighowskie, rozpraszanie
   klasyczne
R рэлеевское рассеяние

R band
A band in the electronic spectra of organic
molecules containing heteroatoms; it
corresponds to the $n-\pi^*$ electronic transitions
and is of much lower intensity than the
K band.
D R-Band

F bande R
P pasmo R
R полоса R, диапазон R

R-branch, positive branch
The higher-frequency lines in the rotational-
vibrational band that result from transitions
in which the rotational quantum number
changes by $+1$ as the vibrational quantum
number increases.
D R-Zweig, positiver Zweig
F branche R
P gałąź R, gałąź dodatnia
R R-ветвь, положительная ветвь

RBS → high-energy ion scattering spectroscopy

RDE → rotating-disk electrode

RE → Raman effect

reactance, X
The imaginary part of the impedance of any
linear part of an electric circuit with
a sinusoidal alternating current.
SI unit: ohm, Ω.
D Blindwiderstand, Reaktanz
F réactance électrique
P reaktancja, opór (elektryczny) bierny
R реактанс, реактивное сопротивление
   (*электрической цепи*)

reaction (*of a solution*)
A characteristic feature of a solution given
by its hydrogen ion exponent (pH). An aqueous
solution is said to be neutral when its pH = 7,
acidic when pH < 7, and basic when pH > 7.
D Reaktion
F réaction
P odczyn
R реакция

reaction layer (*in electrochemistry*)
The space in the vicinity of an electrode in
which, due to the decrease in concentration of
the electroactive substance, new quantities of
the electroactive substance are formed from
a substance which does not participate directly
in the electrode reaction.
D Reaktionsschicht
F couche réactionnelle
P warstwa reakcyjna
R реакционный слой

reaction overpotential
An overpotential of an electrode process
caused by a slow, as compared with the charge
transfer step, chemical reaction that precedes
or follows the charge transfer.
D Reaktionsüberspannung
F surtension de réaction
P nadpotencjał reakcyjny, nadpotencjał
   kinetyczny
R перенапряжение реакции, химическое
   перенапряжение, реакционное
   перенапряжение

reaction-rate constant, k
A constant coefficient at a given temperature
and pressure characteristic for a given reaction,
occuring in the kinetic equation of the reaction
$v = kc^n$, where v — reaction rate,
c — concentration of a substrate, n — order of
reaction.
D Geschwindigkeitskonstante der Reaktion
F constante de vitesse de réaction
P stała szybkości reakcji
R константа скорости реакции

reactive sputtering, chemical sputtering
The sputtering process realized with a beam
of reactive ions or reactive low-pressure
plasma.
D reaktive Zerstäubung
F pulvérisation réactive
P rozpylanie (jonowe) reaktywne
R реактивное распыление

readability (*of indications of measuring
instrument*)
D Ablesbarkeit
F appréciation de lecture
P odczytywalność
R ...

reagent
A chemical substance of defined purity grade
used in the laboratory.
D Reagens, Reagent
F réactif
P odczynnik
R реактив, реагент

rearrangement ion (*in mass spectrometry*)
An electrically charged dissociation product,
involving a molecular or parent ion, in which
atoms or groups of atoms have transferred
from one portion of the molecule or molecular
moiety to another during the fragmentation
process.
D Rearrangemention
F réarrangement ionique
P jon przegrupowany
R перегруппировочный ион

recording spectrophotometer
A fully automatic spectrophotometer, most
frequently double-beam, equipped with
a recorder which gives a record of the entire
spectrum over a given wavelength range.
D registrierendes Spektralphotometer
F spectrophotomètre enregistreur
P spektrofotometr rejestrujący
R регистрирующий спектрофотометр

recovery factor, degree of extraction,
extraction efficiency, R
The fraction or percentage (R_A) of the total
quantity of a substance A extracted (usually
into the organic solvent phase) under specified
conditions. $R_A = Q_A/(Q_A)'$, where $(Q_A)'$ and
Q_A — initial and final quantities of
substance A.

Remark: The term extractability is not recommended. If
an aqueous solution is extracted with n successive portions
of organic solvent, the ratio of the volumes of the phases
being $V_o/V_w = r$ in each case, the recovery factor for
a particular substance is given by
$$R_n = 1 - (rD_c + 1)^{-n}$$
where D_c — concentration distribution ratio. If $n = r = 1$,
$R_1 = D_c/(1 + D_c)$.
D Spurenelementausbeute
F rendement de l'extraction
P współczynnik odzysku
R фактор извлечения

recrystallization
Repeated crystallization of a material from
a fresh solvent to obtain a product of greater
purity.
D Rekristallisation
F recristallisation
P rekrystalizacja, przekrystalizowanie
R перекристаллизация

redistilled water
Water that has been twice or repeatedly
distilled.
D redestilliertes Wasser
F eau redistillée
P woda redestylowana
R повторно дистиллированная вода,
бидистиллят

redox electrode, oxidation-reduction electrode
An electrode (half-cell), consisting of
a material chemically inert (usually gold,
platinum, or carbon) which is immersed in
a solution containing both the oxidized and
reduced states of a reversible oxidation-
reduction system, e.g. Fe^{3+}, $Fe^{2+}_{(aq)}|Pt_{(s)}$.
The electrode develops a potential proportional
to the ratio of the activities of the oxidized and
reduced forms.
D Redoxelektrode, Reduktions-Oxidations-
-Elektrode
F électrode d'oxydo-réduction, électrode redox
P elektroda redoks(owa)
R окислительно-восстановительный электрод,
редокс-электрод

redox indicator → oxidation-reduction indicator

redox ion exchanger
An ion exchanger in which reversible redox
pairs have been introduced by complex
formation, sorption or as counter-ions.
D Redoxionenaustauscher
F échangeur d'ions redox
P wymieniacz jonowy redoks
R редоксионообменник

redox polymer
An organic polymer containing functional
groups which can be reversibly reduced or
oxidized.
Remark: The term electron exchanger may be used as
a synonym.

D Redoxpolymer
F polymère redox
P polimer redoks
R редоксполимер

redox potential → oxidation-reduction potential

redox reaction
The electron exchange between the reductant (electron donor) and the oxidant (electron acceptor), taking place according to the following scheme

$$n_2 \text{Red}_1 + n_1 \text{Ox}_2 \rightleftarrows n_2 \text{Ox}_1 + n_1 \text{Red}_2$$

where n_2 − number of electrons supplied by the reductant, n_1 − number of electrons accepted by the oxidant.
D Redoxreaktion
F réaction d'oxydo-réduction, réaction redox
P reakcja redoks
R реакция окисления-восстановления

redox system
The system in which an identical element exists in two different states of oxidation, e.g. the system $\text{MnO}_4^-/\text{Mn}^{2+}$

$$\text{MnO}_4^- + 8\text{H}^+ + 5e \rightleftarrows \text{Mn}^{2+} + 4\text{H}_2\text{O}$$

D Redoxsystem
F système redox
P układ redoks
R система редокс

redox titration → oxidation-reduction titration

redox titration curve
A plot describing the redox potential of a system as a function of the volume of a standard oxidant (or reductant) solution added.
D Redox-Titrationskurve
F courbe de titrage redox
P krzywa miareczkowania redoks
R кривая титрования редокс

red shift → bathochromic shift

reduced mobile phase velocity → reduced velocity of the mobile phase

reduced parameters, reduced variables, X_r
Dimensionless quantities defined as the ratios of the state parameters X to the corresponding criticalparameters X_c of the given substance $X_r = X/X_c$.
D reduzierte Größen, reduzierte Parameter
F paramètres réduits
P parametry zredukowane, wielkości zredukowane, zmienne zredukowane
R приведённые величины, приведённые переменные

reduced plate height, h_r
The ratio of the height equivalent to a theoretical plate h to the average diameter d_p

of solid particles of column packing
$h_r = h/d_p$.
D reduzierte Trennstufenhöhe
F hauteur de plateaux réduite
P wysokość półki zredukowana
R приведённая высота эквивалентная теоретической тарелке

reduced variables → reduced parameters

reduced velocity of the mobile phase, reduced mobile phase velocity, v
A dimensionless quantity given by

$$v = u d_p \, D_M^{-1}$$

where D_M — diffusion coefficient of a solute in the mobile phase, u − interstitial velocity of the mobile phase, d_p − average diameter of solid particles of column packing.
D reduzierte Geschwindigkeit der mobilen Phase
F vitesse réduite de la phase mobile
P prędkość przepływu (międzyziarnowa) zredukowana fazy ruchomej
R приведённая скорость потока подвижной фазы

reducer → reductant

reducing agent → reductant

reductant, reducing agent, reducer
A substance donating its electrons to another (being reduced) substance, called the oxidant, which gains electrons; thus in this process of electron transfer the reductant is oxidized.
D Reduktionsmittel
F réducteur, agent réducteur, désoxydant
P reduktor, środek redukujący
R восстановитель

reduction
A chemical reaction involving the transfer of electron(s) from one molecule (atom or ion) to another.
D Reduktion
F réduction
P redukcja
R восстановление

reductometric titration
Oxidation-reduction titration in which the oxidant is titrated with a standard solution of a reductant. During the titration reduction of the substance being determined occurs.
D reduktometrische Titration
F titrage par réduction
P miareczkowanie reduktometryczne
R восстановительное титрование

reductometry
The determination of the substance by titration with a standard solution of reductant.

D Reduktionsanalyse
F réductométrie
P reduktometria
R редуктометрия

reference band (*in spectroscopy*)
The spectral band of an auxiliary sample of
known frequency, superimposed on the
investigated spectrum in order to obtain a
reference point on the frequency scale.
D Bezugsband, Vergleichsband
F bande de référence
P pasmo odniesienia
R полоса сравнения

reference electrode
An electrode (half-cell) which maintains
a virtually invariant potential under the
conditions prevailing in electrochemical
measurement, and when a negligible current
passes through it; the electrode serves for
the observation or control of the potential of
the working or indicator electrode.
D Referenzelektrode, Referenz-Elektrode,
   Bezugselektrode
F électrode de référence, électrode de
   comparaison
P elektroda porównawcza, elektroda
   odniesienia
R электрод сравнения

reference material
A substance with precisely defined properties
used for the calibration of a piece of apparatus,
or for testing a measuring method.
D Vergleichssubstanz, Bezugssubstanz
F matériau de référence
P substancja odniesienia
R эталон

reference method
An analytical method characterized by an
appropriate precision and accuracy, used for
the estimation of analytical results obtained
by other methods.
D Vergleichsmethode
F méthode de comparaison
P metoda odniesienia
R метод сравнения

reference sample (*in absorptiometry*)
A solvent, blank solution, or reference
solution which is used to set the instrument.
The reference material is placed in the light
path and the instrument is set to show 0%
transmittance when no light passes to the
detector and 100% transmittance with the
shutter open and light on. In some cases air is
used as the reference.
D Vergleichssubstanz, Bezugssubstanz
F matériau de référence
P odnośnik
R эталонное вещество

reference solution (*in spectrophotometry*)
A solution with the same solvent as in the
sample, containing the analyte, and possibly
some concomitants, in known concentrations.
D Vergleichslösung, Bezugslösung
F solution de référence
P roztwór odniesienia
R раствор сравнения

reflected wave method (*in EPR spectroscopy*)
The method in which the paramagnetic
resonance absorption is detected by the change
in power reflected from the cavity resonator
containing the sample of the substance under
investigation.
D zurückgestrahlte Mikrowellen-Methode
F méthode de micro-onde réfléchie
P metoda fali odbitej
R метод отражённой волны

reflecting grating → reflection grating

reflection grating, reflecting grating
A diffraction grating which reflects
polychromatic light in the form of a spectrum.
D Reflexionsgitter
F réseau par réflexion, réseau de réflexion
P siatka odbijająca, siatka odbiciowa
R отражательная решётка

**reflection high-energy electron diffraction,
RHEED**
The surface sensitive diffraction technique in
which use is made of electrons of energy from
10 to 50 keV for probing a solid surface at
a glancing angle.
D Beugung schneller Elektronen in Reflexion
F diffraction d'électrons à grande vitesse par
   réflexion
P dyfrakcja odbitych elektronów szybkich
R дифракция быстрых электронов
   в отраженном пучке

refraction
The change of the direction of propagation
of any wave, such as an electromagnetic or
sound wave, when it passes from one medium
into another in which it has a different velocity.
D Brechung, Refraktion
F réfraction
P załamanie
R преломление, рефракция

refractive index, absolute refractive index, n
The ratio of the phase velocity of radiation in
free space (in vacuum) to the phase velocity of
this radiation in the given medium. The
absolute value of the refractive index for air
is 1.000 29, so the values of the refractive
index measured in air are close to absolute
values.

Remark: The expression $n_D{}^{20}$ means that the refractive
index was determined at 20°C by using the yellow D line
in the sodium emission spectrum ($\lambda = 589.3$ nm).

D Brechungsindex, absoluter Brechungsindex
F indice de réfraction, indice de réfraction absolu
P współczynnik załamania, współczynnik załamania absolutny
R абсолютный показатель преломления

refractometer
An instrument used to measure the refractive index of a substance by one of several ways, such as measurement of the refraction produced by a prism, measurement of the critical angle, observation of the interference pattern produced by passing light through the substance, and measurement of the relative permittivity of the substance.

D Refraktometer, Brechungsmesser
F réfractomètre
P refraktometr
R рефрактометр

refractometry
Measurements of the refractive index of a substance by using an instrument known as refractometer; also a method of chemical analysis.

D Refraktometrie, Brechzahlbestimmung
F réfractométrie
P refraktometria
R рефрактометрия

regression → regression function

regression analysis
Methods applied to investigate statistical dependences. The regression analysis, among others, makes possible the estimation of regression coefficients, the testing of regression function significance, and the determination of confidence intervals for the regression function and its coefficients.

D Regressions analyse
F analyse de régression
P analiza regresji
R регрессионный анализ

regression coefficient
Parameter of a regression function, which if unknown is estimated by using methods of regression analysis.

D Regressionskoeffizient
F coefficient de régression
P współczynnik (funkcji) regresji
R коэффициент регрессии

regresion (function)
A function describing the relationship between explanatory variables $X_1, X_2, ..., X_n$ of a process and the expected value of a random variable Y which is the dependent variable (process outcome) in this process.
$E(Y|X_1 = x_1, X_2 = x_2, ..., X_n = x_n) =$
$= f(x_1, x_2, ..., x_n, b_1, b_2, ..., b_k)$ where
$b_1, b_2, ..., b_k$ — regression coefficients;
a simplified notation is in common use
$\hat{Y} = f(x_1, x_2, ..., x_n, b_1, b_2, ..., b_k)$.

D Regression, Regressionsfunktion
F régression, function de régression
P regresja, funkcja regresji
R регрессия, функция регрессии

regular packed column → packed column

rejection region, critical region
A part of the sample space. If the sample point falls into the rejection region, the null hypothesis has to be rejected.

D Ablehnungsbereich, kritischer Bereich
F région de rejet, région critique
P obszar odrzuceń, obszar krytyczny
R область отбрасывания, критическая область

relative activity of substance B, activity of substance B, a_B
The effective concentration of substance B in a two- or multicomponent chemical system defined as

$$a_B = \exp[(\mu_B - \mu_B^\ominus)/RT]$$

where μ_B — chemical potential of substance B in the given system, $\mu_B^\ominus$ — chemical potential of substance B in its standard state, R — gas constant, T — thermodynamic temperature; it may be expressed in mole fractions (rational activity), molalities (molal activity), or molarities (molar activity).

D (relative) Aktivität der Substanz B, Aktivität eines Stoffes B
F activité (relative) du constituant B
P aktywność składnika B (względna)
R (относительная) активность вещества B

relative atomic mass, A_r
The ratio of the average mass per atom of a specified isotopic composition of an element to 1/12 of the mass of an atom of the nuclide ^{12}C, e.g. $A_r(Cl) = 35.453$.

D relative Atommasse eines Elementes
F masse atomique relative
P masa atomowa (względna)
R относительная атомная масса

relatvie electrode potential → electrode potential

relative error
The error, usually in percent, defined by the formula: $\varepsilon_R = A/\mu$ where A — absolute error, μ — true value of the measured quantity. In analytical practice, the arithmetic mean $\bar{x}$ is used instead of the true value μ.
Remark: The true value μ is known only when the standard sample is analysed.

D relativer Fehler
F erreur relative
P błąd względny
R относительная ошибка

relative front → R_f value

relative molecular mass, M_r
The ratio of the average mass per molecule of
a specified isotopic composition of a substance
to 1/12 of the mass of an atom of the nuclide
^{12}C, e.g. (KCl) = 74.56.

D relative Molekülmasse einer Substanz
F masse moléculaire relative
P masa cząsteczkowa (względna)
R относительная молекулярная масса

relative permittivity, ε_r
The ratio of the permittivity of a system to the
permittivity of vacuum.

Remark: The term dielectric constant is not recommended.

D relative Permittivität, Dielektrizitätszahl,
    (relative) Dielektrizitätskonstante
F permittivité relative, constante diélectrique
P przenikalność (elektryczna) względna, stała
    dielektryczna
R относительная диэлектрическая
    проницаемость, диэлектрическая
    постоянная

relative refractive index, n_{12}
The ratio of the phase velocity of radiation in
two media equal to the ratio of the sines of the
angles that the incident and refracted beam
make with the normal to the interface of
these media at the point of incidence.

D relativer Brechungsindex
F indice de réfraction relatif
P współczynnik załamania względny
R относительный показатель преломления

relative retention, $r_{1,2}$, $r_{A/B}$, $r_{i,s}$
The ratio of the adjusted retention volume
(or time) of the substance under study (1)
to that of the reference compound (2),
e.g. of a standard substance, determined under
identical conditions

$$r_{1,2} = \frac{V'_{R1}}{V'_{R2}} = \frac{V_{N1}}{V_{N2}} = \frac{V_{g1}}{V_{g2}} = \frac{t'_{R1}}{t'_{R2}}$$

where V'_R — adjusted retention volume,
V_N — net retention volume, V_g — specific
retention volume, and t'_R — adjusted retention
time.

D relative Retention
F (grandeur de) rétention relative
P retencja względna
R относительное удерживание

relative standard deviation → coefficient of
variation

relative standard deviation of the sample →
empirical variation coefficient

relative supersaturation
The quantity expressed by the formula

$(Q-S)/S$

where Q — concentration of the substance in
solution prior to precipitation,

S — concentration of the substance in the
saturated solution after precipitation.
D relative Übersättigung, relative
    Löslichkeitserhohung
F sursaturation relative
P przesycenie względne
R относительное пересыщение

relaxation
The tendency of the pertubed system to
acquire a stationary state.
D Relaxation
F relaxation
P relaksacja
R релаксация

relaxation effect → relaxation-time effect

relaxation methods → transient methods

relaxation(-time) effect, asymmetry effect
Deformation of an ionic atmosphere when the
electric field is applied to the electrolyte
solution. A central ion and its ionic
atmosphere move in opposite directions; thus
the ionic atmosphere becomes smaller in
front of the ion and becomes greater behind
the ion.
D Relaxationseffekt, Asymmetrieeffekt,
    Effekt der Relaxationszeit
F effet d'asymétrie, effet du temps de
    relaxation
P efekt relaksacyjny, efekt asymetrii
R релаксационный эффект, релаксационное
    торможение

repeatability
The term meaning the precision of a method
expressed as the agreement attainable between
the independent determinations run by single
analyst using the same apparatus and technique.
D Wiederholbarkeit
F répétabilité
P powtarzalność
R повторяемость

replacement titration, displacement titration
Indirect titration in which substituent A being
determined displaces from an appropriate
compound the equivalent amount of
constituent B which is then directly titrated.
Displacement titration is applied mainly in
compleximetry.
D Substitutionstitration, Verdrängungstitration
F titrage par déplacement
P miareczkowanie podstawieniowe,
    miareczkowanie substytucyjne
R вытеснительное титрование

reprecipitation
The dissolution of a precipitate and its
repeated precipitation in order to remove
contaminents.

D Umfällung
F reprécipitation
P strącanie powtórne, strącanie dwukrotne
R переосаждение

reproducibility
The precision of a method expressed as the agreement between the results obtained by different persons in different laboratories, or in the same laboratory but at different times, provided the individual results refer to the same product and method.

D Reproduzierbarkeit
F reproductibilité
P odtwarzalność
R воспроизводимость

reproducibility error *See* instrumental error
D Reproduzierbarkeitsfehler
F erreur de reproductibilité
P błąd odtwarzania
R ошибка воспроизведения

residual current (*in electrochemistry*)
The current which flows through the indicator electrode, equal to the sum of the capacity current and the current arising from the reduction or oxidation of impurities.

D Reststrom, Grundstrom
F courant résiduel
P prąd szczątkowy, prąd resztkowy
R остаточный ток

residual variance (*in regression analysis*)
The variance of deviations of the observed values from their estimators calculated on the basis of the regression equation given by the formula

$$\sigma^2_{y-\hat{y}} = \frac{\sum\limits_{i=1}^{N} (y_i - \hat{y}_i (x_1, x_2, ..., x_p))^2}{N-K}$$

where N — number of observations,
K — number of regression coefficients inclusive of the intercept, y_i — ith observed value,
$\hat{y}_i$ — estimator of the ith observed value.

D Reststreuung
F variance residuelle
P wariancja resztowa
R остаточная дисперсия

resistance, R
In an electric circuit with a constant current, the scalar quantity defined as the ratio of the voltage applied to a linear element of the circuit to the electric current flowing through that element. In an electric circuit with a sinusoidal alternating current, the real part of the impedance of an linear element of this circuit. SI unit: ohm, Ω.

D Resistanz, elektrischer Widerstand,
   Wirkwiderstand
F résistance électrique

P rezystancja, opór elektryczny
R электрическое сопротивление, активное
   сопротивление (*электрической цепи*)

resistance overpotential
An ohmic drop of the potential at the working electrode surface occurring during the current flow, caused by the low conductivity of the layers which are formed at (or disappear from) this electrode surface.

D Widerstandsüberspannung
F surtention de résistance
P nadpotencjał oporowy
R перенапряжение сопротивления

resistance to mass transfer → mass transfer term

resistivity (*formerly called specific resistance*), ρ
The resistance per unit length of the substance taken for unit crosssection. The reciprocal of conductivity. SI unit: ohm metre, Ω m.

D specifischer elektrischer Widerstand,
   Resistivität
F résistivité, résistance spécifique
P rezystywność, opór elektryczny właściwy
R удельное электрическое сопротивление

resolution (*of a photographic emulsion*)
The ability to separate two close points or close lines; it is limited by the graininess of the emulsion.

D Auflösungsvermögen
F pouvoir résolvant, résolution
P zdolność rozdzielcza
R разрешающая сила, разрешающая
   способность

resolution (*of a spectrometer*)
The ability of a spectrometer to resolve the radiation of close wavelengths or energies, expressed by the respective formulae

$$R = \frac{\lambda}{\Delta\lambda}, \qquad R = \frac{E}{\Delta E}$$

D Auflösungsvermögen
F pouvoir de résolution, résolvance
P zdolność rozdzielcza
R разрешающая сила, разрешающая
   способность

resolving time (*in radiometry*)
The smallest time interval which must elapse between two consecutive ionizing events or signal pulses, in order that the measuring device be capable of recording them separately.

D Auflösungszeit
F temps de résolution
P czas rozdzielczy
R разрешающее время

resonance capture
Capture of neutrons of resonance energy.

D Resonanzeinfang
F capture par résonance

P wychwyt rezonansowy
R резонансный захват

resonance fluorescence, resonant fluorescence
Fluorescence occurring when the wavelengths
of the absorbed radiation in the exciting beam
and those of the re-emitted fluorescent
radiation are identical.
D Resonanz-Fluoreszenz
F fluorescence de résonance
P fluorescencja rezonansowa
R резонансная флуоресценция

resonance neutrons
Neutrons of kinetic energy in the range in
which the resonance neutron capture takes
place. The energy limits are arbitrary, generally
from 1 to 1000 eV.
D Resonanzneutronen
F neutrons de résonance
P neutrony rezonansowe
R резонансные нейтроны

resonance peak
A peak on the curve relating the crosssection
of a nuclear reaction to the energy of the
bombarding particle (excitation function); it
corresponds to the resonance absorption.
D Resonanzspitze
F pic de résonance
P pik rezonansowy
R резонансный максимум, резонансный пик

resonance Raman effect, RRE, resonance
Raman scattering, RRS
Raman effect in which the exciting frequency
of the Raman source of radiation is selected
to lie near to or in the region of electronic
absorption of a molecule, and thus to cause
great enhancement of the Raman spectrum.
D Resonanz-Raman-Effekt
F effet Raman de résonance
P efekt Ramana rezonansowy, rozpraszanie
  ramanowskie rezonansowe
R резонансное комбинационное рассеяние
  света

resonance Raman scattering → resonance
Raman effect

resonant cavity → cavity resonator

' resonant fluorescence → resonance
fluorescence

response surface
Geometric pattern of a function presenting
the relationship between factors controlled in
a process and the value observed in this
process (response of the process).
D Wirkungsfläche
F surface de réponse
P powierzchnia odpowiedzi
R поверхность отклика

response time (of chromatographic detector)
The time elapsed between a solute passing
into the detector cell and the pen starting to
move on the chart recorder. It is determined
by three factors: the speed of response of
detection process, the detector dead volume
and the time delay of the electronic equipment.
D Ansprechzeit
F temps de réponse
P czas opóźnienia odpowiedzi
R время отклика детектора

rest mass
The mass of a particle in a system when it is
at rest.
D Ruh(e)masse
F masse au repos
P masa spoczynkowa
R масса покоя

rest point (of a balance)
The position of the pointer with respect to
the pointer scale when the motion of the beam
has ceased.
D Rehepunkt
F position de repos
P punkt równowagi
R точка равновесия

retardation factor → R_f value

retarding field analyser, RFA, retarding
potential analyser, RPA
The concentric four-grid analyser of electron
energy applied especially in the low-energy
electron diffraction equipment. The current
flux is controlled by a retarding potential V_r
which admits only the electrons with energy
greater than eV_r.
D Bremsfeldanalysator
F analyseur par champ retardeur, analyseur
  par potentiel retardeur
P analizator z polem opóźniającym
R анализатор задерживающего потенциала

retarding potential analyser → retarding field
analyser

retention index, Kováts index, I_i
(in chromatography)
The multiplied by a factor of 100 number of
carbon atoms in the molecule of a hypothetical
n-alkane which has the same value of the
adjusted retention time (volume) as the
component being examined, under the given
conditions. Mathematically this can be
expressed by the equation.

$$I_i = 100 \left[\frac{\log X_i - \log X_z}{\log X_{z+1} - \log X_z} + z \right]$$

where X — adjusted retention time or volume,
subscript i refers to the component under
examination, and z — number of carbon
atoms in the molecule of the n-alkane emerging
from the chromatographic bed just before the
component of interest.

D (Kováts-)Retentionsindex
F indice de rétention, indice de Kováts
P indeks retencji, indeks Kowacza
R индекс удерживания (логарифмический),
индекс Ковача

retention temperature (*in temperature
programmed chromatography*)
The column temperature at which the
concentration of the component under
examination in the eluate reaches
maximum value.
D Retentionstemperatur
F température de rétention
P temperatura retencji
R температура удерживания

retention time → total retention time

retention time of the mobile phase → mobile
phase hold-up time

retention volume, total retention volume,
uncorrected retention volume, V_R
(*in chromatography*)
The volume of the eluent required to elute
the component under study from the
chromatographic column, measured from the
injection of the sample to the moment the
concentration of the specified component
in the eluate assumes its maximum, and
is given by

$$V_R = t_R F_c$$

where t_R — total retention time and
F_c — volumetric flow rate of the mobile
phase.
Remark: In liquid chromatography, also known as peak
elution volume; defined as the volume of eluent entering
the column from the start of elution to the appearance
of the peak maximum.
D Gesamtretentionsvolumen
F volume de rétention total, volume de
rétention (non corrigé)
P objętość retencji całkowita, objętość retencji
(niepoprawiona)
R общий удерживаемый объём, объём
удерживания, неисправленный
удерживаемый объём

retrograde extraction → back-extraction

reversed-phase chromatography, RPC
A kind of liquid chromatography in which
the mobile phase is more polar than the
stationary phase.
D Umkehrphasen-Chromatographie, reverse
Phasenchromatographie
F chromatographie en phases inversées
P chromatografia z odwróconymi fazami
R хроматография с обращёнными фазами,
обращённо-фазная хроматография

reversed-phase partition chromatography →
extraction chromatography

reverse isotope dilution analysis
A method of isotope dilution analysis in which
a substance labelled with an isotopic (generally
radioactive) tracer is determined by adding
the same substance with a natural isotopic
composition.
D umgekehrte Isotopenverdünnungsanalyse
F analyse par (la méthode de) dilution
isotopique inverse
P analiza metodą rozcieńczenia izotopowego
odwrotnego
R анализ методом обратного изотопного
разбавления

reversible cell → reversible galvanic cell

reversible coagulation
The coagulation in which the product (usually
a gel) can be converted back into the sol state,
e.g. the coagulation of lyophilic colloids by the
action of an electrolytic solution.
D reversible Koagulation
F coagulation réversible
P koagulacja odwracalna
R обратимая коагуляция

reversible electrode
An electrode (half-cell) in the state of
thermodynamic equilibrium or in the state of
meta-stable equilibrium. Slow irreversible
processes taking place at this electrode do not,
under given conditions, influence the measured
electrode potential determined by a fast and
reversible electrode reaction.
D reversible Elektrode, umkehrbare Elektrode
F électrode réversible
P elektroda odwracalna, elektroda
równowagowa
R обратимый электрод

reversible electrode reaction
The electrode reaction in which the slowest
step is the transport of an electroactive
substance to the electrode surface. The ratio
of concentrations of the oxidized and reduced
forms is described by the Nernst equation.
In thermodynamics, the reversible electrode
reaction is defined as the electrode reaction
which is not accompanied by an increase of
entropy of the system.
Remark: In electrochemistry, for practical reasons this
type of reaction is considered to be the electrode reaction
consisting of two simultaneously proceeding, but
in opposite direction partial electrode reactions. The
products of the first reaction are identical with the
substrates of the second one and vice-versa. The ratio of
their reaction rates lies within 0.1 – 10.
D reversible Durchtrittsreaktion
F réaction d'électrode réversible
P reakcja elektrodowa odwracalna
R обратимая электродная реакция

reversible (galvanic) cell
A galvanic cell in which all the electrode
reactions which occur at each electrode are
thermodynamically reversible.

D reversible (galvanische) Zelle
F pile réversible, cellule (galvanique) réversible
P ogniwo (galwaniczne) odwracalne
R обратимый (гальванический) элемент,
  обратимая (гальваническая) цепь

reversible redox indicator
The redox system whose reduced and oxidized
forms are differently coloured and in which the
transitions between these forms are reversible.
D reversibler Redoxindikator
F indicateur d'oxydo-réduction réversible
P wskaźnik redoks odwracalny
R обратимый редокс-индикатор, обратимый
  окислительно-восстановительный
  индикатор

RFA→ retarding field analyser

rf plasma → radiofrequency plasma

rf polarography → radio-frequency polarography

RHEED → reflection high-energy electron
diffraction

ribbed funnel
A funnel for rapid filtration made of glass,
porcelain, plastic or metal.
D Rippentrichter, Analysentrichter für
  schnelle Filtration
F entonnoir cannelé
P lejek karbowany
R рифлёная воронка, ребристая воронка

right-handed circularly polarized light,
d circularly polarized light
Circularly polarized light in which the electrical
component moves clockwise.
D rechts-zirkular polarisiertes Licht
F lumière à polarisation circulaire droite
P światło spolaryzowane kołowo w prawo
R свет поляризованный по кругу вправо,
  правополяризованный по кругу свет

rigid matrix electrode
A term recommended in the IUPAC
nomenclature to signify an ion-selective
electrode in which the sensing membrane is
a thin piece of glass. *See* glass membrane
electrode.
D ...
F électrode (à membrane sélective) à matrice
  rigide
P elektroda (jonoselektywna) z membraną
  sztywną, elektroda (jonoselektywna)
  z matrycą sztywną
R электрод с жёсткой матрицей

ring chromatography → circular
chromatography

rocking vibrations
A type of bending vibrations in which the
structural unit swings back and forth in the
symmetry plane of the molecule.

D Schaukelschwingungen, ρ-Schwingungen,
  Pendelschwingungen
F vibrations de tangage, vibrations de rotation
P drgania (deformacyjne płaskie) wahadłowe,
  drgania kołyszące
R маятниковые колебания

roentgenogram → X-ray pattern

Ross filters → balanced filters

rotatability
A property of the experimental design
consisting in that the variance of the estimator
of the regression function in any given point
of the factor space depends only on the
distance (in the factor space) from this point
to the centre point of the experiment.
D Drehbarkeit
F ...
P rotatabilność
R ротатабельность

rotatable design
An experimental design satisfying the
rotatability criterion.
D drehbarer Versuchsplan
F ...
P plan rotatabilny, plan o symetrii obrotowej
R ротатабельный план

rotating-disk electrode, RDE (*in electrochemistry*)
An electrode consisting of a disk of electrode
material set in an inert material such as
Teflon which is rotated about its axis at
a known constant velocity.
D rotierende Scheibenelektrode
F électrode à disque tournant, électrode
  tournante à disque
P elektroda dyskowa wirująca
R вращающийся дисковый электрод

rotating-disk electrode, rotrode (*in arc and
spark spectroscopy*)
A vertical disk electrode which rotates
continuously and is partially immersed in the
sample solution. The electrode provides
a continuous supply of the liquid sample to
the discharge space usually crystalline graphite
disks are used.
D rotierende Scheibenelektrode,
  Graphiträdchen-Elektrode
F électrode à disque tournant, électrode
  tournante à disque, électrode rotative,
  rotrode
P elektroda (dyskowa) wirująca
R вращающийся дисковый электрод

rotating-platinum-wire electrode amperometry
See amperometry with one indicator electrode
D Amperometrie mit rotierender
  Platindraht-Elektrode
F ampérométrie à une électrode de platine
  tournante

P amperometria z wirującą elektrodą
platynową
R амперометрия с вращающимся платиновым
проволочным электродом

rotating-ring disk electrode, RRDE
A modification of the rotating disk electrode
in which a disk of electrode material is
surrounded by an annulus of an insulator and
by an annulus of an electric conductor making
the indicator electrode for monitoring the
intermediates of products of the electrode
reaction proceeding at the disk electrode.
D rotierende Scheibenelektrode mit Ring
F ...
P elektroda wirująca dyskowa z pierścieniem
R вращающийся дисковый электрод с кольцом

rotating-wire electrode (*in electrochemistry*)
An electrode consisting of a piece of a metal
wire, usually platinum, which is rotated about
its axis at a known constant velocity used as
an indicator or working electrode.
D rotierende Drahtelktrode
F ...
P elektroda wirująca drutowa
R вращающийся проволочный электрод

rotational spectrum
A molecular spectrum corresponding to
transitions between the rotational energy levels
of the same vibrational and electron state of
a molecule; a pure rotational spectrum appears
in the far-infrared and microwave regions, and
consists of equally spaced lines.
D Rotationsspektrum
F spectre de rotation
P widmo rotacyjne
R вращательный спектр

**rotation of the plane (of polarization) of
polarized light** *See* optical rotation
D Drehung der Polarisationsebene des Lichtes
F rotation du plan de polarisation de la
lumière
P skręcenie płaszczyzny polaryzacji światła
R вращение плоскости поляризации света

rotatory dispersion → optical rotatory
dispersion

rotrode → rotating disk electrode (*in arc and
spark spectroscopy*)

roughness factor, *R*
The ratio of the actual surface area available
for adsorption to the geometrical area of the
plane surface.
D Rauhigkeitszahl, Rauheitszahl
F facteur de rugosité
P współczynnik szorstkości
R фактор шероховатости

routine analysis
Analysis effected repeatedly consisting of
given sequence of operations, e.g. in
production control.
D Routineanalyse, Reihenanalyse
F analyse de routine, analyse courante
P analiza rutynowa
R серийный анализ

RPA → retarding field analyser

RPC → reversed-phase chromatography

RRDE → rotating-ring disk electrode

RRE → resonance Raman effect

RRS → resonance Raman effect

RS → Raman effect

RSD → coefficient of variation

RU → raies ultimes

Rutherford backscattering spectroscopy →
high-energy ion scattering spectroscopy

rydberg, Ry
The atomic unit of energy equal to the product
of the Rydberg constant *R*, the Planck
constant *h*, and the velocity of light in
vacuum *c*,
$1\ Ry = R\,h\,c = 2.179\ 9072 \times 10^{-18}$ J.
D Rydberg
F rydberg
P rydberg
R ридберг

Rydberg constant, *R*
The constant of dimension of wavenumber,
used in atomic and molecular spectroscopy

$$R = \frac{\mu_0^2\,m_e\,e^4 c^3}{8h^3} =$$

$$= (1.097\ 3731 \pm 0.000\ 0003) \times 10^7\ m^{-1}$$

where μ_0 — magnetic permittivity in vacuum,
m_e — electron rest mass, e — elementary
charge, c — velocity of light in vacuum,
h — Planck's constant.
D Rydberg-Konstante
F constante de Rydberg
P stała Rydberga
R постоянная Ридберга

S

SAES → Auger-electron microscopy

safety head, splash head
An adapter preventing centrainment of drops
of a liquid by steam.

D Tropfenfänger
F dévésiculeur, ampoule de garde
P łapacz kropel
R каплеуловитель

salt bridge
A device used to obtain electrical contact
between two half-cells of an electrochemical
cell without mixing the electrolytes, and to
reduce the liquid-junction potential between
the electrolytes; usually, an inverted glass
U-tube filled with the required solution of the
salt (the ions of which have approximately
the same transference numbers, e.g. KCl and
KNO_3).

D Stromschlüssel, Salzbrücke, Elektrolytbrücke
F pont électrolytique, siphon électrolytique
P mostek elektrolityczny, klucz elektrolityczny
R электролитический мостик, солевой
мостик, электролитический сифон

salt effect, uncommon ion effect
The increase of solubility of the precipitate
due to the presence of foreign ions in the
solution.

D Salzeffekt
F effet de sel
P efekt obcych jonów, efekt solny
R солевой эффект

salt error
The error in the determination of the end
point of titration with indicators due to a high
concentration of neutral salts in the solution.

D Salzfehler
F erreur de sel
P błąd solny
R солевая ошибка

salt form of an ion exchanger
The ionic form of an ion exchanger in which
the counter-ions are neither hydrogen nor

hydroxide ions. The kind of counter-ions
defines the form of ion exchanger, e.g. sodium
form, chloride form, Fe(II)-form.

D salzige Form des Ionenaustauschers
F forme saline d'un échangeur d'ions
P forma solna wymieniacza jonów
R солевая форма ионообменника

salting-out
Improving the extraction of a substance by the
addition of an electrolyte to the aqueous phase.

D Aussalzen
F relargage, salaison
P wysalanie
R высаливание

salting-out chromatography
A chromatographic technique used for the
separation of non-electrolytes in which an ion
exchanger acts as the stationary phase.
The addition of an inorganic salt to the
aqueous mobile phase results in the increase
of the distribution coefficients of the
non-electrolytes to be separated.

D Aussalzchromatographie
F chromatographie avec effet de sel
P chromatografia z wysalaniem,
chromatografia wysalająca
R хроматография с высаливанием,
высаливающая хроматография

sample
A part of a population (representative of the
population from which it was taken) being
a subject of investigation with respect to
a particular attribute in order to draw
conclusions about the distribution of the
value of that attribute in the population;
e.g. a series of measurements of the same
quantity carried out by the same method.

D Stichprobe
F échantillon
P próba, próbka, zbiorowość próbna
R выборка

sample function → statistic

sample injector, injection port
(*in chromatography*)
A device by means of which a liquid or
gaseous sample is introduced directly into the
chromatographic column or is swept into the
column by the mobile phase.

D Probengeber, Probeninjektor
F injecteur d'echantillon
P dozownik próbki
R дозатор образца

sample median, median, *m*
The central value of a sample; when the
sample elements are ordered according to their
magnitude, the sample median takes the
values: $m = x_{(n+1)/2}$ when the number of
elements *n* is odd and $m = (x_{n/2} + x_{n/2+1})/2$
when *n* is an even number.

D Stichprobenmedian
F médiane de l'échantillon
P mediana z próby
R (выборочная) медиана

sample path length → path length

sample size (*in statistics*)
The number of elements in a sample.
D Umfang der Stichprobe, Stichprobenumfang
F effectif de l'échantillon, taille de l'échantillon
P licz(eb)ność próby
R объём выборки

sample splitter, inlet splitter (*in gas chromatography*)
A device, used to apply very small samples to capillary columns in which the homogeneous sample is split into two portions (in a known ratio), the smaller of which is introduced into the column and the major part is allowed to escape.
D Probenteiler
F diviseur d'échantillon
P dzielnik próbki
R делитель пробы

sampling boat → combustion boat

sampling error
The error of estimation of a given feature of population due to non-representative sampling. This error may be either random (depending on the amount of sample and on the distribution of the investigated feature in the population) or systematic (dependent on the method of sampling).
D Probenahmefehler
F erreur d'échantillonnage
P błąd pobrania próby
R ошибка выборочного обследования

sampling variance
The variance described by the equation

$$s^2 = \frac{1}{n-1} \sum_{n-1}^{n} (x_i - \bar{x})^2$$

where s — standard deviation of the sample, x_i — value of the ith element of the sample, $\bar{x}$ — arithmetic mean of the sample, n — sample size. The sampling variance s^2 is an estimator of the population variance σ^2.
D Stichprobenstreuung
F dispersion de l'échantillon
P wariancja z próby
R выборочная дисперсия

Sandell's sensitivity index, S
The number of micrograms of the determinand per cubic centimetre of a solution that gives an absorbance of 0.001 per path-length of one centimetre used for evaluation of the sensitivity of a spectrophotometric method developed by Sandell. The sensitivity index is expressed n µg cm^{-2}.

D Sandell-Empfindlichkeitsindex
F indice de sensibilité de Sandell
P współczynnik Sandella
R показатель чувствительности Санделля

Sand equation
The equation describing the relationship between the transition time τ and the bulk concentration c_i^0 of an electroactive substance at constant current I flowing through the elecrode under the conditions of linear semi-infinite diffusion

$$\tau = 0.25\pi n^2 F^2 A^2 D_i I^{-2} (c_i^0)^2$$

where n — number of electrons exchanged in the electrode reaction, F — Faraday constant, A — surface area of the electrode, D_i — diffusion coefficient of the electroactive substance.
D Sand-Gleichung
F équation de Sand
P równanie Sanda
R уравнение Санда

sandwich chamber → S-chamber

SARISA → surface analysis by resonance ionization of sputtered atoms

saturated-calomel electrode, SCE
A calomel electrode, in which the chloride solution is saturated with potassium chloride.
D gesättigte Kalomelelektrode
F électrode au calomel saturée
P elektroda kalomelowa nasycona
R насыщенный каломельный электрод

saturated chamber (*in chromatography*)
A large volume chamber lined with filter paper to increase the evaporation surface area and to ensure uniform saturation of the atmosphere with eluent vapours.
D gesättigte Kammer
F chambre à atmosphère saturée
P komora nasycona
R насыщенная камера

saturated design
An experimental design in which the number of experimental points is equal to the number of investigated factors (and sometimes their interactions).
D gesättigter Versuchsplan
F plan saturé
P plan nasycony
R насыщенный план

saturated solution
A solution in which the concentration of the solute is the same as that in a solution being in equilibrium with the undissolved solute at given temperature and pressure.
D gesättigte Lösung
F solution saturée
P roztwór nasycony
R насыщенный раствор

saturation
The state of a saturated solution.
D Sättigung
F saturation
P nasycenie
R насыщение

saturation activity
Activity reached after an infinitely long time
of activation, practically after several half-life
periods.
D Sättigungsaktivität
F activité à saturation
P aktywność nasycenia
R активность насыщения

saturation factor, growth factor
The ratio of the activity measured after
a specified lapse of time to the saturation
activity measured under the same conditions.
D Zeitfaktor, Wachstumsfaktor
F facteur de saturation
P współczynnik nasycenia
R фактор насыщения

scale error, calibration error *See* instrumental
error
D Kalibrierungsfehler, Eichfehler
F erreur d'étalonnage
P błąd wzorcowania
R калибровочная ошибка

scanning Auger-electron spectroscopy →
Auger-electron microscopy

scanning electron microscope, SEM
The electron microscope in which the sample
surface is scanned by the electron beam of
a diameter of the order of 0.1 µm what
causes the emission of secondary electrons and
characteristic X-rays, generated at the point of
collision of the primary electron beam with the
sample surface, enabling investigation of the
surface processes, real structures and surface
topography. The measurement of X-ray
intensity permits determination of the
distribution of concentration of elements on
the surface analysed.
D Rasterelektronenmikroskop
F microscope électronique à balayage
P mikroskop elektronowy skaningowy,
  mikroskop elektronowy rastrowy
R растровый электронный микроскоп

scattering
The deflection of incident particles (electrons,
photons, etc.) due to collisions with other
particles in their path. Scattering is said to be
elastic or inelastic, depending on the type of
interaction.
D Streuung
F diffusion
P rozpraszanie
R рассеяние

scavenger (*in radiochemistry*)
A substance used to remove from solution, by
adsorption or coprecipitation, a large fraction
of one or more radionuclides. The scavenger is
usually a solid with a well developed surface.
D Scavenger, Spülmittel
F scavenger, entraîneur, épurateur
P zmiatacz
R изоморфный носитель

SCE → saturated-calomel electrode

S-chamber, sandwich chamber
(*in chromatography*)
The arrangement of two plates: the lower
one — a thin layer plate and the upper one — an
uncoated support plate placed over the top of
the former. The lower edge of the so formed
flat chamber (sandwich) is dipped in a solvent
which ascends the plate.
D Sandwich-Kammer
F chambre sandwich
P komora typu sandwicz, komora typu S
R сэндвич-камера

Schellbach burette
A burette with a blue ribbon on a white
background which simplifies burette readings
and considerably reduces the paralax error.
D Bürette mit Schellbachstreifen, Bürette nach
  Schellbach
F burette de Schellbach
P biureta Schellbacha
R бюретка с полоской

Schöniger method
A rapid method for determination of sulfur,
halogens and other elements involving the
combustion of the analysed organic substance
in a sealed flask, previously filled with oxygen.
The flask contains also a solution absorbing
the gaseous combustion products, which are
subsequently titrated.
D Schöniger-Methode
F méthode de Schöniger
P metoda Schönigera
R метод Шенигера

Schöniger stopper
A ground glass stopper whose lower end is
drawn into a bar to which a sample holder is
attached. It is designed to stopper the
oxygen-filled conical flask in which
mineralization by the Schöniger methods is
performed.
D Schöniger-Stopfen
F bouchon de Schöniger, bouchon pour la fiole
  de Schöniger
P korek Schönigera
R пробка Шенигера

scintillation counter
A radiation counter in which the light flashes
produced in a scintillator by ionizing radiation
are subsequently converted into electrical
pulses by means of a photomultiplier.

D Szintillationszähler
F compteur à scintillation
P licznik scyntylacyjny
R сцинтилляционный счётчик

scintillation detector
A radiation detector using a medium in which
a burst of luminescence radiation is produced
along the path of an ionizing particle.
D Szintillationsdetektor
F détecteur à scintillation
P detektor scyntylacyjny
R сцинтилляционный детектор

scintillation spectrometer
A measuring assembly incorporating
a scintillation detector and a pulse amplitude
analyser, used for determining the energy
spectrum of certain types of radiation.
D Szintillationsspektrometer
F spectromètre à scintillation
P spektrometr scyntylacyjny
R сцинтилляционный спектрометр

scissoring vibrations
A type of bending vibrations in which two
atoms bonded to a central atom move toward
or away from each other with deformation of
the bond angle.
D Scherenschwingungen, Beugenschwingungen
F vibrations de cisaillement
P drgania (deformacyjne) nożycowe
R ножничные колебания

screen analysis → sieve analysis

screened indicator
A mixture of an indicator (acid-base or other
type) and a suitable neutral dyestuff chosen
so as to screen the unwanted parts of the
visible range spectrum transmitted by the
indicator in one of its forms.
D Mischindikator
F indicateur avec effet d'écran
P wskaźnik z tłem
R внутренний фильтр

scrubbing
The process of removing impurities from the
separated phase containing the main
substance (i.e. from the extract or
back-extract).
D Waschen
F épuration
P oczyszczanie
R промывка

secondary battery → battery

secondary cell → electrical accumulator

secondary combustion zone, secondary reaction
zone, outer zone (*of a flame*)
The zone formed at the edge of the flame in
which the hot gas comes into contact with
the surrounding atmosphere.

D Außenkegel, äußerer Kegel, sekundäre
  Reaktionszone
F zone de combustion secondaire, zone de
  diffusion, zone externe, cône externe,
  panache
P obszar dyfuzyjny, stożek zewnętrzny
R зона вторичного сгорания, внешняя зона

secondary electrode → auxiliary electrode

secondary interfering reaction
A nuclear interfering reaction induced by
secondary particles produced in the sample or
its surroundings as a result of nuclear
reactions effected by the stream of primary
bombarding particles.
D ...
F interférence de deuxième catégorie
P reakcja jądrowa przeszkadzająca wtórna
R вторичная конкурирующая реакция

secondary ion imaging mass spectroscopy,
SIIMS, secondary-ion microscopy, SIM
The technique involving ion bombardment of
the solid surface in order to determine the
secondary ion patterns of element or compound
distribution in the surface region by means of
the mass analysis of the secondary ions.
D Sekundär-Ionen-Mikroskopie
F microscopie d'ions secondaires
P mikroskopia jonów wtórnych
R ионная микроскопия

secondary ion mass spectroscopy, SIMS
The branch of mass spectroscopy in which
a small spot on the surface of a sample is
bombarded by a beam of primary ions and
the positive and negative secondary ions
sputtered from the surface are analysed
subsequently by means of conventional mass
spectrometry methods.
D Sekundärionen-Massenspektroskopie
F spectroscopie de masse d'ions secondaires
P spektroskopia mas jonów wtórnych
R масс-спектроскопия вторичных ионов

secondary-ion microscopy → secondary ion
imaging mass spectroscopy

secondary reaction zone → secondary
combustion zone

secondary standard (*in titrimetric analysis*)
A substance used for standardization, whose
content of the active agent has been determined
against a primary standard.
D sekundäre Urtitersubstanz
F substance étalon secondaire
P substancja podstawowa wtórna
R вторичный эталон

secondary X-radiation, secondary X-rays
The fluorescent and scattered primary
X-radiations.

Remark: In the narrower sense, the term is used as a synonym of the X-ray fluorescence radiation.

D sekundäre Röntgenstrahlen
F rayons X secondaires
P promieniowanie rentgenowskie wtórne
R вторичное рентгеновское излучение

secondary X-rays → secondary X-radiation

second-class conductor → ionic conductor

second-derivative potentiometric titration
The potentiometric titration in which the second derivative d^2E/dV^2, where E is the potential of the indicator electrode and V is the volume of added titrant, is measured.

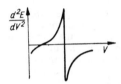

V — volume of a titrant

D potentiometrische Titration mit zweiter Ableitung
F titrage potentiomètrique doublement dérivé
P miareczkowanie potencjometryczne różniczkowe drugiej pochodnej
R потенциометрическое титрование с регистрацией второй производной

second-harmonic ac polarography
The higher harmonic ac polarography in which only the component of the sinusoidal alternating current corresponding to the second harmonic is recorded.

D ac-Polarographie mit zweiter Harmonischer
F polarographie à tension alternative surimposée par détection du second harmonique
P polarografia zmiennoprądowa sinusoidalna drugiej harmonicznej
R переменно-токовая полярография второй гармоники

second order design
An experimental design on the basis of which the response surface can be approximated by a second order polynomial.

D Versuchsplan 2. Ordnung
F plan de second ordre
P plan drugiego rzędu
R план второго порядка

second overtone (band) *See* overtone band

D zweite Oberschwingung
F second harmonique
P nadton drugi, pasmo nadtonowe drugie
R второй обертон

second Wien effect → dissociation field effect

secular (radioactive) equilibrium
A stationary equilibrium between a parent radioactive nuclide and its daughternuclide in the case when the half-life of the parent is very long, as compared with that of the daughter. In the stationary equilibrium state the following relation is satisfied: $N_1 \lambda_1 = N_2 \lambda_2 = \dots = N_i \lambda_i$ where N_i — number of nuclei of the ith member of the radioactive series, λ_i — disintegration constant of this member.

D (radioaktives) Dauer-Gleichgewicht, säkulares Gleichgewicht
F équilibre (radioactif) séculaire
P równowaga (promieniotwórcza) trwała, równowaga wiekowa
R вековое (радиоактивное) равновесие

sedimentation
Precipitation and settling of particles of the dispersed solid phase under the influence of gravitational or centrifugal forces. Sedimentation is used to separate solids from liquids.

D Sedimentation, Sedimentbildung, Abscheiden
F sédimentation
P sedymentacja
R седиментация

sedimentation analysis
The separation of particles (grains) of disintegrated solids based on the differences in sedimentation rates in the given medium, in order to determine the percentage of the particular grain-size fractions.

D Sedimentationsanalyse
F analyse de sédimentation, analyse sédimentaire
P analiza sedymentacyjna
R седиментационный анализ, седиментометрический анализ

sedimentation balance
A balance for the determination of the grain size of homogeneous loose materials. This balance makes possible the determination of the gravitational sedimentation rate of suspensions, e.g. the Figurowski balance.

D Sedimentationswaage
F balance de sédimentation
P waga sedymentacyjna
R седиментационные весы

sedimentation potential, electrophoretic potential, Dorn effect
Difference of electric potentials generated between the top and the bottom of a column in which particles of a suspension are settling.

D Sedimentationspotential, Dorn-Effekt
F potentiel de sédimentation, effet Dorn
P potencjał sedymentacji, efekt Dorna
R потенциал седиментации, потенциал оседания, эффект Дорна

segregation
The differentiation of the composition of an alloy on solidification due to differences in the freezing points and densities of the particular components.
D Seigerung, Saigerung
F ségrégation
P segregacja
R серрегация

Seidel transformation
A mathematical transformation of the emulsion calibration curve into a straight linear form according to the formula:
$P = \log(1-T) - \log T$, where P — transformed value and T — transmittance.
D Seidel-Transformation
F transformation de Seidel
P transformacja Seidla
R преобразование Зейделя

selection rules
Rules which govern radiative transitions between the quantum mechanical states of a system.
D Auswahlregeln
F régles de sélection
P reguły wyboru
R правила отбора

selective elution
An elution procedure in which a selective eluent is used, e.g. a complexing agent that forms stable, non-sorbable complexes with a group of the components to be separated, but affects the adsorbability of other components only to a negligible extent.
D selektive Elution
F élution sélective
P elucja selektywna
R селективное элюирование

selective ion exchanger
An ion exchanger containing functional groups which show distinct preference with respect to certain ions.
D selektiver Ionenaustauscher
F échangeur d'ions sélectif
P wymieniacz jonowy selektywny
R селективный ионообменник

selective reaction
The reaction of a selective reagent conducted under condition that ensure its selective action (e.g. appropriate pH of the solution, masking the interfering ions.).
D selektive Reaktion
F réaction sélective
P reakcja selektywna
R селективная реакция, избирательная реакция

selective reagent
A reagent reacting under given conditions with a limited number of ions or compounds.

D selektives Reagens
F réactif sélectif
P odczynnik selektywny
R селективный реагент

selectivity coefficient, $k_{A/B}$, k_B^A (*in ion exchange chromatography*)
The equilibrium coefficient obtained by formal application of the law of mass action to the ion exchange process and characterizing quantitatively the relative preference of an ion exchanger to select one ion rather than another one, e.g. for the ion exchange reaction:

$$2\overline{Na}^+ + Mg^{2+} \rightleftarrows \overline{Mg}^{2+} + 2Na^+$$

$$k_{Mg/Na} = \frac{[\overline{Mg}^{2+}][Na^+]^2}{[\overline{Na}^+]^2[Mg^{2+}]}$$

where the barred symbols refer to the ion exchanger phase. For exchangers of ions differing as regards their charges, the numerical value of $k_{A/B}$ depends on the choice of the concentration scales for the ion exchanger and the solution. Concentration units used must therefore be clearly stated.
D Selektivitätskoeffizient
F coefficient de sélectivité
P współczynnik selektywności
R коэффициент селективности

selectivity of a reagent
The ability of a reagent to enter into reaction under given conditions with a selected group of ions or compounds.
D Selektivität eines Reagens
F sélectivité d'un réactif
P selektywność odczynnika
R селективность реагента, избирательность реагента

self-absorption
Absorption of an electromagnetic or corpuscular radiation by the emitter.
D Selbstabsorption
F autoabsorption
P autoabsorpcja
R самопоглощение

self-absorption (*in emission spectroscopy*)
Decrease of the intensity of radiation of the wavelength corresponding to the central part of the spectral line, due to its selective absorption in the cold outer vapours surrounding the hot core of the excitation source.
D Selbstabsorption
F autoabsorption, réabsorption
P autoabsorpcja
R самопоглощение

self-dissociation → autoprotolysis

self electrode (*in arc and spark spectroscopy*)
An electrode made of the material being analysed.

D Selbstemissionselektrode
F électrode auto-émettrice, électrode soumise
 à l'analyse
P elektroda z badanego materiału
R самоэмитирующий электрод

self-reversal (*of a spectral line*)
A case of self-absorption, when a line is self-
absorbed to such an extent that the peak or
central wavelength intensity is less than that
at the wings or non-central wavelengths. In the
extreme case, the intensity at the wavelength
centre may become so weak that practically
only the wings remain, giving the appearance
of two fuzzy lines.
D Selbstumkehr
F renversement (d'une raie)
P odwrócenie (linii)
R самообращение линии

self-shielding
The lowering of the flux density in the inner
part of a sample due to absorption in the
outer layers of the sample.
D Selbstabschirmung
F autoprotection
P samoosłanianie
R самоэкранирование

SEM → scanning electron microscope

semiconductor detector, solid-state detector
Detector of ionizing radiation which transforms
a photon or particle energy to an electrical
pulse in a semiconductor material containing
the polarized *p-n* junction, usually of silicon or
germanium one.
D Halbleiterdetektor
F détecteur semi-conducteur
P detektor półprzewodnikowy
R полупроводниковый детектор

semimicrochemical balance
A laboratory balance of weighing capacity up
to 100 g and sensitivity of 10^{-4} g.
D Halbmikrowaage
F balance semi-microanalytique
P waga półmikroanalityczna
R полумикроаналитические весы

semiquantitative analysis
The approximative determination of the
quantitative composition of the sample.
D halbquantitative Analyse
F analyse semi-quantitative
P analiza półilościowa
R полуколичественный анализ

sensitive volume of a detector
The volume of that part of a radiation
detector from which an output signal may
originate.

D empfindliches Volumen des Detektors
F volume sensible d'un détecteur
P objętość czynna detektora
R чувствительный объём датчика

sensitivity (*of analytical method*)
The ratio of the analytical signal increment to
the corresponding increment of
concentration (or contents) of the determinand.
D Empfindlichkeit
F sensibilité
P czułość
R чувствительность

sensitivity → sensitivity of analytical reaction

sensitivity curve of balance
The curve characterizing the sensitivity of
a balance (the number of scale divisions
per 1 mg) as a function of the load.
D ...
F courbe de sensibilité de la balance
P krzywa czułości wagi
R кривая чувствительности весов

sensitivity of a balance → balance sensitivity

sensitivity of analytical reaction
The property of a reaction determined by the
smallest quantity of a substance which can be
detected by that reaction.
D Empfindlichkeit von analytischer Reaktion
F sensibilité d'une réaction analytique
P czułość reakcji analitycznej
R чувствительность аналитической реакции

sensitized ion-selective electrode
An electrochemical sensor for potentiometric
determinations. It consists of the ion-selective
electrode proper and a system which transforms
the component to be determined or separates
(isolates) it from the sample. *See* enzyme-
substrate electrode and gas-sensing electrode.
D ...
F électrode à membrane sélective sensibilisée
P elektroda jonoselektywna uczulana
R сенсибилизированный ионоселективный
 электрод, активированный ионоселективный
 электрод

sensory analysis
The estimation of a quality effected by senses
(sight, smell, taste and touch) by a special
team under conditions that ensure the accuracy
and precision of the results obtained.
D sensorische Analyse, Sinnesprüfung
F analyse sensorielle
P analiza sensoryczna
R сенсорный анализ

separated flame
A flame which is protected mechanically from
direct contact with the surrounding
atmosphere. This is done by placing a tube on

the top of the burner round the flame, and the secondary-combustion zone is thus separated from the primary zone.

D abgetrennte Flamme
F flamme séparée
P płomień rozdzielony
R разделённое пламя

separation
The isolation of a single component or a group of components from other substances present in the sample.

D Trennung
F séparation
P oddzielanie
R отделение

separation factor, $\alpha_{A/B}$
The ratio of the distribution coefficient of solutes A and B in a given medium at a given temperature

$$\alpha_{A/B} = (D_c)_A/(D_c)_B \quad \text{for} \quad (D_c)_A > (D_c)_B$$

where $(D_c)_A$ and $(D_c)_B$ — concentration distribution ratio of substance A and B, respectively.

D Trennfaktor
F facteur de séparation
P współczynnik rozdzielenia
R коэффициент распределения, фактор разделения

separation temperature (*in chromatography*)
The temperature of the chromatographic bed.
Remark: In column chromatography the term column temperature is often used.

D Trenntemperatur
F température de séparation
P temperatura rozdzielania
R температура разделения

Sephadex G
The trade name for hydrophilic, cross-linked polydextran gels which are used as the stationary phases in gel-permeation chromatography, and are available with different exclusion limits so that an optimum grade may be selected for a given application.

D Sephadex G
F Sephadex G
P Sefadeks G
R сефадекс G

sequential analysis
A statistical method in which the number of necessary experiments is not determined in advance. After each single experiment a test is made to find whether the experimental series is to be continued or not.

D Sequentialanalyse, Sequenzanalyse
F analyse séquentielle
P analiza sekwencyjna
R последовательный анализ

sequential test
Method of hypothesis testing in which experimental runs are continued until the obtained results allow either to accept hypothesis tested with the probability β of the error of second kind, or to reject it with the probability α of the error of first kind. The admissible probabilities α and β are set up before starting the experiment.

D sequentieller Test
F test séquentiel
P test sekwencyjny
R последовательный критерий

series of spectral lines → spectral series

SFC → supercritical fluid chromatography

SHE → standard hydrogen electrode

sheathed flame → shielded flame

shielded flame, sheathed flame
A flame which is protected from direct contact with the surrounding atmosphere by a sheath of inert gas that emerges from openings at the rim of the burner top.

D abgeschirmte Flamme
F flame gainée
P płomień osłonięty
R защищённое пламя

short analysis, proximate analysis
Analysis in which only the main constituent or the most important constituents in a sample are determined.

D Kurzanalyse
F analyse rapide
P analiza skrócona
R сокращённый анализ

short-lived radioisotope
A radioisotope with a short half-life (usually shorter than several hours).

D kurzlebiges Isotop
F radioisotope de période courte
P izotop krótkożyciowy
R короткоживущий изотоп

short-wavelength limit (*of an* X-*ray spectrum*)
In the continuous spectrum of an X-ray tube, a sharp lower limit of the wavelength corresponding to the maximum of the X-ray tube voltage.

D Grenzwellenlänge, Kurzwellengrenze
F valeur critique de longueur d'onde, limite de Duane-Hunt
P granica krótkofalowa
R коротковолновая граница непрерывного спектра

side-reaction coefficient, α
A quantity characterizing the degree of progress of reactions other than the main

reaction of the given component; usually these are the interfering reactions

$$\alpha_{M(L)} = [M']/[M] = 1 + \sum_i \beta_i [L]^i$$

where $[M']$ — concentration of component M which does not undergo the main reaction, $[M]$ — concentration of the free component M (e.g. aquacomplexes $M_{(ag)}^{2+}$), $[L]$ — concentration of the substance forming with M compound ML_i (e.g. a ligand which forms complexes with metal ions), β — concentration overall stability constant of ML_s (e.g. stability constant of complexes). Occasionally this coefficient is used as the reciprocal of the molar fraction of substance M, which does not react with component L.

D Nebenreaktions-Koeffizient
F ...
P współczynnik reakcji ubocznych
R коэффициент побочной реакции

sieve analysis, screen analysis
The mechanical separation of particles of the disintegrated material, effected on standard screens to determine the percentage of the particular grain-size fractions.

D Siebanalyse
F analyse granulométrique par tamisage
P analiza sitowa
R ситовый анализ

signal-noise ratio → signal-to-noise ratio

signal-(to-) noise ratio, S/N
The ratio of the signal value S to the noise level N. A signal is assumed to exist when its level is double that of the noise, i.e. $S/N \geqslant 2$.

D Signal-Rausch-Verhältnis
F rapport signal-bruit
P iloraz sygnału i szumu, stosunek sygnał-szum
R отношение сигнал-шум

sign convention, Stockholm (sign) convention, IUPAC convention
The convention of the sign of the electromotive force and electrode (half-cell) potential values in all forms of reversible galvanic cells, according to which for the cell represented as, e.g.,

$$M_{(s)}|Zn_{(s)}|Zn_{(aq)}^{2+}||Cu_{(aq)}^{2+}|Cu_{(s)}^{2+}|M'_{(s)}$$

the electromotive force is positive if the cell reaction from left to right is spontaneous, i.e. the positive charges flow through the cell from M to M′

$$Zn_{(s)} + Cu_{(aq)}^{2+} + 2e_{(M')} \rightarrow Zn_{(aq)}^{2+} + Cu_{(s)} + 2e_{(M)}$$

where M and M′ are identical metallic terminals attached to the left- and right-hand electrodes, respectively. The electromotive force of a galvanic cell E is defined as

$$E = -\Delta G/nF$$

where ΔG — change of Gibbs free energy of the cell reaction, n — number of electrons flowing through the cell from M′ to M, and F — Faraday constant, which is equal to the difference of potentials between the right- and left-hand electrodes. For the reversed representation of the cell, E is negative. The potential of an electrode (half-cell), under open circuit conditions, is regarded as the electromotive force of a galvanic cell in which the electrode on the left is a standard hydrogen electrode and that on the right is the electrode in question. The sign of the right-hand electrode (RHE) is the same as that of the electromotive force.

D Stockholmer Konvention
F convention de Stockholm
P konwencja sztokholmska, konwencja o znakach, konwencja IUPAC
R конвенция о знаках, Стокгольмское соглашение, Стокгольмская конвенция

significance level
The probability of rejection of the tested hypothesis under the assumption that this hypothesis is true (i.e. the probability of error of the first kind). In practice, the significance level is usually assumed as 0.01, 0.02 or 0.05.

D Signifikanzniveau
F niveau de signification, seuil de signification
P poziom istotności
R уровень значимости

significance test
A statistical test enabling (on the basis of a random sample) either to reject the tested hypothesis (with the risk of committing a determined error by the significance level) or to conclude that there is no reason for rejecting the hypothesis.

D Signifikanztest
F test de signification
P test istotności
R критерий значимости

SIIMS → secondary ion imaging mass spectroscopy

Si(Li) detector → lithium-drifted silicon detector

silver electrode
An electrode of the first kind, consisting of a silver wire in contact with Ag^+ ions in solution. The half-cell is represented as $Ag_{(aq)}^+|Ag_{(s)}$; it is used both in potentiometry and electrogravimetry.

D Silberelektrode
F électrode d'argent
P elektroda srebrna
R серебряный электрод

silver-silver chloride electrode
An electrode of the second kind, consisting of
a strip or disc of silver coated with a thin
deposit of silver chloride, immersed in a
chloride solution of known concentration.
The half-cell is represented as
$Cl^-_{(aq)} |AgCl_{(s)}, Ag_{(s)}$; it is used as a reference
electrode or an indicator electrode for
chloride ions.
D Silber-Silberchlorid-Elektrode, Silber-
   Chlorsilber-Elektrode,
   Silberchloridelektrode
F électrode à chlorure d'argent, électrode
   argent-chlorure d'argent
P elektroda chlorosrebrowa
R хлор(о)серебряный электрод

silylating reagent → silylation reagent

silylation
A process of replacing an active proton in
a molecule of a chemical compound by the
silyl group H_3Si— (or more often by its
derivatives, e.g. $(CH_3)_3$ Si—).
D Silylierung
F silylation
P sililowanie
R силилирование

silylation reagent, silylating reagent
A reagent which converts the free hydroxyl
groups on the support or adsorbent surface
into inactive silyl groups, and thus reduces the
support or adsorbent polarity.
D Silylierungsreagens
F réactif à silylation
P odczynnik sililujący
R силилирующий реактив

SIM → secondary ion imaging mass
spectroscopy

simple electrode reaction
The electrode reaction consisting of only one
elementary step.
D einfache Elektrodenreaktion
F réaction d'électrode simple
P reakcja elektrodowa prosta
R простая электродная реакция

simplex
A geometrical figure which has $n + 1$
vertices when a responce is being optimized
with respect to n factors. For example, for
two factors the simplex is a traingle.
D Simplex
F simplex
P sympleks
R симплекс

simplex design
An experimental design consisting of points
hwich are the vertices of a regular simplex.

D Simplex-Plan
F plan symplexe
P plan sympleksowy
R симплекс-план

SIMS → secondary ion mass spectroscopy

simultaneous determination
The determination by a given method of more
than one constituents in an analysed sample;
e.g. the determination of calcium and
magnesium by titration with EDTA in the
presence of eriochromic black.
D Simultanbestimmung
F détermination simultanée
P oznaczanie jednoczesne
R одновременное определение

simultaneous nucleation
The nucleation of more than one compound
simultaneously.
D gleichzeitige Keimbildung
F nucléation simultanée
P zarodkowanie jednoczesne
R совместное образование зародышей

simultaneous techniques
The simultaneous investigation of a sample by
combined techniques, e.g. by thermogravimetry
and differential thermal analysis.
D simultane Untersuchungsmethoden,
   gleichzeitige Untersuchungen
F techniques simultanées
P techniki jednoczesne
R одновременные методы

single-beam spectrophotometer
A spectrophotometer in which the reference
material and the investigated sample are placed
successively in the same monochromatic
beam of radiation.
D Einstrahlspektralphotometer
F spectrophotomètre monofaisceau
P spektrofotometr jednowiązkowy
R однолучевой спектрофотометр

single-comparator method → monostandard
method

single escape peak *See* pair escape peak
D ein-Quant-Escape-Peak, ein-Quant-Escape-
   -Linie
F pic de premier échappement
P pik ucieczki pojedynczej
R пик одиночного вылета, пик вылета
   одного кванта

single-sweep oscillographic polarography →
single-sweep polarography

single-sweep polarography
A modification of polarography in which
a linearly changing in time potential, applied
to the indicator electrode, is swept and a single

current-potential curve is recorded directly during the life of a drop.

Remark: The synonyms: single-sweep oscillographic polarography and cathode-ray polarography are not recommended.

If the sweep is so fast that the change of the surface area of the electrode is negligible, the term dropping electrode amperometry with linear potential sweep is recommended.

 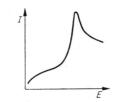

D Single-sweep-Polarographie, Single-sweep-Oszillopolarographie, Katodenstrahlpolarographie
F polarographie à simple balayage, polarographie oscillographique à simple balayage, polarographie sur écran cathodique
P polarografia z pojedynczą zmianą potencjału
R полярография с однократной развёрткой потенциала, осциллографическая полярография с однократной развёрткой потенциала, катодно-лучевая полярография

singlet line (*in atomic spectroscopy*)
A spectral line corresponding to an electronic transition between two energy levels belonging to two atomic singlet terms.

D Singulettlinie, Einfachlinie
F singlet, singulet
P singlet widmowy
R синглет, одиночная линия

SIRIS → surface analysis by resonance ionization of sputtered atoms

skeletal vibrations
Vibrations that occur between various groups of atoms in linear and non-linear molecules.

D Gerüstschwingungen
F vibrations de squelette
P drgania szkieletowe
R скелетные колебания

slit (*of an optical instrument*)
An aperture, limiting the geometrical width of a spectral beam, through which the radiation enters or emerges from an optical instrument.

D Spalt
F fente
P szczelina
R щель

slot-burner (*in flame spectroscopy*)
A burner whose outlet orifice for the gas mixture has the shape of a slot.

D Schlitzbrenner
F brûleur à fente
P palnik szczelinowy
R щелевая горелка

slowing down (*of neutrons*)
Decreasing the neutron energy by elastic collisions with atomic nuclei of the medium.

D Abbremsung, Verlangsamung
F ralentissement, modération
P spowalnianie neutronów
R замедление нейтронов

slow neutrons
Neutrons having kinetic energies from zero to about 1000 eV.

D langsame Neutronen
F neutrons lents
P neutrony powolne
R медленные нейтроны

slurry packing (*in high performance liquid chromatography*)
The wet packing procedure consisting in pumping a suspension of fine particles (e.g. $5-10$ μm) of the sorbent, ion exchange resin, etc. into a column at high pressure.

D Einschlämmtechnik, Slurry-Verfahren
F remplissage (de la colonne) par une suspension
P napełnianie kolumn zawiesiną
R суспензионное заполнение колонки, заполнение колонки суспензией

SME → streaming mercury electrode

SNMS → glow discharge mass spectroscopy

soap chromatography
Ion-pair chromatography in which an ionic detergent is the source of hydrophobic counter-ions.

D ...
F chromatographie avec utilisation de détergents
P chromatografia z użyciem detergentów jonowych
R ...

soft X-ray appearance-potential spectroscopy → appearance-potential spectroscopy

soft X-ray photoelectron spectroscopy → X-ray photoelectron spectroscopy

sol
A suspension of colloidal particles of one substance in another. Hydrosol is a dispersion in water, and aerosol is a dispersion in air.

D Sol
F sol
P zol, roztwór koloidalny
R золь

solid addition funnel → powder funnel

solid core support → superficially porous support

solid electrode
An electrode prepared from a substance or a mixture of substances in the solid state and immersed in an electrolyte; used in potentiometry (e.g. crystal membrane electrode) and in current methods (e.g. rotating disc electrode).
D Festkörper-Elektrode
F électrode solide
P elektroda stała
R твёрдый электрод, твердотельный электрод

solid ion exchanger
A solid containing ions exachngeable for other ions of the same sign present in the solution in which the ion exchanger is considered to be insoluble, e.g. clays, zeolites, inorganic metal oxides and salts as well as synthetic polymers and polycondensates containing ionogenic groups, i.e. ion exchange resins and cellulose derivatives.
D fester Ionenaustauscher
F échangeur d'ions solide
P jonit
R ионит

solid solution → mixed crystal

solid-state detector → semiconductor detector

solid-state membrane
An ion-selective membrane produced from a single- or polycrystalline compound, sparingly soluble salt, or a homogeneous mixture of these compounds. An active electrode material may also be dispersed in an inert support.
D Festkörpermembran
F membrane solide
P membrana stała
R твердотельная мембрана, твёрдая мембрана

solid-state membrane electrode → solid-state membrane ion-selective electrode

solid-state membrane (ion-selective) electrode
An ion-selective electrode with a solid-state membrane.
D Festkörpermembran-Elektrode, ionenselektive Elektrode mit fester Membran, ionenselektive Festkörperelektrode
F électrode à membrane (sélective) solide
P elektroda (jonoselektywna) z membraną stałą, elektroda membranowa stała
R твёрдый мембранный электрод, (ионоселективный) электрод с твёрдой мембраной

solid support (*in chromatography*), chromatographic support
A solid (usually inert and porous) on which the liquid stationary phase is deposited or to which certain active groups or organic radicals involved in the separation process are attached.
Remark: The term carrier in this sense is not recommended.
D fester Träger
F support solide, support de chromatographie
P nośnik
R твёрдый носитель

solid volume (*in chromatography*)
The volume occupied by the solid support in the column.
Remark: In gas chromatography also the volume occupied by the active solid in the column.
D Volumen der festen Phase, Festkörpervolumen
F volume du solide
P objętość fazy stałej
R объём твёрдого наполнителя, объём твёрдой фазы, объём твёрдого тела

Soller collimator, Soller slit
A collimator made of parallel metallic or metal-plated closely spaced blades to form slits, used in the wavelength dispersive X-ray fluorescence analysis.
D Soller-Kollimator
F collimateur de Soller
P kolimator Sollera, szczelina Sollera
R диафрагма Соллера, щель Соллера

Soller slit → Soller collimator

solubility
The ability or tendency of a solid, liquid or gaseous substance to form homogeneous systems with other substances. The solubilities are given under specified conditions of temperature and pressure as the concentration of the saturated solution.
D Löslichkeit
F solubilité
P rozpuszczalność
R растворимость

solubility parameter, δ
The parameter characterizing the polarity of a liquid, $\delta = (E/V)^{1/2}$, where E — cohesive energy and V_m — molar volume of liquid. At low vapour pressures the solubility parameter corresponds to the energy of vaporization per cubic centimetre.
D Löslichkeitsparameter
F paramètre de solubilité
P parametr rozpuszczalności
R параметр растворимости

solubility product, K_{so}
The product of the activities of ions of a sparingly soluble salt in saturated solution at constant temperature and pressure, written as

$$K_{so} = a_+^{v+} a_-^{v-}$$

where a_+ and a_- — activities of cations

and anions, respectively, v_+ and v_- — numbers of cations and anions formed from one molecule of a salt.

D Löslichkeitsprodukt
F produit de solubilité
P iloczyn rozpuszczalności
R произведение растворимости

solubilization chromatography
A chromatographic technique in which the (aqueous) mobile phase contains substances which increase the solubility of non-electrolytes in water and thus decrease their distribution coefficients in the ion exchange — solution system.

D Solubilisationschromatographie
F chromatographie par solubilisation
P chromatografia solubilizacyjna
R солюбилизационная хроматография

solute
The minor component of a solution which is regarded as having been dissolved by the solvent.

D gelöster Stoff, Gelöstes
F soluté
P substancja rozpuszczona, solut
R растворённое вещество

solute transport detector → transport detector

solution
A homogeneous liquid phase comprising at least two different substances.

D Lösung
F solution
P roztwór
R раствор

solvatochromism
Shift of the spectral absorption band caused by the solvent.

D Solvatochromie
F solvatochromisme
P solwatochromia
R сольватохромия

solvent
A liquid (usually the major component of a solution) which is used to dissolve a solute or solutes.

D Lösungsmittel, Lösemittel
F solvant
P rozpuszczalnik
R растворитель

solvent demixing, dehomogenisation of a mobile phase
Frontal chromatography of mixed eluents of different elution strength resulting in formation of zones having different qualitative and quantitative composition due to selective sorption of the components of the mixture along the bed.

D Lösungsmittelentmischung
F démixtion du solvant
P odmieszanie rozpuszczalnika
R расслаивание растворителя, расслоение растворителя

solvent front → mobile phase front

solvent migration distance → mobile phase distance

solvent programming → gradient elution

solvent regain → weight swelling in solvent

solvent strength, ε^0
According to concept introduced by Snyder, a quantity, characterizing the ability of a solvent to elute the components from a polar adsorbent (alumina) in adsorption chromatography. In this classification all solvents have a value of ε^0 within a range of 0.00 for pentane to 1.00 for acetic acid.

D Lösungsmittelstärke, Fließmittelstärke
F force d'élution du solvant
P siła rozpuszczalnika, moc rozpuszczalnika
R элюирующая сила растворителя, элюирущая способность растворителя

sorbate
The substance being sorbed.

D Sorbat, Sorptiv
F sorbat
P sorbat
R сорбат

sorbent, sorbing agent
The substance capable of sorbing another substance (known as the sorbate).

D Sorbens, Sorptionsmittel
F sorbant
P sorbent
R сорбент

sorbing agent → sorbent

sorption
The process of absorption, adsorption or chemisorption of a substance (sorbate) by another substance (sorbent).

D Sorption
F sorption
P sorpcja
R сорбция

sorption isotherm
A relationship between the concentration of a given substance in the sorbent and its concentration in the external solution (liquid, gas) at equilibrium, under specified conditions, and at constant temperature.

D Sorptionsisotherme
F isotherme de sorption
P izoterma sorpcji
R изотерма сорбции

source (*in the structure of a field effect transistor*)
One of the two current electrodes, usually grounded.
D Quelle
F source
P źródło
R источкник

source-detector geometry → counting geometry

Soxhlet apparatus, Soxhlet extractor
A laboratory apparatus for continuous extraction of solid substances with a cold liquid. A solvent, heated in a flask, circulates continuously over the material placed in a paper thimble and carries away its soluble constituents.
D Extraktionsapparat nach Soxhlet
F extracteur de Soxhlet
P aparat ekstrakcyjny Soxhleta
R аппарат Сокслета

Soxhlet extractor → Soxhlet apparatus

spark → electrical spark

spark ionization → spark source ionization

spark ionization source, spark source (*in mass spectrometry*)
An ion source in which the sample molecules are ionized by a high-voltage, high-frequency spark generated between two electrodes made of the sample itself (if the sample is conducting) or of a mixture of graphite and the material to be examined.
D Funkenquelle
F source à étincelle(s)
P źródło iskrowe
R искровой источник

spark source → spark ionization source

spark (source) ionization
Ionization produced by a spark discharge occurring between two electrodes.
Remark: It is recommended to drop the word source from this term.
D Funkenquellen-Ionisation
F ionisation par une source à étincelle(s)
P wzbudzenie iskrowe, jonizacja źródłem iskrowym
R ионизация в искровом источнике

spark source mass spectrometry, SSMS
A method of mass spectrometry in which the ionization of the sample is carried out by generating a high-voltage, high frequency spark or discharge between two electrodes of the material under examination. Quantitative analysis of impurities in inorganic materials as low as 10 ppb can be carried out by this procedure.

D Funkenquelle-Massenspektrometrie
F spectrométrie de masse (à source) à étincelle(s)
P spektrometria mas ze wzbudzeniem iskrowym
R масс-спектрометрия с искровым источником

spark spectrum
An emission spectrum of an element obtained by exciting the element with an electric spark; it is composed mainly of the emission lines of singly-ionized atoms, but also of the spectral lines of multiply-ionized and neutral atoms.
D Funkenspektrum
F spectre d'étincelles
P widmo iskrowe
R искровой спектр

specific absorption coefficient → specific decadic absorption coefficient

specific activity, a
The activity per unit mass of a pure radioisotope, or the activity of a radioisotope in a material per unit mass of that material; given in disintegrations per second per kg.
D spezifische Aktivität
F activité spécifique
P aktywność właściwa
R удельная активность

specific adsorption
Adsorption of a particular constituent from a multi-component system effected preferentially to the other components.
D spezifische Adsorption
F adsorption spécifique
P adsorpcja specyficzna
R специфическая адсорбция

specific conductance → electrical conductivity

specific conductance → electrolytic conductivity

specific (decadic) absorption coefficient, absorptivity, a
The measure of the efficiency with which molecules absorb radiation, defined from the Beer's law as the absorbance A divided by the product of the sample path length b (in centimetres) traversed by the radiation and the concentration c (in grams per litre) of the absorbing substance B: $a = A/bc_B$.
Remark: The terms absorbancy index, extinction coefficient, specific extinction are not recommended.
D spezifischer (dekadischer) Absorptionskoeffizient, Extinktionkoeffizient
F coefficient d'absorption specifique
P współczynnik absorpcji właściwy, współczynnik ekstynkcji, absorpcja właściwa
R удельный (десятичный) показатель поглощения, абсорбируемость, коэффициент поглощения, показатель экстинкции, коэффициент экстинкции

217 spectral band

specific elution
An elution procedure in which a specific eluent is used, e.g. a complexing agent that yields stable, non-sorbable complexes with one component only.
D spezifische Elution
F élution spécifique
P elucja specyficzna
R специфическое элюирование

specific extinction → specific decadic absorption coefficient

specific gravity bottle → pycnometer

specificity of a reagent
The property of a reagent to react with only one ion or compound under specified conditions.
D Spezifität eines Reagens
F spécificité d'un réactif
P specyficzność odczynnika
R специфичность реагента

specific reaction
The reaction of a specific reagent carried out under conditions that ensure its specificity (e.g. an appropriate pH of the solution, masking of interfering ions).
D spezifische Reaktion
F réaction spécifique
P reakcja specyficzna
R специфическая реакция, специфичная реакция

specific reagent
A reagent reacting solely with a particular ion (or compound), thus permitting the detection (or determination) of the latter in the presence of other ions (compounds).
D spezifisches Reagens
F réactif spécifique
P odczynnik specyficzny
R специфический реагент

specific refraction, specific refractivity, *r*
The relationship between the refractive index of a substance *n* at any definite wavelength, and its density ρ; usually given by the formula of Lorentz and Lorentz

$$r = \frac{n^2-1}{n^2+2} \cdot \frac{1}{\rho}$$

D spezifische Refraktion
F réfraction spécifique
P refrakcja właściwa
R удельная рефракция

specific refractivity → specific refraction

specific resistance → resistivity

specific retention volume, V_g (*in chromatography*)
The net retention volume V_N, per gram of the

stationary liquid phase, active solid or solvent-free gel at 0°C and is given by:

$$V_g = \frac{V_N}{w_L} \cdot \frac{273,15}{T}$$

where w_L — mass of the stationary liquid phase, active solid, or solvent-free gel and T — absolute temperature of the column.
Remark: In liquid chromatography, the adjusted and net retention volumes are identical; hence, the specific retention volume may be calculated by using the adjusted retention volume.
D spezifisches Retentionsvolumen
F volume de rétention spécifique
P objętość retencji właściwa
R удельный удерживаемый объём, удельный объём удерживания

specific rotation, specific rotatory power, $[\alpha]_\lambda^t$
The measure of rotation exhibited by an quid, optically active substance. For a pure liquid the specific rotation is given by the equation

$$[\alpha]_\lambda^t = \frac{\alpha}{b\rho}$$

where α — observed optical rotation in degrees, b — layer thickness in decimetres, ρ — density, t — temperature of the measurement, and λ — wavelength used in the measurement. For a solution, the corresponding equation is

$$[\alpha]_\lambda^t = \frac{100\alpha}{bc}$$

where c — concentration of solute in grams per 100 ml of solution.
D spezifische Drehung
F rotation spécifique, pouvoir rotatoire spécifique
P skręcalność właściwa
R удельное вращение

specific rotatory power → specific rotation

specific surface area (*of an adsorbent*)
The total surface area per unit mass of an adsorbent, in m²/g.
D spezifische Oberfläche
F surface spécifique, aire spécifique
P powierzchnia właściwa
R удельная поверхность, специфическая поверхность

specpure reagent → spectrally pure reagent

spectral band
Grouped together spectral lines emitted by a molecule.
D Spektralbande
F bande spectrale
P pasmo widmowe
R спектральная полоса

spectral line, spectrum line
A definite single line corresponding to
a particular wavelength, characteristic of an
element in the atomic state.
D Spektrallinie
F raie spectrale
P linia spektralna, linia widmowa
R спектральная линия

spectral(ly) pure reagent, specpure reagent
A reagent provided with a certificate
indicating the spectral lines of impurities
detectable under specified conditions of
excitation.
D spektrographisch reines Reagens,
spektralreines Reagens, spektroskopisch
reines Reagens
F réactif spectroscopiquement pur
P odczynnik spektralnie czysty
R спектрально чистый реактив

spectral order
The consecutive number of a given interference
maximum of a specified spectral line starting
from the radiation beam directly reflected, or
transmitted (called the zero order).
D Beugungsordnung
F ordre d'un spectre
P rząd widma
R порядок спектра

spectral pure reagent → spectrally pure reagent

spectral series, series of spectral lines
A group of lines in an atomic absorption or
emission spectrum of atoms or ions, in which
the lines all arise from transitions from or to
the energy level characterized by the same
value of the principal quantum number.
D Spektralserie
F série spectrale
P seria widmowa
R спектральная серия

spectroanalysis → spectrochemical analysis

spectrochemical analysis, spectroanalysis,
spectroscopic analysis, spectrum analysis
Qualitative identification and quantitative
determination of the constituents or
components of a sample investigated by
spectral methods.
Remark: In the Russian and Polish literature the term
spectrochemical analysis is also used to denote the
emission spectrochemical analysis with a preliminary
chemical treatment of the sample investigated.
D spektrochemische Analyse Spektralanalyse
F analyse spectrochimique, analyse spectrale,
analyse spectroscopique
P analiza spektrochemiczna, analiza spektralna
analiza widmowa
R спектрохимический анализ,
спектроскопический анализ, спектральный
анализ

spectrochemical buffer
A substance added to a sample
with the purpose to reduce the influence of
interfering components on the emission or
absorption of radiation by the analyte.
D spektrochemischer Puffer
F tampon spectrochimique
P bufor spektrochemiczny
R спектрохимический буфер

spectrochemical carrier
A substance which when added to a spectral
sample gives rise to a gas which can help to
transport the vapour of the sample material
into the excitation region of the source, e.g.
carbon in an air atmosphere when carbon
dioxide is formed.
D Spektralträger
F entraîneur spectrochimique
P nośnik spektroskopowy
R спектрохимический носитель

spectrochemical source → excitation source

spectrofluorimeter
An instrument for measuring the fluorescence
over a wide range of wavelengths, which
consists of a radiation source, a sample
cuvette, a detector, and a pair of
monochromators. The excitation
monochromator selects only the specific
excitation wavelength and the emission
monochromator selects only the emission
wavelengths. Both kinds of fluorescence
spectra, excitation and emission, may be
observed and recorded.
D Spektralfluorimeter
F spectrofluorimètre, spectrofluoromètre
P spektrofluorymetr
R спектрофлуориметр, спектрофлуорометр

spectrofluorimetry
A group of methods of studying of the
intensity of fluorescent radiation as a function
of wavelength.
Remark: This term is often used as synonym of
fluorimetric analysis utilizing a spectrofluorimeter for
measurements.
D Spektrofluorimetrie
F spectrofluorimétrie
P spektrofluorymetria
R спектрофлуориметрия,
спектрофлуорометрия

spectrogram
A photographic or graphical record of
a spectrum.
D Spektrogram
F spectrogramme
P spektrogram
R спектрограмма

spectrograph
An instrument with an entrance slit and
dispersing device that produces a spectrum and

records it on a photographic plate. The
radiant flux passing through the optical system
is integrated over time, and the quantity
recorded is a function of radiant energy.

D Spektrograf
F spectrographe
P spektrograf
R спектрограф

spectrographic analysis
Spectrochemical analysis based on the
investigation of the emission spectra recorded
on a photographic plate.

D spektrografische Analyse
F analyse spectrographique
P analiza spektrograficzna
R спектрографический анализ

spectrometer → optical spectrometer

spectrometry
A branch of spectroscopy concerned with the
measurement of spectra.

D Spektrometrie
F spectrométrie
P spektrometria
R спектрометрия

spectrophotometer
A spectrometer with the associated equipment,
which gives the ratio or the function of the
ratio, of the radiant flux of two electromagnetic
beams as a function of spectral wavelength.
These two beams may be separated in time
and space. Spectrophotometers operate in
various regions of the electromagnetic
spectrum, e.g. ultraviolet, visible and infrared.

D Spektralphotometer, Spektrophotometer
F spectrophotomètre
P spektrofotometr
R спектрофотометр

spectrophotometric titration
Any titration in which the titrant, a reactant,
or reaction product absorbs radiation and has
a sufficiently large molar absorptivity. During
titration the absorbance of the solution is
measured after each addition of the titrant
and the absorbance is plotted versus the volume
of the titrant added. The titration curve
consists, if the reaction is complete, of two
straight lines intersecting at the equivalence
point.

D spektrophotometrische Titration
F titrage spectrophotométrique
P miareczkowanie spektrofotometryczne
R спектрофотометрическое титрование

spectrophotometry
A group of methods dealing with the
investigation of the relationship between the
intensity of electromagnetic radiation
interacting with matter (emission, absorption,
reflection, scattering) and the radiation
wavelength.

D Spektralphotometrie
F spectrophotométrie
P spektrofotometria
R спектрофотометрия

spectropolarimeter
A device used to measure the optical rotatory
dispersion.

D Spektralpolarimeter, Spektropolarimeter
F spectropolarimètre
P spektropolarymetr, polarymetr spektralny
R спектрополяриметр

spectropolarimetry
Methods of measuring the optical rotatory
dispersion, using a spectropolarimeter; also
a method of chemical analysis based on
spectropolarimetric measurements, mainly
used for structure determination of the
optically active substances.

D Spektralpolarimetrie
F spectropolarimétrie
P spektropolarymetria
R спектрополяриметрия

spectroscope
An optical instrument consisting of an
entrance slit, collimator, prism or grating,
telescope or objective lens, which formed
spectra for visual examination.

D Spektroskop
F spectroscope
P spektroskop
R спектроскоп

spectroscopic analysis → spectrochemical
analysis

spectroscopy
A branch of physics and chemistry concerned
with the production, measurement and
interpretation of the energy spectra
(electromagnetic or particle) arising from
either emission or absorption of radiant
energy or particles from a substance upon
bombardment by electromagnetic radiation,
electrons, neutrons, protons, ions or neutrals
or upon heating or excitement by an electrical
or magnetic field; used to investigate nuclear,
atomic, molecular, or solid-state structure, and
also in chemical analysis.

D Spektroskopie
F spectroscopie
P spektroskopia
R спектроскопия

spectrum (*of radiant energy*)
A display or graph of intensity of radiation
(particles, photons, acoustic radiation, etc.)
versus frequency, wavelength or related
quantity.

D Spektrum
F spectre
P widmo
R спектр

spectrum analysis → spectrochemical analysis

spectrum line → spectral line

spectrum of electromagnetic radiation → electromagnetic spectrum

speed of light in vacuum → velocity of light in vacuum

spherical aberration
An aberration of lenses and mirrors with spherical surfaces, due to the fact that rays striking the lens or mirror at the points close to the edges are focused nearer than those being close to the axis. Thus, a point object is not focused as a point image, but as a small disc.
D sphärische Aberration, sphärische Abweichung
F aberration de sphéricité, aberration sphérique
P aberracja sferyczna
R сферическая аберрация

splash head → safety head

spontaneous electrogravimetry → internal electrogravimetry

spot (*in planar chromatography*)
A zone of circular or oval shape.
D Fleck
F tache, spot
P plamka
R пятно

spot broadening *See* peak broadening
D Fleckenverbreiterung
F étalement de tache
P rozmycie plamy
R расширение пятна

spot plate → spotting plate

spot test analysis
A method of qualitative analysis which consists in carrying out the chemical reaction in a drop of the solution analysed, usually on a porcelaine plate, filter paper, or on the sample surface. The appearance of an appropriate coloration, precipitate, or gas bubbles proves the presence of the expected constituent.
D Tüpfelanalyse
F analyse à la goutte, analyse à la touche, essai par touches
P analiza kroplowa
R капельный анализ

spot(ting) plate
D Tüpfelplatte
F plaque à godets (pour essai par touches)
P płytka do analizy kroplowej
R капельная пластинка, пластинка с углублениями

spreader, thin-layer spreader (*in thin-layer chromatography*)
A device used to coat support plates with a thin uniform layer of a sorbent suspension.
D Streichgerät, Dünnschicht-Streicher
F étaleur, dispositif d'étalement
P powlekacz
R прибор для нанесения тонких слоев

spring balance
A balance which determines the mass of an object by measuring the deformation of the balance spring.
D Federwaage
F balance à ressort
P waga sprężynowa
R пружинные весы

sputtered neutrals mass spectroscopy → glow discharge mass spectroscopy

sputtering
The removal of surface atoms as neutrals or positive or negative ions to the gas phase by bombardment with high-energy particles.
D Zerstäubung, Sputtering-Prozeß
F pulvérisation
P rozpylanie
R распыление

sputtering yield, Y
The number of sputtered atoms per one incident particle; it is a function of sample composition, energy and kind of incident particle, angle of incidence of the primary beam, and of crystal orientation.
D Zerstäubungsergiebigkeit, Sputtering-Ausbeute
F rendement de la pulvérisation, richesse de la pulvérisation
P współczynnik wydajności rozpylania
R коэффициент распыления, выход распыления

sputter-initiated resonance ionization spectroscopy → surface analysis by resonance ionization of sputtered atoms

square-wave polarography, sw-polarography
The alternating current polarography in which a square-wave voltage with a frequency of $200-250$ Hz and an amplitude of $5-50$ mV, is used. The square-wave current component of the current I_{sw} is measured as a function of the direct potential E_{dc} of the indicator electrode. The current is recorded at the end part of every pulse and just before the next one.

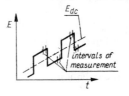
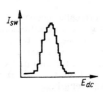

D Rechteckwellen-Polarographie, Square-wave-
Polarographie
F polarographie à créneaux de potentiel,
polarographie à tension carrée,
polarographie à onde carrée
P polarografia zmiennoprądowa prostokątna
R полярография с прямоугольным
напряжением, квадратно-волновая
полярография

SSMS → spark source mass spectrometry

staircase polarography
The polarography in which the potential
applied to the indicator electrode changes
stepwise and is synchronized with the drop
time. The current is usually measured at the
end of the lifetime of a drop.

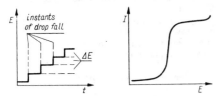

D Treppenstufen-Polarographie
F polarographie pas à pas
P polarografia schodkowa
R полярография со ступенчатой развёрткой
потенциала

standard addition method, known addition
method
The method used in various instrumental
analytical techniques for the analyte
concentration ranges in which the analytical
signal is proportional to concentration. It
consists in measuring the analytical signal of
the analyte before and after addition of
a known amount of it. The analyte content is
determined graphically or by calculation.
D Standardadditionsmethode, Methode der
bekannten Zusätze
F méthode d'étalonnage par ajouts dosés,
méthode à étalon interne, méthode des
additions connues
P metoda dodawania wzorca, metoda
dodatków
R метод добавления стандарта, метод
добавок

standard cell
A primary cell whose EMF is accurately
known and remains sufficiently constant.
D Standardelement, Normalelement
F pile étalon, élément étalon
P ogniwo standardowe, ogniwo normalne
R стандартный элемент, нормальный
элемент, эталонный элемент

standard conditions, normal conditions
The temperature of 273.15 K (0°C) and the
pressure of 1013.25 hPa (1 atm).

D Standardzustände, Normzustände
F conditions normales
P warunki standardowe, warunki normalne
R стандартные условия, нормальные условия

standard deviation (*of a random variable or
probability distribution*), σ
One of the measures of the dispersion of
a probability distribution expressed as the
positive square root of the variance of
distribution.
D Standardabweichung
F écart-type
P odchylenie standardowe
R стандартное отклонение

standard deviation of a sample, *s*
Statistic describing the dispersion of a sample
around its mean value, being a square root of
the sampling variance.
D empirische Standardabweichung
F écart type empirique, écart-type de l'épreuve
P ochylenie standardowe z próby
R выборочное стандартное отклонение

standard deviation of the mean, $s_{\bar{x}}$
The quantity defined by the formula

$$s_{\bar{x}} = \frac{s}{\sqrt{n}}$$

where *s* — standard deviation of a sample,
n — size of sample.
D Standardabweichung des Mittelwertes
F écart-type de la moyenne de l'epreuve
P odchylenie standardowe średniej z próby
R среднее стандартное отклонение

standard (electrode) potential, normal
(electrode) potential, $E°$
The potential of an electrode (half-cell) in
which the components which contribute to the
electrode potential are in their ground states.
D Standard-Elektrodenpotential,
Standardpotential, Normalpotential
F potentiel standard d'électrode, potentiel
normal d'électrode
P potencjał (elektrody) standardowy, potencjał
(elektrody) normalny
R стандартный (электродный) потенциал,
нормальный (электродный) потенциал

standard electromotive force, normal
electromotive force, standard EMF
The EMF of a cell in which all substances
participating in the cell reaction are in the
ground state.
D Standard-EMK, Normal-EMK
F F.e.m. normale, force électromotrice normale
P siła elektromotoryczna standardowa, siła
elektromotoryczna normalna
R стандартная электродвижущая сила,
стандартная э.д.с.

standard EMF → standard electromotive force

standard hydrogen electrode, SHE, normal
hydrogen electrode, NHE
The hydrogen-gas electrode in which the
partial pressure of electrolytic hydrogen gas
is equal to 1 atm (1 013.25 hPa), and the
activity of hydrogen ions in the solution is
equal to unit activity. The electrode potential
is then said to be standard (normal) and is
arbitrarily assigned the value of zero at any
temperature. Used as a standard reference
electrode.

D Standard-Wasserstoffelektrode, Normal-
   Wasserstoffelektrode
F électrode standard à hydrogène, électrode
   à hydrogène normale, électrode normale
   à hydrogène
P elektroda wodorowa standardowa, elektroda
   wodorowa normalna
R стандартный водородный электрод,
   нормальный водородный электрод

standardization
The procedure carried out to determine the
titre of a standard solution, or the molar
concentration of a constituent in the titrant.

D Titerstellung
F étalonnage
P mianowanie, nastawianie miana
R стандартизация

standardized method
The method of analysis given in an obligatory
standard.

D Standardmethode
F méthode normalisée
P metoda znormalizowana
R стандартный метод, унифицированный
   метод

standardized normal distribution
Normal distribution with a zero mean and unit
standard deviation.

D Standardnormalverteilung, standardisierte
   Normalverteilung, Null-Eins-Verteilung
F distribution normale réduite
P rozkład normalny standardowy, rozkład
   normalny zmiennej standardyzowanej
R стандартизированное нормальное
   распределение

standardized random variable
A random variable with expected value equal
to zero and variance equal to one.
Remark: If X is the random variable with expected value
μ and variance σ^2, then the standardized random variable
Z can be obtained using the transformation $Z = \dfrac{X-\mu}{\sigma}$

D standardisierte Zufallsgröße
F variable aléatoire réduite
P zmienna losowa standardyzowana
R стандартизированная случайная величина

standard matching solution
A solution whose characteristic feature is well
specified (e.g. its colour, or turbidity), used as

a reference for matching with an examined
solution.

D Vergleichslösung
F solution témoin
P roztwór porównawczy
R контрольный раствор

standard potential → standard electrode
potential

**standard rate constant of electrode
reaction,** $k°$
The rate constant of an electrode reaction,
characterizing the rate at which a given
electrode reaction proceeds at standard
potential.

D Standardkonstante der Elektrodenreaktion
F constante de vitesse normale de la reaction
   d'électrode
P stała szybkości (reakcji elektrodowej)
   standardowa
R стандартная константа скорости
   электродной реакции

standard reference solution (*in titrimetric
analysis*)
The solution applied for standardization of
other solutions, prepared from a standard
substance, or standardized by any other
method.

D Vergleichsstandardlösung
F solution étalon de référence
P roztwór odniesienia
R образцовый раствор

standard sample
A substance with a strictly defined chemical
composition (given in the manufacturer's
certificate), determined by a large number of
analyses carried out in various laboratories.
The standard sample is used in comparative
analyses, estimation of analytical methods, etc.

D Standardprobe, Normalprobe
F échantillon type, échantillon de comparaison
P wzorzec analityczny
R эталон, нормаль

standard series
An ordered set of references, differing as
regards the physical properties or
concentration of the components.

D Eichprobenserie
F série des étalons
P skala wzorców
R комплект эталонов

standards for elemental analysis
Organic compounds of the structure and
elemental composition similar to that of the
compound to be analysed. Such test compounds
should be stable and possess a high degree of
purity.

D Testsubstanzen für Elementaranalyse
F substances-types pour l'analyse

P wzorzec do analizy elementarnej
R стандартные вещества для элементного анализа

standard solution
A solution having strictly defined properties, e.g. pH, concentration of ions, colour, titre, conductivity, etc.
D Standardlösung
F solution étalon
P roztwór wzorcowy
R стандартный раствор

standard substance (*in titrimetric analysis*)
A substance of known composition, reacting in titration in a defined way (under given conditions), used as a reference in titrimetric analysis. Such a substance must be non-hygroscopis, easy to purify and dry.
D Urtitersubstanz, Urtiter
F substance étalon, substance de base
P substancja podstawowa
R основное вещество

starting line, starting point (*in planar chromatography*)
The line or point on a sheet of paper or on a thin layer plate where the sample to be chromatographed is applied. Also a point on a chromatogram corresponding to the time at which the sample was applied and the development started.
D Startlinie, Startpunkt
F ligne de départ, point de départ
P linia startu, punkt startu
R стартовая линия, стартовая точка

starting point → starting line

stationary electrode process
The electrode process which runs at a constant rate and constant potential of the indicator electrode with no detectable changes of the parameters of the system.
D ...
F processus d'électrode à l'état stationnaire
P proces elektrodowy stacjonarny
R стационарный электродный процесс

stationary-electrode voltammetry → chronoamperometry with linear potential sweep

stationary liquid phase, liquid stationary phase, liquid phase (*in chromatography*)
A liquid of low volatility at the operating temperature held on a solid support, employed in gas-liquid and liquid-liquid chromatography as the stationary phase.
D stationäre Flüssigphase, flüssige (stationäre) Phase
F phase (stationnaire) liquide
P faza (nieruchoma) ciekła, faza stacjonarna ciekła
R неподвижная жидкая фаза, жидкая фаза

stationary liquid volume, (total) pore volume, interstitial volume, V_i (*in permeation chromatography*)
The volume of the liquid mobile phase that is stationary in the internal pores of the column packing material.
D Porenvolumen, Gesamtporenvolumen
F volume de liquide stationnaire, volume poreux
P objętość cieczy unieruchomionej
R объём пор, объём подвижной застойной фазы

stationary methods → steady-state methods

stationary phase (*in chromatography*)
An active solid or a liquid held on a solid support through which the mobile phase passes, and on which the separation takes place.
Remark: The solid support itself should not be identified with the stationary phase.
D stationäre Phase, nichtmobile Phase
F phase stationnaire, phase immobile
P faza nieruchoma, faza stacjonarna
R неподвижная фаза, стационарная фаза

stationary-phase fraction, ε_S (*in chromatography*)
The volume of the stationary phase V_s, divided by the bed volume X, $\varepsilon_S = V_s/X$.
D Anteil der stationären Phase
F fraction (volumique) de la phase stationnaire
P objętość fazy nieruchomej ułamkowa
R доля неподвижной фазы

stationary-phase volume, V_s (*in chromatography*)
The volume of the stationary liquid phase or active solid or gel in the column.
Remark: The volume of any solid support is not included.
D Volumen der stationären Phase
F volume de la phase stationnaire
P objętość fazy nieruchomej
R объём неподвижной фазы

statistic, sample function
A value calculated on the base of a sample being a comprehensive characteristic of the sample; any function of the measurements or observations.
D Statistik, Stichprobenfunktion
F statistique, fonction des observations de l'échantillon
P statystyka, funkcja próby
R статистика, выборочная функция

statistical hypothesis
Any supposition about the unknown distribution of a random variable.
The statistical hypothesis may concern either the unknown value of the population parameter or the unknown type of the distribution function.

D statistische Hypothese
F hypothèse statistique
P hipoteza statystyczna
R статистическая гипотеза

statistical probability
The relative frequency of events in a given
statistical material treated as an approximation
of the theoretical probability.

D statistische Wahrscheinlichkeit
F probabilité statistique
P prawdopodobieństwo statystyczne
R статистическая вероятность

statistical quality control
Quality control which uses statistical methods
of investigation of the production quality in
a manufacturing process, acceptance inspection
or utilization of products. The aim of
statistical quality control is to keep the quality
level within limits determined by random
causes only. The reduction of the variability
of quality to the random variation means that
the practical aim of statistical quality control is
to trace and to eliminate the systematic
variation of quality or at least to minimize it.

D statistische Qualitätskontrolle
F contrôle statistique de la qualité
P statystyczna kontrola jakości
R статистический контроль качества

statistical test → test

statistics
Mathematical theory related to probability
theory useful for making decisions in the case
of uncertainty.

D mathematische Statistik
F statistique mathématique
P statystyka matematyczna
R математическая статистика

steady-state methods, stationary methods (*in
electrochemistry*)
A number of electroanalytical methods in
which the current flowing through the
indicator electrode or the potential of this
electrode does not change in time, or changes
very slowly and reflects these changes in the
concentration of an electroactive substance at
the electrode surface.

D stationäre Methoden
F méthodes stationnaires
P metody stacjonarne
R стационарные методы

step filter, step weakener
A quartz (occasionally glass) plate decreasing
the intensity of the transmitted radiation in
a stepwise manner.

D Stufenfilter
F filtre à échelons
P filtr stopniowy, osłabiacz stopniowy
R ступенчатый фильтр, ступенчатый
    ослабитель

step height (*on an integral chromatogram*)
The distance, perpendicular to the time or
volume axis, by which the baseline is shifted
during the record of a step. *See* integral
chromatogram.

D Stufenhöhe (in einem Integral-
    Chromatogramm)
F hauteur de palier (sur chromatogramme
    intégral)
P wysokość stopnia (chromatogramu całkowego)
R высота ступени (на интегральной
    хроматограмме)

step (*on an integral chromatogram*)
A shift of the baseline perpendicular to the
retention time (volume) axis corresponding to
the appearance in the eluate of the successive
component of a mixture. *See* integral
chromatogram.

D Stufe (in einem Integral-Chromatogramm)
F palier (sur un chromatogramme intégral)
P stopień (chromatogramu całkowego)
R ступень (на интегральной хроматограмме)

step weakener → step filter

stepwise electrode reaction
The electrode reaction which consists of
a series of intermediate elementary steps.

D stufenweise Elektrodenreaktion
F ...
P reakcja elektrodowa złożona, reakcja
    elektrodowa wieloetapowa
R сложная электродная реакция

stepwise elution
An elution procedure in which several eluents
of different qualitative and/or quantitative
composition and arranged in gradually
increasing order of elution strength are used in
succession to elute groups or single
components of the sample in separate
fractions.

D stufenweise Elution
F élution par gradient de composition en
    escalier, élution par paliers, élution par
    étage
P elucja stopniowana, elucja ze skokową
    zmianą składu
R ступенчатое элюирование

stepwise formation constant → stepwise
stability constant

stepwise regression
A procedure of choosing (from among many
possible) the best regression equation consisting
of successive adjoining of particular variables
to the equation (and sometimes their removing
from the equation) until all variables having a
significant influence on the observed value
(experimental outcome) are included in the
regression equation; the significance of
variables is tested by the F-test.

D schrittweise Regression
F régression pas à pas
P regresja krokowa
R постепенная регрессия

stepwise stability constant, stepwise formation
constant, K_n
The equilibrium constant of the successive
reaction of a complex formation
$ML_{n-1} + L \leftrightarrows ML_n$. Under conditions
ensuring stability of the activity coefficients
it is given by the equation

$$K_n = \frac{[ML_n]}{[ML_{n-1}][L]}$$

where $[ML_n]$ and $[ML_{n-1}]$ — concentrations of
the complexes at equilibrium, $[L]$ —
concentration of the ligand at equilibrium.
D ...
F constante de formation successive
P stała trwałości kolejna, stała trwałości
   stopniowa, stała tworzenia kompleksu
   kolejna
R ступенчатая константа образования

sticking probability, adsorption probability,
adsorption coefficient, s
The ratio of the number of collisions leading
to adsorption to the total number of collisions
of gaseous molecules with the surface. For
a given system, the sticking probability is
a function of the surface temperature, surface
coverage and surface roughness.
D Haftwahrscheinlichkeit
F probabilité de fixation, probabilité de collage
P prawdopodobieństwo przylgnięcia,
   prawdopodobieństwo adsorpcji
R коэффициент прилипания

stimulated emission
The process in which a photon colliding with
an excited atom causes emission of a second
photon of the same energy as the first.
D stimulierte Emission, induzierte Emission
F emission stimulée, emission induite
P emisja wymuszona
R индуцированное излучение, вынужденное
   излучение

stimulated Raman effect
The Raman effect produced by the high
intensity light pulses emitted from the pulsed
ruby laser. The Raman scattering intensity is
by many orders of magnitude greater than that
which occurs during the normal Raman
effect.
D stimulierter Raman-Effekt
F effet Raman stimulé
P efekt Ramana stymulowany
R вынужденное комбинационное рассеяние
   света, вынужденный эффект Рамана

stimulus threshold (*in sensory analysis*)
The lowest intensity of an impulse that
causes a sensation.

D absoluter Schwellenwert
F seuil absolu
P próg absolutny
R абсолютный порог ощущения

stirred-mercury-pool amperometry *See*
amperometry with one indicator electrode
D Amperometrie mit gerührter Quecksilber-
   Pool-Elektrode
F ampérométrie à une électrode de mercure
P amperometria z makroelektrodą rtęciową
R амперометрия с перемешиваемым
   ртутным макроэлектродом

stochastic variable → random variable

Stockholm convention → sign convention

Stockholm sign convention → sign convention

stoichiometric end-point → equivalence-point
(*of titration*)

Stokes band → Stokes lines

Stokes law
The wavelength of luminescence excited by
a radiation is always longer than that of the
exciting radiation.
D Stokessches Gesetz
F régle de Stokes
P prawo Stokesa
R закон Стокса

Stokes lines, Stokes band
In the Raman spectrum, the Raman lines
occurring on the longer-wavelength (low-
frequency) side of the exciting line and
corresponding to vibrational energy transitions
in the scattering molecule in which the final
vibrational level, to which the excited molecule
drops back, is higher than the initial
vibrational level.
D Stokessche Linien
F raies de Stokes, bande Stokes
P linie stokesowskie, pasmo stokesowskie
R стоксовы линии

storage battery → battery

storage cell → electrical accumulator

streaming mercury electrode, SME
An electrode made by a stream of mercury
flowing under pressure through a capillary
immersed in the solution being investigated.
The electrode is characterized by having
a considerably greater surface area than
mercury drop electrodes. Used in polarography
or in measurements of the potential of zero
charge, depending on the construction.
D strömende Quecksilberelektrode,
   Quecksilberstrahlelektrode

F électrode à jet de mercure
P elektroda rtęciowa strumieniowa
R струйчатый ртутный электрод

streaming potential
The potential diffrence arising between opposite sides of a porous solid material or between the two ends of a capillary when a liquid is forced through a plug of finely divided material or through a capillary tube.

D Strömungspotential
F potentiel d'écoulement
P potencjał przepływu
R потенциал течения

stretching vibrations, valence vibrations
Vibrations which imply a rhythmical movement of the bonded atoms back and forth along the bond axis.

D Valenzschwingungen, ν-Schwingungen, Streckschwingungen
F vibrations de valence, vibrations d'extension
P drgania walencyjne, drgania rozciągające
R валентные колебания

stripping → back-extraction

stripping analysis → electrochemical stripping analysis

stripping solution
The solution (usually aqueous, sometimes water alone) used for extracting a substance from the extract (usually organic).

D Mittel zum Strippen, Mittel zum Rückextrahieren
F solution d'extraction en retour
P roztwór do reekstrakcji
R реэкстрагент

stripping voltammetry
A group of analytical techniques in which the substance to be analysed is first concentrated electrochemically at the surface of an electrode (or in a mercury amalgam) and then is stripped of, again electochemically. Voltammetric methods are classed as anodic or cathodic, depending on the polarity of the electrode during stripping.

D Inversvoltametrie
F voltampérométrie avec redissolution
P woltamperometria inwersyjna, woltamperometria z zatężaniem, woltamperometria stripingowa
R инверсионная вольтамперометрия

Student's distribution, *t*-distribution
Probability distribution of the continuous random variable *t* with the probability density function defined by the formula

$$f(t) = \frac{\Gamma\left[(k+1)/2\right]}{\Gamma\left(k/2\right)\sqrt{\pi k}} \left(1 + \frac{t^2}{k}\right)^{-(k+1)/2}$$

for $-\infty < t < +\infty$ and for k degrees of freedom. When k is large enough ($k > 30$) the

t-distribution may be approximated by a normal distribution; the *t*-distribution finds application, e.g. for testing hypotheses about mean values.

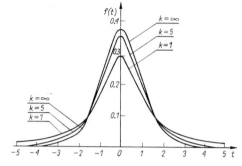

D Student-Verteilung, *t*-Verteilung
F distribution de Student, distribution de *t*
P rozkład *t*(-Studenta)
R распределение Стьюдента

Student's test, *t*-test
The test of significance used for verification of the hypothesis concerning the unknown expected value of a random variable, having a normal distribution and unknown variance; also used for comparison of unknown expected values of two random variables, both of them having the normal distribution.

D Student-Test, *t*-Test
F test de Student, test *t*
P test Studenta, test *t*
R критерий Стьюдента

submicro-analysis
Analysis in which the mass of the investigated sample lies in the range of 10^{-4} to 10^{-3} g.

Remark: The submicro-analysis partly encompasses the microgram methods (from 10^{-6} to 10^{-3} g).

D Submikroanalyse
F submicroanalyse
P submikroanaliza
R микроанализ

substance
Any chemical compound (complex substance), or chemical element in its free state (simple substance).

D Substanz
F substance
P substancja
R вещество

(+)-substance → dextrarotatory substance

(−)-substance → laevorotatory substance

d-substance → dextrarotatory substance

l-substance → laevorotatory substance

substoichiometric isotope dilution analysis
A method of istotope dilution analysis in
which the final isotopic abundance is estimated
from the amount of the nuclide present in
a given quantity of the relevant element
isolated from the sample. This quantity is
smaller than the total amount of that element
present in the sample.
D substöchiometrische
  Isotopenverdünnnungsanalyse
F analyse par (la méthode de) dilution
  isotopique substoechiométrique
P analiza metodą rozcieńczenia izotopowego
  substechiometrycznego
R субстехиометрический метод в изотопном
  разбавлении

successive titration, consecutive titration
The direct sequential titration of the
constituents of a solution (without
preliminary separation).
D Sequentialtitration
F ...
P miareczkowanie sukcesywne
R ...

sum peak
A peak in a γ-ray spectrum registered due to
the simultaneous photoelectric absorption of
two γ-quanta of different energies. It is
generally present in γ-ray spectra of
radioisotopes that emit γ-radiation in cascade.
D Summen-Peak
F pic de somme
P pik sumy
R пик суммарной энергии, суммарный пик

supercritical fluid chromatography, SFC
High pressure gas chromatography conducted
at temperatures higher than the critical
temperature of the carrier gas.
D superkritische Fluidchromatographie,
  Fluid-Chromatographie
F chromatographie à fluide supercritique
P chromatografia w obszarze nadkrytycznym
R надкритическая флюидная хроматография,
  сверхкритическая флюидная хроматография

superficially porous support, porous layer
support, pellicular support, solid core support,
controlled surface porosity support
A support which consists of a solid, impervious
core, usually glass, on the surface of which
a thin porous layer (outer shell) is formed.
D Träger mit poröser Oberfläche, oberflächlich
  poröses Trägermaterial, Pellicular-
  Trägermaterial
F support superficiellement poreux, support
  à couche mince poreuse, support
  pelliculaire
P nośnik powierzchniowo porowaty, nośnik
  z cienką warstwą porowatą na powierzchni,
  nośnik pelikularny

R поверхностно-пористый носитель,
  пелликулярный носитель, плёночный
  носитель, носитель с контролируемой
  поверхностной пористостью

supersaturated solution
A metastable solution whith a concentration
of solute higher than that of a saturated
solution at the same temperature and pressure.
D übersättigte Lösung
F solution sursaturée
P roztwór przesycony
R пересыщенный раствор

supersaturation
The state when a solution is supersaturated.
D Übersättigung
F sursaturation
P przesycenie
R пересыщение

supersaturation ratio
The square root of the ratio of the ionic
activity products of a supersaturated solution
to the ionic activity products of a saturated
solution, under identical conditions.
D Übersättigungsverhltnis
F rapport de sursaturation
P stosunek przesycenia
R отношение пересыщения

supporting electrode (*in arc and spark
spectroscopy*)
An electrode made of a material different
from that of the sample being investigated
serving for introduction of the sample into
the discharge gap.
D Trägerelektrode, Hilfselektrode
F électrode porte-échantillon
P elektroda pomocnicza
R электрод-носитель

supporting electrolyte, indifferent electrolyte
An electrolyte added, in a large excess with
respect to the electroactive constituents (at
least 100:1), to a solution being investigated
in order to decrease the effect of migration of
electroactive ions in an electric field and to
reduce the resistance of the solution.
D indifferenter Elektrolyt, Leitsalz,
  Leitelektrolyt, Grundelektrolyt,
  Zusatzelektrolyt
F électrolyte indifférent, électrolyte support,
  électrolyte de base
P elektrolit podstawowy
R индифферентный электролит, фоновый
  электролит

support plate (*in thin-layer chromatography*)
The plate (e.g. glass plate) that supports a thin
layer of active solid.
D Trägerplatte, Trägerunterlage
F plaque support
P płytka nośna
R пластинка-подложка, несущая пластинка

surface (*in solid state physics and chemistry*)
The outermost region of a solid, comprising
from 1 to 3 monoatomic layers, differing in
properties from the bulk.
D Oberfläche
F surface
P powierzchnia
R поверхность

surface-active agent → surfactant

surface analysis
The determination of the composition,
electronic and crystallographic structures,
dynamic and thermodynamic properties of the
outermost layer of a solid and the species
adsorbed on it.
D Oberflächenanalyse
F analyse de surface
P analiza powierzchni
R анализ поверхности

**surface analysis by resonance ionization of
sputtered atoms, SARISA,** sputter-initiated
resonance ionization spectroscopy, SIRIS
A very sensitive technique involving the
sputtering of surface atoms with a pulsed ion
beam and analysis of these atoms when
ionized in the gas phase by laser resonance.
The method permits quantitative
determination of trace amounts at the level
of 10^{-11} ppm of the sputtered monolayer.
D ...
F analyse de surface par ionisation résonante
d'atomes projetés
P analiza powierzchni metodą rezonansowej
jonizacji rozpylonych atomów
R анализ поверхности резонансной
ионизацией распыленных атомов

surface barrier detector
A semiconductor detector of charged particles,
made of *n*-type silicon whose surface has been
allowed to oxidize in order to bring about the
formation of a *p*-layer.
D Oberflächensperrschichtzähler
F détecteur à barrière de surface
P detektor z barierą powierzchniową
R поверхностно-барьерный
полупроводниковый детектор

surface charge density, Q
The electric charge divided by the surface area
of the interface (electrode) as given by the
Lippmann equation:

$$Q = (\partial\gamma/\partial E)_{T, \, p, \, \mu_i}... \text{ (in C m}^{-2})$$

where γ − interfacial tension, E − potential
of the electrode with respect to the reference
electrode, T − thermodynamic temperature,
p − external pressure, μ_i − set of
chemical potentials which are held constant.
Remark: The physical charge density (in C m^{-2}) believed
to occur on either side of the electrical double layer is

recognized as the free charge density on the interface, σ.
It depends upon the model assumed for the interface. For
an ideal polarized electrode $\sigma = Q$.
D Flächenladungsdichte
F densité de charge superficielle
P gęstość ładunku powierzchniowego
R поверхностная плотность (электрического)
заряда

surface coverage, degree of surface coverage, θ
The ratio of the number of adsorption sites
actually occupied by the adsorbate to the total
number of sites available for adsorption.
D Oberflächenbedeckung
F couvercle superficiel, degré de recouvrement
de la surface
P pokrycie powierzchni, stopień pokrycia
powierzchni
R поверхностное покрытие, степень
заполнения поверхности

surface electric potential of phase β, χ^β
The electric potential of phase β connected
with the dipolar distribution of charge on the
surface of phase β.
D elektrisches Oberflächenpotential der Phase β
F potentiel électrique de surface de la phase β
P potencjał (elektryczny) powierzchniowy
fazy β
R поверхностный электрический потенциал
фазы β

surface ionization
Ionization of an atom or molecule on
interaction with a solid surface. Ionization
occurs only when the relationship between the
work function of the surface, the temperature
of the surface, and the ionization energy of
the atom or molecule is appropriate.
D Oberflächenionisierung, Oberflächenionisation
F ionisation superficielle
P jonizacja powierzchniowa
R поверхностная ионизация

surface science
A branch of physics and chemistry concerned
the state of the solid surface and phenomena
occurring on it when in contact with a gaseous
or liquid environment.
D ...
F science des surfaces
P nauka o powierzchni
R ...

surface sensitive techniques
The group of spectroscopic, microscopic and
diffraction methods which provide information
on the state of the outermost layer of a solid
and the surface species on it. The measurements
with these techniques, with the exception of
ellipsometry and SERS, must be performed
under high or ultra high vacuum.

D ...
F techniques d'étude des surfaces
P metody analizy powierzchni
R методы исследования поверхности

surfactant, surface-active agent

A chemical compound which when dissolved
in water or in any other solvent lowers the
surface tension of the latter.
D grenzflächenaktiver Stoff, Surfaktant,
oberflächenaktiver Stoff, Tensid
F surfactant, substance à activité interfaciale,
tenside, substance tensio-active, agent
tensio-actif
P substancja powierzchniowoczynna
R поверхностно-активное вещество

susceptance, *B*
The imaginary part of the admittance of
a linear part of an electric circuit with
a sinusoidal alternating current.
SI unit: siemens, S.
D Blindleitwert, Suszeptanz
F susceptance (électrique)
P susceptancja, przewodność (elektryczna)
bierna
R реактивная проводимость (*электрической
цепи*)

susceptometry
A branch of high-frequency conductometry
concerned with measurements of the
susceptance.
D Suszeptometrie
F susceptométrie
P susceptometria
R сусцептометрия

swelling
The process of binding a liquid by a gel or
solid, connected with an increase of volume
of the latter.
D Quellung
F gonflement
P pęcznienie
R набухание

swelling pressure
The pressure occurring when a gel or solid is
swelling.
D Quellungsdruck
F pression de gonflement
P ciśnienie pęcznienia
R давление набухания

sw-polarography → square-wave polarography

SXAPS → appearance-potential spectroscopy

SXPS → X-ray photoelectron spectroscopy

symmetrical vibrations
Vibrations of atoms in atomic groups
occurring symmetrically.

D symmetrische Schwingungen
F vibrations symétriques
P drgania symetryczne
R симметричные колебания

systematic analysis → systematic (qualitative)
analysis

systematic error
The deviation of the mean value of a series of
measurements from the true value occurring
permanently when the measurement is carried
out by a particular method or with a given
instrument.
D systematischer Fehler
F erreur systématique
P błąd systematyczny
R систематическая ошибка, систематическая
п огрешность

systematic (qualitative) analysis
The qualitative analysis in which groups of
ions are successively separated by group
reagents; the particular ions of a group are
identified after elimination of the interfering
ions.
D systematische (qualitative) Analyse
F analyse systématique
P analiza (jakościowa) systematyczna
R систематический (качественный) анализ

T

Tafel's equation
The relationship between the electroactivation overpotential η_e and the partial current density j

$$|\eta_e| = a + b \log |j|$$

where

$$a = -\frac{(\ln 10)\, v\, RT}{\alpha\, nF} \log j_0, \qquad b = \frac{(\ln 10)\, v\, RT}{\alpha\, nF}$$

v — stoichiometric number of the transfer reaction, n — number of electrons exchanged in the electrode reaction, α — (cathodic or anodic) charge transfer coefficient, j_0 — density of the exchange current.

D Tafel-Gleichung
F équation de Tafel, relation de Tafel
P równanie Tafela
R формула Тафеля, уравнение Тафеля

tailing
Asymmetry of a chromatographic peak such that the ascending part of the peak, the front, is steeper than the descending part of the peak, the rear. In planar chromatography, the apperance of a diffused zone behind the spot.

D Tailing
F traînée (située sur le pic après son sommet)
P rozmycie tylnej części piku/pasma/plamy, tworzenie ogonów
R образование хвоста

take-off angle
The angle between the surface of an analysed sample and the axis of a collimator, limiting the beam of excited radiation, or the detector axis.

D Abnahmewinkel, Austrittwinkel
F angle d'émergence, angle de prise
P kąt wyjścia
R угол выхода

target (*in radiochemistry*)
A substance or an object exposed to the action of nuclear radiation or bombarded with accelerated ions or elementary particles. During these processes specific nuclear reactions occur in the target.

D Target, Auffänger, Treffplatte
F cible
P tarcza
R мишень

Tast polarography
The polarography in which the current is recorded in the final part of drop life time.

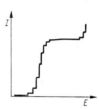

polarization curve recorded under conditions of the dc polarography

D Tast-Polarographie
F polarographie en systéme Tast
P polarografia prądu maksymalnego
R Таст-полярография

TCD → thermal conductivity detector

TD → thermodilatometry

TELS → transmission energy loss spectroscopy

temperature coefficient
The slope of the curve presenting the dependence of the cell EMF variation with temperature at a constant extent of reaction and under constant pressure.

D Temperaturquotient
F coefficient de température
P współczynnik temperaturowy
R температурный коэффициент

temperature gradient chromatography → chromathermography

temperature-programmed chromatography, programmed temperature gas chromatography, PTGC.
A chromatographic technique, commonly used in gas chromatography, in which the temperature of the column is varied during a part (or the whole) of the separation process according to a predetermined program.

D temperaturprogrammierte Chromatographie
F chromatographie avec programmation de la température
P chromatografia z programowaniem temperatury,
R хроматография с программированием температуры

temperature programmed desorption, TPD
The desorption process of adsorbed species induced by increasing the adsorbent

temperature at a prescribed rate, usually linear or hyperbolic. This process is used to determine the respective surface states of adsorbed species, surface coverage and the kinetic parameters of the desorption equation, i.e. its order and activation energy.

D ...
F désorption thermique programmée
P desorpcja termoprogramowana, termodesorpcja programowana
R температурно-программированная десорбция, термодесорбция

temporary hardness
The hardness of water due to the presence of hydrogencarbonates of calcium and magnesium, which on boiling decompose with precipitation of calcium carbonate and partly magnesium carbonate. Temporary hardness is the difference between the total and permanent hardness of water.

D temporäre Härte, vorübergehende Härte
F dureté temporaire
P twardość przemijająca
R временная жёсткость

tensammetry → nonfaradaic admittance measurement

test (*in chemical analysis*)
A standardized procedure, carried out to establish the presence (or absence) of a substance, or a group of substances in the examined sample. The procedure may involve the observation of colour change, formation (or disappearance) of precipitate, gas or smell evolution or indication of an alytic instrument. The test may have a semi-quantitative character.

D Prüfung, Test
F épreuve, essai
P próba, test
R проба, испытание

test, statistical test
A procedure leading to the decision of accepting or rejecting a given statistical hypothesis.

D (statistischer) Test
F test (statistique)
P test (statystyczny)
R (статистический) критерий

t-**test** → Student's test

test electrode → indicator electrode

test for goodness of fit
A statistical test used for verification of hypotheses concerning the shape (type of function) of the probability distribution. Tests for goodness of fit are used, e.g. to compare the goodness of fit between empirical and theoretical distributions and to compare the goodness of agreement between two or more empirical distributions.

D Anpassungstest
F test de validité de l'ajustement
P test zgodności
R критерий согласия

test paper → indicator paper

TG → thermogravimetry

thalamid electrode, thallium amalgam-thallous chloride electrode
An electrode (half-cell) consisting of thallium amalgam (40%) covered with thallous chloride, in contact with a saturated solution of potassium chloride. The half-cell is represented by $KCl_{(sat)} \mid TlCl_{(s)}$, $Tl(Hg)_{(l)}$; used as a reference electrode.

D Thalamid-Elektrode, Thalliumamalgam-Thallium(I)-chlorid-Elektrode
F ...
P elektroda talamidowa
R таламидный электрод, таллиевоамальгамный-хлорталлиевый электрод

thallium amalgam-thallous chloride electrode → thalamid electrode

theoretical end-point → equivalence-point (*of titration*)

theoretical plate height → height equivalent to a theoretical plate

theoretical plate number → number of theoretical plates

theoretical specific capacity → total specific capacity

theory of double layer
The theory which describes the model of the electrical doube layer at the electrode-electrolyte interface. According to this theory, on the solution side (close to the electrode) a compact layer (Helmholtz layer) is formed, and next to it the diffuse layer (Gouy-Chapman layer) extends into the bulk of the solution.

D Theorie der elektrochemische Doppelschicht
F théorie de la couche double
P teoria budowy warstwy podwójnej
R теория строения двойного электрического слоя

thermal analysis
A term covering a group of techniques in which a physical property of a substance is measured as a function of temperature whilst the substance is subjected to a controlled temperature programme.

Remark: The term thermography is not recommended.

D thermische Analyse
F analyse thermique
P analiza termiczna
R термический анализ

thermal conductivity detector, TCD katharometer
A detector which measures the difference
between the thermal conductivity of the pure
carrier gas and that containing the sample
components.
D Wärmeleitfähigkeits-Meßzelle,
  Katharometer
F détecteur á conductibilité thérmique,
  catharomètre
P detektor termokonduktometryczny,
  detektor konduktywności cieplnej, detektor
  przewodnictwa cieplnego, katarometr
R термокондуктометрический детектор,
  термокондуктометрическая ячейка,
  катарометр

thermal curve
A term relating to any curve obtained from
any method of thermal analysis.
D therm(oanalyt)ische Kurve
F courbe thermoanalytique, courbe
  thermométrique, thermogramme
P krzywa termiczna
R термоаналитическая кривая, термограмма

thermal desorption
Desorption taking place at elevated
temperatures.
D thermische Desorption
F désorption thermique
P desorpcja termiczna
R термическая десорбция

thermal detector → thermal radiation detector

thermal dissociation
The decomposition of a molecule taking place
at elevated temperatures.
D thermische Dissoziation
F dissociation thermique
P dysocjacja termiczna
R термическая диссоциация

thermal emission ion source, thermal ionization
source (*in mass spectrometry*)
An ion source consisting of a filament
assembly with one or more filaments, on
which the solid sample is placed. The rate of
ion formation on the heated surface of the
filament is inversely proportional to the
ionization potential of the element.
D Thermoionisations-Ionenquelle
F source (d'ions) par ionisation thermique
P źródło jonów przez jonizację termiczną
R термоионизационный ионный источник

thermal ionization
Ionization of an atom or molecule when it
interacts with a heated surface or is in a
gaseous environment at high temperatures.
D thermische Ionisation, Thermoionisation
F ionisation thermique
P jonizacja termiczna, termojonizacja
R термическая ионизация, термоионизация

thermal ionization source → thermal emission
ion source

thermal neutrons
Neutrons in thermal equilibrium with the
atoms or molecules of the substance through
which they are passing.
D thermische Neutronen
F neutrons thermiques
P neutrony termiczne
R термические нейтроны

thermal (radiation) detector
A detector in which radiation produces
a heating effect that alters some physical
property of the instrument.
D Wärmedetektor
F détecteur thermique
P detektor termiczny
R термодетектор

thermobalance
A device for continuous weighing of the
investigated sample during its heating or
cooling.
D Thermowaage
F thermobalance
P termowaga
R термовесы

thermocouple, thermoelement (*in IR
spectroscopy*)
A thermal detector usually made of a small
piece of blackened gold foil welded to the tips
of two wire conductors made of different
metals chosen so as to give a large
electromotive force when heated.
D Thermopaar, Thermoelementpaar
F couple thermoélectrique, thermocouple
P termopara
R термопара, термоэлемент

thermodiffusion potential
The difference in the inner potentials due to
the temperature gradient in a solution.
D Thermodiffusionspotential
F potentiel de termodiffusion
P potencjał termodyfuzyjny
R термодиффузионный потенциал

thermodilatometry, TD
The method of thermal analysis in which the
changes in dimension(s) of a substance are
measured as a function of temperature.
D Dilatometrie
F dilatométrie
P termodylatometria
R дилатометрия

thermoelement → thermocouple

thermogalvanic cell → galvanic thermocell

thermography → thermal analysis

thermogravimetry, TG
A technique whereby the mass of a substance,
in an environment heated or cooled at
a controlled rate, is recorded as a function of
time or temperature.
D Thermogravimetrie
F thermogravimétrie
P termograwimetria
R термогравиметрия

thermomanometric analysis
The detection of the gaseous products
liberated in thermal decomposition, and their
determination effected by measuring the
pressure gain in a sealed system.
D thermomanometrische Analyse,
   Thermomanometrie
F analyse thermomanométrique
P analiza termomanometryczna
R термоманометрический анализ

thermometric titration, enthalpimetric titration
The titration controlled by a thermosensitive
device (thermistor, thermoelement,
thermometer) located in the reaction medium
which is in a thermally insulated vessel.
D thermometrische Titration,
   enthalpometrische Titration, Enthalpie-
   Titration
F titrage thermométrique
P miareczkowanie termometryczne,
   miareczkowanie entalpimetryczne
R энтальпиметрическое титрование

thermovolumetric analysis
The detection of the gaseous products liberated
in thermal decomposition, and their
determination effected by measuring the
increase of gas volume in a sealed system.
D thermo-gasvolumetrische Analyse,
   Thermogasvolumetrie, Thermovolumetrie
F analyse thermovolumétrique
P analiza termowolumetryczna
R термогазоволюметрический анализ

thick specimen, infinite thickness specimen
(*in X-ray fluorescence analysis*)
A sample of such a thickness that its increase
does not affect the intensity of the excited
X-rays.
D (unendlich) dicke Probe
F échantillon épais
P próbka gruba, próbka o grubości
   nieskończonej, próbka o grubości nasycenia
R грубый образец

thick target
A target of such a thickness that the energy
loss or absorption of the incident particles or
photons traversing it is appreciable.
D dickes Target
F cible épaisse
P tarcza gruba
R толстая мишень

thin-film specimen (*in X-ray fluorescence
analysis*)
A sample that exhibits a linear dependence
of the intensity of the characteristic X-rays
of a given element on its content in the sample.
The matrix effects in such specimens are
practically negligible.
D dünne Probe
F échantillon mince, spécimen mince
P próbka cienka
R тонкий образец

thin-layer chromatography, TLC
Chromatography in which the stationary phase
is a thin layer of sorbent (adsorbent, ion
exchanger, gel) spread on a support plate,
e.g. glass plate, plastic sheet.
D Dünnschichtchromatographie
F chromatographie sur couche mince
P chromatografia cienkowarstwowa,
   chromatografia w cienkich warstwach
R тонкослойная хроматография,
   хроматография в тонком слое

thin-layer spreader → spreader

thin target
A target of such thickness that the energy
loss or absorption of the incident particles or
photons traversing it is negligible.
D dünnes Target
F cible mince
P tarcza cienka
R тонкая мишень

three-sigma rule
The rule that nearly all values obtained in
a sample lie within the interval $\bar{x}-3s$, $\bar{x}+3s$,
where $\bar{x}$ — arithmetic mean of the sample,
s — standard deviation of the sample.
D Drei-Sigma-Regel
F règle des trois sigmas
P reguła trzech sigm
R правило трёх сигм

threshold energy
The limiting kinetic energy of an incident
particle or energy of an incident photon
below which a specified process cannot take
place.
D Schwellenenergie
F seuil d'énergie
P energia progowa
R пороговая энергия

time drift → drift

time of flight, TOF (*in mass spectrometry*)
The time it takes an accelerated ion to travel
along a fixed horizontal path in the field-free
region of a time-of-flight mass spectrometer.
D Laufzeit, Flugzeit
F temps de vol
P czas przelotu
R время полёта

time-of-flight mass spectrometer, TOF
spectrometer
A mass spectrometer which produces the ions
from the sample in pulses and sorts them out
into bunches of different mass-to-charge ratios
according to the time required for the ions to
travel along a fixed horizontal path in a field-
free region after they have all been initially
given the same kinetic energy, or the same
pulse. The ions of lowest mass arrive to the
detector first.

D Flugzeitmassenspektrometer,
   Laufzeitspektrometer
F spectromètre de masse à temps de vol
P spektrometr mas czasu przelotu
R времяполётный масс-спектрометр

time-resolved spectrum
A spectrum which shows the variation of
radiation intensity over very short time
intervals, e.g. microseconds or miliiseconds.
Usually for investigating successive phases of
the excitation cycle.

D zeitlaufgelöstes Spektrum
F spectre avec résolution dans le temps
P widmo rozdzielone w czasie
R спектр с разрешением во времени

titanometry
The determination of oxidising substances
by titration with a standard solution of
titanium (III) salt.

D Titanometrie
F ...
P tytanometria
R титанометрия

titer → titre

titrant
The solution containing the active agent with
which a titration is made.

D Titrans, Titersubstanz
F solution titrante, titrant
P titrant
R титрант

titrant generation → elektrochemical generation
of titrant

titration
The process of determining substance A by
adding to its solution portions of substance B
(almost always as a standardized solution)
with provision for recognizing the point at
which all of A has reacted, thus allowing the
amount of A to be found from the known
amount of B added up to this point, the
reacting ratio of A to B being known from
stoichiometry or from other data. The reverse
process − portionwise addition of A to B −
is seldom applied, except in for standardization
titrations.

D Titration, Titrierung
F titrage
P miareczkowanie
R титрование

titration curve
A curve illustrating the characteristic
quantity (e.g. potential, absorbance, electric
current, etc.) measured during titration as
a function of the volume of a titrant added, or
the fraction titrated.

D Titrationskurve
F courbe de titrage
P krzywa miareczkowania
R кривая титрования

titration error
The difference between the volume of titrant
necessary for end-point and volume sufficient
for equivalence-point to be reached.

D Titrierfehler, Titrationsfehler,
   Indikationsfehler
F erreur de titrage
P błąd miareczkowania
R ошибка титрования

titre, titer, T
The reacting strength of a standard solution,
usually given as the mass of the titrated
substance equivalent to 1 cm^3 of the standard
solution. Unit: gram per millilitre, g ml^{-1}.
Remark: The term titre must not be used to denote the
volume of titrant consumed in a particular titration.

D Titer
F titre
P miano
R титр

titrimetric analysis
A technique of quantitative analysis based on
titration; the determination of a substance is
usually effected from the measured volume of
the titrant.
Remark: The term volumetric analysis is not recommended.

D titrimetrische Analyse, Titrieranalyse,
   volumetrische Analyse
F analyse titrimétrique, titrimétrie, volumétrie
P analiza miareczkowa, analiza objętościowa
R титриметрический анализ, объёмный
   анализ

titrimetric conversion factor
The factor giving the amount (usually mass) of
the titrated substance corresponding to unit
amount (usually cm^3) of the standard
solution.

D Faktor der Titration
F facteur de conversion titrimétrique
P współczynnik przeliczeniowy miareczkowania
R титриметрический фактор пересчёта

TLC → thin-layer chromatography

TMCS → trimethylchlorosilane

TOF → time of flight

TOF spectrometer → time-of-flight mass spectrometer

torsion balance
A helical coil balance in which the determination of mass is effected from the torsional or bending moment of a resilient element, e.g. a quarz filament coil.
D Torsionswaage
F balance de torsion
P waga torsyjna
R торзионные весы

total absorption peak → full energy peak

total analysis → complete analysis

total consumption burner → direct-injection burner

total (internal) reflection
Reflection of an incident ray of light back into the initial medium at an interface between this medium and one that has a lower refractive index.
D Totalreflexion
F réflexion totale
P całkowite wewnętrzne odbicie
R полное внутреннее отражение

total liquid volume, V_t (*in permeation chromatography*)
The total volume of the mobile phase in the chromatographic system. It is the sum of the stationary liquid volume V_i, the interstitial volume V_o, and the extra-column volume V_{ext},

$$V_t = V_i + V_o + V_{ext}$$

The total liquid volume represents the total retention volume of all solutes, the molecules of which are smaller than the smallest pores of the column packing.
D Gesamtelutionsvolumen
F volume de liquide total
P objętość cieczy całkowita
R полный объём растворителя в колонке, полный объём подвижной фазы

totally porous support
A support of high specific surface, possessing pores in its whole structure, e.g. diatomaceous earth, porous silica beads.
D poröser Träger
F support poreux
P nośnik porowaty
R пористый носитель

total pore volume → stationary liquid volume

total reflection → total internal reflection

total retention time, uncorrected retention time, retention time, t_R
The time interval from the moment of injection of a sample into the column to the appearance of the peak maximum of the given component in the eluate.
Remark: In liquid chromatography the total retention time should be defined as the time elapsed between the start of elution and the appearance of the peak maximum.
D Gesamtretentionszeit, Bruttoretentionszeit, Retentionszeit
F temps de rétention (total), temps de rétention non corrigé, temps de rétention brut
P czas retencji (całkowity), czas retencji niepoprawiony
R (общее) время удерживания, неисправленное время удерживания

total retention volume → retention volume

total specific capacity, theoretical specific capacity, Q_o (*of an ion exchanger*)
The number of milliequivalents of ionogenic groups per 1 g of the dry ion exchanger. If not otherwise stated, the mass of the exchanger should refer to the H^+-form of the cation exchanger and Cl^--form of the anion exchanger.
D totale Ionenaustauschkapazität, Gesamtkapazität
F capacité totale spécifique, capacité théorique spécifique
P zdolność wymienna całkowita, pojemność wymienna właściwa całkowita
R теоретическая удельная ёмкость

total water hardness
The total content of calcium and magnesium ions in water given in $mval/dm^3$, or degrees of (water) hardness.
D Gesamthärte
F dureté totale
P twardość wody ogólna, twardość wody całkowita
R общая жёсткость воды

TPD → temperature programmed desorption

trace analysis
The detection and/or determination of the constituents in a sample present in quantities lower than 100 ppm.
D Spurenanalyse
F analyse de traces
P analiza śladowa
R анализ следов

trace carrier → trace collector

trace collector, trace scavenger, trace carrier
A solid substance added to or formed in a solution to collect a micro- or macro-component.

D Spurenfänger, Spurensammler, Kollektor
F collecteur (de traces), entraîneur
P kolektor śladów, nośnik śladów
R коллектор (следов)

trace constituent
A constituent whose content in a sample is
less than 0.01%.
D Spurenbestandteil
F trace
P składnik śladowy
R следовой компонент

trace scavenger → trace collector

transfer coefficient → charge-transfer
coefficient

transference number of ionic species B →
transport number of ionic species B

transfer pipette, one mark pipette
A graduated glass tube for measuring out
a fixed quantity of a liquid.
D Vollpipette mit einer Marke, Pipette auf
  Abstrich
F pipette courante à un trait
P pipeta z jedną kreską
R пипетка с одной меткой

transient equilibrium → transient radioactive
equilibrium

transient methods, relaxation methods
(*in electrochemistry*)
A number of electroanalytical methods in
which the current flowing through the
indicator electrode, or the potential of this
electrode, varies in time, and reflects changes
in the concentration of an electroactive
substance at the electrode surface.
D Relaxationsmethoden
F méthodes de relaxation
P metody relaksacyjne
R релаксационные методы

transient (radioactive) equilibrium
The stationary state between a radioactive
parent nuclide and its daughter in the case
when the half-life of the parent nuclide is by
several to several tens times longer than that
of the daughter.
D (radioaktives) Übergangsgleichgewicht,
  laufendes (radioaktives) Gleichgewicht
F équilibre (radioactif) transitoire
P równowaga (promieniotwórcza) przejściowa
R переходное (радиоактивное) равновесие

transition interval (*of a visual indicator*)
The concentration range of hydrogen ion,
metal ion, or other species over which the eye
is able to perceive a variation in hue, colour
intensity, fluorescence, or other property of
a visual indicator arising from the varying

ratio of the two conjugate forms of the
indicator. The range is usually given in terms
of the negative decadic logarithm of
concentration (e.g. pH). For an oxidation-
reduction indicator the range of oxidation-
reduction potential corresponds to the
transition interval.
D Umschlagsintervall
F intervalle de virage, intervalle de transition
P obszar przejścia
R интервал перехода

transition time, τ (*in chronopotentiometry*)
The time from the moment when the
electrolysis is started to the moment when the
concentration of the electroactive substance
at the electrode surface decreases
approaching zero.
D Transitionszeit, Übergangszeit
F temps de transition
P czas przejścia
R переходное время, время перехода

transmission → internal transmittance

transmission energy loss spectroscopy, TELS,
characteristic energy loss spectroscopy, CELS
A method involving the transmission of an
electron beam of energy from 20 to 60 keV
through a thin sample (about 100 nm thick),
in order to determine the energy losses of
primary electrons due to ionization and
excitation of atoms in the sample.
D Transmissions-Elektronenenergieverlust-
  Spektroskopie
F spektroskopie de transmission de pertes
  d'énergie
P spektroskopia transmisyjna strat energii
  elektronów
R спектроскопия потерь проходящей энергии,
  спектроскопия характеристических потерь
  энергии

transmission factor, transmittance, τ
The ratio of the radiant or luminous flux
transmitted by a substance to the flux incident
upon that substance.
D Transmissionsfaktor, Durchlässigkeit
F facteur de transmission, transmittance
P transmitancja
R коэффициент пропускания, пропускание

transmission grating
A diffraction grating produced on
a transparent base through which the incident
radiation is transmitted.
D Transmissionsgitter, Durchlaßgitter
F réseau par transmission
P siatka transmisyjna, siatka przepuszczająca
R прозрачная решётка

transmittance → transmission factor

transmittancy → internal transmittance

transmitted wave method (*in EPR spectroscopy*)
The method in which the paramagnetic resonance absorption is detected by the change of power transmitted through the cavity resonator containing the substance under investigation.

D durchgelassene Mikrowellen-Methode
F méthode de micro-onde transmise
P metoda fali przechodzącej
R метод проходящей волны

transport detector, phase transformation detector, solute transport detector (*in chromatography*)
The detector in which the liquid column effluent is fed continuously to a transport system, such as a moving belt, wire or chain, and after evaporation of the volatile components of the mobile phase the residue of less volatile components is burned to carbon dioxide and water. The carbon dioxide is next reduced in a hydrogen stream over a nickel catalyst to methane which is then passed with a gas stream to the flame ionization detector, the response of which is proportional to the number of carbon atoms in the solute molecule.

D Transportdetektor
F détecteur á transport du soluté
P detektor transportowy
R детектор транспортного типа

transport number of ionic species B,
transference number of ionic species B, t_B
The fraction of the current carried by a given ionic species B, given by the equation

$$t_B = \frac{|z_B|\, u_B\, c_B}{\sum_i |z_i|\, u_i\, c_i}$$

where z — charge number, u — electric mobility, and c — actual concentration of the ionic species.

D Überführungszahl der Ionensorte B
F nombre de transport de l'ion B
P liczba transportu jonu B, liczba przenoszenia jonu B
R транспортное число иона B, число переноса иона B

trend → drift

triangular-wave polarography
A modification of polarography in which, during the life time of one drop, one triangular-wave potential pulse is applied to the indicator electrode. Slopes of the linear

 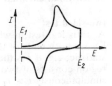

sections of the potential-time curve (*a* and *b*) may be equal or unequal.

D Dreieckwellen-Polarographie, Dreiecksspannungspolarographie
F polarographie à balayage triangulaire
P polarografia z falą trójkątną
R полярография с треугольной развёрткой потенциала

triangular-wave voltammetry
The voltammetry under conditions of a diffusion mass transport to/from the surface of the indicator electrode with a non-renewable surface, when the potential of this electrode is varied linearly in time from a value E_1 to E_2 and back to E_1. The slopes of the two linear segments of the potential-time curve (*a* and *b*) may be different.

Remark: When *a* and *b* are equal, the term cyclic voltammetry is sometimes used.

 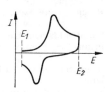

D Dreieckwellen-Voltammetrie, Dreiecksspannungsvoltammetrie
F voltampérométrie à balayage triangulaire
P woltamperometria z falą trójkątną
R вольтамперометрия с треугольной развёрткой потенциала

trimethylchlorosilane, TMCS
A silylation reagent used in gas chromatography either for deactivation of supports and adsorbents or for derivatization of sample components.

D Trimethylchlorsilan
F triméthylchlorosilane
P trimetylochlorosilan
R триметилхлорсилан

triple ions → ion triplet

tritium target
A metal target covered with a thin layer of titanium or zirconium with tritium adsorbed on it, used in fast neutron generators.

D Tritium-Target, Tritiumauffänger
F cible tritiée, cible de tritium
P tarcza trytowa
R тритиевая мишень, мишень из трития

true degree of dissociation
The fraction of the overall number of molecules of an electrolyte which are dissociated at a given concentration of the electrolyte.

D wahrer Dissoziationsgrad
F degré de dissociation réel
P stopień dysocjacji rzeczywisty
R истинная степень электролитической диссоциации

true (half-) width
The spectral line width resulting from the
physical properties of the systems emitting or
absorbing radiation, but free from the
influence of the instrumental factors; it is
usually presented as the true half-width $\Delta v^t_{1/2}$.
D wahre Halbwertsbreite
F largeur vraie (à mi-absorption)
P szerokość linii spektralnej rzeczywista
R истинная ширина спектральной линии

true width → true half-width

tunable laser
A laser in which the frequency of the output
radiation can be tuned over a part or all the
ultraviolet, visible or infrared regions of the
spectrum.
D abstimmbarer Laser
F laser accordable
P laser przestrajalny
R перестраиваемый лазер

tungsten filament lamp → tungsten
incandescent lamp

tungsten incandescent lamp, tungsten filament
lamp
The incandescent lamp which consists of
a tungsten filament in a glass enclosure, usually
employed as a continuous source of radiation
for the visible, near-infrared and near-ultraviolet
regions. The operating temperature is about
2850 K.
D Wolframbandlampe, Wolframdraht-
Glühlampe
F lampe à filament au tungstène, lampe à tube
au tungstène
P lampa wolframowa
R лампа (накаливания) с вольфрамовой
нитью, лампа с вольфрамовой спиралью

turbidimeter
A device that measures the loss of intensity of
a light beam as it passes through a suspension
of particles large enough to scatter the light.
D Turbidimeter, Trübungsmesser
F turbidimètre
P turbidymetr
R турбидиметр

turbidimetric analysis → turbidimetry

turbidimetric titration, heterometric titration
Precipitation titration in which the end-point
is determined from the variation of the logarithm
of the ratio of the intensity of incident ray to
the intensity of the non-scattered ray passing
through the titrated solution.
D heterometrische Titration, turbidimetrische
Titration
F titrage turbidimétrique
P miareczkowanie turbidymetryczne,
miareczkowanie heterometryczne
R турбидиметрическое титрование

turbidimetry, turbidimetric analysis
A method of quantitative analysis which
consists in the measurement of the change of
intensity of the light beam transmitted caused
by the turbidity of the solution due to the
presence of a solid or colloidal suspension.
D Turbidimetrie
F turbidimétrie
P turbidymetria, analiza turbidymetryczna
R турбидиметрия

turbulent flame
A flame produced when fuel and oxidant gases
are mixed above the burner orifices due to
their turbulent motion.
D turbulente Flamme
F flamme turbulente
P płomień turbulentny
R турбулентное пламя

twisting vibrations,
A type of bonding vibrations in which the
structural unit rotates back and forth around
the bond connecting it to the rest of the
molecule.
D Torsionsschwingungen, τ-Schwingungen,
Drillschwingungen
F vibrations de torsion
P drgania (deformacyjne niepłaskie) skręcające
R торзионные колебания, крутильные
колебания

two-colour indicator
An indicator exhibiting two different colours,
one on each side of the transition interval.
D Zweifarbigerindikator
F indicateur bicolore, indicateur deux couleurs
P wskaźnik dwubarwny
R двухцветный индикатор

two-dimensional chromatography, two-
dimensional development
A technique applied in paper or thin-layer
chromatography, in which the chromatogram
is developed successively in two directions
usually at right angles to each other, and in
two different solvent systems.
D zweidimensionale Chromatographie,
zweidimensionale Entwicklung
F chromatographie bidimensionnelle,
chromatographie á deux dimensions
P chromatografia dwukierunkowa, technika
dwukierunkowa
R двумерная хроматография

two-dimensional development → two-
dimensional chromatography

two-sided test
A statistical test in which the rejection region
lies on both ends of the set of values of the test
statistic; e.g. if t is the statistic, then this
region consists of two regions for which
simultaneously $t > t_1$ and $t < t_0$.

D zweiseitiger Test
F test bilatéral
P test dwustronny
R двухстронний критерий

Tyndall effect, Tyndall scattering
Scattering of light which is due to the
interaction of light with small particles of the
dispersed phase rather than with molecules.
D Tyndall-Effekt
F effet Tyndall
P efekt Tyndalla
R эффект Тиндаля

Tyndall scattering → Tyndall effect

type I error → error of first kind

type II error → error of second kind

type I model → fixed effects model

type II model → random effects model

U

ultra-high vacuum
The vacuum conditions in which the gas
pressure in the system is kept below 10^{-8} Pa.
D Ultrahochvakuum, Höchstvakuum
F ultra vide, vide ultrapoussé
P ultrapróżnia, próżnia ultrawysoka
R сверхвысокий вакуум

ultramicro analysis
Analysis of a sample whose mass does not
exceed 10^{-4} g; the minimum amount of the
sample is not specified.
Remark: The ultramicro analysis encompasses micro-
(10^{-6} to 10^{-3} g), nano- (10^{-9} to 10^{-6} g), and
picogram (10^{-12} to 10^{-9} g) methods of analysis.

D Ultramikroanalyse
F ultramicroanalyse
P ultramikroanaliza
R ультрамикроанализ

ultramicro balance
A balance having a permissible load of several
hundred of milligrams for weighing with in
the sensitivity range of 0.01 to 0.1 µg.
D Ultramikrowaage
F ultramicrobalance, balance ultra-
  microchimique
P ultramikrowaga
R ультрамикровесы

ultrasonic nebulizer
A nebulizer in which the sample is fed to, or
caused to flow over a surface which vibrates at
an ultrasonic frequency.
D Ultraschallnebelblaser
F nébuliseur à ultrasons
P nebulizer ultradźwiękowy
R ультразвуковой распылитель

ultraviolet (light), ultraviolet radiation, UV
Electromagnetic radiation with wavelengths
ranging from 10 to 380 nm.
D Ultraviolett(strahlung)
F ultraviolet, radiation ultraviolette
P nadfiolet, ultrafiolet, promieniowanie
  nadfioletowe
R ультрафиолет(овое излучение)

Entries:

ultraviolet photoelectron spectroscopy,
ultraviolet photoemission spectroscopy, UPS
A type of photoelectron spectroscopy which utilizes ultraviolet photons of energy up to 40.8 eV in order to observe the changes in the valence-band region. These changes may be also observed for surfaces covered with adsorbed layers and may be related to the electronic structure of the surface-adsorbate bond.

D UV-Photoelektronenspektroskopie
F spectroscopie des photoélectrons (excités par rayons), spectroscopie des photoélectrons induits par radiation UV
P spektroskopia fotoelektronów wybijanych promieniowaniem z zakresu UV
R ультрафиолетовая фотоэлектронная спектроскопия, УФ-фотоэлектронная спектроскопия

ultraviolet photoemission spectroscopy → ultraviolet photoelectron spectroscopy

ultraviolet radiation → ultraviolet light

ultraviolet region
The part of electromagnetic spectrum with wavelengths ranging from 10 to 380 nm; this range begins at the short wavelength limit of visible light and overlaps the wavelength of long X-rays.

D Ultraviolett-Bereich
F région ultraviolette
P zakres nadfioletu
R ультрафиолетовая область

ultraviolet spectrum, UV spectrum
Electromagnetic spectrum in the ultraviolet wavelength region with bands corresponding to transitions between different electronic energy levels in atoms or molecules; either emission or absorption one.

D Ultraviolett-Spektrum
F spectre ultraviolet
P widmo w nadfiolecie, widmo UV
R ультрафиолетовый спектр, спектр УФ

umpire analysis
The analysis carried out to determine the composition of a given sample, when the results obtained in different laboratories are inconsistant. The result obtained from the umpire analysis carried out by the institution accepted by both parties for arbitration, are not open to question.

D Schiedsanalyse
F analyse d'arbitrage
P analiza arbitrażowa, analiza rozjemcza
R арбитражный анализ

unbiased estimator
An estimator with the expected value equal to the true value of the estimated parameter, irrespectively of the size of the sample used for calculation of the estimator.

D erwartungstreue Schätzfunktion, unverzerrte Schätzfunktion
F estimateur sans biais
P estymator nieobciążony
R несмещённая оценка

uncommon ion effect → salt effect

uncorrected retention time → total retention time

uncorrected retention volume → retention volume

undamped balance
A balance with the beam supported by a centrally located prism; the equipoise (equilibrium point) is established when the excursions of the pointer towards either side of the mid-point of the scales are of equal amplitude.

D ungedämpfte Waage, periodische Waage, schwingende Waage
F balance périodique
P waga periodyczna
R периодические весы

under-exposure region (*of a photographic emulsion*)
The lower part of the emulsion characteristic curve corresponding to the region of insufficient exposure.

D Unterexpositionsgebiet
F région de sous-exposition
P obszar niedoświetleń
R область недодержек

unified atomic mass unit, atomic mass unit, u
A unit of mass not belonging to the International System of Units:
1 u = 1.661×10^{-27} kg.

D (vereinheitlichte) atomare Masseneinheit
F unité de masse atomique
P jednostka masy atomowej
R (унифицированная) атомная единица

unit sample
The part of a bulk material which is composed of all the primary samples taken from one package.

D Einzelprobe
F échantillon unitaire
P próbka jednostkowa
R индивидуальная проба

universal indicator, multiple range indicator
A mixture of several acid-base indicators, changing continuously its colour in a broad pH range, used for rough pH determination.

D Universalindikator
F indicateur universel
P wskaźnik uniwersalny
R универсальный индикатор, комбинированный индикатор

unpaired electron
The electron on singly occupied atomic or
molecular orbitals responsible for the
paramagnetic effect in atoms or molecules.

D ungepaartes Elektron
F électron célibataire, électron non apparié
P elektron niesparowany
R неспаренный электрон

unresolved peak
A composite peak representing two or more
unseparated or partially separated components.

D nichtaufgelöster Peak
F pic non résolu
P pik nierozdzielony
R неразрешённый пик

unsaturated chamber (*in chromatography*)
A large volume chamber which is not
presaturated with the eluent vapour and the
paper or thin layer plate is placed in it
immidiately after its filling with the eluent.

D ungesättigte Kammer
F chambre à atmosphère insaturée
P komora nienasycona
R ненасыщенная камера

unsaturated Weston cell
The Weston cell in which a solution of
cadmium sulfate, saturated at temperature $4°C$
(and unsaturated at ambient temperature) is
used as the electrolyte; it is characterized by
a very small temperature coefficient of the
EMF.

D ungesättigtes Weston-Element
F pile de Weston insaturée
P ogniwo Westona nienasycone
R ненасыщенный элемент Вестона

Unterzaucher method for oxygen determination
A method for direct determination of oxygen
involving the pyrolysis of organic substance
at high temperature, in pure nitrogen
atmosphere; the resulting gaseous product,
containing the whole oxygen present in the
compound analysed, is converted into water
and carbon dioxide. Both these compounds
are subsequently passed over carbon at
a temperature of $1120°C$ to form carbon
monoxide only which is next oxidized with
anhydroiodic acid. Liberated iodine is
determined by titration according to the
Winkler-Leipert method.

D Unterzaucher-Methode der
  Sauerstoffbestimmung
F méthode d'Unterzaucher de dosage
  d'oxygène
P metoda Unterzauchera oznaczania tlenu
R метод Унтерзаухера определения кислорода

UPS → ultraviolet photoelectron spectroscopy

U-tube
A type of absorption tube.

D U-Rohr
F tube absorbeur en U
P U-rurka
R U-образная трубка

UV → ultraviolet light

UV spectrum → ultraviolet spectrum

vacuum desiccator
A desiccator used for drying and keeping materials under reduced pressure.
D Vakuumexsikkator
F dessiccateur sous vide
P eksykator próżniowy
R вакуум-эксикатор

vacuum ultraviolet, VUV
The part of ultraviolet radiation with wavelengths shorter than about 200 nm.
D Vakuumultraviolett, Vakuum-UV, Schumann-Ultraviolett
F ultraviolet à vide
P nadfiolet próżniowy, nadfiolet Schumanna
R вакуумный ультрафиолет

val → gram equivalent

valence level electron spectroscopy
The spectroscopy of electrons displaced from the outer electronic levels.
D Elektronenspektroskopie der Valenzniveaus
F spectroscopie d'électrons déplacés des couches de valence d'atome
P spektroskopia elektronów walencyjnych
R спектроскопия валентных электронов

valence vibrations → stretching vibrations

R_B **value**, R_s **value** (*in planar chromatography*)
The ratio of the distance travelled by the centre of a zone of the solute of interest to the distance simultaneously travelled by the centre of the zone of the reference standard B (both measured from the starting line).
D R_B-Wert
F valeur R_B
P wartość R_B
R величина R_B

R_f **value**, retardation factor, relative front (*in planar chromatography*)
The ratio of the distance travelled by the centre of a zone of a solute of interest to the distance simultaneously travelled by the mobile phase front (both measured from the starting line).

Remark: The symbol R_F is also used. The value of R_f is always smaller than unity and it is usually given up to two decimal digits. In order to simplify this notation often the so-called hR_f value, equal to 100 R_f is used.
D R_f-Wert
F valeur R_f
P wartość R_f
R величина R_f

R_f' **value**, corrected R_f value (*in planar chromatography*)
The R_f value multiplied by a correction factor, which provides for the formation of the "false" front of the mobile phase that is the result of preadsorption of vapours from the chamber atmosphere.
D R_f'-Wert, korrigierter R_f-Wert
F valeur R_f'
P wartość R_f', wartość R_f skorygowana
R исправленная величина R_f

R_M **value** (*in planar chromatography*)
A quantity which depends linearly on the free energy of transfer of a substance from one phase to the other, and is calculated from the R_f value by means of the equation

$$R_M = \log\left(\frac{1}{R_f} - 1\right)$$

In a given chromatographic system, R_M is an additive value for the particular elements of the molecular structure of a chemical compound.
D R_M-Wert
F valeur R_M
P wartość R_M
R величина R_M

R_M' **value**, corrected R_M value (*in planar chromatography*)
The quantity expressed by the equation

$$R_M' = \log\left(\frac{1}{R_f'} - 1\right)$$

where R_f' — corrected R_f value.
D R_M'-Wert, korrigierter R_M-Wert
F valeur R_M'
P wartość R_M', wartość R_M skorygowana
R исправленная величина R_M

R_s **value** → R_B value

value of a division (*of a balance*)
The inverse of the sensitivity of a balance, expressed in mass units per scale division, e.g. 0.2 mg per division.
D Masse pro Skalenteil
F valeur d'une division
P wartość działki skali
R цена деления шкалы

van Deemter equation
The equation representing the dependence of
the height equivalent to a theoretical plate h on
the interstitial velocity of the mobile phase u,

$$h = A + \frac{B}{u} + Cu$$

where A, B, and C are the plate height
contributions from eddy diffusion, molecular
diffusion, and mass transfer effects, respectively.

D van-Deemter-Gleichung
F équation de van Deemter
P równanie van Deemtera
R уравнение ван Деемтера

van der Waals adsorption → physisorption

variable take-off angle method (*in X-ray
fluorescence analysis*)
A method for the elimination of matrix
effects based on the measurement of the
intensity of excited characteristic X-rays in an
analyte for different take-off angles.

D Veränderlichabnahmewinkel-Methode
F méthode d'angle de prise variable
P metoda zmiennego kąta wyjścia
R метод переменного угла выхода

variance, σ^2
The measure of dispersion of a probability
distribution defined by the formula

$$\sigma^2 = E\{(X - E\{X\})^2\}$$

where E — expected value of a random
variable, X — random variable.

D Streuung, Varianz, Dispersion
F variance, dispersion
P wariancja
R дисперсия

variance propagation
The principle for calculation of the variance of
indirectly measured variable X which is
a function $X = f(X_1, X_2, ..., X_n)$ of some
stochastically independent variables
$X_1, X_2, ..., X_n$, according to the formula

$$\sigma^2(X) = \sum_{i=1}^{n} \left(\frac{\partial X}{\partial X_i}\right)^2 \sigma^2(X_i)$$

This principle is used for estimation of the
maximum absolute error.

D Fehlerfortpflanzungsgesetz
F loi de sommation des erreurs
P prawo sumowania błędu, prawo rozwijania
   błędu, prawo przenoszenia błędów
R распространение дисперсии

Večeřa catalyst
Specially prepared mixture of oxides of Co(II)
and Co(III), often supported on appropriate
carriers (e.g. asbestos, corundum), used as an
oxidizer in the combustion tubes.

D Večeřa-Katalysator
F catalyseur de Večeřa
P katalizator Večeřy, utleniacz Večeřy
R катализатор Вэчержи

velocity of light in vacuum, speed of light in
vacuum, c
The velocity of propagation of electromagnetic
waves in vacuum, a physical constant of the
value $c = 2.997\,92458 \times 10^8$ m s^{-1}

D Lichtgeschwindigkeit im Vakuum
F vitesse de la lumière dans le vide
P prędkość światła w próżni
R скорость света в вакууме

vibrating capacitor → vibrating condenser

vibrating condenser, vibrating capacitor
A condenser with a vibrating plate (the reference
electrode), the other plate (the investigated
electrode) being immobile; the device serves
to monitor changes in the surface potential.
In ultra-high vacuum, it is used as a sensitive
indicator of the cleanness of a surface.

D Schwingkondensator
F condensateur à lame vibrante
P kondensator wibrujący, kondensator
   dynamiczny
R вибрирующий конденсатор, динамический
   конденсатор

vibrational-rotational spectrum
A molecular spectrum corresponding to
transitions between the rotational energy levels
which occur simultaneously with the transitions
from one vibrational energy level to another
within the same electronic level of the molecule;
the spectrum consists of rotation-vibration
bands and appears in the near- and middle-
infrared region.

D Rotations-Schwingungs-Spektrum
F spectre de vibration-rotation
P widmo oscylacyjno-rotacyjne
R вращательно-колебательный спектр

vibrational spectroscopy
A group of the surface sensitive techniques in
which an electric field, electron beam or
infrared light is used to raise an adsorbate to
the permitted vibration state. This makes it
possible to determine the molecular structure
of the adsorbate and to characterize the
solid-adsorbate bonds.

D Schwingungs-Spektroskopie
F spectroscopie vibrationnelle
P spektroskopia wibracyjna
R колебательная спектроскопия

vibrational spectrum
A molecular spectrum corresponding to
transitions between two vibrational energy
levels belonging to the same electronic level of
a non-rotating molecule, and appearing in the
infrared region.

D Schwingungsspektrum
F spectre de vibration
P widmo oscylacyjne
R колебательный спектр

vibration band (*in infrared spectroscopy*)
The band corresponding to the vibrational transition which causes a simultaneous change in the rotational energy.
D Schwingungsband
F bande de vibration
P pasmo oscylacyjne
R колебательная полоса

vibration-rotation band
A band in vibrational-rotational spectrum, made up of a series of closely spaced lines of definite frequency due to transitions between rotational energy levels which occur simultaneously with the transition between two different vibrational energy levels of the same electronic energy level. Every vibrational transition gives a band o this type.
D Rotations-Schwingungsbande
F bande de vibration-rotation
P pasmo oscylacyjno-rotacyjne
R вращательно-колебательная полоса

vibrations (*in molecular spectroscopy*)
The periodical movements of the fragments of molecules, atoms or groups of atoms; also the oscillations of a whole molecule with respect to its equilibrium position.
D Schwingungen
F vibrations
P drgania, oscylacje
R колебания

VIS → visible region

visible radiation → light

visible region, VIS
The range of the electromagnetic spectrum with wavelengths ranging from 380 to 780 nm; the range to which the human eye is sensitive.
D sichtbares Gebiet, sichtbarer Bereich
F région visible
P zakres widzialny
R видимая область

visible spectrum
Electromagnetic spectrum (either emission or absorption) in the visible wavelength region with lines or bands corresponding to transitions between different electronic energy levels in atoms or molecules.
D sichtbares Spektrum
F spectre visible
P widmo widzialne
R видимый спектр

visual colorimetry
A method of analysis in which the eye is used to estimate the intensity of colour of a sample solution as compared to the colour intensity of a standard solution of the substance being determined.
D visuelle Farbmessung, subjektive Farbmessung
F colorimétrie visuelle
P kolorymetria wizualna
R визуальная колориметрия

visual indicator → indicator

visual titration
Titration in which the end point is visually determined (by the change of colour, by appearance of fluorescence, etc.).
D visuelle Titration
F titrage visuel
P miareczkowanie wizualne
R визуальное титрование

void volume → interstitial volume

volatilization
Conversion of a chemical substance from the liquid or solid state to a gaseous or vapour state by application of heat or by reducing pressure.
D Verflüchtigung
F volatilisation
P ulatnianie się
R испарение

Volhard method for determination of halides
A method of determination of halides in acidic medium, consisting in adding an excess of standard silver nitrate solution which is subsequently back-titrated with a standard ammonium thiocyanate solution in the presence of ferric ions as indicator.
D Volhardsche Bestimmung der Halogenide
F méthode de Volhard (de dosage des halogénures)
P metoda Volharda oznaczania halogenków
R метод Фольгарда определения галогенидов

Volhard method for determination of manganese
The determination of manganese consisting in oxidizing manganese(II) to manganese(IV) with a titrant containing permanganate in a neutral medium

$$2MnO_4^- + 3Mn^{2+} + 2H_2O \rightarrow 5MnO_2\downarrow + 4H^+$$

D Volhard-Wolff-Methode
F méthode de Volhard de dosage du manganèse
P metoda Volharda oznaczania manganu
R метод Фольгарда определения марганца

voltage → electric potential difference

voltaic cell → galvanic cell

voltammetric analysis *See* voltammetry
D voltammetrische Analyse
F analyse voltammétrique
P analiza woltamperometryczna
R вольтамперометрический анализ

voltammetry
A group of electrochemical methods based on measurements of the dependence of the electric current I flowing through the indicator electrode on the potential E of this electrode under conditions of forced known changes in the potential or voltage of the cell.
D Voltammetrie
F voltam(péro)métrie
P woltamperometria
R вольтамперометрия

voltammogram → polarization curve

Volta potential → outer electric potential of phase β

Volta pontential difference, $\Delta_\alpha^\beta \psi$
The difference in the outer electrical potentials of two phases.

$$\Delta_\alpha^\beta \psi = \psi^\beta - \psi^\alpha$$

D Differenz der Volta-Potentiale, Volta-Potentialdifferenz, Voltaspannung, Voltasche Kontaktspannung
F différence de potentiel de Volta, tension de Volta
P różnica potencjałów Volty, napięcie Volty, różnica potencjałów kontaktowa
R разность Вольта-потенциалов, напряжение Вольта, контактная разность потенциалов

volume capacity, Q_v (*of an ion exchanger*)
The number of milliequivalents of ionogenic groups per 1 cm^3 (true volume) of the swollen ion exchanger (the ionic form of the ion exchanger and the medium must be specified).
D Ionenaustauscher-Volumenkapazität
F capacité volumique
P zdolność wymienna objętościowa, pojemność wymienna właściwa objętościowa
R объёмная ёмкость

volume capacity of ion exchanger bed → bed volume capacity

volume distribution coefficient → distribution coefficient D_v

volume fraction (of substance B), ϕ_B
The ratio of the volume V_B of a specified constituent B to the volume V_G of a given system containing that constituent

$$\phi_B = \frac{V_B}{V_G}$$

Unit: litre per litre, $1\,l^{-1}$, % (V/V).

D Volumenanteil, Volumengehalt,
F fraction en volume
P ułamek objętościowy
R объёмная доля

volume of the column tube → column volume

volume swelling ratio
The ratio of the volume of a swollen gel or solid to the true volume of the dry gel or solid (e.g. ion exchanger).
D Quellungsvolumensquotient
F rapport (de volume) de gonflement, gonflement relatif
P iloraz pęcznienia objętościowy
R относительная объёмная набухаемость

volumetric analysis → titrimetric analysis

volumetric flask
A flask with a narrow neck on which the rated capacity of the vessel is marked.
D Meßkolben
F fiole jaugée, flacon volumétrique
P kolba pomiarowa
R измерительная колба

volumetric flask with graduated neck
D Meßkolben mit graduiertem Hals
F fiole jaugée avec col gradué
P kolba pomiarowa z szyjką kalibrowaną
R измерительная колба с градуированным горлом

volumetric flow rate F_a, mobile phase volumetric flow rate F_A
The flow rate of the mobile phase in $cm^3\ min^{-1}$ at a given outlet pressure and ambient temperature.
D Volumenfluß F_a der mobilen Phase, Volumenströmungsgeschwindigkeit F_a
F débit d'écoulement F_a, débit volumique F_a de la phase mobile
P prędkość przepływu objętościowa fazy ruchomej
R объёмная скорость потока F_a, объёмная скорость F_a подвижной фазы

volumetric flow rate F_c, mobile phase volumetric flow rate F_c
The flow rate of the mobile phase in $cm^3\ min^{-1}$ at a given outlet pressure and column temperature

$$F_c = F_a \frac{T_c}{T_a}$$

where T_c − column temperature, T_a − ambient temperature, F_a − volumetric flow rate of the mobile phase at a given outlet pressure and ambient temperature.
D Volumenfluß F_c (der mobilen Phase), Volumenströmungsgeschwindigkeit F_c
F débit d'écoulement F_c, débit volumique F_c de la phase mobile

P prędkość przepływu objętościowa
skorygowana fazy ruchomej
R объёмная скорость потока F_c, объёмная
скорость F_c подвижной фазы

volumetric gas analysis
The determination of a constituent (or
constituents) of gaseous mixtures based on the
volume changes under constant pressure and
temperature.
D volumetrische Gasanalyse, Gasvolumetrie
F analyse volumétrique des gaz
P analiza gazowa objętościowa
R газоволюметрический анализ,
газоволюметрия, объёмный газовый анализ

VUV → wacuum ultraviolet

W

wagging vibrations,
A type of bending vibration in which the
structural unit swings back and forth in
a plane perpendicular to the symmetry plane
of the molecule; no change takes place in the
angle between the bonds of the vibrating group.
D Nickschwingungen, ω-Schwingungen,
Kippschwingungen
F vibrations de roulis, vibrations de
balancement
P drgania wachlarzowe
R веерообразные колебания, веерные
колебания

washing of precipitate
The treating of a separated precipitate with a
washing solution.
D Auswaschen des Niederschlages,
Niederschlagsauswaschen
F lavage du précipité
P przemywanie osadu
R промывка осадка

washing solution
A solution for washing precipitates. Generally
it is an aqueous solution of substances
diminishing the solubility of crystalline
precipitates, or a solution of electrolyte
inhibiting peptization of colloidal precipitates.
D Waschflüssigkeit
F solution de lavage
P roztwór przemywający
R промывной раствор

water hardness
The property of water due mainly to the
presence of calcium and magnesium ions which
form an insoluble precipitate (boiler scale)
and inhibit lathering of soap solution.
D Härte des Wassers
F dureté de l'eau
P twardość wody
R жёсткость воды

water memory effect (*in elemental analysis*)
An effect manifested by the erroneous results
(too low, or too high) of hydrogen content in

the samples analysed, caused by chemical
reagents present in the combustion tube.
These reagents may retain water present in
the system, or given up trace amounts of
water previously occluded by them. This effect
is often observed when compounds with
various hydrogen contents are analysed one
after another.
D ...
F effet mémoire relatif à l'eau
P efekt pamięci wody
R ...

water of crystallization
The water occurring in hydrates in a
stoichiometric ratio, having a definite location
in the crystalline lattice, and bound directly
to the particular anions or cations; it is
removed stepwise upon heating.
D Kristallwasser
F eau de cristallisation
P woda krystalizacyjna
R кристаллизационная вода

wavelength, λ
The distance measured along the line of
propagation of a wave, between two nearest
points which are in the same phase in the
wave.
SI units: nanometre, nm, micrometre, µm,
centimetre, cm, metre, m.
D Wellenlänge
F longueur d'onde
P długość fali
R длина волны

**wavelength dispersive X-ray fluorescence
analysis**
A method of the X-ray fluorescence analysis
based on measurement of the wavelength of
the X-ray emission spectrum; the separation
of X-rays according to their wavelength is
achieved with a crystal of known interplanar
spacing.
D wellenlängendispersive
Röntgenfluoreszenzanalyse
F analyse par fluorescence X avec
dispersion des ondes
P analiza fluorescencyjna rentgenowska
z dyspersją fal
R рентгенофлуоресцентный анализ с волновой
дисперсией

wave number, $\sigma, \tilde{v}$
The reciprocal of the wavelength.
SI unit: cm^{-1}.
Remark: This term is sometimes incorrectly substituted by
the term frequency, due to the relationship: $\sigma = v/c$, where
c — velocity of the electromagnetic wave motion and
v — frequency.
D Wellenzahl
F nombre d'ondes
P liczba falowa
R волновое число

weighed amount, weighed sample
A quantity of a substance weighed out for
analysis with a definnit accuracy.
D Einwaage
F pesée
P odważka
R навеска

weighed sample → weighed amount

weighing bottle
A stoppered glass or quartz vessel for weighing,
drying and keeping samples of solids.
D Wägeglas, Wägegefäß
F flacon à tare, vase à peser
P naczynko wagowe
R весовой стаканчик, сосуд для взвешивания

weighing scoop
D Wägeschiffchen
F main pour poudres
P szufelka do odważania substancji
R ложка для взятия навески

weight swelling in solvent, solvent regain, w_s
The quantity of solvent (in grams) taken up by
one gram of dry solid or gel (e.g. ion
exchanger) after the swelling equilibrium has
been reached; when water is the solvent, the
symbol w_{H_2O} is used.
D ...
F gonflement dans un solvant donné
P pęcznienie masowe w rozpuszczalniku
R весовая набухаемость в растворителе

weight titration
A titration in which the amount of titrant is
determined by the measurement of its mass.
D gravimetrische Titration, Wägetitration
F titrage par pesée
P miareczkowanie wagowe
R весовое титрование

well-type counter → well-type scintillation
counter

well-type (scintillation) counter
A scintillation counter whose scintillator has
a cylindrical hole to accomodate a sample of
a small size; this permits to measure the
activity of a sample at a solid angle of 4π.
D Zählrohr mit Probenkanal
F scintillateur à puits, compteur à canal
P licznik, studzienkowy, licznik wnękowy
R счётчик с каналом (для образцов)

Weston normal cell → Weston standard cell

Weston standard cell, Weston normal cell
A standard cell schematically represented as:
$Cd(Hg)_{(l)}$ (12.5%) | $CdSO_4$ 8/3 H_2O sat.sol.|
|$Hg_2SO_{4(s)}$ | $Hg_{(l)}$
and having a very well reproducible and
invariant in time EMF.

D Westonsches Normalelement,
 Weston(-Normal)-Element
F pile standard Weston, pile de Weston, pile
 étalon Weston, élément Weston standard
P ogniwo Westona standardowe, ogniwo
 Westona normalne
R стандартный элемент Вестона,
 нормальный элемент Вестона, эталонный
 элемент Вестона

wet method
The method of analysis of a solid sample after
passing it into solution.
D nasses Verfahren
F procédé par voie humide
P metoda mokra
R мокрый путь (анализа)

Wheatstone bridge
The arrangement of resistances used for the
measurement of an unknown resistance R_x
by comparison with known resistors R_1, R_2
and R_3; under balanced conditions no current
flows through the measuring instrument G
(e.g. galvanometer) and the resistance R_x can
be determined from the following equation

$$R_x = R_3 \; \frac{R_1}{R_2}$$

D Wheatstonesche Brücke
F pont de Wheatstone
P mostek Wheatstone'a
R мостик Уитстона

white light
Any one of a variety of spectral energy
distributions producing the same colour
sensation as the average noon sunlight.
D weißes Licht
F lumière blanche
P światło białe
R белый свет

Wien's effect, first Wien effect, field strength
effect
An increase in the molar conductivity of
a strong electrolyte when measurements are
made at a high electric field strength
($10^4 - 10^5$ V/cm) due to the partial
disappearance of the ionic atmosphere.
D erster Wien-Effekt, Feldstärkeeffekt
F effet Wien
P efekt Wiena, efekt natężenia pola
 elektrycznego
R эффект Вина

window counter, end window counter
A Geiger-Müller counter with a window
penetrable to β- or α-radiation.
D Fensterzählrohr
F compteur à fenêtre
P licznik okienkowy
R счётчик с окном

working electrode
An electrode (half-cell) used for the
determination of the concentration of an
electroactive substance by measuring the
current flowing through this electrode.
Processes occurring at the working electrode
change significantly the bulk composition of
the solution being investigated.
D Arbeitselektrode
F électrode de travail
P elektroda pracująca
R рабочий электрод

X

XAES → X-ray induced Auger electron spectroscopy

XAS → X-ray absorption spectroscopy

XES → X-ray emission spectroscopy

XPS → X-ray photoelectron spectroscopy

X-radiation, X-rays
The electromagnetic radiation of wavelengths (approximately) from 10 to 10 000 pm.
D Röntgenstrahlung, Röntgenstrahlen
F rayonnement X, rayons X
P promieniowanie rentgenowskie, promieniowanie X, promieniowanie Röntgena
R рентгеновское излучение, рентгеновские лучи

X-radiation intensity
The number of photons recorded by the detector in a given solid angle energy range in a unit time.
D Röntgenstrahlintensität
F intensité de rayonnement X
P natężenie promieniowania rentgenowskiego
R интенсивность рентгеновских лучей

X-ray absorption spectroscopy, XAS
A branch of spectroscopy which deals with the investigation of X-ray absorption spectra.
D Röntgenabsorptionsspektroskopie
F spectroscopie d'absorption des rayons X
P spektroskopia absorpcyjna promieniowania rentgenowskiego
R абсорбционная рентгеновская спектроскопия

X-ray characteristic spectrum
The discrete X-ray spectrum corresponding to the electron transitions in the inner electronic shells of atoms.
D charakteristisches Röntgenspektrum
F spectre de rayons X caractéristiques
P widmo charakterystyczne rentgenowskie
R рентгеновский характеристический спектр

X-ray diffraction, XRD
The elastic and coherent scattering and interference of X-rays in a crystalline substance.
D Beugung der Röntgenstrahlen, Röntegenbeugung, Röntgendiffraktion
F diffraction des rayons X
P dyfrakcja promieniowania rentgenowskiego
R дифракция рентгеновских лучей

X-ray diffraction analysis
Determination of the crystal structure from X-ray diffraction data: the angular positions and intensities of the interference peaks observed on the diffraction pattern. It is also used for identification of crystalline substances (phase analysis) and for quantitative determination of their composition.
D Röntgendiffraktionsanalyse
F analyse par diffraction des rayons X
P analiza dyfrakcyjna rentgenowska
R рентгенодифракционный анализ

X-ray diffractometer
An instrument used for measuring the angular positions and intensities of the interference peaks produced by X-rays diffracted in a crystalline medium.
D Röntgendiffraktometer, Röntgenbeugungsgerät
F diffractomètre à rayons X
P dyfraktometr rentgenowski
R рентгеновский дифрактометр

X-ray electron appearance potential spectroscopy
Probing of a solid surface with X-ray beam in order to detect the threshold of the appearance of electrons emitted from the solid.
D ...
F spectroscopie d'apparition d'électrons excités par rayons X
P spektroskopia potencjału pojawiania się elektronów wybijanych promieniowaniem rentgenowskim
R спектроскопия внешнего потенциала электронов индуцированных рентгеновским излучением

X-ray emission spectroscopy, XES
A branch of spectroscopy dealing with the investigation of the X-ray emission spectra of atoms and molecules.
D Röntgenemissionsspektroskopie
F spectroscopie d'émission des rayons X
P spektroskopia emisyjna promieniowania rentgenowskiego
R эмиссионная рентгеновская спектроскопия

X-ray escape peak
A peak in the spectra of γ- or X-radiation caused by the escape of an X-ray photon emitted by an atom of the detector material after photoelectron emission.

D Röntgen-Escape-Peak
F pic de fuite des photons
P pik ucieczki promieniowania rentgenowskiego
R рентгеновский пик утечки

X-ray filter
A thin layer of material, usually metallic,
changing the spectral composition of X-rays
which pass through it, used to decrease the
intensity of interfering radiation or for
monochromatisation of the continuous
radiation.
D Röntgenfilter
F filtre pour rayons X
P filtr promieniowania rentgenowskiego
R фильтр рентгеновского излучения

X-ray fluorescence, XRF
The emission of X-ray fluorescent radiation
due to the transition of an ion from the excited
to the ground state. Such an ion may be
formed by the ejection of an electron from the
inner shell of an atom by a photon of
primary X radiation.
Remark: In chemical analysis the term X-ray fluorescence
analysis is used.
D Röntgenfluoreszenz
F fluorescence X
P fluorescencja rentgenowska
R рентгенофлуоресценция, рентгеновская
флуоресценция

X-ray fluorescence analysis, XRFA,
X-ray fluorimetry
A non destructive physical method used for
chemical analyses of solids and liquids based
upon X-ray fluorescence.
D Röntgenfluoreszenz-Analyse
F analyse par fluorescence X
P analiza fluorescencyjna rentgenowska
R рентгеноспектральный флуоресцентный
анализ

X-ray fluorescence radiation, fluorescent
X-rays
The X-ray characteristic radiation excited
within a sample by the primary X-rays.
D Röntgenfluoreszenzstrahlung
F rayonnement de fluorescence X
P promieniowanie fluorescencyjne
rentgenowskie
R рентгенофлуоресцентное излучение,
флуоресцентное рентгеновское излучение

X-ray fluorescence spectroscopy
A branch of spectroscopy dealing with the
investigation of the X-ray fluorescence spectra.
D Röntgenfluoreszenzspektroskopie
F spectroscopie de fluorescence X
P spektroskopia fluorescencyjna
promieniowania rentgenowskiego
R рентгеновская флуоресцентная
спектроскопия

X-ray fluorescence spectrum
The characteristic X-ray spectrum excitated
within a sample by primary X radiation.
D Röntgenfluoreszenzspektrum
F spectre de fluorescence X
P widmo fluorescencyjne rentgenowskie
R рентгеновский флуоресцентный спектр

X-ray fluorescence yield, ω
The ratio of the number of emitted photons of
characteristic radiation belonging to the
given series to the total number of atoms
excited to this series (e.g. K level for K series).
It is also used as the measure of the probability
of fluorescent X-ray emission.
D Röntgen-Fluoreszenzausbeute
F rendement de fluorescence (de rayons) X
P wydajność fluorescencji rentgenowskiej
R выход рентгеновской флуоресценции,
отдача рентгеновской флуоресценции

X-ray fluorimetry → X-ray fluorescence
analysis

X-ray focusing spectrometer
The X-ray spectrometer with wavelength
dispersion, having a curved analysing crystal
focusing the diffracted radiation at the
circumference of the focal circle, where the
detector is located.
D fokussierendes Kristallspektrometer
F spectromètre des rayons X avec focalisation
P spektrometr rentgenowski ogniskujący
R фокусирующий рентгеновский спектрометр

X-ray goniometer
A device for setting the angle between the
beam of analysed X-rays and the crystal plane
(glancing angle), the crystal plane and the
detector (take-off angle) and for accurate
measurement of these angles.
D Röntgengoniometer
F goniomètre à rayons X
P goniometr rentgenowski
R рентгеновский гониометр

X-ray induced Auger electron spectroscopy,
XAES
A type of Auger-electron spectroscopy
involving soft X-ray radiation as the primary
beam for the excitation of Auger electrons.
D Auger-Elektronen-Spektroskopie mit
Röntgenstrahl-Anregung
F spectroscopie d'électrons Auger excités par
rayons X
P spektroskopia elektronów Augera
wybijanych promieniowaniem rentgenowskim
R спектроскопия оже-электронов
индуцированных рентгеновским
излучением

X-ray oscillating-crystal pattern
The X-ray diffraction pattern obtained by the
oscillating-crystal method.

D Röntgenkonusaufnahme,
  Röntgenschwenkaufbnahme
F diagramme de cristal oscillant
P rentgenogram kryształu wahanego,
  dyfraktogram kryształu wahanego
R рентгенограмма колебания

X-ray pattern, roentgenogram
The X-ray diffraction spectrum produced by
a single crystal or powder sample recorded
either on photographic film or on a paper tape
(when a counter-diffractometer is used). The
positions and intensities of the diffraction
peaks recorded contain information on the
crystal structure of the sample, or its phase
composition.

D Röntgenaufnahme, Röntgenogramm
F spectrogramme X, image radiographique X
P rentgenogram, dyfraktogram
R рентгенограмма

X-ray photoelectron spectroscopy, X-ray
photoemission spectroscopy, XPS, electron
spectroscopy for chemical analysis, ESCA,
soft X-ray photoelectron spectroscopy, SXPS
The spectroscopy of electrons ejected from
atoms on irradiation with X-rays. The spectra
provide information on the chemical
composition of the irradiated sample.

D Röntgen-Photoelektronen-Spektroskopie,
  Elektronenspektroskopie für die chemische
  Analyse
F spectroscopie des photoélectrons X,
  spectroscopie des photoélectrons induits par
  rayonnement X, spectroscopie d'électrons
  pour analyse chimique, spectroscopie
  photoélectronique des rayons X
P spektroskopia fotoelektronów wybijanych
  promieniowaniem rentgenowskim,
  spektroskopia elektronów dla celów analizy
  chemicznej
  рентгеновская фотоэлектронная
R спектроскопия, электронная спектроскопия
  для химического анализа, спектроскопия
  рентгеновских фотоэлектронов

X-ray photoemission spectroscopy → X-ray
photoelectron spectroscopy

X-ray powdered-crystal pattern
The X-ray pattern obtained by the Debye-
Scherrer-Hull method; the diffraction pattern,
in the form of concentric rings, is obtained by
exposing the powdered sample to a beam of
monochromatic X-rays. When a counter-
diffractometer is used, the pattern contains
intenference peaks at angular positions
characteristic for the crystal structure of the
crystalline sample investigated.

D Debye-Scherrer-Aufnahme, Pulveraufnahme
F diagramme de poudre en rayons X
P rentgenogram proszkowy, dyfraktogram
  proszkowy, debajogram
R дебаеграмма, порошкограмма

X-ray rotating-crystal pattern
The X-ray pattern obtained when a monocrystal sample rotates about a selected
crystallographic axis.

D Röntgendrehkristallaufnahme
F diagramme de cristal tournant des rayons X
P rentgenogram warstwicowy, rentgenogram
  kryształu obracanego, dyfraktogram
  warstwicowy, dyfraktogram obrotowy
R рентгенограмма вращения кристалла

X-rays → X-radiation

X-ray spectrochemical analysis
The determination of the chemical composition
of the specimem and some physical and
chemical properties of its component elements
carried out by X-ray emission and/or
absorption spectroscopic analysis.

D Röntgenspektralanalyse
F analyse spectrochimique X
P analiza spektralna rentgenowska
R рентгеновский спектрохимический анализ

X-ray spectrometer
The instrument in which diffraction on crystal
planes of known spacing is used to determine
unknown wavelengths; useful for the study of
the chemical composition of samples.

D Röntgenspektrometer
F spectromètre à rayons X
P spektrometr rentgenowski
R рентгеновский спектрометр

X-ray tube
A vacuum tube generating X-rays when
high-speed electrons undergo deceleration or
eject electrons from the metallic anode; it
emits a discrete and/or continuous spectrum.

D Röntgenröhre
F tube à rayons X
P lampa rentgenowska
R рентгеновская трубка

XRD → X-ray diffraction

XRF → X-ray fluorescence

XRFA → X-ray fluorescence analysis

Z

Zeeman effect
Splitting of the energy levels of an atom or molecule by external magnetic field, involving the splitting of their spectral lines.
D Zeeman-Effekt
F effet de Zeeman
P efekt Zeemana
R явление Зеемана

Zeeman levels
Sublevels arising from splitting of the energetic level of an atom, ion, molecule or paramagnetic nucleus in the magnetic field.
D Zeeman-Niveaus
F sous-niveaux Zeeman
P poziomy zeemanowskie
R зеемановские уровни

Zeisel reaction
The process in which the alkoxy group reacts with boiling hydroiodic acid to form the corresponding alkyl iodide

$$ROR' + HI \leftrightarrows ROH + R'I$$

It is used for determination of alkoxy groups.
D Zeisel-Reaktion
F réaction de Zeisel
P reakcja Zeisela
R реакция Цейзеля

zeolites
Crystalline aluminosilicates having either ion exchange and sorptive properties. This enables the zeolites to act as ion exchangers or as molecular sieves.
D Zeolithe
F zéolithes
P zeolity
R цеолиты

zero branch → Q branch

zero-current potentiometric titration → potentiometric titration

zero-current potentiometry → potentiometry

zero point of the scale
The rest point of the properly adjusted balance with no load on the pans and the rider (or chain) in the zero position.
D Nullpunkt der Zeigerskala
F zéro de l'échelle
P punkt zerowy skali
R нуль шкалы

zero-point potential → potential at the point of zero charge

zeta-potential → electrokinetic potential

zone, band (*in chromatography*)
A region in a chromatographic column or layer where one or more components of the sample are located.
D Zone, Bande
F zone, bande
P strefa, pasmo
R зона, полоса

zwitterion → ampholyte ion

DEUTSCHES WÖRTERVERZEICHNIS

A

Abbrand *m* burn-up
Abbremsung *f* slowing down
Abdampfen *n* evaporation
Abdampfen *n* **der Trennflüssigkeit** bleeding
Aberration *f* optical aberration
abgeschirmte Flamme *f* shielded flame
abgeschwächte Totalreflexion *f* attenuated total
reflectance
abgeschwächte Totalreflexions-Spektrum *n*
attenuated total reflectance spectrum
abgetrennte Flamme *f* separated flame
Ablauf *m* effluent
Ablehnungsbereich *m* rejection region
Ablesbarkeit *f* readability
Abnahmewinkel *m* take-off angle
Abscheiden *n* sedimentation
Abscheidungspotential *n* deposition potential
absolute Detektorempfindlichkeit *f* absolute
detector sensitivity
absolute Dielektrizitätskonstante *f* permittivity
absoluter Brechungsindex *m* refractive index
absoluter Fehler *m* absolute error
absoluter Schwellenwert *m* stimulus threshold
absolutes Elektrodenpotential *n* absolute
electrode potential
Absolutfehler *m* absolute error
Absorbanz *f* absorbance
Absorbat *n* absorbate
Absorbens *n* absorbent
Absorber *m* absorber
absorptiometrische Gasanalyse *f*
absorptiometric gas analysis
Absorption *f* **1.** absorption **2.** absorption
(*of radiation*)
Absorptionseffekt *m* absorption effect
Absorptionsfaktor *m* absorptance
Absorptionsfilter *n, m* absorption filter
Absorptionsgrad *m* absorptance
Absorptionskante *f* absorption edge
Absorptions-Kantenfilter *npl* balanced filters
Absorptionskoeffizient *m* specific decadic
absorption coefficient
Absorptionskurve *f* absorption curve
Absorptionsküvette *f* absorption cell
Absorptionsmittel *n* absorbent
Absorptions-Molekülspektroskopie *f* molecular
absorption spectroscopy
Absorptionsquerschnitt *m* absorption cross
section
Absorptionsröhrchen *n* absorption tube
Absorptionsspektralanalyse *f* absorption
spectrochemical analysis

Absorptionsspektralphotometrie *f* absorption
spectrophotometry
Absorptionsspektroskopie *f* absorption
spectroscopy
Absorptionsspektrum *n* absorption spectrum
Absorptiv *n* absorbate
absteigende Chromatographie *f* descending
development
absteigende Entwicklung *f* descending
development
absteigende Methode *f* descending development
abstimmbarer Laser *m* tunable laser
Acidimetrie *f* acidimetry
acidimetrische Titration *f* acidimetric
titration
ac-Polarographie *f* **höherer Harmonischer**
higher-harmonic ac polarography
ac-Polarographie *f* **mit zweiter Harmonischer**
second harmonic ac polarography
Additivität *f* additivity law
Adhäsion *f* adhesion
Admittanz *f* admittance
Adsorbat *n* adsorbate
Adsorbens *n* adsorbent
Adsorbensaktivierung *f* activation of an
adsorbent
Adsorbensaktivität *f* adsorbent activity
Adsorber *m* adsorbent
Adsorption *f* adsorption
Adsorptionschromatographie *f* adsorption
chromatography
Adsorptionsindikator *m* adsorption indicator
Adsorptionsisobare *f* adsorption isobar
Adsorptionsisostere *f* adsorption isostere
Adsorptionsisotherme *f* adsorption isotherm
Adsorptionsmittel *n* adsorbent
Adsorptionsstrom *m* adsorption current
Adsorptionsstufe *f* adsorption wave
Adsorptiv *n* adsorbate
Affinität *f* affinity
Affinitätschromatographie *f* affinity
chromatography
Agarose *f* agarose
Agglomeration *f* agglomeration
Aggregat *n* aggregate
Aggregation *f* aggregation
Akkumulator *m* electrical accumulator
aktiver Festkörper *m* active solid
aktiver Versuch *m* active experiment
Aktivierung *f* activation
Aktivierung *f* **mit Ladungsträger** charged-
particle activation
Aktivierungsanalyse *f* activation analysis

Aktivierungsanalyse *f* **mit geladenen Kernteilchen** charged-particle activation analysis

Aktivierungsanalyse *f* **mit γ-Quanten** photoactivation analysis

Aktivierungsanalyse *f* **mit schnellen Ionen** charged-particle activation analysis

Aktivierungsenergie *f* activation energy

Aktivierungsquerschnitt *m* activation cross section

Aktivierungsüberspannung *f* activation overpotential

Aktivität *f* activity

Aktivität *f* **der Substanz B** relative activity of substance B

Aktivität *f* **des Elektrolytes B** activity of electrolyte B

Aktivität *f* **eines Stoffes B** relative activity of substance B

Aktivitätskoeffizient *m* **der Substanz B** activity coefficient of substance B

Aktivitätsskale *f* **nach Brockmann** Brockmann scale of activity

Aktivitätsverminderung *f* radioactive cooling

Aktor *m* actor

Akzeptorlösungsmittel *n* protophilic solvent

Alkalimetrie *f* alkalimetry

alkalimetrische Titration *f* alkalimetric titration

Alkoholometrie *f* alcoholometry

Altern *n* ageing

Alternativhypothese *f* alternative hypothesis

Altern *n* **eines Niederschlages** ageing of a precipitate

Alterung *f* ageing

Alterung *f* **der Sole** ageing of sols

Amalgam *n* amalgam

Amalgamelektrode *f* amalgam electrode

Amalgampolarographie *f* amalgam polarography

amerikanische Härte *f* American degree of water hardness

amerikanischer Härtegrad *m* American degree of water hardness

Amperometrie *f* amperometry

Amperometrie *f* **mit einer Indikator-Elektrode** amperometry with one indicator electrode

Amperometrie *f* **mit einer polarisierbaren Elektrode** amperometry with one indicator electrode

Amperometrie *f* **mit gerührter Quecksilber--Pool-Elektrode** stirred-mercury-pool amperometry

Amperometrie *f* **mit rotierender Platindraht--Elektrode** rotating-platinum-wire electrode amperometry

Amperometrie *f* **mit zwei Indikator-Elektroden** amperometry with two indicator electrodes

Amperometrie *f* **mit zwei polarisierbaren Elektroden** amperometry with two indicator electrodes

amperometrische Analyse *f* amperometric analysis

amperometrische Indikation *f* amperometric titration with one indicator electrode

amperometrische Titration *f* amperometric titration with one indicator electrode

amperometrische Titration *f* **einer Quecksilbertropfelektrode** amperometric titration with a dropping mercury electrode

amperometrische Titration *f* **mit einer polarisierbaren Elektrode** amperometric titration with one indicator electrode

amperometrische Titration *f* **mit zwei Indikator-Elektroden** amperometric titration with two indicator electrodes

amperometrische Titration *f* **mit zwei polarisierbaren Elektroden** amperometric titration with two indicator electrodes

amperostatische Methode *f* galvanostatic method

amphiprotisches Lösungsmittel *n* amphiprotic solvent

Ampholyt *m* ampholyte

ampholytisches Lösungsmittel *n* amphiprotic solvent

amphoterer Elektrolyt *m* ampholyte

amphoterer Ionenaustauscher *m* amphoteric ion exchanger

amphoterer Stoff *m* ampholyte

amphoteres Ion *n* ampholyte ion

Analysator *m* analyser

Analysatorkristall *m* analysing crystal

Analyse *f* **durch Kernstrahlenabsorption** nuclear radiation absorptiometry

Analyse *f* **durch Neutronenabbremsung** analysis by neutron slowing-down

Analysenlinie *f* analytical line

Analysen-Probe *f* analytical sample

Analysenstrecke *f* analytical gap

Analysentrichter *m* **für schnelle Filtration** ribbed funnel

Analysenverfahren *n* analytical method

Analysenwaage *f* analytical balance

analytische Chemie *f* analytical chemistry

analytische Gewichte *npl* analytical weights

analytische Reaktion *f* analytical reaction

analytischer Faktor *m* analytical factor

analytisches Signal *n* analytical signal

analytische Wellenlänge *f* analytical wavelength

Anhydrone *n* Anhydrone

Anionenaustausch *m* anion exchange

Anionenaustauscher *m* anion exchanger

Anionenaustauschharz *n* anion-exchange resin

Ankergruppen *fpl* ionogenic groups

Anode *f* anode

anodische Inversvoltammetrie *f* anodic stripping voltammetry

Anolyt *m* anolyte

anomale Dispersion *f* anomalous dispersion

Anpassungstest *m* test for goodness of fit

χ^2-Anpassungstest *m* chi-squared test for goodness of fit

Anregung *f* excitation

Anregungsfunktion *f* excitation function

Anregungsmonochromator *m* excitation monochromator

Anregungsquelle *f* excitation source

Anreicherungsfaktor *m* enrichment factor

Anreicherungsmethode *f* concentration method

Ansprechwahrscheinlichkeit *f* **des Detektors** intrinsic detector efficiency.

Ansprechzeit *f* response time

Anteil *m* der stationären Phase stationary phase
fraction
Antikoinzidenzverfahren *n* anticoincidence
technique
Antimonelektrode *f* antimony electrode
anti-Stokessche Linien *fpl* anti-Stokes lines
antisymmetrische Schwingungen *fpl*
antisymmetrical vibrations
Anwendungsbereich *m* der Waage range of
applicability of a balance
Anzahl *f* der Freiheitsgrade number of degrees
of freedom
aperiodische Waage *f* damped balance
Apperancepotential-Spektroskopie *f*
appearance-potential spectroscopy
aprotisches Lösungsmittel *n* aprotic solvent
Aquametrie *f* aquametry
äquieluotrope Reihe *f* equieluotropic series
Äquivalentgrenzleitfähigkeit *f* eines
Elektrolyten limiting equivalent conductivity
of electrolyte
Äquivalentgrenzleitfähigkeit *f* eines Ions
limiting equivalent conductivity of ionic
species B
Äquivalentleitfähigkeit *f* bei der
Grenzverdünnung limiting equivalent
conductivity of electrolyte
Äquivalentleitfähigkeit *f* bei unendlicher
Verdünnung limiting equivalent conductivity
of electrolyte
Äquivalentleitfähigkeit *f* eines Elektrolyten
equivalent conductivity of electrolyte
Äquivalentleitfähigkeit *f* eines Elektrolyten bei
der Grenzverdünnung limiting equivalent
conductivity of electrolyte
Äquivalentleitfähigkeit *f* eines Elektrolyten bei
unendlicher Verdünnung limiting equivalent
coductivity of an electrolyte
Äquivalentleitfähigkeit *f* eines Ions equivalent
conductivity of ionic species B
Äquivalenz-Faktor *m* der Komponente B
equivalence factor of component B
Äquivalenz-Faktor *m* in den
Neutralisationsreaktionen equivalence factor
acid-base reactions
Äquivalenz-Faktor *m* in den Redoxreaktionen
equivalence factor for redox reactions
Äquivalenzpunkt *m* equivalence-point
Aräometer *n* hydrometer
Arbeitselektrode *f* working electrode
Arbeitsspannung *f* burning voltage
Argentometrie *f* argentimetry
argentometrische Titration *f* argentimetric
titration
Argon-Detektor *m* argon ionization detector
Argon-Ionisationsdetektor *m* argon ionization
detector
arithmetisches Mittel *n* arithmetic mean
Arrhenius-Ostwald-Theorie *f* der
elektrolytischen Dissoziation Arrhenius' and
Ostwald's theory of electrolytic dissociation
Ascarit *n* Ascarite
Assoziationsgrad *m* der Ionen association degree
of ions
Assoziationstheorie *f* der Ionen von Bjerrum
Bjerrum's theory of ionic association

Astigmatismus *m* astigmatism
Asymmetrieeffekt *m* relaxation-time effect
Asymmetriefehler *m* coma
asymmetrische Schwingungen *fpl* antisymmetrical
vibrations
Atom-Absorptions-Spektroskopie *f* atomic
absorption spectroscopy
atomare Masseneinheit *f* unified atomic mass
unit
atomarer Schwächungskoeffizient *m* atomic
attenuation coefficient
atomarer Wirkungsquerschnitt *m* atomic
cross section
Atom-Emissions-Spektroskopie *f* atomic
emission spectroscopy
Atom-Fluoreszenz-Spektroskopie *f* atomic
fluorescence spectroscopy
Atomisator *m* atomizer
Atomisierung *f* atomization
Atomisierungseinrichtung *f* atomizer
Atomisierungsgrad *m* efficiency of atomization
Atomrefraktion *f* atomic refraction
Atomrumpf *m* atomic core
Atomspektralanalyse *f* atomic spectrochemical
analysis
Atomspektroskopie *f* atomic spectroscopy
Atomspektrum *n* atomic spectrum
ATR-Methode *f* attenuated total reflectance
Aufeinandertürmen *n* pile-up
Auffänger *m* target
Aufladungs-Effekt *m* charging effect
Auflösen *n* dissolution
Auflösungsanalyse *f* electrochemical stripping
analysis
Auflösungsvermögen *n* 1. resolution (*of
a photographic emulsion*) 2. resolution (*of
a spectrometer*) 3. energy resolution of
a gamma spectrometer
Auflösungsvermögen *n* des Kristalls crystal
resolving power
Auflösungszeit *f* resolving time
Aufschliessenverfahren *n* fusion method
Aufschluß *m* digestion
Aufschluß *m* in der Paar-Bombe Paar bomb
fusion
Aufschlußverfahren *n* fusion method
Aufspaltung *f* branching decay
aufsteigende Chromatographie *f* ascending
development
aufsteigende Entwicklung *f* ascending
development
aufsteigende Methode *f* ascending development
Auftragegerät *n* applicator
Auger-Effekt *m* Auger effect
Auger-Elektronenmikroskopie *f* Auger-electron
microscopy
Auger-Elektronenspektroskopie *f* Auger-
electron spectroscopy
Auger-Elektronen-Spektroskopie *f* mit
Röntgenstrahl-Anregung X-ray induced
Auger electron spectroscopy
Auger-Übergang *m* Auger effect
Ausbrand *m* burn-up
Ausgangsnuklid *n* parent nuclide
Ausreißer *m* outlier

Aussalzchromatographie f salting-out chromatography
Aussalzen n salting-out
Ausschluß m exclusion
Ausschlußgrenze f exclusion limit
Außen-Indikator m external indicator
Außenkegel m secondary combustion zone
äußere Helmholtz-Schicht f Helmholtz double layer
äußerer Indikator m external indicator
äußerer Kegel m secondary combustion zone
äußerer Standard m external standard
äußeres elektrisches Potential n der Phase β outer electric potential of phase β
Ausströmgeschwindigkeit f des Quecksilbers an der Kapillare flow-rate of mercury
Austauschadsorption f exchange adsorption
austauschaktive Ankergruppen fpl ionogenic groups
Austauschermatrix f ion exchanger matrix
Austauschstrom m exchange current
Austrittwinkel m take-off angle
Auswahlregeln fpl selection rules
Auswaschen n des Niederschlages washing of precipitate
automatische Kolbenbürette f piston automatic burette
automatische Kohlenstoff-, Wasserstoff- und Stickstoffbestimmung f automatic determination of carbon, hydrogen and nitrogen
automatische Pipette f automatic zero pipette
automatische Waage f automatic balance
Autoprotolyse f autoprotolysis
Autoradiogramm n autoradiograph
Auxochrom n auxochrome
auxochrome Gruppe f auxochrome
av-Polarographie f alternating-voltage polarography
axiale Molekulardiffusion f molecular diffusion term
axiale Streudiffusion f eddy diffusion term
axiale Wirbeldiffusion f eddy diffusion term

B

balancierte Viskositätsmethode f balanced-density slurry packing
Bande f zone
Bandengruppe f band group
Bandenkante f band head
Bandenkopf m band head
Bandenspektrum n band spectrum
Bandensystem n band group
Bandverbreiterung f band broadening
Barn n barn
basische Form f des Anionenaustauschers base form of an anion exchanger
basisches Lösungsmittel n protophilic solvent
Basislinie f base-line
Basislinie-Methode f base-line method
Basislinienabweichung f base-line drift
Basislinienänderung f base-line drift
Basisliniendrift f base-line drift
Basispeak m base peak
bathochrome Gruppe f bathochrome

bathochrome Verschiebung f bathochromic shift
B Band n B band
Becquerel n becquerel
Beersches Gesetz n Beer's law
Belastung f load
Beleuchtungsstärke f illuminance
Bereichschätzung f interval estimation
Berg m peak
Bergbasis f peak base
Bergfläche f peak area
Bernoullische Verteilung f binomial distribution
bestimmte Substanz f determinand
Bestimmung f determination
Bestimmungsgrenze f limit of determination
Bestrahlung f radiant exposure
Bestrahlungsdauer f radiant exposure
Bestrahlungsstärke f irradiance
Bettvolumen n bed volume
Beugenschwingungen fpl scissoring vibrations
Beugung f der Röntgenstrahlen X-ray diffraction
Beugung f langsamer Elektronen low-energy electron diffraction
Beugung f langsamer Ionen low-energy ion diffraction
Beugungsanalyse f diffractometry
Beugung f schneller Elektronen high-energy electron diffraction
Beugung f schneller Elektronen in Reflexion reflection high-energy diffraction
Beugungsgitter n diffraction grating
Beugungsordnung f spectral order
bevorzugte Zerstäubung f preferential sputtering
Bezugsband n reference band
Bezugselektrode f reference electrode
Bezugslinie f internal reference line
Bezugslösung f reference solution
Bezugssubstanz f 1. reference material 2. marker 3. reference sample
Biamperometrie f amperometry with two indicator electrodes
biamperometrische Titration f amperometric titration with two indicator electrodes
bifunktioneller Ionenaustauscher m bifunctional ion exchanger
Binomialverteilung f binomial distribution
Biolumineszenz f bioluminescence
Bipotentiometrie f controlled-current potentiometry with two indicator electrodes
bipotentiometrische Titration f controlled-current potentiometric titration with two indicator electrodes
Biuretreaktion f biuret reaction
Bjerrumsche Theorie f der Ionenassoziation Bjerrum's theory of ionic association
Blasenzähler m bubble gauge
bleibende Härte f permanent hardness
Blindleitwert m susceptance
Blindlösung f blank solution
Blindprobe f blank test
Blindreaktion f blank test
Blindtitration f blank titration
Blindversuch m indicator blank
Blindwert m indicator blank
Blindwiderstand m reactance

Blitzpyrolyse f flash pyrolysis
Blitzverbrennung f flash combustion
Block m block
Blockierung f des Metallindikators blocking of
metal indicator
BN-Kammer f BN-chamber
Bogenspektrum n arc spectrum
Bolometer n bolometer
Bombenkalorimeter n bomb calorimeter
Boraxperle f borax bead
Boraxsalzperle f borax bead
Bornsche Solvatation-Theorie f Born's theory
of solvation
Bornsche Theorie f Born's theory of solvation
Bouguer-Lambert Beersches Gesetz n Beer's law
Bouguer-Lambertsches Gesetz n Lambert's law
Braggsche Gleichung f Bragg's law
Braggsches Reflexionsgesetz n Bragg's law
Brechung f refraction
Brechungsindex m refractive index
Brechungsmesser m refractometer
Brechzahlbestimmung f refractometry
Bremsfeldanalysator m retarding field analyser
Bremsstrahlung f bremsstrahlung
Brenner m burner
Brenner-Niederwieser-Kammer f BN-chamber
Brennspannung f burning voltage
Bromatometrie f bromatometry
Bruchstückbildung f fragmentation
Bruchstückion n fragment ion
Bruttoretentionszeit f total retention time
Büchner-Nutsche f Büchner funnel
Büchner-Trichter m Büchner funnel
Bündelung f collimation
Bunte-Bürette f Bunte gas burette
Bürette f burette
Bürette f amtlich geprüft certified burette
Bürette f mit automatischer Nullpunkteinstellung
automatic zero burette
Bürette f mit Schellbachstreifen Schellbach
burette
Bürette f nach Daffert Daffert automatic zero
burette
Bürette f nach Schellbach Schellbach burette
Butler-Volmer-Gleichung f Butler-Volmer
equation
Bypaß-Injektor m by-pass injector
Bypaß-Probengeber m by-pass injector

C

Carbonathärte f carbonate hardness
Carbowax Carbowax
Carius Aufschluß m Carius method
Carius Methode f Carius method
Castaingsche Mikrosonde f electron microprobe
X-ray analyser
Celite Celite
Centigramm-Methode f centigram method
Cerenkov-Strahlung f Cerenkov radiation
Cerimetrie f cerimetry
cerimetrische Titration f cerimetric titration
charakteristische Frequenz f characteristic
frequency
charakteristische Röntgenstrahlung f
characteristic X-radiation

charakteristisches Röntgenspektrum n X-ray
charakteristic spectrum
Chelat n chelate compound
Chelatharz n chelating resin
Chelat-Ionenaustauscher m chelating ion
exchanger
Chelatkomplex m chelate compound
Chelatligand m chelating agent
Chelatring m chelate ring
Chelatverbindung f chelate compound
Chemiionisation f chemiionization
Chemilumineszenz f chemiluminescence
Chemilumineszenzindikator m chemiluminescent
indicator
chemische Adsorption f chemisorption
chemische Aktivierungsanalyse f destructive
activation analysis
chemische Analyse f chemical analysis
chemische Ionisation f chemical ionization
chemischer Effekt m chemical effect
chemisches Äquivalent n chemical equivalent
chemisches Coulometer n chemical coulometer
chemisches Potential n der Substanz B chemical
potential of species B in phase α
chemische Verschiebung f chemical shift
chemische Zelle f chemical cell
chemisch gebundene stationäre Phase f
chemically bonded stationary phase
chemisch reines Reagens n chemically pure
reagent
Chemisorption f chemisorption
Chinhydronelektrode f quinhydrone electrode
chromatische Aberration f chromatic aberration
Chromatogramm n chromatogram
Chromatogrammsentwicklung f development of
a chromatogram
Chromatograph m chromatograph
Chromatographie f chromatography
chromatographieren chromatograph
Chromatographiesäule f chromatographic
column
chromatographische Analyse f chromatographic
analysis
chromatographische Kammer f chromatographic
chamber
chromatographische Säule f chromatographic
column
chromatographisches Bett n chromatographic
bed
Chromatometrie f chromatometry
Chromatopolarographie f chromatopolarography
Chromophor m chromophore
chromophore Gruppe f chromophore
Chromosorb m chromosorb
Chronoamperometrie f chronoamperometry
Chronoamperometrie f mit doppelter
Potentialstufe double-potential-step
chronoamperometry
Chronoamperometrie f mit linearem
Potentialanstieg chronoamperometry with
linear potential sweep
Chronoamperometrie f mit linearem Potential/
Spannungsanstieg an der
Quecksilbertropfelektrode dropping
electrode chronoamperometry with linear
potential/voltage sweep

Chronocoulometrie *f* chronocoulometry
Chronocoulometrie *f* mit doppelter
  Potentialstufe double-potential-step
  chronocoulometry
Chronopotentiometrie *f* chronopotentiometry
Chronopotentiometrie *f* mit linearem
  Stromanstieg chronopotentiometry with
  linear current sweep
Chronopotentiometrie *f* mit linearem Strom-
  Zeit-Verlauf chronopotentiometry with
  . linear current sweep
Chronopotentiometrie *f* mit programmiertem
  Strom programmed-current
  chronopotentiometry
Chronopotentiometrie *f* mit Stromstufen
  current-step chronopotentiometry
Chronopotentiometrie *f* mit überlagertem
  Wechselstrom chronopotentiometry with
  superimposed alternating current
Chronopotentiometrie *f* mit umgekehrtem
  Strom current-reversal chronopotentiometry
Chronopotentiometrie *f* mit unterbrochenem
  Strom current-cessation chronopotentiometry
Circulardichroismus *m* circular dichroism
Circular-Entwicklung *f* circular chromatography
Clarksche Methode *f* Clark's method
Coionen *mpl* co-ions
Compton-Effekt *m* Compton effect
Compton-Kante *f* Compton edge
Compton-Streuung *f* Compton effect
Comptonverteilung *f* Compton continuum
Cottrell-Gleichung *f* Cottrell equation
Coulogravimetrie *f* coulogravimetry
Coulombmeter *n* chemical coulometer
Coulometer *n* coulometer
Coulometrie *f* coulometry
Coulometrie *f* bei konstantem Potential
  controlled-potential coulometry
Coulometrie *f* bei konstantem Strom
  controlled-current coulometry
Coulometrie *f* mit gesteuertem Potential
  controlled-potential coulometry
Coulometrie *f* mit tropfender Elektrode
  droppig electrode coulometry
coulometrische Analyse *f* coulometric analysis
coulometrische Titration *f* controlled-current
  coulometry
coulostatische Methode *f* coulostatic method
Craig-Verteilung *f* countercurrent distribution
CT-Band *n* charge-transfer band
Curie *n* curie
curvilineare Regression *f* curvilinear regression
cyclische Chronopotentiometrie *f* cyclic
  chronopotentiometry
cyclische Chronopotentiometrie *f* mit
  Stromumkehr cyclic current-reversal
  chronopotentiometry
cyclische Dreieckspannungspolarographie *f*
  cyclic triangular-wave polarography
cyclische Dreieckspannungsvoltammetrie *f*
  cyclic triangular-wave voltammetry
cyclische Dreieckwellen-Polarographie *f* cyclic
  triangular-wave polarography
cyclische Dreieckwellen-Voltammetrie *f* cyclic
  triangular-wave voltammetry

cyclische Stromstufen-Chronopotentiometrie
  f cyclic current-step chronopotentiometry
cyclische Stromumkehr-Chronopotentiometrie
  f cyclic current-reversal chronopotentiometry

D

Dampf-Entladungslampe *f* metal-vapour lamp
Dampfraumanalyse *f* headspace analysis
Dämpfungswaage *f* damped balance
Dauer-Gleichgewicht *n* secular radioactive
  equilibrium
dc-Polarographie *f* direct-current polarography
dead-stop Titration *f* amperometric titration
  with two indicator electrodes
Deaktivierung *f* deactivation
Debye-Falkenhagen-Effekt *m* Debye-
  Falkenhagen's effect
Debye-Hückel-Brönsted-Gleichung *f* Debye-
  Hückel-Brönsted equation
Debye-Hückel-Onsagersches Grenzgesetz *n*
  Debye-Hückel-Onsager's limiting law for
  conductivity
Debye-Hückelsche Gleichung *f* Debye-Hückel
  equation
Debye-Hückelsche Grenzbeziehung *f* Debye-
  Hückel limiting law
Debye-Hückelsches Grenzgesetz *n* Debye-
  Hückel limiting law
Debye-Hückelsche Theorie *f* der starken
  Elektrolyte Debye-Hückel's theory of strong
  electrolytes
Debye-Hückel-Theorie *f* Debye-Hückel's
  theory of strong electrolytes
Debyescher Radius *m* radius of ionic
  atmosphere
Debye-Scherrer-Aufnahme *f* X-ray powdered-
  crystal pattern
Deformationsschwingungen *fpl* bending
  vibrations
deionisiertes Wasser *n* deionized water
Dekametrie *f* dielectrometry
dekametrische Titration *f* dielectrometric
  titration
Dekantation *f* decantation
Dekontamination *f* decontamination
Dekontaminationsfaktor *m* decontamination
  factor
Demaskierungsmittel *n* demasking agent
Demodulationspolarographie *f* demodulation
  polarography
Depolarisator *m* depolarizer
Derivatisierung *f* derivatization
Derivativ-Dilatometrie *f* derivative
  thermodilatometry
derivative Chronopotentiometrie *f* derivative
  chronopotentiometry
derivative Polarographie *f* derivative
  polarography
derivative potentiometrische Titration *f*
  derivative potentiometric titration
derivative Puls-Polarographie *f* derivative pulse
  polarography
derivative Voltammetrie *f* derivative
  voltammetry

Derivativ-Polarographie f derivative
polarography
Derivatograph m derivatograph
Desaktivierung f deactivation
Desorption f desorption
destilliertes Wasser n distilled water
Detektor m detector
Detektorempfindlichkeit f detector sensitivity
Deuteriumlampe f deuterium discharge lamp
deutsche Härte f German degree of water
hardness
deutscher Härtegrad m German degree of water
hardness
Dextran Blau n Blue Dextran
Diatomeenerde f diatomaceous earth
Dichromatometrie f chromatometry
Dichtemittel n mode
Dicke f **der Ionenwolke** radius of ionic
atmosphere
dicke Probe f thick specimen
dickes Target n thick target
Dielektrizitätskonstante f **1.** permittivity
2. relative permittivity
Dielektrizitätszahl f relative permittivity
Dielektrometrie f dielectrometry
dielektrometrische Analyse f dielectrometric
analysis
dielektrometrische Titration f dielectrometric
titration
Dielkometrie f dielectrometry
dielkometrische Titration f dielectrometric
titration
Differentialamperometrie f differential
amperometry
Differentialchromatogramm n differential
chromatogram
Differentialdetektor m differential detector
Differentialdilatometrie f differential
thermodilatometry
differentiale thermische Analyse f differential
thermal analysis
Differentialpolarographie f differential
polarography
Differentialpotentiometrie f differential
potentiometry
differentialpotentiometrische Indikation f
differential potentiometric titration
differentialpotentiometrische Titration f
differential potentiometric titration
Differentialpulspolarographie f differential pulse
polarography
Differential-Scanning-Kalorimetrie f differential
scanning calorimetry
Differential-Spektrophotometrie f differential
spectrophotometry
Differentialthermoanalyse f differential thermal
analysis
Differentialthermogravimetrie f derivative
thermogravimetry
Differentialvoltammetrie f differential
voltammetry
**differentielle anodische/kathodische
Pulsinversvoltammetrie** f differential pulse
anodic/cathodic stripping voltammetry
differentielle Kapazität f **der Doppelschicht**
differential capacitance of double layer

differentielle Pulspolarographie f differential
pulse polarography
Differenz-Amperometrie f differential
amperometry
Differenz f **der elektrischen Potentiale** electric
potential difference
Differenz f **der elektrischen Potentiale einer
galvanischen Zelle** electric potential
difference of a galvanic cell
Differenz f **der Galvani-Potentiale** Galvani
potential difference
Differenz f **der Volta-Potentiale** Volta potential
difference
differenzierender Effekt m differentiating effect
differenzierendes Lösungsmittel n
differentiating solvent
Differenzierung f differentiating effect
Differenz-Polarographie f differential
polarography
Differenz-Potentiometrie f differential
potentiometry
differenzpotentiometrische Titration f
differential potentiometric titration
Differenz-Puls-Polarographie f differential pulse
polarography
Differenz-Voltammetrie f differential
voltammetry
Diffraktion f **des Lichtes** light diffraction
Diffraktionsgitter n diffraction grating
Diffraktometrie f diffractometry
diffuse Doppelschicht f diffuse double layer
diffuse Schicht f diffuse double layer
Diffusion f diffusion
Diffusionsflamme f diffusion flame
Diffusionsgrenzstrom m limiting diffusion current
Diffusionskoeffizient m diffusion coefficient
Diffusions-Kontrolle f diffusion controlled
process
Diffusionskonstante f diffusion coefficient
Diffusionspotential n diffusion potential
Diffusionsschicht f diffusion layer
Diffusionsstrom m diffusion current
Diffusionsüberspannung f diffusion overpotential
Dilatometer n dilatometer
Dilatometrie f thermodilatometry
Dimethyldichlorsilan n dimethyldichlorsilane
direkte Isotopenverdünnungsanalyse f direct
isotope dilution analysis
direktes Verfahren n direct method
direkte Titration f direct titration
Direktverfahren n direct method
direktzerstäubender Brenner m direct-injection
burner
diskontinuierliches Spektrum n discontinuous
spectrum
diskrete Variable f discrete random variable
diskrete Zufallsgröße f discrete random
variable
Dispersion f **1.** dispersion **2.** dispersion (*in
statistics*) **3.** variance
Dispersion f **des Lichtes** light dispersion
Dispersionsanalyse f analysis of variance
Dispersionseffekt m Debye-Falkenhagen's effect
Dispersionseffekt m **der Leitfähigkeit** Debye-
Falkenhagen's effect
Dissoziation f dissociation

Dissoziationsfeldeffekt *m* dissociation field effect
Dissoziationsgrad *m* **nach Arrhenius** apparent
　degree of dissociation
DK-Metrie *f* dielectrometry
Donnan-Gleichgewicht *n* membrane equilibrium
Donnanpotential *n* membrane potential
Donorlösungsmittel *n* protogenic solvent
Doppelpotentialstufenchronoamperometrie *f*
　double-potential-step chronoamperometry
Doppelpotentialstufenchronocoulometrie *f*
　double-potential-step chronocoulometry
Doppelschicht *f* electrochemical double layer
doppelte Isotopenverdünnungsanalyse *f* double
　isotope dilution analysis
Doppeltonpolarographie *f* double-tone
　polarography
Dorn-Effekt *m* sedimentation potential
Dosierschleife *f* by-pass injector
Drahtüberzug-Elektrode *f* coated wire ion-
　selective electrode
Drain *n* drain
Drechsel-Waschflasche *f* Drechsel gas washing
　bottle
drehbarer Versuchsplan *m* rotatable design
Drehbarkeit *f* rotatability
Drehung *f* **der Polarisationsebene des Lichtes**
　rotation of the plane of polarization of
　polarized light
Dreieckspannungspolarographie *f* triangular-
　wave polarography
Dreieckspannungsvoltammetrie *f* triangular-
　wave voltammetry
Dreieckwellen-Polarographie *f* triangular-wave
　polarography
Dreieckwellen-Voltammetrie *f* triangular-wave
　voltammetry
Drei-Sigma-Regel *f* three-sigma rule
Drift *f* **eines Potentials** drift of a potential
Drillschwingungen *fpl* twisting vibrations
Druckgradient-Korrekturfaktor *m* pressure-
　gradient correction factor
Dumas Methode *f* Dumas method
Dumas-Pregl-Methode *f* Dumas-Pregl method
Dunkelstrom *m* dark current
dünne Probe *f* thin-film specimen
dünnes Target *n* thin target
Dünnschichtchromatographie *f* thin-layer
　chromatography
Dünnschicht-Streicher *m* spreader
Durchbruch *m* breakthrough
Durchbruchskurve *f* breakthrough curve
durchflußprogrammierte Chromatographie *f* flow-
　programmed chromatography
Durchflußzählrohr *n* propotional gas-flow
　counter
durchgelassene Mikrowellen-Methode *f*
　transmitted wave method
Durchlaßgitter *n* transmission grating
Durchlässigkeit *f* transmission factor
Durchschnittsprobe *f* average sample
Durchstrahlungsverfahren *n* **mit**
　Quantenstrahlung gamma-ray absorptiometry
Durchtrittsfaktor *m* charge-transfer coefficient
Durchtrittsüberspannung *f* activation
　overpotential
Dynode *f* dynode

E

ebullioskopische Bestimmung *f* **der Molmasse**
　ebullioscopic determination of the molar mass
Eddy-Diffusionsterm *m* eddy diffusion term
Effekt *m* **der Relaxationszeit** relaxation-time
　effect
Effekt *m* **elektrostatischer Aufladungen** effect of
　electrostatic charges
effektive Reichweite *f* effective depth
effektiver Radius *m* **der Ionenatmosphäre**
　radius of ionic atmosphere
effektiver Radius *m* **der Ionenwolke** radius of
　ionic atmosphere
effektives Probevolumen *n* effective layer
effektive theoretische Bodenzahl *f* number of
　effective plates
effektive Trennstufenhöhe *f* height equivalent
　to an effective plate
effektive Trennstufenzahl *f* number of
　effective plates
Effluent *m* effluent
Eichfehler *m* scale error
Eichkurve *f* analytical curve
Eichprobenserie *f* standard series
einfache Elektrodenreaktion *f* simple electrode
　reaction
einfache Isotopenverdünnungsanalyse *f* direct
　isotope dilution analysis
Einfachlinie *f* singlet line
einfarbiger Indikator *m* one-colour indicator
Einflußkoeffizientenverfahren *n* empirical
　coefficient method
ein-Quant-Escape-Linie *f* single escape peak
ein-Quant-Escape-Peak *m* single escape peak
Einschlämmtechnik *f* slurry packing
einseitiger Test *m* one-sided test
Einstrahlspektralphotometer *n* single-beam
　spectrophotometer
Einwaage *f* weighed amount
Einzelordnung *f* order of reaction with respect
　to a given substance
Einzelprobe *f* unit sample
Einzelteilchen *n* elementary entity
elastischer Stoß *m* elastic collision
elastische Streuung *f* elastic scattering
elektrische Beweglichkeit *f* **der Ionen B** electric
　mobility of ion B
elektrische Doppelschicht *f* electrochemical
　double layer
elektrische Leitfähigkeit *f* 1. electrical
　conductivity 2. *früher* conductance
elektrische Leitung *f* electrical conduction
elektrische Potentialdifferenz *f* electric
　potential difference
elektrische Potentialdifferenz *f* **einer**
　galvanischen Zelle electric potential difference
　of a galvanic cell
elektrischer Leiter *m* electrical conductor
elektrischer Leitwert *m* conductance
elektrischer Strom *m* electric current
elektrischer Widerstand *m* resistance
elektrisches Oberflächenpotential *n* **der Phase** *β*
　surface electric potential of phase *β*
elektrische Spannung *f* electric potential
　difference

elektrische Stromdichte *f* current density
elektrische Stromstärke *f* electric current
elektrische Waage *f* electromechanical balance
Elektrizitätsleitung *f* electrical conduction
Elektroanalyse *f* electroanalysis
Elektrochemie *f* electrochemistry
elektrochemische Analyse *f* electroanalysis
elektrochemische Doppelschicht *f*
    electrochemical double layer
elektrochemische Erzeugung *f* des Reagenzes
    electrochemical generation of a titrant
elektrochemischer Durchtrittsfaktor *m* charge
    transfer coefficient
elektrochemische Reagenzerzeugung *f*
    elektrochemical generation of a titrant
elektrochemischer Sensor *m* electrochemical
    sensor
elektrochemische Spannungsreihe *f*
    electrochemical series
elektrochemisches Potential *n* electrochemical
    potential of ionic component B in phase α
elektrochemische Wertigkeit *f* electrochemical
    equivalent
elektrochemische Zelle *f* electrochemical cell
Elektrode *f* 1. electrode 2. electrode (*in*
    electrochemistry)
Elektrode *f* dritter Art electrode of the third
    kind
Elektrode *f* erster Art electrode of the first kind
Elektrode *f* mit elektroneutralem
    Ladungsüberträger neutral-carrier membrane
    ion-selective electrode
Elektrode *f* mit flüssiger
    Ionenaustauschermembran liquid ion-exchange
    membrane electrode
Elektrode *f* mit neutralem Ligand neutral-
    carrier membrane ion-selective electrode
Elektrodenabstand *m* analytical gap
elektrodenlose Entladungslampe *f* electrodeless
    discharge lamp
Elektrodenpotential *n* electrode potential
Elektrodenprozeß *m* electrode process
Elektrodenreaktion *f* electrode reaction
Elektrode *f* zweiter Art electrode of the second
    kind
Elektrodialyse *f* electrodialysis
Elektroendoosmose *f* electro-endoosmosis
Elektrographie *f* electrography
Elektrogravimetrie *f* electrogravimetry
Elektrogravimetrie *f* bei konstanter
    Stromstärke constant-current
    electrogravimetry
Elektrogravimetrie *f* mit gesteuertem
    Potential controlled-potential
    electrogravimetry
Elektrogravimetrie *f* mit konstantem Strom
    constant-current electrogravimetry
Elektrogravimetrie *f* mit kontrolliertem
    Potential controlled-potential
    electrogravimetry
elektrokapillare Erscheinungen *fpl*
    electrocapillary phenomena
Elektrokapillareffekte *mpl* electrocapillary
    phenomena
Elektrokapillarität *f* electrocapillary
    phenomena

elektrokinetische Effekte *mpl* electrokinetic
    phenomena
elektrokinetische Erscheinungen *fpl*
    electrokinetic phenomena
elektrokinetisches Potential *n* electrokinetic
    potential
Elektrolumineszenz *f* electroluminescence
Elektrolyse *f* electrolysis
Elektrolyse *f* bei konstanter Stromstärke
    constant-current electrolysis
Elektrolyse *f* bei kontrolliertem Potential
    controlled-potential electrolysis
Elektrolyse *f* mit konstantem Strom constant-
    current electrolysis
Elektrolysezelle *f* electrolytic cell
Elektrolysierzelle *f* electrolytic cell
Elektrolyt *m* electrolyte
Elektrolytbrücke *f* salt bridge
elektrolytische Abscheidung *f* electrodeposition
elektrolytische Dissoziation *f* electrolytic
    dissociation
elektrolytische Fällung *f* electrodeposition
elektrolytische Leitfähigkeit *f* 1. electrolytic
    conductivity 2. *früher* conductance of
    electrolyte
elektrolytische Leitung *f* ionic conduction
elektrolytischer Leiter *m* ionic conductor
elektrolytischer Leitwert *m* conductance of
    electrolyte
elektrolytisches Abscheiden *n* electrodeposition
    of metals
elektrolytische Thermozelle *f* galvanic thermocell
elektrolytische Zelle *f* electrolytic cell
elektromagnetisches Spektrum *n*
    electromagnetic spectrum
elektromagnetische Strahlung *f* electromagnetic
    radiation
elektromotorische Kraft *f* einer galvanischen
    Zelle electromotive force of a galvanic cell
Elektron-Capture-Detektor *m* electron capture
    detector
Elektronenaustauscher *m* electron exchanger
Elektronenband *n* electronic band
Elektronenbeschuß-Ionenquelle *f* electron-
    impact ion source
Elektronen-Donator-Akzeptor-Komplex *m*
    charge-transfer complex
Elektroneneinfangdetektor *m* electron capture
    detector
Elektronenemission *f* electron emission
Elektronenenergieverlustspektroskopie *f*
    electron-energy loss spectroscopy
Elektronenleiter *m* electronic conductor
Elektronenleitung *f* electronic conduction
Elektronenresonanzspektroskopie *f* electron
    paramagnetic resonance spectroscopy.
Elektronen-Rotationsschwingungsspektrum *n*
    electronic-vibration-rotation spectrum
Elektronenspektroskopie *f* 1. electron
    spectroscopy 2. electronic spectroscopy
Elektronenspektroskopie *f* der innersten
    atomaren Niveaus core level electron
    spectroscopy
Elektronenspektroskopie *f* der
    Valenzniveaus valence level electron
    spectroscopy

Elektronenspektroskopie *f* **für die chemische Analyse** X-ray photoelectron spectroscopy

Elektronenspektrum *n* electronic spectrum

Elektronenspin-Resonanz *f* electron paramagnetic resonance

Elektronen-Spin-Resonanzspektroskopie *f* electron paramagnetic resonance spectroscopy

Elektronenstoß-Ionenquelle *f* electron-impact ion source

Elektronenstoßionisation *f* electron-impact ionization

elektronenstrahlinduzierte Desorption *f* electron stimulated desorption

elektronenstrahlinduzierte Desorption *f* **von Ionen** electron stimulated desorption of ions

elektronenstrahlinduzierte Desorption *f* **von Neutralteilchen** electron stimulated desorption of neutrals

Elektronenstrahl-Mikroanalysator *m* electron microprobe X-ray analyser

$\pi - \pi^*$**-Elektronenübergang** *m* $\pi - \pi^*$ electronic transition

$\sigma - \pi^*$**-Elektronenübergang** *m* $\sigma - \pi^*$ electronic transition

$n - \pi^*$**-Elektronenübergang** *m* $n - \pi^*$ electronic transition

$n - \sigma^*$**-Elektronenübergang** *m* $n - \sigma^*$ electronic transition

Elektronenvervielfacher *m* electron multiplier

Elektronenvolt *n* electronvolt

Elektron-Kern-Doppelresonanz *f* electron nuclear double resonance

Elektroosmose *f* electro-endoosmosis

Elektropherogramm *n* electrophorogram

Elektrophorese *f* electrophoresis

elektrophoretischer Effekt *m* electrophoretic effect

Elektrowaage *f* electromechanical balance

Elementaranalyse *f* **1.** elemental analysis **2.** elemental analysis (*of organic compounds*)

Elementarereignis *n* elementary event

Elementarindividuum *n* elementary entity

Elementarreaktion *f* elementary reaction

Elementarstufe *f* elementary step

Ellipsometrie *f* ellipsometry

elliptische Polarisation *f* elliptical polarization

elliptisch polarisiertes Licht *n* elliptically polarized light

Eluant *m* eluent

Eluat *n* eluate

Eluent *m* eluent

eluotrope Reihe *f* eluotropic series

Elution *f* elution

Elutionsanalyse *f* elution chromatography

Elutionsbande *f* peak (*in chromatography*)

Elutionschromatographie *f* elution chromatography

Elutionskurve *f* elution curve

Elutionsmittel *n* eluent

Elutriation *f* elutriation

Emaniermethode *f* emanometric method

Emissions-Molekülspektroskopie *f* molecular emission spectroscopy

Emissionsmonochromator *m* emission monochromator

Emissionsspektralanalyse *f* emission spectrochemical analysis

Emissionsspektroskopie *f* emission spectroscopy

Emissionsspektrum *n* emission spectrum

EMK *f* **der Zelle** electromotive force of a galvanic cell

empfindliches Volumen *n* **des Detektors** sensitive volume of a detector

Empfindlichkeit *f* sensitivity (*of analytical method*)

Empfindlichkeit *f* **der Waage** balance sensitivity

Empfindlichkeitsbereich *m* latitude

Empfindlichkeit *f* **von analytischer Reaktion** sensitivity of analytical reaction

empirischer Variationskoeffizient *m* empirical variation coefficient

empirische Standardabweichung *f* standard deviation of a sample

empirische Verteilung *f* empirical distribution

energiedispersive Röntgenfluoreszenzanalyse *f* energy dispersive X-ray fluorescence analysis

Energieauflösungsvermögen *n* energy resolution of a gamma spectrometer

englische Härte *f* English degree of water hardness

englischer Härtegrad *m* English degree of water hardness

Entaktivierung *f* deactivation

Enthalpiemetrie *f* enthalpimetry

Enthalpie-Titration *f* thermometric titration

enthalpometrische Titration *f* thermometric titration

entionisiertes Wasser *n* deionized water

Entladungslampe *f* discharge lamp

Entladungspolarographie *f* incremental-charge polarography

Entseuchung *f* decontamination

Entseuchungsgrad *m* decontamination factor

Entwicklung *f* **1.** development **2.** development (*in photography*)

Entwicklung *f* **des Chromatogramms** development of a chromatogram

Entwicklungskammer *f* chromatographic chamber

Enzym-Elektrode *f* enzyme-substrate electrode

Epikadmiumneutronen *npl* epicadmium neutrons

epithermische Neutronen *npl* epithermal neutrons

Erfassungsgrenze *f* limit of detection

Erhitzungskurve *f* heating curve

erlaubter Übergang *m* allowed transition

Erregerfrequenz *f* exciting frequency

Erregerlinie *f* **1.** exciting line **2.** Rayleigh line

Erregungs-Quelle *f* excitation source

Erregungs-Spektrum *n* excitation spectrum

Erscheinungspotential-Spektroskopie *f* appearance-potential spectroscopy

erste Oberschwingung *f* first overtone band

erster Wien-Effekt *m* Wien's effect

erwartungstreue Schätzfunktion *f* unbiased estimator

Erwartungswert *m* expected value

erweiterte Nernst-Gleichung *f* Nikolsky equation

Escape-Peak *m* pair escape peak

Evaporieren *n* evaporation

Everhart-Thornley-Detektor *m* Everhart-Thornley detector

Exoelektronenspektroskopie *f* exoelectron
  spectroscopy
Exponentialverteilung *f* exponential
  distribution
Exponierungszeit *f* exposure time
Exposition *f* radiant exposure
Expositionszeit *f* exposure time
Exsikkator *m* desiccator
Extinktion *f* absorbance
Extinktionskoeffizient *m* specific decadic
  absorption coefficient
Extrakt *m* extract
Extrakteur *m* extractor
Extraktion *f* extraction
Extraktionsapparat *m* nach Soxhlet Soxhlet
  apparatus
Extraktionsausbeute *f* in % extraction
  percentage
Extraktionschromatographie *f* extraction
  chromatography
Extraktionsgrad *m* in % extraction percentage
Extraktionsindikator *m* extraction indicator
Extraktionskoeffizient *m* concentration
  distribution ratio (*in extraction*)
Extraktionskonstante *f* extraction constant
Extraktionskurve *f* extraction curve
Extraktionsmittel *n* extractant
Extraktionsreagent *n* extracting agent
extraktive Titration *f* extractive titration

F

Faktor *m* factor
Faktor-Analyse *f* factor analysis
Faktor *m* der Titration titrimetric conversion
  factor
Faktoren-Analyse *f* factor analysis
Fällung *f* precipitation
Fällung *f* aus homogener Lösung precipitation
  from homogeneous solution
Fällungsindikator *m* precipitation indicator
Fällungsmittel *n* precipitating agent
Fällungstitration *f* precipitation titration
Fällungstitrationskurve *f* precipitation titration
  curve
Faltenfilter *n*, *m* fluted filter
Faraday-Konstante *f* Faraday constant
Faradaysche Gesetze *npl* Faraday's laws
Faradaysche Gleichrichtung *f* high-level
  faradaic rectification
Faradayscher Strom *m* faradaic current
Faradaysche Zahl *f* Faraday constant
Farbenabweichung *f* chromatic aberration
Farbstofflaser *m* dye laser
Farbtonverschiebung *f* bathochromic shift
Federwaage *f* spring balance
Fehler *m* erster Art error of first kind
Fehlerfortpflanzungsgesetz *n* variance
  propagation
Fehler *m* zweiter Art error of second kind
Fehlingsche Lösung *f* Fehling's reagent
Feinstruktur *f* fine structure
Felddesorption *f* field desorption
Felddissoziation *f* dissociation field effect
Feldeffekttransistor *m* field effect transistor
Feldionenquelle *f* field ionization ion source

Feldionisation *f* field ionization
Feldstärkeeffekt *m* Wien's effect
Feldsweep-Methode *f* field-sweep method
Fensterzählrohr *n* window counter
Fermi-Resonanz *f* Fermi resonance
fernes Infrarot *n* far infrared
fernes Ultrarot *n* far infrared
fernes Ultraviolett *n* far ultraviolet
feste Lösung *f* mixed crystal
fester Ionenaustauscher *m* solid ion exchanger
fester Träger *m* solid support
Festionen *npl* fixed ions
Festkörper-Elektrode *f* solid electrode
Festkörpermembran *f* solid-state membrane
Festkörpermembran-Elektrode *f* solid-state
  membrane ion-selective electrode
Festkörpervolumen *n* solid volume
Ficksche Gesetze *npl* Fick's laws
Filterfluorimeter *n* filter fluorimeter
Filternutsche *f* nach Büchner Büchner funnel
Filtrat *n* filtrate
Filtration *f* filtration
Filtrieren *n* filtration
Fingerabdruck *m* fingerprint region
Fishersche Verteilung *f* F-distribution
Fixierungspaar *n* fixation pair
Flächenladungsdichte *f* surface charge density
Flamme *f* flame
Flammen-Absorptions-Spektroskopie *f* flame
  absorption spectroscopy
Flammen-Atom-Absorptions-Spektroskopie *f*
  flame atomic absorption spectroscopy
Flammen-Atom-Emissions-Spektroskopie *f*
  flame atomic emission spectroscopy
Flammen-Atom-Fluoreszenz-Spektroskopie *f*
  flame atomic fluorescence spectroscopy
Flammen-Emissions-Spektroskopie *f* flame
  emission spectroscopy
Flammenfärbung *f* flame coloration
Flammen-Fluoreszenz-Spektroskopie *f* flame
  fluorescence spectroscopy
Flammenionisationsdetektor *m* flame ionization
  detector
flammenlose Atomisierungseinrichtung *f*
  electrothermal atomizer
Flammenphotometer *n* flame photometer
flammenphotometrischer Detektor *m* flame
  photometric detector
Flammenspektralphotometer *n* flame
  spectrophotometer
Flammenspektrophotometer *n* flame
  spectrophotometer
Flash-filament-Methode *f* flash desorption
Flash-Pyrolyse *f* flash pyrolysis
Fleck *m* spot
Fleckenverbreiterung *f* spot broadening
Fließmittelfront *f* mobil phase front
Fließmittelstärke *f* solvent strength
Fließmittelwanderungsstrecke *f* mobile phase
  distance
Flockung *f* coagulation
Flockungskonzentration *f* coagulation
  concentration
Flockungsschwellenwert *m* coagulation
  concentration

Flockungswert *m* coagulation concentration
Fluid-Chromatographie *f* supercritical fluid chromatography
Flugzeit *f* time of flight
Flugzeitmassenspektrometer *n* time-of flight mass spectrometer
Fluoreszenz *f* fluorescence
Fluoreszenzanalyse *f* fluorimetry
Fluoreszenz-Emissionsspektrum *n* fluorescence emission spectrum
Fluoreszenz-Erregungsspektrum *n* fluorescence excitation spectrum
Fluoreszenzindikator *m* fluorescent indicator
Fluoreszenz-Molekülspektroskopie *f* molecular fluorescence spectroscopy
Fluoreszenz-Spektralanalyse *f* fluorescence spectrochemical analysis
Fluoreszenz-Spektroskopie *f* fluorescence spectroscopy
Fluoreszenzspektrum *n* fluorescence spectrum
Fluoreszenztitration *f* fluorimetric titration
Fluorimeter *n* fluorimeter
Fluorimetrie *f* fluorimetry
fluorimetrische Titration *f* fluorimetric titration
Flußdichte *f* der Teilchen oder Photonen flux density of particles or photons
Flüssigaustauscher-Elektrode *f* liquid ion-exchange membrane electrode
Flüssigchromatographie *f* liquid chromatography
flüssige Ionenaustauschermembran *f* liquid ion exchanger membrane
flüssige Phase *f* stationary liquid phase
flüssiger Ionenaustauscher *m* liquid ion exchanger
flüssige stationäre Phase *f* stationary liquid phase
Flüssig-Fest-Chromatographie *f* liquid-solid chromatography
Flüssig-Flüssig-Chromatographie *f* liquid-liquid chromatography
Flüssig-Flüssig-Extraktion *f* liquid-liquid extraction
Flüssig-Flüssig-Verteilung *f* liquid-liquid distribution
Flüssig-Gel-Chromatographie *f* liquid-gel chromatography
Flüssigkeitschromatographie *f* liquid chromatography
Flüssigkeits-Festkörper-Chromatographie *f* liquid-solid chromatography
Flüssigkeits-Flüssigkeits-Chromatographie *f* liquid-liquid chromatography
Flüssigkeitsgrenzflächenpotential *n* liquid-junction potential
Flüssigkeitspotential *n* liquid-junction potential
Flüssigkeitsvolumen *n* liquid phase volume
Flüssigmembran *f* liquid membrane
Flüssigmembran-Elektrode *f* liquid membrane ion-selective electrode
Flußmittel *n* flux
flußprogrammierte Chromatographie *f* flow-programmed chromatography
fokussierendes Kristallspektrometer *n* X-ray focusing spectrometer
Fokussierkreis *m* focusing circle

formales Potential *n* conditional electrode potential
Formalität *f* formality
Fotoelektron *n* photoelectron
Fotopolarographie *f* photopolarography
Fragmentierung *f* fragmentation
Fragmentierungsmuster *n* fragmentation pattern
Fragmention *n* fragment ion
Fraktion *f* fraction
fraktionierte Fällung *f* fractional precipitation
Fraktionssammler *m* fraction collector
französische Härte *f* French degree of water hardness
französischer Härtegrad *m* French degree of water hardness
Frequenz *f* frequency
Frequenzsweep-Methode *f* frequency-sweep method
Frontalanalyse *f* frontal chromatography
Frontalchromatographie *f* frontal chromatography
Front *f* der mobilen Phase mobile phase front
Fronting *n* fronting
F-Test *m* F-test
Fundamentalparameter-Verfahren *n* fundamental parameter method
Funken *m* electrical spark
Funkenquelle *f* spark ionization source
Funkenquelle-Massenspektrometrie *f* spark source mass spectrometry
Funkenquellen-Ionisation *f* spark source ionization
Funkenspektrum *n* spark spectrum
funktionelle Analyse *f* functional group analysis
funktionelle Gruppe *f* functional group
funktionelle Gruppenanalyse *f* functional group analysis
Funktionsverstärker *m* operational amplifier
F-Verteilung *f* F-distribution

G

Galvani-Potential *n* inner electric potential of phase β
Galvani-Potentialdifferenz *f* Galvani potential difference
galvanische Kette *f* galvanic cell
galvanisches Element *n* galvanic cell
galvanische Zelle *f* galvanic cell
Galvani-Spannung *f* Galvani potential difference
Galvanometer *n* galvanometer
Galvanostat *m* galvanostat
galvanostatische Coulometrie *f* controlled-current coulometry
galvanostatische Methode *f* galvanostatic method
Gamma-Wert *m* gamma
Gasabsorptiometrie *f* absorptiometric gas analysis
Gasanalyse *f* gas analysis
Gasanalysenapparat *m* nach Orsat Orsat gas analyser
Gasbürette *f* nach Bunte Bunte gas burette
Gaschromatographie *f* gas chromatography

gasdurchlässige Membran *f* gas-permeable
   membrane
Gaselektrode *f* gas electrode
Gasentwicklungsanalyse *f* evolved gas detection
Gasentwicklungsapparat *m* nach Kipp Kipp gas
   generator
Gas-Fest-Chromatographie *f* gas-solid
   chromatography
Gas-Festkörper-Chromatographie *f* gas-solid
   chromatography
Gasfilter *n, m* filter tube
Gas-Flüssig-Chromatographie *f* gas-liquid
   chromatography
Gas-Flüssigkeits-Chromatographie *f* gas-liquid
   chromatography
Gasgravimetrie *f* gravimetric gas analysis
Gasmanometrie *f* manometric gas analysis
gaspermeable Membran *f* gas-permeable
   membrane
Gaspipette *f* gas sampling pipette
Gassammelröhre *f* gas sampling pipette
gas-sensitive Elektrode *f* gas sensing electrode
Gas-Sensor *m* gas sensing electrode
Gastitrimetrie *f* gas titration analysis
Gasverstärkung *f* gas amplification
Gasvolumetrie *f* volumetric gas analysis
Gasvolumetrie *f* und Gasmanometrie *f*
   gasometric analysis
Gaswaschflasche *f* mit Glasfritte gas washing
   bottle with sintered head
Gaswaschflasche *f* nach Drechsel Drechsel gas
   washing bottle
Gaswaschflasche *f* nach Friedrichs Friedrichs
   gas washing bottle
Gate *n* gate
Gaußsche Verteilung *f* normal distribution
gebundene Ionen *npl* fixed ions
gedämpfte Waage *f* damped balance
gefüllte Säule *f* packed column
Gegenelektrode *f* 1. auxiliary electrode (*in
   electrochemistry*) 2. counter electrode (*in arc
   and spark spectroscopy*)
Gegenionen *npl* counter-ions
Gegenstromextraktion *f* countercurrent
   extraktion
Gegenstromverteilung *f* countercurrent
   distribution
Gehalt *m* content
Geiger-Müller-Zähler *m* Geiger-Müller counter
   tube
Geiger-Müller-Zählrohr *n* Geiger-Müller counter
   tube
Geiger-Zähler *m* Geiger-Müller counter tube
Geister *mpl* ghosts
gekoppelte Säulen *fpl* coupled columns
gekuppelte Untersuchungsmethode *fpl* coupled
   simultaneous techniques
Gel *n* gel
Gelchromatographie *f* gel chromatography
Gelfiltration *f* gel filtration
gelöster Stoff *m* solute
Gelöstes *n* solute
Gel-Permeationschromatographie *f* gel-
   permeation chromatography
gemuffelter Ofen *m* muffle furnace
Geometriefaktor *m* geometry factor

geometrisches Mittel *n* geometric mean
geometrisches Säulenvolumen *n* column volume
gepackte Säule *f* packed column
Gerätefehler *m* instrumental error
Gerüstschwingungen *fpl* skeletal vibrations
Gesamtanalyse *f* complete analysis
Gesamtelutionsvolumen *n* total liquid volume
Gesamtenergiepeak *m* full energy peak
Gesamthärte *f* total water hardness
Gesamtkapazität *f* total specific capacity
Gesamtordnung *f* der Reaktion order of
   reaction
Gesamtporenvolumen *n* stationary liquid
   volume
Gesamtretentionsvolumen *n* retention volume
Gesamtretentionszeit *f* total retention time
gesättigte Kalomelelektrode *f* saturated-calomel
   electrode
gesättigte Kammer *f* saturated chamber
gesättigte Lösung *f* saturated solution
gesättigter Versuchsplan *m* saturated design
Geschwindigkeit *f* im Hohlraumbereich
   interstitial velocity
Geschwindigkeit *f* im Zwischenraum
   interstitial velocity
Geschwindigkeitskonstante *f* der
   Elektrodenreaktion rate constant of
   electrode reaction
Geschwindigkeitskonstante *f* der Reaktion
   reaction-rate constant
getrockneter Ionenaustauscher *m* absolutely dry
   ion exchanger
Gewichtsanalyse *f* gravimetric analysis
Glanzbildner *m* brightening agent
Glanzwinkel *m* 1. glancing angle 2. blaze angle
   (*of a diffraction grating*)
Glanzwinkel-Wellenlänge *f* blaze wavelength
Glaselektrode *f* glass membrane electrode
Gläser *npl* mit kontrollierter Porosität
   controlled-porosity glass
Glasfiltertiegel *m* crucible with sintered disk
Glasfiltertrichter *m* filter funnel with sintered
   disk
Glaskohlenelektrode *f* glassy carbon electrode
Glaskohlenstoffelektrode *f* glassy carbon
   electrode
Glasmembranelektrode *f* glass membrane
   electrode
Glastrichter *m* glass funnel
gleicharmige Hebelwaage *f* laboratory balance
Gleichgewichtspotential *n* equilibrium electrode
   potential
Gleichstrombogen *m* direct current arc
Gleichstrompolarographie *f* direct current
   polarography
gleichzeitige Keimbildung *f* simultaneous
   nucleation
gleichzeitige Untersuchungen *fpl* simultaneous
   techniques
Glimmentladung *f* glow discharge
Globar *m* Globar
Glühen *n* des Filters incineration of filter paper
Glühen *n* des Niederschlages ignition of
   precipitate
Glühschiffchen *n* combustion boat
GM-Zählrohr *n* Geiger-Müller counter tube

Golay-Detektor *m* Golay pneumatic detector
Golay-Zelle *f* Golay pneumatic detector
Gooch-Tiegel *m* Gooch crucible
Gouy-Chapman-Schicht *f* diffuse double layer
Gradation *f* γ gamma
Gradationskurve *f* emulsion calibration curv
Gradient *m* **des Flusses der Teilchen oder Photonen** flux gradient of particles or photons
Gradientelution *f* gradient elution
Gradientenelution *f* gradient elution
Gradientpackung *f* gradient layer
Gradientschicht *f* gradient layer
Grammäquivalent *n* gram equivalent
Gramm-Molekül *n* gram molecule
Gran-Methode *f* Gran's plot
Graphit-Küvette *f* graphite cuvette
Graphiträdchen-Elektrode *f* rotating-disk electrode
Graphitrohrküvette *f* graphite cuvette
Graphitrohrofen *m* graphite-tube furnace
Graufilter *n*, *m*, neutral-density filter
Gravimetrie *f* gravimetric analysis
gravimetrische Titration *f* weight titration
Grenzfläche *f* interface
grenzflächenaktiver Stoff *m* surfactant
Grenzkonzentration *f* concentration limit
Grenzstrom *m* limiting current
Grenzwellenlänge *f* short-wavelength limit
grober Fehler *m* gross error
Grundbande *f* fundamental vibration band
Grundelektrolyt *m* supporting electrolyte
Grundgesamtheit *f* population
Grundkörper *m* **des Ionenaustauschers** ion exchanger matrix
Grundlinie *f* base-line
Grundschwingungsbande *f* fundamental vibration band
Grundstrom *m* residual current
Gruppe *f* analytical group
Gruppenanalyse *f* functional group analysis
Gruppenreagens *n* group reagent
Gruppe *f* **von Elementen** analytical group

H

Haber-Luggin-Kapillare *f* Luggin-Haber capillary
Haftwahrscheinlichkeit *f* sticking probability
Halbleiterdetektor *m* semiconductor detector
Halbmikroanalyse *f* mesoanalysis
Halbmikrowaage *f* semimicrochemical balance
Halbpeakpotential *n* half-peak potential
halbquantitative Analyse *f* semiquantitative analysis
Halbstufenpotential *n* half-wave potential
Halbwellenpotential *n* half-wave potential
Halbwertsbreite *f* half-intensity width
Halbwertsbreite *f* **des Peak** width at half-height
Halbwertsdicke *f* half-thickness
Halbwertschicht *f* half-thickness
Halbwertszeit *f* half-life period
Halbzelle *f* electrode (*in electrochemistry*)
handbetätigtes Spektralphotometer *n* non-recording spectrophotometer

hängende Quecksilbertropfelektrode *f* hanging mercury drop electrode
Härte *f* **des Wassers** water hardness
Hartmannblende *f* Hartmann diaphragm
Hauptbestandteil *m* major constituent
Headspace-Technik *f* headspace analysis
heißes Laboratorium *n* hot laboratory
heiße Zelle *f* hot cell
Heißwassertrichter *m* funnel heater
Heizdraht-Methode *f* flash desorption
Heizmantel *m* **für Trichter** funnel heater
Heiztisch *m* **nach Kofler** Kofler micro melting point apparatus
Helmholtzsche Doppelschicht *f* compact layer
Helmholtz-Schicht *f* compact layer
Hendersonsche Gleichung *f* Henderson's equation
heterogene Keimbildung *f* heterogeneous nucleation
heterogene Membran *f* heterogeneous membrane
heterogene Membran-Elektrode *f* heterogeneous membrane ion-selective electrode
heterometrische Titration *f* turbidimetric titration
Heteropolysäure *f* heteropolyacid
Hexamethyldisilazan *n* hexamethyldisilazane
hf-Polarographie *f* radio-frequency polarography
Hg-Niveaudifferenz *f* height of mercury column
Hilfselektrode *f* **1.** auxiliary electrode (*in electrochemistry*) **2.** supporting electrode (*in arc and spark spectroscopy*)
Histogramm *n* histogram
hochauflösendes kernmagnetisches Resonanzspektrum *n* high-resolution nuclear magnetic resonance spectrum
Hochdruckflüssigchromatographie *f* high-performance liquid chromatography
Hochdruckflüssigkeitschromatographie *f* high-performance liquid chromatography
Hochfrequenzkonduktometrie *f* high-frequency conductometry
hochfrequenzkonduktometrische Titration *f* high-frequency conductometric titration
Hochfrequenzplasma *n* radiofrequency plasma
Hochfrequenz-Polarographie *f* radio-frequency polarography
Hochfrequenztitration *f* high-frequency conductometric titration
Hochleistungs-Flüssigchromatographie *f* high-performance liquid chromatography
Hochleistungs-Flüssigkeitschromatographie *f* high-performance liquid chromatography
Höchstvakuum *n* ultra-high vacuum
Höhenäquivalent *n* **eines effektiven theoretischen Bodens** height equivalent to an effective plate
Höhenäquivalent *n* **eines theoretischen Bodens** height equivalent to a theoretical plate
Hohlgitter *n* concave grating
Hohlkatodenentladung *f* hollow-cathode discharge
Hohlkatodenlampe *f* hollow-cathode lamp
Hohlraumanteil *m* interstitial fraction
Hohlraumresonator *m* cavity resonator

Hohlraumvolumen *n* interstitial volume
homogene Keimbildung *f* homogeneous
    nucleation
homogene Membran *f* homogeneous membrane
homogene Membran-Elektrode *f* homogeneous
    membrane ion-selective electrode
homogenes Verteilungsgesetz *n* homogeneous
    distribution law
homologe Linien *fpl* homologous lines
Hookesches Gesetz *n* Hooke's law
Hopkalit *m* Hopcalite
hydrodynamische Voltammetrie *f* hydrodynamic
    voltammetry
Hydrogel *n* hydrogel
hydroskopisches Wasser *n* hydroscopic water
Hydrosol *n* hydrosol
hydrostatische Waage *f* hydrostatic balance
hyperchromer Effekt *m* hyperchromic effect
Hyperfeinstruktur *f* hyperfine structure
Hypersorption *f* hypersorption
hypochromer Effekt *m* hypochromic effect
hypsochrome Gruppe *f* hypsochrome
hypsochrome Verschiebung *f* hypsochromic shift
Hysterese *f* hysteresis
Hysteresis *f* hysteresis

I

ideale nicht polarisierbare Elektrode *f* ideal
    non-polarizable electrode
idealetpolarisierbare Elektrode *f* ideal
    polarizable electrode
Identifizierung *f* identification
Identitätsreaktion *f* identification reaction
Ilkovič-Gleichung *f* Ilkovič equation
Immune-Elektrophorese *f* immunoelectrophoresis
Impedanz *f* impedance
Impulsanhäufung *f* pile-up
Impulsdesorption *f* flash desorption
Impulshöhenanalysator *m* pulse amplitude
    analyser
Impulshöhenanalyse *f* pulse amplitude analysis
indifferenter Elektrolyt *m* supporting electrolyte
indifferentes Lösungsmittel *n* aprotic solvent
Indikationsfehler *m* titration error
Indikator-Elektrode *f* indicator electrode
Indikatorexponent *m* indicator exponent
Indikatorfehler *m* indicator error
Indikatorpapier *n* indicator paper
indirekte Methode *f* indirect method
indirektes Verfahren *n* indirect method
indirekte Titration *f* indirect titration
induktivgekoppeltes Plasma *n* inductively
    coupled plasma
Induktor *m* inductor
induzierte Emission *f* stimulated emission
induzierte Reaktion *f* induced reaction
Inertpeak *m* air peak
Inflexionspunkte *mpl* inflection points
Infrarot *n* infrared
Infrarotgebiet *n* infrared region
Infrarotspektrum *n* infrared spectrum
Infusorienerde *f* diatomaceous earth
inkohärente Streuung *f* **von Photonen** Compton
    effect
Innen-Indikator *m* internal indicator

Innenkegel *m* primary combustion zone
Innenstandard *m* **1.** internal standard **2.**
    internal standard (*in spectrochemical
    analysis*)
Innenstandardlinie *f* internal reference line
innere Bezugselektrode *f* internal reference
    electrode
innere Durchlässigkeit *f* internal transmittance
Innere *n* **einer Lösung** bulk of a solution
innere Elektrolyse *f* internal electrolysis
innerer Indikator *m* internal indicator
innerer Kegel *m* primary combustion zone
innerer Standard *m* **1.** internal standard
    2. internal standard (*in spectrochemical
    analysis*)
inneres elektrisches Potential *n* **der Phase** β
    inner electric potential of phase β
inneres Komplexsalz *n* neutral chelate
inneres Photoansprechvermögen *n* intrinsic
    photopeak efficiency
inneres Transmissionsvermögen *n* internal
    transmittance
Instrumentalanalyse *f* instrumental analysis
Instrumentenanalyse *f* instrumental analysis
Instrumentenfehler *m* instrumental error
Integralchromatogramm *n* integral
    chromatogram
Integraldetektor *m* integral detector
integrale Kapazität *f* **der Doppelschicht** integral
    capacitance of double layer
Interelementanregung *f* enhancement effect
Interferenz-Filter *n, m* interference filter
Interferometer *n* optical interferometer
Interferometrie *f* interferometry
interne Elektrogravimetrie *f* internal
    electrogravimetry
Interpolation *f* interpolation
inverse derivative potentiometrische Titration *f*
    inverse derivative potentiometric titration
inverse Titration *f* inverse titration
Inversvoltammetrie *f* stripping voltammetry
Iodatometrie *f* iodatometry
iodimetrische Titration *f* iodimetric titration
Iodometrie *f* iodimetry
Iodzahl *f* iodine number
Ionenäquivalentleitfähigkeit *f* equivalent
    conductivity of ionic species B
Ionenäquivalentleitfähigkeit *f* **bei unendlicher
    Verdünnung** limiting equivalent conductivity
    of ionic species B
Ionenassoziationssystem *n* ion-association system
Ionenatmosphäre *f* ionic atmosphere
Ionenausschluß *m* ion exclusion
Ionenausschlußchromatographie *f* ion-exclusion
    chromatography
Ionenaustausch *m* ion exchange
Ionenaustauschchromatographie *f* ion-exchange
    chromatography
Ionenaustauscher *m* ion exchanger
Ionenaustauscher-Gesamtkapazität *f* total
    specific capacity
Ionenaustauscherharz *n* ion-exchange resin
Ionenaustauscher-Membrane *f* ion-exchange
    membrane
Ionenaustauscher-Volumenkapazität *f* volume
    capacity

Ionenaustauschisotherme *f* ion-exchange
isotherm
Ionenaustauschmembrane *f* ion-exchange
membrane
Ionenbeweglichkeit *f* electric mobility of ion B
Ionendrilling *n* ion triplet
Ionenentladung *f* discharge of ion
Ionenimplantation *f* ion implantation
Ionenleiter *m* ionic conductor
Ionenleitung *f* ionic conduction
Ionenneutralisierungsspektroskopie *f* ion
neutralization spectroscopy
Ionenpaarchromatographie *f* ion-pair partition
chromatography
Ionenquelle *f* ionization source
Ionenrefraktion *f* ionic refraction
ionenselektive Elektrode *f* ion-selective
electrode
ionenselektive Elektrode *f* **mit fester Membran**
solid-state membrane ion-selective electrode
ionenselektive Elektrode *f* **mit flüssigem**
Ionenaustauscher liquid ion-exchange
membrane electrode
ionenselektive Elektrode *f* **mit flüssiger**
Membran liquid ion-selective electrode
ionenselektive Festkörperelektrode *f* solid-state
membrane ion-selective electrode
ionenselektive Glasmembranelektrode *f* glass
membrane electrode
ionenselektive Membran *f* ion-selective
membrane
ionenselektive Transistorelektrode *f* ion-
sensitive field effect transistor
ionensensitive Elektrode *f* ion-selective
electrode
Ionenspektrum *n* ionic spectrum
ionenspezifische Elektrode *f* ion-selective
electrode
Ionenstärke *f* **der Lösung** ionic strength of
a solution
Ionenstärke-Gesetz *n* **von Lewis und Randall**
Lewis-Randall's ionic strength law
Ionenstoßdesorption *f* ion impact desorption
ionenstrahl-induzierte Auger-
-Elektronenspektroskopie *f* ion-induced
Auger electron spectroscopy
Ionenstreuungsspektroskopie *f* low-energy ion
scattering spectroscopy
Ionentriplett *n* ion triplet
Ionenwanderung *f* migration of ions
Ionenwolke *f* ionic atmosphere
Ionisationsdetektor *m* ionization detector
Ionisationskammer *f* ionization chamber
ionisierende Strahlung *f* ionizing radiation
ionogene Gruppen *fpl* ionogenic groups
irreversible Elektrode *f* irreversible electrode
irreversible Koagulation *f* irreversible
coagulation
irreversibler Redoxindikator *m* irreversible
redox indicator
IR-Spektrum *n* infrared spectrum
Isochrome *f* isochrome
isoelektrischer Punkt *m* isoelectric point
isokratische Elution *f* isocratic elution
Isopolysäure *f* isopolyacid
isosbestischer Punkt *m* isosbestic point

Isotachophorese *f* isotachophoresis
Isotherme *f* isotherm
Isotopenaustausch *m* isotope exchange
Isotopeneffekt *m* isotope effect
Isotopenhäufigkeit *f* abundance
Isotopen-Verdünnungsanalyse *f* isotope dilution
analysis
Isotopenwirkungsquerschnitt *m* isotopic cross
section
isotoper Indikator *m* isotopic tracer
isotoper Träger *m* isotopic carrier
isotopische Neutronenquelle *f* isotopic neutron
source
isotopischer Träger *m* isotopic carrier

J

Johann-Kristall *m* Johann crystal
Johannson-Kristall *m* Johannson crystal

K

Kadmium-Verhältnis *n* cadmium ratio
Kalibrierungsfehler *m* scale error
Kaliumbromidscheibe *f* potassium bromide disk
Kaliumbromid-Tablette *f* potassium bromide
disk
Kalkhärte *f* calcium hardness
Kalomelektrode *f* calomel electrode
Kalorimeterbombe *f* bomb calorimeter
Kalousek-Polarographie *f* Kalousek
polarography
kalte Neutronen *npl* cold neutrons
Kammersättigung *f* chamber saturation
Kapazitätsfaktor *m* mass distribution ratio
Kapazitätsstrom *m* capacity current
Kapillarchromatographie *f* open-tube
chromatography
Kapillaren *fpl* capillary tubes
Kapillarsäule *f* capillary column
Karl-Fischer-Lösung *f* Karl Fischer reagent
katalytische Analyse *f* kinetic analysis
katalytische Reaktion *f* catalytic reaction
katalytischer Strom *m* catalytic current
katalytische Verbrennung *f* catalytic combustion
Katharometer *m* thermal conductivity detector
Kathode *f* cathode
Katholyt *m* catholyte
Kationenaustausch *m* cation exchange
Kationenaustauscher *m* cation exchanger
Kationenaustauscherharz *n* cation-exchange
resin
Katode *f* cathode
Katodenschicht *f* cathode layer
Katodenstrahlpolarographie *f* single-sweep
polarography
katodische Metallabscheidung *f*
electrodeposition of metals
Katodolumineszenz *f* cathodoluminescence
Katolyt *m* catholyte
Kayser *n* kayser
K-Band *n* K band
KBr-Preßling *m* potassium bromide disk
KBr-Preßlingtechnik *f* pellet technique
KBr-Preßtechnik *f* pellet technique
Keim *m* nucleus

Keimbildung *f* nucleation
Keimbildungsgeschwindigkeit *f* rate of
  nucleation
Kelvin-Methode *f* Kelvin method
Kernbruchstücke *npl* fission fragments
kernmagnetische Resonanz *f* nuclear magnetic
  resonance
kernmagnetisches Resonanz-Spektrometer *n*
  nuclear magnetic resonance spectrometer
Kernreinheit *f* nuclear purity
Kernresonanzspektroskopie *f* nuclear magnetic
  resonance spectroscopy
Kernspaltung *f* fission
Kernstrahlungsspektrum *n* nuclear radiation
  spectrum
Kette *f* **mit Überführung** cell with transference
Kette *f* **ohne Überführung** cell without
  transference
Kieselgur *f* diatomaceous earth
kinetische Analyse *f* kinetic analysis
kinetische Kontrolle *f* kinetic controlled process
kinetische Reaktionsordnung *f* order of reaction
kinetischer Strom *m* kinetic current
Kippscher Apparat *m* Kipp gas generator
Kippscher Gasapparat *m* Kipp gas generator
Kippschwingungen *fpl* wagging vibrations
Kjeldahl-Methode *f* Kjeldahl method
klassische Wechselstrompolarographie *f*
  conventional alternating-current
  polarography
Klemmenspannung *f* 1. cell voltage 2. electric
  potential difference of a galvanic cell
Knallgasverbrennung *f* oxy-hydrogen flame
  method
Koagulation *f* coagulation
kohärente Streuung *f* coherent scattering
Kohlepasteelektrode *f* carbon-paste electrode
Kohlrauschsche Regel *f* **der unabhängigen
  Ionenwanderung** Kohlrausch's additive law
Kohlrausches Gesetz *n* **von der unabhängigen
  Ionenwanderung** Kohlrausch's additive law
Koinzidenzverfahren *n* coincidence technique
Kolben *m* **für jodometrische Bestimmung** iodine
  flask
Kollektor *m* trace collector
Kollimation *f* collimation
Kollimator *m* collimator
kolloider Niederschlag *m* colloid precipitate
Kolorimeter *n* colorimeter
Kolorimeterzylinder *m* **nach Hehner** Hehner
  measuring cylinder
Kolorimeterzylinder *m* **nach Neßler** Nessler
  measuring cylinder
Kolorimetrie *f* colorimetric analysis
kolorimetrische Analyse *f* colorimetric analysis
kolorimetrische Titration *f* colorimetric titration
Kombinationsbande *f* combination band
kombinierte Elektrode *f* combination electrode
kombinierte Säulen *fpl* coupled columns
kombinierte Untersuchungsmethoden *fpl* 1.
  combined techniques 2. multiple techniques
Komma *n* coma
Komparatormethode *f* monostandard method
Komparatortechnik *f* monostandard method
Kompensationsmethode *f* **von Poggendorff**
  Poggendorff's compensations method

Komplexbildungstitration *f* complexus
  titration
komplexer Scheinleitwert *m* admittance
komplexer Scheinwiderstand *m* impedance
Kompleximetrie *f* complexometry
Komplexometrie *f* complexometry
komplexometrische Titration *f* complexometric
  titration
komplexometrische Titrationskurve *f*
  compleximetric titration curve
Komplexonen *npl* complexones
Kondensatorstrom *m* capacity current
Konduktanz *f* conductance
Konduktimetrie *f* conductometry
Konduktivität *f* electrical conductivity
Konduktometrie *f* conductometry
konduktometrische Analyse *f* conductometric
  analysis
konduktometrische Indikation *f* conductometric
  titration
konduktometrische Titration *f* conductometric
  titration
Konduktor *m* electrical conductor
Konfidenzgrenzen *fpl* confidence limits
Konfidenzintervall *n* confidence intervall
Konfidenzniveau *n* confidence level
Konformation *f* conformation
Konformationsanalyse *f* conformational analysis
konjugierte Reaktion *f* conjugate reaction
Konkavgitter *n* concave grating
konkreter Versuchsplan *m* exact design
konsistente Schätzfunktion *f* consistent
  estimator
konstante *f* **Masse** constant mass
Konstellation *f* conformation
kontinuierliche Extraktion *f* continuous
  extraction
kontinuierliches Spektrum *n* continuous
  spectrum
Kontrastfaktor *m* gamma
Kontrollkarte *f* control chart
Kontrollprobe *f* check test
Kontrolltitration *f* control titration
Konvektion *f* convection
konvektive Chronoamperometrie *f* convective
  chronoamperometry
konvektive Chronocoulometrie *f* convective
  chronocoulometry
konventionelle Analyse *f* conventional analysis
konventionelle Geschwindigkeitskonstante *f* **der
  Elektrodenreaktion** conditional rate constant
  of electrode reaction
Konzentration *f* **eines Stoffes B** amount-of-
  substance concentration of substance B
Konzentrationskette *f* concentration cell
Konzentrationsprofil *n* depth profile
Konzentrationsüberspannung *f* concentration
  overpotential
Konzentrationsverteilungsverhältnis *n* 1.
  concentration distribution ratio (*in
  chromatography*) 2. concentration
  distribution ratio (*in extraction*)
Konzentrationszelle *f* concentration cell
koordinierende Gruppe *f* coordinating group
Körbl-Katalysator *m* Körbl catalyst
Korngrößeeffekt *m* grain size effect

korpuskulare Strahlung f corpuscular radiation
Korrektion f dilution-correction factor
Korrekturfaktor m **für den Druckabfall** pressure-gradient correction factor
Korrelation f correlation
Korrelationskoeffizient m coefficient of correlation
Korrelationsmatrix f correlation matrix
korrigierter Selektivitätskoeffizient m corrected selectivity coefficient
korrigierter R_f **–Wert** m R'_f value
korrigierter R_M **–Wert** m R'_M value
korrigiertes Retentionsvolumen n corrected retention volume
Korrosionspotential n corrosion potential
Kovarianz f covariance
Kovarianzmatrix f covariance matrix
Kováts-Retentionsindex m retention index
Kraftkonstante f force constant
Kratergrube f crater
kristalliner Niederschlag m crystalline precipitate
Kristallisation f crystallization
Kristallisationsüberspannung f crystallization overpotential
Kristallisierschale f crystallizing dish
Kristallmembran f crystalline membrane
Kristallmembran-Elektrode f crystalline membrane ion-selective electrode
kristalloskopische Mikroanalyse f microcrystalloscopic analysis
Kristallwasser n water of crystallization
kritische Bestimmung f critical determination
kritischer Bereich m rejection region
kritisches Übersättigungsverhältnis n critical supersaturation ratio
kryoskopische Bestimmung f **der Molmasse** cryoscopic determination of the molar mass
Kupellation f cupellation
Kupfercoulometer n copper coulometer
Kurzanalyse f short analysis
kurzlebiges Isotop n short-lived radioisotope
Kurzwellengrenze f short-wavelength limit

L

Ladestrom m capacity current
Ladungsaustauschband n change-transfer band
Ladungs-Durchtrittsreaktion f charge-transfer step
Ladungsinkrement-Polarographie f incremental-charge polarography
Ladungsstrom m capacity current
Ladungsstufenpolarographie f incremental-charge polarography
Ladungs-Transferreaktion f charge-transfer step
Ladungsübertragungskomplex m charge-transfer complex
Lambert-Beersches Gesetz n Beer's law
Lambertsches Gesetz n Lambert's law
laminare Flamme f laminar flame
langlebiges Isotop n long-lived radioisotope
langsame Neutronen npl slow neutrons
Laser m laser
Laser-Fluoreszenzspektroskopie f laser fluorescence spectroscopy

Laser-Raman-Spektroskopie f laser Raman spectroscopy
lateinisches Quadrat n latin square
laufendes Gleichgewicht n transient radioactive equilibrium
laufendes radioaktives Gleichgewicht n transient radioactive equilibrium
Laufmittelfront f mobile phase front
Laufstrecke f **der mobilen Phase** mobile phase distance
Laufzeit f time of flight
Laufzeitspektrometer n time-of-flight mass spectrometer
Laugung f leaching
Leclanché-Element n Leclanché cell
Leclanché-Zelle f Leclanché cell
Leerrohrverbrennung f empty tube combustion
Leerversuch m blank test
Leitelelektrolyt m supporting electrolyte
Leiter m **erster Klasse** electronic conductor
Leiter m **1. Ordnung** electronic conductor
Leiter m **2. Ordnung** ionic conductor
Leiter m **zweiter Klasse** ionic conductor
Leitfähigkeit f conductivity
Leitfähigkeitsmesser m conductometer
Leitisotop n isotopic tracer
Leitsalz n supporting electrolyte
letzte Linien fpl raies ultimes
Leuchtdichte f luminance
Lewis-Sargentsche Gleichung f Lewis-Sargent's relation
Lewitsch-Gleichung f Levich equation
Licht n light
Lichtbeugung f light diffraction
Lichtbogen m electrical arc
Lichtbrechung f light refraction
Lichtenergie f luminous energy
Lichtfilter n, m light filter
Lichtgeschwindigkeit f **im Vakuum** velocity of light in vacuum
Lichtintensität f luminous intensity
Lichtmenge f luminous energy
Lichtquelle f light source
Lichtstärke f luminous intensity
Lichtstreuung f light scattering
Lichtstrom m luminous flux
Lichtzerstreuung f light scattering
Liebig-Methode f Liebig's method
Ligandenaustauschchromatographie f ligand-exchange chromatography
lineare Dispersion f linear dispersion
lineare Polarisation f plane polarization
linearer Absorptionskoeffizient m linear decadic absorption coefficient
linearer dekadischer Absorptionskoeffizient m linear decadic absorption coefficient
lineare Regression f linear regression
linearer Schwachungskoeffizient m linearer attenuation coefficient
Linearisierung f linearization
linear polarisiertes Licht n plane-polarized light
Linienbreite f line width
Linienkoinzidenz f line coincidence
Linienpaar n line pair
Linienspektrum n line spectrum

linksdrehende Substanz *f* laevorotatory substance
links-zirkular polarisiertes Licht *n* left-handed
  circularly polarized light
lithiumgedrifteter Germaniumdetektor *m*
  lithium-drifted germanium detector
lithiumgedrifteter Siliciumdetektor *m* lithium-
  drifted silicon detector
Löcherleitung *f* hole conduction
logarithmische Normalverteilung *f* logarithmic-
  normal distribution
logarithmisches Verteilungsgesetz *n* logarithmic
  distribution law
Lokalanalyse *f* local analysis
Löschung *f* quenching
Lösemittel *n* solvent
Lösen *n* dissolution
Löslichkeit *f* solubility
Löslichkeitskonstante *f* solubility product
Löslichkeitsparameter *m* solubility parameter
Löslichkeitsprodukt *n* solubility product
Lösung *f* solution
Lösungsmittel *n* solvent
Lösungsmittelentmischung *f* solvent demixing
Lösungsmittelfront *f* mobile phase front
Lösungsmittelgemisch *n* mixed solvent
Lösungsmittelstärke *f* solvent strength
Lötrohrprobe *f* blow-pipe test
Löckenleitung *f* hole conduction
Luftberg *m* air peak
Luftpeak *m* air peak
Luftspalt-Elektrode *f* air-gap electrode
Luggin-Kapillare *f* Luggin-Haber capillary
Lumineszenz *f* luminescence

M

Macht *f* eines Tests power of a test
Magnesiahärte *f* magnesium hardness
magnetische Ablenkung *f* magnetic deflection
magnetische Kernresonanzspektroskopie *f*
  nuclear magnetic resonance spectroscopy
magnetische Resonanz *f* magnetic resonance
Makroanalyse *f* macroanalysis
Makrokomponente *f* macro-component
makroporöser Ionenaustauscher *m* macroporous
  ion exchanger
makroskopischer Wirkungsquerschnitt *m*
  macroscopic cross section
Manganometrie *f* manganometry
manometrische Gasanalyse *f* manometric gas
  analysis
manometrische Verfahren *npl* manometric
  methods
markiertes Volumenmeßgefäß *n* designated
  volume
markierte Verbindung *f* labelled compound
Maskierungsmittel *n* masking agent
Maskierungsreagens *n* masking agent
Maskierung *f* von Ionen masking of ions
Masse-Ladung-Verhältnis *n* mass-to-charge
  ratio
Massenanalysator *m* mass analyser
Massenanteil *m* mass fraction of substance B
Massengehalt *n* mass fraction of substance B
Massenkonzentration *f* eines Stoffes B
  concentration of substance B

Massenschwächungskoeffizient *m* mass
  attenuation coefficient
Massenspektrograph *m* mass spectrograph
Massenspektrometer *n* mass spectrometer
Massenspektrometrie *f* mass spectrometry
Massenspektroskop *n* mass spectroscope
Massenspektroskopie *f* mass spectrometry
Massenspektroskopie *f* nachionisierter
  Neutralteilchen an Oberflächen glow
  discharge mass spectroscopy
Massenspektrum *n* mass spectrum
Massenspektrum *n* negativer Ionen negative-ion
  mass spectrum
Massenspektrum *n* positiver Ionen positive-ion
  mass spectrum
Massenübergangswiderstand *m* mass transfer
  term
Massenverteilungsverhältnis *n* mass
  distribution ratio
Masse *f* pro Skalenteil value of a division
mathematische Statistik *f* statistics
mathematische Stichprobe *f* random sample
Matrix *f* matrix
Matrix *f* des Ionenaustauschers ion exchanger
  matrix
Matrixeffekt *m* 1. matrix effect 2. matrix
  effect (*in X-ray fluorescence analysis*)
Maximum-Likelihood-Methode *f* maximum
  likelihood method
Mehrfachelektrode *f* complex electrode
mehrfache Regression *f* multiple regression
Mehrkomponentenanalyse *f* multiple analysis
Membran-Elektrode *f* membrane ion-selective
  electrode
Membrangleichgewicht *n* membrane equilibrium
Membran *f* mit elektroneutralem
  Ladungsüberträger neutral-carrier membrane
Membranpotential *n* membrane potential
Mengenverhältnis *n* quantitative composition
Merkurimetrie *f* mercurimetry
merkurimetrische Titration *f* mercurimetric
  titration
Merkurometrie *f* mercurometry
merkurometrische Titration *f* mercurometric
  titration
Meßkolben *m* volumetric flask
Meßkolben *m* mit graduiertem Hals volumetric
  flask with graduated neck
Meßpipette *f* graduated pipette
Messung *f* des nichtfaradayschen Leitwerts
  measurement of nonfaradaic admittance
Meßzylinder *m* measuring cylinder
Metallchelat *n* chelate compound
Metallelektrode *f* metal-metal ion electrode
Metallfluoreszenzindikator *m* metallofluorescent
  indicator
Metallindikator *m* metallochromic indicator
Metallindikatoren *mpl* metal indicators
Metallionenelektrode *f* metal-metal ion electrode
Metallionenexponent *m* pM
metallischer Leiter *m* electronic conductor
Metall-Metallion-Elektrode *f* metal-metal ion
  electrode
Metalloxidelektrode *f* oxide electrode
Metallpuffer *m* metal buffer
metastabiler Abbau *m* metastable decomposition

metastabiler Peak *m* metastable ion peak
metastabiles Gleichgewicht *n* **einer Elektrode**
metastable equilibrium of an electrode
metastabiles Ion *n* metastable ion
Methode *f* **der bekannten Zusätze** standard
addition method
Methode *f* **der β-Durchstrahlung** beta-
particle absorptiometry
Methode *f* **der Isotopenverdünnung** isotope
dilution analysis
Methode *f* **der kleinen Schwingungen** method of
short swings
Methode *f* **der kleinsten Quadrate** least-squares
method
Methode *f* **der Zufallsbalance** random balance
method
Methode *f* **des hängenden Tropfens** hanging-drop
method
Methode *f* **des inneren Standards** internal-
standard method
Methode *f* **von Covell** Colvell's method
methodischer Fehler *m* error of method
Migrationsstrom *m* migration current
Migration *f* **von Ionen** migration of ions
Mikroanalyse *f* microanalysis
mikroanalytische Methode *f* milligram method
mikroanalytisches Reagens *n* microanalytical
reagent
Mikroazotometer *n* micronitrometer
Mikrobestandteil *m* micro-component
Mikrobürette *f* microburette
mikrochemische Waage *f* microchemical balance
Mikrocoulometrie *f* polarographic coulometry
Mikrokomponente *f* micro-component
Mikrokristalloskopie *f* microcrystalloscopic
analysis
Mikromethode *f* milligram method
Mikrophotometer *n* microphotometer
Mikropipette *f* micropipette
mikroskopischer Wirkungsquerschnitt *m*
microscopic cross section
Mikrospuren *fpl* microtraces
Mikrowellen *fpl* microwaves
Mikrowellengebiet *n* microwave region
Mikrowellen-Plasma *n* microwave plasma
Mikrowellen-Spektroskopie *f* microwave
spectroscopy
Mikrowellenspektrum *n* microwave spectrum
Mikrozirkularchromatographie *f* microcircular
chromatography
Millersche Indizes *mpl* Miller indices
Milliardstel *n* parts per billion
Millicoulometrie *f* polarographic coulometry
Millionstel *n* parts per million
Mineralisation *f* mineralization
mineralogischer Effekt *m* mineralogical effect
Mischbett *n* mixed bed
Mischindikator *m* 1. mixed indicator
2. screened indicator
Mischkammerbrenner *m* premixed gas burner
Mischkristall *m* mixed crystal
Mischpotential *n* mixed polyelectrode potential
Mitfällung *f* coprecipitation
Mittelwert *m* **der Spannweite** mid-range
Mittel *n* **zum Rückextrahieren** stripping
solution

Mittel *n* **zum Strippen** stripping solution
mittlere Abweichung *f* mean deviation
mittlere Aktivität *f* **eines Elektrolyten** mean
activity of electrolyte
mittlere Ionenaktivität *f* mean activity of
electrolyte
mittlere Ionenkonzentration *f* mean
concentration of electrolyte
mittlerer Aktivitätskoeffizient *m* **der Elektrolyte**
mean activity coefficient of electrolyte
mittlerer Ionenaktivitätskoeffizient *m* mean
activity coefficient of electrolyte
mittlerer Ionendurchmesser *m* mean ionic
diameter
mittleres Infrarot *n* middle-infrared
mittleres Ultrarot *n* middle-infrared
mittlere Trägergasgeschwindigkeit *f* **im**
Hohlraumbereich mean interstitial velocity of
the carrier gas
Mixtur-Versuchsplan *m* mixture design
mobile Phase *f* mobile phase
Mobilvolumen *n* mobil phase hold-up volume
Mobilzeit *f* mobile phase hold-up time
Modalwert *m* mode
Mode *m* mode
Modell *n* **I** fixed effects model
Modell *n* **II** random effects model
Modell *n* **mit festen Effekten** fixed effects model
Modell *n* **mit zufälligen Effekten** random
effects model
modifizierter aktiver Festkörper *m* modified
active solid
Modulationspolarographie *f* modulation
polarography
Mohrsche Methode *f* Mohr's method
Mol *n* **1.** gram molecule **2.** mole
Molalität *f* molality
molare Drehung *f* molar rotation
molare Grenzleitfähigkeit *f* **eines Elektrolyten**
limiting molar conductivity
molare Grenzleitfähigkeit *f* **eines Ions** limiting
molar conductivity of ionic species B
molare Ionengrenzleitfähigkeit *f* limiting molar
conductivity of ionic species B
molare Ionenleitfähigkeit *f* molar conductivity
of ionic species B
molare Ionenleitfähigkeit *f* **bei unendlicher**
Verdünnung limiting molar conductivity of
ionic species B
molare Konzentration *f* amount-of-substance
concentration of substance B
molare Leitfähigkeit *f* **bei unendlicher**
Verdünnung limiting molar conductivity
of electrolyte
molare Leitfähigkeit *f* **eines Elektrolyten**
molar conductivity of electrolyte
molare Leitfähigkeit *f* **eines Elektrolyten bei**
unendlicher Verdünnung limiting molar
conductivity of electrolyte
molare Leitfähigkeit *f* **eines Ions** molar
conductivity of ionic species B
molare Masse *f* molar mass
molarer Absorptionskoeffizient *m* molar
absorptivity
molarer dekadischer Absorptionskoeffizient *m*
molar absorptivity

molarer Extinktionskoeffizient *m* molar
absorptivity
Molekulardiffusion *f* molecular diffusion term
Molekulardiffusionsterm *m* molecular diffusions
term
molekulare spektrochemische Analyse *f*
molecular spectrochemical analysis
Molekularsiebchromatographie *f* molecular-sieve
chromatography
Molekularsiebe *npl* molecular sieves
Molekularstrahl *m* molecular beam
Molekülion *n* molecular ion
Molekülpeak *m* molecular peak
Molekülradiation *f* molecular radiation
Molekülschwingungen *fpl* molecular vibrations
Molekülspektroskopie *f* molecular spectroscopy
Molekülspektrum *n* molecular spectrum
Molmasse *f* molar mass
Molrefraktion *f* molar refraction
monochromatische Strahlung *f* monochromatic
radiation
Monochromator *m* monochromator
monofunktioneller Ionenaustauscher *m*
monofunctional ion exchanger
Monoschicht *f* monolayer
Moseleysches Gesetz *n* Moseley law
Mößbauer-Effekt *m* Mössbauer effect
Mößbauer-Spektroskopie *f* Mössbauer
spectroscopy
Muffelofen *m* muffle furnace
Mull-Verfahren *n* mull technique
multiple Regression *f* multiple regression
multiple Regressionsmethode *f* multiple
regression method
multipler Korrelationskoeffizient *m* coefficient
of multiple correlation
Multiplett *n* multiplet
Multisweep-Polarographie *f* multisweep
polarography
Murexidreaktion *f* murexide reaction
Mutterion *f* parent ion
Mutternuklid *n* parent nuclide

N

Nachfällung *f* postprecipitation
Nachweis *m* detection
Nachweisgrenze *f* limit of detection
Nachweisreaktion *f* identification reaction
nahes Infrarot *n* near infrared
nahes Ultrarot *n* near infrared
nahes Ultraviolett *n* near ultraviolet
Nanogrammethode *f* nanogram method
Nanospuren *fpl* nanotraces
nasses Verfahren *n* wet method
natürliche Isotopenhäufigkeit *f* natural isotopic
abundance
natürliche Konvektion *f* natural convection
natürliche Linienbreite *f* natural line-width
Nebelblaser *m* nebulizer
Nebenbestandteil *m* minor constituent
Nebenreaktions-Koeffizient *m* side-reaction
coefficient
negativer Zweig *m* P-branch
negatives Ion *n* negative ion

Nephelometer *n* nephelometer
Nephelometrie *f* nephelometry
nephelometrische Titration *f* nephelometric
titration
Nernst-Gleichung *f* Nernst equation
Nernst-Lampe *f* Nernst glower
Nernstsche Neigung *f* Nernstian slope
Nernstscher Verteilungskoeffizient *m*
distribution constant (*in extraction*)
Nernstscher Verteilungssatz *m* Nernst
distribution law
Nernst-Stift *m* Nernst glower
Neßlers Reagens *f* Nessler reagent
Nettoretentionsvolumen *n* net retention volume
Nettoretentionszeit *f* net retention time
Neutralfilter *n*, *m* neutral-density filter
Neutralisation *f* neutralization
Neutralisationsanalyse *f* acidimetry and
alkalimetry
Neutralisationskurve *f* neutralization titration
curve
Neutralisationstitration *f* acid-base titration
Neutronenabsorptionsmethode *f* neutron
absorptiometry
Neutronenabsorptionsverfahren *n* neutron
absorptiometry
Neutronenaktivierung *f* neutron activation
Neutronenaktivierungsanalyse *f* neutron
activation analysis
Neutronenausbeute *f* neutron output
Neutronendetektor *m* neutron detector
Neutronengenerator *m* neutron generator
Neutronenspektrum *n* neutron spectrum
Neutronenthermalisierung *f* neutron
thermalization
Nichrom-Spirale *f* coil of Nichrome wire
nichtaufgelöster Peak *m* unresolved peak
Nichtcarbonathärte *f* non-carbonate hardness
nichtfaradaysche Admittanzmessung *f*
measurement of non-faradaic admittance
nichtisotoper Träger *m* non-isotopic carrier
nichtlineare Regression *f* non-linear regression
nichtmobile Phase *f* stationary phase
nichtparametrischer Test *m* non-parametric test
nichtstationäre
Temperaturgradientenchromatographie *f*
chromathermography
nicht umkehrbare Elektrode *f* irreversible
electrode
Nickschwingungen *fpl* wagging vibrations
Niederschlag *m* precipitate
Niederschlagsauswaschen *n* washing of
precipitate
Niederschlagsmembran-Elektrode *f*
precipitate-based ion-selective electrode
Nitrometer *n* nitrometer
nivellierender Effekt *m* levelling effect
nivellierendes Lösungsmittel *n* levelling solvent
Nivellierung *f* levelling effect
NMR-Spektroskopie *f* nuclear magnetic
resonance spectroscopy
nominelle lineare Flußgeschwindigkeit *f* nominal
linear flow
normale Dispersion *f* normal dispersion
Normalelement *n* standard cell
Normal-EMK *f* standard electromotive force

normale Puls-Polarographie *f* pulse polarography
normaler Raman-Effekt *m* Raman effect
Normalgleichungen *fpl* normal equations
Normalität *f* normality
Normalpotential *n* standard electrode potential
Normalprobe *f* standard sample
Normalschwingungen *fpl* normal vibrations
Normalverteilung *f* normal distribution
Normal-Wasserstoffelektrode *f* standard hydrogen electrode
Normzustände *mpl* standard conditions
Nujol Nujol
Nujol-Technik *f* mull technique
nukleare Reinheit *f* nuclear purity
Nulladungspotential *n* potential at the point of zero charge
Nulleffekt *m* background
Null-Eins-Verteilung *f* standardized normal distribution
Nullhypothese *f* null hypotesis
Null-Linie *f* base-line
Nullpunkt *m* der Zeigerskale zero point of the scale
Nullstrom-Potentiometrie *f* potentiometry
Nullstrom-potentiometrische Titration *f* potentiometric titration
Nullwert *m* background
Nullzweig *m* Q-branch
numerometrische Titration *f* numerical titration

O

Oberfläche *f* surface
oberflächenaktiver Stoff *m* surfactant
Oberflächenanalyse *f* surface analysis
Oberflächenbedeckung *f* surface coverage
Oberflächenionisation *f* surface ionization
Oberflächenionisierung *f* surface ionization
Oberflächensperrschichtzähler *m* surface barrier detector
oberflächlich poröses Trägermaterial *n* superficially porous support
Oberschwingung *f* overtone band
Oberschwingungsbande *f* overtone band
Obertonpolarographie *f* higher-harmonic ac polarography
Obertonpolarographie *f* mit Phasengleichrichtung higher-harmonic ac polarography with phase-sensitive rectification
OC-Funktion *f* operating characteristic function
Odorimetrie *f* odorimetry
offene röhrenartige Säule *f* open-tube column
Okklusion *f* occlusion
oleophiler Ionenaustauscher *m* oleophilic ion exchanger
Onsagersches Grenzgesetz *n* Debye-Hückel-Onsager's limiting law for conductivity
Operationscharakteristik *f* operating characteristic function
Operationsverstärker *m* operational amplifier
Optimalitätskriterium *n* eines Versuchsplanes criterion of optimality of experimental design
Optimierung *f* optimization

optisch aktive Substanz *f* optically active substance
optisch-akustische Spektroskopie *f* photoacoustic spectroscopy
optische Achse *f* optical axis
optische Aktivität *f* optical activity
optische Bank *f* optical bench
optische Dichte *f* absorbance
optische Drehung *f* optical rotation
optische Reinheit *f* optical purity
optische Rotationsdispersion *f* optical rotatory dispersion
optisches Drehvermögen *n* optical rotation
optisches Filter *n* optical filter
optisches Interferometer *n* optical interferometer
optische Spektroskopie *f* optical spectroscopy
optische Spektroskopie *f* von Glimmentladung an Oberflächen glow-discharge optical emission spectroscopy
optisches Spektrum *n* optical spectrum
organische Elementaranalyse *f* elemental analysis (*of organic compounds*)
organischer Ionenaustauscher *m* ion-exchange resin
organoleptischer Test *m* organoleptic assessment
Orsat-Apparat *m* Orsat gas analyser
orthogonaler Versuchsplan *m* orthogonal design
Ostwald-Reifung *f* Ostwald ripening
Ostwaldsches Verdünnungsgesetz *n* Ostwald dilution law
Ostwald-Viskosimeter *n* Ostwald viscometer
Oszillationsebene *f* plane of vibration
oszillographische Polarographie *f* oscillopolarography
oszillographischer Polarograph *m* oscillopolarograph
Oszillometrie *f* high-frequency conductometry
oszillometrische Indikation *f* high-frequency conductometric titration
oszillometrische Titration *f* high-frequency conductometric titration
Oszillopolarographie *f* oscillopolarography
Oxidans *n* oxidant
Oxidation *f* oxidation
Oxidationsmittel *n* oxidant
Oxidations-Reduktionspotential *n* oxidation-reduction potential
Oxidations-Reduktions-Titration *f* oxidation-reduction titration
Oxidimetrie *f* oxidimetry
oxidimetrische Titration *f* oxidimetric titration

P

Papierchromatographie *f* paper chromatography
Papierfilter *n, m* filter paper
paramagnetische Elektronenresonanz *f* electron paramagnetic resonance
Parameter *m* parameter
Parnas-Wagner-Apparatur *f* Parnas-Wagner apparatus
Partie *f* bulk of material
partielle Elektrodenreaktion *f* partial electrode reaction

partieller anodisch/katodischer Strom *m* partial anodic/cathodic current

partikel-induzierte Röntgenemissionsspektroskopie *f* particle induced X-ray emission

passiver Versuch *m* passive experiment

Passivierung *f* einer Elektrode passivation of electrode

Peak *m* peak

Peakauflösung *f* peak resolution

Peakbasis *f* peak base

Peakbreite *f* an der Basislinie peak width at base

Peakbreite *f* in halber Höhe peak width at half-height

Peak *m* eines metastabilen Ions metastable ion peak

Peakelutionsvolumen *n* peak elution volume

Peakfläche *f* peak area

Peakhöhe *f* peak height

Peakkapazität *f* peak capacity

Peakmaximum *n* peak maximum

Peakpotential *n* peak potential

Peakstrom *m* peak current

Peakasymmetriefaktor *m* asymmetry factor of a peak

Peakasymmetriekoeffizient *m* asymmetry factor of a peak

Peak-to-total-Verhältnis *n* peak-to-Compton ratio

Peakverbreiterung *f* peak broadening

Pehametrie *f* pH-metry

Pellicular-Ionenaustauscher *m* pellicular ion exchanger

Pellicular-Trägermaterial *n* superficially porous support

Pendelschwingungen *fpl* rocking vibrations

Peptisation *f* peptization

periodische Extraktion *f* periodic extraction

periodische Waage *f* undament balace

permanente Härte *f* permanet hardness

Permanganometrie *f* manganometry

Permeationschromatographie *f* permeation chromatography

Permeationsselektivität permselectivity

Permittivität *f* permittivity

Permselektivität *f* permselectivity

persönlicher Fehler *m* operative error

pH pH

pH-Anzeigegerät *n* pH-meter

Phasenanalyse *f* phase analysis

Phasenverhältnis *n* phase ratio

pH-Elektrode *f* pH electrode

pH-Glaselektrode *f* glass pH electrode

pH-Indikator *m* acid-base indicator

pH-Messer *m* pH-meter

pH-Meter *n* pH-meter

pH-Metrie *f* pH-metry

Phosphoreszenz *f* phosphorescence

Phosphorescenz-Emissionsspektrum *n* phosphorescence emission spectrum

Phosphoreszenz-Erregungsspektrum *n* phosphorescence excitation spectrum

Phosphorimetrie *f* phosphorimetry

Phosphorsalzperle *f* microcosmic salt bead

Photoaktivierung *f* photoactivation

Photoansprechvermögen *n* absolute photopeak efficiency

Photoanteil *m* photofraction

Photodesorption *f* photodesorption

Photodetektor *m* photodetector

Photodiode *f* photodiode

Photoeffekt *m* photoelectric effect

photoelektrischer Strom *m* photocurrent

photoelektrisches Strahlungsmeßgerät *n* photodetector

Photoelektron *n* photoelectron

Photoelektronenlinie *f* photoelectron line

Photoelektronenmikroskopie *f* photoelectron microscopy

Photoelektronenspektroskopie *f* photoelectron spectroscopy

Photoionisation *f* photoionization

Photoionisations-Ionenquelle *f* photoionization ion source

Photoionisierung *f* photoionization

Photokolorimeter *n* photocolorimeter

Photoleiter *m* photoconductor

Photolinie *f* photoelectric peak

Photolumineszenz *f* photoluminescence

Photometer *m* photometer

Photometrie *f* photometry

photometrischer Detektor *m* flame photometric detector

photometrische Titration *f* photometric titration

Photon *n* photon

Photoneutron *n* photoneutron

Photopeak *m* photoelectric peak

Photoplatte *f* photographic plate

Photopolarographie *f* photopolarography

Photostrom *m* photocurrent

Phototransistor *m* phototransistor

Photovervielfacher *m* photomultiplier tube

Photozelle *f* photocell

pH-Wert *m* pH

physikalische Adsorption *f* physisorption

Picospuren *fpl* picotraces

Pik *m* peak

pile-up-Effekt *m* pile-up

Pipette *f* pipette

Pipette *f* auf Abstrich transfer pipette

Plangitter *n* plane grating

Plasma *n* plasma

Plasma-Anodisierung *f* plasma anodization

Plasma-Ätzung *f* plasma etching

Plasmabrenner *m* plasma jet

Plasmafackel *f* plasma torch

Plasma-Reinigung *f* plasma cleaning

Plasma-Veraschung *f* plasma ashing

Plateau *n* des Zählrohres plateau of a counter

Platinasbest *m* platinized asbestos

Platinelektrode *f* platinum electrode

Platin-Gasruß *m* platinized carbon

Platinieren *n* der Elektroden platinization of electrodes

Platin-Kohle *f* platinized carbon

pneumatischer Nebelblaser *m* pneumatic nebulizer

Poggendorffsche Kompensationsmethode *f* Poggendorff's compensation method

pOH pOH

Poisson-Verteilung *f* Poisson distribution
Polarimeter *n* polarimeter
Polarimetrie *f* polarimetry
Polarisation *f* der Elektrode polarization of electrode
Polarisation *f* des Lichtes polarization of light
Polarisationsebene *f* plane of polarization
Polarisationskurve *f* polarization curve
Polarisationsstromtitration *f* amperometric titration with two indicator electrodes
polarisiertes Licht *n* polarized light
Polarogramm *n* polarogram
Polarograph *m* polarograph
Polarographie *f* polarography
Polarographie *f* mit linearer Stromabtastung current-scanning polarography
Polarographie *f* mit linearem Strom-Zeit-Verlauf current-scanning polarography
Polarographie *f* mit überlagerter periodischer Spannung alternating-current polarography
polarographische Analyse *f* polarographic analysis
polarographische Chronoamperometrie *f* polarographic chronoamperometry
polarographische Coulometrie *f* polarographic coulometry
polarographische Maxima *npl* polarographic maxima
polarographische Stromspannungskurve *f* polarogram
polarographische Stufe *f* polarographic wave
polarographische Titration *f* polarographic titration
polarographische Welle *f* polarographic wave
Polarometrie *f* polarimetry
polarometrische Titration *f* polarometric titration
polychromatische Strahlung *f* polychromatic radiation
Polyelektrolyt *m* polyelectrolyte
polyfunktioneller Ionenaustauscher *m* polyfunctional ion exchanger
Polysäure *f* polyacid
Porapak Porapak
Porenvolumen *n* stationary liquid volume
poröser Träger *m* totally porous support
positiver Zweig *m* R-branch
positives Ion *n* positive ion
potentialbestimmende Reaktion *f* 1. electrode reaction 2. potential-determining reaction
potentialkontrollierte Coulometrie *f* controlled-potential coulometry
potentialkontrollierte coulometrische Titration *f* controlled-potential coulometry
potentialkontrollierte Elektrogravimetrie *f* controlled-potential electrogravimetry
Potentialstufenchronocoulometrie *f* chronocoulometry
Potentiometer *n* potentiometer
Potentiometrie *f* potentiometry
Potentiometrie *f* mit gesteuertem Strom controlled-current potentiometry with one indicator electrode
Potentiometrie *f* mit gesteuertem Strom und zwei Indikatorelektroden controlled-current potentiometry with two indicator electrodes

potentiometrische Analyse *f* potentiometric analysis
potentiometrische coulometrische Titration *f* potentiometric coulometric titration
potentiometrische Indikation *f* potentiometric titration
potentiometrische Titration *f* potentiometric titration
potentiometrische Titration *f* bei konstantem Strom mit zwei polarisierbaren Elektroden controlled-current potentiometric titration with two indicator electrodes
potentiometrische Titration *f* mit gesteuertem Strom controlled-current potentiometric titration
potentiometrische Titration *f* mit zweiter Ableitung second-derivative potentiometric titration
potentiometrische Titration *f* mit gesteuertem Strom und zwei Indikatorelektroden controlled-current potentiometric titration with two indicator electrodes
Potentiostat *m* potentiostat
potentiostatische Coulometrie *f* controlled-potential coulometry
potentiostatische Impulsmethode *f* potentiostatic method
potentiostatische Methode *f* potentiostatic method
praktische spezifische Ionenaustauscherkapazität *f* practical specific capacity
Präzipitation *f* precipitation
Präzision *f* precision
Präzision *f* der Waage precision of a balance
Präzision *f* der Wägung precision of a weighing
Präzisions-Nullstrom-Potentiometrie *f* differential potentiometry
Pregl-Verfahren *n* der Elementarmikroanalyse Pregl procedure
Primärelement *n* primary cell
primäre Reaktionszone *f* primary combustion zone
primäre Röntgenstrahlen *mpl* primary X-radiation
primäre Urtitersubstanz *f* primary standard
Primärzelle *f* primary cell
Prisma *n* optical prism
pro-analytisches Reagens *n* analytical reagent
Probenahmefehler *m* sampling error
Probengeber *m* sample injector
Probeninjektor *m* sample injector
Probenteiler *m* sample splitter
prompte Neutronen *npl* prompt neutrons
prompte *γ*-Strahlung *f* prompt gamma radiation
Proportionalgaszähler *m* proportional gas-flow counter
Proportionalzähler *m* proportional counter tube
Proportionalzählrohr *n* proportional counter tube
protolytisches Lösungsmittel *n* protolytic solvent
Protonenresonanz *f* proton magnetic resonance
protonophiles Lösungsmittel *n* protophilic solvent
prozentualer Fehler *m* percentage error
Prozeßanalyse *f* process analysis

Prüfung *f* test (*in chemical analysis*)
Prüfung *f* der Gewichte calibration of weights
Puffer *m* buffer solution
Puffergemisch *n* buffer solution
Pufferlösung *f* buffer solution
Puls-Polarographie *f* pulse polarography
Pulveraufnahme *f* X-ray powdered-crystal
  pattern
Pulvertrichter *m* powder funnel
Punktschätzung *f* point estimation
Pyknometer *n* pycnometer
Pyrolyse *f* pyrolysis
Pyrolyser *m* pyrolyzer
P-Zweig *m* P-branch

Q

Quadrupol *m* quadrupole
Quadrupolmassenspektrometer *n* quadrupole
  mass analyser
Quadrupolrelaxation *f* quadrupol relaxation
qualitative Analyse *f* qualitative analysis
qualitative organische Analyse *f* qualitative
  analysis of organic compounds
qualitative Zusammensetzung *f* qualitative
  composition
γ-Quant *n* gamma quantum
Quantenausbeute *f* counting efficiency
Quantil *n* quantile of a probability distribution
Quantil *n* der Ordnung *p* quantile of a
  probability distribution
quantitative Analyse *f* quantitative analysis
quantitative Fällungsreaktion *f* quantitative
  precipitation
quantitative organische Analyse *f* quantitative
  analysis of organic compounds
Quecksilber(II)-oxid-Elektrode *f* mercury-
  mercuric oxide electrode
Quecksilber-Quecksilber(I)-chlorid-Elektrode
  f calomel electrode
Quecksilberstrahlelektrode *f* streaming mercury
  electrode
Quecksilber (I)-sulfat-Elektrode *f* mercury-
  mercurous sulphate electrode
Quecksilbertropfelektrode *f* dropping-mercury
  electrode
Quecksilberzelle *f* mercury dry cell
Quelle *f* source
Quellung *f* swelling
Quellungsdruck *m* swelling pressure
Quellungsvolumensquotient *m* volume swelling
  ratio
Quotient *m* Masse/Ladung mass-to-charge
  ratio
Q-Zweig *m* Q-branch

R

Radiation *f* radiation
radioaktiver Indikator *m* radioactive tracer
radioaktiver Zerfall *m* radioactive decay
radioaktives Dauer-Gleichgewicht *n* secular
  radioactive equilibrium
radioaktives Indikatorverfahren *n* radioactive
  tracer method
radioaktives Kryptonat *n* radioactive kryptonate

radioaktives Leitisotop *n* radioactive tracer
radioaktives Reagens *n* radioactive reagent
radioaktives Übergangsgleichgewicht *n* transient
  radioactive equilibrium
radioaktive Umwandlung *f* radioactive decay
Radioaktivität *f* radioactivity
radiochemische Analyse *f* radiochemical
  analysis
radiochemische Ausbeute *f* radiochemical yield
radiochemische Reinheit *f* radiochemical purity
radiochemische Trennung *f* radiochemical
  separation
Radiochromatographie *f* radiochromatography
Radioelektrophorese *f* radioelectrophoresis
radioexchange Analyse *f* radio-exchange method
Radiofrequenz-Polarographie *f* radio-frequency
  polarography
radioisotopische Röntgenfluoreszenzanalyse *f*
  radioisotope-excited X-ray fluorescence
  analysis
Radiolumineszenz *f* radioluminescence
Radiometrie *f* radiometry
radiometrische Analyse *f* radiometric analysis
radiometrische Titration *f* radiometric titration
Radioreagens *n* radioactive reagent
Radius *m* der Ionenatmosphäre radius of ionic
  atmosphere
Radius *m* der Ionenwolke radius of ionic
  atmosphere
Raman-Effekt *m* Raman effect
Raman-Frequenz-Verschiebung *f* Raman
  frequency shift
Raman-Linien *fpl* Raman lines
Raman-Spektroskopie *f* Raman spectroscopy
Raman-Spektrum *n* Raman spectrum
Ramman-Strahlung *f* Raman effect
Raman-Verschiebung *f* Raman frequency shift
Randles-Ševčik-Gleichung *f* Randles-Ševčik
  equation
Randomisierung *f* randomization
Rasterelektronenmikroskop *n* scanning electron
  microscope
Rast-Methode *f* Rast method
Rauheitszahl *f* roughness factor
Rauhigkeitszahl *f* roughness factor
Rauschen *n* noise level
Rauschpegel *m* noise level
Rayleigh-Linie *f* Rayleigh line
Rayleighsche Streuung *f* Rayleigh scattering
Rayleigh-Streuung *f* Rayleigh scattering
R-Band *n* R-band
Reagens *n* reagent
Reagens *n* zur Analyse analytical reagent
Reagent *n* reagent
Reagenzpapier *n* indicator paper
Reaktanz *f* reactance
Reaktion *f* reaction
Reaktionsordnung *f* order of reaction
Reaktionsschicht *f* reaction layer
Reaktionsüberspannung *f* reaction
  overpotential
reaktive Zerstäubung *f* reactive sputtering
Rearrangemention *n* rearrangement ion
Rechenverstärker *m* operational amplifier
Rechteckwellen-Polarographie *f* square-wave
  polarography

rechtsdrehende Substanz *f* dextrarotatory
substance
rechts-zirkular polarisiertes Licht *n* right-
handed circularly polarized light
redestilliertes Wasser *n* redistilled water
Redoxelektrode *f* redox electrode
Redoxindikator *m* oxidation-reduction indicator
Redoxionenaustauscher *m* redox ion exchanger
Redoxpolymer *n* redox polymer
Redoxpotential *n* oxidation-reduction potential
Redoxpuffer *m* oxidation-reduction buffer
Redoxreaktion *f* redox reaction
Redoxsystem *n* redox system
Redoxtitration *f* oxidation-reduction titration
Redox-Titrationskurve *f* redox titration curve
Reduktion *f* reduction
Reduktionsanalyse *f* reductometry
Reduktionsmittel *n* reductant
Reduktions-Oxidations-Elektrode *f* redox
elektrode
reduktometrische Titration *f* reductometric
titration
reduzierte Geschwindigkeit *f* der mobilen Phase
reduced velocity of the mobile phase
reduzierte Größen *fpl* reduced parameters
reduzierte Parameter *mpl* reduced parameters
reduzierte Retentionszeit *f* adjusted retention
time
reduziertes Retentionsvolumen *n* adjusted
retention volume
reduzierte Trennstufenhöhe *f* reduced plate
height
Referenz-Elektrode *f* reference electrode
Referenzelektrode *f* reference electrode
Reflektions-Absorptions-Infrarotspektroskopie
f infrared reflectance-absorption spectroscopy
Reflexionsgitter *n* reflection grating
Refraktion *f* refraction
Refraktometer *n* refractometer
Refraktometrie *f* refractometry
registrierendes Spektralphotometer *n* recording
spectrophotometer
Regression *f* regression function
Regressionsanalyse *f* regression analysis
Regressionsfunktion *f* regression function
Regressionskoeffizient *m* regression coefficient
Reihenanalyse *f* routine analysis
reines Reagens *n* pure reagent
Reinheit *f* des Reagens purity of a reagent
Rekristallisation *f* recrystallization
relative Aktivität *f* der Substanz B relative
activity of substance B
relative Atommasse *f* eines Elementes relative
atomic mass
relative Dielektrizitätskonstante *f* relative
permittivity
relative empirische Standardabweichung *f*
empirical variation coefficient
relative Löslichkeitserhohung *f* relative
supersaturation
relative Molekülmasse *f* einer Substanz
relative molecular mass
relative Permittivität *f* relative
permittivity
relativer Brechungsindex *m* relative refractive
index

relative Retention *f* relative retention
relativer Fehler *m* relative error
relatives Elektrodenpotential *n* electrode
potential
relative Standardabweichung *f* coefficient of
variation
relative Übersättigung *f* relative supersaturation
Relaxation *f* relaxation
Relaxationseffekt *m* relaxation-time effect
Relaxationsmethoden *fpl* transient methods
Reproduzierbarkeit *f* reproducibility
Reproduzierbarkeitsfehler *m* reproducibility error
Resistanz *f* resistance
Resistivität *f* resistivity
Resonanzdetektor *m* resonance detector
Resonanzeinfang *m* resonance capture
Resonanz-Fluoreszenz *f* resonance fluorescence
Resonanzmonochromator *m* resonance detector
Resonanzneutronen *npl* resonance neutrons
Resonanz-Raman-Effekt *m* resonance Raman
effect
Resonanzspitze *f* resonance peak
Resonator *m* cavity resonator
Reststreuung *f* residual variance
Reststrom *m* residual current
Retentionsindex *m* retention index
Retentionstemperatur *f* retention temperature
Retentionszeit *f* total retention time
reverse Phasenchromatographie *f* reversed-phase
chromatography
reversible Durchtrittsreaktion *f* reversible
electrode reaction
reversible Elektrode *f* reversible electrode
reversible galvanische Zelle *f* reversible
galvanic cell
reversible Koagulation *f* reversible coagulation
reversibler Redoxindikator *m* reversible redox
indicator
reversible Zelle *f* reversible galvanic cell
rf-Polarographie *f* radio-frequency
polarography
Richtigkeit *f* accuracy
Ringchromatographie *f* circular chromatography
Rippentrichter *m* ribbed funnel
Röntgenabsorptionsspektroskopie *f* X-ray
absorption spectroscopy
Röntgenaufnahme *f* X-ray pattern
Röntgenbeugung *f* X-ray diffraction
Röntgenbeugungsgerät *n* X-ray diffractometer
Röntgendiffraktion *f* X-ray diffraction
Röntgendiffraktionsanalyse *f* X-ray diffraction
analysis
Röntgendiffraktometer *n* X-ray diffractometer
Röntgendrehkristallaufnahme *f* X-ray rotating-
crystal pattern
Röntgenemissionsspektroskopie *f* X-ray emission
spectroscopy
Röntgen-Escape-Peak *m* X-ray escape peak
Röntgenfilter *n*, *m* X-ray filter
Röntgenfluoreszenz *f* X-ray fluorescence
Röntgenfluoreszenz-Analyse *f* X-ray
fluorescence analysis
Röntgen-Fluoreszenzausbeute *f* X-ray
fluorescence yield
Röntgenfluoreszenzspektroskopie *f* X-ray
fluorescence spectroscopy

Röntgenfluoreszenzspektrum *n* X-ray
  fluorescence spectrum
Röntgenfluoreszenzstrahlung *f* X-ray
  fluorescence radiation
Röntgengoniometer *n* X-ray goniometer
Röntgenkonusaufnahme *f* X-ray oscillating-
  crystal pattern
Röntgenogramm *n* X-ray pattern
Röntgen-Photoelektronen-Spektroskopie *f* X-ray
  photoelectron spectroscopy
Röntgenprimärstrahlung *f* primary X-radiation
Röntgenröhre *f* X-ray tube
Röntgenschwenkaufnahme *f* X-ray oscillating-
  crystal pattern
Röntgenspektralanalyse *f* X-ray
  spectrochemical analysis
Röntgenspektrometer *n* X-ray spectrometer
Röntgenstrahlen *mpl* X-radiation
Röntgenstrahlintensität *f* X-radiation intensity
Röntgenstrahlung *f* X-radiation
Rotations-Schwingungsbande *f* vibration-
  rotation band
Rotations-Schwingungs-Spektrum *n* vibrational-
  rotational spectrum
Rotationsspektrum *n* rotational spectrum
rotiendere Drahtelektrode *f* rotating-wire
  electrode
rotierende Scheibenelektrode *f* 1. rotating-disk
  electrode (*in electrochemistry*) 2. rotating-
  disk electrode (*in arc and spark spectroscopy*)
rotierende Scheibenelektrode *f* mit Ring
  rotating-ring disk electrode
Routineanalyse *f* routine analysis
Rowland-Kreis *m* focusing circle
Rückextraktion *f* back-extraction
Rückhalteträger *m* hold-back carrier
Rückschüttelung *f* back-extraction
Rückspülung *f* backflushing
Rückstreukoeffizient *m* backscattering factor
Rückstreumaximum *n* backscatter peak
Rückstreu-Peak *m* backscatter peak
Rückstreuung *f* backscattering
Rücktitration *f* back-titration
Ruhepotential *n* open-circuit electrode potential
Ruhepunkt *m* rest point
Ruhmasse *f* rest mass
Rumpfniveau-Elektronenspektroskopie *f* core
  level electron spectroscopy
Rydberg *n* rydberg
Rydberg-Konstante *f* Rydberg constant
R-Zweig *m* R-branch

S

Saigerung *f* segregation
säkulares Gleichgewicht *n* secular radioactive
  equilibrium
Salzbrücke *f* salt bridge
Salzeffekt *m* salt effect
Salzfehler *m* salt error
salzige Form *f* des Ionenaustauschers salt form
  of an ion exchanger
Sammeln *n* collection
Sammelprobe *f* gross sample
Sammler *m* electrical accumulator

Sandell-Empfindlichkeitsindex *m* Sandell's
  sensitivity index
Sand-Gleichung *f* Sand equation
Sandwich-Kammer *f* S-chamber
Sättigung *f* saturation
Sättigung *f* der Kammer chamber saturation
Sättigungsaktivität *f* saturation activity
saure Form *f* des Kationenaustauschers acid
  form of a cation exchanger
saure Gruppe *f* acidic group
saures Lösungsmittel *n* protogenic solvent
Sauerstoffelektrode *f* oxygen gas electrode
Säulenchromatographie *f* column
  chromatography
Säulenfüllung *f* column packing
Säulenpackung *f* column packing
Säulentemperatur *f* column temperature
Säulenvolumen *n* column volume
Saülenvorvolumen *n* extra-column volume
Säulenwirksamkeit *f* column performance
Säure-Base-Indikator *m* acid-base indicator
Säure-Base-Reaktion *f* acid-base reaction
Säure-Base-Titration *f* acid-base titration
Säure-Base-Titrationskurve *f* neutralization
  titration curve
Scavenger *m* scavenger
Schätzfunktion *f* estimator
Schaukelschwingungen *fpl* rocking vibrations
scheinbare Halbwertsbreite *f* apparent half-
  width
Scheinleitwert *m* admittance
Scheinwiderstand *m* impedance
Scherenschwingungen *fpl* scissoring vibrations
Schichtäquilibrierung *f* layer equilibration
Schichtdicke *f* path length
Schiedsanalyse *f* umpire analysis
Schiffchen *n* combustion boat
Schleier *m* fog
Schleierschwärzung *f* fog
Schlichtzbrenner *m* slot-burner
Schmelzen *n* melting
Schmelzpunktröhrchen *npl* melting-point
  capillaries
schnelle Flüssigchromatographie *f* high-
  performance liquid chromatography
schnelle Flüssigkeitschromatographie *f* high-
  performance liquid chromatography
schnelle Neutronen *npl* fast neutrons
Schöniger-Methode *f* Schöniger method
Schöniger-Stopfen *m* Schöniger stopper
schrittweise Regression *f* step-wise regression
Schumann-Ultraviolett *n* vacuum ultraviolet
Schwächungskoeffizient *m* attenuation coefficient
Schwärzung *f* blackening
Schwärzungskurve *f* emulsion characteristic
  curve
Schwellenenergie *f* threshold energy
schwingende Waage *f* undamped balance
Schwingkondensator *m* vibrating condenser
Schwingungen *fpl* vibrations
v-Schwingungen *fpl* stretching vibrations
ρ-Schwingungen *fpl* rocking vibrations
τ-Schwingungen *fpl* twisting vibrations
ω-Schwingungen *fpl* wagging vibrations
Schwingungen *fpl* aus der Ebene out-of-plane
  vibrations

Schwingungen *fpl* **in der Ebene** in-plane vibrations
Schwingungsband *n* vibration band
Schwingungsfrequenz *f* frequency
Schwingungsmethode *f* method of swings
Schwingungsspektroskopie *f* vibrational spectroscopy
Schwingungsspektrum *n* vibrational spectrum
Schwingungszahl *f* frequency
Sedimentation *f* sedimentation
Sedimentationsanalyse *f* sedimentation analysis
Sedimentationspotential *n* sedimentation potential
Sedimentationswaage *f* sedimentation balance
Sedimentbildung *f* sedimentation
Seidel-Transformation *f* Seidel transformation
Seigerung *f* segregation
Sektorfeldmassenspektrometer *m* magnetic-deflection mass spectrometer
Sekundäranregung *f* enhancement effect
Sekundarbatterie *f* battery
Sekundärelektronenvervielfacher *n* photomultiplier tube
Sekundärelement *n* electrical accumulator
sekundäre Reaktionszone *f* secondary combustion zone
sekundäre Röntgenstrahlen *mpl* secondary X-radiation
sekundäre Störreaktion *f* interfering second order reaction
sekundäre Urtitersubstanz *f* secondary standard
Sekundärionen-Massenspektroskopie *f* secondary ion mass spectroscopy
Sekundär-Ionen-Mikroskopie *f* secondary ion imaging mass spectroscopy
Sekundärreaktion *f* induced reaction
Selbstabschirmung *f* self-shielding
Selbstabsorption *f* **1.** self-absorption **2.** self-absorption (*in emission spectroscopy*)
Selbstemissionselektrode *f* self-electrode
Selbstumkehr *f* self-reversal
selektive Elution *f* selective elution
selektive Reaktion *f* selective reaction
selektiver Ionenaustauscher *m* selective ion exchanger
selektives Reagens *n* selective reagent
Selektivität *f* **eines Reagens** selectivity of a reagent
Selektivitätskoeffizient *m* selectivity coefficient
Selectivitätskonstante *f* potentiometric selectivity coefficient
sensorische Analyse *f* sensory analysis
Sephadex G Sephadex G
Sequentialanalyse *f* sequential analysis
Sequentialtitration *f* successive titration
sequentieller Test *m* sequential test
Sequenzanalyse *f* sequential analysis
sichtbarer Bereich *m* visible region
sichtbares Gebiet *n* visible region
sichtbares Licht *n* light
sichtbares Spektrum *n* visible spectrum
sichtbare Strahlung *f* light
Sickerelektrode *f* porous cup electrode
Siebanalyse *f* sieve analysis
Siedenpunktbestimmer *m* boiling point apparatus
Signal-Rausch-Verhältnis *n* signal-to-noise ratio

Signifikanzniveau *n* significance level
Signifikanztest *m* significance test
Silberchloridelektrode *f* silver-silver chloride electrode
Silber-Chlorsilber-Elektrode *f* silver-silver chloride electrode
Silberelektrode *f* silver electrode
Silber-Silberchlorid-Elektrode *f* silver-silver chloride electrode
Silitheizstab *m* Globar
Silitstift *m* Globar
Silylierung *f* silylation
Silylierungsreagens *n* silylation reagent
Simplex *n* simplex
Simplex-Plan *m* simplex design
Simultanbestimmung *f* simultaneous determination
simultane diskontinuierliche Untersuchungsmethoden *fpl* discontinuous simultaneous techniques
simultane gekuppelte Untersuchungsmethoden *fpl* coupled simultaneous techniques
simultane kontinuierliche Untersuchungsmethoden *fpl* continuous simultaneous techniques
simultane Untersuchungsmethoden *fpl* simultaneous techniques
Single-sweep-Oszillopolarographie *f* single-sweep polarography
Single-sweep-Polarographie *f* single-sweep polarography
Singulettlinie *f* singlet line
Sinnesprüfung *f* sensory analysis
δ-**Skala** *f* delta scale
Slurry-Verfahren *n* slurry packing
Sol *n* sol
Soller-Kollimator *m* Soller collimator
Solubilisationschromatographie *f* solubilization chromatography
Solvatochromie *f* solvatochromism
Solvensfront *f* mobile phase front
Sorbat *n* sorbate
Sorbens *n* sorbent
Sorption *f* sorption
Sorptionsmittel *n* sorbent
Sorptionsisotherme *f* sorption isotherm
Sorptiv *n* sorbate
Spalt *m* slit
Spaltausbeute *f* fission yield
Spaltneutronen *npl* fission neutrons
Spaltprodukte *npl* fission fragments
Spaltung *f* fission
Spannweite *f* range
Spektralanalyse *f* spectrochemical analysis
Spektralbande *f* spectral band
Spektralfluorimeter *n* spectrofluorimeter
Spektrallinie *f* spectral line
Spektralphotometer *n* spectrophotometer
Spektralphotometrie *f* spectrophotometry
Spektralpolarimeter *n* spectropolarimeter
Spektralpolarimetrie *f* spectropolarimetry
spektralreines Reagens *n* spectrally pure reagent
Spektralserie *f* spectral series
Spektralträger *m* spectrochemical carrier
spektrochemische Analyse *f* spectrochemical analysis

spektrochemischer Puffer *m* spectrochemical buffer
Spektrofluorimetrie *f* spectrofluorimetry
Spektrograf *m* spectrograph
spektrographische Analyse *f* spectrographic analysis
Spektrogramm *n* spectrogram
spektrographisch reines Reagens *n* spectrally pure reagent
Spektrometer *n* optical spectrometer
γ-Spektrometer *n* gamma-ray spectrometer
Spektrometrie *f* spectrometry
Spektrophotometer *n* spectrophotometer
Spektrophotometrie *f* spectrophotometry
spektrophotometrische Titration *f* spectrophotometric titration
Spektropolarimeter *n* spectropolarimeter
Spektroskop *n* spectroscope
Spektroskopie *f* spectroscopy
Spektroskopie *f* **hochenergetischer rückgestreuter Ionen** high-energy ion scattering spectroscopy
spektroskopisch reines Reagens *n* spectrally pure reagent
Spektrum *n* spectrum
Sperrschichtelement *n* photovoltaic cell
Sperrschichtphotozelle *f* photovoltaic cell
spezifische Adsorption *f* specific adsorption
spezifische Aktivität *f* specific activity
spezifische Drehung *f* specific rotation
spezifische Elution *f* specific elution
spezifische Leitfähigkeit *f* **1.** electrical conductivity **2.** electrolytic conductivity
spezifische Oberfläche *f* specific surface area
spezifischer Absorptionskoeffizient *m* specific decadic absorption coefficient
spezifischer dekadischer Absorptionskoeffizient *m* specific decadic absorption coefficient
spezifische Reaktion *f* specific reaction
spezifische Refraktion *f* specific refraction
spezifischer elektrischer Widerstand *m* resistivity
spezifisches Reagens *n* specific reagent
spezifisches Retentionsvolumen *n* specific retention volume
Spezifität *f* **eines Reagens** specifity of a reagent
sphärische Aberration *f* spherical aberration
sphärische Abweichung *f* spherical aberration
Spitzenpotential *n* peak potential
Spitzenstrom *m* peak current
spontane Elektrogravimetrie *f* internal electrogravimetry
Spülmittel *n* scavenger
Spurenanalyse *f* trace analysis
Spurenbestandteil *m* trace constituent
Spurenelementausbeute *f* recovery factor
Spurenfänger *m* trace collector
Spurensammler *m* trace collector
Sputtering-Ausbeute *f* sputtering yield
Sputtering-Prozeß *m* sputtering
Square-wave-Polarographie *f* square-wave polarography
Standardabweichung *f* standard deviation
Standardabweichung *f* **des Mittelwertes** standard deviation of the mean
Standardadditionsmethode *f* standard addition method

Standard-Elektrodenpotential *n* standard electrode potential
Standardelement *n* standard cell
Standard-EMK *f* standard electromotive force
standardisierte Normalverteilung *f* standardized normal distribution
standardisierte Zufallsgröße *f* standardized random variable
Standardkonstante *f* **der Elektrodenreaktion** standard rate constant of electrode reaction
Standardlösung *f* standard solution
Standardmethode *f* standarized method
Standardnormalverteilung *f* standardized normal distribution
Standardpotential *n* standard electrode potential
Standardprobe *f* standard sample
Standard-Wasserstoffelektrode *f* standard hydrogen electrode
Standardzustände *mpl* standard conditions
Startlinie *f* starting line
Startpunkt *m* starting line
stationäre Flüssigphase *f* stationary liquid phase
stationäre Methoden *fpl* steady-state methods
stationäre Quecksilbertropfelektrode *f* hanging mercury drop electrode
stationäre Phase *f* stationary phase
Statistik *f* statistic
statistische Hypothese *f* statistical hypothesis
statistische Qualitätskontrolle *f* statistical quality control
statistischer Test *m* test
statistische Versuchsplanung *f* experimental design
statistische Wahrscheinlichkeit *f* statistical probability
stetige Variable *f* continuous random variable
stetige Zufallsgröße *f* continuous random variable
Stichprobe *f* **1.** sample **2.** primary sample
Stichprobenfunktion *f* statistic
Stichprobenmedian *f* sample median
Stichprobenstreuung *f* sampling variance
Stichprobenumfang *m* sample size
stimulierte Emission *f* stimulated emission
stimulierter Raman-Effekt *m* stimulated Raman effect
stöchiometrische Konzentration *f* analytical concentration
stöchiometrischer Faktor *m* analytical factor
stöchiometrischer Punkt *m* equivalence-point
Stockholmer Konvention *f* sign convention
Stoffaustausch *m* mass exchange
Stoffmenge *f* amount of substance
Stoffmengenanteil *m* mole fraction of substance B
Stoffmengenkonzentration *f* amount-of-substance concentration of substance B
Stokessche Linien *fpl* Stokes lines
Stokessches Gesetz *n* Stokes law
störende Ionen *npl* interfering ions
störende Kernreaktion *f* interfering nuclear reaction
störende Substanz *f* interfering substance
γ-Strahl-Spektrometer *n* gamma-ray spectrometer
Strahlendetektor *m* radiation detector

Strahlenmessung *f* radiometry
Strahlennachweis *m* detection of radiation
Strahlenschwächung *f* attenuation of radiation
Strahlenzähler *m* radiation counter
Strahlenzählrohr *n* radiation counter
Strahler *m* emitter
Strahlung *f* radiation
γ-**Strahlung** *f* gamma radiation
Strahlungsdetektor *m* radiation detector
Strahlungsenergie *f* radiant energy
Strahlungsfluß *m* radiant flux
Strahlungsleistung *f* radiant flux
Strahlungsquelle *f* radiation source
Strahlungsuntergrund *m* background radiation
Streckschwingungen *fpl* stretching vibrations
Streichgerät *n* spreader
Streubreite *f* range
Streudiffusion *f* eddy diffusion term
Streuung *f* **1.** dispersion (*in statistics*)
  2. variance **3.** scattering
Streuungsmatrix *f* covariance matrix
Strippen *n* back-extraction
Strom-Abtast-Polarographie *f* current scanning
  polarography
Stromausbeute *f* current efficiency
strömende Quecksilberelektrode *f* streaming
  mercury electrode
stromkontrollierte Coulometrie *f* controlled-
  current coulometry
stromkontrollierte Coulometrie *f* **mit**
  potentiometrischer Endpunktsbestimmung
  controlled-current coulometry with
  potentiometric end-point detection
stromkontrollierte Potentiometrie *f* controlled-
  current potentiometry with one indicator
  electrode
stromkontrollierte Potentiometrie *f* **mit zwei**
  Indicator-Elektroden controlled-current
  potentiometry with two indicator electrodes
stromkontrollierte potentiometrische Titration
  f controlled-current potentiometric titration
stromkontrollierte potentiometrische Titration
  f **mit zwei Indikator-Elektroden** controlled-
  current potentiometric titration with two
  indicator electrodes
stromprogrammierte Chronopotentiometrie *f*
  programmed-current chronopotentiometry
Stromschlüssel *m* salt bridge
Strom-Spannungskurve *f* polarization curve
Stromstufen-Chronopotentiometrie *f* current-
  step chronopotentiometry
Stromumkehr-Chronopotentiometrie *f* current-
  reversal chronopotentiometry
Strömungspotential *n* streaming potential
strömungsprogrammierte Chromatographie *f*
  flow-programmed chromatography
Stromunterbrechungschronopotentiometrie *f*
  current-cessation chronopotentiometry
Student-Test *m* Student's test
Student-Verteilung *f* Student's distribution
Stufe *f* **in einem Integral-Chromatogramm** step on
  an integral chromatogram
Stufenfilter *n, m* step filter
Stufenhöhe *f* step height
Stufenhöhe *f* **in einem Integral-Chromatogramm**
  step height on an integral chromatogram

stufenweise Elektrodenreaktion *f* stepwise
  electrode reaction
stufenweise Elution *f* stepwise elution
subjektive Farbmessung *f* visual colorimetry
Submikroanalyse *f* submicro-analysis
Submikromethode *f* picogram method
Substanz *f* substance
(+)-Substanz *f* dextrarotatory substance
(−)-Substanz *f* laevorotatory substance
d-**Substanz** *f* dextrarotatory substance
l-**Substanz** *f* laevorotatory substance
Substitutionstitration *f* replacement titration
substöchiometrische Isotopenverdünnungsanalyse
  f substoichiometric isotope dilution analysis
Summen-Peak *m* sum peak
superkritische Fluidchromatographie *f*
  supercritical fluid chromatography
Surfaktant *m* surfactant
Suszeptanz *f* susceptance
Suszeptometrie *f* susceptometry
symmetrische Schwingungen *fpl* symmetrical
  vibrations
systematische Analyse *f* systematic
  qualitative analysis
systematische qualitative Analyse *f* systematic
  qualitative analysis
systematischer Fehler *m* systematic error
Szintillations-β-Spektrometer *n* beta
  scintillation spectrometer
Szintillationsdetektor *m* scintillations detector
Szintillationsspektrometer *n* scintillation
  spectrometer
Szintillationszähler *m* scintillation counter

T

Tafel-Gleichung *f* Tafel's equation
Tailing *n* tailing
Target *n* target
Tast-Polarographie *f* Tast polarography
technische Waage *f* chemical balance
Teilstromprobengeber *m* by-pass injector
temperaturprogrammierte Chromatographie *f*
  temperature-programmed chromatography
Temperaturquotient *m* temperature coefficient
temporäre Härte *f* temporary hardness
Tensammetrie *f* measurement of non-faradaic
  admittance
Tensid *n* surfactant
Test *m* **1.** test (*in chemical analysis*) **2.** test
t-**Test** *m* Student's test
Test-Elektrode *f* indicator electrode
Testsubstanzen *fpl* **für Elementaranalyse**
  standards for elemental analysis
Thalamid-Elektrode *f* thalamid electrode
Thalliumamalgam-Thallium(I)-chlorid-Elektrode
  f thalamid electrode
theoretische Bodenzahl *f* number of
  theoretical plates
theoretischer Endpunkt *m* equivalence-point
theoretischer Endpunkt *m* **der Titration**
  equivalence-point
theoretische Trennstufenzahl *f* number of
  theoretical plates
Theorie *f* **der elektrochemische Doppelschicht**
  theory of double layer

Theorie *f* von Debye und Hückel Debye-
Hückel's theory of strong electrolytes
thermische Analyse *f* thermal analysis
thermische Desorption *f* thermal desorption
thermische Dissoziation *f* thermal dissociation
thermische Ionisation *f* thermal ionization
thermische Kurve *f* thermal curve
thermische Neutronen *npl* thermal neutrons
thermoanalytische Kurve *f* thermal curve
Thermochromatographie *f* chromathermography
Thermodiffusionspotential *n* thermodiffusion
potential
thermodynamische Verteilungskonstante *f*
partition constant
thermoelektrischer Atomisator *m*
electrothermal atomizer
Thermoclement *n* thermogalvanic cell
Thermoelementpaar *n* thermocouple
thermogalvanische Zelle *f* thermogalvanic cell
Thermogasvolumetrie *f* thermovolumetric
analysis
thermo-gasvolumetrische Analyse *f*
thermovolumetric analysis
Thermogravimetrie *f* thermogravimetry
Thermoionisation *f* thermal ionization
Thermoionisations-Ionenquelle *f* thermal
emission ion source
Thermomanometrie *f* thermomanometric
analysis
thermomanometrische Analyse *f*
thermomanometric analysis
thermometrische Titration *f* thermometric
titration
Thermopaar *n* thermocouple
Thermovolumetrie *f* thermovolumetric analysis
Thermowaage *f* thermobalance
Tiefenprofil *n* depth profile
Tiegel *m* mit Deckel crucible with cover
Tintometer *n* colorimeter
Titanometrie *f* titanometry
Titer *m* titre
Titerstellung *f* standardization
Titersubstanz *f* titrant
Titrans *n* titrant
Titration *f* titration
Titration *f* in nichtwässriger Lösung non-aqueous
titration
Titration *f* mit zwei Indikatoren double-indicator
titration
Titrationsendpunkt *m* end-point of titration
Titrationsfehler *m* titration error
Titrationsgrad *m* per cent titrated
Titrationskurve *f* titration curve
Titrieranalyse *f* titrimetric analysis
Titrierfehler *m* titration error
Titrierung *f* titration
titrimetrische Analyse *f* titrimetric analysis
Tochterion *n* daughter ion
Tochternuklid *n* daughter nuclide
Torsionschwingungen *fpl* twisting vibrations
Torsionswaage *f* torsion balance
totale Ionenaustauschkapazität *f* total specific
capacity
Totalkonzentration *f* analytical concentration
Totalreflexion *f* total internal reflection
Totvolumen *n* extra-column volume

Totvolumen *n* des Detektors intensitive volume
of a detector
Totzeit *f* 1. dead time (*in chromatography*)
2. dead time (*of a device*))
Träger *m* carrier
Trägerdestillation *f* carrier distillation
Trägerelektrode *f* supporting electrode
Trägergas *n* carrier gas
Träger *m* mit poröser Oberfläche superficially
porous support
Trägerplatte *f* support plate
Trägersubstanz *f* carrier
Tragerunterlage *f* support plate
Tragfähigkeit *f* capacity
Transitionszeit *f* transition time
Transmissions-Elektronenenergieverlust-
-Spektroskopie *f* transmission energy loss
spectroscopy
Transmissionsfaktor *m* transmission factor
Transmissionsgitter *n* transmission grating
Transportdetektor *m* transport detector
Treffplatte *f* target
Trend *m* drift
Trennfaktor *m* separation factor
Trennkammer *f* chromatographic chamber
Trennkapillare *f* capillary column
Trennleistung *f* der Säule column performance
Trennsäule *f* chromatographic column
Trennstufenhöhe *f* height equivalent to a
theoretical plate
Trenntemperatur *f* separation temperature
Trennung *f* separation
Treppenstufen-Polarographie *f* staircase
polarography
Trimethylchlorsilan *n* trimethylchlorsilane
Tripelionen *npl* ion-triplets
Tritiumauffänger *m* tritium target
Tritium-Target *n* tritium target
trockenes Verfahren *n* dry method
Trockenmittel *n* desiccant
Trockenpackung *f* dry-packing
Trocknen *n* des Niederschlages drying of
precipitate
Trogkammer *f* chromatographic chamber
Tropfenabstand *m* drop time
tropfende Quecksilberelektrode *f* dropping-
mercury electrode
Tropfenfänger *m* safety head
Tropfenzähler *m* dropper
Tropfer *m* dropper
Tropfpipette *f* dropper
Tropfzeit *f* drop time
Trübungsmesser *m* turbidimeter
Tscherenkow-Strahlung *f* Cerenkov radiation
Tschugajews Reagens *n* Chugaev reagent
Tulpe *f* crucible holder
Tüpfelanalyse *f* spot test analysis
Tüpfelplatte *f* spotting plate
Turbidimeter *n* turbidimeter
Turbidimetrie *f* turbidimetry
turbidimetrische Titration *f* turbidimetric titration
turbulente Flamme *f* turbulent flame
Tynndall-Effekt *m* Tyndall effect
Tyndallometrie *f* nephelometry
tyndallometrische Titration *f* nephelometric
titration

U

Überexpositionsgebiet *n* over-exposure region
überführungsfreie Zelle *f* cell without
  transference
Überführungszahl *f* der Ionensorte B transport
  number of ionic species B
Übergangsgleichgewicht *n* transient
  radioactive equilibrium
Übergangszeit *f* transition time
übersättigte Lösung *f* supersaturated solution
Übersättigung *f* supersaturation
Übersättigungsverhältnis *n* supersaturation
  ratio
Überspannung *f* overpotential
Überspannung *f* des Wasserstoffs hydrogen
  overpotential
Uhrzeit *f* clock time
Ultrahochvakuum *n* ultra-high vacuum
Ultramikroanalyse *f* ultramicro analysis
Ultramikromethode *f* microgram method
Ultramikrowaage *f* ultramicro balance
Ultrarot *n* infrared
Ultrarotgebiet *n* infrared region
Ultraschallnebelblaser *m* ultrasonic nebulizer
Ultraspektrum *n* infrared spectrum
Ultraviolett *n* ultraviolet light
Ultraviolett-Bereich *m* ultraviolet region
Ultraviolett-Spektrum *n* ultraviolet spectrum
Ultraviolettstrahlung *f* ultraviolet light
Umfang *m* der Stichprobe sample size
Umfällung *f* reprecipitation
umgekehrte Isotopenverdünnungsanalyse *f*
  reverse isotope dilution analysis
umkehrbare Elektrode *f* reversible electrode
Umkehrphasen-Chromatographie *f* reversed-
  phase chromatography
Umkehrphasen-Verteilungschromatographie *f*
  extraction chromatography
Umschlagsintervall *n* transition interval
Umschlagsintervall *n* des Farbindikators colour
  change interval
Umwandlungsfolge *f* decay chain
unelastische Diffraktion *f* der energiearmen
  Elektronen inelastic low energy electron
  diffraction
unelastischer Stoß *m* inelastic collision
unelastische Streuung *f* inelastic scattering
unendlich dicke Probe *f* thick specimen
unendlich dicke Schicht *f* infinitely thick layer
unendlich dünne Schicht *f* infinitely thin layer
ungedämpfte Waage *f* undamped balance
ungepaartes Elektron *n* unpaired electron
ungesättigte Kammer *f* unsaturated chamber
ungesättigtes Weston-Element *n* unsaturated
  Weston cell
Universalbombe *f* combustion bomb
Universalindikator *m* universal indicator
Unterexpositionsgebiet *n* under-exposure region
Untergrund *m* background
Untergrundkompensator *m* background
  corrector
Untersuchung *f* der flüchtigen
  Zersetzungsprodukte evolved gas analysis
unterthermische Neutronen *npl* cold neutrons

Unterzaucher-Methode *f* der
  Sauerstoffbestimmung Unterzaucher method
  for oxygen determination
unverzerrte Schätzfunktion *f* unbiased
  estimator
U-Rohr *n* U-tube
Urtiter *m* standard substance
Urtitersubstanz *f* standard substance
UV-Photoelektronenspektroskopie *f* ultraviolet
  photoelectron spectroscopy

V

Vakuumexsikkator *m* vacuum desiccator
Vakuumultraviolett *n* vacuum ultraviolet
Vakuum-UV *n* vacuum ultraviolet
Valenzschwingungen *fpl* stretching vibrations
van-Deemter-Gleichung *f* van Deemter equation
van-der-Waalssche Adsorption *f* physisorption
Varianz *f* variance
Varianzanalyse *f* analysis of variance
Variationsbreite *f* range
Variationskoeffizient *m* coefficient of variation
Večeřa-Katalysator *m* Večeřa catalyst
Vektorpolarograph *m* vector polarograph
Veränderlichabnahmewinkel-Methode *f*
  variable take-off angle method
Veraschen *n* ashing technique
Veraschung *f* ashing technique
Verästelung *f* branching decay
verbotener Übergang *m* forbidden transition
Verbrennung *f* combustion
Verbrennung *f* im leeren Rohr empty tube
  combustion
Verbrennungsanalyse *f* combustion analysis
Verbrennungsbombe *f* bomb calorimeter
Verbrennungsofen *m* combustion furnace
Verbrennungsrohr *n* combustion tube
Verbrennungsschiffchen *n* combustion boat
Verdrängung *f* displacement
Verdrängungsanalyse *f* displacement
  chromatography
Verdrängungschromatographie *f* displacement
  chromatography
Verdrängungstitration *f* replacement titration
Verdünnungseffekt *m* dilution effect
Verdünnungsgrenze *f* dilution limit
Verdünnungskorrektion *f* dilution-correction
  factor
Verdünnungsmittel *n* diluent
Verdunstung *f* evaporation
vereinheitlichte atomare Masseneinheit *f*
  unified atomic mass unit
Verfahrensfehler *m* error of method
Verflüchtigung *f* volatilization
Vergleichsband *n* reference band
Vergleichslösung *f* 1. comparison solution
  2. reference solution 3. standard matching
  solution
Vergleichsmethode *f* reference method
Vergleichsstandardlösung *f* standard reference
  solution
Vergleichssubstanz *f* 1. reference material
  2. reference sample
Verkohlung *f* des Filters charring of filter
  paper

Verlangsamung *f* slowing down
Vernebeln *n* nebulization
Vernebelung *f* nebulization
Vernetzung *f* cross-linking
Verstärkungsreaktion *f* amplification reaction
Versuch *m* experiment
Versuchsbereich *m* experimental region
Versuchsmatrix *f* design matrix
Versuchsplan *m* exact design
Versuchsplan *m* **1. Ordnung** first order design
Versuchsplan *m* **2. Ordnung** second order design
Versuchsplanung *f* experimental design
Versuchspunkt *m* experimental point
χ-Verteilung *f* chi-squared distribution
t-Verteilung *f* Stundent's distribution
Verteilungschromatographie *f* partition
  chromatography
Verteilungsdichte *f* probability density function
Verteilungsfunktion *f* distribution function
Verteilungsisotherme *f* **1.** distribution
  isotherm **2.** partition isotherm
Verteilungskoeffizient *n* **1.** partition
  coefficient (*in chromatography*)
  2. concentration distribution ratio
  (*in chromatography*) **3.** concentration
  distribution ratio (*in extraction*)
  4. distribution constant (*in extraction*)
Verteilungskoeffizient *m* D_g distribution
  coefficient D_g
Verteilungskoeffizient *m* D_s distribution
  coefficient D_s
Verteilungskoeffizient *m* D_v volume distribution
  coefficient D_v
Verteilungskonstante *f* **1.** distribution
  constant (*in chromatography*) **2.** distribution
  constant (*in extraction*)
Verteilungsverhältnis *n* **1.** concentration
  distribution ratio (*in chromatography*)
  2. concentration distribution ratio (*in
  extraction*)
Vertrauensgrenzen *fpl* confidence limits
Vertrauensintervall *n* confidence interval
Verunreinigung *f* **des Niederschlages**
  contamination of a precipitate
Verzerrung *f* **der Schätzfunktion** bias of
  estimator
verzögerte Neutronen *npl* delayed neutrons
Verzögerung *f* **des Stoffaustausches** mass
  transfer terms
Vibrationsebene *f* plane of vibration
visuelle Farbmessung *f* visual colorimetry
visueller Indikator *m* indicator
visuelle Titration *f* visual titration
Volhardsche Bestimmung *f* **der Halogenide**
  Volhard method for determination of halides
Volhard-Wolff-Methode *f* Volhard method for
  determination of manganese
Vollpipette *f* **mit einer Marke** transfer pipette
vollständiger faktorieller Versuchsplan *m*
  complete factorial design
Voltammetrie *f* voltammetry
Voltammetrie *f* **an stationärer Elektrode**
  chronoamperometry with linear potential
  sweep
Voltammetrie *f* **mit anodischer Auflösung**
  anodic stripping voltammetry

Voltammetrie *f* **mit linearem Spannungsanstieg**
  chronoamperometry with linear potential
  sweep
voltammetrische Analyse *f* voltammetric
  analysis
voltammetrische Indikation *f* controlled-current
  potentiometric titration
voltammetrische Indikation *f* **mit zwei
  Indikatorelektroden** controlled-current
  potentiometric titration with two indicator
  electrodes
voltammetrische Kurve *f* polarization curve
Volta-Potential *n* outer electric potential of
  phase β
Volta-Potentialdifferenz *f* Volta potential
  difference
Voltasche Kontaktspannung *f* Volta potential
  difference
Voltaspannung *f* Volta potential difference
Volumenanteil *m* volume fraction of
  substance B
Volumenbruch *m* volume fraction of
  substance B
Volumen *n* **der festen Phase** solid volume
Volumen *n* **der flüssigen Phase** liquid phase
  volume
Volumen *n* **der stationären Phase** stationary-
  phase volume
Volumenfluß *m* F_a **der mobilen Phase** volumetric
  flow rate F_a
Volumenfluß *m* F_c **der mobilen Phase**
  volumetric flow rate F_c
Volumengehalt *m* volume fraction of
  substance B
Volumenströmungsgeschwindigkeit *f* F_a
  volumetric flow rate F_a
Volumenströmungsgeschwindigkeit *f* F_c
  volumetric flow rate F_c
volumetrische Analyse *f* titrimetric analysis
volumetrische Gasanalyse *f* volumetric gas
  analysis
Vorbrennungszeit *f* pre-arc period
Vorfunkzeit *f* pre-spark period
Vorstoß *m* **für Filtertiegel** crucible holder
vorübergehende Härte *f* temporary hardness

W

Wachstumsfaktor *m* saturation factor
Wägefläschchen *n* pycnometer
Wägegefäß *n* weighing bottle
Wägeglas *n* weighing bottle
Wägegläschen *n* **für hygroskopische Substanzen**
  piggie
Wägepipette *f* **nach Lunge-Rey** Lunge-Rey
  weighing pipette
Wägeschiffchen *n* weighing scoop
Wägetitration *f* weight titration
wahre Halbwertsbreite *f* true half-width
wahrer Dissoziationsgrad *m* true degree of
  dissociation
wahre Zeit *f* live time
Wahrscheinlichkeit *f* probability
Wahrscheinlichkeitsdichte *f* probability density
  function
Wahrscheinlichkeitsrechnung *f* probability theory

Wahrscheinlichkeitstheorie *f* probability theory
Wahrscheinlichkeitsverteilung *f* probability distribution
Wärmedetektor *m* thermal radiation detector
Wärmeleitfähigkeits-Meßzelle *f* thermal conductivity detector
Waschen *n* scrubbing
Waschflüssigkeit *f* **1.** wash liquid **2.** washing solution
wasserähnliches Lösungsmittel *n* hydroxylic solvent
Wasserstoffelektrode *f* hydrogen-gas electrode
Wasserstoff-Entladungs-Lampe *f* hydrogen discharge lamp
Wasserstoffionenexponent *m* pH
Wechselspannungschronopotentiometrie *f* alternating-voltage chronopotentiometry
Wechselspannungspolarographie *f* alternating-voltage polarography
Wechselstrombogen *m* alternating current arc
Wechselstrom-Chronopotentiometrie *f* alternating-current chronopotentiometry
Wechselstrompolarographie *f* alternating-current polarography
Wechselstrompolarographie *f* **höherer Harmonischer** higher-harmonic ac polarography
Wechselstrompolarographie *f* **höherer Harmonischer mit phasenempfindlicher Gleichrichtung** higher-harmonic ac polarography with phase-sensitive rectification
weißes Licht *n* white light
Wellenlänge *f* wavelength
wellenlängendispersive Röntgenfluoreszenzanalyse *f* wavelength dispersive X-ray fluorescence analysis
Wellenzahl *f* wave number
Wendepunkte *mpl* inflection points
Wendepunkte *mpl* **des Peak** inflection points of a peak
R_B-**Wert** *m* R_B value
R_f-**Wert** *m* R_f value
R_f'-**Wert** *m* R_f' value
R_M-**Wert** *m* R_M value
R_M'-**Wert** *m* R_M' value
Weston-Element *n* Weston standard cell
Weston-Normal-Element *n* Weston standard cell
Westonsches Normalelement *n* Weston standard cell
Wheatstonesche Brücke *f* Wheatstone bridge
Widerstandskapazität *f* cell constant of a conductivity cell
Widerstandskonstante *f* cell constant of a conductivity cell
Widerstandsüberspannung *f* resistance overpotential
Wiederholbarkeit *f* repeatability
Winkeldispersion *f* angular dispersion
Winkeldispersion *f* **des Analysatorkristalls** angular dispersion of an analysing crystal
Wirbeldiffusion *f* eddy diffusion term
Wirkleitwert *m* conductance
wirksamste Schätzfunktion *f* most efficient estimator

Wirkungsfläche *f* response surface
Wirkungsquerschnitt *m* **1.** cross section **2.** activation cross section
Wirkwiderstand *m* resistance
Wolframbandlampe *f* tungsten incandescent lamp
Wolframdraht-Glühlampe *f* tungsten incandescent lamp

Z

Zählausbeute *f* counting efficiency
Zahl *f* **der Freiheitsgrade** number of degrees of freedom
Zahl *f* **der theoretischen Böden** number of theoretical plates
Zahl *f* **der theoretischen Stufen** number of theoretical plates
2π-**Zähler** *m* 2π-counter
4π-**Zähler** *m* 4π-counter
Zählgeometrie *f* counting geometry
Zählrate *f* counting rate
Zählrohrcharakteristik *f* plateau of a counter
Zählröhre *f* counter tube
Zählrohr *n* **mit Probenkanal** well-type scintillation counter
Zeeman-Effekt *m* Zeeman effect
Zeeman-Niveaus *npl* Zeeman levels
Zeisel-Reaktion *f* Zeisel reaction
Zeitfaktor *m* saturation factor
zeitlaufgelöstes Spektrum *n* time-resolved spectrum
Zelle *f* **mit Überführung** cell with transference
Zelle *f* **ohne Überführung** cell without transference
Zellkonstante *f* cell constant of a conductivity cell
Zellspannung *f* electric potential difference of a galvanic cell
Zentrifugieren *n* centrifuging
Zeolithe *mpl* zeolites
Zerfallsgesetz *n* law of radioactive decay
Zerfallskurve *f* decay curve
Zerfallsreihe *f* decay chain
Zerfallsschema *n* decay scheme
Zersetzungspotential *n* decomposition potential
Zersetzungsspannung *f* decomposition voltage
Zerstäubung *f* sputtering
Zerstäubungsergiebigkeit *f* sputtering yield
Zerstäubungsgrad *m* efficiency of nebulization
zerstörungsfreie Aktivierungsanalyse *f* non-destructive activation analysis
zerstörungsfreie Analyse *f* non-destructive analysis
Zetapotential *n* electrokinetic potential
Zirkularchromatographie *f* circular chromatography
zirkulare Doppelbrechung *f* circular birefringence
zirkulare Polarisation *f* circular polarization
Zirkularpolarisation *f* circular polarization
zirkular polarisiertes Licht *n* circularly polarized light
Zone *f* zone
zufälliges Ereignis *n* random event
zufällige Variable *f* random variable

zufällige Veränderliche *f* random variable
Zufallsbalancemethode *f* random balance method
Zufallsfehler *m* random error
Zufallsstichprobe *f* random sample
Zufallsvariable *f* random variable
Zufallszahlentabellen *fpl* random number tables
zurückgestrahlte Mikrowellen-Methode *f*
    reflected wave method
Zurückspülung *f* backflushing
zusammengesetzter Versuchsplan *m* composite
    design
Zusatzelektrolyt *m* supporting electrolyte
zweidimensionale Chromatographie *f* two-
    dimensional chromatography
zweidimensionale Entwicklung *f* two-
    dimensional chromatography

Zweifarbigerindikator *m* two-colour indicator
Zweiphasentitration *f* phase titration
zwei-Quanten-Escape-Linie *f* double escape peak
zwei-Quanten-Escape-Peak *m* double escape peak
zweiseitiger Test *m* two-sided test
Zweistrahlspektralphotometer *n* double-beam
    spectrophotometer
zweite Oberschwingung *f* second overtone band
zweiter Wien-Effekt *m* dissociation field effect
Zwischenraumanteil *m* interstitial fraction
Zwischenraumvolumen *n* interstitial volume
Zwischenzone *f* interzonal region
Zwitterion *n* ampholyte ion
Zylinder-Spiegelanalysator *m* cylindrical
    mirror analyser

INDEX FRANCAIS

A

aberration *f* optical aberration
aberration *f* chromatique chromatic aberration
aberration *f* de sphéricité spherical aberration
aberration *f* sphérique spherical aberration
abondance *f* isotopique abundance
abondance *f* isotopique naturelle natural
   isotopic abundance
absorbance *f* absorbance
absorbant *m* absorbent
absorbat *m* absorbate
absorbeur *m* absorber
absorption *f* absorption
absorption *f* de résonance resonance absorption
accumulateur *m* electrical accumulator
acidimétrie *f* acidimetry
acido-alcalimétrie *f* acidimetry and alkalimetry
activation *f* activation
activation *f* d'un adsorbant activation of an
   adsorbent
activation *f* par neutrons neutron activation
activation *f* par porteurs électrisés
   charged-particle activation
activité *f* activity
activité *f* à saturation saturation activity
activité *f* de l'électrolyte B activity of
   electrolyte B
activité *f* du constituant B relative activity of
   substance B
activité *f* d'un adsorbant adsorbent activity
activité *f* moyenne d'un électrolyte mean
   activity of electrolyte
activité *f* nucléaire activity
activité *f* optique optical activity
activité *f* relative du constituant B relative
   activity of substance B
activité *f* spécifique specific activity
additivité *f* additivity law
adhérence *f* adhesion
adhésion *f* adhesion
admittance *f* complexe admittance
admittance *f* électrique admittance
adsorbant *m* adsorbent
adsorbat *m* adsorbate
adsorbendum *m* adsorbate
adsorpt *m* adsorbate
adsorption *f* adsorption
adsorption *f* chimique chemisorption
adsorption *f* d'échange exchange adsorption
adsorption *f* de van der Waals physisorption
adsorption *f* physique physisorption
adsorption *f* spécifique specific adsorption
affinité *f* chimique affinity

agarose *f* agarose
agent *m* brillanteur brightening agent
agent *m* chélatant chelating agent
agent *m* de rétention hold-back carrier
agent *m* de séchage desiccant
agent *m* extractant extracting agent
agent *m* masquant masking agent
agent *m* oxydant oxidant
agent *m* réducteur reductant
agent *m* tensio-actif surfactant
agglomération *f* agglomeration
agrégat *m* aggregate
agrégation *f* aggregation
aire *f* du pic peak area
aire *f* spécifique specific surface area
alcalimétrie *f* alkalimetry
alcoométrie *f* alcoholometry
amalgame *m* amalgam
amiante *m* platiné platinized asbestos
amiante *m* sodé Ascarite
ampérométrie *f* amperometry
ampérométrie *f* à deux électrodes indicatrices
   amperometry with two indicator electrodes
ampérométrie *f* à une électrode de mercure
   stirred-mercury-pool amperometry
ampérométrie *f* à une électrode de platine
   tournante rotating-platinum-wire electrode
   amperometry
ampérométrie *f* à une électrode indicatrice
   amperometry with one indicator electrode
ampérométrie *f* à une électrode polarisée
   amperometry with one indicator electrode
ampérométrie *f* avec deux électrodes polarisées
   amperometry with two indicator electrodes
ampérométrie *f* différentielle differential
   amperometry
amphion *m* ampholyte ion
ampholyte *m* 1. amphiprotic solvent
   2. ampholyte
amplificateur *m* opérationnel operational
   amplifier
amplification *f* gazeuse gas amplification
ampoule *f* de garde safety head
analyse *f* absorptiométrique des gaz
   absorptiometric gas analysis
analyse *f* à la goutte spot test analysis
analyse *f* à la touche spot test analysis
analyse *f* ampérométrique amperometric analysis
analyse *f* calorimétrique différentielle
   differential scanning calorimetry
analyse *f* canonique canonical analysis
analyse *f* centigrammique mesoanalysis
analyse *f* chimique chemical analysis

analyse *f* **chromatographique** chromatographic analysis
analyse *f* **colorimétrique** colorimetric analysis
analyse *f* **conductométrique** conductometric analysis
analyse *f* **conformationnelle** conformational analysis
analyse *f* **coulométrique** coulometric analysis
analyse *f* **courante** routine analysis
analyse *f* **d'amplitude** pulse amplitude analysis
analyse *f* **d'arbitrage** umpire analysis
analyse *f* **de dispersion** analysis of variance
analyse *f* **de head-space** headspace analysis
analyse *f* **d'élution** elution chromatography
analyse *f* **de phase** phase analysis
analyse *f* **de plate-forme** process analysis
analyse *f* **de régression** regression analysis
analyse *f* **de routine** routine analysis
analyse *f* **de sédimentation** sedimentation analysis
analyse *f* **des gaz** gas analysis
analyse *f* **des gaz émis** evolved gas analysis
analyse *f* **destructive par activation** destructive activation analysis
analyse *f* **de surface** surface analysis
analyse *f* **de surface par ionisation résonante d'atomes projetés** surface analysis by resonance ionization of sputtered atoms
analyse *f* **de tête espace** headspace analysis
analyse *f* **de traces** trace analysis
analyse *f* **de variance** analysis of variance
analyse *f* **diélectrométrique** dielectrometric analysis
analyse *f* **d'usage** conventional analysis
analyse *f* **électrochimique** electroanalysis
analyse *f* **élémentaire** elemental analysis
analyse *f* **élémentaire organique** elemental analysis (*of organic compounds*)
analyse *f* **enthalpique différentielle** differential scanning calorimetry
analyse *f* **factorielle** factor analysis
analyse *f* **fonctionnelle** functional group analysis
analyse *f* **fractionnée** fractional analysis
analyse *f* **frontale** frontal chromatography
analyse *f* **granulométrique par tamisage** sieve analysis
analyse *f* **gravimétrique des gaz** gravimetric gas analysis
analyse *f* **instrumentale** instrumental analysis
analyse *f* **locale** local analysis
analyse *f* **manométrique des gaz** manometric gas analysis
analyse *f* **microcristalloscopique** microcrystalloscopic analysis
analyse *f* **non destructive** non-destructive analysis
analyse *f* **non destructive par activation** non-destructive activation analysis
analyse *f* **ordinaire** conventional analysis
analyse *f* **organique qualitative** qualitative analysis of organic compounds
analyse *f* **organique quantitative** quantitative analysis of organic compounds
analyse *f* **par absorption de rayonnement nucléaire** nuclear radiation absorptiometry
analyse *f* **par absorption de neutrons** neutron absorptiometry

analyse *f* **par absorption des particules** β beta-particle absorptiometry
analyse *f* **par absorption des rayons** γ gamma-ray absorptiometry
analyse *f* **par activation** activation analysis
analyse *f* **par activation aux particules chargées** charged-particle activation analysis
analyse *f* **par activation de neutrons** neutron activation analysis
analyse *f* **par activation instrumentale** non-destructive activation analysis
analyse *f* **par combustion** combustion analysis
analyse *f* **par diffraction** diffractometry
analyse *f* **par diffraction des rayons X** X-ray diffraction analysis
analyse *f* **par dilution isotopique** isotope dilution analysis
analyse *f* **par dilution isotopique double** double isotope dilution analysis
analyse *f* **par dilution isotopique inverse** reverse isotope dilution analysis
analyse *f* **par dilution isotopique simple** direct isotope dilution analysis
analyse *f* **par dilution isotopique substoechiométrique** substoichiometric isotope dilution analysis
analyse *f* **par fluorescence X** X-ray fluorescence analysis
analyse *f* **par fluorescence X avec dispersion de l' énergie** energy dispersive X-ray fluorescence analysis
analyse *f* **par fluorescence X avec dispersion des ondes** wavelength dispersive X-ray fluorescence analysis
analyse *f* **par fluorescence excitée par radio-isotope** radioisotope-excited X-ray fluorescence analysis
analyse *f* **par groupements fonctionnels** functional group analysis
analyse *f* **par la méthode de dilution isotopique double** double isotope dilution analysis
analyse *f* **par la méthode de dilution isotopique inverse** reverse isotope dilution analysis
analyse *f* **par la méthode de dilution isotopique simple** direct isotope dilution analysis
analyse *f* **par la méthode de dilution isotopique substoechiométrique** substoichiometric isotope dilution analysis
analyse *f* **par les groupements fonctionnels** functional group analysis
analyse *f* **par photoactivation** photoactivation analysis
analyse *f* **par ralentissement de neutrons** analysis by neutron slowing-down
analyse *f* **par redissolution** electrochemical stripping analysis
analyse *f* **polarographique** polarographic analysis
analyse *f* **pondérale** gravimetric analysis
analyse *f* **pondérale des gaz** gravimetric gas analysis
analyse *f* **potentiométrique** potentiometric analysis
analyse *f* **pyrognostique** 1. pyrochemical analysis 2. blow-pipe test

analyse *f* **pyrognostique au chalumeau**
  blow-pipe test
analyse *f* **qualitative** qualitative analysis
analyse *f* **quantitative** quantitative analysis
analyse *f* **radiochimique** radiochemical analysis
analyse *f* **radiométrique** radiometric analysis
analyse *f* **rapide** short analysis
analyse *f* **sédimentaire** sedimentation analysis
analyse *f* **semi-quantitative** semiquantitative
  analysis
analyse *f* **sensorielle** sensory analysis
analyse *f* **séquentielle** sequential analysis
analyse *f* **spectrale** spectrochemical analysis
analyse *f* **spectrochimique** spectrochemical
  analysis
analyse *f* **spectrochimique atomique** atomic
  spectrochemical analysis
analyse *f* **spectrochimique d'absorption**
  absorption spectrochemical analysis
analyse *f* **spectrochimique de fluorescence**
  fluorescence spectrochemical analysis
analyse *f* **spectrochimique d'émission** emission
  spectrochemical analysis
analyse *f* **spectrochimique moléculaire** molecular
  spectrochemical analysis
analyse *f* **spectrochimique X** X-ray
  spectrochemical analysis
analyse *f* **spectrographique** spectrographic
  analysis
analyse *f* **spectrophotométrique simultanée**
  multiple analysis
analyse *f* **spectroscopique** spectrochemical
  analysis
analyse *f* **systématique** systematic qualitative
  analysis
analyse *f* **thermique** thermal analysis
analyse *f* **thermique différentielle** differential
  thermal analysis
analyse *f* **thermomanométrique**
  thermomanometric analysis
analyse *f* **thermovolumétrique**
  thermovolumetric analysis
analyse *f* **titrimétrique** titrimetric analysis
analyse *f* **titrimétrique des gaz** gas titrimetric
  analysis
analyse *f* **totale** complete analysis
analyseur *m* analyser
analyseur *m* **à miroir cylindrique** cylindrical
  mirror analyser
analyseur *m* **d'amplitudes** pulse amplitude
  analyser
analyseur *m* **de champ magnétique**
  magnetic-deflection mass spectrometer
analyseur *m* **de masse** mass analyser
analyseur *m* **de masse quadripolaire** quadrupole
  mass analyser
analyseur *m* **par champ retardeur** retarding
  field analyser
analyseur *m* **par potentiel retardeur** retarding
  field analyser
analyse *f* **voltammétrique** voltammetric analysis
analyse *f* **volumétrique des gaz** volumetric gas
  analysis
analyte *m* analyte
angle *m* **d'émergence** take-off angle
angle *m* **de miroitement** blaze angle

angle *m* **de prise** take-off angle
angle *m* **de réflexion** glancing angle
Anhydrone *f* Anhydrone
anode *f* anode
anodisation *f* **par plasma** plasma anodization
anolyte *m* anolyte
anti-entraîneur *m* hold-back carrier
appareil *m* **à distiller de Kjeldahl** Kjeldahl
  apparatus
appareil *m* **de détermination du point d'ébullition**
  boiling point apparatus
appareil *m* **de Kipp** Kipp gas generator
appareil *m* **de Parnas et Wagner**
  Parnas-Wagner apparatus
appareil *m* **d'Orsat** Orsat gas analyser
applicateur *m* applicator
appréciation *f* **de lecture** readability
aquamétrie *f* aquametry
arc *m* **à courant alternatif** alternating current arc
arc *m* **à courant continu** direct current arc
arc *m* **électrique** electrical arc
aréomètre *m* hydrometer
argentométrie *f* argentimetry
Ascarite *f* Ascarite
ascorbinométrie *f* ascorbimetry
astigmatisme *m* astigmatism
atmosphère *f* **ionique** ionic atmosphere
atomisation *f* atomization
atomiseur *m* atomizer
atomiseur *m* **électrothermique** electrothermal
  atomizer
atténuation *f* **de rayonnement** attenuation of
  radiation
auto-absorption *f* 1. self-absorption
  2. self-absorption (*in emission spectroscopy*)
autoprotection *f* self-shielding
autoprotolyse *f* autoprotolysis
autoradiogramme *m* autoradiograph
autoradiographe *m* autoradiograph
auxochrome *m* auxochrome
axe *m* **optique** optical axis

B

balance *f* **à amortisseurs** damped balance
balance *f* **analytique** analytical balance
balance *f* **à ressort** spring balance
balance *f* **automatique** automatic balance
balance *f* **de laboratoire** laboratory balance
balance *f* **de sédimentation** sedimentation
  balance
balance *f* **de torsion** torsion balance
balance *f* **électromécanique** electromechanical
  balance
balance *f* **hydrostatique** hydrostatic balance
balance *f* **microchimique** microchemical balance
balance *f* **périodique** undamped balance
balance *f* **semi-microanalytique**
  semimicrochemical balance
balance *f* **ultra-microchimique** ultramicro
  balance
banc *m* **chauffant Kofler** Kofler micro melting
  point apparatus
banc *m* **Kofler** Kofler micro melting point
  apparatus
banc *m* **d'optique** optical bench

bande f zone
bande f **antistokes** anti-Stokes lines
bande f B B band
bande f **de combinaison** combination band
bande f **d'elution** peak (*in chromatography*)
bande f **de référence** reference band
bande f **de transfert de charge** charge-transfer
  band
bande f **de vibration** vibration band
bande f **de vibration-rotation** vibration-rotation
  band
bande f **électronique** electronic band
bande f **fondamentale** fundamental vibration
  band
bande f **harmonique** overtone band
bande f K K band
bande f R R band
bande f **spectrale** spectral band
bande f **Stokes** Stokes lines
barn m barn
base f **du pic** peak base
batterie f battery
becquerel m becquerel
biais m **de l'estimateur** bias of estimator
biampérométrie f amperometry with two
  indicator electrodes
bioluminescence f bioluminescence
bipotentiométrie f controlled-current
  potentiometry with two indicator electrodes
biréfringence f **circulaire** circular birefringence
blanc m blank solution
bloc m block
blocage m **d'un métal indicateur** blocking of
  metal indicator
bolomètre m bolometer
bombe f **au peroxyde de sodium** combustion
  bomb
bombe f **calorimétrique** bomb calorimeter
bouchon m **de Schöniger** Schöniger stopper
bouchon m **pour la fiole de Schöniger** Schöniger
  stopper
branche f P P-branch
branche f Q Q-branch
branche f R R-branch
brillanteur m brightening agent
bromatométrie f bromatometry
broyage m **dans le nujol** mull technique
bruit m noise level
bruit m **de fond** background
brûleur m burner
brûleur m **à consommation totale**
  direct-injection burner
brûleur m **à fente** slot-burner
brûleur m **à injection directe** direct-injection
  burner
brûleur m **à mélange préalable** premixed gas
  burner
brûleur m **à plasma** plasma jet
burette f burette
burette f **à piston à lecture numérique** piston
  automatic burette
burette f **à zéro automatique** automatic zero
  burette
burette f **à zéro automatique de Daffert**
  Daffert automatic zero burette

burette f **de Bunte** Bunte gas burette
burette f **de Schellbach** Schellbach burette
burette f **vérifiée avec certificat d'étalonnage**
  certified burette

C

calcul m **des probabilités** probability theory
capacité f capacity
capacité f **de fixation d'un lit d'échangeur**
  d'ions breakthrough capacity
  of an ion exchanger bed
capacité f **de pics** peak capacity
capacité f **de séparation exprimée en noremb**
  de pics peak capacity
capacité f **différentielle de la couche double**
  differential capacitance of double layer
capacité f **du lit de résine** bed volume capacity
capacité f **intégrale de la couche double**
  integral capacitance of double layer
capacité f **spécifique pratique** practical specific
  capacity
capacité f **théorique spécifique** total specific
  capacity
capacité f **totale spécifique** total specific
  capacity
capacité f **volumique** volume capacity
capillaire m **de Luggin** Luggin-Haber capillary
capteur m **électrochimique** electrochemical
  sensor
capture f **par résonance** resonance capture
caractéristique f **d'uu compteur** plateau of
  a counter
carbonisation d'un filtre f charring of filter
Carbowax m Carbowax
carré m **latin** latin square
carte f **de contrôle** control chart
catalyseur m **de Körbl** Körbl catalyst
catalyseur m **de Večeřa** Večeřa catalyst
catharomètre m thermal conductivity detector
cathode f cathode
cathodoluminescence f cathodoluminescence
catholyte m catholyte
cavité f **résonante** cavity resonator
Célite f Celite
cellule f **à couche d'arrêt** photovoltaic cell
cellule f **à haute activité** hot cell
cellule f **de graphite** graphite cuvette
cellule f **d'électrolyse** electrolytic cell
cellule f **électrolytique** electrolytic cell
cellule f **galvanique** galvanic cell
cellule f **galvanique réversible** reversible
  galvanic cell
cellule f **photoélectrique** photocell
cellule f **photovoltaïque** photovoltaic cell
cellule f **pneumatique de Golay** Golay
  pneumatic detector
cellule f **réversible** reversible galvanic cell
cellule f **thermoélectrique** galvanic thermocell
centre m **de l'intervalle de variation** mid-range
centrifugation f centrifuging
cercle m **de focalisation** focusing circle
cérimétrie f cerimetry
chaleur f **rayonnante** infrared radiation
chambre f **à atmosphère insaturée** unsaturated
  chamber

chambre *f* **à atmosphère saturée** saturated chamber

chambre *f* **BN** BN-chamber

chambre *f* **de Brenner et Niederwieser** BN-chamber

chambre *f* **de développement** chromatographic chamber

chambre *f* **d'ionisation** ionization chamber

chambre *f* **sandwich** S-chamber

charbon *m* **platiné** platinized carbon

charge *f* load

chauffe-entonnoir *m* funnel heater

chélate *m* chelate compound

chimie *f* **analytique** analytical chemistry

chimiionisation *f* chemiionization

chimiluminescence *f* chemiluminescence

chimioluminescence *f* chemiluminescence

chimisorption *f* chemisorption

choc *m* **élastique** elastic collision

choc *m* **inélastique** inelastic collision

chromathermographie *f* chromathermography

chromatogramme *m* chromatogram

chromatogramme *m* **différentiel** differential chromatogram

chromatogramme *m* **intégral** integral chromatogram

chromatographe *m* chromatograph

chromatographie *f* chromatography

chromatographie *f* **à deux dimensions** two-dimensional chromatography

chromatographie *f* **à écoulement programmé** flow-programmed chromatography

chromatographie *f* **à fluide supercritique** supercritical fluid chromatography

chromatographie *f* **ascendante** ascending development

chromatographie *f* **avec effet de sel** salting-out chromatography

chromatographie *f* **avec programmation de la température** temperature-programmed chromatography

chromatographie *f* **avec programmation du débit** flow-programmed chromatography

chromatographie *f* **avec utilisation de détergents** soap chromatography

chromatographie *f* **bidimensionnelle** two-dimensional chromatography

chromatographie *f* **circulaire** circular chromatography

chromatographie *f* **d'adsorption** adsorption chromatography

chromatographie *f* **d'affinité** affinity chromatography

chromatographie *f* **d'échange d'ions** ion-exchange chromatography

chromatographie *f* **d'élution** elution chromatography

chromatographie *f* **de paires d'ions** ion-pair partition chromatography

chromatographie *f* **de partage** partition chromatography

chromatographie *f* **de partage en phases inversées** extraction chromatography

chromatographie *f* **de perméation** permeation chromatography

chromatographie *f* **descendante** descending development

chromatographie *f* **en phase gazeuse** gas chromatography

chromatographie *f* **en phase liquide** liquid chromatography

chromatographie *f* **en phase liquide à haute performance** high-performance liquid chromatography

chromatographie *f* **en phase liquide rapide** high-performance liquid chromatography

chromatographie *f* **en phase liquide sous haute pression** high-performance liquid chromatography

chromatographie *f* **en phases inversées** reversed-phase chromatography

chromatographie *f* **gaz-liquide** gas-liquid chromatography

chromatographie *f* **gaz-solide** gas-solid chromatography

chromatographie *f* **liquide-gel** liquid-gel chromatography

chromatographie *f* **liquide-liquide** liquid-liquid chromatography

chromatographie *f* **liquide-solide** liquid-solid chromatography

chromatographie *f* **microcirculaire** microcircular chromatography

chromatographie *f* **par adsorption** adsorption chromatography

chromatographie *f* **par déplacement** displacement chromatography

chromatographie *f* **par échange de ligands** ligand-exchange chromatography

chromatographie *f* **par échange d'ions** ion-exchange chromatography

chromatographie *f* **par élution** elution chromatography

chromatographie *f* **par exclusion d'ions** ion-exclusion chromatography

chromatographie *f* **par extraction** extraction chromatography

chromatographie *f* **par filtration de gel** gel filtration

chromatographie *f* **par formation de paires d'ions** ion-pair partition chromatography

chromatographie *f* **par perméation** permeation chromatography

chromatographie *f* **par perméation sur gel** gel-permeation chromatography

chromatographie *f* **par solubilisation** solubilization chromatography

chromatographier *v* chromatograph

chromatographie *f* **sur colonne** column chromatography

chromatographie *f* **sur colonne capillaire** open-tube chromatography

chromatographie *f* **sur couche mince** thin-layer chromatography

chromatographie *f* **sur gel** gel chromatography

chromatographie *f* **sur papier** paper chromatography

chromatographie *f* **sur tamis moléculaires** molecular-sieve chromatography

chromatopolarographie *f* chromatopolarography

chromométrie *f* chromatometry

chromophore *m* chromophore
Chromosorb *m* Chromosorb
chronoampérométrie *f* chronoamperometry
chronoampérométrie *f* **à double échelon de potentiel** double-potential-step chronoamperometry
chronoampérométrie *f* **avec balayage linéaire du potentiel** chronoamperometry with linear potential sweep
chronoampérométrie *f* **avec balayage linéaire du potentiel à une électrode à gouttes** dropping electrode chronoamperometry with linear potential/voltage sweep
chronoampérométrie *f* **convective** convective chronoamperometry
chronoampérométrie *f* **polarographique** polarographic chronoamperometry
chronocoulométrie *f* chronocoulometry
chronocoulométrie *f* **à double échelon de potentiel** double-potential-step chronocoulometry
chronocoulométrie *f* **à échelon de potentiel** chronocoulometry
chronocoulométrie *f* **convective** convective chronocoulometry
chronopotentiométrie *f* chronopotentiometry
chronopotentiométrie *f* **à courant alternatif** alternating-current chronopotentiometry
chronopotentiométrie *f* **à courant alternatif surimposé** chronopotentiometry with superimposed alternating current
chronopotentiométrie *f* **à courant programmé** programmed-current chronopotentiometry
chronopotentiométrie *f* **à échelon de courant** current-step chronopotentiometry
chronopotentiométrie *f* **à inversion de courant** current-reversal chronopotentiometry
chronopotentiométrie *f* **à inversion cyclique de courant** cyclic current-reversal chronopotentiometry
chronopotentiométrie *f* **à tension alternative** alternating-voltage chronopotentiometry
chronopotentiométrie *f* **avec arrêt de courant** current-cessation chronopotentiometry
chronopotentiométrie *f* **avec balayage linéaire du courant** chronopotentiometry with linear current sweep
chronopotentiométrie *f* **cyclique** cyclic chronopotentiometry
chronopotentiométrie *f* **cyclique à échelon de courant** cyclic current-step chronopotentiometry
chronopotentiométrie *f* **dérivée** derivative chronopotentiometry
cible *f* target
cible *f* **de tritium** tritium target
cible *f* **épaisse** thick target
cible *f* **mince** thin target
cible *f* **tritiée** tritium target
circuit *m* **d'essais** interlaboratory research
coagulation *f* coagulation
coagulation *f* **irréversible** irreversible coagulation
coagulation *f* **réversible** reversible coagulation
coefficient *m* **d'absorption linéaire** linear decadic absorption coefficient
coefficient *m* **d'absorption molaire** molar absorptivity

coefficient *m* **d'absorption spécifique** specific decadic absorption coefficient
coefficient *m* **d'activité du constituant B** activity coefficient of substance B
coefficient *m* **d'atténuation** attenuation coefficient
coefficient *m* **d'atténuation atomique** atomic attenuation coefficient
coefficient *m* **d'atténuation linéaire** linear attenuation coefficient
coefficient *m* **d'atténuation massique** mass attenuation coefficient
coefficient *m* **de corrélation** coefficient of correlation
coefficient *m* **de corrélation multiple** coefficient of multiple correlation
coefficient *m* **de diffusion** diffusion coefficient
coefficient *m* **de distribution 1.** concentration distribution ratio (*in chromatography*) **2.** concentration distribution ratio (*in extraction*)
coefficient *m* **de distribution de concentration** concentration distribution ratio (*in chromatography*)
coefficient *m* **de distribution D_g** distribution coefficient D_g
coefficient *m* **de distribution D_s** distribution coefficient D_s
coefficient *m* **de distribution D_v** distribution coefficient D_v
coefficient *m* **de distribution en volume** distribution coefficient D_v
coefficient *m* **de distribution massique** mass distribution ratio
coefficient *m* **de partage 1.** distribution constant (*in extraction*) **2.** distribution constant (*in chromatography*) **3.** partition coefficient (*in gas chromatography*)
coefficient *m* **de régression** regression coefficient
coefficient *m* **de sélectivité** selectivity coefficient
coefficient *m* **de sélectivité corrigé** corrected selectivity coefficient
coefficient *m* **de sélectivité potentiométrique** potentiometric selectivity coefficient
coefficient *m* **de température** temperature coefficient
coefficient *m* **de transfert** charge-transfer coefficient
coefficient *m* **de variabilité** coefficient of variation
coefficient *m* **de variation empirique** empirical variation coefficient
coefficient *m* **d'extraction** concentration distribution ratio (*in extraction*)
coefficient *m* **d'ionisation** apparent degree of dissociation
coefficient *m* **moyen d'activité ionique** mean activity coefficient of electrolyte
coïncidence *f* **de raies** line coincidence
co-ions *mpl* co-ions
collecteur *m* trace collector
collecteur *m* **de fractions** fraction collector
collecteur *m* **de traces** trace collector
collection *f* collection
collimateur *m* collimator
collimateur *m* **de Soller** Soller collimator

collimation *f* collimation
collision *f* élastique elastic collision
collision *f* inélastique inelastic collision
cologarithme *m* de l'activité des ions hydrogènes
pH
colonne *f* à tube ouvert open-tube column
colonne *f* capillaire capillary column
colonne *f* chromatographique chromatographic
column
colonne *f* garnie classique packed column
colonne *f* régulièrement remplie packed column
colonne *f* remplie packed column
colonne *f* tubulaire ouverte open-tube column
coloration *f* de flamme flame coloration
colorimètre *m* colorimeter
colorimétrie *f* colorimetric analysis
colorimétrie *f* visuelle visual colorimetry
coma *f* coma
combinaison *f* marquée labelled compound
combustion *f* combustion
combustion *f* catalytique catalytic combustion
combustion *f* du filtre incineration of filter paper
combustion *f* éclair flash combustion
combustion *f* en tube vide empty tube
combustion
combustion *f* totale par plasma plasma ashing
complexe *m* de transfert de charge
donor-acceptor complex
complexe *m* interne neutral chelate
complexométrie *f* compleximetry
complexones *mpl* complexones
composé *m* marqué labelled compound
composition *f* du projet mixture design
composition *f* qualitative qualitative composition
composition *f* quantitative quantitative
composition
compte-bulles *m* bubble gauge
compte-gouttes *m* dropper
compteur *m* 2π 2π-counter
compteur *m* 4π 4π-counter
compteur *m* à canal well-type scintillation
counter
compteur *m* à fenêtre window counter
compteur *m* à scintillation scintillation counter
compteur *m* de rayonnement radiation counter
compteur *m* GM Geiger-Müller counter tube
compteur *m* proportionnel proportional counter
tube
compteur *m* proportionnel de courant gazeux
proportional gas-flow counter
concentration *f* analytique analytical
concentration
concentration *f* du constituant B amount-of-
substance concentration of substance B
concentration *f* en masse du constituant B
mass concentration of substance B
concentration *f* en quantité de matière du
constituant B amount-of-substance
concentration of substance B
concentration *f* ionique moyenne mean
concentration of electrolyte
concentration *f* limite concentration limit
concentration *f* molaire amount-of-substance
concentration of substance B
condensateur *m* à lame vibrante vibrating
condenser

conditions *fpl* normales standard conditions
conductance *f* d'un électrolyte conductance of
an electrolyte
conductance *f* électrique conductance
conductance *f* électrique d'un électrolyte
conductance of electrolyte
conductance *f* spécifique electrical conductivity
conductance *f* spécifique d'un électrolyte
electrolytic conductivity
conducteur *m* de deuxième espèce ionic conductor
conducteur *m* de première espèce electronic
conductor
conducteur *m* de seconde classe ionic conductor
conducteur *m* électrique electrical conductor
conducteur *m* électrolytique ionic conductor
conducteur *m* électronique electronic conductor
conducteur *m* ionique ionic conductor
conducteur *m* métallique electronic conductor
conductibilité *f* conductivity
conductimétrie *f* conductometry
conduction *f* électrique electrical conduction
conduction *f* électrolytique ionic conduction
conduction *f* électronique electronic conduction
conduction *f* ionique ionic conduction
conduction *f* par des électrons electronic
conduction
conduction *f* par des ions ionic conduction
conduction *f* par des lacunes hole conduction
conduction *f* par des trous hole conduction
conductivimètre *m* conductometer
conductivité *f* conductivity
conductivité *f* d'un électrolyte electrolytic
conductivity
conductivité *f* électrique electrical conductivity
conductivité *f* électrolytique electrolytic
conductivity
conductivité *f* équivalente à une dilution infinie
limiting equivalent conductivity of electrolyte
conductivité *f* équivalente d'un électrolyte
equivalent conductivity of electrolyte
conductivité *f* équivalente d'un électrolyte
à une dilution infinie limiting equivalent
conductivity of electrolyte
conductivité *f* équivalente limite d'un électrolyte
limiting equivalent conductivity of electrolyte
conductivité *f* ionique équivalente equivalent
conductivity of ionic species B
conductivité *f* ionique équivalente à une
dilution infinie limiting equivalent conductivity
of ionic species B
conductivité *f* ionique équivalente limite limiting
equivalent conductivity of ionic species B
conductivité *f* ionique molaire molar
conductivity of ionic species B
conductivité *f* ionique molaire à une dilution
infinie limiting molar conductivity of ionic
species B
conductivité *f* ionique molaire limite limiting
molar conductivity of ionic species B
conductivité *f* molaire à une dilution infinie
limiting molar conductivity of electrolyte
conductivité *f* molaire d'un électrolyte molar
couductivity of electrolyte
conductivité *f* molaire d'un électrolyte à une
dilution infinie limiting molar conductivity
of electrolyte

conductivité *f* **molaire limite d'un électrolyte** limiting molar conductivity of electrolyte
conductométrie *f* conductometry
conductométrie *f* **haute fréquence** high-frequency conductometry
cône *m* **externe** secondary combustion zone
cône *m* **interne** primary combustion zone
conformation *f* conformation
consommation *f* burn-up
constante *f* **de diffusion** diffusion coefficient
constante *f* **de distribution 1.** distribution constant (*in chromatography*) **2.** distribution constant (*in extraction*)
constante *f* **de Faraday** Faraday constant
constante *f* **de force** force constant
constante *f* **de formation successive** stepwise stability constant
constante *f* **de cellule** cell constant of a conductivity cell
constante *f* **de partage** partition constant
constante *f* **de partage thermodynamique** parition constant
constante *f* **de Rydberg** Rydberg constant
constante *f* **de sélectivité** potentiometric selectivity coefficient
constante *f* **de vitesse de la réaction d'électrode** rate constant of electrode reaction
constante *f* **de vitesse de réaction** reaction-rate constant
constante *f* **d'extraction** extraction constant
constante *f* **de vittesse normale de la réaction d'électrode** standard rate constant of electrode reaction
constante *f* **diélectrique** relative electric permittivity
constante *f* **du courant de diffusion** limiting current constant
constante *f* **globale de formation** cumulative stability constant
constituant *m* **majeur** major constituent
constituant *m* **mineur** minor constituent
contamination *f* **d'un précipité** contamination of a precipitate
contre-balayage *m* backflushing
contre-électrode *f* **1.** auxiliary electrode (*in electrochemistry*) **2.** counter electrode (*in arc and spark spectroscopy*)
contre-ions *mpl* counter-ions
contrôle *m* **statistique de la qualité** statistical quality control
convection *f* convection
convection *f* **naturelle** natural convection
convention *f* **de Stockholm** sign convention
coprécipitation *f* coprecipitation
corps *m* **électro-actif** electroactive substance
correcteur *m* **d'absorption non spécifique** background corrector
correcteur *m* **de fond** background corrector
correction *f* **d'indicateur** indicator blank
corrélation *f* correlation
couche *f* **active** effective layer
couche *f* **altérée** altered layer
couche *f* **cathodique** cathode layer
couche *f* **de demi-absorption** half-thickness
couche *f* **de diffusion** diffusion layer

couche *f* **de Gouy** diffuse double layer
couche *f* **diffuse** diffuse double layer
couche *f* **double electrochemical** double layer
couche *f* **double électrique** electrochemical double layer
couche *f* **double électrochimique** electrochemical double layer
couche *f* **infiniment épaisse** infinitely thick layer
couche *f* **infiniment mince** infinitely thin layer
couche *f* **réactionnelle** reaction layer
couche *f* **rigide d'Helmholtz** compact layer
coulogravimétrie *f* coulogravimetry
coulomètre *m* coulometer
coulomètre *m* **à cuivre** copper coulometer
coulomètre *m* **chimique** chemical coulometer
coulométrie *f* coulometry
coulométrie *f* **à courant contrôlé** controlled-current coulometry
coulométrie *f* **à courant imposé** controlled-current coulometry
coulométrie *f* **à courant imposé avec détection potentiométrique du point de fin de titrage** controlled-current coulometry with potentiometric end-point detection
coulométrie *f* **à intensité de courant constante** controlled-current coulometry
coulométrie *f* **à potentiel constant** controlled-potential coulometry
coulométrie *f* **à potentiel contrôlé** controlled-potential coulometry
coulométrie *f* **à une électrode à gouttes** dropping electrode coulometry
coulométrie *f* **polarographique** polarographic coulometry
coulométrie *f* **potentiostatique** controlled-potential coulometry
coupellation *f* cupellation
couplage *m* **de colonnes** coupled columns
couple *f* **de raies** line pair
couple *m* **thermoélectrique** thermocouple
courant *m* **capacitif** capacity current
courant *m* **catalytique** catalytic current
courant *m* **cinétique** kinetic current
courant *m* **d'adsorption** adsorption current
courant *m* **d'échange** exchange current
courant *m* **de charge** capacity current
courant *m* **de crête** peak current
courant *m* **de diffusion** diffusion current
courant *m* **de migration** migration current
courant *m* **d'obscurité** dark current
courant *m* **électrique** electric current
courant *m* **faradaïque** faradaic current
courant *m* **global** net current
courant *m* **limite** limiting current
courant *m* **maximum** peak current
courant *m* **mesuré** observed current
courant *m* **observé** observed current
courant *m* **partiel anodique/cathodique** partial anodic/cathodic current
courant *m* **photoélectrique** photocurrent
courant *m* **résiduel** residual current
courant *m* **total** net current
courbe *f* **analytique** analytical curve
courbe *f* **caractéristique 1.** operating characteristic function **2.** emulsion characteristic curve

courbe *f* caractéristique d'une émulsion
emulsion characteristic
courbe *f* courant-tension polarization curve
courbe *f* d'absorption 1. absorption curve
2. absorption curve (*in magnetic resonance*)
courbe *f* d'échauffement heating curve
courbe *f* de décroissance decay curve
courbe *f* de désintégration decay curve
courbe *f* de Hurter et Driffield emulsion
characteristic curve
courbe *f* d'élution elution curve
courbe *f* de noircissement emulsion
characteristic curve
courbe *f* de noircissement d'une émulsion
emulsion characteristic curve
courbe *f* de percée breakthrough curve
courbe *f* de polarisation polarization curve
courbe *f* de sensibilité de la balance
sensitivity curve of balance
courbe *f* d'étalonnage analytical curve
courbe *f* d'étalonnage d'émulsion emulsion
calibration curve
courbe *f* de titrage titration curve
courbe *f* de titrage acido-basique
neutralization titration curve
courbe *f* de titrage complexométrique
complexometric titration curve
courbe *f* de titrage de précipitation
precipitation titration curve
courbe *f* de titrage redox redox titration curve
courbe *f* d'extraction extraction curve
courbe *f* intensité-potentiel polarization curve
courbe *f* OC operating characteristic function
courbe *f* thermoanalytique thermal curve
courbe *f* thermométrique thermal curve
courbe *f* voltampérométrique polarization curve
couvercle *m* superficiel surface coverage
covariance *f* covariance
cratère *m* crater
creuset *m* crucible
creuset *m* de Gooch Gooch crucible
creuset *m* filtrant crucible with sintered disk
cristal *m* analyseur analysing crystal
cristal *m* de Johann Johann crystal
cristal *m* de Johannson Johannson crystal
cristallisation *f* crystallization
cristallisoir *m* crystallizing dish
cristal *m* mixte mixed crystal
curie *m* curie
cuve *f* absorbante absorption cell
cuve *f* à chromatographie chromatographic
chamber
cuve *f* à électrolyse electrolytic cell
cuve *f* de développement chromatographic
chamber
cuvette *f* de graphite graphite cuvette
cycle *m* chélaté chelate ring

D

débit *m* d'écoulement F_a volumetric flow
rate F_a
débit *m* d'écoulement F_c volumetric flow
rate F_c
débit *m* du mercure par le capillaire flow rate
of mercury

débit *m* volumique F_a de la phase mobile
volumetric flow rate F_a
débit *m* volumique F_c de la phase mobile
volumetric flow rate F_c
décantation *f* decantation
décapage *m* par plasma plasma etching
décharge *f* à cathode creuse hollow-cathode
discharge
décharge *f* luminescente glow discharge
décomposition *f* métastable metastable
decomposition
décontamination *f* decontamination
décroissance *f* radioactive radioactive decay
degré *m* d'association des ions association
degree of ions
degré *m* de dissociation apparent degree of
dissociation
degré *m* de dissociation réel real degree of
dissociation
degré *m* de recouvrement de la surface surface
coverage
degré *m* d'ionisation apparent degree of
dissociation
degré *m* hydrotimétrique allemand German
degree of water hardness
degré *m* hydrotimétrique américain American
degree of water hardness
degré *m* hydrotimétrique anglais English
degree of water hardness
degré *m* hydrotimétrique francais French
degree of water hardness
demi-cellule *f* electrode (*in electrochemistry*)
demi-largeur *f* half-intensity width
demi-pile *f* electrode (*in electrochemistry*)
démixtion *f* du solvant solvent demixing
densité *f* de charge superficielle surface charge
density
densité *f* de courant current density
densité *f* de flux flux density of particles or
photons
densité *f* de flux de particules ou de photons
flux density of particles or photons
densité *f* de probabilité probability density
function
densité *f* lumineuse luminance
densité *f* optique 1. absorbance 2. blackening
densitomètre *m* microphotometer
déplacement *m* displacement
déplacement *m* bathochrome bathochromic shift
déplacement *m* chimique chemical shift
déplacement *m* de la ligne de base base-line
drift
déplacement *m* des raies Raman Raman
frequency shift
déplacement *m* en fréquence des raies Raman
Raman frequency shift
déplacement *m* hypsochrome hypsochromic shift
dépolarisant *m* depolarizer
dépolariseur *m* depolarizer
dérivatisation *f* derivatization
dérivatographe *m* derivatograph
dérive *f* de la ligne de base base-line drift
dérive *f* d'un potentiel drift of a potential
désactivation *f* 1. deactivation
2. decontamination
désintégration *f* radioactive radioactive decay

désorption *f* desorption
désorption *f* de champ field desorption
désorption *f* de particules neutres par choc d'électrons electron stimulated desorption of neutrals
désorption *f* de particules neutres par impact d'électrons electron stimulated desorption of neutrals
désorption *f* d'ions par choc d'électrons electron stimulated desorption of ions
désorption *f* d'ions par impact d'électrons electron stimulated desorption of ions
désorption *f* par choc d'électrons electron stimulated desorption
désorption *f* par choc d'ions ion impact desorption
désorption *f* par impact d'électrons electron stimulated desorption
désorption *f* par impact d'ions ion impact desorption
désorption *f* thermique thermal desorption
désorption *f* thermique programmée temperature programmed desorption
désoxydant *m* reductant
desséchant *m* desiccant
dessiccateur *m* desiccator
dessiccateur *m* sous vide vacuum desiccator
détecteur *m* detector
détecteur *m* à argon argon ionization detector
détecteur *m* à barrière de surface surface barrier detector
détecteur *m* à capture d'électrons electron capture detector
détecteur *m* à conductibilité thermique thermal conductivity detector
détecteur *m* à ionisation ionization detector
détecteur *m* à ionisation d'argon argon ionization detector
détecteur *m* à ionisation de flamme flame ionization detector
détecteur *m* à photométrie de flamme flame photometric detector
détecteur *m* à résonance resonance detector
détecteur *m* à scintillation scintillation detector
détecteur *m* à transport du soluté transport detector
détecteur *m* d'argon argon ionization detector
détecteur *m* de Everhart et Thornley Everhart-Thornley detector
détecteur *m* de neutrons neutron detector
détecteur *m* de rayonnement radiation detector
détecteur *m* différentiel differential detector
détecteur *m* intégral integral detector
détecteur *m* photoélectrique photodetector
détecteur *m* photométrique flame photometric detector
détecteur *m* semi-conducteur semiconductor detector
détecteur *m* semi-conducteur Ge(Li) lithium-drifted germanium detector
détecteur *m* semi-conducteur Si(Li) lithium-drifted silicon detector
détecteur *m* thermique thermal radiation detector
détection *f* detection
détection *f* de rayonnement detection of radiation

détection *f* des gaz émis evolved gas detection
détermination *f* cryoscopique de la masse molaire cryoscopic determination of the molar mass
détermination *f* ébullioscopique de la masse molaire ebullioscopic determination of the molar mass
détermination *f* simultanée simultaneous determination
développement *m* 1. development 2. development (*in photography*)
développement *m* ascendant ascending development
développement *m* d'un chromatogramme development of a chromatogram
développement *m* radial circular chromatography
développement *m* vertical descendant descending development
dévésiculeur *m* safety head
déviation *f* magnétique magnetic deflection
diagramme *m* de cristal oscillant X-ray oscillating-crystal pattern
diagramme *m* de cristal tournant des rayons X X-ray rotating-crystal pattern
diagramme *m* d'électrophorèse electrophorogram
diagramme *m* de poudre en rayons X X-ray powdered-crystal pattern
diaphragm *m* de Hartmann Hartmann diaphragm
dichroïsme *m* circulaire circular dichroism
diélcométrie *f* dielectrometry
diélectrométrie *f* dielectrometry
différence *f* de potentiel electric potential difference
différence *f* de potentiel de Galvani Galvani potential difference
différence *f* de potentiel d'une cellule galvanique electric potential difference of a galvanic cell
différence *f* de potentiel de Volta Volta potential difference
différence *f* de potentiel entre deux points electric potential difference
diffraction *f* de la lumière light diffraction
diffraction *f* d'électrons à basse énergie low-energy electron diffraction
diffraction *f* d'électrons à grande vitesse high-energy electron diffraction
diffraction *f* d'électrons à grande vitesse par réflexion reflection high-energy electron diffraction
diffraction *f* d'électrons de faible énergie low-energy electron diffraction
diffraction *f* d'électrons lents low-energy electron diffraction
diffraction *f* des rayons X X-ray diffraction
diffraction *f* d'ions lents low-energy ion diffraction
diffraction *f* inélastique des électrons de faible énergie inelastic low-energy electron diffraction
diffractomètre *m* à rayons X X-ray diffractometer
diffractométrie *f* diffractometry
diffusion *f* 1. diffusion 2. scattering
diffusion *f* cohérente coherent scattering
diffusion *f* de Compton Compton effect

diffusion *f* de la lumière light scattering
diffusion *f* de Rayleigh Rayleigh scattering
diffusion *f* élastique elastic scattering
diffusion *f* en retour backscattering
diffusion *f* incohérente Compton effect
diffusion *f* inélastique inelastic scattering
diffusion *f* longitudinale molecular diffusion term
diffusion *f* moléculaire molecular diffusion term
diffusion *f* moléculaire axiale molecular
    diffusion term
diffusion *f* turbulente eddy diffusion term
diffusivité *f* diffusion coefficient
digestion *f* digestion
dilatomètre *m* dilatometer
dilatométrie *f* thermodilatometry
dilatométrie *f* dérivée derivative
    thermodilatometry
dilatométrie *f* différentielle differential
    thermodilatometry
diluant *m* diluent
diméthyldichlorosilane *m* dimethyldichlorosilane
discontinuité *f* d'absorption absorption edge
dispersion *f* 1. dispersion 2. dispersion
    (*in statistics*) 3. variance
dispersion *f* angulaire angular dispersion
dispersion *f* angulaire d'une cristal analyseur
    angular dispersion of an analysing crystal
dispersion *f* anormale anomalous dispersion
dispersion *f* de conductance
    Debye-Falkenhagen's effect
dispersion *f* de la lumière light dispersion
dispersion *f* de l'échantillon sampling variance
dispersion *f* de surface par faisceau moléculaire
    molecular beam surface scattering
dispersion *f* linéaire linear dispersion
dispersion *f* normale normal dispersion
dispersion *f* réactive par faisceau moléculaire
    molecular beam reactive scattering
dispersion *f* rotatoire optical rotatory dispersion
dispersion *f* statistique dispersion (*in statistics*)
dispositif *m* d'étalement spreader
dissociation *f* dissociation
dissociation *f* électrolytique electrolytic
    dissociation
dissociation *f* thermique thermal dissociation
dissolution *f* dissolution
distance *f* de la phase mobile mobile phase
    distance
distance *f* de migration du solvant mobile phase
    distance
distillation *f* avec un entraîneur carrier
    distillation
distribution *f* à contre-courant countercurrent
    distribution
distribution *f* binomiale binomial distribution
distribution *f* Compton Compton continuum
distribution *f* de χ^2 chi-squared distribution
distribution *f* de Bernoulli binomial distribution
distribution *f* de Fisher F-distribution
distribution *f* de Laplace-Gauss normal
    distribution
distribution *f* de Poisson Poisson distribution
distribution *f* de probabilité probability
    distribution
distribution *f* de Student Student's distribution
distribution *f* de *t* Student's distribution

distribution *f* empirique empirical distribution
distribution *f* exponentielle exponential
    distribution
distribution *f* liquide-liquide liquid-liquid
    distribution
distribution *f* logarithmico-normale
    logarithmic-normal distribution
distribution *f* normale normal distribution
distribution *f* normale réduite standardized
    normal distribution
diviseur *m* d'échantillon sample splitter
domaine *m* d'utilisation d'une balance range of
    applicability of a balance
dominante *f* mode
dosage *m* determination
dosage *m* automatique du carbone, de
    l'hydrogène et de l'azote automatic
    determination of carbon, hydrogen and
    nitrogen
dosage *m* critique critical determination
dosages *mpl* manométriques manometric
    methods
drain *m* drain
durée *f* de vie d'une goutte drop time
dureté *f* calcicité calcium hardness
dureté *f* de l'eau water hardness
dureté *f* magnésienne magnesium hardness
dureté *f* permanente 1. non-carbonate hardness
    2. permanent hardness
dureté *f* temporaire 1. carbonate hardness
    2. temporary hardness
dureté *f* totale total water hardness
dynode *f* − dynode

E

eau *f* de cristallisation water of crystallization
eau *f* déionisée deionized water
eau *f* déminéralisée deionized water
eau *f* distillée distilled water
eau *f* non combinée hygroscopic water
eau *f* redistillée redistilled water
écart *m* moyen mean deviation
écart-type *m* standard deviation
écart-type *m* de la moyenne de l'épreuve
    standard deviation of the mean
écart-type *m* de l'épreuve standard deviation
    of a sample
écart-type *m* empirique standard deviation of
    a sample
écart-type *m* relatif coefficient of variation
échange *m* d'anions anion exchange
échange *m* de cations cation exchange
échange *m* de masse mass exchange
échange *m* d'ions ion exchange
échange *m* isotopique isotope exchange
échangeur *m* d'anions anion exchanger
échangeur *m* de cations cation exchanger
échangeur *m* d'électrons electron exchanger
échangeur *m* d'ions ion exchanger
échangeur *m* d'ions à macropores macroporous
    ion exchanger
échangeur *m* d'ions amphotère amphoteric ion
    exchanger
échangeur *m* d'ions bifonctionnel bifunctional
    ion exchanger

échangeur *m* **d'ions chélateur** chelating ion exchanger
échangeur *m* **d'ions difonctionnel** bifunctional ion exchanger
échangeur *m* **d'ions liquide** liquid ion exchanger
échangeur *m* **d'ions macroporeux** macroporous ion exchanger
échangeur *m* **d'ions monofonctionnel** monofunctional ion exchanger
échangeur *m* **d'ions oléophile** oleophilic ion exchanger
échangeur *m* **d'ions pelliculaire** pellicular ion exchanger
échangeur *m* **d'ions polyfonctionnel** polyfunctional ion exchanger
échangeur *m* **d'ions redox** redox ion exchanger
échangeur *m* **d'ions sec** absolutely dry ion exchanger
échangeur *m* **d'ions sélectif** selective ion exchanger
échangeur *m* **d'ions solide** solid ion exchanger
échantillon *m* sample
échantillon *m* **aléatoire** random sample
échantillon *m* **brut** primary sample
échantillon *m* **de comparaison** standard sample
échantillon *m* **épais** thick specimen
échantillon *m* **global** gross sample
échantillon *m* **mince** thin-film specimen
échantillon *m* **moyen** average sample
échantillon *m* **pastillé** potassium bromide disk
échantillon *m* **pour laboratoire** average sample
échantillon *m* **type** standard sample
échantillon *m* **unitaire** unit sample
échelle *f* δ delta scale
échelle *f* **d'activité de Brockmann** Brockmann scale of activity
éclairement *m* **énergétique** irradiance
éclairement *m* **lumineux** illuminance
effectif *m* **de l'échantillon** sample size
effet *m* **Auger** Auger effect
effet *m* **bathochrome** bathochromic shift
effet *m* **chimique** chemical effect
effet *m* **d'absorption** absorption effect
effet *m* **d'asymétrie** relaxation-time effect
effet *m* **Debye-Falkenhagen** Debye-Falkenhagen's effect
effet *m* **de charge** charging effect
effet *m* **de Compton** Compton effect
effet *m* **de dilution** dilution effect
effet *m* **de matrice 1.** matrix effect **2.** matrix effect (*in X-ray fluorescence analysis*)
effet *m* **des charges électrostatiques** effect of electrostatic charges
effet *m* **de sel** salt effect
effet *m* **de Zeeman** Zeeman effect
effet *m* **d'exaltation** enhancement effect
effet *m* **dissociant du champ électrique** dissociation field effect
effet *m* **Dorn** sedimentation potential
effet *m* **d'un ion commun** common ion effect
effet *m* **du temps de relaxation** relaxation-time effect
effet *m* **égalisant** levelling effect
effet *m* **électrophorétique** electrophoretic effect
effet *m* **granulométrique** grain size effect
effet *m* **hyperchromique** hyperchromic effect

effet *m* **hypochromique** hypochromic effect
effet *m* **isotopique** isotope effect
effet *m* **mémoire relatif à l'eau** water memory effect
effet *m* **minéralogique** mineralogical effect
effet *m* **Mössbauer** Mössbauer effect
effet *m* **photoélectrique** photoelectric effect
effet *m* **Raman** Raman effect
effet *m* **Raman de résonance** resonance Raman effect
effet *m* **Raman stimulé** stimulated Raman effect
effet *m* **sélectif** differentiating effect
effet *m* **Tyndall** Tyndall effect
effet *m* **Wien** Wien's effect
efficacité *f* **absolue du pic d'absorption totale** absolute full energy peak efficiency
efficacité *f* **de comptage** counting efficiency
efficacité *f* **du détecteur** intrinsic detector efficiency
efficacité *f* **d'une colonne** column performance
efficacité *f* **photoélectrique absolue** absolute photopeak efficiency
efficacité *f* **photoélectrique intrinsèque** intrinsic photopeak efficiency
effluent *m* effluent
élargissement *m* **du pic** peak broadening
électro-analyse *f* electroanalysis
électrocapillarité *f* electrocapillary phenomena
électrochimie *f* electrochemistry
électrode *f* **1.** electrode **2.** electrode (*in electrochemistry*)
électrode *f* **à amalgame** amalgam electrode
électrode *f* **à bulle d'air** air-gap electrode
électrode *f* **à chlorure d'argent** silver-silver chloride electrode
électrode *f* **à disque tournant 1.** rotating-disk electrode (*in electrochemistry*) **2.** rotating-disk electrode (*in arc and spark spectroscopy*)
électrode *f* **à échangeur d'ions liquide** liquid ion-exchange membrane electrode
électrode *f* **à gaz** gas electrode
électrode *f* **à goutte de mercure pendante** hanging mercury drop electrode
électrode *f* **à gouttes de mercure** dropping-mercury electrode
électrode *f* **à hydrogène** hydrogen-gas electrode
électrode *f* **à hydrogène normale** standard hydrogen electrode
électrode *f* **à jet de mercure** streaming mercury electrode
électrode *f* **à matrice rigide** rigid matrix electrode
électrode *f* **à membrane cristalline** crystalline membrane ion-selective electrode
électrode *f* **à membrane hétérogène** heterogeneous membrane ion-selective electrode
électrode *f* **à membrane homogène** homogeneous membrane ion-selective electrode
électrode *f* **à membrane liquide** liquid membrane ion-selective electrode
électrode *f* **à membrane sélective 1.** ion-selective electrode **2.** membrane ion-selective electrode
électrode *f* **à membrane sélective à échangeur d'ions liquide** liquid ion-exchange membrane electrode

électrode *f* à membrane sélective à matrice
    rigide rigid matrix electrode
électrode *f* à membrane sélective cristalline
    crystalline membrane ion-selective electrode
électrode *f* à membrane sélective de verre glass
    membrane electrode
électrode *f* à membrane sélective hétérogène
    heterogeneous membrane ion-selective
    electrode
électrode *f* à membrane sélective homogène
    homogeneous membrane ion-selective
    electrode
électrode *f* à membrane sélective liquide liquid
    membrane ion-selective electrode
électrode *f* à membrane sélective sensibilisée
    sensitized ion-selective electrode
électrode *f* à membrane sélective solide
    solid-state membrane ion-selective
    electrode
électrode *f* à membrane solide solid-state
    membrane ion-selective electrode
électrode *f* à oxygène oxygen gas electrode
électrode *f* à quinhydrone quinhydrone
    electrode
électrode *f* argent-chlorure d'argent
    silver-silver chloride electrode
électrode *f* à substrat enzymatique
    enzyme-substrate electrode
électrode *f* à sulfate mercureux
    mercury-mercurous sulphate electrode
électrode *f* au calomel calomel electrode
électrode *f* au calomel saturée saturated-calomel
    electrode
électrode *f* auto-émettrice self-electrode
électrode *f* auxiliaire auxiliary electrode
électrode *f* combinée combination electrode
électrode *f* cristalline crystalline membrane
    ion-selective electrode
électrode *f* d'antimoine antimony electrode
électrode *f* d'antimoine-oxyde antimonieux
    antimony electrode
électrode *f* d'argent silver electrode
électrode *f* de comparaison reference electrode
électrode *f* de deuxième espèce electrode of the
    second kind
électrode *f* de graphite vitrifié glassy carbon
    electrode
électrode *f* de mesure indicator electrode
électrode *f* de platine platinum electrode
électrode *f* de première espèce electrode of the
    first kind
électrode *f* de premier ordre electrode of the
    first kind
électrode *f* de référence reference electrode
électrode *f* de référence interne internal
    reference electrode
électrode *f* de second ordre electrode of the
    second kind
électrode *f* de travail working electrode
électrode *f* de troisième espèce electrode of the
    third kind
électrode *f* de troisième ordre electrode of the
    third kind
électrode *f* de verre glass membrane electrode
électrode *f* de verre indicatrice des ions
    hydrogène glass pH electrode

électrode *f* d'oxydo-réduction redox electrode
électrode *f* idéalement non-polarisable ideal
    non-polarisable electrode
électrode *f* idéalement polarisable ideal
    polarizable electrode
électrode *f* impolarisable ideal non-polarisable
    electrode
électrode *f* indicatrice indicator electrode
électrode *f* indicatrice de gaz gas sensing
    electrode
électrode *f* indicatrice de pH pH electrode
électrode *f* indicatrice d'ions à transistor
    à effet de champ ion-selective field effect
    transistor
électrode *f* irréversible irreversible electrode
électrode *f* mercure-oxyde mercurique
    mercury-mercuric oxide electrode
électrode *f* mercure-sulfate mercureux
    mercury-mercurous sulfate electrode
électrode *f* métal-ion métallique metal-metal
    ion electrode
électrode *f* métallique metal-metal ion
    electrode
électrode *f* métal-oxyde métallique oxide
    electrode
électrode *f* multiple complex electrode
électrode *f* normale à hydrogène standard
    hydrogen electrode
électrode *f* poreuse porous cup electrode
électrode *f* porte-échantillon supporting
    electrode
électrodéposition *f* electrodeposition
électrodéposition *f* de métaux electrodeposition
    of metals
électrode *f* redox redox electrode
électrode *f* réversible reversible electrode
électrode *f* rotative rotating-disk electrode
    (*in arc and spark spectroscopy*)
électrode *f* sélective indicatrice d'ions
    ion-selective electrode
électrode *f* solide solid electrode
électrode *f* soumise à l'analyse self-electrode
électrode *f* spécifique ion-selective electrode
électrode *f* standard à hydrogène standard
    hydrogen electrode
électrode *f* totalement polarisable ideal
    polarizable electrode
électrode *f* tournante à disque 1. rotating-disk
    electrode (*in electrochemistry*) 2. rotating-disk
    electrode (*in arc and spark spectroscopy*)
électrodialyse *f* electrodialysis
électroendoosmose *f* electro-endoosmosis
électrographie *f* electrography
électrogravimétrie *f* electrogravimetry
électrogravimétrie *f* à potentiel contrôlé
    controlled-potential electrogravimetry
électrogravimétrie *f* à tension d'électrode
    contrôlée controlled-potential
    electrogravimetry
électrogravimétrie *f* à tension d'électrolyse
    constante constant-current electrogravimetry
électrogravimétrie *f* interne internal
    electrogravimetry
électrogravimétrie *f* spontanée internal
    electrogravimetry
électroluminescence *f* electroluminescence

électrolyse *f* electrolysis
électrolyse *f* **à courant constant**
  constant-current electrolysis
électrolyse *f* **à potentiel contrôlé**
  controlled-potential electrolysis
électrolyse *f* **à tension d'électrode constante**
  controlled-potential electrolysis
électrolyse *f* **à tension d'électrolyse constante**
  constant-current electrolysis
électrolyse *f* **interne** internal electrolysis
électrolyse *f* **potentiostatique** controlled-potential
  electrolysis
électrolyseur *m* electrolytic cell
électrolyte *m* electrolyte
électrolyte *m* **de base** supporting electrolyte
électrolyte *m* **indifférent** supporting electrolyte
électrolyte *m* **polimérique** polyelectrolyte
électrolyte *m* **support** supporting electrolyte
électromigration *f* migration of ions
électron *m* **célibataire** unpaired electron
électron *m* **non apparié** unpaired electron
électron-volt *m* electronvolt
électro-osmose *f* electro-endoosmosis
électrophorèse *f* electrophoresis
électrophorogramme *m* electrophorogram
élément *m* **de Leclanché** Leclanché cell
élément *m* **étalon** standard cell
élément *m* **primaire** primary cell
élément *m* **secondaire** electrical accumulator
élément *m* **Weston standard** Weston standard cell
ellipsométrie *f* ellipsometry
éluant *m* eluent
éluat *m* eluate
élution *f* elution
élution *f* **isochratique** isocratic elution
élution *f* **par étage** stepwise elution
élution *f* **par gradient de composition** gradient
  elution
élution *f* **par gradient de composition en
  escalier** stepwise elution
élution *f* **par paliers** stepwise elution
élution *f* **sélective** selective elution
élution *f* **spécifique** specific elution
élutriation *f* elutriation
émetteur *m* emitter
émission *f* **d'électrons** electron emission
émission *f* **des rayons X par bombardement
  ionique** ion induced X-ray spectroscopy
émission *f* **induite** stimulated emission
émission *f* **stimulée** stimulated emission
empilement *m* **d'impulsions** pile-up
empreinte *f* **digitale** fingerprint region
enceinte *f* **étanche** hot cell
énergie *f* **d'activation** activation energy
énergie *f* **rayonnante** radiant energy
enthalpimétrie *f* enthalpimetry
entité *f* **élementaire** elementary entity
entonnoir *m* **allonge pour creuset filtrant**
  crucible holder
entonnoir *m* **à plaque de verre fritté** filter
  funnel with sintered disk
entonnoir *m* **à poudre** powder funnel
entonnoir *m* **à solide** powder funnel
entonnoir *m* **cannelé** ribbed funnel
entonnoir *m* **chauffant** funnel heater

entonnoir *m* **de Büchner** Büchner funnel
entonnoir *m* **en verre** glass funnel
entraînement *m* collection
entraîneur *m* 1. carrier 2. scavenger 3. trace
  collector
entraîneur *m* **isotopique** isotopic carrier
entraîneur *m* **non isotopique** non-isotopic carrier
entraîneur *m* **spectrochimique** spectrochemical
  carrier
entrode *f* analytical gap
épaisseur *f* **del'échantillon** path length
épaisseur-moitié *f* half-thickness
épreuve *f* test (*in chemical analysis*)
éprouvette *f* **graduée** measuring cylinder
épurateur *m* scavenger
épuration *f* scrubbing
équation *f* **de Cottrell** Cottrell equation
équation *f* **de Debye-Hückel** Debye-Hückel
  equation
équation *f* **de Debye-Hückel-Brönsted**
  Debye-Hückel-Brönsted equation
équation *f* **de Ilkovič** Ilkovič equation
équation *f* **de Levich** Levich equation
équation *f* **de Lewis et Sargent** Lewis-Sargent's
  relation
équation *f* **de Nernst** Nernst equation
équation *f* **de Nernst modifiée** Nikolsky
  equation
équation *f* **de Randles** Randles-Ševčik equation
équation *f* **de Sand** Sand equation
équation *f* **de Tafel** Tafel's equation
équation *f* **de van Deemter** van Deemter
  equation
équation *f* **d'Henderson** Henderson's equation
équation *f* **d'Onsager** Debye-Hückel-Onsager's
  limiting law for conductivity
équations *fpl* **normales** normal equations
équilibrage *m* **de la couche** layer equilibration
équilibre *m* **de Donnan** membrane equilibrium
équilibre *m* **de membrane** membrane
  equilibrium
équilibre *m* **métastable d'une électrode**
  metastable equilibrium of an electrode
équilibre *m* **radioactif séculaire** secular
  radioactive equilibrium
équilibre *m* **radioactif transitoire** transient
  radioactive equilibrium
équilibre *m* **séculaire** secular radioactive
  equilibrium
équilibre *m* **transitoire** transient radioactive
  equilibrium
équivalent *m* **chimique** chemical equivalent
équivalent *m* **électrochimique** electrochemical
  equivalent
équivalent-gramme *m* gram equivalent
erreur *f* **aberrante** gross error
erreur *f* **absolue** absolute error
erreur *f* **aléatoire** random error
erreur *f* **commise par l'opérateur** operative
  error
erreur *f* **d'échantillonnage** sampling error
erreur *m* **de méthode** error of method
erreur *f* **de première espèce** error of first kind
erreur *f* **de reproductibilité** reproducibility
  error

erreur *f* **de seconde espèce** error of second kind
erreur *f* **de sel** salt error
erreur *f* **d'étalonnage** scale error
erreur *f* **de titrage** titration error
erreur *f* **d'indicateur** indicator error
erreur *f* **d'instrument** instrumental error
erreur *f* **en pourcentage** percentage error
erreur *f* **relative** relative error
erreur *f* **systématique** systematic error
espace *m* **factoriel** experimental region
espace *m* **inter-électrode** analytical gap
espèce *f* **électro-active** electroactive substance
espérance *f* **mathématique** expected value
essai *m* test (*in chemical analysis*)
essai *m* **à blanc** blank test
essai *m* **à blanc de l'indicateur** indicator blank
essai *m* **de contrôle** check test
essai *m* **organoleptique** organoleptic assessment
essai *m* **par touches** spot test analysis
estimateur *m* estimator
estimateur *m* **convergent** consistent estimator
estimateur *m* **le plus efficient** most efficient estimator
estimateur *m* **sans biais** unbiased estimator
estimation *f* **intervalle** interval estimation
estimation *f* **par intervalle** interval estimation
estimation *f* **ponctuelle** point estimation
étalement *m* **de bande** band broadening
étalement *m* **de tache** spot broadening
étaleur *m* spreader
étalon *m* **externe** external standard
étalon *m* **interne** 1. internal standard 2. internal standard (*in spectrochemical analysis*)
étalonnage *m* standardization
étape *f* **du processus** elementary step
étendue *f* range
étincelle *f* electrical spark
évaporation *f* evaporation
événement *m* **aléatoire** random event
événement *m* **élémentaire** elementary event
exactitude *f* accuracy
examen *m* **de contrôle** check test
exciccateur *m* desiccator
excitation *f* excitation
exclusion *f* exclusion
exclusion *f* **d'ions** ion exclusion
expérience *f* experiment
expérience *f* **active** active experiment
expérience *f* **passive** passive experiment
exposant *m* **de l'indicateur** indicator exponent
exposition *f* radiant exposure
exposition *f* **énergétique** radiant exposure
extinction *f* absorbance
extractant *m* 1. extractant 2. extracting agent
extracteur *m* extractor
extracteur *m* **de Soxhlet** Soxhlet apparatus
extraction *f* extraction
extraction *f* **continue** continuous extraction
extraction *f* **en contre-courant** countercurrent extraction
extraction *f* **en retour** back-extraction
extraction *f* **liquide-liquide** liquid-liquid extraction
extraction *f* **périodique** periodic extraction
extrait *m* extract

F

facteur *m* factor
facteur *m* **d'absorption** absorptance
facteur *m* **d'analyse** analytical factor
facteur *m* **d'asymétrie du pic** asymmetry factor of a peak
facteur *m* **de capacité** mass distribution ratio
facteur *m* **de compressibilité** pressure-gradient correction factor
facteur *m* **de contraste** gamma
facteur *m* **de conversion titrimétrique** titrimetric conversion factor
facteur *m* **de correction de compressibilité** pressure-gradient correction factor
facteur *m* **de correction du gradient de pression** pressure-gradient correction factor
facteur *m* **de décontamination** decontamination factor
facteur *m* **de diffusion en retour** backscattering factor
facteur *m* **de dilution** dilution-correction factor
facteur *m* **de dissociation** apparent degree of dissociation
facteur *m* **de géométrie** geometry factor
facteur *m* **de James et Martin** pressure-gradient correction factor
facteur *m* **d'enrichissement** enrichment factor
facteur *m* **d'équivalence dans les réactions acide-base** equivalence factor in acid-base reactions
facteur *m* **d'équivalence dans les réactions de précipitation et de complexation** equivalence factor in precipitation and complex formation reactions
facteur *m* **d'équivalence dans les réactions d'oxydo-réduction** equivalence factor in redox reactions
facteur *m* **d'équivalence du réactif B** equivalence factor of component B
facteur *m* **de rugosité** roughness factor
facteur *m* **de saturation** saturation factor
facteur *m* **de sélectivité** potentiometric selectivity coefficient
facteur *m* **de séparation** separation factor
facteur *m* **de transmission** transmission factor
facteur *m* **de transmission interne** internal transmittance
faisceau *m* **moléculaire** molecular beam
fantômes *mpl* ghosts
F.e.m. **d'une pile** electromotive force of a galvanic cell
F.e.m. **normale** standard electromotive force
fente *f* slit
filament *m* **de Nernst** Nernst glower
filament *m* **de nichrome** coil of Nichrome wire
filtrat *m* filtrate
filtration *f* filtration
filtration *f* **sur gel** gel filtration
filtre *m* **à échelons** step filter
filtre *m* **à gaz** filter tube
filtre *m* **à plis** fluted filter
filtre *m* **d'absorption** absorption filter
filtre *m* **de lumière** light filter
filtre *m* **d'interférence** interference filter
filtre *m* **interférentiel** interference filter

filtre *m* **neutre** neutral-density filter
filtre *m* **optique** optical filter
filtre *m* **plissé** fluted filter
filtre *m* **pour rayons X** X-ray filter
filtres *mpl* **balancés** balanced filters
filtres *mpl* **équilibrés** balanced filters
fiole *f* **à iode** iodine flask
fiole *f* **jaugée** volumetric flask
fiole *f* **jaugée avec col gradué** volumetric flask
  with graduated neck
fission *f* fission
fission *f* **nucléaire** fission
flacon *m* **à tare** weighing bottle
flacon *m* **compte-gouttes** dropper
flacon *m* **d'iode** iodine flask
flacon *m* **laveur à plaque frittée** gas washing
  bottle with sintered head
flacon *m* **laveur de Drechsel** Drechsel gas
  washing bottle
flacon *m* **laveur de Friedrichs** Friedrichs gas
  washing bottle
flacon *m* **volumétrique** volumetric flask
flamme *f* flame
flamme *f* **de diffusion** diffusion flame
flamme *f* **gainée** shielded flame
flamme *f* **laminaire** laminar flame
flamme *f* **séparée** separated flame
flamme *f* **turbulente** turbulent flame
floculation *f* coagulation
fluorescence *f* fluorescence
fluorescence *f* **de résonance** resonance
  fluorescence
fluorescence *f* **X** X-ray fluorescence
fluorimètre *m* fluorimeter
fluorimètre *m* **à filtre** filter fluorimeter
fluorimétrie *f* fluorimetry
flux *m* flux
flux *m* **énérgétique** radiant flux
flux *m* **lumineux** luminous flux
fonction *f* **de distribution** distribution function
fonction *f* **de régression** regression function
fonction *f* **des observations de l'échantillon**
  statistic
fonction *f* **d'excitation** excitation function
fond *m* **Compton** Compton continuum
fond *m* **de rayonnement** background radiation
fondant *m* flux
force *f* **d'élution du solvant** solvent strength
force *f* **électromotrice d'une cellule galvanique**
  electromotive force of a galvanic cell
force *f* **électromotrice normale** standard
  electromotive force
force *f* **ionique** ionic strength of a solution
force *f* **ionique d'une solution** ionic strength of
  a solution
formalité *f* formality
forme *f* **acide d'un échangeur de cations** acid
  form of a cation exchanger
forme *f* **basique d'un échangeur d'anions** base
  form of an anion exchanger
forme *f* **saline d'un échangeur d'ions** salt form
  of an ion exchanger
formule *f* **de Bragg** Bragg's law
formule *f* **de Debye-Hückel** Debye-Hückel
  equation
formule *f* **de Nernst** Nernst equation

formule *f* **de Nikolski** Nikolsky equation
formule *f* **d'Henderson** Henderson's equation
four *m* **à moufle** muffle furnace
four *m* **à tube de carbone** graphite-tube furnace
four *m* **à tube de graphite** graphite-tube furnace
four *m* **à tubes** combustion furnace
four *m* **graphite** graphite-tube furnace
fraction *f* fraction
fraction *f* **de dissociation** apparent degree of
  dissociation
fraction *f* **de la phase stationnaire**
  stationary-phase fraction
fraction *f* **en masse** mass fraction of substance B
fraction *f* **en volume** volume fraction of
  substance B
fraction *f* **interstitielle** interstitial fraction
fraction *f* **volumique de la phase stationnaire**
  stationary-phase fraction
fragmentation *f* fragmentation
fragments *mpl* **de fission** fission fragments
fréquence *f* frequency
fréquence *f* **caractéristique** characteristic
  frequency
fréquence *f* **excitatrice** exciting frequency
front *m* **Compton** Compton edge
front *m* **de la phase mobile** mobile phase front
front *m* **de solvant** mobile phase front
front *m* **diffus** fronting
front *m* **diffus situé sur le pic avant le sommet**
  fronting
fuite *f* **de la phase stationnaire** bleeding
fusion *f* melting
fusion *f* **en bombe de Parr** Parr bomb fusion

G

galvanomètre *m* galvanometer
galvanostat *m* galvanostat
gamma *m* gamma
garnissage *m* **de colonne** column packing
gazométrie *f* gasometric analysis
gaz *m* **porteur** carrier gas
gaz *m* **vecteur** carrier gas
gel *m* gel
générateur *m* **de neutrons** neutron generator
générateur *m* **secondaire** electrical accumulator
génération *f* **électrochimique d'un réactif**
  electrochemical generation of a titrant
géométrie *f* **de comptage** counting geometry
germe *m* nucleus
ghosts *mpl* ghosts
globar *m* globar
gonflement *m* swelling
gonflement *m* **dans un solvant donné** weight
  swelling in solvent
gonflement *m* **relatif** volume swelling ratio
goniomètre *m* **à rayons X** X-ray goniometer
gradient *m* **de composition de la couche mince**
  gradient layer
gradient *m* **de composition du remplissage**
  gradient layer
gradient *m* **de flux de particules ou de photons**
  flux gradient of particles or photons
gradient *m* **d'élution** gradient elution
grandeur *f* **de rétention relative** relative retention
gravimétrie *f* gravimetric analysis

grillage *m* d'un précipité ignition of precipitate
grille *f* gate
groupe *m* analytical group
groupe *m* **analytique** analytical group
groupe *m* **chromophique** chromophore
groupe *m* **coordonnant** coordinating group
groupe *m* **fonctionnel** functional group
groupement *m* **acide** acidic group
groupement *m* **bathochrome** bathochrome
groupement *m* **fonctionnel** functional group
groupement *m* **hypsochrome** hypsochrome
groupements *mpl* **ionogènes** ionogenic groups

H

harmonique *m* overtone band
hauteur *f* de la colonne de mercure height of
   mercury column
hauteur *f* de palier step height
hauteur *f* de palier sur chromatogramme
   intégral step height
hauteur *f* de plateaux réduite reduced plate
   height
hauteur *f* du pic peak height
hauteur *f* équivalente à un plateau effectif height
   equivalent to an effective plate
hauteur *f* équivalente à un plateau théorique
   height equivalent to a theoretical plate
hauteur *f* équivalente à un plateau théorique
   effectif height equivalent to an effective plate
hétéropolyacide *m* heteropolyacid
hexaméthyldisilazane *m* hexamethyldisilazane
histogramme *m* histogram
Hopcalite *f* Hopcalite
hydrogel *m* hydrogel
hydromètre *m* hydrometer
hydrosol *m* hydrosol
hypersorption *f* hypersorption
hypothèse *f* alternative alternative hypothesis
hypothèse *f* nulle null hypothesis
hypothèse *f* statistique statistical hypothesis
hystérésis *f* hysteresis

I

identification *f* identification
image *f* radiographique X X-ray pattern
immunoélectrophorèse *f* immunoelectrophoresis
impédance *f* complexe impedance
impédance *f* électrique impedance
implantation *f* d'ions ion implantation
impureté *f* d'un précipité contamination of
   a precipitate
inactivation *f* deactivation
indicateur *m* acide-base acid-base indicator
indicateur *m* avec effet d'écran screened
   indicator
indicateur *m* bicolore two-colour indicator
indicateur *m* chimiluminescent chemiluminescent
   indicator
indicateur *m* d'adsorption adsorption indicator
indicateur *m* de pH acid-base indicator
indicateur *m* deux couleurs two-colour
   indicator
indicateur *m* d'oxydo-réduction
   oxidation-reduction indicator

indicateur *m* d'oxydo-réduction irréversible
   irreversible redox indicator
indicateur *m* d'oxydo-réduction réversible
   reversible redox indicator
indicateur *m* externe external indicator
indicateur *m* interne internal indicator
indicateur *m* isotopique isotopic tracer
indicateur *m* métallochrome metallochromic
   indicator
indicateur *m* mixte mixed indicator
indicateur *m* monocolore one-colour indicator
indicateur *m* par chimiluminescence
   chemiluminescent indicator
indicateur *m* par extraction extraction indicator
indicateur *m* par fluorescence fluorescent
   indicator
indicateur *m* par métallofluorescence
   metallofluorescent indicator
indicateur *m* par précipitation precipitation
   indicator
indicateur *m* radioactif radioactive tracer
indicateur *m* redox oxidation-reduction
   indicator
indicateur *m* universel universal indicator
indicateur *m* visuel indicator
indice *m* de Kováts retention index
indice *m* de réfraction refractive index
indice *m* de réfraction absolu refractive index
indice *m* de réfraction relatif relative refractive
   index
indice *m* de rétention retention index
indice *m* de sensibilité de Sandell Sandell's
   sensitivity index
indice *m* d'iode iodine number
indices *mpl* de Miller Miller indices
infrarouge *m* infrared radiation
infrarouge *m* lointain far infrared
infrarouge *m* moyen middle infrared
infrarouge *m* proche near infrared
injecteur *m* à dérivation by-pass injector
injecteur *m* avec by-pass by-pass injector
injecteur *m* d'échantillon sample injector
intensité *f* de courant électrique electric current
intensité *f* de rayonnement X X-radiation
   intensity
intensité *f* lumineuse luminous intensity
interface *f* interface
interférence *f* de deuxième catégorie secondary
   interfering reaction
interféromètre *m* optique optical interferometer
interférométrie *f* interferometry
interpolation *f* interpolation
intervalle *m* de confiance confidence interval
intervalle *m* de transition transition interval
intervalle *m* de virage 1. colour change interval
   2. transition interval
iodimétrie *f* iodimetry
iodométrie *f* iodimetry
ion *m* ampholyte ampholyte ion
ion *m* amphotère ampholyte ion
ion *m* fils daughter ion
ion *m* fragment fragment ion
ion *m* hermaphrodite ampholyte ion
ionisation *f* chimique chemical ionization
ionisation *f* par champ field ionization

ionisation *f* **par impact électronique**
electron-impact ionization
ionisation *f* **par une source à étincelles** spark
source ionization
ionisation *f* **superficielle** surface ionization
ionisation *f* **thermique** thermal ionization
ion *m* **métastable** metastable ion
ion *m* **moléculaire** molecular ion
ion *m* **négatif** negative ion
ion-parent *m* parent ion
ion *m* **positif** positive ion
ions *mpl* **fixes** fixed ions
ions *mpl* **interférants** interfering ions
ions *mpl* **retenus** fixed ions
irradiance *f* irradiance
isobare *f* **d'adsorption** adsorption isobar
isochrome *f* isochrome
isopolyacide *m* isopolyacid
isostère *f* **d'adsorption** adsorption isostere
isotachophorèse *f* isotachophoresis
isotherme *f* isotherm
isotherme *f* **d'adsorption** adsorption isotherm
isotherme *f* **d'échange d'ions** ion-exchange
isotherm
isotherme *f* **de distribution** distribution isotherm
isotherme *f* **de partage** partition isotherm
isotherme *f* **de sorption** sorption isotherm

J

jet *m* **de plasma** plasma jet
justesse *f* accuracy

K

kayser *m* kayser
kieselguhr *m* diatomaceous earth
kryptonate *m* **radioactif** radioactive kryptonate

L

laboratoire *m* **chaud** hot laboratory
laboratoire *m* **de haute activité** hot laboratory
lampe *f* **à cathode creuse** hollow-cathode lamp
lampe *f* **à décharge** discharge lamp
lampe *f* **à deutérium** deuterium discharge lamp
lampe *f* **à excitation haute fréquence sans
électrode** electrodeless discharge lamp
lampe *f* **à filament au tungstène** tungsten
incandescent lamp
lampe *f* **à hydrogène** hydrogen discharge lamp
lampe *f* **à incandescence** incandescent lamp
lampe *f* **à tube au tungstène** tungsten
incandescent lamp
lampe *f* **à vapeur de métal** metal-vapour lamp
lampe *f* **Nernst** Nernst glower
largeur *f* **à mi-intensité** half-intensity width
largeur *f* **apparente à mi-absorption** apparent
half-width
largeur *f* **d'une raie** line width
largeur *f* **du pic à demi-hauteur** peak width at
half-height
largeur *f* **du pic à la base** peak width at base
largeur *f* **du pic à mi-hauteur** peak width at
half-height

largeur *f* **naturelle d'une raie spectrale** natural
line-width
largeur *f* **vraie** true half-width
largeur *f* **vraie à mi-absorption** true half-width
laser *m* laser
laser *m* **accordable** tunable laser
laser *m* **à colorant** dye laser
latitude *f* **d'exposition** latitude
lavage *m* **du précipité** washing of precipitate
lessivage *m* leaching
ligne *f* **de base** base-line
ligne *f* **de départ** starting line
ligne *f* **de Rayleigh** Rayleigh line
limite *f* **de décèlement** limit of detection
limite *f* **de détection** limit of detection
limite *f* **de détermination** limit of determination
limite *f* **de dilution** dilution limit
limite *f* **de dosage** limit of determination
limite *f* **de Duane-Hunt** short-wavelength limit
limite *f* **d'exclusion** exclusion limit
limites *fpl* **de confiance** confidence limits
linéarisation *f* linearization
lit *m* **chromatographique** chromatographic bed
lit *m* **mixte** mixed bed
loi *f* **de Beer** Beer's law
loi *f* **de Bouguer** Lambert's law
loi *f* **de Bouguer-Beer** Bouguer-Lambert-Beer
law
loi *f* **de Bouguer-Lambert** Lambert's law
loi *f* **de Bragg** Bragg's law
loi *f* **de désintégration radioactive** law of
radioactive decay
loi *f* **de dilution d'Ostwald** Ostwald dilution law
loi *f* **de distribution** Nernst distribution law
loi *f* **de distribution homogène** homogeneous
distribution law
loi *f* **de distribution logarithmique** logarithmic
distribution law
loi *f* **de Hooke** Hooke's law
loi *f* **de Kohlrausch** Kohlrausch's additive law
loi *f* **de Lambert** Lambert's law
loi *f* **de Lambert-Beer** Bouguer-Lambert-Beer
law
loi *f* **de l'indépendance de la migration des ions**
Kohlrausch's additive law
loi *f* **de Moseley** Moseley law
loi *m* **de Nernst** Nernst distribution law
loi *f* **de partage de Nernst** Nernst distribution
law
loi *f* **de sommation des erreurs** variance
propagation
loi *f* **limite de Debye et Hückel** Debye-Hückel
limiting law
loi *f* **limite de Debye' Hückel et Onsager**
Debye-Hückel-Onsager's limiting law for
conductivity
lois *fpl* **de Faraday** Faraday's laws
lois *fpl* **de Fick** Fick's laws
longueur *f* **d'onde** wavelength
longueur *f* **d'onde analytique** analytical
wavelength
longueur *f* **d'onde de blaze** blaze wavelength
longueur *f* **du trajet d'absorption** path length
lumière *f* light
lumière *f* **à polarisation circulaire** circularly
polarized light

lumière f **à polarisation circulaire droite**
right-handed circularly polarized light
lumière f **à polarisation circulaire gauche**
left-handed circularly polarized light
lumière f **à polarisation elliptique** elliptically
polarized light
lumière f **à polarisation rectiligne** plane-
polarized light
lumière f **blanche** white light
lumière f **polarisée** polarized light
lumière f **polarisée circulairement** circularly
polarized light
lumière f **polarisée elliptiquement** elliptically
polarized light
lumière f **rectilignement polarisée**
plane-polarized light
luminance f luminance
luminescence f luminescence

M

macro-analyse f macroanalysis
macroconstituant m macro-component
main f **pour poudres** weighing scoop
manganométrie f manganometry
marqueur m marker
masquage m **d'ions** masking of ions
masse f **atomique relative** relative atomic mass
masse f **au repos** rest mass
masse f **constante** constant mass
masse f **molaire** molar mass
masse f **moléculaire relative** relative molecular
mass
matériau m **de référence** 1. reference material
2. reference sample
matrice f matrix
matrice f **d'échangeur d'ions** ion exchanger
matrix
matrice f **de corrélation** correlation matrix
matrice f **de covariance** covariance matrix
matrice f **de dispersion** covariance matrix
maxima mpl **polarographiques** polarographic
maxima
maximum m **du pic** peak maximum
maximums mpl **polarographiques** polarographic
maxima
mécanisme m **de fragmentation** fragmentation
pattern
médiane f **de l'échantillon** sample median
mélange m **tampon** buffer solution
membrane f **à échangeur d'ions liquide** liquid
ion exchanger membrane
membrane f **cristalline** crystalline membrane
membrane f **échangeuse d'ions** ion-exchange
membrane
membrane f **hétérogène** heterogeneous
membrane
membrane f **homogène** homogeneous membrane
membrane f **liquide** liquid membrane
membrane f **perméable aux gaz** gas-permeable
membrane
membrane f **sélective** ion-selective membrane
membrane f **solide** solid-state membrane
mémoire f **de l'électrode** hysteresis
mercurimétrie f mercurimetry
mercurométrie f mercurometry

mésoanalyse f mesoanalysis
mesure f **d'admittance non faradaïque**
measurement of nonfaradaic admittance
métal-indicateurs mpl metal indicators
méthode f **à étalon interne** standard addition
method
méthode f **à ligne de base** base-line method
méthode f **analytique** analytical method
méthode f **ascendante** ascending development
méthode f **centigrammique** centigram method
méthode f **coulostatique** coulostatic method
méthode f **d'analyse cinétique** kinetic analysis
méthode f **d'angle de prise variable** variable
take-off angle method
méthode f **d'anticoïncidences** anticoincidence
technique
méthode f **de balayage de fréquence**
frequency-sweep method
méthode f **de balayage du champ magnétique**
field-sweep method
méthode f **de calcination** ashing technique
méthode f **de Carius** Carius method
méthode f **d'échange des réactifs radioactifs**
radio-exchange method
méthode f **de Clark** Clark's method
méthode f **de combustion en chalumeau
oxhydrique** oxy-hydrogen flame method
méthode f **de comparaison** reference method
méthode f **de compensation de Poggendorff**
Poggendorff's compensation method
méthode f **de comptage de coïncidences**
coincidence technique
méthode f **de concentration** concentration
method
méthode f **de Covell** Covell's method
méthode f **de dégagement de radioactivité**
radio-release method
méthode f **de dilution isotopique** isotope
dilution analysis
méthode f **de Dumas** Dumas method
méthode f **de Dumas-Pregl** Dumas-Pregl
method
méthode f **de fusion** fusion method
méthode f **de Gran** Gran's plot
méthode f **de Kelvin** Kelvin method
méthode f **de Kjeldahl** Kjeldahl method
méthode f **de l'étalon interne** internal-standard
method
méthode f **de Liebig** Liebig's method
méthode f **d'émanation** emanometric method
méthode f **de marquage radioactif** radioactive
tracer method
méthode f **de micro-onde réflechie** reflected
wave method
méthode f **de micro-onde transmise**
transmitted wave method
méthode f **de Mohr** Mohr's method
méthode f **d'enrichissement** concentration
method
méthode f **de Rast** Rast method
méthode f **de régression multiple** multiple
regression method
méthode f **des additions connues** standard
addition method
méthode f **descendante** descending development
méthode f **de Schöniger** Schöniger method

méthode *f* des **coefficients d'influence** empirical
coefficient method
méthode *f* des **moindres carrés** least-squares
method
méthode *f* des **paramètres fondamentaux**
fundamental parameter method
méthode *f* des **radioéléments traceurs**
radioactive tracer method
méthode *f* d'**étalonnage par ajouts dosés**
standard addition method
méthode *f* de **Volhard** Volhard method for
determination of halides
méthode *f* de **Volhard de dosage des halogénures**
Volhard method for determination of halides
méthode *f* de **Volhard de dosage du manganèse**
Volhard method for determination of
manganese
méthode *f* d'**incinération** ashing technique
méthode *f* **directe** direct method
méthode *f* d'**oscillations** method of swings
méthode *f* du **maximum de vraisemblance**
maximum likelihood method
méthode *f* d'**Unterzaucher de dosage d'oxygène**
Unterzaucher method for oxygen
determination
méthode *f* **galvanostatique** galvanostatic method
méthode *f* **indirecte** indirect method
méthode *f* **microgrammique** microgram method
méthode *f* **milligrammique** milligram method
méthode *f* **nanogrammique** nanogram method
méthode *f* **normalisée** standardized method
méthode *f* **picogrammique** picogram method
méthode *f* **potentiostatique** potentiostatic
method
méthodes *fpl* de **relaxation** transient methods
méthodes *fpl* **stationnaires** steady-state methods
micro-analyse *f* microanalysis
microazotimètre *m* micronitrometer
microburette *f* microburette
microcomposant *m* micro-component
microconstituant *m* micro-component
microcoulométrie *f* polarographic coulometry
microcristalloscopie *f* microcrystalloscopic
analysis
microméthode *f* milligram method
micronitromètre *m* micronitrometer
micro-ondes *fpl* microwaves
microphotomètre *m* microphotometer
micropipette *f* micropipette
microscope *m* **électronique à balayage** scanning
electron microscope
microscopie *f* d'**électrons Auger** Auger-electron
microscopy
microscopie *f* des **photoélectrons** photoelectron
microscopy
microscopie *f* d'**ions secondaires** secondary ion
imaging mass spectroscopy
microsonde *f* de **Castaing électronique** electron
microprobe X-ray analyser
microsonde *f* **électronique** electron microprobe
X-ray analyser
microtraces *fpl* microtraces
migration *f* **bathochrome** bathochromic shift
migration *f* des **ions** migration of ions
migration *f* **hypsochrome** hypsochromic shift
mi-largeur *f* half-intensity width

millicoulométrie *f* polarographic coulometry
minéralisation *f* mineralization
mobilité *f* de **l'ion B** electric mobility of ion **B**
mobilité *f* **électrique de l'ion B** electric
mobility of ion **B**
mobilité *f* **ionique** electric mobility of ion **B**
mode *m* mode
modèle *m* **I** fixed effects model
modèle *m* **II** random effects model
modèle *m* à **effets aléatoires** random effects
model
modèle *m* à **effets fixes** fixed effects model
modération *f* slowing down
molalité *f* molality
molarité *f* amount-of-substance concentration
of substance **B**
mole *f* **1.** gram molecule **2.** mole
molécule *f* **chélatée** chelate compound
molécule-gramme *m* gram molecule
monochromateur *m* monochromator
monochromateur *m* d'**émission** emission
monochromator
monochromateur *m* d'**excitation** excitation
monochromator
monocouche *f* monolayer
mouvement *m* **propre** background
moyenne *f* **arithmétique** arithmetic mean
moyenne *f* **géométrique** geometric mean
multiplet *m* multiplet
multiplicateur *m* **électronique** electron
multiplier
mûrissement *m* d'**Ostwald** Ostwald ripening

N

nacelle *f* à **combustion** combustion boat
nanotraces *fpl* nanotraces
nébulisation *f* nebulization
nébuliseur *m* nebulizer
nébuliseur *m* à **ultrasons** ultrasonic nebulizer
nébuliseur *m* **pneumatique** pneumatic nebulizer
néphélomètre *m* nephelometer
néphélométrie *f* nephelometry
nettoyage *m* **par plasma** plasma cleaning
neutralisation *f* neutralization
neutrons *mpl* de **fission** fission neutrons
neutrons *mpl* de **résonance** resonance neutrons
neutrons *mpl* **épicadmiques** epicadmium
neutrons
neutrons *mpl* **épicadmium** epicadmium neutrons
neutrons *mpl* **épithermiques** epithermal neutrons
neutrons *mpl* **froids** cold neutrons
neutrons *mpl* **instantanés** prompt neutrons
neutrons *mpl* **lents** slow neutrons
neutrons *mpl* **rapides** fast neutrons
neutrons *mpl* **retardés** delayed neutrons
neutrons *mpl* **thermiques** thermal neutrons
nitromètre *m* nitrometer
niveau *m* de **bruit** noise level
niveau *m* de **confiance** confidence level
niveau *m* de **signification** significance level
noircissement *m* blackening
nombre *m* de **plateaux effectifs** number of
effective plates
nombre *m* de **plateaux théoriques** number of
theoretical plates

nombre *m* de plateaux théoriques effectifs number of effective plates

nombre *m* des degrés de liberté number of degrees of freedom

nombre *m* de transport de l'ion B transport number of ionic species B

nombre *m* d'ondes wave number

normalité *f* normality

nuage *m* ionique ionic atmosphere

nucléation *f* nucleation

nucléation *f* hétérogène heterogeneous nucleation

nucléation *f* homogène homogeneous nucleation

nucléation *f* simultanée simultaneous nucleation

nuclide *m* père parent nuclide

nujol *m* Nujol

O

occlusion *f* occlusion

odorimétrie *f* odorimetry

optimisation *f* optimization

ordre *m* d'une réaction order of reaction with respect to a particular substance

ordre *m* d'un spectre spectral order

ordre *m* global d'une réaction order of reaction

ordre *m* partiel d'une réaction order of reaction with respect to a given substance

oscillométrie *f* high-frequency conductometry

oscillopolarographe *m* oscillopolarograph

oxydant *m* oxidant

oxydation *f* oxidation

oxydimétrie *f* oxidimetry

P

paire *f* de raies de fixation fixation pair

palier *m* step on an integral chromatogram

palier *m* sur un chromatogramme intégral step on an integral chromatogram

panache *m* secondary combustion zone

papier-filtre *m* filter paper

papier *m* réactif indicator paper

paramètre *m* parameter

paramètre *m* de solubilité solubility parameter

paramètres *mpl* réduits reduced parameters

parent *m* nucléaire parent nuclide

partie *f* de matériau bulk of material

parties *fpl* par million parts per million

passivation *f* d'une électrode passivation of an electrode

pastillage *m* pellet technique

pastille *f* de bromure de potassium potassium bromide disk

pente *f* nernstienne Nernstian slope

peptisation *f* peptization

percée *f* breakthrough

performance *f* d'une colonne column performance

période *f* radioactive half-life period

perle *f* au borax borax bead

perle *f* de métaphosphate de sodium microcosmic salt bead

perméation *f* sélective permselectivity

permittivité *f* permittivity

permittivité *f* absolue permittivity

permittivité *f* relative relative permittivity

pesée *f* weighed amount

pèse-substance *m* forme cochonnet piggie

pH pH

phase *f* immobile stationary phase

phase *f* liquide stationary liquid phase

phase *f* mobile mobile phase

phase *f* stationnaire stationary phase

phase *f* stationnaire greffée chemically bonded stationary phase

phase *f* stationnaire liquide stationary liquid phase

phénomènes *mpl* électrocapillaires electrocapillary phenomena

phénomènes *mpl* électrocinétiques electrokinetic phenomena

pH-mètre *m* pH-meter

pH-metrie *f* pH-metry

phosphorescence *f* phosphorescence

phosphorimétrie *f* phosphorimetry

photoactivation *f* photoactivation

photocellule *f* à couche d'arrêt photovoltaic cell

photocolorimètre *m* photocolorimeter

photoconducteur *m* photoconductor

photodésorption *f* photodesorption

photodétecteur *m* photodetector

photodiode *f* photodiode

photoélectron *m* photoelectron

photofraction *f* photofraction

photoionisation *f* photoionization

photo-ligne *f* photoelectric peak

photoluminescence *f* photoluminescence

photomètre *m* photometer

photomètre *m* de flamme flame photometer

photométrie *f* photometry

photométrie *f* de flamme flame emission spectroscopy

photomultiplicateur *m* photomultiplier tube

photon *m* photon

photoneutron *m* photoneutron

photopic *m* photoelectric peak

photopile *f* photovoltaic cell

photopolarographie *f* photopolarography

phototransistor *m* phototransistor

physisorption *f* physisorption

pic *m* peak

pic *m* d'absorption totale full energy peak

pic *m* de base base peak

pic *m* d'échappement pair escape peak

pic *m* de deuxième échappement double escape peak

pic *m* de fuite pair escape peak

pic *m* de fuite des photons X-ray escape peak

pic *m* de l'air air peak

pic *m* de premier échappement single escape peak

pic *m* de résonance resonance peak

pic *m* de rétrodiffusion backscatter peak

pic *m* de somme sum peak

pic *m* d'ion métastable metastable ion peak

pic *m* métastable metastable ion peak

pic *m* moléculaire molecular peak

picnomètre *m* pycnometer

pic *m* non résolu unresolved peak

picotraces *fpl* picotraces

pic *m* photoélectrique photoelectric peak

piégeage *m* **mécanique** mechanical entrapment
pile *f* **à mercure** mercury dry cell
pile *f* **avec transport** cell with transference
pile *f* **chimique** chemical cell
pile *f* **de concentration** concentration cell
pile *f* **de Leclanché** Leclanché cell
pile *f* **de Weston** Weston standard cell
pile *f* **de Weston insaturée** unsaturated Weston cell
pile *f* **électrique** galvanic cell
pile *f* **électrochimique** electrochemical cell
pile *f* **étalon** standard cell
pile *f* **étalon Weston** Weston standard cell
pile *f* **galvanique** galvanic cell
pile *f* **primaire** primary cell
pile *f* **réversible** reversible galvanic cell
pile *f* **sans transport** cell without transference
pile *f* **sèche à mercure** mercury dry cell
pile *f* **secondaire** electrical accumulator
pile *f* **standard Weston** Weston standard cell
pile *f* **thermoélectrique** galvanic thermocell
pipette *f* pipette
pipette *f* **à zéro automatique** automatic zero pipette
pipette *f* **courante à un trait** transfer pipette
pipette *f* **graduée** graduated pipette
pipette *f* **gravimétrique de Lunge-Rey** Lunge-Rey weighing pipette
pipette *f* **pour les gaz** gas sampling pipette
plan *m* **de polarisation** plane of polarization
plan *m* **de premier degré** first order design
plan *m* **de vibration** plane of vibration
plan *m* **d'expérience** exact design
planification *f* **d'expérience** experimental design
plan *m* **de premier ordre** first order design
plan *m* **de second ordre** second order design
plan *m* **d'experience complete** complete factorial design
plan *m* **factoriel complet** complete factorial design
plan *m* **orthogonal** orthogonal design
plan *m* **saturé** saturated design
plan *m* **symplexe** simplex design
plaque *f* **à godets** spotting plate
plaque *f* **à godets pour essai par touches** spotting plate
plaque *f* **photographique** photographic plate
plaque *f* **support** support plate
plasma *m* plasma
plasma *m* **à couplage inductif** inductively coupled plasma
plasma *m* **à haute fréquence** radiofrequency plasma
plasma *m* **induit par haute fréquence** inductively coupled plasma
plasma *m* **micro-onde** microwave plasma
plateau *m* **du compteur** plateau of a counter
platinisation *f* **des électrodes** platinization of electrodes
pM pM
pOH pOH
poids *mpl* **analytiques** analytical weights
point *m* **de départ** starting line
point *m* **de fin de titrage** end-point of titration
point *m* **d'équivalence** equivalence-point

point *m* **d'expérience** experimental point
point *m* **isobestique** isosbestic point
point *m* **iso-électrique** isoelectric point
point *m* **isopotentiel** isopotential point
points *mpl* **d'inflexion** inflection points of a peak
points *mpl* **d'inflexion du pic** inflection points of a peak
point *m* **stœchiométrique** equivalence-point
point *m* **théorique de fin de titrage** equivalence-point
polarimètre *m* polarimeter
polarimétrie *f* polarimetry
polarisation *f* **circulaire** circular polarization
polarisation *f* **de la lumière** polarization of light
polarisation *f* **d'électrode** polarization of electrode
polarisation *f* **elliptique** elliptical polarization
polarisation *f* **rectiligne** plane polarization
polarogramme *m* polarogram
polarographe *m* polarograph
polarographie *f* polarography
polarographie *f* **à balayage linéaire de courant** current-scanning polarography
polarographie *f* **à balayages multiples** multisweep polarography
polarographie *f* **à balayage triangulaire** triangular-wave polarography
polarographie *f* **à balayage triangulaire cyclique** cyclic triangular-wave polarography
polarographie *f* **à courant alternatif imposé** alternating voltage polarography
polarographie *f* **à créneaux de potentiel** square-wave polarography
polarographie *f* **à double tonalité** double-tone polarography
polarographie *f* **à impulsions** pulse polarography
polarographie *f* **à incrément de charge** incremental-charge polarography
polarographie *f* **à onde carrée** square-wave polarography
polarographie *f* **à sauts de charge** incremental-charge polarography
polarographie *f* **à simple balayage** single-sweep polarography
polarographie *f* **à tension alternative** 1. conventional alternating-current polarography 2. alternating-current polarography
polarographie *f* **à tension alternative surimposée par détection d'harmonique supérieur** higher-harmonic ac polarography
polarographie *f* **à tension alternative surimposée par détection d'harmonique supérieur et redressement de phase** higher-harmonic ac polarography with phase sensitive rectification
polarographie *f* **à tension alternative surimposée par détection du second harmonique** second-harmonic ac polarography
polarographie *f* **à tension carrée** square-wave polarography
polarographie *f* **à tension périodique surimposée** alternating-current polarography
polarographie *f* **à tension sinusoïdale surimposée** conventional alternating-current polarography

polarographie *f* **aux radiofréquences** radio-frequency polarography
polarographie *f* **avec modulation** modulation polarography
polarographie *f* **cc** direct current polarography
polarographie *f* **de Kalousek** Kalousek polarography
polarographie *f* **dérivée** derivative polarography
polarographie *f* **dérivée à impulsions** derivative pulse polarography
polarographie *f* **DF** demodulation polarography
polarographie *f* **différentielle** differential polarography
polarographie *f* **différentielle à impulsions** differential pulse polarography
polarographie *f* **en courant continu** direct current polarography
polarographie *f* **en système Tast** Tast polarography
polarographie *f* **oscillographique** oscillopolarography
polarographie *f* **oscillographique à simple balayage** single-sweep polarography
polarographie *f* **par décharge** incremental-charge polarography
polarographie *f* **par démodulation faradaïque** demodulation polarography
polarographie *f* **pas à pas** staircase polarography
polarographie *f* **RF** radio-frequency polarography
polarographie *f* **sur écran cathodique** single-sweep polarography
polyacide *m* polyacid
polyélectrolyte *m* polyelectrolyte
polymère *m* **redox** redox polymer
pont *m* **de Wheatstone** Wheatstone bridge
pont *m* **électrolytique** salt bridge
population *f* population
population *f* **générale** population
Porapak *m* Porapak
porode *f* porous cup electrode
porosité *f* **interparticulaire** interstitial fraction
porteur *m* carrier
porteur *m* **isotopique** isotopic carrier
pose *f* exposure time
position *f* **de repos** rest point
postprécipitation *f* postprecipitation
potentiel *m* **absolu d'électrode** absolute electrode potential
potentiel *m* **apparent d'electrode** conditional electrode potential
potentiel *m* **chimique du constituant B dans la phase** α chemical potential of species B in phase α
potentiel *m* **de charge au zéro** potential at the point of zero charge
potentiel *m* **de corrosion** corrosion potential
potentiel *m* **de décomposition** decomposition potential
potentiel *m* **d'écoulement** streaming potential
potentiel *m* **de demi-onde** half-wave potential
potentiel *m* **de demi-palier** half-wave potential
potentiel *m* **de demi-vague** half-wave potential
potentiel *m* **de diffusion** diffusion potential
potentiel *m* **de Galvani** inner electric potential of phase β

potentiel *m* **de jonction liquide** liquid-junction potential
potentiel *m* **d'électrode** electrode potential
potentiel *m* **d'électrode à l'équilibre** equilibrium electrode potential
potentiel *m* **de membrane** membrane potential
potentiel *m* **de précipitation** deposition potential
potentiel *m* **d'équilibre** equilibrium electrode potential
potentiel *m* **de repos** open-circuit electrode potential
potentiel *m* **de sédimentation** sedimentation potential
potentiel *m* **de thermodiffusion** thermodiffusion potential
potentiel *m* **d'oxydo-réduction** oxidation-reduction potential
potentiel *m* **électrique de surface de la phase** β surface electric potential of phase β
potentiel *m* **électrique extérieur** outer electric potential of phase β
potentiel *m* **électrique externe de la phase** β outer electric potential of phase β
potentiel *m* **électrique intérieur** inner electric potential of phase β
potentiel *m* **électrique interne de la phase** β inner electric potential of phase β
potentiel *m* **électrochimique de l'espèce ionique B dans la phase** α electrochemical potential of ionic component B in phase α
potentiel *m* **électrocinétique** electrokinetic potential
potentiel *m* **en circuit ouvert** open-circuit electrode potential
potentiel *m* **formel** conditional electrode potential
potentiel *m* **formel d'électrode** conditional electrode potential
potentiel *m* **mixte** mixed polyelectrode potential
potentiel *m* **normal d'électrode** standard electrode potential
potentiel *m* **redox** oxidation-reduction potential
potentiel *m* **relatif d'électrode** electrode potential
potentiel *m* **standard d'électrode** standard electrode potential
potentiel *m* **Volta** outer electric potential of phase β
potentiel *m* **zêta** electrokinetic potential
potentiomètre *m* potentiometer
potentiométrie *f* potentiometry
potentiométrie *f* **à courant imposé** controlled-current potentiometry with one indicator electrode
potentiométrie *f* **à courant imposé à deux électrodes indicatrices** controlled-current potentiometry with two indicator electrodes
potentiométrie *f* **à courant nul** potentiometry
potentiométrie *f* **à intensité constante** controlled-current potentiometry with one indicator electrode
potentiométrie *f* **à intensité nulle** potentiometry
potentiométrie *f* **de précision par compensation** differential potentiometry
potentiométrie *f* **différentielle** differential potentiometry

potentiostat *m* potentiostat
pourcentage *m* **de titrage** per cent titrated
pourcentage *m* **d'extraction** extraction percentage
pouvoir *m* **de résolution** resolution
pouvoir *m* **de résolution de cristal** crystal resolving power
pouvoir *m* **résolvant** resolution
pouvoir *m* **rotatoire naturel** optical rotation
pouvoir *m* **rotatoire spécifique** specific rotation
précipitation *f* precipitation
précipitation *f* **en milieu homogène** precipitation from homogeneous solution
précipitation *f* **fractionnée** fractional precipitation
précipitation *f* **homogène** precipitation from homogeneous solution
précipitation *f* **quantitative** quantitative precipitation
précipité *m* precipitate
précipité *m* **colloïdal** colloid precipitate
précipité *m* **cristallin** crystalline precipitate
précision *f* precision
précision *f* **d'une balance** precision of a balance
précision *f* **d'une pesée** precision of a weighing
prélèvement *m* **élémentaire** primary sample
première harmonique *m* first overtone band
pression *f* **de gonflement** swelling pressure
prise *f* **d'essai** analytical sample
prisme *m* optical prism
probabilité *f* probability
probabilité *f* **de collage** sticking probability
probabilité *f* **de fixation** sticking probability
probabilité *f* **statistique** statistical probability
procédé *m* **à sec** dry method
procédé *m* **de Pregl de microanalyse élémentaire** Pregl procedure
procédé *m* **par voie humide** wet method
procédé *m* **par voie sèche** dry method
processus *m* **contrôlé par diffusion** diffusion controlled process
processus *m* **contrôlé par le transfert de charge** electroactivation control
processus *m* **contrôlé par une réaction chimique** kinetic controlled process
processus *m* **d'électrode** electrode process
processus *m* **d'électrode à l'état stationnaire** stationary electrode process
processus *m* **de transfert de charge** charge-transfer step
produit *m* **de filiation** daughter nuclide
produit *m* **de solubilité** solubility product
profil *m* **de profondeur** depth profile
profondeur *f* **effective** effective depth
programmation *f* **de débit** flow-programmed chromatography
programmation *f* **de solvant** gradient elution
puissance *f* **d'un test** power of a test
puissance *f* **rayonnante** radiant flux
pulvérisation *f* sputtering
pulvérisation *f* **préférentielle** preferential sputtering
pulvérisation *f* **réactive** reactive sputtering
pureté *f* **d'un réactif** purity of a reagent
pureté *f* **nucléaire** nuclear purity
pureté *f* **optique** optical purity

pureté *f* **radiochimique** radiochemical purity
pycnomètre *m* pycnometer
pyrolyse *f* pyrolysis
pyrolyse *f* **par flash** flash pyrolysis
pyrolyseur *m* pyrolyzer

Q

quadripole *m* quadrupole
quadrupole *m* quadrupole
quantilé *m* **de distribution de probabilité** quantile of a probability distribution
quantile *m* **d'orde** *p* **de distribution de probabilité** quantile of a probability distribution
quantité *f* **de lumière** luminous energy
quantité *f* **de matière** amount of substance
quantum *m* γ gamma quantum
quenching *m* quenching

R

radiation *f* radiation
radiation *f* **électromagnétique** electromagnetic radiation
radiation *f* **moléculaire** molecular radiation
radiation *f* **monochromatique** monochromatic radiation
radiation *f* **polychromatique** polychromatic radiation
radiation *f* **ultraviolette** ultraviolet light
radiation *f* **visible** light
radioactivité *f* radioactivity
radiochromatographie *f* radiochromatography
radioélectrophorèse *f* radioelectrophoresis
radioisotope *m* **à longue période** long-lived radioisotope
radioisotope *m* **de période courte** short-lived radioisotope
radioisotope *m* **de période longue** long-lived radioisotope
radioluminescence *f* radioluminescence
radiométrie *f* radiometry
raie *f* **d'analyse** analytical line
raie *f* **de l'élément de référence interne** internal reference line
raie *f* **de l'étalon interne** internal reference line
raie *f* **de Rayleigh** Rayleigh line
raie *f* **excitatrice** 1. exciting line 2. Rayleigh line
raies *fpl* **antistokes** anti-Stokes lines
raies *fpl* **de Raman** Raman lines
raies *fpl* **de Stokes** Stokes lines
raies *fpl* **homologues** homologous lines
raie *f* **spectrale** spectral line
raies *fpl* **persistantes** raies ultimes
raies *fpl* **ultimes** raies ultimes
raie *f* **zéro** Q-branch
ralentissement *m* slowing down
ramification *f* branching decay
randomisation *f* randomization
rapport *m* **cadmique** cadmium ratio
rapport *m* **de distribution** 1. concentration distribution ratio (*in chromatography*) 2. concentration distribution ratio (*in extraction*)

rapport *m* de distribution de concentration
1. concentration distribution ratio
(*in chromatography*) **2.** concentration
distribution ratio (*in extraction*)
rapport *m* de gonflement volume swelling ratio
rapport *m* de partage mass distribution ratio
rapport *m* de phase phase ratio
rapport *m* de sursaturation supersaturation ratio
rapport *m* de sursaturation critique critical
supersaturation ratio
rapport *m* de volume de gonflement volume
swelling ratio
rapport *m* du pic photoélectrique au Compton
peak-to-Compton ratio
rapport *m* masse/charge mass-to-charge ratio
rapport *m* masse sur charge mass-to-charge
ratio
rapport *m* signal-bruit signal-to-noise ratio
rayon *m* de l'atmosphère ionique radius of ionic
atmosphere
rayon *m* ionique mean ionic diameter
rayonnement *m* γ gamma radiation
rayonnement *m* calorifique infrared radiation
rayonnement *m* corpusculaire corpulscular
radiation
rayonnement *m* de Cerenkov Cerenkov radiation
rayonnement *m* de fluorescence X X-ray
fluorescence radiation
rayonnement *m* de freinage bremsstrahlung
rayonnement *m* électromagnétique
electromagnetic radiation
rayonnement *m* infrarouge infrared radiation
rayonnement *m* γ instantané prompt gamma
radiation
rayonnement *m* ionisant ionizing radiation
rayonnement *m* visible light
rayonnement *m* X X-radiation
rayonnement *m* X caractéristique
characteristic X-radiation
rayons *mpl* X X-radiation
rayons *mpl* X primaires primary X-radiation
rayons *mpl* X secondaires secondary X-radiation
réabsorption *f* self-absorption
réactance *f* électrique reactance
réactif *m* reagent
réactif *m* analytique analytical reagent
réactif *m* à silylation silylation reagent
réactif *m* chimiquement pur chemically pure
reagent
réactif *m* de Fehling Fehling's reagent
réactif *m* de Fischer Karl Fischer reagent
réactif *m* de groupe group reagent
réactif *m* démasquant demasking agent
réactif *m* de Nessler Nessler reagent
réactif *m* de Tschugaeff Chugaev reagent
réactif *m* micro-analytique microanalytical
reagent
réactif *m* précipitant precipitating agent
réactif *m* pur pure reagent
réactif *m* radiométrique radioactive reagent
réactif *m* sélectif selective reagent
réactif *m* spécifique specific reagent
réactif *m* spectroscopiquement pur spectrally
pure reagent
réaction *f* reaction
réaction *f* acide-base acid-base reaction

réaction *f* analytique analytical reaction
réaction *f* catalytique catalytic reaction
réaction *f* conjuguée conjugate reaction
réaction *f* d'amplification amplification
reaction
réaction *f* de la murexide murexide reaction
réaction *f* d'électrode electrode reaction
réaction *f* d'électrode irréversible irreversible
electrode reaction
réaction *f* d'électrode réversible reversible
electrode reaction
réaction *f* d'électrode simple simple electrode
reaction
réaction *f* de Zeisel Zeisel reaction
réaction *f* d'identification identification reaction
réaction *f* d'oxydo-réduction redox reaction
réaction *f* du biuret biuret reaction
réaction *f* élémentaire elementary reaction
réaction *f* nucléaire d'interférence interfering
nuclear reaction
réaction *f* partielle partial electrode reaction
réaction *f* redox redox reaction
réaction *f* secondaire induced reaction
réaction *f* sélective selective reaction
réaction *f* spécifique specific reaction
réarrangement *m* ionique rearrangement ion
recristallisation *f* recrystalization
redressement *m* faradaïque de haut niveau high-
level faradaic rectification
réducteur *m* reductant
réduction *f* reduction
réductométrie *f* reductometry
réflexion *f* totale total internal reflection
réflexion *f* totale atténuée attenuated total
reflectance
réfraction *f* refraction
réfraction *f* atomique atomic refraction
réfraction *f* de la lumière light refraction
réfraction *f* ionique ionic refraction
réfraction *f* molaire molar refraction
réfraction *f* spécifique specific refraction
réfractivité *f* molaire molar refraction
réfractomètre *m* refractometer
réfractométrie *f* refractometry
refroidissement *m* radioactive cooling
région *f* critique rejection region
région *f* de rejet rejection region
région *f* des micro-ondes microwave region
région *f* de sous-exposition under-exposure
region
région *f* de surexposition over-exposure region
région *f* infrarouge infrared region
région *f* intermédiaire interzonal region
région *f* ultraviolette ultraviolet region
région *f* visible visible region
règle *f* de la force ionique de Lewis et Randall
Lewis-Randall's ionic strength law
règle *f* de Lewis et Randall Lewis-Randall's
ionic strength law
règle *f* de Stokes Stokes law
règle *f* des trois sigmas three-sigma rule
règles *fpl* de sélection selection rules
régression *f* regression function
régression *f* curviligne curvilinear regression
régression *f* curvilinéaire curvilinear regression
régression *f* linéaire linear regression

régression *f* **multiple** multiple regression
régression *f* **non-linéaire** non-linear regression
régression *f* **pas à pas** stepwise regression
relargage *m* salting-out
relation *f* **de Butler-Volmer** Butler-Volmer equation
relation *f* **de cadmium** cadmium ratio
relation *f* **de Tafel** Tafel's equation
relaxation *f* relaxation
relaxation *f* **quadrupolaire** quadrupole relaxation
remplissage *m* **de la colonne** column packing
remplissage *m* **de la colonne par la méthode sèche** dry-packing
remplissage *m* **de la colonne par une suspension** slurry packing
remplissage *m* **de la colonne par une suspension à densité compensée** balanced-density slurry packing
remplissage *m* **par la méthode sèche** dry-packing
remplissage *m* **par une suspension** slurry packing
remplissage *m* **par une suspension à densité compensée** balanced-density slurry packing
rendement *m* **d'atomisation** efficiency of atomization
rendement *m* **de comptage** counting efficiency
rendement *m* **de courant** current efficiency
rendement *m* **de fission** fission yield
rendement *m* **de fluorescence X** X-ray fluorescence yield
rendement *m* **de la nébulisation** efficiency of nebulization
rendement *m* **de la pulvérisation** sputtering yield
rendement *m* **de l'extraction** recovery factor
rendement *m* **de neutrons** neutron output
rendement *m* **du détecteur** intrinsic detector efficiency
rendement *m* **radiochimique** radiochemical yield
renversement *m* self-reversal
renversement *m* **d'une raie** self-reversal
répétabilité *f* repeatability
réponse *f* **d'une électrode** electrode response
réponse *f* **nernstienne** Nernstian electrode response
reprécipitation *f* reprecipitation
reproductibilité *f* reproducibility
réseau *m* **concave** concave grating
réseau *m* **de diffraction** diffraction grating
réseau *m* **de réflexion** reflection grating
réseau *m* **par réflexion** reflection grating
réseau *m* **par transmission** transmission grating
réseau *m* **plan** plane grating
résine *f* **chélatante** chelating resin
résine *f* **échangeuse d'anions** anion-exchange resin
résine *f* **échangeuse de cations** cation-exchange resin
résine *f* **échangeuse d'ions** ion-exchange resin
résine *f* **pelliculaire** pellicular ion exchanger
résistance *f* **au transfert de masse** mass transfer term
résistance *f* **électrique** resistance
résistance *f* **spécifique** resistivity
résistivité *f* resistivity
résolution *f* resolution

résolution *f* **de détection de l'énergie du photopic** energy resolution of a gamma spectrometer
résolution *f* **de deux pics** peak resolution
résolvance *f* resolution
résonance *f* **de spin électronique** electron paramagnetic resonance
résonance *f* **Fermi** Fermi resonance
résonance *f* **magnétique** magnetic resonance
résonance *f* **magnétique nucléaire** nuclear magnetic resonance
résonance *f* **magnétique nucléaire du proton** proton magnetic resonance
résonance *f* **paramagnétique électronique** electron paramagnetic resonance
résonance *f* **paramagnétique électronique double** electron nuclear double resonance
résultat *m* **aberrant** outlier
rétention *f* **relative** relative retention
réticulation *f* cross-linking
rétrodiffusion *f* backscattering
richesse *f* **de la pulvérisation** sputtering yield
richesse *f* **en isotopes** abundance
rotation *f* **du plan de polarisation de la lumière** rotation of the plane of polarization of polarized light
rotation *f* **moléculaire** molar rotation
rotation *f* **spécifique** specific rotation
rotrode *f* rotating-disk electrode (*in arc and spark spectroscopy*)
RPE-spectroscopie *f* electron paramagnetic resonance spectroscopy
rydberg *m* rydberg

S

salaison *m* salting-out
saturation *f* saturation
saturation *f* **de la chambre** chamber saturation
saturation *f* **de la chambre de développement** chamber saturation
scavenger *m* scavenger
schéma *m* **de désintégration** decay scheme
science *f* **des surfaces** surface science
scintillateur *m* **à puits** well-type scintillation counter
séchage *m* **d'un précipité** drying of precipitate
second harmonique *m* second overtone band
section *f* **efficace** cross section
section *f* **efficace atomique** atomic cross section
section *f* **efficace d'absorption** absorption cross section
section *f* **efficace d'activation** activation cross section
section *f* **efficace isotopique** isotopic cross section
section *f* **efficace macroscopique** macroscopic cross section
section *f* **efficace microscopique** microscopic cross section
sédimentation *f* sedimentation
ségrégation *f* segregation
sein *m* **d'une solution** bulk of a solution
sélectivité *f* **d'un réactif** selectivity of a reagent

semiconducteur *m* photoélectrique photoconductor
sensibilité *f* sensitivity
sensibilité *f* absolue d'un détecteur absolute detector sensitivity
sensibilité *f* d'une réaction analytique sensitivity of analytical reaction
sensibilité *f* d'un détecteur detector sensitivity
sensibilité *f* d'une balance balance sensitivity
séparation *f* separation
séparation *f* radiochimique radiochemical separation
Sephadex *m* G Sephadex G
série *f* de désintégrations decay chain
série *f* des étalons standard series
série *f* des potentiels electrochemical series
série *f* électrochimique electrochemical series
série *f* éluotropique eluotropic series
série *f* équiéluotropique equieluotropic series
série *f* spectrale spectral series
seuil *m* absolu stimulus threshold
seuil *m* d'énergie threshold energy
seuil *m* de signification significance level
signal *m* d'absorption absorption curve (*in magnetic, resonance*)
signal *m* de l'analyte analytical signal
signal *m* de photoélectrons photoelectron line
silylation *f* silylation
simplex *m* simplex
singlet *m* singlet line
singulet *m* singlet line
siphon *m* de Haber et Luggin Luggin-Haber capillary
siphon *m* électrolytique salt bridge
sol *m* sol
solide *m* actif active solid
solide *m* activé active solid
solide *m* activé modifié modified active solid
solubilité *f* solubility
soluté *m* solute
solution *f* solution
solution *f* de comparaison comparison solution
solution *f* de lavage washing solution
solution *f* de référence reference solution
solution *f* d'extraction en retour stripping solution
solution *f* étalon standard solution
solution *f* étalon de référence standard reference solution
solution *f* saturée saturated solution
solution *f* solide mixed crystal
solution *f* sursaturée supersaturated solution
solution *f* tampon buffer solution
solution *f* tampon redox oxidation-reduction buffer
solution *f* témoin standard matching solution
solution *f* titrante titrant
solvant *m* solvent
solvant *m* acide protogenic solvent
solvant *m* aprotique aprotic solvent
solvant *m* basique protophilic solvent
solvant *m* capable de différencier des acides ou des bases differentiating solvent
solvant *m* d'extraction extractant
solvant *m* différenciant differentiating solvent
solvant *m* hydroxylé hydroxylic solvent

solvant *m* inerte aprotic solvent
solvant *m* mixte mixed solvent
solvant-niveleur *m* levelling solvent
solvant *m* protolytique protolytic solvent
solvant *m* proton-accepteur protophilic solvent
solvant *m* proton-donneur protogenic solvent
solvant *m* protophile protophilic solvent
solvatochromisme *m* solvatochromism
sorbant *m* sorbent
sorbat *m* sorbate
sorption *f* sorption
source *f* source
source *f* à émission de champ field ionization ion source
source *f* à étincelle spark ionization source
source *f* à étincelles spark ionization source
source *f* à impact d'électrons electron-impact ion source
source *f* à impact électronique electron-impact ion source
source *f* de lumière 1. light source 2. light source (*in atomic absorption and atomic fluorescence spectroscopy*)
source *f* d'énergie rayonnante radiation source
source *f* de rayonnement radiation source
source *f* d'excitation excitation source
source *f* d'ions ionization source
source *f* d'ions à impact électronique electron-impact ion source
source *f* d'ions par ionisation thermique thermal emission ion source
source *f* d'ions par photoionisation photoionization ion source
source *f* isotopique de neutrons isotopic neutron source
source *f* par ionisation thermique thermal emission ion source
sous-niveaux *mpl* Zeeman Zeeman levels
spécificité *f* d'un réactif specificity of a reagent
spécimen *m* mince thin-film specimen
spectranalyse *f* d'absorption absorption spectrochemical analysis
spectre *m* spectrum
spectre *m* atomique atomic spectrum
spectre *m* avec résolution dans le temps time-resolved spectrum
spectre *m* continu continuous spectrum
spectre *m* d'absorption 1. absorption spectrum 2. excitation spectrum
spectre *m* d'arc arc spectrum
spectre *m* de bande band spectrum
spectre *m* de fluorescence fluorescence spectrum
spectre *m* de fluorescence X X-ray fluorescence spectrum
spectre *m* de masse mass spectrum
spectre *m* de masse d'ions négatifs negative-ion mass spectrum
spectre *m* de masse d'ions positifs positive-ion mass spectrum
spectre *m* de micro-ondes microwave spectrum
spectre *m* d'émission emission spectrum
spectre *m* d'émission de fluorescence fluorescence emission spectrum
spectre *m* d'émission de phosphorescence phosphorescence emission spectrum
spectre *m* de neutrons neutron spectrum

spectre *m* de radiation nucléaire nuclear radiation spectrum
spectre *m* de raies line spectrum
spectre *m* de rayons X caractéristique X-ray characteristic spectrum
spectre *m* de réflexion totale atténuée attenuated total reflectance spectrum
spectre *m* de résonance magnétique nucléaire à haute résolution high-resolution nuclear magnetic resonance spectrum
spectre *m* de rotation rotational spectrum
spectre *m* d'étincelles spark spectrum
spectre *m* de vibration vibrational spectrum
spectre *m* de vibration-rotation vibrational-rotational spectrum
spectre *m* d'excitation excitation spectrum
spectre *m* d'excitation de fluorescence fluorescence excitation spectrum
spectre *m* d'excitation de phosphorescence phosphorescence excitation spectrum
spectre *m* discontinu discontinuous spectrum
spectre *m* électromagnétique electromagnetic spectrum
spectre *m* électronique electronic spectrum
spectre *m* électronique de rotation-vibration electronic-vibration-rotation spectrum
spectre *m* infrarouge infrared spectrum
spectre *m* ionique ionic spectrum
spectre *m* moléculaire molecular spectrum
spectre *m* Mössbauer Mössbauer spectrum
spectre *m* optique optical spectrum
spectre *m* Raman Raman spectrum
spectre *m* ultraviolet ultraviolet spectrum
spectre *m* visible visible spectrum
spectrofluorimètre *m* spectrofluorimeter
spectrofluoromètre *m* spectrofluorimeter
spectrofluorimétrie *f* spectrofluorimetry
spectrogramme *m* spectrogram
spectrogramme *m* X X-ray pattern
spectrographe *m* spectrograph
spectrographe *m* de masse mass spectrograph
spectromètre *m* optical spectrometer
spectromètre *m* à rayons X X-ray spectrometer
spectromètre *m* à scintillation scintillation spectrometer
spectromètre *m* à scintillations β beta scintillation spectrometer
spectromètre *m* de masse mass spectrometer
spectromètre *m* de masse à temps de vol time-of-flight mass spectrometer
spectromètre *m* de résonance magnétique nucléaire nuclear magnetic resonance spectrometer
spectromètre *m* des rayons γ gamma-ray spectrometer
spectromètre *m* des rayons X avec focalisation X-ray focusing spectrometer
spectromètre *m* γ gamma-ray spectrometer
spectromètre *m* non-enregistreur non-recording spectrophotometer
spectrométrie *f* spectrometry
spectrométrie *f* de masse mass spectrometry
spectrométrie *f* de masse à étincelles spark source mass spectrometry
spectrométrie *f* de masse à source à étincelles spark source mass spectrometry

spectrophotomètre *m* spectrophotometer
spectrophotomètre *m* à double faisceau double-beam spectrophotometer
spectrophotomètre *m* à flamme flame spectrophotometer
spectrophotomètre *m* bifaisceau double-beam spectrophotometer
spectrophotomètre *m* enregistreur recording spectrophotometer
spectrophotomètre *m* monofaisceau single-beam spectrophotometer
spectrophotométrie *f* spectrophotometry
spectrophotométrie *f* d'absorption absorption spectrophotometry
spectrophotométrie *f* différentielle differential spectrophotometry
spectropolarimètre *m* spectropolarimeter
spectropolarimétrie *f* spectropolarimetry
spectroscope *m* spectroscope
spectroscope *m* de masse mass spectroscope
spectroscopie *f* spectroscopy
spectroscopie *f* atomique atomic spectroscopy
spectroscopie *f* d'absorption absorption spectroscopy
spectroscopie *f* d'absorption atomique atomic absorption spectroscopy
spectroscopie *f* d'absorption atomique de flamme flame atomic absorption spectroscopy
spectroscopie *f* d'absorption de flamme flame absorption spectroscopy
spectroscopie *f* d'absorption des rayons X X-ray absorption spectroscopy
spectroscopie *f* d'absorption moléculaire molecular absorption spectroscopy
spectroscopie *f* d'apparition d'électrons excités par rayons X X-ray electron appearance potential spectroscopy
spectroscopie *f* de diffusion des ions low-energy ion scattering spectroscopy
spectroscopie *f* de dispersion d'ions de haute énergie high-energy ion scattering spectroscopy
spectroscopie *f* de flamme par absorption flame absorption spectroscopy
spectroscopie *f* de fluorescence fluorescence spectroscopy
spectroscopie *f* de fluorescence atomique atomic fluorescence spectroscopy
spectroscopie *f* de fluorescence atomique de flamme flame atomic fluorescence spectroscopy
spectroscopie *f* de fluorescence de flamme flame fluorescence spectroscopy
spectroscopie *f* de fluorescence moléculaire molecular fluorescence spectroscopy
spectroscopie *f* de fluorescence X X-ray fluorescence spectroscopy
spectroscopie *f* d'électrons Auger Auger-electron spectroscopy
spectroscopie *f* d'électrons Auger excités par bombardement ionique ion-induced Auger electron spectroscopy
spectroscopie *f* d'électrons Auger excités par rayons X X-ray induced Auger electron spectroscopy
spectroscopie *f* d'électrons déplacés des couches de valence d'atome valence level electron spectroscopy

spectroscopie *f* d'électrons déplacés des niveaux de coeur d'atome core level electron spectroscopy

spectroscopie *f* d'électrons pour analyse chimique X-ray photoelectron spectroscopy

spectroscopie *f* de masse mass spectroscopy

spectroscopie *f* de masse d'ions secondaires secondary ion mass spectroscopy

spectroscopie *f* de masse sous plasma glow discharge mass spectroscopy

spectroscopie *f* de micro-ondes microwave spectroscopy

spectroscopie *f* d'émission emission spectroscopy

spectroscopie *f* d'émission atomique atomic emission spectroscopy

spectroscopie *f* d'émission atomique de flamme flame atomic emission spectroscopy

spectroscopie *f* d'émission de flamme flame emission spectroscopy

spectroscopie *f* d'émission des rayons X X-ray emission spectroscopy

spectroscopie *f* d'émission des rayons X excités par particules particle induced X-ray emission

spectroscopie *f* d'émission moléculaire molecular emission spectroscopy

spectroscopie *f* d'émission optique sous plasma glow-discharge optical emission spectroscopy

spectroscopie *f* de pertes d'énergie d'électrons electron-energy loss spectroscopy

spectroscopie *f* de résonance de spin électronique electron paramagnetic resonance spectroscopy

spectroscopie *f* de résonance magnétique électronique electron paramagnetic resonance spectroscopy

spectroscopie *f* de résonance magnétique nucléaire nuclear magnetic resonance spectroscopy

spectroscopie *f* de résonance paramagnétique électronique electron paramagnetic resonance spectroscopy

spectroscopie *f* des photoélectrons photoelectron spectroscopy

spectroscopie *f* des photoélectrons excités par rayons UV ultraviolet photoelectron spectroscopy

spectroscopie *f* des photoélectrons induits par radiation UV ultraviolet photoelectron spectroscopy

spectroscopie *f* des photoélectrons induits par rayonnement X X-ray photoelectron spectroscopy

spectroscopie *f* des photoélectrons UV ultraviolet photoelectron spectroscopy

spectroscopie *f* des photoélectrons X X-ray photoelectron spectroscopy

spectroscopie *f* de transmission de pertes d'énergie transmission energy loss spectroscopy

spectroscopie *f* d'exoélectrons exoelectron spectroscopy

spectroscopie *f* du rayonnement monochromatique de freinage bremsstrahlung isochrome spectroscopy

spectroscopie *f* électronique 1. electron spectroscopy 2. electronic spectroscopy

spectroscopie *f* infrarouge d'absorption-réflexion infrared reflectance-absorption spectroscopy

spectroscopie *f* moléculaire molecular spectroscopy

spectroscopie *f* Mössbauer Mössbauer spectroscopy

spectroscopie *f* optique optical spectroscopy

spectroscopie *f* opto-acoustique photoacoustic spectroscopy

spectroscopie *f* par fluorescence à laser laser fluorescence spectroscopy

spectroscopie *f* par neutralisation des ions ion neutralization spectroscopy

spectroscopie *f* photoélectronique des rayons X X-ray photoelectron spectroscopy

spectroscopie *f* Raman Raman spectroscopy

spectroscopie *f* Raman à excitation laser laser Raman spectroscopy

spectroscopie *f* Raman à laser laser Raman spectroscopy

spectroscopie *f* vibrationnelle vibrational spectroscopy

sphère *f* d'activité mean ionic diameter

spot *m* spot

statistique *f* statistic

statistique *f* mathématique statistics

structure *f* fine fine structure

structure *f* hyperfine hyperfine structure

submicroanalyse *f* submicro-analysis

substance *f* substance

(+)-substance *f* dextrarotatory substance

(−)-substance *f* laevorotatory substance

d-substance *f* dextrarotatory substance

l-substance *f* laevorotatory substance

substance *f* à activité interfaciale surfactant

substance *f* absorbée absorbate

substance *f* de base standard substance

substance *f* déterminée determinand

substance *f* dextrogyre dextrarotatory substance

substance *f* dosée determinand

substance *f* électro-active electroactive substance

substance *f* étalon standard substance

substance *f* étalon primaire primary standard

substance *f* étalon secondaire secondary standard

substance *f* fille daughter nuclide

substance *f* interférente interfering substance

substance *f* lévogyre laevorotatory substance

substance *f* optiquement active optically active substance

substances-types *fpl* pour l'analyse standards for elemental analysis

substance *f* tensio-active surfactant

superposition *f* d'impulsions pile-up

support *m* à couche mince poreuse superficially porous support

support *m* de chromatographie solid support

support *m* pelliculaire superficially porous support

support *m* poreux totally porous support

support *m* solide solid support

support *m* superficiellement poreux superficially porous support

surface *f* surface

surface *f* de réponse response surface

surface *f* du pic peak area

surface *f* spécifique specific surface area
surfactant *m* surfactant
sursaturation *f* supersaturation
sursaturation *f* relative relative supersaturation
surtension *f* overpotential
surtension *f* d'activation activation
 overpotential
surtension *f* de concentration concentration
 overpotential
surtension *f* de cristallisation crystallization
 overpotential
surtension *f* de diffusion diffusion overpotential
surtension *f* de l'hydrogène hydrogen
 overpotential
surtension *f* de réaction reaction overpotential
surtension *f* de résistance resistance
 overpotential
surtension *f* de résistance ohmique pseudo-
 resistance overpotential
surtension *f* de transfert activation overpotential
survoltage *m* overpotential
susceptance *f* susceptance
susceptance *f* électrique susceptance
susceptométrie *f* susceptometry
système *m* de bandes band group
système *m* d'équations normales normal
 equations
système *m* redox redox system

T

tables *fpl* de nombres aléatoires random
 number tables
tache *f* spot
taille *f* de'l échantillon sample size
tamis *mpl* moléculaires molecular sieves
tampon *m* buffer solution
tampon *m* spectrochimique spectrochemical
 buffer
taux *m* de comptage counting rate
technique *f* de pastillage pellet technique
technique *f* du filament chauffé flash desorption
techniques *fpl* associées coupled simultaneous
 techniques
techniques *fpl* combinées combined techniques
techniques *fpl* d'étude des surfaces surface
 sensitive techniques
techniques *fpl* multiples multiple techniques
techniques *fpl* simultanées simultaneous
 techniques
techniques *fpl* simultanées associées coupled
 simultaneous techniques
techniques *fpl* simultanées continues
 continuous simultaneous techniques
température *f* de la colonne column temperature
température *f* de rétention retention
 temperature
température *f* de séparation separation
 temperature
temps *m* actif live time
temps *m* de goutte drop time
temps *m* de pose exposure time
temps *m* de pré-étincelage pre-spark period
temps *m* de préflambage pre-arc period
temps *m* de réponse response time
temps *m* de résolution resolving time

temps *m* de rétention total retention time
temps *m* de rétention brut total retention time
temps *m* de rétention de l'air mobile phase
 hold-up time
temps *m* de rétention d'un composé non-retenu
 mobile phase hold-up time
temps *m* de rétention net net retention time
temps *m* de rétention non corrigé total
 retention time
temps *m* de rétention réduit adjusted retention
 time
temps *m* de rétention total total retention time
temps *m* de transition transition time
temps *m* de vol time of flight
temps *m* mort 1. dead time (*of a device*)
 2. dead time (*in chromatography*)
temps *m* réel clock time
tendance *f* drift
teneur *f* content
teneur *f* isotopique abundance
tensamétrie *f* measurement of non-faradaic
 admittance
tenside *m* surfactant
tension *f* chimique d'une cellule galvanique
 electromotive force of a galvanic cell
tension *f* de décomposition decomposition
 voltage
tension *f* de fonctionnement burning voltage
tension *f* de Galvani Galvani potential difference
tension *f* d'électrode à l'équilibre equilibrium
 electrode potential
tension *f* de Volta Volta potential difference
tension *f* d'oxydo-réduction oxidation-reduction
 potential
tension *f* électrique electric potential difference
tension *f* électrique d'une cellule galvanique
 electric potential difference of a galvanic cell
tension *f* électrique relative d'électrode
 electrode potential
tension *f* redox oxidation-reduction potential
terre *f* de diatomées diatomaceous earth
terre *f* d'infusoires diatomaceous earth
test *m* test
test *m* bilatéral two-sided test
test *m* de signification significance test
test *m* de Student Student's test
test *m* de validité de l'ajustement test for
 goodness of fit
test *m* χ^2 de validité de l'ajustement chi-squared
 test for goodness of fit
test *m* F F-test
test *m* F de Snedecor F-test
test *m* non paramétrique non-parametric test
test *m* séquentiel sequential test
test *m* statistique test
test *m* t Student's test
test *m* unilatéral one-sided test
tête *f* de bande band head
théorie *f* d'Arrhenius-Ostwald Arrhenius'
 and Ostwald's theory of electrolytic
 dissociation
théorie *f* de la couche double theory of double
 layer
théorie *f* de la dissociation électrolytique
 d'Arrhenius-Ostwald Arrhenius' and
 Ostwald's theory of electrolytic dissociation

théorie *f* **de la solvatation de Born** Born's theory of solvation

théorie *f* **de l'association des ions de Bjerrum** Bjerrum's theory of ionic association

théorie *f* **des électrolytes forts de Debye et Hückel** Debye-Hückel's theory of strong electrolytes

théorie *f* **des probabilités** probability theory

thermalisation *f* **des neutrons** neutron thermalization

thermobalance *f* thermobalance

thermochromatographie *f* chromatography

thermocouple *m* thermocouple

thermogramme *m* thermal curve

thermogravimétrie *f* thermogravimetry

thermogravimétrie *f* **dérivée** derivative thermogravimetry

thermogravimétrie *f* **isobare** isobaric mass-change determination

thermogravimétrie *f* **isotherme** isothermal mass-change determination

titrage *m* titration

titrage *m* **à arrêt net** amperometric titration with two indicator electrodes

titrage *m* **à blanc** blank titration

titrage *m* **acide-base** acid-base titration

titrage *m* **acidimétrique** acidimetric titration

titrage *m* **alcalimétrique** alkalimetric titration

titrage *m* **ampérométrique** amperometric titration with one indicator electrode

titrage *m* **ampérométrique à deux électrodes indicatrices** amperometric titration with two indicator electrodes

titrage *m* **ampérométrique à une électrode à gouttes de mercure** amperometric titration with a dropping mercury electrode

titrage *m* **ampérométrique à une électrode indicatrice** amperometric titration with one indicator electrode

titrage *m* **ampérométrique à une électrode polarisée** amperometric titration with one indicator electrode

titrage *m* **argentimétrique** argentimetric titration

titrage *m* **avec deux indicateurs** double-indicator titration

titrage *m* **biampérométrique** amperometric titration with two indicator electrodes

titrage *m* **bipotentiométrique** controlled-current potentiometric titration with two indicator electrodes

titrage *m* **cérimétrique** cerimetric titration

titrage *m* **colorimétrique** colorimetric titration

titrage *m* **complexométrique** compleximetric titration

titrage *m* **conductométrique** conductometric titration

titrage *m* **conductométrique haute fréquence** high-frequency conductometric titration

titrage *m* **coulométrique** controlled-current coulometry

titrage *m* **coulométrique à potentiel contrôlé** controlled-potential coulometry

titrage *m* **coulométrique potentiométrique** potentiometric coulometric titration

titrage *m* **de contrôle** control titration

titrage *m* **de phase** phase titration

titrage *m* **dielcométrique** dielectrometric titration

titrage *m* **diélectrométrique** dielectrometric titration

titrage *m* **d'iodate** iodate titration

titrage *m* **direct** direct titration

titrage *m* **en milieu non aqueux** non-aqueous titration

titrage *m* **en présence de plusieurs phases** phase titration

titrage *m* **en retour** back-titration

titrage *m* **extracteur** extractive titration

titrage *m* **fluorimétrique** fluorimetric titration

titrage *m* **gazométrique** gasometric titration

titrage *m* **indirect** indirect titration

titrage *m* **inverse** inverse titration

titrage *m* **iodométrique** iodimetric titration

titrage *m* **mercurimétrique** mercurimetric titration

titrage *m* **mercurométrique** mercurometric titration

titrage *m* **néphélométrique** nephelometric titration

titrage *m* **numérique** numerical titration

titrage *m* **oscillométrique** high-frequency conductometric titration

titrage *m* **par déplacement** replacement titration

titrage *m* **par oxydation** oxidimetric titration

titrage *m* **par oxydo-réduction** oxidation-reduction titration

titrage *m* **par pesée** weight titration

titrage *m* **par précipitation** precipitation titration

titrage *m* **par réduction** reductometric titration

titrage *m* **photométrique** photometric titration

titrage *m* **polarographique** polarographic titration

titrage *m* **polarométrique** polarometric titration

titrage *m* **potentiométrique** potentiometric titration

titrage *m* **potentiométrique à courant imposé** controlled-current potentiometric titration

titrage *m* **potentiométrique à courant imposé à deux électrodes indicatrices** controlled-current potentiometric titration with two indicator electrodes

titrage *m* **potentiométrique à courant nul** potentiometric titration

titrage *m* **potentiométrique à intensité constante** controlled-current potentiometric titration

titrage *m* **potentiométrique dérivé** derivative potentiometric titration

titrage *m* **potentiométrique dérivé inverse** inverse derivative potentiometric titration

titrage *m* **potentiométrique différentiel** differential potentiometric titration

titrage *m* **potentiométrique doublement dérivé** second-derivative potentiometric titration

titrage *m* **radiométrique** radiometric titration

titrage *m* **spectrophotométrique** spectrophotometric titration

titrage *m* **thermométrique** thermometric titration
titrage *m* **turbidimétrique** turbidimetric titration
titrage *m* **visuel** visual titration
titrant *m* titrant
titre *m* titre
titrimétrie *f* titrimetric analysis
torche *f* **à plasma** plasma torch
trace *f* trace constituent
traceur *m* **isotopique** isotopic tracer
traceur *m* **radioactif** radioactive tracer
traînée *f* tailing
traînée *f* **située sur le pic après son sommet** tailing
transformation *f* **de Seidel** Seidel transformation
transistor *m* **à effet de champ** field effect transistor
transistor *m* **à effet de champ sensible au gaz** gas-sensitive field effect transistor
transition *f* **électronique** $\pi - \pi^*$ $\pi - \pi^*$ electronic transition
transition *f* **électronique** $\sigma - \pi^*$ $\sigma - \pi^*$ electronic transition
transition *f* **électronique** $n - \pi^*$ $n - \pi^*$ electronic transition
transition *f* **électronique** $n - \sigma^*$ $n - o^*$ electronic transition
transition *f* **interdite** forbidden transition
transition *f* **permise** allowed transition
transitométrie *f* chronopotentiometry
transmittance *f* transmission factor
transmittance *f* **interne** internal transmittance
trébuchet *m* chemical balance
triméthylchlorosilane *m* trimethylchlorosilane
triplet *m* **ionique** ion triplet
tronc *m* **d'atome** atomic core
tube *m* **absorbeur** absorption tube
tube *m* **absorbeur en U** U-tube
tube *m* **à combustion** combustion tube
tube *m* **à rayons X** X-ray tube
tube *m* **à tare avec pieds** piggie
tube *m* **compteur** counter tube
tube *m* **compteur de Geiger-Müller** Geiger-Müller counter tube
tube *m* **de graphite** graphite-tube furnace
tube *m* **de Nessler** Nessler measuring cylinder
tube *m* **de Nessler pour colorimétrie** Nessler measuring cylinder
tube *m* **de rayonnement** radiation counter
tube *m* **d'Hehner** Hehner measuring cylinder
tube *m* **échantillonneur de gaz** gas sampling pipette
tube *m* **photomultiplicateur** photomultiplier tube
tubes *mpl* **capillaires** capillary tubes
tubes *mpl* **capillaires pour mesurer le point de fusion** melting-point capillaries
turbidimètre *m* turbidimeter
turbidimétrie *f* turbidimetry

U

unité *f* **de masse atomique** unified atomic mass unit
ultramicroanalyse *f* ultramicro analysis
ultramicrobalance *f* ultramicro balance
ultramicrométhode *f* microgram method
ultra vide *m* ultra-high vacuum

ultraviolet *m* ultraviolet light
ultraviolet *m* **à vide** vacuum ultraviolet
ultraviolet *m* **lointain** far ultraviolet
ultraviolet *m* **proche** near ultraviolet

V

vague *f* **polarographique** polarographic wave
valence-gramme *m* gram equivalent
valeur *f* **critique de longueur d'onde** short-wavelength limit
valeur *f* **de coagulation** coagulation concentration
valeur *f* **d'une division** value of a division
valeur *f* R_B R_B value
valeur *f* R_f R_f value
valeur *f* R_f' R_f' value
valeur *f* R_M R_M value
valeur *f* R_M' R_M' value
vanne *f* **à boucle d'injection** by-pass injector
variable *f* **aléatoire** random variable
variable *f* **aléatoire continue** continuous random variable
variable *f* **aléatoire discrète** discrete random variable
variable *f* **aléatoire réduite** standardized random variable
variance *f* variance
variance *f* **résiduelle** residual variance
vase *m* **à peser** weighing bottle
vérification *f* **des poids** calibration of weights
verre *m* **de porosité contrôlée** controlled-porosity glass
vibrations *fpl* vibrations
vibrations *fpl* **antisymétriques** antisymmetrical vibrations
vibrations *fpl* **dans le plan** in-plane vibrations
vibrations *fpl* **de balancement** wagging vibrations
vibrations *fpl* **de cisaillement** scissoring vibrations
vibrations *fpl* **de déformation** bending vibrations
vibrations *fpl* **de rotation** rocking vibrations
vibrations *fpl* **de roulis** wagging vibrations
vibrations *fpl* **de squelette** skeletal vibrations
vibrations *fpl* **de tangage** rocking vibrations
vibrations *fpl* **de torsion** twisting vibrations
vibrations *fpl* **de valence** streching vibrations
vibrations *fpl* **d'extension** streching vibrations
vibrations *fpl* **hors de plan** out-of-plane vibrations
vibrations *fpl* **moléculaires** molecular vibrations
vibrations *fpl* **normales** normal vibrations
vibrations *fpl* **symétrique** symmetrical vibrations
vide *m* **ultrapoussé** ultra-high vacuum
vieillissement *m* ageing of a precipitate
vieillissement *m* **des colloïdes** ageing of colloids
viscosimètre *m* **d'Ostwald** Ostwald viscometer
vitesse *f* **d'écoulement du mercure** flow rate of mercury
vitesse *f* **d'écoulement linéaire nominale** nominal linear flow
vitesse *f* **de la lumière dans le vide** velocity of light in vacuum
vitesse *f* **de nucléation** rate of nucleation

vitesse *f* **de sortie du mercure** flow rate of mercury
vitesse *f* **interstitielle** interstitial velocity
vitesse *f* **moyenne interstitielle du gaz-vecteur** mean interstitial velocity of the carrier gas
vitesse *f* **réduite de la phase mobile** reduced velocity of the mobile phase
voile *m* fog
volatilisation *f* volatilization
voltammétrie *f* voltammetry
voltampérométrie *f* voltammetry
voltampérométrie *f* **à balayage linéaire** chronoamperometry with linear potential sweep
voltampérométrie *f* **à balayage triangulaire** triangular-wave voltammetry
voltampérométrie *f* **à balayage triangulaire cyclique** cyclic triangular-wave voltammetry
voltampérométrie *f* **à une électrode stationnaire** chronoamperometry with linear potential sweep
voltampérométrie *f* **avec redissolution** stripping voltammetry
voltampérométrie *f* **avec redissolution anodique** anodic stripping voltammetry
voltampérométrie *f* **dérivée** derivative voltammetry
voltampérométrie *f* **différentielle** differential voltammetry
voltampérométrie *f* **hydrodynamique** hydrodynamic voltammetry
volume *m* **de la colonne** column volume
volume *m* **de la phase liquide** liquid phase volume
volume *m* **de la phase stationnaire** stationary-phase volume
volume *m* **de liquide stationnaire** stationary liquid volume
volume *m* **de liquide total** total liquid volume
volume *m* **d'élution du pic** peak elution volume
volume *m* **de phase liquide** liquid phase volume
volume *m* **de rétention** retention volume
volume *m* **de rétention absolu** net retention volume
volume *m* **de rétention corrigé** corrected retention volume

volume *m* **de rétention de l'air** mobile phase hold-up volume
volume *m* **de rétention d'un composé non-retenu** mobile phase hold-up volume
volume *m* **de rétention limite** corrected retention volume
volume *m* **de rétention net** net retention volume
volume *m* **de rétention non corrigé** retention volume
volume *m* **de rétention réduit** adjusted retention volume
volume *m* **de rétention spécifique** specific retention volume
volume *m* **de rétention total** retention volume
volume *m* **du liquide** liquid phase volume
volume *m* **du lit** bed volume
volume *m* **d'un détecteur** sensitive volume of a detector
volume *m* **du remplissage** bed volume
volume *m* **du solide** solid volume
volume *m* **géométrique de la colonne** column volume
volume *m* **hors de la colonne** extra-column volume
volume *m* **interstitiel** interstitial volume
volume *m* **mort** extra-column volume
volume *m* **nominal** designated volume
volume *m* **poreux** stationary liquid volume
volume *m* **sensible d'un détecteur** insensitive volume of a detector
volumétrie *f* titrimetric analysis

Z

zéolithes *fpl* zeolites
zéro *m* **de l'échelle** zero point of the scale
zone *f* zone
zone *f* **de combustion primaire** primary combustion zone
zone *f* **de diffusion** secondary combustion zone
zone *f* **de solarisation** over-exposure region
zone *f* **externe** secondary combustion zone
zone *f* **interne** primary combustion zone
zwitterion *m* ampholyte ion

INDEKS POLSKI

A

aberracja *f* optical aberration
aberracja *f* chromatyczna chromatic aberration
aberracja *f* sferyczna spherical aberration
absorbancja *f* absorbance
absorbat *m* absorbate
absorbent *m* absorbent
absorber *m* absorber
absorpcja *f* absorption
absorpcja *f* rezonansowa resonance absorption
absorpcja *f* właściwa specific decadic absorption coefficient
abundancja *f* abundance
acydymetria *f* acidimetry
adhezja *f* adhesion
admitancja *f* admittance
adsorbat *m* adsorbate
adsorbent *m* adsorbent
adsorpcja *f* adsorption
adsorpcja *f* chemiczna chemisorption
adsorpcja *f* fizyczna physisorption
adsorpcja *f* siłami van der Waalsa physisorption
adsorpcja *f* specyficzna specific adsorption
adsorpcja *f* van der Waalsa physisorption
adsorpcja *f* wymienna exchange adsorption
agaroza *f* agarose
aglomeracja *f* agglomeration
agregacja *f* aggregation
agregat *m* aggregate
aktor *m* actor
aktywacja *f* activation
aktywacja *f* adsorbentu activation of an adsorbent
aktywacja *f* neutronowa neutron activation
aktywacja *f* za pomocą cząstek naładowanych charged-particle activation
aktywność *f* activity
aktywność *f* adsorbentu adsorbent activity
aktywność *f* elektrolitu B activity of electrolyte B
aktywność *f* jonowa średnia mean activity of electrolyte
aktywność *f* nasycenia saturation activity
aktywność *f* optyczna optical activity
aktywność *f* promieniotwórcza activity
aktywność *f* składnika B względna relative activity of substance B
aktywność *f* średnia elektrolitu mean activity of electrolyte
aktywność *f* właściwa specific activity
akumulator *m* electrical accumulator
akwametria *f* aquametry
alkacymetria *f* acidimetry and alkalimetry
alkalimetria *f* alkalimetry

alkoholometria *f* alcoholometry
amalgamat *m* amalgam
amfolit *m* ampholyte
amperometria *f* amperometry
amperometria *f* różnicowa differential amperometry
amperometria *f* z dwiema elektrodami spolaryzowanymi amperometry with two indicator electrodes
amperometria *f* z jedną elektrodą spolaryzowaną amperometry with one indicator electrode
amperometria *f* z makroelektrodą rtęciową stirred-mercury-pool amperometry
amperometria *f* z wirującą elektrodą platynową rotating-platinum-wire electrode amperometry
amperostat *m* galvanostat
analit *m* analyte
analiza *f* aktywacyjna activation analysis
analiza *f* aktywacyjna destrukcyjna destructive activation analysis
analiza *f* aktywacyjna instrumentalna non-destructive activation analysis
analiza *f* aktywacyjna neutronowa neutron activation analysis
analiza *f* aktywacyjna niedestrukcyjna non--destructive activation analysis
analiza *f* aktywacyjna za pomocą cząstek naładowanych charged-particle activation analysis
analiza *f* amperometryczna amperometric analysis
analiza *f* amplitudy impulsów pulse amplitude analysis
analiza *f* arbitrażowa umpire analysis
analiza *f* całkowita complete analysis
analiza *f* chemiczna chemical analysis
analiza *f* chromatograficzna chromatographic analysis
analiza *f* czołowa frontal chromatography
analiza *f* czynnikowa factor analysis
analiza *f* dielektrometryczna dielectrometric analysis
analiza *f* dyfrakcyjna diffractometry
analiza *f* dyfrakcyjna rentgenowska X-ray diffraction analysis
analiza *f* elektrochemiczna electroanalysis
analiza *f* elementarna związków organicznych elemental analysis (*of organic compounds*)
analiza *f* fazowa phase analysis
analiza *f* fazy gazowej nad roztworem headspace analysis
analiza *f* fluorescencyjna fluorescence analysis
analiza *f* fluorescencyjna rentgenowska X-ray fluorescence analysis

analiza *f* fluorescencyjna rentgenowska
  radioizotopowa radioisotope-excited X-ray
  fluorescence analysis
analiza *f* fluorescencyjna rentgenowska
  z dyspersją energii energy dispersive X-ray
  fluorescence analysis
analiza *f* fluorescencyjna rentgenowska
  z dyspersją fal wavelength dispersive X-ray
  fluorescence analysis
analiza *f* fotoaktywacyjna photoactivation
  analysis
analiza *f* frakcyjna fractional analysis
analiza *f* gazometryczna gasometric analysis
analiza *f* gazowa gas analysis
analiza *f* gazowa absorpcjometryczna
  absorptiometric gas analysis
analiza *f* gazowa manometryczna manometric
  gas analysis
analiza *f* gazowa objętościowa volumetric gas
  analysis
analiza *f* gazowa wagowa gravimetric gas
  analysis
analiza *f* grawimetryczna gravimetric analysis
analiza *f* ilościowa quantitative analysis
analiza *f* ilościowa związków organicznych
  quantitative analysis of organic compounds
analiza *f* instrumentalna instrumental analysis
analiza *f* inwersyjna electrochemical stripping
  analysis
analiza *f* jakościowa qualitative analysis
analiza *f* jakościowa systematyczna systematic
  qualitative analysis
analiza *f* jakościowa związków organicznych
  qualitative analysis of organic compounds
analiza *f* kanoniczna canonical analysis
analiza *f* kinetyczna kinetic analysis
analiza *f* kolorymetryczna colorimetric analysis
analiza *f* konduktometryczna conductometric
  analysis
analiza *f* konformacyjna conformational analysis
analiza *f* konwencjonalna conventional analysis
analiza *f* kroplowa spot test analysis
analiza *f* kulometryczna coulometric analysis
analiza *f* lokalna local analysis
analiza *f* metodą rozcieńczenia izotopowego
  isotope dilution analysis
analiza *f* metodą rozcieńczenia izotopowego
  odwrotnego reverse isotope dilution analysis
analiza *f* metodą rozcieńczenia izotopowego
  podwójnego double isotope dilution analysis
analiza *f* metodą rozcieńczenia izotopowego
  prostego direct isotope dilution analysis
analiza *f* metodą rozcieńczenia izotopowego
  substechiometrycznego substoichiometric
  isotope dilution analysis
analiza *f* miareczkowa titrimetric analysis
analiza *f* miareczkowa gazów gas titration
  analysis
analiza *f* międzyoperacyjna process analysis
analiza *f* mikrokrystaliczna microcrystalloscopic
  analysis
analiza *f* mikrokrystaloskopowa
  microcrystalloscopic analysis
analiza *f* niedestrukcyjna non-destructive
  analysis
analiza *f* nieniszcząca non-destructive analysis

analiza *f* objętościowa titrimetric analysis
analiza *f* pierwiastkowa elemental analysis
analiza *f* pirochemiczna pyrochemical analysis
analiza *f* polarograficzna polarographic analysis
analiza *f* potencjometryczna potentiometric
  analysis
analiza *f* powierzchni surface analysis
analiza *f* powierzchni metodą rezonansowej
  jonizacji rozpylonych atomów surface analysis
  by resonance ionization of sputtered atoms
analiza *f* półilościowa semiquantitative analysis
analiza *f* przez spalanie combustion analysis
analiza *f* radiochemiczna radiochemical analysis
analiza *f* radiometryczna radiometric analysis
analiza *f* regresji regression analysis
analiza *f* rozjemcza umpire analysis
analiza *f* ruchowa process analysis
analiza *f* rutynowa routine analysis
analiza *f* sedymentacyjna sedimentation analysis
analiza *f* sekwencyjna sequential analysis
analiza *f* sensoryczna sensory analysis
analiza *f* sitowa sieve analysis
analiza *f* skrócona short analysis
analiza *f* spektralna spectrochemical analysis
analiza *f* spektralna absorpcyjna absorption
  spectrochemical analysis
analiza *f* spektralna atomowa atomic
  spectrochemical analysis
analiza *f* spektralna cząsteczkowa molecular
  spectrochemical analysis
analiza *f* spektralna emisyjna emission
  spectrochemical analysis
analiza *f* spektralna fluorescencyjna fluorescence
  spectrochemical analysis
analiza *f* spektralna rentgenowska X-ray
  spectrochemical analysis
analiza *f* spektralna rentgenowska ze
  wzbudzaniem cząstkami naładowanymi particle
  induced X-ray emission
analiza *f* spektrochemiczna spectrochemical
  analysis
analiza *f* spektrograficzna spectrographic
  analysis
analiza *f* stripingowa electrochemical stripping
  analysis
analiza *f* systematyczna systematic qualitative
  analysis
analiza *f* śladowa trace analysis
analiza *f* termiczna thermal analysis
analiza *f* termiczna różnicowa differential
  thermal analysis
analiza *f* termomanometryczna
  thermomanometric analysis
analiza *f* termowolumetryczna thermovolumetric
  analysis
analiza *f* turbidymetryczna turbidimetry
analiza *f* umowna conventional analysis
analiza *f* wagowa gravimetric analysis
analiza *f* wariancji analysis of variance
analiza *f* według grup funkcyjnych functional
  group analysis
analiza *f* widmowa spectrochemical analysis
analiza *f* wieloskładnikowa multiple analysis
analiza *f* woltamperometryczna voltammetric
  analysis
analiza *f* wydzielanego gazu evolved gas analysis

analiza *f* wysokości impulsów pulse amplitude
    analysis
analizator *m* analyser
analizator *m* kwadrupolowy quadrupole mass
    analyser
analizator *m* mas mass analyser
analizator *m* wysokości impulsów pulse
    amplitude analyser
analizator *m* z polem opóźniającym retarding
    field analyser
analizator *m* z sektorem magnetycznym
    magnetic-deflection mass spectrometer
Anhydron *m* Anhydrone
anionit *m* anion exchanger
anoda *f* anode
anodowanie *n* plazmowe plasma anodization
anolit *m* anolyte
aparat *m* do oznaczania temperatury wrzenia
    boiling point apparatus
aparat *m* ekstrakcyjny Soxhleta Soxhlet
    apparatus
aparat *m* Kippa Kipp gas generator
aparat *m* Kjeldahla Kjeldahl apparatus
aparat *m* Orsata Orsat gas analyser
aparat *m* Parnasa i Wagnera Parnas-Wagner
    apparatus
aplikator *m* applicator
areometr *m* hydrometer
argentometria *f* argentimetry
Askaryt *m* Ascarite
askorbinometria *f* ascorbimetry
astygmatyzm *m* astigmatism
atmosfera *f* jonowa ionic atmosphere
atomizacja *f* atomization
atomizer *m* atomizer
atomizer *m* bezpłomieniowy electrothermal
    atomizer
atomizer *m* elektrotermiczny electrothermal
    atomizer
auksochrom *m* auxochrome
autoabsorpcja *f* 1. self-absorption 2. self-
    absorption (*in emission spectroscopy*)
autodysocjacja *f* autoprotolysis
autojonizacja *f* autoprotolysis
autoprotoliza *f* autoprotolysis
autoradiogram *m* autoradiograph
azbest *m* platynowany platinized asbestos
azotometr *m* nitrometer

B

badania *npl* międzylaboratoryjne interlaboratory
    research
barn *m* barn
barwienie *n* płomienia flame coloration
bateria *f* battery
batochrom *m* bathochrome
bekerel *m* becquerel
biamperometria *f* amperometry with two indi-
    cator electrodes
bioluminescencja *f* bioluminescence
bipotencjometria *f* controlled-current
    potentiometry with two indicator electrodes
biureta *f* burette
biureta *f* automatyczna tłokowa piston
    automatic burette

biureta *f* Buntego Bunte gas burette
biureta *f* legalizowana certified burette
biureta *f* Schellbacha Schellbach burette
biureta *f* z automatycznym nastawianiem zera
    automatic zero burette
biureta *f* z automatycznym nastawianiem zera
    Dafferta Daffert automatic zero burette
blok *m* block
blokowanie *n* metalowskaźnika blocking of
    metal indicator
błąd *m* analityka operative error
błąd *m* bezwzględny absolute error
błąd *m* drugiego rodzaju error of second kind
błąd *m* gruby gross error
błąd *m* instrumentalny instrumental error
błąd *m* losowy random error
błąd *m* metody error of method
błąd *m* miareczkowania titration error
błąd *m* odtwarzania reproducibility error
błąd *m* pierwszego rodzaju error of first kind
błąd *m* pobrania próby sampling error
błąd *m* popełniony przez analityka operative error
błąd *m* procentowy percentage error
błąd *m* przypadkowy random error
błąd *m* przyrządu instrumental error
błąd *m* solny salt error
błąd *m* systematyczny systematic error
błąd *m* wskaźnika indicator error
błąd *m* względny relative error
błąd *m* wzorcowania scale error
bolometr *m* bolometer
bomba *f* do mineralizacji combustion bomb
bomba *f* kalorymetryczna bomb calorimeter
bramka *f* gate
bromianometria *f* bromatometry
bufor *m* pH buffer solution
bufor *m* pM metal buffer
bufor *m* redoks oxidation-reduction buffer
bufor *m* spektrochemiczny spectrochemical buffer

C

całkowite wewnętrzne odbicie *n* total internal
    reflection
Carbowax *m* Carbowax
Celit *m* Celite
cerometria *f* cerimetry
chelat *m* wewnętrzny neutral chelate
chemia *f* analityczna analytical chemistry
chemijonizacja *f* chemiionization
chemiluminescencja *f* chemiluminescence
chemisorpcja *f* chemisorption
chmura *f* jonowa ionic atmosphere
chromatermografia *f* chromathermography
chromatograf *m* chromatograph
chromatografia *f* chromatography
chromatografia *f* adsorpcyjna adsorption
    chromatography
chromatografia *f* bibułowa paper
    chromatography
chromatografia *f* cieczowa liquid
    chromatography
chromatografia *f* cieczowa szybka
    high-performance liquid chromatography
chromatografia *f* cieczowa wysokociśnieniowa
    high-performance liquid chromatography

chromatografia *f* cieczowa wysokosprawna high-performance liquid chromatography
chromatografia *f* cienkowarstwowa thin-layer chromatography
chromatografia *f* czołowa frontal chromatography
chromatografia *f* dwukierunkowa two-dimensional chromatography
chromatografia *f* ekskluzyjna exclusion chromatography
chromatografia *f* elucyjna elution chromatography
chromatografia *f* gazowa gas chromatography
chromatografia *f* jonitowa ion-exchange chromatography
chromatografia *f* jonowo-asocjacyjna ion-pair partition chromatography
chromatografia *f* jonowo-ekskluzyjna ion-exclusion chromatography
chromatografia *f* jonowymienna ion-exchange chromatography
chromatografia *f* kapilarna open-tube chromatography
chromatografia *f* kolumnowa column chromatography
chromatografia *f* krążkowa circular chromatography
chromatografia *f* mikrokrążkowa microcircular chromatography
chromatografia *f* na bibule paper chromatography
chromatografia *f* na kolumnach o otwartym przekroju open-tube chromatography
chromatografia *f* na kolumnach otwartych open-tube chromatography
chromatografia *f* permeacyjna permeation chromatography
chromatografia *f* podziałowa partition chromatography
chromatografia *f* podziałowa z odwróconymi fazami extraction chromatography
chromatografia *f* powinowactwa affinity chromatography
chromatografia *f* przez rugowanie displacement chromatography
chromatografia *f* rugująca displacement chromatography
chromatografia *f* sitowo-molekularna molecular-sieve chromatography
chromatografia *f* solubilizacyjna solubilization chromatography
chromatografia *f* w cienkich warstwach thin-layer chromatography
chromatografia *f* w obszarze nadkrytycznym supercritical fluid chromatography
chromatografia *f* w układzie ciecz-ciało stałe liquid-solid chromatography
chromatografia *f* w układzie ciecz-ciecz liquid-liquid chromatography
chromatografia *f* w układzie ciecz-żel liquid-gel chromatography
chromatografia *f* w układzie gaz-ciało stałe gas-solid chromatography
chromatografia *f* w układzie gaz-ciecz gas-liquid chromatography

chromatografia *f* wstępująca ascending development
chromatografia *f* wysalająca salting-out chromatography
chromatografia *f* z odwróconymi fazami reversed-phase chromatography
chromatografia *f* z programowaniem prędkości przepływu flow-programmed chromatography
chromatografia *f* z programowaniem temperatury temperature-programmed chromatography
chromatografia *f* z użyciem detergentów jonowych soap chromatography
chromatografia *f* z wymianą ligandów ligand-exchange chromatography
chromatografia *f* z wysalaniem salting-out chromatography
chromatografia *f* zstępująca descending development
chromatografia *f* żelowa gel chromatography
chromatografia *f* żelowo-permeacyjna gel-permeation chromatography
chromatografować *v* chromatograph
chromatogram *m* chromatogram
chromatogram *m* całkowy integral chromatogram
chromatogram *m* różnicowy differential chromatogram
chromatopolarografia *f* chromatopolarography
chromianometria *f* chromatometry
chromofor *m* chromophore
Chromosorb *m* Chromosorb
chronoamperometria *f* chronoamperometry
chronoamperometria *f* konwekcyjna convective chronoamperometry
chronoamperometria *f* polarograficzna polarographic chronoamperometry
chronoamperometria *f* z podwójną zmianą potencjału double-potential-step chronoamperometry
chronoamperometria *f* z podwójnym skokiem potencjału double-potential-step chronoamperometry
chronokulometria *f* chronocoulometry
chronokulometria *f* konwekcyjna convective chronocoulometry
chronokulometria *f* ze skokiem potencjału chronocoulometry
chronokulometria *f* z podwójną zmianą potencjału double-potential-step chronocoulometry
chronokulometria *f* z podwójnym skokiem potencjału double-potential-step chronocoulometry
chronopotencjometria *f* chronopotentiometry
chronopotencjometria *f* cykliczna cyclic chronopotentiometry
chronopotencjometria *f* cykliczna z odwróceniem kierunku prądu cyclic current-reversal chronopotentiometry
chronopotencjometria *f* cykliczna ze skokową zmianą prądu cyclic current-step chronopotentiometry
chronopotencjometria *f* pierwszej pochodnej derivative chronopotentiometry
chronopotencjometria *f* różniczkowa derivative chronopotentiometry

chronopotencjometria *f* wielocykliczna cyclic
chronopotentiometry
chronopotencjometria *f* z liniowo narastającym
prądem chronopotentiometry with linear
current sweep
chronopotencjometria *f* z nałożonym prądem
zmiennym chronopotentiometry with
superimposed alternating current
chronopotencjometria *f* z odwróceniem
kierunku prądu current-reversal
chronopotentiometry
chronopotencjometria *f* z programowanym
prądem programmed-current
chronopotentiometry
chronopotencjometria *f* z wyłączeniem prądu
current-cessation chronopotentiometry
chronopotencjometria *f* ze skokową zmianą
prądu current-step chronopotentiometry
chronopotencjometria *f* zmiennonapięciowa
alternating-voltage chronopotentiometry
chronopotencjometria *f* zmiennoprądowa
alternating-current chronopotentiometry
chronowoltamperometria *f* chronoamperometry
with linear potential sweep
chronowoltamperometria *f* cykliczna cyclic
triangular-wave voltammetry
ciśnienie *n* pęcznienia swelling pressure
cylinder *m* kolorymetryczny Hehnera Hehner
measuring cylinder
cylinder *m* kolorymetryczny Nesslera Nessler
measuring cylinder
cylinder *m* pomiarowy measuring cylinder
cylindryczny analizator *m* zwierciadlany
cylindrical mirror analyser
czas *m* martwy 1. dead time (*of a device*) 2. dead
time (*in chromatography*)
czas *m* naświetlenia exposure time
czas *m* opóźnienia odpowiedzi response time
czas *m* przediskrzenia pre-spark period
czas *m* przedpalenia pre-arc period
czas *m* przejścia transition time
czas *m* przelotu time of flight
czas *m* retencji absolutny net retention time
czas *m* retencji całkowity total retention time
czas *m* retencji niepoprawiony total retention
time
czas *m* retencji składnika nie zatrzymywanego
w kolumnie mobile phase hold-up time
czas *m* retencji zredukowany adjusted retention
time
czas *m* rozdzielczy resolving time
czas *m* trwania kropli drop time
czas *m* zegarowy clock time
czas *m* żywy live time
cząstka *f* elementary entity
częstość *f* frequency
częstość *f* charakterystyczna characteristic
frequency
częstość *f* grupowa characteristic frequency
częstotliwość *f* frequency
częstotliwość *f* wzbudzająca exciting frequency
czoło *n* rozpuszczalnika mobile phase front
czujnik *m* elektrochemiczny electrochemical
sensor
czujnik *m* gazowy selektywny gas sensing
electrode

czujnik *m* tlenowy Clark oxygen-sensing probe
czułość *f* 1. sensitivity 2. sensitivity of
analytical reaction
czułość *f* detektora detector sensitivity
czułość *f* detektora absolutna absolute detector
sensitivity
czułość *f* detektora bezwzględna absolute
detector sensitivity
czułość *f* reakcji analitycznej sensitivity of
analytical reaction
czułość *f* wagi balance sensitivity
czynnik *m* factor
czynnik *m* chelatujący chelating agent
czynność *f* optyczna optical activity
czystość *f* jądrowa nuclear purity
czystość *f* odczynnika purity of a reagent
czystość *f* optyczna optical purity
czystość *f* radiochemiczna radiochemical purity

D

debajogram *m* X-ray powdered-crystal pattern
dekantacja *f* decantation
dekontaminacja *f* decontamination
dekstran *m* błękitny Blue Dextran
depolaryzator *m* depolarizer
derywatograf *m* derivatograph
derywatyzacja *f* derivatization
desorpcja *f* desorption
desorpcja *f* błyskawiczna flash desorption
desorpcja *f* cząstek obojętnych wywołana
bombardowaniem elektronowym electron
stimulated desorption of neutrals
desorpcja *f* jonów wywołana bombardowaniem
elektronowym electron stimulated
desorption of ions
desorpcja *f* natychmiastowa flash desorption
desorpcja *f* polowa field desorption
desorpcja *f* stymulowana jonowo ion impact
desorption
desorpcja *f* termiczna thermal desorption
desorpcja *f* termoprogramowana temperature
programmed desorption
desorpcja *f* wywołana bombardowaniem
elektronowym electron stimulated desorption
desorpcja *f* wywołana bombardowaniem
jonowym ion impact desorption
destylacja *f* nośnikowa carrier distillation
detekcja *f* promieniowania detection of radiation
detektor *m* detector
detektor *m* argonowy argon detector
detektor *m* całkowy integral detector
detektor *m* Everharta i Thornleya
Everhart-Thornley detector
detektor *m* fotoelektryczny photodetector
detektor *m* fotometryczny płomieniowy flame
photometric detector
detektor *m* germanowo-litowy lithium-drifted
germanium detector
detektor *m* Golaya Golay pneumatic detector
detektor *m* jonizacyjny ionization detector
detektor *m* jonizacyjny argonowy argon
ionization detector
detektor *m* jonizacyjny płomieniowy flame
ionization detector

detektor *m* **konduktywności cieplnej** thermal conductivity detector
detektor *m* **krzemowo-litowy** lithium-drifted silicon detector
detektor *m* **neutronów** neutron detector
detektor *m* **płomieniowo-fotometryczny** flame photometric detector
detektor *m* **płomieniowo-jonizacyjny** flame ionization detecotr
detektor *m* **półprzewodnikowy** semiconductor detector
detektor *m* **promieniowania** radiation detector
detektor *m* **przewodnictwa cieplnego** thermal conductivity detector
detektor *m* **rezonansowy** resonance detector
detektor *m* **różnicowy** differential detector
detektor *m* **scyntylacyjny** scintillation detector
detektor *m* **termiczny** thermal radiation detector
detektor *m* **termokonduktometryczny** thermal conductivity detector
detektor *m* **transportowy** transport detector
detektor *m* **wychwytu elektronów** electron capture detector
detektor *m* **z barierą powierzchniową** surface barrier detector
dezaktywacja *f* deactivation
diafragma *f* **Hartmanna** Hartmann diaphragm
dichroizm *m* **kołowy** circular dichroism
dielektrometria *f* dielectrometry
dimetylodichlorosilan *m* dimethyldichlorosilane
długość *f* **fali** wavelength
długość *f* **fali analityczna** analytical wavelength
długość *f* **fali błysku** blaze wavelength
dojrzewanie *n* **osadu** Ostwald ripening
dokładność *f* accuracy
dominanta *f* mode
dozownik *m* **bocznikowy** by-pass injector
dozownik *m* **dwustrumieniowy** by-pass injector
dozownik *m* **próbki** sample injector
dren *m* drain
drgania *npl* vibrations
drgania *npl* **antysymetryczne** antisymmetrical vibrations
drgania *npl* **cząsteczkowe** molecular vibrations
drgania *npl* **deformacyjne** bending vibrations
drgania *npl* **deformacyjne niepłaskie** twisting vibrations
drgania *npl* **deformacyjne nożycowe** scissoring vibrations
drgania *npl* **deformacyjne płaskie wahadłowe** rocking vibrations
drgania *npl* **kołyszące** rocking vibrations
drgania *npl* **niepłaskie** out-of-plane vibrations
drgania *npl* **normalne** normal vibrations
drgania *npl* **nożycowe** scissoring vibrations
drgania *npl* **płaskie** in-plane vibrations
drgania *npl* **rozciągające** stretching vibrations
drgania *npl* **skręcające** twisting vibrations
drgania *npl* **symetryczne** symmetrical vibrations
drgania *npl* **szkieletowe** skeletal vibrations
drgania *npl* **wachlarzowe** wagging vibrations
drgania *npl* **wahadłowe** rocking vibrations
drgania *npl* **walencyjne** stretching vibrations
drgania *npl* **zginające** bending vibrations
droga *f* **rozwijania** mobile phase distance
drugi efekt *m* **Wiena** dissociation field effect

drugi nadton *m* second overtone
dryf *m* drift
dryf *m* **czasowy** drift
dryf *m* **linii podstawowej** base-line drift
dryf *m* **potencjału** drift of a potential
duchy *mpl* ghosts
dwójłomność *f* **kołowa** circular birefringence
dyfrakcja *f* **elektronów o dużych energiach** high energy electron diffraction
dyfrakcja *f* **elektronów o małych energiach** low energy electron diffraction
dyfrakcja *f* **elektronów o małych energiach rozproszonych nieelastycznie** inelastic low energy electron diffraction
dyfrakcja *f* **elektronów powolnych** low energy electron diffraction
dyfrakcja *f* **elektronów szybkich** high energy electron diffraction
dyfrakcja *f* **jonów o małych energiach** low energy ion diffraction
dyfrakcja *f* **odbitych elektronów szybkich** reflection high energy electron diffraction
dyfrakcja *f* **promieniowania rentgenowskiego** X-ray diffraction
dyfrakcja *f* **światła** light diffraction
dyfraktogram *m* X-ray pattern
dyfraktogram *m* **kryształu wahanego** X-ray oscillating-crystal pattern
dyfraktogram *m* **obrotowy** X-ray rotating-crystal pattern
dyfraktogram *m* **proszkowy** X-ray powdered-crystal pattern
dyfraktogram *m* **warstwicowy** X-ray rotating-crystal pattern
dyfraktometr *m* **rentgenowski** X-ray diffractometer
dyfraktometria *f* diffractometry
dyfuzja *f* diffusion
dyfuzja *f* **molekularna** molecular diffusion term
dyfuzja *f* **molekularna osiowa** molecular diffusion term
dyfuzja *f* **podłużna** molecular diffusion term
dyfuzja *f* **wirowa** eddy diffusion term
dyfuzja *f* **wirowa osiowa** eddy diffusion term
dylatometr *m* dilatometer
dynoda *f* dynode
dysocjacja *f* dissociation
dysocjacja *f* **elektrolityczna** electrolytic dissociation
dysocjacja *f* **pod wpływem pola elektrycznego** dissociation field effect
dysocjacja *f* **termiczna** thermal dissociation
dyspersja *f* dispersion
dyspersja *f* **anomalna** anomalous dispersion
dyspersja *f* **kątowa** angular dispersion
dyspersja *f* **kątowa kryształu analizującego** angular dispersion of an analysing crystal
dyspersja *f* **liniowa** linear dispersion
dyspersja *f* **normalna** normal dispersion
dyspersja *f* **skręcalności optycznej** optical rotatory dispersion
dyspersja *f* **statystyczna** dispersion
dyspersja *f* **światła** light dispersion
dystans *m* **rozwijania** mobile phase distance
dystrybuanta *f* distribution funktion
dzielnik *m* **próbki** sample splitter

E

efekt *m* absorpcji absorption effect
efekt *m* asymetrii relaxation-time effect
efekt *m* Augera Auger effect
efekt *m* batochromowy bathochromic shift
efekt *m* chemiczny chemical effect
efekt *m* Comptona Compton effect
efekt *m* Debye'a i Falkenhagena
  Debye-Falkenhagen's effect
efekt *m* Dorna sedimentation potential
efekt *m* dysocjacyjny pola elektrycznego
  dissociation field effect
efekt *m* elektroforetyczny electrophoretic effect
efekt *m* elektrostatyczny effect of electrostatic
  charges
efekt *m* fotoelektryczny photoelectric effect
efekt *m* granulacji grain size effect
efekt *m* hiperchromowy hyperchromic effect
efekt *m* hipochromowy hypochromic effect
efekt *m* hipsochromowy hypsochromic shift
efekt *m* izotopowy isotope effect
efekt *m* ładowania elektrycznego charging effect
efekt *m* matrycy 1. matrix effect 2. matrix
  effect (*in X-ray fluorescence analysis*)
efekt *m* mineralogiczny mineralogical effect
efekt *m* Mössbauera Mössbauer effect
efekt *m* natężenia pola elektrycznego Wien's effect
efekt *m* obcych jonów salt effect
efekt *m* pamięci wody water memory effect
efekt *m* Ramana Raman effect
efekt *m* Ramana rezonansowy resonance
  Raman effect
efekt *m* Ramana stymulowany stimulated
  Raman effect
efekt *m* relaksacyjny relaxation-time effect
efekt *m* rozcieńczania dilution effect
efekt *m* różnicujący differentiating effect
efekt *m* składników towarzyszących matrix effect
efekt *m* solny salt effect
efekt *m* Tyndalla Tyndall effect
efekt *m* uziarnienia grain size effect
efekt *m* Wiena Wien's effect
efekt *m* wspólnego jonu common ion effect
efekt *m* wyrównujący levelling effect
efekt *m* wzmocnienia enhancement effect
efekt *m* Zeemana Zeeman effect
efekt *m* ziarnistości grain size effect
efekty *mpl* elektrokapilarne electrocapillary
  phenomena
efekty *mpl* elektrokinetyczne electrokinetic
  phenomena
ekskluzja *f* exclusion
ekskluzja *f* jonów ion exclusion
eksperyment *m* experiment
eksperyment *m* bierny passive experiment
eksperyment *m* czynny active experiment
eksperyment *m* planowany active experiment
esktrahent *m* extractant
ekstrakcja *f* extraction
ekstrakcja *f* ciągła continuous extraction
ekstrakcja *f* ciecz-ciało stałe leaching
ekstrakcja *f* ciecz-ciecz liquid-liquid extraction
ekstrakcja *f* okresowa periodic extraction
ekstrakcja *f* przeciwprądowa countercurrent
  extraction

ekstrakcja *f* w układzie ciecz-ciało stałe leaching
ekstrakcja *f* w układzie ciecz-ciecz liquid-liquid
  extraction
ekstrakt *m* extract
ekstraktor *m* extractor
ekstynkcja *f* absorbance
eksykator *m* desiccator
eksykator *m* próżniowy vacuum desiccator
elektroanaliza *f* electroanalysis
elektroanaliza *f* chemiczna electroanalysis
elektrochemia *f* electrochemistry
elektroda *f* 1. electrode 2. electrode, half cell
  (*in electrochemistry*)
elektroda *f* amalgamatowa amalgam electrode
elektroda *f* antymonowa antimony electrode
elektroda *f* chinhydrynowa quinhydrone
  electrode
elektroda *f* chlorosrebrowa silver-silver chloride
  electrode
elektroda *f* doskonale niepolaryzowalna ideal
  non-polarizable electrode
elektroda *f* doskonale polaryzowalna ideal
  polarizable electrode
elektroda *f* drugiego rodzaju electrode of the
  second kind
elektroda *f* dyskowa wirująca 1. rotating disk
  electrode (*in arc and spark spectroscopy*)
  2. rotating disk electrode (*in electrochemistry*)
elektroda *f* enzymatyczna enzyme-substrate
  electrode
elektroda *f* gazowa gas electrode
elektroda *f* gazowa selektywna gas sensing
  electrode
elektroda *f* jonoselektywna ion-selective electrode
elektroda *f* jonoselektywna membranowa
  membrane ion-selective electrode
elektroda *f* jonoselektywna szklana glass
  membrane electrode
elektroda jonoselektywna uczulona
  sensitized ion-selective electrode
elektroda *f* jonoselektywna z matrycą sztywną
  rigid matrix electrode
elektroda *f* jonoselektywna z membraną ciekłą
  liquid membrane ion-selective electrode
elektroda *f* jonoselektywna z membraną
  heterogeniczną heterogeneous membrane
  ion-selective electrode
elektroda *f* jonoselektywna z membraną
  homogeniczną homogeneous membrane
  ion-selective electrode
elektroda *f* jonoselektywna z membraną
  krystaliczną crystalline membrane
  ion-selective electrode
elektroda *f* jonoselektywna z membraną stałą
  solid-state membrane ion-selective electrode
elektroda *f* jonoselektywna z membraną
  sztywną rigid matrix electrode
elektroda *f* jonoselektywna z membraną
  z nośnikiem obojętnym neutral-carrier
  membrane ion-selective electrode
elektroda *f* jonoselektywna z membraną
  z wymieniaczem jonowym ciekłym liquid
  ion-exchange membrane electrode
elektroda *f* jonoselektywna z powleczonym
  drutem coated wire ion-selective electrode

elektroda *f* jonoselektywna z trudno
rozpuszczalnym materiałem aktywnym
precipitate-based ion-selective electrode
elektroda *f* jonoselektywna ze stałym kontaktem
all-solid-state ion-selective electrode
elektroda *f* jonospecyficzna ion-selective
electrode
elektroda *f* kalomelowa calomel electrode
elektroda *f* kalomelowa nasycona
saturated-calomel electrode
elektroda *f* kombinowana combination electrode
elektroda *f* membranowa membrane
ion-selective electrode
elektroda *f* membranowa ciekła liquid membrane
ion-selective electrode
elektroda *f* membranowa heterogeniczna
heterogeneous membrane ion-selective
electrode
elektroda *f* membranowa homogeniczna
homogeneous membrane ion-selective
electrode
elektroda *f* membranowa krystaliczna
crystalline membrane ion-selective electrode
elektroda *f* membranowa stała solid-state
membrane ion-selective electrode
elektroda *f* membranowa szklana glass
membrane electrode
elektroda *f* membranowa z nośnikiem
obojętnym neutral-carrier membrane
ion-selective electrode
elektroda *f* membranowa z wymieniaczem
jonowym ciekłym liquid ion-exchange
membrane electrode
elektroda *f* metalowa metal-metal ion electrode
elektroda *f* mieszana complex electrode
elektroda *f* nieodwracalna irreversible electrode
elektroda *f* nierównowagowa irreversible
electrode
elektroda *f* odniesienia reference electrode
elektroda *f* odniesienia wewnętrzna internal
reference electrode
elektroda *f* odwracalna reversible electrode
elektroda *f* pehametryczna pH electrode
elektroda *f* pH pH electrode
elektroda *f* pierwszego rodzaju electrode of the
first kind
elektroda *f* platynowa platinum electrode
elektroda *f* pomocnicza 1. auxiliary electrode
(*in electrochemistry*) 2. supporting electrode
(*in arc and spark spectroscopy*)
elektroda *f* porowata porous cup electrode
elektroda *f* porównawcza reference electrode
elektroda *f* porównawcza wewnętrzna internal
reference electrode
elektroda *f* pracująca working electrode
elektroda *f* przesączalna porous cup electrode
elektroda *f* redoksowa redox electrode
elektroda *f* równowagowa reversible electrode
elektroda *f* rtęciowa kapiąca dropping-mercury
electrode
elektroda *f* rtęciowa kroplowa kapiąca
dropping-mercury electrode
elektroda *f* rtęciowa kroplowa wisząca hanging
mercury drop electrode
elektroda *f* rtęciowa strumieniowa streaming
mercury electrode

elektroda *f* siarczanowo-rtęciowa
mercury-mercurous sulfate electrode
elektroda *f* srebrna silver electrode
elektroda *f* stała solid electrode
elektroda *f* szklana glass membrane electrode
elektroda *f* szklana wodorowa glass pH electrode
elektroda *f* talamidowa thalamid electrode
elektroda *f* tlenkowa oxide electrode
elektroda *f* tlenkowortęciowa mercury-mercuric
oxide electrode
elektroda *f* tlenowa oxygen gas electrode
elektroda *f* tlenowa Clarka Clark oxygen-sensing
probe
elektroda *f* trzeciego rodzaju electrode of the
third kind
elektroda *f* wirująca rotating disk electrode
elektroda *f* wirująca drutowa rotating-wire
electrode
elektroda *f* wirująca dyskowa z pierścieniem
rotating-ring disk electrode
elektroda *f* wodorowa hydrogen-gas electrode
elektroda *f* wodorowa normalna standard
hydrogen electrode
elektroda *f* wodorowa standardowa standard
hydrogen electrode
elektroda *f* wskaźnikowa indicator electrode
elektroda *f* z badanego materiału self-electrode
elektroda *f* z matrycą sztywną rigid matrix
electrode
elektroda *f* z membraną ciekłą liquid membrane
ion-selective electrode
elektroda *f* z membraną heterogeniczną
heterogeneous membrane ion-selective
electrode
elektroda *f* z membraną homogeniczną
homogeneous membrane ion-selective
electrode
elektroda *f* z membraną jonoselektywną
membrane ion-selective electrode
elektroda *f* z membraną krystaliczną
crystalline membrane ion-selective electrode
elektroda *f* z membraną stałą solid-state
membrane ion-selective electrode
elektroda *f* z membraną sztywną rigid matrix
electrode
elektroda *f* z membraną z nośnikiem obojętnym
neutral-carrier membrane ion-selective
electrode
elektroda *f* z membraną z wymieniaczem
jonowym ciekłym liquid ion-exchanger
membrane electrode
elektroda *f* z pasty węglowej carbon-paste
electrode
elektroda *f* z przerwą powietrzną air-gap
electrode
elektroda *f* z węgla szklistego glassy carbon
electrode
elektroda *f* z wymieniaczem jonowym ciekłym
liquid ion-exchange membrane electrode
elektroda *f* złożona complex electrode
elektrodializa *f* electrodialysis
elektroendoosmoza *f* electro-endoosmosis
elektroforeza *f* electrophoresis
elektroforogram *m* electrophorogram
elektrografia *f* electrography
elektrograwimetria *f* electrogravimetry

elektrograwimetria f przy stałym prądzie constant-current electrogravimetry
elektrograwimetria f samorzutna internal electrogravimetry
elektrograwimetria f stałoprądowa constant-current electrogravimetry
elektrograwimetria f wewnętrzna internal electrogravimetry
elektrograwimetria f z kontrolowanym potencjałem controlled-potential electrogravimetry
elektrokapilarność f electrocapillary phenomena
elektrolit m electrolyte
elektrolit m amfoteryczny ampholyte
elektrolit m podstawowy supporting electrolyte
elektroliza f electrolysis
elektroliza f przy stałym prądzie constant-current electrolysis
elektroliza f stałoprądowa constant-current electrolysis
elektroliza f wewnętrzna internal electrolysis
elektroliza f z kontrolowanym potencjałem controlled-potential electrolysis
elektrolizer m electrolytic cell
elektroluminescencja f electroluminescence
electron m niesparowany unpaired electron
elektronowolt m electronvolt
elektroosmoza f electro-endoosmosis
elektrowydzielanie n electrodeposition
elektrowydzielanie n metali electrodeposition of metals
elipsometria f ellipsometry
eluant m eluent
eluat m eluate
elucja f elution
elucja f gradientowa gradient elution
elucja f izokratyczna isocratic elution
elucja f selektywna selective elution
elucja f specyficzna specific elution
elucja f stopniowana stepwise elution
elucja f ze skokową zmianą składu stepwise elution
eluent m eluent
elutriacja f elutriation
emisja f elektronu electron emission
emisja f wymuszona stimulated emission
emiter m emitter
endoosmoza f electro-endoosmosis
energia f aktywacji activation energy
energia f progowa threshold energy
entalpimetria f enthalpimetry
estymacja f przedziałowa interval estimation
estymacja f punktowa point estimation
estymator m estimator
estymator m najefektywniejszy most efficient estimator
estymator m nieobciążony unbiased estimator
etap m elementarny elementary step
etap m przeniesienia ładunku charge-transfer step

F

fala f adsorpcyjna adsorption wave
fala f polarograficzna polarographic wave
faza f ciekła stationary liquid phase
faza f nieruchoma stationary phase

faza f nieruchoma ciekła stationary liquid phase
faza f nieruchoma związana chemicznie chemically bonded stationary phase
faza f ruchoma mobile phase
faza f stacjonarna stationary phase
faza f stacjonarna ciekła stationary liquid phase
faza f związana chemicznie chemically bonded stationary phase
filtr m absorpcyjny absorption filter
filtr m gazowy filter tube
filtr m interferencyjny interference filter
filtr m neutralny neutral-density filter
filtr m optyczny optical filter
filtr m promieniowania rentgenowskiego X-ray filter
filtr m stopniowy step filter
filtr m szary neutral-density filter
filtr m świetlny light filter
filtracja f żelowa gel filtration
filtry mpl Rossa balanced filters
filtry mpl zbalansowane balanced filters
filtry mpl zrównoważone balanced filters
flokulacja f coagulation
fluorescencja f fluorescence
fluorescencja f rentgenowska X-ray fluorescence
fluorescencja f rezonansowa resonance fluorescence
fluorymetr m fluorimeter
fluorymetr m z filtrem filter fluorimeter
fluorymetria f fluorimetry
forma f kwasowa wymieniacza kationów acid form of a cation exchanger
forma f solna wymieniacza jonów salt form of an ion exchanger
forma f zasadowa wymieniacza anionów base form of an anion exchanger
formalność f formality
formalność f roztworu formality
fosforescencja f phosphorescence
fosforymetria f phosphorimetry
fotoaktywacja f photoactivation
fotodesorpcja f photodesorption
fotodetektor m photodetector
fotodioda f photodiode
fotoelektron m photoelectron
fotofrakcja f photofraction
fotojonizacja f photoionization
fotokolorymetr m photocolorimeter
fotoluminescencja f photoluminescence
fotometr m photometer
fotometr m płomieniowy flame photometer
fotometria f photometry
fotometria f płomieniowa flame emission spectroscopy
foton m photon
fotoneutron m photoneutron
fotoogniwo n photovoltaic cell
fotopik m photoelectric peak
fotopolarografia f photopolarography
fotopowielacz m photomultiplier tube
fotoprzewodnik m photoconductor
fototranzystor m phototransistor
fragmentacja f fragmentation
fragmenty mpl rozszczepienia fission fragments
frakcja f fraction

front *m* **fazy ruchomej** mobile phase front
front *m* **rozpuszczalnika** mobile phase front
funkcja *f* **OC** operating characteristic function
funkcja *f* **operacyjno-charakterystyczna**
    operating characteristic function
funkcja *f* **próby** statistic
funkcja *f* **regresji** regression
funkcja *f* **wzbudzenia** excitation function

G

galwanometr *m* galvanometer
galwanostat *m* galvanostat
gałąź *f* **dodatnia** R-branch
gałąź *f* **P** P-branch
gałąź *f* **Q** Q-branch
gałąź *f* **R** R-branch
gałąź *f* **ujemna** P-branch
gałąź *f* **zerowa** Q-branch
gaz *m* **nośny** carrier gas
gaz *m* **wymywający** carrier gas
generator *m* **neutronów** neutron generator
generowanie *n* **elektrolityczne titrantu**
    electrochemical generation of a titrant
geometria *f* **liczenia** counting geometry
geometria *f* **pomiaru** counting geometry
gęstościomierz *m* hydrometer
gęstość *f* **ładunku powierzchniowego** surface
    charge density
gęstość *f* **optyczna** 1. absorbance 2. blackening
gęstość *f* **prawdopodobieństwa** probability
    density function
gęstość *f* **prądu** current density
gęstość *f* **strumienia cząstek lub fotonów** flux
    density of particles or photons
globar *m* globar
głębia *f* **roztworu** bulk of a solution
głębokość *f* **czynna** effective depth
głowica *f* **pasma** band head
goniometr *m* **rentgenowski** X-ray goniometer
gradient *m* **strumienia cząstek lub fotonów** flux
    gradient of particles or photons
gramocząsteczka *f* gram molecule
gramorównoważnik *m* gram equivalent
granica *f* **ekskluzji** exclusion limit
granica *f* **faz** interface
granica *f* **krótkofalowa** short-wavelength limit
granica *f* **oznaczalności** limit of determination
granica *f* **wykrywalności** limit of detection
granice *fpl* **ufności** confidence limits
grawimetria *f* gravimetric analysis
grubość *f* **atmosfery jonowej** radius of ionic
    atmosphere
grubość *f* **warstwy absorbującej** path length
grubość *f* **warstwy próbki** path length
grupa *f* **analityczna** analytical group
grupa *f* **chromoforowa** chromophore
grupa *f* **auksochromowa** auxochrome
grupa *f* **batochromowa** bathochrome
grupa *f* **funkcyjna** functional group
grupa *f* **hipsochromowa** hypsochrome
grupa *f* **koordynująca** coordinating group
grupa *f* **kwasowa** acidic group
grupy *fpl* **funkcyjne jonogenne** ionogenic groups
grupy *fpl* **jonogenne** ionogenic groups

H

heksametylodisilazan *m* hexamethyldisilazane
heteropolikwas *m* heteropolyacid
hipersorpcja *f* hypersorption
hipoteza *f* **alternatywna** alternative hypothesis
hipoteza *f* **statystyczna** statistical hypothesis
hipoteza *f* **zerowa** null hypothesis
hipsochrom *m* hypsochrome
histereza *f* hysteresis
histogram *m* histogram
Hopkalit *m* Hopcalite
hydrozol *m* hydrosol
hydrożel *m* hydrogel

I

identyfikacja *f* identyfication
iloczyn *m* **rozpuszczalności** solubility product
iloraz *m* **fazowy** phase ratio
iloraz *m* **pęcznienia objętościowy** volume
    swelling ratio
iloraz *m* **podziału** concentration distribution
    ratio
iloraz *m* **podziału masowy** mass distribution
    ratio
iloraz *m* **podziału stężeniowy** concentration
    distribution ratio
iloraz *m* **sygnału i szumu** signal-to-noise ratio
ilość *f* **energii promienistej** radiant energy
ilość *f* **materii** amount of substance
ilość *f* **światła** luminous energy
immunoelektroforeza *f* immunoelectrophoresis
impedancja *f* impedance
implantacja *f* **jonów** ion implantation
indeks *m* **Kowacza** retention index
indeks *m* **retencji** retention index
induktor *m* inductor
interferometr *m* **optyczny** optical interferometer
interferometria *f* interferometry
interpolacja *f* interpolation
iskra *f* electrical spark
izobara *f* **adsorpcji** adsorption isobar
izochromata *f* isochrome
izopolikwas *m* isopolyacid
izostera *f* **adsorpcji** adsorption isostere
izotachoforeza *f* isotachophoresis
izoterma *f* isotherm
izoterma *f* **adsorpcji** adsorption isotherm
izoterma *f* **podziału** 1. partition isotherm
    2. distribution isotherm
izoterma *f* **sorpcji** sorption isotherm
izoterma *f* **wymiany jonowej** ion-exchange
    isotherm
izotop *m* **długożyciowy** long-lived radioisotope
izotop *m* **krótkożyciowy** short-lived radioisotope

J

jednostka *f* **masy atomowej** unified atomic mass
    unit
jodanometria *f* iodatometry
jodometria *f* iodimetry
jon *m* **cząsteczkowy** molecular ion
jon *m* **dodatni** positive ion

jon *m* dwubiegunowy ampholyte ion
jon *m* fragmentacyjny fragment ion
jon *m* macierzysty parent ion
jon *m* metastabilny metastable ion
jon *m* obojnaczy ampholyte ion
jon *m* pochodny daughter ion
jon *m* przegrupowany rearrangement ion
jon *m* ujemny negative ion
jonit *m* solid ion exchanger
jonit *m* bezwzględnie suchy absolutely dry ion
   exchanger
jonit *m* błonkowaty pellicular ion exchanger
jonit *m* makroporowaty macroporous ion
   exchanger
jonit *m* oleofilowy oleophilic ion exchanger
jonit *m* organiczny ion-exchange resin
jonit *m* pelikularny pellicular ion exchanger
jonit *m* suchy absolutely dry ion exchange
jonizacja *f* chemiczna chemical ionization
jonizacja *f* elektronowa electron-impact
   ionization
jonizacja *f* polem field ionization
jonizacja *f* powierzchniowa surface ionization
jonizacja *f* termiczna thermal ionization
jonizacja *f* źródłem iskrowym spark source
   ionization
jony *mpl* przeszkadzające interfering ions
jony *mpl* związane fixed ions

K

kajzer *m* kayser
kalorymetria *f* skaningowa różnicowa
   differential scanning calorimetry
kapilara *f* Ługgina Luggin-Haber capillary
kapilarki *fpl* do oznaczania temperatury
   topnienia melting-point capillaries
karta *f* kontrolna control chart
katalizator *m* Körbla Körbl catalyst
katalizator *m* Večeřy Večeřa catalyst
katarometr *m* thermal conductivity detector
kationit *m* cation exchanger
katoda *f* cathode
katodoluminescencja *f* cathodoluminescence
katolit *m* catholyte
kąt *m* błysku blaze angle
kąt *m* odbłysku glancing angle
kąt *m* połysku glancing angle
kąt *m* wyjścia take-off angle
kiur *m* curie
klucz *m* elektrolityczny salt bridge
koagulacja *f* coagulation
koagulacja *f* nieodwracalna irreversible
   coagulation
koagulacja *f* odwracalna reversible coagulation
kojony *mpl* co-ions
kolba *f* jodowa iodine flask
kolba *f* pomiarowa volumetric flask
kolba *f* pomiarowa z szyjką kalibrowaną
   volumetric flask with graduated neck
kolektor *m* frakcji fraction collector
kolektor *m* śladów trace collector
kolimacja *f* collimation
kolimator *m* collimator
kolimator *m* Sollera Soller collimator
kolorymetr *m* colorimeter

kolorymetr *m* fotoelektryczny photocolorimeter
kolorymetria *f* colorimetric analysis
kolorymetria *f* wizualna visual colorimetry
kolumna *f* chromatograficzna chromatographic
   column
kolumna *f* kapilarna capillary column
kolumna *f* o otwartym przekroju open-tube
   column
kolumna *f* o otwartym świetle open-tube column
kolumna *f* otwarta open-tube column
kolumna *f* OT open-tube column
kolumna *f* z wypełnieniem klasyczna packed
   column
kolumna *f* z wypełnieniem zwykła packed
   column
kolumny *fpl* łączone coupled columns
koło *n* ogniskujące focusing circle
koma *f* coma
komora *f* BN BN-chamber
komora *f* Brennera i Niederwiesera BN-chamber
komora *f* chromatograficzna chromatographic
   chamber
komora *f* chromatograficzna Brennera
   i Niederwiesera BN-chamber
komora *f* gorąca hot cell
komora *f* jonizacyjna ionization chamber
komora *f* nasycona saturated chamber
komora *f* nienasycona unsaturated chamber
komora *f* typu sandwicz S-chamber
komora *f* typu S S-chamber
komórka *f* fotoelektryczna photocell
komórka *f* fotowoltaiczna photovoltaic cell
kompleks *m* chelatowy chelate compound
kompleks *m* donorowo-akceptorowy
   donor-aceeptor complex
kompleks *m* wewnętrzny neutral chelate
kompleksometria *f* compleximetry
kompleksony *mpl* complexones
kondensator *m* dynamiczny vibrating condenser
kondensator *m* wibrujący vibrating condenser
konduktancja *f* conductance
konduktancja *f* elektrolityczna conductance
   of electrolyte
konduktometr *m* conductometer
konduktometria *f* conductometry
konduktometria *f* wielkiej częstości
   high-frequency conductometry
konduktywność *f* electrical conductivity
konduktywność *f* elektrolityczna electrolytic
   conductivity
konduktywność *f* jonowa molowa molar
   conductivity of ionic species B
konduktywność *f* molowa elektrolitu molar
   conductivity of electrolyte
konduktywność *f* molowa graniczna elektrolitu
   limiting molar conductivity of electrolyte
konduktywność *f* molowa jonowa graniczna
   limiting molar conductivity of ionic species B
kondycjonowanie *n* warstwy layer equilibration
konformacja *f* conformation
kontinuum *n* komptonowskie Compton
   continuum
kontrola *f* dyfuzyjna diffusion controlled
   process
kontrola *f* elektroaktywacyjna electroactivation
   control

kontrola *f* **kinetyczna** kinetic controlled process
konwekcja *f* convection
konwekcja *f* **naturalna** natural convection
konwencja *f* **IUPAC** sign convention
konwencja *f* **o znakach** sign convention
konwencja *f* **sztokholmska** sign convention
korek *m* **Schönigera** Schöniger stopper
korektor *m* **tła** background corrector
korelacja *f* correlation
kowariancja *f* covariance
krater *m* crater
krawędź *f* **absorpcji** absorption edge
krawędź *f* **komptonowska** Compton edge
kroplomierz *m* dropper
kryptonat *m* **promieniotwórczy** radioactive
 kryptonate
krystalizacja *f* crystallization
krystalizator *m* crystallizing dish
kryształ *m* **analizujący** analysing crystal
kryształ *m* **Johanna** Johann crystal
kryształ *m* **Johannsona** Johannson crystal
kryształ *m* **mieszany** mixed crystal
kryterium *n* **optymalności planu eksperymentu**
 criterion of optimality of experimental design
krzywa *f* **absorpcji** absorption curve
krzywa *f* **analityczna** analytical curve
krzywa *f* **charakterystyczna emulsji** emulsion
 characteristic curve
krzywa *f* **czułości wagi** sensitivity curve of
 balance
krzywa *f* **ekstrakcji** extraction curve
krzywa *f* **elucji** elution curve
krzywa *f* **miareczkowania** titration curve
krzywa *f* **miareczkowania alkacymetrycznego**
 neutralization titration curve
krzywa *f* **miareczkowania**
 kompleksometrycznego compleximetric
 titration curve
krzywa *f* **miareczkowania redoks** redox
 titration curve
krzywa *f* **miareczkowania strąceniowego**
 precipitation titration curve
krzywa *f* **ogrzewania** heating curve
krzywa *f* **polarograficzna** polarogram
krzywa *f* **polaryzacyjna** polarization curve
krzywa *f* **prąd-potencjał** polarization curve
krzywa *f* **przebicia** breakthrough curve
krzywa *f* **rozpadu promieniotwórczego** decay
 curve
krzywa *f* **termiczna** thermal curve
krzywa *f* **woltamperometryczna** polarization
 curve
krzywa *f* **wzorcowa** analytical curve
krzywa *f* **wzorcowania emulsji** emulsion
 calibration curve
krzywa *f* **zaczernienia emulsji** emulsion
 characteristic curve
kulograwimetria *f* coulogravimetry
kulometr *m* coulometer
kulometr *m* **chemiczny** chemical coulometer
kulometr *m* **miedziowy** copper coulometer
kulometria *f* coulometry
kulometria *f* **amperostatyczna** controlled-current
 coulometry
kulometria *f* **galwanostatyczna**
 controlled-current coulometry

kulometria *f* **polarograficzna** polarographic
 coulometry
kulometria *f* **potencjostatyczna** controlled-
 potential coulometry
kulometria *f* **z elektrodą kapiącą** dropping
 electrode coulometry
kulometria *f* **z kontrolowanym natężeniem prądu**
 controlled-current coulometry
kulometria *f* **z kontrolowanym potencjałem**
 controlled-potential coulometry
kulometria *f* **z kontrolowanym prądem**
 i potencjometryczną detekcją punktu
 końcowego controlled-current coulometry
 with potentiometric end-point detection
kupelacja *f* cupellation
kuweta *f* **absorpcyjna** absorption cell
kuweta *f* **grafitowa** graphite cuvette
kwadrat *m* **łaciński** latin square
kwadrupol *m* quadrupole
kwant *m* γ gamma quantum
kwantyl *m* **rozkładu prawdopodobieństwa**
 quantile of a probability distribution
kwantyl *m* **rzędu p rozkładu prawdopodobieństwa**
 quantile of a probability distribution

L

laboratorium *n* **gorące** hot laboratory
lampa *f* **deuterowa** deuterium discharge lamp
lampa *f* **rentgenowska** X-ray tube
lampa *f* **wodorowa** hydrogen discharge lamp
lampa *f* **wolframowa** tungsten incandescent
 lamp
lampa *f* **wyładowcza** discharge lamp
lampa *f* **wyładowcza z parami metalu**
 metal-vapour lamp
lampa *f* **z katodą wnękową** hollow-cathode
 lamp
lampa *f* **z wyładowaniem bezelektrodowym**
 electrodeless discharge lamp
laser *m* laser
laser *m* **barwnikowy** dye laser
laser *m* **przestrajalny** tunable laser
lejek *m* **Büchnera** Büchner funnel
lejek *m* **do materiałów sypkich** powder funnel
lejek *m* **do sączenia na gorąco** funnel heater
lejek *m* **karbowany** ribbed funnel
lejek *m* **Schotta** filter funnel with sintered disk
lejek *m* **sitowy** Büchner funnel
lejek *m* **szklany** glass funnel
lejek *m* **z filtrem ze spiekanego szkła** filter
 funnel with sintered disk
liczba *f* **falowa** wave number
liczba *f* **jodowa** iodine number
liczba *f* **półek efektywnych** number of effective
 plates
liczba *f* **półek teoretycznych** number of
 theoretical plates
liczba *f* **półek teoretycznych efektywna** number
 of effective plates
liczba *f* **przenoszenia jonu B** transport number
 of ionic species B
liczba *f* **stopni swobody** number of degrees of
 freedom
liczba *f* **transportu jonu B** transport number
 of ionic species B

liczebność *f* **próby** sample size
licznik *m* **2π** 2π-counter
licznik *m* **4π** 4π-counter
licznik *m* **G-M** Geiger-Müller counter tube
licznik *m* **Geigera i Müllera** Geiger-Müller
counter tube
licznik *m* **okienkowy** window counter
licznik *m* **pęcherzyków** bubble gauge
licznik *m* **promieniowania** radiation counter
licznik *m* **proporcjonalny** proportional counter
licznik *m* **proporcjonalny przepływowy**
proportional gas-flow counter
licznik *m* **scyntylacyjny** scintillation counter
licznik *m* **studzienkowy** well-type scintillation
counter
licznik *m* **wnękowy** well-type scintillation
counter
liczność *f* **materii** amount of substance
liczność *f* **próby** sample size
linearyzacja *f* linearization
linia *f* **analityczna** analytical line
linia *f* **czołowa fazy ruchomej** mobile phase
front
linia *f* **fotoelektronowa** photoelectron line
linia *f* **podstawowa** baseline
linia *f* **rayleighowska** Rayleigh line
linia *f* **spektralna** spectral line
linia *f* **startu** starting line
linia *f* **widmowa** spectral line
linia *f* **wzbudzająca 1.** exciting line **2.** Rayleigh
line
linia *f* **wzorca wewnętrznego** internal reference
line
linie *fpl* **antystokesowskie** anti-Stokes lines
linie *fpl* **homologiczne** homologous lines
linie *fpl* **ostatnie** raies ultimes
linie *fpl* **Ramana** Raman lines
linie *fpl* **stokesowskie** Stokes lines
logarytmiczne **prawo** *n* **podziału** logarithmic
distribution law
luminancja *f* luminance
luminescencja *f* luminescence

Ł

łapacz *m* **kropli** safety head
ława *f* **optyczna** optical bench
łódeczka *f* **do spalań** combustion boat
ługowanie *n* leaching
łuk *m* **elektryczny** electrical arc
łuk *m* **prądu stałego** direct current arc
łuk *m* **prądu zmiennego** alternating current arc

M

macierz *f* **doświadczenia** design matrix
macierz *f* **dyspersji** covariance matrix
macierz *f* **korelacji** correlation matrix
macierz *f* **kowariancji** covariance matrix
macierz *f* **planowania** design matrix
macierz *f* **planu** design matrix
macierz *f* **wejść** design matrix
makroanaliza *f* macroanalysis
makroskładnik *m* macro-component
maksima *f* **polarograficzne** polarographic
maxima

maksimum *n* **piku** peak maximum
manganometria *f* manganometry
marker *m* marker
masa *f* **atomowa** relative atomic mass
masa *f* **atomowa względna** relative atomic mass
masa *f* **cząsteczkowa** relative molecular mass
masa *f* **cząsteczkowa względna** relative
molecular mass
masa *f* **molowa** molar mass
masa *f* **spoczynkowa** rest mass
masa *f* **stała** constant mass
maskowanie *n* **jonów** masking of ions
matryca *f* matrix
mediana *f* **z próby** sample median
membrana *f* **ciekła** liquid membrane
membrana *f* **elektrodowa** ion-selective membrane
membrana *f* **heterogeniczna** heterogeneous
membrane
membrana *f* **homogeniczna** homogeneous
membrane
membrana *f* **jonoselektywna** ion-selective
membrane
membrana *f* **jonowymienna** ion-exchange
membrane
membrana *f* **krystaliczna** crystalline membrane
membrana *f* **przepuszczalna dla gazów**
gas-permeable membrane
membrana *f* **stała** solid-state membrane
membrana *f* **z nośnikiem nienaładowanym**
neutral-carrier membrane
membrana *f* **z nośnikiem obojętnym**
neutral-carrier membrane
membrana *f* **z wymieniaczem jonowym ciekłym**
liquid ion exchanger membrane
merkurometria *f* mercurometry
merkurymetria *f* mercurimetry
metalowskaźniki *mpl* metal indicators
metoda *f* **absorpcji neutronów** neutron
absorptiometry
metoda *f* **absorpcji promieniowania β**
beta-particle absorptiometry
metoda *f* **absorpcji promieniowania γ**
gamma-ray absorptiometry
metoda *f* **absorpcji promieniowania jądrowego**
nuclear radiation absorptiometry
metoda *f* **admitancji niefaradajowskiej**
measurement of nonfaradaic admittance
metoda *f* **amperostatyczna** galvanostatic method
metoda *f* **analityczna** analytical method
metoda *f* **antykoincydencji** anticoincidence
technique
metoda *f* **badania próbki w postaci pastylki**
pellet technique
metoda *f* **badania próbki w postaci zawiesiny**
mull technique
metoda *f* **bezpośrednia** direct method
metoda *f* **bilansu losowego** random balance
method
metoda *f* **Cariusa** Carius method
metoda *f* **centygramowa** centigram method
metoda *f* **Clarka** Clark's method
metoda *f* **Covella** Covell's method
metoda *f* **Craiga** countercurrent distribution
metoda *f* **czołowa** frontal chromatography
metoda *f* **dodatków** standard addition method

metoda *f* dodawania wzorca standard addition method
metoda *f* Dumasa Dumas method
metoda *f* Dumasa i Pregla Dumas-Pregl method
metoda *f* emanacyjna emanometric method
metoda *f* EVOP evolutionary operation
metoda *f* fali odbitej reflected wave method
metoda *f* fali przechodzącej transmitted wave method
metoda *f* galwanostatyczna galvanostatic method
metoda *f* Grana Gran's plot
metoda *f* izochromat rentgenowskiego promieniowania hamowania bremsstrahlung isochrome spectroscopy
metoda *f* Kelvina Kelvin method
metoda *f* Kjeldahla Kjeldahl method
metoda *f* koincydencji coincidence technique
metoda *f* kompensacyjna pomiaru SEM Poggendorff's compensation method
metoda *f* kulostatyczna coulostatic method
metoda *f* Liebiga Liebig's method
metoda *f* linii podstawowej base-line method
metoda *f* małych wahnień method of short swings
metoda *f* mikrogramowa microgram method
metoda *f* miligramowa milligram method
metoda *f* Mohra Mohr's method
metoda *f* mokra wet method
metoda *f* monokomparatorów monostandard method
metoda *f* monostandardów monostandard method
metoda *f* najmniejszej sumy kwadratów least-squares method
metoda *f* najmniejszych kwadratów least-squares method
metoda *f* największej wiarygodności maximum likelihood method
metoda *f* nanogramowa nanogram method
metoda *f* odniesienia reference method
metoda *f* osłabionego całkowitego odbicia attenuated total reflectance
metoda *f* parametrów podstawowych fundamental parameter method
metoda *f* pastylkowania pellet technique
metoda *f* pikogramowa picogram method
metoda *f* Poggendorffa Poggendorff's compensation method
metoda *f* pośrednia indirect method
metoda *f* potencjostatyczna potentiostatic method
metoda *f* przemiatania częstotliwości frequency-sweep method
metoda *f* przemiatania polem field-sweep method
metoda *f* Rasta Rast method
metoda *f* regresji wielokrotnej multiple regression method
metoda *f* rozcieńczenia izotopowego isotope dillution analysis
metoda *f* rugująca displacement chromatography
metoda *f* Schönigera Schöniger method
metoda *f* spowalniania neutronów analysis by neutron slowing-down
metoda *f* standardu wewnętrznego internal-standard method

metoda *f* stapiania fusion method
metoda *f* sucha dry method
metoda *f* Unterzauchera oznaczania tlenu Unterzaucher method for oxygen determination
metoda *f* uwalniania wskaźnika promieniotwórczego radio-release method
metoda *f* Volharda oznaczania halogenków Volhard method for determination of halides
metoda *f* Volharda oznaczania manganu Volhard method for determination of manganese
metoda *f* wahnień method of swings
metoda *f* wiszącej kropli hanging-drop method
metoda *f* wskaźników promieniotwórczych radioactive tracer method
metoda *f* współczynników empirycznych empirical coefficient method
metoda *f* wstępująca ascending development
metoda *f* wymiany promieniotwórczej radio-exchange method
metoda *f* wymiany wskaźnika promieniotwórczego radio-exchange method
metoda *f* wzbogacania concentration method
metoda *f* wzorca wewnętrznego internal-standard method
metoda *f* zagęszczania concentration method
metoda *f* zmiennego kąta wyjścia variable take-off angle method
metoda *f* znaczników promieniotwórczych radioactive tracer method
metoda *f* znormalizowana standardized method
metoda *f* zstępująca descending development
metody *fpl* analizy powierzchni surface sensitive techniques
metody *fpl* manometryczne manometric methods
metody *fpl* relaksacyjne transient methods
metody *fpl* stacjonarne steady-state methods
mezoanaliza *f* mesoanalysis
miano *n* titre
mianowanie *n* standardization
miareczkowanie *n* titration
miareczkowanie *n* acydymetryczne acidimetric titration
miareczkowanie *n* alkacymetryczne acid-base titration
miareczkowanie *n* alkalimetryczne alkalimetric titration
miareczkowanie *n* amperometryczne amperometric titration with one indicator electrode
miareczkowanie *n* amperometryczne z dwiema elektrodami spolaryzowanymi amperometric titration with two indicator electrodes
miareczkowanie *n* amperometryczne z elektrodą rtęciową kroplową amperometric titration with a dropping mercury electrode
miareczkowanie *n* amperometryczne z jedną elektrodą spolaryzowaną amperometric titration with one indicator electrode
miareczkowanie *n* argentometryczne argentimetric titration
miareczkowanie *n* bezpośrednie direct titration
miareczkowanie *n* biamperometryczne amperometric titration with two indicator electrodes

miareczkowanie *n* bipotencjometryczne
controlled-current potentiometric titration
with two indicator electrodes
miareczkowanie *n* cerometryczne cerimetric
titration
miareczkowanie *n* dielektrometryczne
dielectrometric titration
miareczkowanie *n* do punktu martwego
amperometric titration with two indicator
electrodes
miareczkowanie *n* ekstrakcyjne extractive
titration
miareczkowanie *n* entalpimetryczne
thermometric titration
miareczkowanie *n* fazowe phase titration
miareczkowanie *n* fluorymetryczne fluorimetric
titration
miareczkowanie *n* fotometryczne photometric
titration
miareczkowanie *n* gazometryczne gasometric
titration
miareczkowanie *n* heterometryczne
turbidimetric titration
miareczkowanie *n* inwersyjne inverse titration
miareczkowanie *n* jodanometryczne iodate
titration
miareczkowanie *n* jodometryczne iodimetric
titration
miareczkowanie *n* kolorymetryczne colorimetric
titration
miareczkowanie *n* kompleksometryczne
compleximetric titration
miareczkowanie *n* konduktometryczne
conductometric titration
miareczkowanie *n* konduktometryczne wielkiej
częstości high-frequency conductometric
titration
miareczkowanie *n* kontrolne control titration
miareczkowanie *n* kulometryczne
controlled-current coulometry
miareczkowanie *n* kulometryczne
potencjometryczne potentiometric
coulometric titration
miareczkowanie *n* kulometryczne
z kontrolowanym potencjałem
controlled-potential coulometry
miareczkowanie *n* merkurometryczne
mercurometric titration
miareczkowanie *n* merkurymetryczne
mercurimetric titration
miareczkowanie *n* nefelometryczne
nephelometric titration
miareczkowanie *n* numerometryczne numerical
titration
miareczkowanie *n* odwrotne back-titration
miareczkowanie *n* oksydymetryczne oxidimetric
titration
miareczkowanie *n* oscylometryczne
high-frequency conductometric titration
miareczkowanie *n* podstawieniowe replacement
titration
miareczkowanie *n* polarograficzne
polarographic titration
miareczkowanie *n* polarometryczne polarometric
titration
miareczkowanie *n* pośrednie indirect titration

miareczkowanie *n* potencjometryczne
potentiometric titration
miareczkowanie *n* potencjometryczne różnicowe
differential potentiometric titration
miareczkowanie *n* potencjometryczne
różniczkowe derivative potentiometric
titration
miareczkowanie *n* potencjometryczne
różniczkowe drugiej pochodnej
second-derivative potentiometric titration
miareczkowanie *n* potencjometryczne
różniczkowe odwrotne inverse derivative
potentiometric titration
miareczkowanie *n* potencjometryczne
różniczkowe pierwszej pochodnej derivative
potentiometric titration
miareczkowanie *n* potencjometryczne
z kontrolowanym prądem controlled-current
potentiometric titration
miareczkowanie *n* potencjometryczne
z kontrolowanym prądem z dwiema
elektrodami spolaryzowanymi
controlled-current potentiometric titration
with two indicator electrodes
miareczkowanie *n* radiometryczne radiometric
titration
miareczkowanie *n* redoks oxidation-reduction
titration
miareczkowanie *n* redoksymetryczne
oxidation-reduction titration
miareczkowanie *n* reduktometryczne
reductometric titration
miareczkowanie *n* spektrofotometryczne
spectrophotometric titration
miareczkowanie *n* strąceniowe precipitation
titration
miareczkowanie *n* substytucyjne replacement
titration
miareczkowanie *n* sukcesywne successive
titration
miareczkowanie *n* ślepe blank titration
miareczkowanie *n* termometryczne
thermometric titration
miareczkowanie *n* turbidymetryczne
turbidimetric titration
miareczkowanie *n* w roztworze niewodnym
non-aqueous titration
miareczkowanie *n* wagowe weight titration
miareczkowanie *n* wizualne visual titration
miareczkowanie *n* wobec dwóch wskaźników
double-indicator titration
migracja *f* jonów migration of ions
mikroanaliza *f* microanalysis
mikroanaliza *f* elementarna według Pregla
Pregl procedure
mikroanalizator *m* rentgenowski electron
microprobe X-ray analyser
mikroazotometr *m* micronitrometer
mikrobiureta *f* microburette
mikrofale *fpl* microwaves
mikrofotometr *m* microphotometer
mikrokapilarki *fpl* capillary tubes
mikrokulometria *f* polarographic coulometry
mikrometoda *f* milligram method
mikropipetka *f* micropipette
mikroskładnik *m* micro-component

mikroskop *m* elektronowy rastrowy scanning electron microscope

mikroskop *m* elektronowy skaningowy scanning electron microscope

mikroskop *m* Koflera Kofler micro melting point apparatus

mikroskopia *f* elektronów Augera Auger-electron microscopy

mikroskopia *f* fotoelektronów photoelectron microscopy

mikroskopia *f* jonów wtórnych secondary ion imaging mass spectroscopy

mikroskopia *f* skaningowa elektronów Augera Auger-electron microscopy

mikrosonda *f* elektronowa electron microprobe X-ray analyser

mikroślady *mpl* microtraces

milikulometria *f* polarographic coulometry

mineralizacja *f* mineralization

mineralizacja *f* zapłonowa flash combustion

mnożnik *m* analityczny analytical factor

mnożnik *m* przeliczeniowy analytical factor

moc *f* jonowa ionic strength of a solution

moc *f* jonowa roztworu ionic strength of a solution

moc *f* promieniowania radiant flux

moc *f* rozpuszczalnika solvent strength

moc *f* testu power of a test

moda *f* mode

model *m* I fixed effects model

model *m* II random effects model

model *m* losowy random effects model

model *m* stały fixed effects model

mol *m* 1. gram molecule 2. mole

molalność *f* molality

monochromator *m* monochromator

monochromator *m* promieniowania wzbudzającego excitation monochromator

monochromator *f* promieniowania wzbudzonego emission monochromator

monowarstwa *f* monolayer

mostek *m* elektrolityczny salt bridge

mostek *m* Wheatstone'a Wheatstone bridge

multiplet *m* multiplet

N

nachylenie *n* nernstowskie Nernstian slope

naczynko *n* elektrolityczne electrolytic cell

naczynko *n* wagowe weighing bottle

naczynko *n* wagowe leżące piggie

nadfiolet *m* ultraviolet light

nadfiolet *m* bliski near ultraviolet

nadfiolet *m* daleki far ultraviolet

nadfiolet *m* próżniowy vacuum ultraviolet

nadfiolet *m* Schumanna vacuum ultraviolet

nadnapięcie *n* overpotential

nadpotencjał *m* overpotential

nadpotencjał *m* dyfuzyjny diffusion overpotential

nadpotencjał *m* elektroaktywacyjny activation overpotential

nadpotencjał *m* kinetyczny reaction overpotential

nadpotencjał *m* krystalizacji crystallization overpotential

nadpotencjał *m* oporowy resistance overpotential

nadpotencjał *m* pseudo-oporowy pseudo-resistance overpotential

nadpotencjał *m* reakcji activation overpotential

nadpotencjał *m* reakcyjny reaction overpotential

nadpotencjał *m* stężeniowy concentration overpotential

nadpotencjał *m* wydzielania wodoru hydrogen overpotential

nadton *m* overtone band

nadton *m* drugi second overtone band

nadton *m* pierwszy first overtone band

nakładanie *n* się impulsów pile-up

nakładanie *n* się linii line coincidence

nanoślady *mpl* nanotraces

napełnianie *n* kolumn na sucho dry-packing

napełnianie *n* kolumn zawiesiną slurry packing

napełnianie *n* kolumn zawiesiną o wyrównanej gęstości balanced-density slurry packing

napięcie *n* elektryczne electric potential difference

napięcie *n* Galvaniego Galvani potential difference

napięcie *n* ogniwa galwanicznego electric potential difference of a galvanic cell

napięcie *n* robocze burning voltage

napięcie *n* rozkładu decomposition voltage

napięcie *n* Volty Volta potential difference

nastawianie *n* miana standardization

nasycenie *n* saturation

nasycenie *n* komory chamber saturation

naświetlenie *n* radiant exposure

natężenie *n* prądu elektrycznego electric current

natężenie *n* napromieniania irradiance

natężenie *n* oświetlenia illuminance

natężenie *n* promieniowania rentgenowskiego X-radiation intensity

nauka *f* o powierzchni surface science

nebulizacja *f* nebulization

nebulizer *m* nebulizer

nebulizer *m* pneumatyczny pneumatic nebulizer

nebulizer *m* ultradźwiękowy ultrasonic nebulizer

nefelometr *m* nephelometer

nefelometria *f* nephelometry

neutrony *mpl* epikadmowe epicadmium neutrons

neutrony *mpl* epitermiczne epithermal neutrons

neutrony *mpl* nadkadmowe epicadmium neutrons

neutrony *mpl* natychmiastowe prompt neutrons

neutrony *mpl* opóźnione delayed neutrons

neutrony *mpl* powolne slow neutrons

neutrony *mpl* prędkie fast neutrons

neutrony *mpl* rezonansowe resonance neutrons

neutrony *mpl* rozszczepieniowe fission neutrons

neutrony *mpl* termiczne thermal neutrons

neutrony *mpl* zimne cold neutrons

nitrometr *m* nitrometer

normalność *f* normality

nośnik *m* 1. carrier 2. solid support (*in chromatography*)

nośnik *m* izotopowy isotopic carrier

nośnik *m* nieizotopowy non-isotopic carrier

nośnik *m* pelikularny superficially porous support

nośnik *m* porowaty totally porous support

nośnik *m* powierzchniowo-porowaty superficially porous support

nośnik *m* spektroskopowy spectrochemical carrier

nośnik *m* śladów trace collector
nośnik *m* z cienką warstwą porowatą na
    powierzchni superficially porous support
nośnik *m* zatrzymujący hold-back-carrier
nośnik *m* zwrotny hold-back carrier
nośność *f* wagi capacity *(of a balance)*
nujol *m* Nujol
nuklid *m* macierzysty parent nuclide
nuklid *m* pochodny daughter nuclide

O

obciążenie *n* load
obciążenie *n* estymatora bias of estimator
obciążenie *n* maksymalne capacity
objętość *f* cieczy całkowita total liquid volume
objętość *f* cieczy unieruchomionej stationary
    liquid volume
objętość *f* czynna detektora sensitive volume
    of a detector
objętość *f* deklarowana designated volume
objętość *f* elucji piku peak elution volume
objętość *f* fazy ciekłej liquid phase volume
objętość *f* fazy nieruchomej stationary-phase-
    volume
objętość *f* fazy nieruchomej ciekłej liquid phase
    volume
objętość *f* fazy nieruchomej ułamkowa
    stationary-phase fraction
objętość *f* fazy stałej solid volume
objętość *f* kolumny column volume
objętość *f* kolumny geometryczna column
    volume
objętość *f* martwa extra-column volume
objętość *f* martwa detektora insensitive volume
    of a detector
objętość *f* międzyziarnowa interstitial volume
objętość *f* międzyziarnowa ułamkowa
    interstitial fraction
objętość *f* pozakolumnowa extra-column volume
objętość *f* retencji retention volume
objętość *f* retencji absolutna net retention
    volume
objętość *f* retencji całkowita retention volume
objętość *f* retencji niepoprawiona retention
    volume
objętość *f* retencji poprawiona corrected
    retention volume
objętość *f* retencji składnika nie
    zatrzymywanego w kolumnie mobile phase
    hold-up volume
objętość *f* retencji skorygowana corrected
    retention volume
objętość *f* retencji właściwa specific retention
    volume
objętość *f* retencji zredukowana adjusted
    retention volume
objętość *f* swobodna interstitial volume
objętość *f* wolna interstitial volume
objętość *f* wyznaczona designated volume
objętość *f* złoża bed volume
obszar *m* dyfuzyjny secondary combustion zone
obszar *m* krytyczny rejection region
obszar *m* niedoświetleń under-exposure region
obszar *m* odrzuceń rejection region
obszar *m* przejścia transition interval

obszar *m* przejściowy interzonal region
obszar *m* prześwietleń over-exposure region
obszar *m* wstępnego spalania primary
    combustion zone
ocena *f* organoleptyczna organoleptic assessment
oczyszczanie *n* scrubbing
oczyszczanie *n* plazmowe plasma cleaning
odchylenie *n* magnetyczne magnetic deflection
odchylenie *n* standardowe standard deviation
odchylenie *n* standardowe średniej z próby
    standard deviation of the mean
odchylenie *n* standardowe względne coefficient
    of variation
odchylenie *n* standardowe względne z próby
    empirical variation coefficient
odchylenie *n* standardowe z próby standard
    deviation of a sample
odchylenie *n* średnie mean deviation
odczyn *m* reaction
odczynnik *m* reagent
odczynnik *m* chelatujący chelating agent
odczynnik *m* chemicznie czysty chemically pure
    reagent
odczynnik *m* Czugajewa Chugaev reagent
odczynnik *m* czysty pure reagent
odczynnik *m* czysty do analizy analytical reagent
odczynnik *m* demaskujący demasking agent
odczynnik *m* do mikroanalizy microanalytical
    reagent
odczynnik *m* ekstrahujący extracting agent
odczynnik *m* Fehlinga Fehling's reagent
odczynnik *m* grupowy group reagent
odczynnik *m* Karla Fischera Karl Fischer
    reagent
odczynnik *m* maskujący masking agent
odczynnik *m* mikroanalityczny microanalytical
    reagent
odczynnik *m* Nesslera Nessler reagent
odczynnik *m* promieniotwórczy radioactive
    reagent
odczynnik *m* selektywny selective reagent
odczynnik *m* sililujący silylation reagent
odczynnik *m* specyficzny specific reagent
odczynnik *m* spektralnie czysty spectrally pure
    reagent
odczynnik *m* strącający precipitating agent
odczytywalność *f* readability
oddzielanie *n* separation
odkażanie *n* decontamination
odmiareczkowanie *n* nadmiaru back-titration
odmieszanie *n* rozpuszczalnika solvent demixing
odnośnik *m* reference sample
odorometria *f* odorimetry
odparowanie *n* evaporation
odpowiedź *f* elektrody electrode response
odpowiedź *f* elektrody nernstowska Nerstian
    electrode response
odpowiedź *f* potencjałowa elektrody electrode
    response
odtwarzalność *f* reproducibility
odważka *f* weighed amount
odważniki *mpl* analityczne analytical weights
odwirowanie *n* centrifuging
odwrócenie *n* w linii self-reversal
ogniwo *n* bez przenoszenia cell without
    transference

ogniwo *n* **chemiczne** chemical cell
ogniwo *n* **elektrochemiczne** electrochemical cell
ogniwo *n* **fotoelektryczne** photovoltaic cell
ogniwo *n* **galwaniczne** galvanic cell
ogniwo *n* **galwaniczne odwracalne** reversible
 galvanic cell
ogniwo *n* **Leclanchégo** Leclanché cell
ogniwo *n* **normalne** standard cell
ogniwo *n* **odwracalne** reversible galvanic cell
ogniwo *n* **pierwotne** primary cell
ogniwo *n* **rtęciowe** mercury dry cell
ogniwo *n* **standardowe** standard cell
ogniwo *n* **stężeniowe** concentration cell
ogniwo *n* **termoelektryczne** galvanic thermocell
ogniwo *n* **Westona nienasycone** unsaturated
 Weston cell
ogniwo *n* **Westona normalne** Weston standard
 cell
ogniwo *n* **Westona standardowe** Weston
 standard cell
ogniwo *n* **wtórne** accumulator
ogniwo *n* **z przenoszeniem** cell with transference
okluzja *f* occlusion
okluzja *f* **molekularna** occlusion
okres *m* **połowicznego zaniku** half-life period
okres *m* **półrozpadu** half-life period
okres *m* **półtrwania** half-life period
oksydymetria *f* oxidimetry
olej *m* **parafinowy** Nujol
opór *m* **bierny** reactance
opór *m* **elektryczny** resistance
opór *m* **elektryczny bierny** reactance
opór *m* **elektryczny pozorny** impedance
opór *m* **elektryczny właściwy** resistivity
opór *m* **pozorny** impedance
opór *m* **przenoszenia masy** mass transfer term
optymalizacja *f* optimization
osad *m* precipitate
osad *m* **bezpostaciowy** colloid precipitate
osad *m* **koloidalny** colloid precipitate
osad *m* **krystaliczny** crystalline precipitate
oscylacje *npl* vibrations
oscylometria *f* high-frequency conductometry
oscylopolarograf *m* oscillopolarograph
oscylopolarografia *f* oscillopolarography
osłabiacz *m* **stopniowy** step filter
osłabienie *n* **promieniowania** attenuation of
 radiation
oszacowanie *n* estimator
oś *f* **optyczna** optical axis
oznaczalność *f* limit of determination
oznaczanie *n* determination
oznaczanie *n* **automatyczne węgla, wodoru
 i azotu** automatic determination of carbon,
 hydrogen and nitrogen
oznaczanie *n* **izobarycznych zmian masy**
 isobaric mass-change determination
oznaczanie *n* **izotermicznych zmian masy**
 isothermal mass-charge determination
oznaczanie *n* **jednoczesne** simultaneous
 determination
oznaczanie *n* **masy molowej ebuliometryczne**
 ebullioscopic determination of the molar mass
oznaczanie *n* **masy molowej kriometryczne**
 cryoscopic determination of the molar mass
oznaczenie *n* **krytyczne** critical determination

P

palnik *m* burner
palnik *m* **plazmowy** plasma jet
palnik *m* **szczelinowy** slot-burner
palnik *m* **z bezpośrednim wtryskiem**
 direct-injection burner
palnik *m* **ze wstępnym mieszaniem** premixed
 gas burner
papierek *m* **wskaźnikowy** indicator paper
para *f* **linii** line pair
para *f* **linii analityczna** line pair
para *f* **linii kontrolna** fixation pair
parametr *m* parameter
parametr *m* **rozpuszczalności** solubility
 parameter
parametry *mpl* **zredukowane** reduced parameters
partia *f* **produktu** bulk of material
pasmo *n* zone
pasmo *n* **antystokesowskie** anti-stokes lines
pasmo *n* **B** B band
pasmo *m* **CT** charge-transfer band
pasmo *n* **elektronowe** electronic band
pasmo *n* **fundamentalne** fundamental vibration
 band
pasmo *n* **K** K band
pasmo *n* **kombinacyjne** combination band
pasmo *n* **nadtonowe** overtone band
pasmo *n* **nadtonowe drugie** second overtone band
pasmo *n* **nadtonowe pierwsze** first overtone band
pasmo *n* **odniesienia** reference band
pasmo *n* **oscylacyjne** vibration band
pasmo *n* **oscylacyjno-rotacyjne**
 vibration-rotation band
pasmo *n* **podstawowe** fundamental vibration
 band
pasmo *n* **przeniesienia ładunku** charge-transfer
 band
pasmo *n* **R** R band
pasmo *n* **stokesowskie** Stokes lines
pasmo *n* **widmowe** spectral band
pastylka *f* **z bromku potasu** potassium bromide
 disk
pastylkowanie *n* pellet technique
pasywacja *f* **elektrody** passivation of an
 electrode
pehametr *m* pH-meter
pehametria *f* pH-metry
pehastat *m* pH-stat
peptyzacja *f* peptyzacja
perła *f* **boraksowa** borax bead
perła *f* **fosforanowa** microcosmic salt bead
pęcznienie *n* swelling
pęcznienie *n* **masowe w rozpuszczalniku** weight
 swelling in solvent
pH *m* pH
pH-stat *m* pH-stat
piec *m* **do spalań** combustion furnace
piec *m* **grafitowy** graphite-tube furnace
piec *m* **muflowy** muffle furnace
pierścień *m* **chelatowy** chelate ring
pik *m* 1. peak 2. peak (*in chromatography*)
pik *m* **całkowitej absorpcji** full energy peak
pik *m* **cząsteczkowy** molecular peak
pik *m* **główny** base peak
pik *m* **jonu metastabilnego** metastable ion peak

pik *m* **molekularny** molecular peak
pik *m* **nierozdzielony** unresolved peak
pik *m* **podstawowy** base peak
pik *m* **powietrza** air peak
pik *m* **rezonansowy** resonance peak
pik *m* **rozpraszania wstecznego** backscatter peak
pik *m* **rozpraszania zwrotnego** backscatter peak
pik *m* **sumy** sum peak
pik *m* **ucieczki pary** pair escape peak
pik *m* **ucieczki pary elektronowej** pair escape peak
pik *m* **ucieczki podwójnej** double escape peak
pik *m* **ucieczki pojedynczej** single escape peak
pik *m* **ucieczki promieniowania rentgenowskiego** X-ray escape peak
piknometr *m* pycnometer
pikoślady *mpl* picotraces
pipeta *f* pipeta
pipeta *f* **do gazu** gas sampling pipette
pipeta *f* **wagowa Lungego i Reya** Lunge-Rey weighing pipette
pipeta *f* **wielomiarowa** graduated pipette
pipeta *f* **z automatycznym nastawianiem zera** automatic zero pipette
pipeta *f* **z jedną kreską** transfer pipette
pipeta *f* **z podziałką** graduated pipette
pipetka *f* **wkraplająca** dropper
piroliza *f* pyrolysis
piroliza *f* **błyskowa** flash pyrolysis
pirolizer *m* pyrolyser
plamka *f* spot
plan *m* **całkowitego eksperymentu czynnikowego** complete factorial design
plan *m* **czynnikowy pełny** complete factorial design
plan *m* **czynnikowy ułamkowy** fractional factorial design
plan *m* **dla mieszanin** mixture design
plan *m* **drugiego rzędu** second order design
plan *m* **eksperymentu** exact design
plan *m* **nasycony** saturated design
plan *m* **ortogonalny** orthogonal design
plan *m* **o symetrii obrotowej** rotatable design
plan *m* **pierwszego rzędu** first order design
plan *m* **rotatabilny** rotatable design
plan *m* **sympleksowy** simplex design
planowanie *n* **eksperymentalne** experimental design
planowanie *n* **ewolucyjne** evolutionary operation
plateau *n* **licznika** plateau of a counter
platynowanie *n* **elektrod** platinization of electrodes
plazma *f* plasma
plazma *f* **mikrofalowa** microwave plasma
plazma *f* **o częstości radiowej** radiofrequency plasma
plazma *f* **sprzężona indukcyjnie** inductively coupled plasma
plazmotron *m* **łukowy** plasma jet
płaszczyzna *f* **drgań** plane of vibration
płaszczyzna *f* **polaryzacji** plane of polarization
płomień *m* flame
płomień *m* **dyfuzyjny** diffusion flame
płomień *m* **laminarny** laminar flame
płomień *m* **osłonięty** shielded flame

płomień *m* **rozdzielony** separated flame
płomień *m* **turbulentny** turbulent flame
płuczka *f* **Drechsela** Drechsel gas washing bottle
płuczka *f* **Friedrichsa** Friedrichs gas washing bottle
płuczka *f* **z bełkotką ze spiekanego szkła** gas washing bottle with sintered head
płyta *f* **fotograficzna** photographic plate
płytka *f* **do analizy kroplowej** spotting plate
płytka *f* **nośna** support plate
pM *m* pM
podczerwień *f* infrared radiation
podczerwień *f* **bliska** near infrared
podczerwień *f* **daleka** far infrared
podczerwień *f* **średnia** middle-infrared
podgrzewacz *m* **do lejków** funnel heater
podstawa *f* **piku** peak base
pOH *m* pOH
pojemność *f* **oporowa naczynka** cell constant
pojemność *f* **pikowa** peak capacity
pojemność *f* **warstwy podwójnej całkowa** integral capacitance of double layer
pojemność *f* **warstwy podwójnej różniczkowa** differential capacitance of double layer
pojemność *f* **wymienna robocza złoża jonitu** breakthrough capacity of an ion exchanger bed
pojemność *f* **wymienna właściwa całkowita** total specific capacity
pojemność *f* **wymienna właściwa objętościowa** volume capacity
pojemność *f* **wymienna właściwa objętościowa złoża jonitu** bed volume capacity
pojemność *f* **wymienna właściwa praktyczna** practical specific capacity
pojemność *f* **wymienna właściwa praktyczna względem określonego jonu** practical specific capacity
pojemność *f* **względem pików** peak capacity
pokrycie *n* **powierzchni** surface coverage
polarograf *m* polarograph
polarografia *f* polarography
polarografia *f* **amalgamatowa** amalgam polarography
polarografia *f* **amalgamatów** amalgam polarography
polarografia *f* **cykliczna z falą trójkątną** cyclic triangular-wave polarography
polarografia *f* **częstości radiowych** radio-frequency polarography
polarografia *f* **demodulacyjna** demodulation polarography
polarografia *f* **dudnieniowa** double-tone polarography
polarografia *f* **impulsowa** pulse polarography
polarografia *f* **impulsowa normalna** pulse polarography
polarografia *f* **Kalouska** Kalousek polarography
polarografia *f* **klasyczna** direct current polarography
polarografia *f* **kulostatyczna** incremental-charge polarography
polarografia *f* **modulacyjna** modulation polarography
polarografia *f* **pierwszej pochodnej** derivative polarography

342

polarografia *f* prądu maksymalnego Tast polarography
polarografia *f* pulsowa pulse polarography
polarografia *f* pulsowa normalna pulse polarography
polarografia *f* pulsowa pochodna derivative pulse polarography
polarografia *f* pulsowa różnicowa differential pulse polarography
polarografia *f* pulsowa różniczkowa derivative pulse polarography
polarografia *f* rf radio-frequency polarography
polarografia *f* rozładowania incremental-charge polarography
polarografia *f* różnicowa differential polarography
polarografia *f* różniczkowa derivative polarography
polarografia *f* schodkowa staircase polarography
polarografia *f* stałoprądowa direct current polarography
polarografia *f* z falą trójkątną triangular-wave polarography
polarografia *f* z prądem narastającym liniowo current-scanning polarography
polarografia *f* ze zmianą potencjału pojedynczą single-sweep polarography
polarografia *f* ze zmianą potencjału wielokrotną multisweep polarography
polarografia *f* zmiennonapięciowa alternating-voltage polarography
polarografia *f* zmiennoprądowa alternating-current polarography
polarografia *f* zmiennoprądowa prostokątna square-wave polarography
polarografia *f* zmiennoprądowa sinusoidalna conventional alternating-current polarography
polarografia *f* zmiennoprądowa sinusoidalna drugiej harmonicznej second-harmonic ac polarography
polarografia *f* zmiennoprądowa sinusoidalna wyższych harmonicznych higher-harmonic ac polarography
polarografia *f* zmiennoprądowa sinusoidalna wyższych harmonicznych z fazoczułym prostowaniem higher-harmonic ac polarography with phase sensitive rectification
polarogram *m* polarogram
polarymetr *m* polarimeter
polarymetr *m* spektralny spectropolarimeter
polarymetria *f* polarimetry
polaryzacja *f* elektrody polarization of electrode
polaryzacja *f* eliptyczna elliptical polarization
polaryzacja *f* kołowa circular polarization
polaryzacja *f* liniowa plane polarization
polaryzacja *f* światła polarization of light
polielektrolit *m* polyelectrolyte
polikwas *m* polyacid
polimer *m* redoks redox polymer
polowe źródło *n* jonów field ionization ion source
poprawka *f* na rozcieńczenie dilution-correction factor
poprawka *f* wskaźnika indicator blank
populacja *f* population
populacja *f* generalna population

Porapak *m* Porapak
potencjał *m* absolutny absolute electrode potential
potencjał *m* bezwzględny absolute electrode potential
potencjał *m* chemiczny chemical potential
potencjał *m* cieczowy liquid-junction potential
potencjał *m* Donnana membrane potential
potencjał *m* dyfuzyjny diffusion potential
potencjał *m* elektrochemiczny electrochemical potential
potencjał *m* elektrody electrode potential
potencjał *m* elektrody absolutny absolute electrode potential
potencjał *m* elektrody bezwzględny absolute electrode potential
potencjał *m* elektrody formalny conditional electrode potential
potencjał *m* elektrody normalny standard electrode potential
potencjał *m* elektrody równowagowy equilibrium electrode potential
potencjał *m* elektrody spoczynkowy open-circuit electrode potential
potencjał *m* elektrody standardowy standard electrode potential
potencjał *m* elektrody warunkowy conditional electrode potential
potencjał *m* elektrody względny electrode potential
potencjał *m* elektrokinetyczny electrokinetic potential
potencjał *m* elektryczny powierzchniowy fazy *β* surface electric potential of phase *β*
potencjał *m* elektryczny wewnętrzny fazy *β* inner electric potential of phase *β*
potencjał *m* elektryczny zewnętrzny fazy *β* outer electric potential of phase *β*
potencjał *m* formalny conditional electrode potential
potencjał *m* korozyjny corrosion potential
potencjał *m* ładunku zerowego potential at the point of zero charge
potencjał *m* membranowy membrane potential
potencjał *m* mieszany mixed polyelectrode potential
potencjał *m* normalny standard electrode potential
potencjał *m* piku peak potential
potencjał *m* powierzchniowy fazy *β* surface electric potential of phase *β*
potencjał *m* półfali half-wave potential
potencjał *m* półpiku half-peak potential
potencjał *m* przepływu streaming potential
potencjał *m* przeponowy membrane potential
potencjał *m* redoks oxidation-reduction potential
potencjał *m* rozkładu decomposition potential
potencjał *m* równowagowy equilibrium electrode potential
potencjał *m* sedymentacji sedimentation potential
potencjał *m* spoczynkowy open-circuit electrode potential
potencjał *m* standardowy standard electrode potential

potencjał *m* termodyfuzyjny thermodiffusion potential
potencjał *m* warunkowy conditional electrode potential
potencjał *m* wydzielania deposition potential
potencjał *m* zeta electrokinetic potential
potencjometr *m* potentiometer
potencjometria *f* potentiometry
potencjometria *f* różnicowa differential potentiometry
potencjometria *f* z kontrolowanym prądem controlled-current potentiometry with one indicator electrode
potencjometria *f* z kontrolowanym prądem z dwiema elektrodami spolaryzowanymi controlled-current potentiometry with two indicator electrodes
potencjostat *m* potentiostat
powielacz *m* elektronowy electron multiplier
powielacz *m* fotoelektronowy photomultiplier tube
powierzchnia *f* surface
powierzchnia *f* odpowiedzi response surface
powierzchnia *f* piku peak area
powierzchnia *f* rozdziału faz interface
powierzchnia *f* właściwa specific surface area
powinowactwo *n* chemiczne affinity
powlekacz *m* spreader
powtarzalność *f* repeatability
poziom *m* istotności significance level
poziom *m* szumu noise level
poziom *m* ufności confidence level
poziomy *mpl* zeemanowskie Zeeman levels
półogniwo *n* electrode
prawa *npl* Faradaya Faraday's laws
prawa *npl* Ficka Fick's law
prawdopodobieństwo *n* probability
prawdopodobieństwo *n* adsorpcji sticking probability
prawdopodobieństwo *n* przylgnięcia sticking probability
prawdopodobieństwo *n* statystyczne statistical probability
prawo *n* addytywności absorbancji additivity law
prawo *n* Beera 1. Beer's law 2. Bouguer-Lambert-Beer law
prawo *n* Bouguera i Beera Bouguer-Lambert-Beer law
prawo *n* Bouguera i Lamberta Lambert's law
prawo *n* Bragga Bragg law
prawo *n* Debye'a i Hückela graniczne Debye-Hückel limiting law
prawo *n* Doernera i Hoskinsa logarithmic distribution law
prawo *n* Hooke'a Hooke's law
prawo *n* Kohlrauscha Kohlrausch's additive law
prawo *n* Lamberta Lambert's law
prawo *n* Lamberta i Beera Bouguer-Lambert-Beer law
prawo *n* Moseleya Moseley law
prawo *n* niezależnego ruchu jonów Kohlrauscha Kohlrausch's additive law
prawo *n* podziału Nernst distribution law
prawo *n* podziału homogeniczne homogeneous distribution law

prawo *n* podziału Nernsta Nernst distribution law
prawo *n* przenoszenia błędów variance propagation
prawo *n* rozcieńczeń Ostwalda Ostwald's dilution law
prawo *n* rozpadu promieniotwórczego law of radioactive decay
prawo *n* rozwijania błędu variance propagation
prawo *n* Stokesa Stokes law
prawo *n* sumowania błędu variance propagation
prażenie *n* osadu ignition of precipitate
prąd *m* adsorpcyjny adsorption current
prąd *m* ciemny dark current
prąd *m* cząstkowy anodowy/katodowy partial anodic/cathodic current
prąd *m* dyfuzyjny diffusion current
prąd *m* elektryczny electric current
prąd *m* faradajowski faradaic current
prąd *m* fotoelektryczny photocurrent
prąd *m* graniczny limiting current
prąd *m* katalityczny catalytic current
prąd *m* kinetyczny kinetic current
prąd *m* mierzony observed current
prąd *m* migracyjny migration current
prąd *m* obserwowany observed current
prąd *m* piku peak current
prąd *m* pojemnościowy capacity current
prąd *m* resztkowy residual current
prąd *m* szczątkowy residual current
prąd *m* wymiany exchange current
prąd *m* wypadkowy net current
precyzja *f* precision
precyzja *f* wagi precision of a balance
precyzja *f* ważenia precision of a weighing
prędkość *f* przepływu liniowa nominalna nominal linear flow
prędkość *f* przepływu międzyziarnowa interstitial velocity
prędkość *f* przepływu międzyziarnowa średnia gazu nośnego mean interstitial velocity of the carrier gas
prędkość *f* przepływu międzyziarnowa zredukowana fazy ruchomej reduced velocity of the mobile phase
prędkość *f* przepływu objętościowa fazy ruchomej volumetric flow rate F_a
prędkość *f* przepływu objętościowa skorygowana fazy ruchomej volumetric flow rate F_c
prędkość *f* przepływu zredukowana fazy ruchomej reduced velocity of the mobile phase
prędkość *f* światła w próżni velocity of light in vacuum
procent *m* ekstrakcji extraction percentage
procent *m* zmiareczkowania per cent titrated
proces *m* elektrodowy electrode process
proces *m* elektrodowy stacjonarny stationary electrode process
proces *m* kontrolowany dyfuzyjnie diffusion controlled process
proces *m* kontrolowany kinetycznie kinetic controlled process
proces *m* kontrolowany szybkością przeniesienia ładunku electroactivation control
profil *m* koncentracji depth profile
profil *m* stężenia depth profile

programowanie n rozpuszczalnika gradient elution
promieniowanie n radiation
promieniowanie n γ gamma radiation
promieniowanie n γ natychmiastowe prompt gamma radiation
promieniowanie n cząsteczkowe molecular radiation
promieniowanie n Czerenkowa Cerenkov radiation
promieniowanie n elektromagnetyczne electromagnetic radiation
promieniowanie n fluorescencyjne rentgenowskie X-ray fluorescence radiation
promieniowanie n hamowania bremsstrahlung
promieniowanie n jonizujące ionizing radiation
promieniowanie n korpuskularne corpuscular radiation
promieniowanie n monochromatyczne monochromatic radiation
promieniowanie n nadfioletowe ultraviolet light
promieniowanie n podczerwone infrared radiation
promieniowanie n polichromatyczne polychromatic radiation
promieniowanie n rentgenowskie X-radiation
promieniowanie n rentgenowskie charakterystyczne characteristic X-radiation
promieniowanie n rentgenowskie pierwotne primary X-radiation
promieniowanie n rentgenowskie wtórne secondary X-radiation
promieniowanie n rentgenowskie wzbudzające primary X-radiation
promieniowanie n Röntgena X-radiation
promieniowanie n widzialne light
promieniowanie n X X-radiation
promień m atmosfery jonowej radius of ionic atmosphere
promień m Debye'a radius of ionic atmosphere
prostowanie n faradajowskie wysokopoziomowe high-level faradaic rectification
próba f 1. sample 2. test (in chemical analysis)
próba f dmuchawkowa blow-pipe test
próba f kontrolna check test
próba f losowa random sample
próba f ślepa blank test
próba f ślepa wskaźnika indicator blank
próba f zerowa blank test
próbka f sample
próbka f analityczna analytical sample
próbka f cienka thin-film specimen
próbka f gruba thick specimen
próbka f jednostkowa unit sample
próbka f laboratoryjna średnia average sample
próbka f o grubości nasycenia thick specimen
próbka f o grubości nieskończonej thick specimen
próbka f ogólna gross sample
próbka f pierwotna primary sample
próg m absolutny stimulus threshold
próg m absorpcji absorption edge
próżnia f ultrawysoka ultra-high vacuum
pryzmat m optical prism
przebicie n breakthrough
przeciwelektroda f counter electrode

przeciwjony mpl counter-ions
przedział m daktyloskopowy fingerprint region
przedział m ufności confidence interval
przejście n dozwolone allowed transition
przejście n elektronowe typu $\pi - \pi^*$ $\pi - \pi^*$ electronic transition
przejście n elektronowe typu $\sigma - \pi^*$ $\sigma - \pi^*$ electronic transition
przejście n elektronowe typu $n - \pi^*$ $n - \pi^*$ electronic transition
przejście n elektronowe typu $n - \sigma^*$ $n - \sigma^*$ electronic transition
przejście n wzbronione forbidden transition
przekrój m czynny cross section
przekrój m czynny atomowy atomic cross section
przekrój m czynny izotopowy isotopic cross section
przekrój m czynny makroskopowy macroscopic cross section
przekrój m czynny mikroskopowy microscopic cross section
przekrój m czynny na absorpcję absorption cross section
przekrój m czynny na aktywację activation cross section
przekrystalizowanie n recrystallization
przemiana f promieniotwórcza radioactive decay
przemywanie n osadu washing of precipitate
przenikalność f elektryczna permittivity
przenikalność f elektryczna bezwzględna permittivity
przenikalność f elektryczna względna relative permittivity
przenikalność f względna relative permittivity
przerwa f analityczna analytical gap
przerwa f międzyelektrodowa analytical gap
przesącz m filtrate
przesłona f Hartmanna Hartmann diaphragm
przestrzeń f czynnikowa experimental region
przesunięcie n batochromowe bathochromic shift
przesunięcie n chemiczne chemical shift
przesunięcie n czerwone bathochromic shift
przesunięcie n częstotliwości ramanowskie Raman frequency shift
przesunięcie n hipsochromowe hypsochromic shift
przesunięcie n niebieskie hypsochromic shift
przesunięcie n ramanowskie Raman frequency shift
przesycenie n supersaturation
przesycenie n względne relative supersaturation
przewodnictwo n dziurowe hole conduction
przewodnictwo n elektrolityczne 1. conductance of electrolyte 2. ionic conduction
przewodnictwo n elektrolityczne właściwe electrolytic conductivity
przewodnictwo n elektronowe electronic conduction
przewodnictwo n elektryczne 1. conductance 2. electrical conduction
przewodnictwo n elektryczne właściwe electrical conductivity
przewodnictwo n jonowe ionic conduction
przewodnictwo n jonowe molowe molar conductivity of ionic species B

przewodnictwo *n* **molowe elektrolitu** molar conductivity of electrolyte
przewodnictwo *n* **molowe graniczne elektrolitu** limiting molar conductivity of electrolyte
przewodnictwo *n* **molowe jonowe graniczne** limiting molar conductivity of ionic species B
przewodnictwo *n* **równoważnikowe elektrolitu** equivalent conductivity of electrolyte
przewodnictwo *n* **równoważnikowe graniczne elektrolitu** limiting equivalent conductivity of electrolyte
przewodnictwo *n* **równoważnikowe jonowe** equivalent conductivity of ionic species B
przewodnictwo *n* **równoważnikowe jonowe graniczne** limiting equivalent conductivity of ionic species B
przewodnik *m* **elektrolityczny** ionic conductor
przewodnik *m* **elektronowy** electronic conductor
przewodnik *m* **elektryczny** electrical conductor
przewodnik *m* **jonowy** ionic conductor
przewodnik *m* **metaliczny** electronic conductor
przewodność *f* conductivity
przewodność *f* **bierna** susceptance
przewodność *f* **elektrolityczna właściwa** electrolytic conductivity
przewodność *f* **elektryczna** conductance
przewodność *f* **elektryczna bierna** susceptance
przewodność *f* **elektryczna czynna** conductance
przewodność *f* **elektryczna pozorna** admittance
przewodność *f* **elektryczna właściwa** electrical conductivity
przewodność *f* **pozorna** admittance
punkt *m* **doświadczalny** experimental point
punkt *m* **izoelektryczny** isoelectric point
punkt *m* **izopotencjałowy** izopotential point
punkt *m* **izozbestyczny** isosbestic point
punkt *m* **końcowy miareczkowania** end-point of titration
punkt *m* **końcowy teoretyczny** equivalence-point
punkt *m* **równowagi** rest point
punkt *m* **równoważnikowy** equivalence-point
punkt *m* **startu** starting line
punkt *m* **stechiometryczny** equivalence-point
punkt *m* **zerowy skali** zero point of the scale
punkty *mpl* **przegięcia** inflection points
punkty *mpl* **przegięcia piku** inflection points

R

rachunek *m* **prawdopodobieństwa** probability theory
radioaktywność *f* radioactivity
radiochromatografia *f* radiochromatography
radioelektroforeza *f* radioelectrophoresis
radioluminescencja *f* radioluminescence
radiometria *f* radiometry
randomizacja *f* randomization
rdzeń *m* **atomowy** atomic core
reakcja *f* **amplifikacji** amplification reaction
reakcja *f* **analityczna** analytical reaction
reakcja *f* **biuretowa** biuret reaction
reakcja *f* **charakterystyczna** identification reaction
reakcja *f* **elektrodowa** electrode reaction

reakcja *f* **elektrodowa cząstkowa** partial electrode reaction
reakcja *f* **elektrodowa nieodwracalna** irreversible electrode reaction
reakcja *f* **elektrodowa odwracalna** reversible electrode reaction
reakcja *f* **elektrodowa prosta** simple electrode reaction
reakcja *f* **elektrodowa wieloetapowa** stepwise electrode reaction
reakcja *f* **elektrodowa złożona** stepwise electrode reaction
reakcja *f* **elementarna** elementary reaction
reakcja *f* **indukowana** induced reaction
reakcja *f* **jądrowa przeszkadzająca** interfering nuclear reaction
reakcja *f* **jądrowa przeszkadzająca wtórna** secondary interfering reaction
reakcja *f* **katalityczna** catalytic reaction
reakcja *f* **kwas-zasada** acid-base reaction
reakcja *f* **mureksydowa** murexide reaction
reakcja *f* **potencjałotwórcza** potential-determining reaction
reakcja *f* **prosta** elementary reaction
reakcja *f* **redoks** redox reaction
reakcja *f* **selektywna** selective reaction
reakcja *f* **specyficzna** specific reaction
reakcja *f* **sprzężona** conjugate reaction
reakcja *f* **wymiany ładunku** charge-transfer step
reakcja *f* **Zeisela** Zeisel reaction
reaktancja *f* reactance
redukcja *f* reduction
reduktometria *f* reductometry
reduktor *m* reductant
reekstrakcja *f* back-extraction
refrakcja *f* **atomowa** atomic refraction
refrakcja *f* **jonowa** ionic refraction
refrakcja *f* **molowa** molar refraction
refrakcja *f* **właściwa** specific refraction
refraktometr *m* refractometer
refraktometria *f* refractometry
regresja *f* regression
regresja *f* **krokowa** step-wise regression
regresja *f* **krzywoliniowa** curvilinear regression
regresja *f* **liniowa** linear regression
regresja *f* **nieliniowa** non-linear regression
regresja *f* **wielokrotna** multiple regression
regresja *f* **wieloraka** multiple regression
reguła *f* **siły jonowej Lewisa i Randalla** Lewis-Randall's ionic strength law
reguła *f* **trzech sigm** three-sigma rule
reguły *fpl* **wyboru** selection rules
rekrystalizacja *f* recrystallization
relaksacja *f* relaxation
relaksacja *f* **kwadrupolowa** quadrupole relaxation
rentgenogram *m* X-ray pattern
rentgenogram *m* **kryształu obracanego** X-ray rotating-crystal pattern
rentgenogram *m* **kryształu wahanego** X-ray oscillating-crystal pattern
rentgenogram *m* **proszkowy** X-ray powdered-crystal pattern
rentgenogram *m* **warstwicowy** X-ray rotating-crystal pattern
retencja *f* **względna** relative retention

rezonans *m* **elektronowo-jądrowy podwójny** electron nuclear double resonance
rezonans *m* **Fermiego** Fermi resonance
rezonans *m* **magnetyczny** magnetic resonance
rezonans *m* **magnetyczny elektronowy** electron paramagnetic resonance
rezonans *m* **magnetyczny jądrowy** nuclear magnetic resonance
rezonans *m* **magnetyczny protonowy** proton magnetic resonance
rezonans *m* **paramagnetyczny elektronowy** electron paramagnetic resonance
rezonans *m* **spinowy elektronowy** electron paramagnetic resonance
rezonator *m* cavity resonator
rezystancja *f* resistance
rezystywność *f* resistivity
rotatabilność *f* rotatability
rozcieńczalnik *m* diluent
rozcieńczenie *n* **graniczne** dilution limit
rozdział *m* **przeciwprądowy** countercurrent distribution
rozdzielanie *n* **radiochemiczne** radiochemical separation
rozdzielanie *n* **w układzie ciecz-ciecz** liquid-liquid distribution
rozkład *m* χ^2 chi-squared distribution
rozkład *m* **Bernoullego** binomial distribution
rozkład *m* **dwumianowy** binomial distribution
rozkład *m* **empiryczny** empirical distribution
rozkład *m* **F-Snedecora** F-distribution
rozkład *m* **F** F-distribution
rozkład *m* **Gaussa i Laplace'a** normal distribution
rozkład *m* **logarytmiczno-normalny** logarithmic-normal distribution
rozkład *m* **normalny** normal distribution
rozkład *m* **normalny standardowy** standardized normal distribution
rozkład *m* **normalny zmiennej standaryzowanej** standardized normal distribution
rozkład *m* **Poissona** Poisson distribution
rozkład *m* **prawdopodobieństwa** probability distribution
rozkład *m* **t** Student's distribution
rozkład *m* **t-Studenta** Student's distribution
rozkład *m* **wykładniczy** exponential distribution
rozmycie *n* **pasma** band broadening
rozmycie *n* **piku** peak broadening
rozmycie *n* **plamy** spot broadening
rozmycie *n* **przedniej części piku/pasma/plamy** fronting
rozmycie *n* **tylnej części piku/pasma/plamy** tailing
rozpad *m* **metastabilny** metastable decomposition
rozpad *m* **promieniotwórczy** radioactive decay
rozpad *m* **rozgałęziony** branching decay
rozpowszechnienie *n* **izotopu** abundance
rozpraszanie *n* scattering
rozpraszanie *n* **elastyczne** elastic scattering
rozpraszanie *n* **klasyczne** Rayleigh scattering
rozpraszanie *n* **komptonowskie** Compton effect
rozpraszanie *n* **nieelastyczne** inelastic scattering
rozpraszanie *n* **niespójne** Compton effect
rozpraszanie *n* **niesprężyste** inelastic scattering

rozpraszanie *n* **powierzchniowe wiązki molekularnej** molecular beam surface scattering
rozpraszanie *n* **ramanowskie** Raman effect
rozpraszanie *n* **ramanowskie rezonansowe** resonance Raman effect
rozpraszanie *n* **rayleighowskie** Rayleigh scattering
rozpraszanie *n* **reaktywne wiązki molekularnej** molecular beam reactive scattering
rozpraszanie *n* **spójne** coherent scattering
rozpraszanie *n* **sprężyste** elastic scattering
rozpraszanie *n* **wsteczne** backscattering
rozpraszanie *n* **zwrotne** backscattering
rozpraszanie *n* **światła** light scattering
rozpuszczalnik *m* solvent
rozpuszczalnik *m* **amfiprotyczny** amphiprotic solvent
rozpuszczalnik *m* **amfoteryczny** amphiprotic solvent
rozpuszczalnik *m* **aprotyczny** aprotic solvent
rozpuszczalnik *m* **bierny** aprotic solvent
rozpuszczalnik *m* **czynny** protolytic solvent
rozpuszczalnik *m* **mieszany** mixed solvent
rozpuszczalnik *m* **protolityczny** protolytic solvent
rozpuszczalnik *m* **protono-akceptorowy** protophilic solvent
rozpuszczalnik *m* **protonofilowy** protophilic solvent
rozpuszczalnik *m* **protogenny** protogenic solvent
rozpuszczalnik *m* **różnicujący** differentiating solvent
rozpuszczalnik *m* **wodopodobny** hydroxylic solvent
rozpuszczalnik *m* **wyrównujący** levelling solvent
rozpuszczalnik *m* **zasadowy** protophilic solvent
rozpuszczalność *f* solubility
rozpuszczanie *n* dissolution
rozpylanie *n* sputtering
rozpylanie *n* **jonowe reaktywne** reactive sputtering
rozpylanie *n* **preferencyjne** preferential sputtering
rozpylanie *n* **reaktywne** reactive sputtering
rozpylanie *n* **selektywne** preferential sputtering
rozrzut *m* dispersion
rozstęp *m* range
rozszczepienie *n* **jądrowe** fission
roztwarzanie *n* digestion
roztwór *m* solution
roztwór *m* **buforowy** pH buffer solution
roztwór *m* **do reekstrakcji** stripping solution
roztwór *m* **koloidalny** sol
roztwór *m* **nasycony** saturated solution
roztwór *m* **odniesienia 1.** reference solution (*in spectrophotometry*) **2.** standard reference solution (*in titrimetric analysis*)
roztwór *m* **porównawczy** standard matching solution
roztwór *m* **przemywający** washing solution
roztwór *m* **przesycony** supersaturated solution
roztwór *m* **stały** mixed crystal
roztwór *m* **wzorcowy** standard solution
roztwór *m* **ślepej próby** blank solution
rozwijanie *n* development
rozwijanie *n* **chromatogramu** development of a chromatogram

rozwijanie *n* spływowe descending development
rozwijanie *n* wstępujące ascending development
rozwijanie *n* zstępujące descending development
równania *n* normalne normal equations
równanie *n* Butlera i Volmera Butler-Volmer
equation
równanie *n* Cottrella Cottrell equation
równanie *n* Debye'a Hückela i Brönsteda
Debye-Hückel-Brönsted equation
równanie *n* Debye'a, Hückela i Onsagera
graniczne Debye-Hückel-Onsager's limiting
law for conductivity
równanie *n* Debye'a i Hückela Debye-Hückel
equation
równanie *n* Debye'a i Hückela graniczne
Debye-Hückel limiting law
równanie *n* Hendersona Henderson equation
równanie *n* Ilkoviča Ilkovič equation
równanie *n* Lewicza Levich equation
równanie *n* Lewisa i Sargenta Lewis Sargent's
relation
równanie *n* Nernsta Nernst equation
równanie *n* Nernsta zmodyfikowane Nikolsky
equation
równanie *m* Nikolskiego Nikolsky equation
równanie *n* Onsagera graniczne
Debye-Hückel-Onsager's limiting law for
conductivity
równanie *n* Randlesa i Ševčika Randles-Ševčik
equation
równanie *n* Sanda Sand equation
równanie *n* Tafela Tafel's equation
równanie *n* van Deemtera van Deemter equation
równowaga *f* Donnana membrane equilibrium
równowaga *f* membranowa membrane
equilibrium
równowaga *f* metastabilna elektrody metastable
equilibrium of an electrode
równowaga *f* promieniotwórcza przejściowa
transient radioactive equilibrium
równowaga *f* promieniotwórcza trwała secular
radioactive equilibrium
równowaga *f* przejściowa transient radioactive
equilibrium
równowaga *f* przeponowa membrane equilibrium
równowaga *f* trwała secular radioactive
equilibrium
równowaga *f* wiekowa secular radioactive
equilibrium
równoważnik *m* chemiczny chemical equivalent
równoważnik *m* elektrochemiczny
electrochemical equivalent
różnica *f* potencjałów elektrycznych electric
potential difference
różnica *f* potencjałów elektrycznych ogniwa
galwanicznego electric potential difference
of a galvanic cell
różnica *f* potencjałów Galvaniego Galvani
potential difference
różnica *f* potencjałów kontaktowa Volta
potential difference
różnica *f* potencjałów Volty Volta potential
difference
ruchliwość *f* jonu B electric mobility of ion B
rugowanie *n* displacement
rura *f* do spalań combustion tube

rurka *f* absorpcyjna absorption tube
rurka *f* filtracyjna filter tube
rydberg *m* rydberg
rząd *m* reakcji order of reaction
rząd *m* reakcji względem danej substancji order
of reaction with respect to a given substance
rząd *m* widma spectral order

S

samoosłanianie *n* self-shielding
sączek *m* filter paper
sączek *m* faldowany fluted filter
sączek *m* karbowany fluted filter
sączenie *n* filtration
schemat *m* fragmentacji fragmentation pattern
schemat *m* rozpadu decay scheme
schładzanie *n* radioactive cooling
sedymentacja *f* sedimentation
Sefadeks *m* G Sephadex G
segregacja *f* segregation
selektywność *f* odczynnika selectivity of
a reagent
selektywność *f* permeacji permselectivity
SEM *f* ogniwa electromotive force of a galvanic
cell
seria *f* ekwieluotropowa equieluotropic series
seria *f* eluotropowa eluotropic series
seria *f* widmowa spectral series
siatka *f* dyfrakcyjna diffraction grating
siatka *f* odbiciowa reflection grating
siatka *f* odbijająca reflection grating
siatka *f* płaska plane grating
siatka *f* przepuszczająca transmission grating
siatka *f* transmisyjna transmission grating
siatka *f* wklęsła concave grating
sieciowanie *n* cross-linking
sililowanie *n* silylation
singlet *m* widmowy singlet line
siła *f* elektromotoryczna electromotive force
of a galvanic cell
siła *f* elektromotoryczna normalna standard
electromotive force
siła *f* elektromotoryczna ogniwa galwanicznego
electromotive force of a galvanic cell
siła *f* elektromotoryczna standardowa standard
electromotive force
siła *f* jonowa ionic strength of a solution
siła *f* jonowa roztworu ionic strength of
a solution
siła *f* rozpuszczalnika solvent strength
sita *npl* cząsteczkowe molecular sieves
sita *npl* molekularne molecular sieves
skala *f* δ delta scale
skala *f* aktywności Brockmanna Brockmann
scale of activity
skala *f* wzorców standard series
skład *m* ilościowy quantitative composition
skład *m* jakościowy qualitative composition
składnik *m* główny major constituent
składnik *m* śladowy trace constituent
składnik *m* uboczny minor constituent
skręcalność *f* molowa molar rotation
skręcalność *f* optyczna optical rotation
skręcalność *f* właściwa specific rotation

skręcenie *n* płaszczyzny polaryzacji światła rotation of the plane of polarization of polarized light
solut *m* solute
solwatochromia *f* solvatochromism
sorbat *m* sorbate
sorbent *m* sorbent
sorbent *m* stały active solid
sorbent *m* stały modyfikowany modified active solid
sorpcja *f* sorption
sól *f* wewnątrzkompleksowa neutral chelate
spalanie *n* combustion
spalanie *n* błyskawiczne flash combustion
spalanie *n* katalityczne catalytic combustion
spalanie *n* sączka incineration of filter paper
spalanie *n* w płomieniu tlenowodorowym oxy-hydrogen flame method
spalanie *n* w pustej rurze empty tube combustion
specyficzność *f* odczynnika specificity of a reagent
spektrofluorymetr *m* spectrofluorimeter
spektrofluorymetria *f* spectrofluorimetry
spektrofotometr *m* spectrophotometer
spektrofotometr *m* dwuwiązkowy double-beam spectrophotometer
spektrofotometr *m* jednowiązkowy single-beam spectrophotometer
spektrofotometr *m* płomieniowy flame spectrophotometer
spektrofotometr *m* punktowy non-recording spectrophotometer
spektrofotometr *m* rejestrujący recording spectrophotometer
spektrofotometria *f* spectrophotometry
spektrofotometria *f* absorpcyjna spectrophotometry
spektrofotometria *f* różnicowa differential spectrophotometry
spektrograf *m* spectrograph
spektrograf *m* mas mass spectrograph
spektrogram *m* spectrogram
spektrometr *m* optical spectrometer
spektrometr *m* γ gamma-ray spectrometer
spektrometr *m* jądrowego rezonansu magnetycznego nuclear magnetic resonance spectrometer
spektrometr *m* mas mass spectrometer
spektrometr *m* mas czasu przelotu time-of-flight mass spectrometer
spektrometr *m* mas z sektorem magnetycznym magnetic-deflection mass spectrometer
spektrometr *m* NMR nuclear magnetic resonance spectrometer
spektrometr *m* promieniowania γ gamma-ray spectrometer
spektrometr *m* rentgenowski X-ray spectrometer
spektrometr *m* rentgenowski ogniskujący X-ray focusing spectrometer
spektrometr *m* scyntylacyjny scintillation spectrometer
spektrometr *m* scyntylacyjny β beta scintillation spectrometer
spektrometria *f* spectrometry
spektrometria *f* mas mass spectrometry
spektropolarymetr *m* spectropolarimeter

spektropolarymetria *f* spectropolarimetry
spektroskop *m* spectroscope
spektroskop *m* mas mass spectroscope
spektroskopia *f* spectroscopy
spektroskopia *f* absorpcyjna absorption spectroscopy
spektroskopia *f* absorpcyjna atomowa atomic absorption spectroscopy
spektroskopia *f* absorpcyjna atomowa płomieniowa flame atomic absorption spectroscopy
spektroskopia *f* absorpcyjna cząsteczkowa molecular absorption spectroscopy
spektroskopia *f* absorpcyjna płomieniowa flame absorption spectroscopy
spektroskopia *f* absorpcyjna promieniowania rentgenowskiego X-ray absorption spectroscopy
spektroskopia *f* atomowa atomic spectroscopy
spektroskopia *f* charakterystycznych strat energii elektronów electron-energy loss spectroscopy
spektroskopia *f* cząsteczkowa molecular spectroscopy
spektroskopia *f* egzoelektronów exoelectron spectroscopy
spektroskopia *f* elektronowa electronic spectroscopy
spektroskopia *f* elektronowego rezonansu paramagnetycznego electron paramagnetic resonance spectroscopy
spektroskopia *f* elektronowego rezonansu spinowego electron paramagnetic resonance spectroscopy
spektroskopia *f* elektronów electron spectroscopy
spektroskopia *f* elektronów Augera Auger-electron spectroscopy
spektroskopia *f* elektronów Augera wybijanych bombardowaniem jonowym ion-induced Auger electron spectroscopy
spektroskopia *f* elektronów Augera wybijanych promieniowaniem rentgenowskim X-ray induced Auger electron spectroscopy
spektroskopia *f* elektronów dla celów analizy chemicznej X-ray photoelectron spectroscopy
spektroskopia *f* elektronów rdzenia atomowego core level electron spectroscopy
spektroskopia *f* elektronów towarzyszących neutralizacji jonów ion neutralization spectroscopy
spektroskopia *f* elektronów walencyjnych valence level electron spectroscopy
spektroskopia *f* emisyjna emission spectroscopy
spektroskopia *f* emisyjna atomowa atomic emission spectroscopy
spektroskopia *f* emisyjna atomowa płomieniowa flame atomic emission spectroscopy
spektroskopia *f* emisyjna cząsteczkowa molecular emission spectroscopy
spektroskopia *f* emisyjna płomieniowa flame emission spectroscopy
spektroskopia *f* emisyjna promieniowania rentgenowskiego X-ray emission spectroscopy
spektroskopia *f* fluorescencji wzbudzanej laserowo laser fluorescence spectroscopy

spektroskopia *f* fluorescencyjna fluorescence
spectroscopy
spektroskopia *f* fluorescencyjna atomowa
atomic fluorescence spectroscopy
spektroskopia *f* fluorescencyjna atomowa
płomieniowa flame atomic fluorescence
spectroscopy
spektroskopia *f* fluorescencyjna cząsteczkowa
molecular fluorescence spectroscopy
spektroskopia *f* fluorescencyjna płomieniowa
flame fluorescence spectroscopy
spektroskopia *f* fluorescencyjna promieniowania
rentgenowskiego X-ray fluorescence
spectroscopy
spektroskopia *f* fotoelektronów photoelectron
spectroscopy
spektroskopia *f* fotoelektronów wybijanych
promieniowaniem rentgenowskim X-ray
photoelectron spectroscopy
spektroskopia *f* fotoelektronów wybijanych
promieniowaniem z zakresu UV ultraviolet
photoelectron spectroscopy
spektroskopia *f* jądrowego rezonansu
magnetycznego nuclear magnetic resonance
spectroscopy
spektroskopia *f* mas mass spectrometry
spektroskopia *f* mas jonów wtórnych secondary
ion mass spectroscopy
spektroskopia *f* mas z wyładowaniem
jarzeniowym glow discharge mass
spectroscopy
spektroskopia *f* mas ze wzbudzeniem iskrowym
spark source mass spectroscopy
spektroskopia *f* mikrofalowa microwave
spectroscopy
spektroskopia *f* molekularna molecular
spectroscopy
spektroskopia *f* Mössbauera Mössbauer
spectroscopy
spektroskopia *f* neutralizacji jonów ion
neutralization spectroscopy
spektroskopia *f* NMR nuclear magnetic
resonance spectroscopy
spektroskopia *f* odbiciowo-absorpcyjna
w podczerwieni infrared
reflectance-absorption spectroscopy
spektroskopia *f* optyczna optical spectroscopy
spektroskopia *f* optyczna cząstek wzbudzonych
wyładowaniem jarzeniowym glow-discharge
optical emission spectroscopy
spektroskopia *f* optyczna emisyjna cząstek
wzbudzonych wyładowaniem jarzeniowym
glow-discharge optical emission spectroscopy
spektroskopia *f* optyczno-akustyczna
photoacoustic spectroscopy
spektroskopia *f* potencjału pojawiania
appearance-potential spectroscopy
spektroskopia *f* potencjału pojawiania się
elektronów wybijanych promieniowaniem
rentgenowskim X-ray electron appearance
potential spectroscopy
spektroskopia *f* potencjału wzbudzania
appearance-potential spectroscopy
spektroskopia *f* potencjału zanikania elektronów
disappearance potential spectroscopy

spektroskopia *f* promieniowania rentgenowskiego
wzbudzanego bombardowaniem jonowym ion
induced X-ray spectroscopy
spektroskopia *f* ramanowska Raman
spectroscopy
spektroskopia *f* ramanowska lasera laser
Raman spectroscopy
spektroskopia *f* rozpraszania jonów o małych
energiach low energy ion scattering
spectroscopy
spektroskopia *f* rozpraszania jonów o dużych
energiach high-energy ion scattering
spectroscopy
spektroskopia *f* transmisyjna strat energii
elektronów transmission energy loss
spectroscopy
spektroskopia *f* wibracyjna vibrational
spectroscopy
spiętrzanie *n* impulsów pile-up
spirala *f* z nichromu coil of Nichrome wire
spopielanie *n* plazmowe plasma ashing
spopielanie *n* próbek ashing technique
spowalnianie *n* neutronów slowing down
sprawdzanie *n* odważników calibration of
weights
sprawność *f* kolumny column performance
stała *f* dielektryczna relative permittivity
stała *f* dyfuzji diffusion coefficient
stała *f* ekstrakcji extraction constant
stała *f* Faradaya Faraday constant
stała *f* naczynka konduktometrycznego cell
constant
stała *f* podziału 1. distribution constant
(*in chromatography*) 2. distribution constant
(*in extraction*) 3. partition coefficient (*in gas
chromatography*)
stała *f* podziału termodynamiczna partition
constant
stała *f* prądu granicznego limiting current
constant
stała *f* Rydberga Rydberg constant
stała *f* siłowa force constant
stała *f* siłowa wiązania force constant
stała *f* szybkości formalna conditional rate
constant of electrode reaction
stała *f* szybkości reakcji reaction-rate constant
stała *f* szybkości reakcji elektrodowej rate
constant of electrode reaction
stała *f* szybkości reakcji elektrodowej formalna
conditional rate constant of electrode
reaction
stała *f* szybkości reakcji elektrodowej
standardowa standard rate constant of
electrode reaction
stała *f* szybkości reakcji elektrodowej utleniania
lub redukcji rate constant of electrode
reaction
stała *f* szybkości reakcji elektrodowej
warunkowa conditional rate constant of
electrode reaction
stała *f* szybkości standardowa standard rate
constant of electrode reaction
stała *f* szybkości warunkowa conditional rate
constant of electrode reaction
stała *f* trwałości całkowita cumulative stability
constant

stała *f* trwałości kolejna stepwise stability constant

stała *f* trwałości ogólna cumulative stability constant

stała *f* trwałości stopniowa stepwise stability constant

stała *f* tworzenia kompleksu całkowita cumulative stability constant

stała *f* tworzenia kompleksu kolejna stepwise stability constant

stapianie *n* w bombie Parra Parr bomb fusion

starzenie *n* ageing of a precipitate

starzenie *n* się osadu ageing of a precipitate

starzenie *n* się roztworów koloidalnych ageing of sols

statystyczna kontrola *f* jakości statistical quality control

statystyka *f* statistic

statystyka *f* matematyczna statistics

stężenie *n* analityczne analytical concentration

stężenie *n* graniczne concentration limit

stężenie *n* jonowe średnie mean concentration of electrolyte

stężenie *n* koagulujące coagulation concentration

stężenie *n* masowe składnika B w roztworze mass concentration of substance B

stężenie *n* molowe składnika B w roztworze amount-of-substance concentration of substance B

stężenie *n* średnie elektrolitu mean concentration of electrolyte

stolik *m* Koflera Kofler micro melting point apparatus

stopień *m* asocjacji jonów association degree of ions

stopień *m* chromatogramu całkowego step on an integral chromatogram

stopień *m* dysocjacji pozorny apparent degree of dissociation

stopień *m* dysocjacji według Arrheniusa apparent degree of dissociation

stopień *m* dysocjacji rzeczywisty true degree of dissociation

stopień *m* pokrycia powierzchni surface coverage

stopień *m* twardości wody amerykański American degree of water hardness

stopień *m* twardości wody angielski English degree of water hardness

stopień *m* twardości wody francuski French degree of water hardness

stopień *m* twardości wody niemiecki German degree of water hardness

stosunek *m* fazowy phase ratio

stosunek *m* fotopiku do tła komptonowskiego peak-to-Compton ratio

stosunek *m* kadmowy cadmium ratio

stosunek *m* masa/ładunek mass to charge ratio

stosunek *m* objętości faz phase ratio

stosunek *m* podziału 1. concentration distribution ratio (*in chromatography*) 2. concentration distribution ratio (*in extraction*)

stosunek *m* podziału masowy mass distribution ratio

stosunek *m* podziału stężeniowy 1. concentration distribution ratio (*in chromatography*) 2. concentration distribution ratio (*in extraction*)

stosunek *m* przesycenia supersaturation ratio

stosunek *m* przesycenia krytyczny critical supersaturation ratio

stosunek *m* sygnał-szum signal-to-noise ratio

stożek *m* wewnętrzny primary combustion zone

stożek *m* zewnętrzny secondary combustion zone

strącanie *n* precipitation

strącanie *n* dwukrotne reprecipitation

strącanie *n* frakcjonowane fractional precipitation

strącanie *n* homogeniczne precipitation from homogeneous solution

strącanie *n* ilościowe quantitative precipitation

strącanie *n* następcze postprecipitation

strącanie *n* powtórne reprecipitation

strącanie *n* różnicowe fractional precipitation

strącanie *n* z nośnikiem collection

strącanie *n* z roztworu jednorodnego precipitation from homogeneous solution

strefa *f* zone

struktura *f* nadsubtelna hyperfine structure

struktura *f* subtelna fine structure

strumień *m* promieniowania radiant flux

strumień *m* świetlny luminous flux

submikroanaliza *f* submicro-analysis

substancja *f* substance

(+)-substancja *f* dextrarotatory substance

(−)-substancja *f* laevorotatory substance

d-substncja *f* dextrarotatory substance

l-substancja *f* laevorotatory substance

substancja *f* lewoskrętna laevorotatory substance

substancja *f* elektroaktywna electroactive substance

substancja *f* odniesienia reference material

substancja *f* optycznie czynna optically active substance

substancja *f* oznaczana determinand

substancja *f* podstawowa standard substance

substancja *f* podstawowa pierwotna primary standard

substancja *f* podstawowa wtórna secondary standard

substancja *f* powierzchniowo-czynna surfactant

substancja *f* prawoskrętna dextrarotatory substance

substancja *f* przeszkadzająca interfering substance

substancja *f* rozpuszczona solute

substancja *f* wybłyszczająca brightening agent

susceptancja *f* susceptance

susceptometria *f* susceptometry

suszenie *n* osadu drying of precipitate

sygnał *m* analityczny analytical signal

sympleks *m* simplex

szczelina *f* slit

szczelina *f* Sollera Soller collimator

szereg *m* ekwieluotropowy equieluoctropic series

szereg *m* eluotropowy eluotropic series

szereg *m* napięciowy electrochemical series

szereg *m* potencjałów normalnych electrochemical series

szereg *m* promieniotwórczy decay chain
szerokość *f* linii spektralnej line width
szerokość *f* linii spektralnej naturalna natural
  line-width
szerokość *f* linii spektralnej połówkowa
  half-intensity width
szerokość *f* linii spektralnej pozorna apparent
  half width
szerokość *f* linii spektralnej rzeczywista
  true-half-width
szerokość *f* linii widmowej line width
szerokość *f* piku przy podstawie peak width
  at base
szerokość *f* piku w połowie wysokości peak
  width at half-height
szkielet *m* jonitu ion exchanger matrix
szkło *n* o kontrolowanej porowatości
  controlled-porosity glass
szufelka *f* do odważania substancji weighing
  scoop
szum *m* noise level
szybkość *f* liczenia counting rate
szybkość *f* rozpadu promieniotwórczego activity
szybkość *f* tworzenia zarodków rate of
  nucleation
szybkość *f* zliczania counting rate

Ś

średnia *f* arytmetyczna arithmetic mean
średnia *f* geometryczna geometric mean
średnica *f* jonu efektywna mean ionic diameter
średnica *f* jonu średnia mean ionic diameter
środek *m* redukujący reductant
środek *m* rozstępu mid-range
środek *m* suszący desiccant
środek *m* utleniający oxidant
świadek *m* miareczkowania comparison solution
światło *n* light
światło *n* białe white light
światło *n* spolaryzowane polarized light
światło *n* spolaryzowane eliptycznie elliptically
  polarized light
światło *n* spolaryzowane kołowo circularly
  polarized light
światło *n* spolaryzowane kołowo w lewo
  left-handed circularly polarized light
światło *n* spolaryzowane kołowo w prawo
  right-handed circularly polarized light
światło *n* spolaryzowane liniowo
  plane-polarized light
światłość *f* luminous intensity

T

tablice *fpl* liczb losowych random number tables
tarcza *f* target
tarcza *f* cienka thin target
tarcza *f* gruba thick target
tarcza *f* trytowa tritium target
technika *f* dwukierunkowa two-dimensional
  chromatography
technika *f* kolumn łączonych coupled columns
technika *f* kolumn sprzężonych coupled columns
techniki *fpl* jednoczesne simultaneous
  techniques

techniki *fpl* jednoczesne sprzężone coupled
  simultaneous techniques
techniki *fpl* jednoczesne współdziałające ciągle
  continuous simultaneous techniques
techniki *fpl* jednoczesne współdziałające
  nieciągle discontinuous simultaneous
  techniques
techniki *fpl* kombinowane combined techniques
techniki *fpl* połączone multiple techniques
techniki *fpl* sprzężone coupled simultaneous
  techniques
temperatura *f* kolumny column temperature
temperatura *f* retencji retention temperature
temperatura *f* rozdzielania separation
  temperature
tensametria *f* measurement of nonfaradaic
  admittance
teoria *f* Bjerruma asocjacji jonów Bjerrum's
  theory of ionic association
teoria *f* budowy warstwy podwójnej theory of
  double layer
teoria *f* dysocjacji elektrolitycznej Arrheniusa
  i Ostwalda Arrhenius and Ostwald's theory
  of electrolytic dissociation
teoria *f* elektrolitów mocnych Debye'a
  i Hückela Debye-Hückel's theory of strong
  electrolytes
teoria *f* prawdopodobieństwa probability theory
teoria *f* solwatacji Borna Born's theory of
  solvation
termalizacja *f* neutronów neutron thermalization
termodesorpcja *f* programowana temperature
  programmed desorption
termodylatometria *f* thermodilatometry
termodylatometria *f* różnicowa differential
  thermodilatometry
termodylatometria *f* różniczkowa derivative
  thermodilatometry
termograwimetria *f* thermogravimetry
termograwimetria *f* różniczkowa derivative
  thermogravimetry
termojonizacja *f* thermal ionization
termoogniwo *n* galwaniczne galvanic thermocell
termopara *f* thermocouple
termowaga *f* thermobalance
test *m* 1. test 2. test (*in chemical analysis*)
test *m* dwustronny two-sided test
test *m* F F-test
test *m* istotności significance test
test *m* jednostronny one-sided test
test *m* nieparametryczny non-parametric test
test *m* sekwencyjny sequential test
test *m* Snedecora F-test
test *m* statystyczny test
test *m* Studenta Student's test
test *m* t Student's test
test *m* zgodności test for goodness of fit
test *m* zgodności χ^2 chi-squared test for
  goodness of fit
titrant *m* titrant
tło *n* promieniowania background radiation
tło *n* przyrządu background
tolerancja *f* naświetlenia latitude
topnienie *n* melting
topnik *m* flux
transformacja *f* Seidla Seidel transformation

transmitancja *f* **1.** internal transmittance **2.** transmission factor
tranzystor *m* **polowy** field effect transistor
tranzystor *m* **polowy czuły na gazy** gas-sensitive field effect transistor
tranzystor *m* **polowy czuły na jony** ion-sensitive field effect transistor
trawienie *n* **plazmowe** plasma etching
trend *m* drift
trimetylochlorosilan *m* trimethylchlorosilane
trójka *f* **jonowa** ion-triplet
tulipan *m* crucible holder
turbidymetr *m* turbidimeter
turbidymetria *f* turbidimetry
twardość *f* **magnezowa** magnesium hardness
twardość *f* **niewęglanowa** non-carbonate hardness
twardość *f* **przemijająca** temporary hardness
twardość *f* **stała** permanent hardness
twardość *f* **wapniowa** calcium hardness
twardość *f* **węglanowa** carbonate hardness
twardość *f* **wody** water hardness
twardość *f* **wody ogólna** total water hardness
twardość *f* **wody całkowita** total water hardness
tworzenie *n* **odwróconej komety** fronting
tworzenie *n* **ogonów** tailing
tworzenie *n* **się zarodków** nucleation
tygiel *m* crucible
tygiel *m* **Goocha** Gooch crucible
tygiel *m* **Schotta** crucible with sintered disk
tygiel *m* **z filtrem ze spiekanego szkła** crucible with sintered disk
tytanometria *f* titanometry

U

udźwig *m* **wagi** capacity
ugięcie *n* **światła** light diffraction
układ *m* **jonowo-asocjacyjny** ion-association system
układ *m* **pasm** band group
układ *m* **redoks** redox system
układ *m* **równań normalnych** normal equations
ulatnianie *n* **się** volatilization
ultrafiolet *m* ultraviolet light
ultramikroanaliza *f* ultramicro analysis
ultramikrowaga *f* ultramicro balance
ultrapróżnia *f* ultra-high vacuum
ułamek *m* **masowy składnika B** mass fraction of substance B
ułamek *m* **molowy** mole fraction of substance B
ułamek *m* **objętościowy składnika B** volume fraction of substance B
U-rurka *f* U-tube
utleniacz *m* oxidant
utleniacz *m* **Večeřy** Večeřa catalyst
utlenianie *n* oxidation

V

val *m* gram equivalent

W

waga *f* **analityczna** analytical balance
waga *f* **aperiodyczna** damped balance
waga *f* **automatyczna** automatic balance

waga *f* **elektromechaniczna** electromechanical balance
waga *f* **hydrostatyczna** hydrostatic balance
waga *f* **laboratoryjna** laboratory balance
waga *f* **mikroanalityczna** microchemical balance
waga *f* **periodyczna** undamped balance
waga *f* **półmikroanalityczna** semimicrochemical balance
waga *f* **sedymentacyjna** sedimentation balance
waga *f* **sprężynowa** spring balance
waga *f* **techniczna** chemical balance
waga *f* **tłumikowa** damped balance
waga *f* **torsyjna** torsion balance
waga *f* **uchylna** damped balance
wal *m* gram equivalent
wariancja *f* variance
wariancja *f* **resztowa** residual variance
wariancja *f* **z próby** sampling variance
warstwa *f* **czynna** effective layer
warstwa *f* **dyfuzyjna** diffusion layer
warstwa *f* **Gouya i Chapmana** diffuse double layer
warstwa *f* **gradientowa** gradient layer
warstwa *f* **Helmholtza** compact layer
warstwa *f* **katodowa** cathode layer
warstwa *f* **nieskończenie cienka** infinitely thin layer
warstwa *f* **nieskończenie gruba** infinitely thick layer
warstwa *f* **podwójna** electrochemical double layer
warstwa *f* **podwójna elektrochemiczna** electrochemical double layer
warstwa *f* **podwójna elektryczna** electrochemical double layer
warstwa *f* **podwójna rozmyta** diffuse double layer
warstwa *f* **połowicznego osłabienia** half-thickness
warstwa *f* **połówkowa** half-thickness
warstwa *f* **reakcyjna** reaction layer
warstwa *f* **rozmyta** diffuse double layer
warstwa *f* **sztywna** compact layer
warstwa *f* **zmodyfikowana** altered layer
wartość *f* **absorpcji** absorbance
wartość *f* **działki skali** value of a division
wartość *f* **koagulacyjna** coagulation concentration
wartość *f* **modalna** mode
wartość *f* **oczekiwania** expected value
wartość *f* **progowa koagulacji** coagulation concentration
wartość *f* **przeciętna** expected value
wartość *f* R_B R_B value
wartość *f* R_f R_f value
wartość *f* R_f' R_f' value
wartość *f* R_f **skorygowana** R_f' value
wartość *f* R_M R_M value
wartość *f* R_M' R_M' value
wartość *f* R_M **skorygowana** R_M' value
warunki *mpl* **normalne** standard conditions
warunki *mpl* **standardowe** standard conditions
węgiel *m* **platynowany** platinized carbon
wiązka *f* **molekularna** molecular beam
widmo *n* spectrum
widmo *n* **absorpcyjne** absorption spectrum

widmo *n* atomowe atomic spectrum
widmo *n* charakterystyczne rentgenowskie
X-ray characteristic spectrum
widmo *n* ciągłe continuous spectrum
widmo *n* cząsteczkowe molecular spectrum
widmo *n* elektromagnetyczne electromagnetic
spectrum
widmo *n* elektronowe electronic spectrum
widmo *n* elektronowo-oscylacyjno-rotacyjne
electronic-vibration-rotation spectrum
widmo *n* emisyjne emission spectrum
widmo *n* emisyjne fluorescencyjne fluorescence
emission spectrum
widmo *n* emisyjne fosforescencyjne
phosphorescence emission spectrum
widmo *n* fluorescencji fluorescence spectrum
widmo *n* fluorescencyjne fluorescence spectrum
widmo *n* fluorescencyjne rentgenowskie X-ray
fluorescence spectrum
widmo *n* iskrowe spark spectrum
widmo *n* jonowe ionic spectrum
widmo *n* komptonowskie Compton continuum
widmo *n* liniowe line spectrum
widmo *n* łukowe arc spectrum
widmo *n* magnetycznego rezonansu jądrowego
wysokiej rozdzielczości high-resolution
nuclear magnetic resonance spectrum
widmo *n* mas mass spectrum
widmo *n* mas jonów dodatnich positive-ion
mass spectrum
widmo *n* mas jonów ujemnych negative-ion
mass spectrum
widmo *n* mikrofalowe microwave spectrum
widmo *n* Mössbauera Mössbauer spectrum
widmo *n* neutronów neutron spectrum
widmo *n* nieciągłe discontinuous spectrum
widmo *n* optyczne optical spectrum
widmo *n* oscylacyjne vibrational spectrum
widmo *n* oscylacyjno-rotacyjne
vibrational-rotational spectrum
widmo *n* osłabionego całkowitego odbicia
attenuated total reflectance spectrum
widmo *n* pasmowe band spectrum
widmo *n* promieniowania jądrowego nuclear
radiation spectrum
widmo *n* Ramana Raman spectrum
widmo *n* rotacyjne rotational spectrum
widmo *n* rozdzielone w czasie time-resolved
spectrum
widmo *n* UV ultraviolet spectrum
widmo *n* w nadfiolecie ultraviolet spectrum
widmo *n* w podczerwieni infrared spectrum
widmo *n* widzialne visible spectrum
widmo *n* wzbudzenia excitation spectrum
widmo *n* wzbudzenia fluorescencyjne
fluorescence excitation spectrum
widmo *n* wzbudzenia fosforescencyjne
phosphorescence excitation spectrum
wielkości *fpl* zredukowane reduced parameters
wiskozymetr *m* Ostwalda Ostwald viscometer
włókno *n* Nernsta Nernst glower
wnęka *f* rezonansowa cavity resonator
wnętrze *n* roztworu bulk of a solution
woda *f* dejonizowana deionized water
woda *f* demineralizowana deionized water
woda *f* destylowana distilled water

woda *f* higroskopijna hygroscopic water
woda *f* krystalizacyjna water of crystallization
woda *f* redestylowana redistilled water
woltamperometria *f* voltammetry
woltamperometria *f* cykliczna cyclic
triangular-wave voltammetry
woltamperometria *f* cykliczna z falą trójkątną
cyclic triangular-wave voltammetry
woltamperometria *f* hydrodynamiczna
hydrodynamic voltammetry
woltamperometria *f* inwersyjna stripping
voltammetry
woltamperometria *f* inwersyjna z rozpuszczaniem
anodowym anodic stripping voltammetry
woltamperometria *f* różnicowa differential
voltammetry
woltamperometria *f* różniczkowa derivative
voltammetry
woltamperometria *f* stripingowa stripping
voltammetry
woltamperometria *f* z liniową zmianą
potencjału chronoamperometry with linear
potential sweep
woltamperometria *f* z zatężaniem stripping
voltammetry
wskaźnik *m* adsorpcyjny adsorption indicator
wskaźnik *m* alkacymetryczny acid-base indicator
wskaźnik *m* chemiluminescencyjny
chemiluminescent indicator
wskaźnik *m* dwubarwny two-colour indicator
wskaźnik *m* ekstrakcyjny extraction indicator
wskaźnik *m* fluorescencyjny fluorescent indicator
wskaźnik *m* izotopowy isotopic tracer
wskaźnik *m* jednobarwny one-colour indicator
wskaźnik *m* metalochromowy metallochromic
indicator
wskaźnik *m* metalofluorescencyjny
metallofluorescent indicator
wskaźnik *m* mieszany mixed indicator
wskaźnik *m* pH acid-base indicator
wskaźnik *m* promieniotwórczy radioactive tracer
wskaźnik *m* redoks oxidation-reduction indicator
wskaźnik *m* redoks nieodwracalny irreversible
redox indicator
wskaźnik *m* redoks odwracalny reversible redox
indicator
wskaźnik *m* strąceniowy precipitation indicator
wskaźnik *m* uniwersalny universal indicator
wskaźnik *m* wewnętrzny internal indicator
wskaźnik *m* wizualny indicator
wskaźnik *m* z tłem screened indicator
wskaźnik *m* zewnętrzny external indicator
wskaźniki *mpl* Millera Miller indices
wskaźniki *mpl* płaszczyzny Miller indices
współczynnik *m* absorpcji liniowy linear decadic
absorption coefficient
współczynnik *m* absorpcji molowy molar
absorptivity
współczynnik *m* absorpcji właściwy specific
decadic absorption coefficient
współczynnik *m* aktywności jonowej średni
mean activity coefficient of electrolyte
współczynnik *m* aktywności składnika B
activity coefficient of substance B
współczynnik *m* aktywności średni elektrolitu
mean activity coefficient of electrolyte

współczynnik *m* asymetrii piku asymmetry
factor of a peak
współczynnik *m* dekontaminacji decontamination
factor
współczynnik *m* dyfuzji diffusion coefficient
współczynnik *m* ekstrakcji concentration
distribution ratio (*in extraction*)
współczynnik *m* ekstynkcji specific decadic
absorption coefficient
współczynnik *m* ekstynkcji molowy molar
absorptivity
współczynnik *m* funkcji regresji regression
coefficient
współczynnik *m* gamma gamma
współczynnik *m* kontrastowości gamma
współczynnik *m* korekcyjny gradientu ciśnienia
pressure-gradient correction factor
współczynnik *m* korelacji coefficient of
correlation
współczynnik *m* korelacji wielokrotnej
coefficient of multiple correlation
współczynnik *m* korelacji wielowymiarowej
coefficient of multiple correlation
współczynnik *m* nasycenia saturation factor
współczynnik *m* odzysku recovery factor
współczynnik *m* osłabienia attenuation
coefficient
współczynnik *m* osłabienia atomowy atomic
attenuation coefficient
współczynnik *m* osłabienia liniowy linear
attenuation coefficient
współczynnik *m* osłabienia masowy mass
attenuation coefficient
współczynnik *m* pochłaniania absorptance
współczynnik *m* podziału 1. concentration
distribution ratio (*in chromatography*)
2. concentration distribution ratio
(*in extraction*)
współczynnik *m* podziału D_g distribution
coefficient D_g
współczynnik *m* podziału D_s distribution
coefficient D_s
współczynnik *m* podziału D_v distribution
coefficient D_v
współczynnik *m* podziału objętościowy
distribution coefficient D_v
współczynnik *m* przejścia charge-transfer
coefficient
współczynnik *m* przeliczeniowy miareczkowania
titrimetric conversion factor
współczynnik *m* przeniesienia ładunku
charge-transfer coefficient
współczynnik *m* reakcji ubocznych side-reaction
coefficient
współczynnik *m* regresji regression coefficient
współczynnik *m* rozdzielenia separation factor
współczynnik *m* równoważności składnika B
equivalence factor of component B
współczynnik *m* równoważności w reakcjach
redoks equivalence factor in redox reactions
współczynnik *m* równoważności w reakcjach
strącania i kompleksowania equivalence
factor in precipitation and complex
formation reactions

współczynnik *m* równoważności w reakcjach
zobojętniania equivalence factor in
acid-base reactions
współczynnik *m* Sandella Sandell's sensitivity
index
współczynnik *m* selektywności selectivity
coefficient
współczynnik *m* selektywności poprawiony
corrected selectivity coefficient
współczynnik *m* selektywności potencjometryczny
potentiometric selectivity coefficient
współczynnik *m* szorstkości roughness factor
współczynnik *m* temperaturowy temperature
coefficient
współczynnik *m* wstecznego rozpraszania
backscattering factor
współczynnik *m* wydajności rozpylenia
sputtering yield
współczynnik *m* wzbogacenia enrichment factor
współczynnik *m* załamania refractive index
współczynnik *m* załamania absolutny refractive
index
współczynnik *m* załamania względny relative
refractive index
współczynnik *m* zmienności coefficient of
variation
współczynnik *m* zmienności w próbie empirical
variation coefficient
współjony *mpl* co-ions
współstrącanie *n* coprecipitation
wychwyt *m* rezonansowy resonance capture
wyciąg *m* extract
wyciek *m* effluent
wydajność *f* atomizacji efficiency of atomization
wydajność *f* detektora wewnętrzna intrinsic
detector efficiency
wydajność *f* emisji neutronów neutron output
wydajność *f* fluorescencji rentgenowskiej X-ray
fluorescence yield
wydajność *f* fotopiku absolutna absolute
photopeak efficiency
wydajność *f* fotopiku wewnętrzna intrinsic
photopeak efficiency
wydajność *f* kapilary flow rate of mercury
wydajność *f* liczenia counting efficiency
wydajność *f* liczenia geometryczna geometry
factor
wydajność *f* nebulizacji efficiency of nebulization
wydajność *f* piku całkowitej absorpcji
absolutna absolute full energy peak
efficiency
wydajność *f* piku całkowitej absorpcji
wewnętrzna intrinsic full energy peak
efficiency
wydajność *f* prądowa current efficiency
wydajność *f* produktu rozszczepienia fission
yield
wydajność *f* radiochemiczna radiochemical yield
wydajność *f* zliczania counting efficiency
wydajność *f* zliczania geometryczna geometry
factor
wydzielanie *n* metali elektrolityczne
electrodeposition of metals
wygaszanie *n* quenching
wykluczanie *n* exclusion
wykluczanie *n* jonów ion exclusion

wykładnik *m* jonów wodorotlenowych pOH
wykładnik *m* jonów wodorowych pH
wykładnik *m* wskaźnika indicator exponent
wykrywalność *f* limit of detection
wykrywanie *n* detection
wykrywanie *n* wydzielanego gazu evolved gas detection
wyładowanie *n* jarzeniowe glow discharge
wyładowanie *n* w katodzie wnękowej hollow-cathode discharge
wymiana *f* anionowa anion exchange
wymiana *f* izotopowa isotope exchange
wymiana *f* jonowa ion exchange
wymiana *f* kationowa cation exchange
wymiana *f* masy mass exchange
wymieniacz *m* anionowy anion exchanger
wymieniacz *m* elektronów electron exchanger
wymieniacz *m* jonowy ion exchanger
wymieniacz *m* jonowy amfoteryczny amphoteric ion exchanger
wymieniacz *m* jonowy chelatujący chelating ion exchanger
wymieniacz *m* jonowy ciekły liquid ion exchanger
wymieniacz *m* jonowy dwufunkcyjny bifunctional ion exchanger
wymieniacz *m* jonowy jednofunkcyjny monofunctional ion exchanger
wymieniacz *m* jonowy redoks redox ion exchanger
wymieniacz *m* jonowy selektywny selective ion exchanger
wymieniacz *m* jonowy wielofunkcyjny polyfunctional ion exchanger
wymieniacz *m* jonów ion exchanger
wymieniacz *m* kationowy cation exchanger
wymywanie *n* elution
wymywanie *n* fazy nieruchomej bleeding
wymywanie *n* zwrotne backflushing
wynik *m* nietypowy outlier
wynik *m* odbiegający outlier
wypalenie *n* burn-up
wypełnienie *n* kolumny column packing
wypieranie *n* displacement
wysalanie *n* salting-out
wysokość *f* piku peak height
wysokość *f* półki zredukowana reduced plate height
wysokość *f* równoważna półce efektywnej height equivalent to an effective plate
wysokość *f* równoważna półce teoretycznej height equivalent to a theoretical plate
wysokość *f* równoważna półce teoretycznej efektywnej height equivalent to an effective plate
wysokość *f* słupa rtęci height of mercury column
wysokość *f* stopnia step height
wysokość *f* stopnia chromatogramu całkowego step height
wytrącanie *n* precipitation
wytwarzanie *n* elektrochemiczne titrantu electrochemical generation of a titrant
wywoływanie *n* development
wzbudzalnik *m* excitation source
wzbudzenie *n* excitation

wzbudzenie *n* iskrowe spark source ionization
wzmacniacz *m* operacyjny operational amplifier
wzmocnienie *n* gazowe gas amplification
wzorce *mpl* do analizy elementarnej standards for elemental analysis
wzorzec *m* analityczny standard sample
wzorzec *m* wewnętrzny 1. internal standard 2. internal standard (*in spectrochemical analysis*)
wzorzec *m* zewnętrzny external standard
wzór *m* Lewisa i Sargenta Lewis Sargent's relation

Z

zaczernienie *n* blackening
zadymienie *n* fog
zakres *m* mikrofalowy microwave region
zakres *m* nadfioletu ultraviolet region
zakres *m* podczerwieni infrared region
zakres *m* pomiarowy wagi range of applicability of a balance
zakres *m* ważenia range of applicability of a balance
zakres *m* widzialny visible region
zakres *m* zmiany barwy wskaźnika colour change interval
załamanie *f* refraction
załamanie *n* światła light refraction
zanieczyszczenie *n* osadu contamination of a precipitate
zarodek *m* nucleus
zarodkowanie *n* nucleation
zarodkowanie *n* heterogeniczne heterogeneous nucleation
zarodkowanie *n* homogeniczne homogeneous nucleation
zarodkowanie *n* jednoczesne simultaneous nucleation
zarodkowanie *n* jednorodne homogeneous nucleation
zarodkowanie *n* niejednorodne heterogeneous nucleation
zatrzymywanie *n* mechaniczne mechanical entrapment
zawartość *f* content
zawartość *f* izotopów w naturalnym pierwiastku natural isotopic abundance
zbieranie *n* collection
zbiorowość *f* generalna population
zbiorowość *f* próbna sample
zdarzenie *n* elementarne elementary event
zdarzenie *n* losowe random event
zderzenie *n* elastyczne elastic collision
zderzenie *n* nieelastyczne inelastic collision
zdolność *f* rozdzielcza 1. resolution (*of a photographic emulsion*) 2. resolution (*of a spectrometer*)
zdolność *f* rozdzielcza kryształu crystal resolving power
zdolność *f* rozdzielcza pików peak resolution
zdolność *f* rozdzielcza spektrometru gamma energy resolution of a gamma spectrometer
zdolność *f* wymienna całkowita total specific capacity

zdolność *f* **wymienna do chwili przebicia złoża jonitu** breakthrough capacity of an ion exchanger bed
zdolność *f* **wymienna objętościowa** volume capacity
zdolność *f* **wymienna objętościowa złoża jonitu** bed volume capacity
zdolność *f* **wymienna praktyczna** practical specific capacity
zdolność *f* **wymienna robocza** practical specific capacity
zdolność *f* **wymienna robocza złoża jonitu** breakthrough capacity of an ion exchanger bed
zeolity *mpl* zeolites
zespół *m* **cząstek** aggregate
ziemia *f* **okrzemkowa** diatomaceous earth
zjawiska *npl* **elektrokinetyczne** electrokinetic phenomena
złoże *n* **chromatograficzne** chromatographic bed
złoże *n* **mieszane** mixed bed
zmiatacz *m* scavenger
zmienna *f* **losowa** random variable
zmienna *f* **losowa ciągła** continuous random variable
zmienna *f* **losowa dyskretna** discrete random variable
zmienna *f* **losowa nieciągła** discrete random variable
zmienna *f* **losowa skokowa** discrete random variable
zmienna *f* **losowa standaryzowana** standardized random variable
zmienne *fpl* **zredukowane** reduced parameters
znacznik *m* marker
znacznik *m* **izotopowy** isotopic tracer

znacznik *m* **promieniotwórczy** radioactive tracer
zobojętnianie *n* neutralization
zol *m* sol
zrąb *m* **atomowy** atomic core
zwęglanie *n* **sączka** charring of filter paper
związek *m* **chelatowy** chelate compound
związek *m* **kleszczowy** chelate compound
związek *m* **znaczony** labelled compound

Ź

źródło *n* source
źródło *n* **iskrowe** spark ionization source
źródło *n* **jonów** ionization source
źródło *n* **jonów uzyskanych przez bombardowanie elektronami** electron-impact ion source
źródło *n* **jonów przez fotojonizację** photoionization ion source
źródło *n* **jonów wytwarzane przez jonizację termiczne** thermal emission ion source
źródło *n* **neutronów izotopowe** isotopic neutron source
źródło *n* **plazmowe** plasma torch
źródło *n* **promieniowania** radiation source
źródło *n* **światła** light source
źródło *n* **wzbudzenia** excitation source

Ż

żel *m* gel
żywica *f* **anionowymienna** anion-exchange resin
żywica *f* **blonkowata** pellicular ion exchanger
żywica *f* **chelatująca** chelating resin
żywica *f* **jonowymienna** ion-exchange resin
żywica *f* **kationowymienna** cation-exchange resin
żywica *f* **pelikularna** pellicular ion exchanger

РУССКИЙ УКАЗАТЕЛЬ

А

аберрация *f* optical aberration

абсолютная диэлектрическая проницаемость *f* permittivity

абсолютная ошибка *f* absolute error

абсолютная фотоэффективность *f* absolute photopeak efficiency

абсолютная чувствительность *f* детектора absolute detector sensitivity

абсолютная эффективность *f* пика полного поглощения absolute full energy peak efficiency

абсолютный показатель *m* преломления refractive index

абсолютный порог *m* ощущения stimulus threshold

абсолютный потенциал *m* absolute electrode potential

абсолютный электродный потенциал *m* absolute electrode potential

абсорбат *m* absorbate

абсорбент *m* absorbent

абсорбер *m* absorber

абсорбированное вещество *n* absorbate

абсорбируемость *f* specific decadic absorption coefficient

абсорбтив *m* absorbate

абсорбционная кривая *f* absorption curve

абсорбционная кювета *f* absorption cell

абсорбционная молекулярная спектроскопия *f* molecular absorption spectroskopy

абсорбционная рентгеновская спектроскопия *f* X-ray absorption spectroscopy

абсорбционная спектроскопия *f* absorption spectroscopy

абсорбционная спектроскопия *f* пламени flame absorption spectroscopy

абсорбционная спектрофотометрия *f* absorption spectrophotometry

абсорбционный газоанализатор *m* Orsat gas analyser

абсорбционный газовый анализ *m* absorptiometric gas analysis

абсорбционный спектр *m* absorption spectrum

абсорбционный спектрохимический анализ *m* absorption spectrochemical analysis

абсорбционный фильтр *m* absorption filter

абсорбционный эффект *m* absorption effect

абсорбция *f* 1. absorption 2. absorption (*of radiation*)

автоматическая пипетка *f* automatic zero pipette

автоматическая поршневая бюретка *f* piston automatic burette

автоматические весы *pl* automatic balance

автоматическое определение *n* углерода, водорода и азота automatic determination of carbon, hydrogen and nitrogen

автопротолиз *m* autoprotolysis

авторадиограмма *f* autoradiograph

авторадиограф *m* autoradiograph

агароза *f* agarose

агломерация *f* agglomeration

агрегат *m* aggregate

агрегирование *n* aggregation

адгезия *f* adhesion

аддитивность *f* additivity law

адсорбат *m* adsorbate

адсорбент *m* adsorbent

адсорбтив *m* adsorbate

адсорбционная волна *f* adsorption wave

адсорбционная хроматография *f* adsorption chromatography

адсорбционный индикатор *m* adsorption indicator

адсорбционный ток *m* adsorption current

адсорбция *f* adsorption

акваметрия *f* aquametry

аккумулятор *m* electrical accumulator

активационный анализ *m* activation analysis

активационный анализ *m* с применением заряженных частиц charged-particle activation analysis

активация *f* activation

активация *f* адсорбента activation of an adsorbent

активация *f* заряженными частицами charged-particle activation

активация *f* нейтронами neutron activation

активированный ионоселективный электрод *m* sensitized ion-selective electrode

активная проводимость *f* conductance

активное сопротивление *n* resistance

активное твёрдое вещество *n* active solid

активность *f* activity

активность *f* адсорбента adsorbent activity

активность *f* вещества B relative activity of substance B

активность *f* насыщения saturation activity

активность *f* электролита B activity of electrolyte B

актор *m* actor

акустооптическая спектрометрия *f* optoacoustic spectrometry

алкалиметрическое титрование *n* alkalimetric titration

алкалиметрия *f* alkalimetry
алкоголиметрия *f* alcoholometry
альтернативная гипотеза *f* alternative hypothesis
амальгама *f* amalgam
амальгамная полярография *f* amalgam polarography
амальгамный электрод *m* amalgam electrode
американский градус *m* жёсткости воды American degree of water hardness
амперометрический анализ *m* amperometric analysis
амперометрическое титрование *n* с двумя индикаторными электродами amperometric titration with two indicator electrodes
амперометрическое титрование *n* с двумя поляризуемыми электродами amperometric titration with two indicator electrodes
амперометрическое титрование *n* с капающим ртутным электродом amperometric titration with a dropping mercury electrode
амперометрическое титрование *n* с одним индикаторным электродом amperometric titration with one indicator electrode
амперометрическое титрование *n* с одним поляризуемым электродом amperometric titration with one indicator electrode
амперометрическое титрование *n* с поляризационным током amperometric titration with two indicator electrodes
амперометрия *f* amperometry
амперометрия *f* с вращающимся платиновым проволочным электродом rotating-platinum-wire electrode amperometry
амперометрия *f* с двумя индикаторными электродами amperometry with two indicator electrodes
амперометрия *f* с двумя поляризуемыми электродами amperometry with two indicator electrodes
амперометрия *f* с индикаторным электродом amperometry with one indicator electrode
амперометрия *f* с перемешиваемым ртутным макроэлектродом stirred-mercury-pool amperometry
амперометрия *f* с поляризуемым электродом amperometry with one indicator electrode
амперостат *m* galvanostat
амперостатическая кулонометрия *f* controlled-current coulometry
амперостатический метод *m* galvanostatic method
амплитудный анализатор *m* импульсов pulse amplitude analyser
амфион *m* ampholyte ion
амфипротный растворитель *m* amphiprotic solvent
амфолит *m* ampholyte
амфотерный ион *m* ampholyte ion
амфотерный ионообменник *m* amphoteric ion exchanger
амфотерный растворитель *m* amphiprotic solvent
амфотерный электролит *m* ampholyte

анализ *m* амплитуды импульсов pulse amplitude analysis
анализ *m* выделенных газов evolved gas analysis
анализ *m* методом двойного изотопного разбавления double isotope dilution analysis
анализ *m* методом изотопного разбавления isotope dilution analysis
анализ *m* методом обратного изотопного разбавления reverse isotope dilution analysis
анализ *m* методом прямого изотопного разбавления direct isotope dilution analysis
анализ *m* на замедляющих нейтронах analysis by neutron slowing-down
анализ *m* паровой фазы над жидкостью headspace analysis
анализ *m* по функциональным группам functional group analysis
анализ *m* поверхности surface analysis
анализ *m* поверхности резонансной ионизацией распыленных атомов surface analysis by resonance ionization of sputtered atoms
анализ *m* равновесного пара headspace analysis
анализ *m* сжиганием combustion analysis
анализ *m* следов trace analysis
анализатор *m* analyser
анализатор *m* задерживающего потенциала retarding field analyser
анализатор *m* с цилиндрическим зеркалом cylindrical mirror analyser
аналитическая группа *f* analytical group
аналитическая длина *f* волны analytical wavelength
аналитическая концентрация *f* analytical concentration
аналитическая линия *f* analytical line
аналитическая пара *f* линий line pair
аналитическая проба *f* analytical sample
аналитическая реакция *f* analytical reaction
аналитическая химия *f* analytical chemistry
аналитические весы *pl* analytical balance
аналитические гири *fpl* analytical weights
аналитический метод *m* analytical method
аналитический множитель *m* analytical factor
аналитический промежуток *m* analytical gap
аналитический сигнал *m* analytical signal
аналитический фактор *m* analytical factor
ангидрон *m* Anhydrone
английский градус *m* жёсткости воды English degree of water hardness
анионит *m* anion exchanger
анионный обмен *m* anion exchange
анионообменная смола *f* anion-exchange resin
анионообменник *m* anion exchanger
анод *m* anode
анодная инверсионная вольтамперометрия *f* anodic stripping voltammetry
анолит *m* anolyte
аномальная дисперсия *f* anomalous dispersion
антисимметричные колебания *npl* antisymmetrical vibrations
антистоксовы линии *fpl* anti-Stokes lines
апериодичные весы *npl* damped balance
аппарат *m* Киппа Kipp gas generator

аппарат *m* Сокслета Soxhlet apparatus
аппликатор *m* applicator
апротонный растворитель *m* aprotic solvent
арбитражный анализ *m* umpire analysis
аргентометрическое титрование *n*
    argentimetric titration
аргентометрия *f* argentimetry
аргон-детектор *m* argon ionization detector
аргоновый детектор *m* argon ionization
    detector
аргоновый ионизационный детектор *m* argon
    ionization detector
ареометр *m* hydrometer
асимметричные колебания *npl*
    antisymmetrical vibrations
аскарит *m* Ascarite
аскорбинометрия *f* ascorbimetry
астигматизм *m* astigmatism
атомизатор *m* atomizer
атомизация *f* atomization
атомная единица *f* unified atomic mass unit
атомная рефракция *f* atomic refraction
атомная спектроскопия *f* atomic spectroscopy
атомная эмиссионная спектроскопия *f* atomic
    emission spectroscopy
атомно-абсорбционная спектроскопия *f*
    atomic absorption spectroscopy
атомно-абсорбционная спектроскопия *f*
    пламени flame atomic absorption
    spectroscopy
атомно-флуоресцентная спектроскопия *f*
    atomic fluorescence spectroscopy
атомно-флуоресцентная спектроскопия *f*
    пламени flame atomic fluorescence
    spectroscopy
атомно-эмиссионная спектроскопия *f* пламени
    flame atomic emission spectroscopy
атомное сечение *n* atomic cross section
атомное эффективное сечение *n* atomic
    cross section
атомный коэффициент *m* ослабления atomic
    attenuation coefficient
атомный остов *m* atomic core
атомный спектр *m* atomic spectrum
атомный спектрохимический анализ *m* atomic
    spectrochemical analysis
ауксохром *m* auxochrome
аффинная хроматография *f* affinity
    chromatography
ацидиметрическое титрование *n* acidimetric
    titration
ацидиметрия *f* acidimetry
ацидиметрия и алкалиметрия *f* acidimetry
    and alkalimetry

Б

байпасный дозатор *m* by-pass injektor
балансные фильтры *mpl* balanced filters
барн *m* barn
батарея *f* battery
батохромная группа *f* bathochrome
батохромное смещение *n* bathochromic shift
батохромное смещение *n* окраски
    bathochromic shift
батохромный эффект *m* bathochromic shift

безэлектродная разрядная лампа *f*
    electrodeless discharge lamp
бекерель *m* becquerel
белый свет *m* white light
бесконечно толстый слой *m* infinitely thick
    layer
бесконечно тонкий слой *m* infinitely thin layer
беспламенный атомизатор *m* electrothermal
    atomizer
биамперометрическое титрование *n*
    amperometric titration with two indicator
    electrodes
биамперометрия *f* amperometry with two
    indicator electrodes
биномиальное распределение *n* binomial
    distribution
биолюминесценция *f* bioluminescence
биопотенциометрия *f* controlled-current
    potentiometry with two indicator electrodes
бипотенциометрическое титрование *n*
    controlled-current potentiometric titration
    with two indicator electrodes
биуретовая реакция *f* biuret reaction
бифункциональный ионообменник *m*
    bifunctional ion exchanger
блескообразователь *m* brightening
ближнее инфракрасное излучение *n*
    near-infrared
ближний ультрафиолет *m* near ultraviolet
блок *m* block
блокирование *n* металлоиндикатора blocking
    of metal indicator
БН-камера *f* BN-chamber
болометр *m* bolometer
бомба *f* для сплавления combustion bomb
броматометрия *f* bromatometry
бумажная хроматография *f* paper
    chromatography
бумажный фильтр *m* filter paper
буферный раствор *m* buffer solution
быстрые нейтроны *mpl* fast neutrons
бюретка *f* burette
бюретка *f* с автоматической установкой
    нулевой метки automatic zero burette
бюретка *f* с автоматической установкой по
    Дафферту Daffert automatic zero burette
бюретка *f* с полоской Schellbach burette

В

вакуум-эксикатор *m* vacuum desiccator
вакуумный ультрафиолет *m* vacuum
    ultraviolet
валентные колебания *npl* stretching vibrations
веерные колебания *npl* wagging vibrations
веерообразные колебания *npl* wagging
    vibrations
вековое равновесие *n* secular radioaktive
    equilibrium
вековое радиоактивное равновесие *n* secular
    radioaktive equilibrium
вектор-полярограф *m* vector polarograph
величина *f* R_B R_B value
величина *f* R_f R_f value
величина *f* R_M R_M value
вентильный фотоэлемент *m* photovoltaic cell

вероятностное распределение *n* probability distribution
вероятность *f* probability
весовая доля *f* mass fraction of substance B
весовая набухаемость *f* в растворителе weight swelling in solvent
весовое титрование *n* weight titration
весовой анализ *m* gravimetric analysis
весовой стаканчик *m* weighing bottle
весы *pl* с успокоителями damped balance
P-ветвь *f* P-branch
Q-ветвь *f* Q-branch
R-ветвь *f* R-branch
вещество *n* substance
(+)-вещество *n* dextrarotatory substance
(−)-вещество *n* laevorotatory substance
d-вещество *n* dextrarotatory substance
l-вещество *n* laevorotatory substance
вибрирующий конденсатор *m* vibrating condenser
видимая область *f* visible region
видимая радиация *f* light
видимое излучение *n* light
видимый спектр *m* visible spectrum
визуальная колориметрия *f* visual colorimetry
визуальное титрование *n* visual titration
визуальный индикатор *m* indicator
вискозиметр *m* Оствальда Ostwald viscometer
вихревая диффузия *f* eddy diffusion term
внеплоскостные колебания *npl* out-of-plane vibrations
внешний индикатор *m* external indicator
внешний электрический потенциал *m* фазы *β* outer electric potential of phase *β*
внешний эталон *m* external standard
внешняя зона *f* secondary combustion zone
внутреннее пропускание *n* internal transmittance
внутренний индикатор *m* internal indicator
внутренний коэффициент *m* пропускания internal transmittance
внутренний фильтр *m* screened indicator
внутренний электрический потенциал *m* фазы *β* inner electric potential of phase *β*
внутренний электрод *m* сравнения internal reference electrode
внутренний стандарт *m* 1. internal standard 2. internal standard (*in spectrochemical analysis*)
внутренний электролиз *m* internal electrolysis
внутренний эталон *m* 1. internal standard 2. internal standard (*in spectrochemical analysis*)
внутренняя зона *f* primary combustion zone
внутренняя поглощательная способность *f* absorbance
внутренняя фотоэффективность *f* intrinsic photopeak efficiency
внутренняя часть *f* раствора bulk of a solution
внутренняя электрогравиметрия *f* internal electrogravimetry
внутрикомплексное соединение *n* neutral chelate
вогнутая решётка *f* concave grating
водородная лампа *f* hydrogen discharge lamp

водородное перенапряжение *n* hydrogen overpotential
водородный газовый электрод *m* hydrogen gas electrode
водородный показатель *m* pH
водородный электрод *m* hydrogen-gas electrode
водородселективный стеклянный электрод *m* glass pH electrode
водородчувствительный стеклянный электрод *m* glass pH electrode
возбуждение *n* excitation
возбуждённая линия *f* 1. exciting line 2. Rayleigh line
волновое число *n* wave number
вольта-потенциал *m* outer electric potential of phase *β*
вольтамперограмм *m* polarization curve
вольтамперометрическая кривая *f* polarization curve
вольтамперометрический анализ *m* voltametric analysis
вольтамперометрия *f* voltametry
вольтамперометрия *f* анодным растворением anodic stripping voltametry
вольтамперометрия *f* с линейной развёрткой потенциала chronoamperometry with linear potential sweep
вольтамперометрия *f* с треугольной развёрткой потенциала triangular-wave voltametry
вольтамперометрия *f* со стационарным электродом chronoamperometry with linear potential sweep
воронка *f* Бюхнера Büchner funnel
воронка *f* для горячего фильтрования funnel heater
воронка *f* для порошков powder funnel
воспроизводимость *f* reproducibility
восстановитель *m* reductant
восстановительное титрование *n* reductometric titration
восстановление *n* reduction
восходящая хроматография *f* ascending development
восходящее проявление *n* ascending development
восходящий метод *m* ascending development
вращательно-колебательная полоса *f* vibration-rotation band
вращательно-колебательно-электронный спектр *m* electronic-vibration-rotation spectrum
вращательно-колебательный спектр *m* vibrational-rotational spectrum
вращательный спектр *m* rotational spectrum
вращающийся дисковый электрод *m* rotating-disk electrode
вращающийся дисковый электрод *m* с кольцом rotating-ring disk electrode
вращающийся проволочный электрод *m* rotating-wire electrode
вращение *n* плоскости поляризации света rotation of the plane of polarization of polarized light
временная жёсткость *f* temporary hardness
время *n* жизни капли drop time
время *n* образования одной капли drop time

время *n* **отклика детектора** response time
время *n* **перехода** transition time
время *n* **полёта** time of flight
время *n* **предварительного обжига** pre-arc period
время *n* **предварительного обыскривания** pre-spark period
время *n* **удерживания** total retention time
время *n* **удерживания несорбирующегося вещества** mobile phase hold-up time
время *n* **экспозиции** exposure time
времяпролётный масс-спектрометр *m* time-of-flight mass spectrometer
вспомогательный электрод *m* auxiliary electrode
вторичная конкурирующая реакция *f* secondary interfering reaction
вторичная реакция *f* induced reaction
вторичное рентгеновское излучение *n* secondary X-radiation
вторичный элемент *m* electrical accumulator
вторичный эталон *m* secondary standard
второе приближение *n* **теории Дебая-Гюккеля** Debye-Hückel equation
второй обертон *m* second overtone
второй эффект *m* **Вина** dissociation field effect
вуаль *f* fog
выборка *f* sample
выборочная дисперсия *f* sample variance
выборочная медиана *f* sample median
выборочная функция *f* statistic
выборочное стандартное отклонение *n* standard deviation of a sample
выборочный коэффициент *m* **вариантности** empirical variation coefficient
выгорание *n* burn-up
вымывание *n* **неподвижной фазы** bleeding
вынужденный эффект *m* **Рамана** stimulated Raman effect
вынужденное излучение *n* stimulated emission
вынужденное комбинационное рассеяние *n* **света** stimulated Raman effect
выпаривание *n* evaporation
высаливание *n* salting-out
высаливающая хроматография *f* salting-out chromatography
высокопроизводительная жидкостная хроматография *f* high-performance liquid chromatography
высокоскоростная жидкостная хроматография *f* high-performance liquid chromatography
высокочастотная безэлектродная лампа *f* electrodeless discharge lamp
высокочастотная кондуктометрия *f* high-frequency conductometry
высокочастотная плазма *f* radiofrequency plasma
высокочастотная полярография *f* radio-frequency polarography
высокочастотное кондуктометрическое титрование *n* high-frequency conductometric titration
высокоэффективная жидкостная хроматография *f* high-performance liquid chromatography

высота *f* **пика** peak height
высота *f* **столба ртути** height of mercury column
высота *f* **ступени** step height
высота *f* **ступени на интегральной хроматограмме** step height
высота *f* **эквивалентная теоретической тарелке** height equivalent to a theoretical plate
высота *f* **эквивалентная эффективной теоретической тарелке** height equivalent to an effective plate
высушивание *n* **осадка** drying of precipitate
вытеснение *n* displacement
вытеснительная хроматография *f* displacement chromatography
вытеснительное титрование *n* replacement titration
вытеснительный метод *m* displacement chromatography
выход *m* **нейтронов** neutron output
выход *m* **осколков деления** fission yield
выход *m* **по току** current efficiency
выход *m* **продуктов деления** fission yield
выход *m* **распыления** sputtering yield
выход *m* **рентгеновской флуоресценции** X-ray fluorescence yield
выщелачивание *n* leaching

Г

гамма коэффициент *m* gamma
газ-носитель *m* carrier gas
газ-элюент *m* carrier gas
газо-твердофазная хроматография *f* gas-solid chromatography
газо-твёрдая хроматография *f* gas-solid chromatography
газовая бюретка *f* **Бунте** Bunte gas burette
газовая пипетка *f* gas sampling pipette
газовая хроматография *f* gas chromatography
газовое усиление *n* gas amplification
газоволюметрический анализ *m* volumetric gas analysis
газоволюметрия *f* volumetric gas analysis
газовый анализ *m* gas analysis
газовый электрод *m* gas electrode
газо-жидкостная хроматография *f* gas-liquid chromatography
газометрический анализ *m* gasometric analysis
газометрическое титрование *n* gasometric titration
газопроницаемая мембрана *f* gas-permeable membrane
газочувствительный электрод *m* gas sensing electrode
гальваническая цепь *f* galvanic cell
гальванический термоэлемент *m* galvanic thermocell
гальванический элемент *m* galvanic cell
гальванометр *m* galvanometer
гальваностат *m* galvanostat
гальваностатическая кулонометрия *f* controlled-current coulometry
гальваностатический метод *m* galvanostatic method
гексаметилдисилазан *m* hexamethyldisilazane

гелевая хроматография *f* gel chromatography
гель *m* gel
гель-проникающая хроматография *f* gel-permeation chromatography
гель-фильтрация *f* gel filtration
гель-хроматография *f* gel chromatography
гельмгольцевский двойной слой *m* compact layer
генеральная проба *f* gross sample
генеральная совокупность *f* population
геометрический коэффициент *m* geometry factor
геометрический фактор *m* geometry factor
геометрия *f* измерения counting geometry
геометрия *f* счёта counting geometry
гетерогенная мембрана *f* heterogeneous membrane
гетерогенное образование *n* зародышей heterogeneous nucleation
гетерогенный мембранный электрод *m* heterogeneous membrane ion-selective electrode
гетерополикислота *f* heteropolyacid
гигроскопическая вода *f* hygroscopic water
гидрогель *m* hydrogel
гидродинамическая вольтамперометрия *f* hydrodynamic voltammetry
гидрозоль *m* hydrosol
гидроксильный растворитель *m* hydroxylic solvent
гидростатические весы *pl* hydrostatic balance
гиперсорбция *f* hypersorption
гиперхромный эффект *m* hyperchromic effect
гипохромный эффект *m* hypochromic effect
гипсохромная группа *f* hypsochrome
гипсохромное смещение *n* hypsochromic shift
гипсохромный эффект *m* hypsochromic shift
гистерезис *m* hysteresis
гистограмма *f* histogram
главный компонент *m* major constituent
глобар *m* Globar
глубина *f* раствора bulk of a solution
голова *f* полосы band head
гомогенная мембрана *f* homogeneous membrane
гомогенное образование *n* зародышей homogeneous nucleation
гомогенное осаждение *n* precipitation from homogeneous solution
гомогенный закон *m* распределения homogeneous distribution law
гомогенный мембранный электрод *m* homogeneous membrane ion-selective electrode
гомологические линии *fpl* homologous lines
гопкалит *m* Hopcalite
горелка *f* burner
горелка *f* с полным потребителем direct-injection burner
горелка *f* с предварительным смешением premixed gas burner
горячая камера *f* hot cell
горячая лаболатория *f* hot laboratory
гравиметрический анализ *m* gravimetric analysis
гравиметрический газовый анализ *m* gravimetric gas analysis

гравиметрия *f* gravimetric analysis
градиент *m* потока flux gradient
градиентное заполнение *n* gradient layer
градиентное элюирование *n* gradient elution
градиентный слой *m* gradient layer
градуированная пипетка *f* graduated pipette
грамм-молекула *f* gram molecule
грамм-моль *m* gram molecule
грамм-эквивалент *m* gram equivalent
граница *f* фаз interface
граничный ток *m* limiting current
графитовая кювета *f* graphite cuvette
графитовая трубчатая печь *f* graphite-tube furnace
грубая ошибка *f* gross error
грубый образец *m* thick specimen
группа *f* analytical group
групповой реагент *m* group reagent

Д

давление *n* набухания swelling pressure
далёкий ультрафиолет *m* far ultraviolet
дальнее инфракрасное излучение *n* far infrared
двойное лучепреломление *n* по кругу circular birefringence
двойной слой *m* electrochemical double layer
двойной электрический слой *m* electrochemical double layer
двойной электронный парамагнитный резонанс *m* electron nuclear double resonance
двумерная хроматография *f* two-dimensional chromatography
двухлучевой спектрофотометр *m* double-beam spectrophotometer
двухсторонний критерий *m* two-sided test
двухцветный индикатор *m* two-colour indicator
двухчастотная полярография *f* double-tone polarography
дебаевская длина *f* radius of ionic atmosphere
дебаевсий радиус *m* radius of ionic atmosphere
дебаеграмма *f* X-ray powdered-crystal pattern
дезактивация *f* 1. deactivation 2. decontamination
дезактивирование *n* deactivation
дейтериевая лампа *f* deuterium discharge lamp
декаметрическое титрование *n* dielectrometric titration
декаметрия *f* dielectrometry
декантация *f* decantation
деление *n* fission
деление *n* ядра fission
делитель *m* пробы sample splitter
демаскировочный реактив *m* demasking agent
деминерализованная вода *f* deionized water
демодуляционная полярография *f* demodulation polarography
денситометр *m* microphotometer
деполяризатор *m* depolarizer
дериватограф *m* derivatograph
десорбция *f* desorption
десорбция *f* ионов при электронном облучении electron stimulated desorption of ions
десорбция *f* нейтральных частиц при электронном облучении electron stimulated desorption of neutrals

десорбция *f* полем field desorption

десорбция *f* при ионной бомбардировке ion impact desorption

десорбция *f* при электронном облучении electron stimulated desorption

детектор *m* detector

детектор *m* излучения radiation detector

детектор *m* нейтронов neutron detector

детектор *m* резонансного излучения resonance detector

детектор *m* транспортного типа transport detector

детектор *m* Эверхарта-Торнли Everhart-Thornley detector

детектор *m* электронного захвата electron capture detector

Ge-(Li) детектор *m* lithium-drifted germanium detector

Si-(Li) детектор *m* lithium-drifted silicon detector

деформационные колебания *npl* bending vibrations

деформированный слой *m* altered layer

дзета-потенциал *m* electrokinetic potential

диапазон *m* K K band

диапазон *m* R R band

диатомит *m* diatomaceous earth

диатомитовая земля *f* diatomaceous earth

диафрагма *f* Гартмана Hartmann diaphragm

диафрагма *f* Соллера Soller collimator

дилатометр *m* dilatometer

дилатометрия *f* thermodilatometry

диметилдихлорсилан *m* dimethyldichlorosilane

динамический конденсатор *m* vibrating condenser

динод *m* dynode

дискретная случайная величина *f* discrete random variable

дискретный спектр *m* discontinuous spectrum

дисперсионный анализ *m* analysis of variance

дисперсия *f* 1. dispersion 2. dispersion (*in statistics*) 3. variance

дисперсия *f* оптического вращения optical rotatory dispersion

дисперсия *f* света light dispersion

дисперсия *f* электропроводности Debye-Falkenhagen's effect

диссоциация *f* dissociation

дистиллированная вода *f* distilled water

дифрактометрия *f* diffractometry

дифракционная решётка *f* diffraction grating

дифракционный анализ *m* diffractometry

дифракция *f* быстрых электронов high energy electron diffraction

дифракция *f* быстрых электронов в отраженном пучке reflection high-energy

дифракция *f* медленных ионов low energy ion diffraction

дифракция *f* медленных электронов low energy electron diffraction

дифракция *f* рентгеновских лучей X-ray diffraction

дифракция *f* света light diffraction

дифференциальная дилатометрия *f* differential thermodilatometry

дифференциальная ёмкость *f* двойного слоя 1. differential capacity of double layer 2. differential capacitance

дифференциальная импульсная полярография *f* differential pulse polarography

дифференциальная полярография *f* differential polarography

дифференциальная потенциометрия *f* differential potentiometry

дифференциальная сканирующая калориметрия *f* differential scanning calorimetry

дифференциальная спектрофотометрия *f* differential spectrophotometry

дифференциальная хроматограмма *f* differential chromatogram

дифференциальное потенциометрическое титрование *n* differential potentiometric titration

дифференциальный детектор *m* differential detector

дифференциальный термический анализ *m* differential thermal analysis

дифференцирующий растворитель *m* differentiating solvent

дифференцирующий эффект *m* differentiating effect

диффузионная постоянная *f* diffusion coefficient

диффузионная часть *f* двойного слоя diffuse double layer

диффузионное перенапряжение *n* diffusion overpotential

диффузионное пламя *n* diffusion flame

диффузионный двойной слой *m* diffuse double layer

диффузионный потенциал *m* diffusion potential

диффузионный слой *m* 1. diffuse double layer 2. diffusion layer

диффузионный ток *m* diffusion current

диффузия *f* diffusion

диэлектрическая постоянная *f* relative permittivity

диэлектрическая проницаемость *f* permittivity

диэлектрическое титрование *n* dielectrometric titration

диэлектрометрический анализ *m* dielectrometric analysis

диэлектрометрия *f* dielectrometry

диэлькометрическое титрование *n* dielectrometric titration

диэлькометрия *f* dielectrometry

длина *f* волны wavelength

длина *f* волны блеска blaze wavelength

длина *f* поглощающего слоя path length

доверительные границы *fpl* confidence limits

доверительные пределы *mpl* confidence limits

доверительный интервал *m* confidence interval

доверительный уровень *m* confidence level

дозатор *m* образца sample injector

долгоживущий изотоп *m* long-lived radioisotope

доля *f* неподвижной фазы stationary-phase fraction

доля *f* свободного объёма interstitial fraction

доннановое равновесие *n* membrane equilibrium

донорно-акцепторный комплекс *m* donor-acceptor complex

дополнительные колонки *fpl* coupled columns
допустимый переход *m* allowed transition
дочерний ион *m* daughter ion
дочерний нуклид *m* daughter nuclide
дрейф *m* базисной линии baseline drift
дрейф *m* чулевой линии baseline drift
дрейф *m* потенциала drift of a potential
дуга *f* переменного тока alternating current arc
дуга *f* постоянного тока direct current arc
дуговой спектр *m* arc spectrum
духи *mpl* ghosts
дырочная проводимость *f* hole conduction

Е

естественная конвекция *f* natural convection
естественная ширина *f* спектральной линии
   natural line-width
ёмкостный ток *m* capacity current
ёмкость *f* пика peak capacity
ёмкость *f* слоя ионообменника до проскока
   breakthrough capacity of an ion
   exchanger bed

Ж

жёсткость *f* воды water hardness
живое время *n* live time
жидкая мембрана *f* liquid membrane
жидкая фаза *f* stationary liquid phase
жидкий ионообменник *m* liquid ion exchanger
жидкостная сортировка *f* elutriation
жидкостная хроматография *f* liquid
   chromatography
жидкостная хроматография *f* при высоких
   давлениях high-performance liquid
   chromatography
жидкостно-гелевая хроматография *f*
   liquid-gel chromatography
жидкостно-жидкостная хроматография *f*
   liquid-liquid chromatography
жидкостно-твёрдая хроматография *f*
   liquid-solid chromatography
жидкостный мембранный электрод *m* liquid
   membrane ion-selective electrode
жидкостный потенциал *m* liquid-junction
   potential
жидкость-гелевая хроматография *f*
   liquid-gel chromatography
жидкость-жидкостная хроматография *f*
   liquid-liquid chromatography
жидкость-жидкостная экстракция *f*
   liquid-liquid extraction
жидкость-жидкостное распределение *n*
   liquid-liquid distribution
жидкость-твёрдая хроматография *f*
   liquid-solid chromatography

З

загрязнение *n* осадка contamination of
   a precipitate
закадмиевые нейтроны *mpl* epicadmium
   neutrons
закон аддитивности *m* additivity law
закон *m* Бугера-Ламберта Lambert's law

закон *m* Бугера-Ламберта-Бэра Beer's law
закон *m* Бэра Beer's law
закон *m* Вульфа-Брэгга Bragg law
закон *m* Гука Hooke's law
закон *m* Денера-Госкинса logarithmic
   distribution law
закон *m* ионной силы Льюса и Рендалла
   Lewis-Randall's ionic strength law
закон *m* Кольрауша Kohlrausch's additive law
закон *m* Ламберта Lambert's law
закон *m* Мозли Moseley law
закон *m* независимости движения ионов
   Kohlrausch's additive law
закон *m* радиоактивного распада law of
   radioactive decay
закон *m* разведения Оствальда Ostwald's
   dilution law
закон *m* распределения Nernst distribution law
закон *m* Стокса Stokes law
законы *mpl* Фарадея Faraday's laws
законы *mpl* Фика Fick's laws
замедление *n* нейтронов slowing down
   (*of neutrons*)
запаздывающие нейтроны *mpl* delayed neutrons
заполнение *n* колонки методом
   сбалансированной плотности суспензии
   balanced-density slurry packing
заполнение *n* колонки суспензией slurry packing
заполнение *n* колонок сухим способом
   dry-packing
заполненная колонка *f* packed column
запрещённый переход *m* forbidden transition
зародыш *m* nucleus
затвор *m* gate
защищённое пламя *n* shielded flame
зеемановские уровни *mpl* Zeeman levels
золь *m* sol
зона *f* zone
зона *f* вторичного сгорания secondary
   combustion zone
зона *f* первичного сгорания primary
   combustion zone

И

идеально поляризуемый электрод *m* ideal
   polarizable electrode
идеально поляризующийся электрод *m* ideal
   polarizable electrode
идентификация *f* identification
избирательная реакция *f* selective reaction
избирательность *f* реагента selectivity of
   a reagent
излучатель *m* emitter
излучение *n* radiation
излучение *n* Вавилова-Черенкова Cerenkov
   radiation
измерение *n* нефарадеевского адмиттанса
   measurement of nonfaradaic admittance
измерительная колба *f* volumetric flask
измерительная колба *f* с градуированным
   горлом volumetric flask with graduated neck
измерительный цилиндр *m* measuring cylinder
изобара *f* адсорбции adsorption isobar
изобарное определение *n* изменения массы
   isobaric mass-change determination

изобестическая точка *f* isosbestic point
изократное элюирование *n* isocratic elution
изоморфный носитель *m* scavenger
изополикислота *f* isopolyacid
изопотенциальная точка *f* isopotential point
изостера *f* адсорбции adsorption isostere
изотахофорез *m* isotachophoresis
изотерма *f* isotherm
изотерма *f* адсорбции adsorption isotherm
изотерма *f* ионного обмена ion-exchange
  isotherm
изотерма *f* распределения 1. distribution
  isotherm 2. partition isotherm
изотерма *f* сорбции sorption isotherm
изотермическое определение *n* изменения
  массы isothermal mass-change determination
изотопический источник *m* нейтронов isotopic
  neutron source
изотопное сечение *n* isotopic cross section
изотопное эффективное сечение *n* isotopic
  cross section
изотопный индикатор *m* isotopic tracer
изотопный носитель *m* isotopic carrier
изотопный обмен *m* isotope exchange
изотопный эффект *m* isotope effect
изохрома *f* isochrome
изохромная спектроскопия *f* тормозного
  излучения bremsstrahlung isochrome
  spectroscopy
изоэлектрическая точка *f* isoelectric point
ИК излучение *n* infrared radiation
ИК спектр *m* infrared spectrum
иммуноэлектрофорез *m* immunoelectrophoresis
импеданс *m* impedance
имплантация *f* ионов ion implantation
импульсная десорбция *f* flash desorption
импульсная полярография *f* pulse
  polarography
импульсный пиролиз *m* flash pyrolysis
инактивация *f* deactivation
инверсионная вольтамперометрия *f* stripping
  voltammetry
инверсионное титрование *n* inverse titration
инверсионный анализ *m* electrochemical
  stripping analysis
индекс *m* Ковача retention index
индекс *m* удерживания retention index
индивидуальная проба *f* unit sample
индикатор *m* pH acid-base indicator
индикаторная ошибка *f* indicator error
индикаторная поправка *f* indicator blank
индикаторный электрод *m* indicator electrode
индифферентный электролит *m* supporting
  electrolyte
индуктивно-связанная плазма *f* inductively
  coupled plasma
индуктор *m* inductor
индуцированная реакция *f* induced reaction
индуцированное излучение *n* stimulated
  emission
инструментальная ошибка *f* instrumental
  error
инструментальный активационный анализ *m*
  non-destructive activation analysis
инструментальный анализ *m* instrumental
  analysis

интегральная ёмкость *f* двойного слоя
  integral capacitance of double layer
интегральная хроматограмма *f* chromatogram
интегральный детектор *m* integral detector
интенсивность *f* intensity
интенсивность *f* рентгеновских лучей
  X-radiation intensity
интервал *m* перехода transition interval
интервал *m* перехода окраски colour change
  interval
интерполяция *f* interpolation
интерференционный фильтр *m* interference filter
интерферометр *m* optical interferometer
интерферометрия *f* interferometry
инфракрасная область *f* infrared region
инфракрасная область *f* спектра infrared region
инфракрасная отражательно-абсорбционная
  спектроскопия *f* infrared reflection
  absorption spectroscopy
инфракрасное излучение *n* infrared radiation
инфракрасный спектр *m* infrared spectrum
инфузорная земля *f* diatomaceous earth
иодное число *n* iodine number
иодометрическое титрование *n* iodimetric
  titration
иодометрия *f* iodimetry
ион-парная хроматография *f* ion-pair
  partition chromatography
ион-эксклюзионная хроматография *f* ion-
  exclusion chromatography
ионизационная камера *f* ionization chamber
ионизационный детектор *m* ionization detector
ионизация *f* в искровом источнике spark
  source ionization
ионизация *f* полем field ionization
ионизация *f* электронным ударом electron-
  impact ionization
ионизирующее излучение *n* ionizing radiation
ионит *m* solid ion exchanger
ионная атмосфера *f* ionic atmosphere
ионная микроскопия *f* secondary ion imaging
  mass spectroscopy
ионная подвижность *f* electric mobility of ion B
ионная проводимость *f* ionic conduction
ионная рефракция *f* ionic refraction
ионная сила *f* ionic strength
ионная сила *f* раствора ionic strength of
  a solution
ионное облако *n* ionic atmosphere
ионный источник *m* с электронной
  бомбардировкой electron-impact ion source
ионный обмен *m* ion exchange
ионный проводник *m* ionic conductor
ионный спектр *m* ionic spectrum
ионный тройник *m* ion triplet
ионогенные группы *fpl* ionogenic groups
ионообменная мембрана *f* ion-exchange
  membrane
ионообменная смола *f* ion-exchange resin
ионообменная хроматография *f* ion-exchange
  chromatography
ионообменник *m* ion exchanger
ионоселективная мембрана *f* ion-selective
  membrane
ионоселективный мембранный электрод *m*
  membrane ion-selective electrode

ионоселективный полевой транзистор *m*
ion-sensitive field effect transistor
ионоселективный электрод *m* ion-selective
electrode
ионоселективный электрод *m* с гетерогенной
мембраной heterogeneous membrane
ion-selective electrode
ионоселективный электрод *m* с гомогенной
мембраной homogeneous membrane ion-
selective electrode
ионоселективный электрод *m* с жидкой
мембраной liquid membrane ion-selective
electrode
ионоселективный электрод *m*
с ионообменной мембраной membrane
ion-selective electrode
ионоселективный электрод *m*
с кристаллической мембраной crystalline
membrane ion-selective electrode
ионоселективный электрод *m* с мембраной
из жидкого ионита liquid ion-exchange
membrane electrode
ионоселективный электрод *m* с мембраной
на основе нейтральных переносчиков
neutral-carrier membrane ion-selective
electrode
ионоселективный электрод *m* с осадочной
мембраной precipitate-based ion-selective
electrode
ионоселективный электрод *m* с твёрдой
мембраной solid-state membrane ion-
selective electrode
ионспецифический электрод *m* ion-selective
electrode
исключение *n* exclusion
исключение *n* ионов ion exclusion
искра *f* electrical spark
искровой источник *m* spark ionization source
искровой спектр *m* spark spectrum
испарение *n* volatilization
испарение *n* неподвижной фазы bleeding
исправленная величина *f* R_f R_f value
исправленная величина *f* R_M R_M value
исправленный коэффициент *m* селективности
corrected selectivity coefficient
исправленный объём *m* удерживания
corrected retention volume
исправленный удерживаемый объём *m*
corrected retention volume
испытание *n* 1. test (*in chemical analysis*)
2. experiment
исследование *n* рядом методов multiple
techniques
истинная скорость *f* interstitial velocity
истинная средняя скорость *f* газа-носителя
mean interstitial velocity of the carrier gas
истинная степень *f* электролитической
диссоциации true degree of dissociation
истинная ширина *f* спектральной линии true
half-width
источник *m* source
источник *m* возбуждения excitation source
источник *m* для ионизации электронным
ударом electron-impact ion source
источник *m* излучения radiation source

источник *m* ионизации полем field ionization
ion source
источник *m* ионов ion source
источник *m* света light source
исходный ион *m* parent ion
исходный нуклид *m* parent nuclide
исчисление *n* вероятностей probability theory

К

кадмиевое отношение *n* cadmium ratio
кажущаяся ширина *f* спектральной линии
apparent half width
кайзер *m* kayser
калибровочная кривая *f* analytical curve
калибровочная кривая *f* фотоэмульсии
emulsion calibration curve
калибровочная ошибка *f* scale error
калибровочный график *m* analytical curve
каломельный электрод *m* calomel electrode
калометрическая бомба *f* bomb calorimeter
кальциевая жёсткость *f* calcium hardness
кальцинирование *n* пробы ashing technique
камера *f* Бреннера-Нидервизера BN-chamber
камера *f* для проявления хроматограмм
chromatographic chamber
канонический анализ *m* canonical analysis
капельная пластинка *f* spotting plate
капельница *f* dropper
капельный анализ *m* spot test analysis
капельный ртутный электрод *m*
dropping-mercury electrode
капилляр *m* Луггина Luggin-Haber capillary
капиллярная колонка *f* capillary column
капиллярная хроматография *f* open-tube
chromatography
капилляры *mpl* capillary tubes
капилляры *mpl* для определения точки
плавления melting-point capillaries
каплеуловитель *m* safety head
карбовакс *m* Carbowax
карбонатная жёсткость *f* carbonate hardness
каркас *m* ионита ion exchanger matrix
катализатор *m* Вэчержи Večeřa catalyst
катализатор *m* Кэрбля Körbl catalyst
каталитическая реакция *f* catalytic reaction
каталитический ток *m* catalytic current
каталитическое сожжение *n* catalytic
combustion
катарометр *m* thermal conductivity detector
катафоретический эффект *m* electrophoretic
effect
катионит *m* cation exchanger
катионный обмен *m* cation exchange
катионообменная смола *f* cation-exchange
resin
катионообменник *m* cation exchanger
катод *m* cathode
катодно-лучевая полярография *f* single-sweep
polarography
катодолюминесценция *f* cathodoluminescence
католит *m* catholyte
качественный анализ *m* qualitative analysis
качественный органический анализ *m*
qualitative analysis of organic compounds
качественный состав *m* qualitative composition

квадратно-волновая полярография *f* square-wave polarography

квадруполь *m* quadrupole

квадрупольная релаксация *f* quadrupole relaxation

квадрупольный масс-анализатор *m* quadrupole mass analyser

γ-квант *m* gamma quantum

квантиль *m* quantile of a probability distribution

квантиль *m* вероятностного распределения quantile of a probability distribution

кизельгур *m* diatomaceous earth

кинетическй метод *m* анализа kinetic analysis

кинетический ток *m* kinetic current

кислородный зонд *m* Clark oxygen-sensing probe

кислородный электрод *m* oxygen gas electrode

кислотная форма *f* катионообменника acid form of a cation exchanger

кислотно-основная кривая *f* титрования neutralization titration curve

кислотно-основное титрование *n* acid-base titration

кислотно-основный индикатор *m* acid-base indicator

кислый растворитель *m* protogenic solvent

классическая полярография *f* direct current polarography

классическая полярография *f* переменного тока conventional alternating-current polarography

коагуляция *f* coagulation

ковариантная матрица *f* covariance matrix

ковариация *f* covariance

когерентное рассеяние *n* coherent scattering

коионы *mpl* co-ions

колба *f* для иодометрического титрования iodine flask

колебания *npl* vibrations

колебания *npl* молекул molecular vibrations

колебательная полоса *f* vibration band

колебательная спектроскопия *f* vibrational spectroscopy

колебательный спектр *m* vibrational spectrum

количественное осаждение *n* quantitative precipitation

количественный анализ *m* quantitative analysis

количественный анализ *m* органических соединений quantitative analysis of organic compounds

количественный состав *m* quantitative composition

количество *n* вещества amount of substance

количество *n* освещения radiant exposure

коллектор *m* trace collector

коллектор *m* следов trace collector

коллектор *m* фракций fraction collector

коллиматор *m* collimator

коллимация *f* collimation

коллоидный осадок *m* colloid precipitate

колоночная хроматография *f* column chromatography

колориметр *m* colorimeter

колориметрический анализ *m* colorimetric analysis

колориметрическое титрование *n* colorimetric titration

колориметрия *f* colorimetric analysis

кома *f* coma

комбинационная полоса *f* combination band

комбинационное рассеяние *n* Raman effect

комбинированные методы *mpl* combined techniques

комбинированный индикатор *m* universal indicator

комбинированный электрод *m* combination electrode

компенсационный метод *m* Поггендорфа Poggendorff's compensation method

комплекс *m* с переносом заряда donor acceptor complex

комплексиметрическая кривая *f* титрования compleximetric titration curve

комплексиметрическое титрование *n* compleximetric titration

комплексная проводимость *f* admittance

комплексное сопротивление *n* impedance

комплексометрическая кривая *f* титрования compleximetric titration curve

комплексометрическое титрование *n* compleximetric titration

комплексометрия *f* compleximetry

комплексоны *mpl* complexones

комплект *m* эталонов standard series

компромиссный потенциал *m* mixed polyelectrode potential

комптоновский край *m* Compton edge

комптоновское распределение *n* Compton continuum

комптоновское рассеяние *n* Compton effect

конвективная хроноамперометрия *f* convective chronoamperometry

конвективная хронокулонометрия *f* convective chronocoulometry

конвекция *f* convection

конвенциональный анализ *m* conventional analysis

конвенция *f* о знаках sign convention

конденсаторный ток *m* capacity current

конденсированный двойной слой *m* compact layer

кондуктиметрия *f* conductometry

кондуктометрический анализ *m* conductometric analysis

кондуктометрическое титрование *n* conductometric titration

кондуктометрия *f* conductometry

конечная точка *f* титрования end-point of titration

конкурирующая ядерная реакция *f* interfering nuclear reaction

константа *f* диффузионного тока limiting current constant

константа *f* распределения 1. distribution constant, partition coefficient, K_D 2. partition constant

константа *f* селективности potentiometric selectivity coefficient

константа *f* скорости реакции reaction-rate constant

константа *f* скорости электродной реакции rate constant of electrode reaction

константа *f* экстракции extraction constant

контактная разность *f* потенциалов Volta potential difference

контрольная карта *f* control chart

контрольная проба *f* check test

контрольное титрование *n* control titration

контрольный раствор *m* standard matching solution

конформационный анализ *m* conformational analysis

конформация *f* conformation

концентрационная цепь *f* concentration cell

концентрационное отношение *n* распределения concentration distribution ratio

концентрационное перенапряжение *n* concentration overpotential

концентрационный элемент *m* concentration cell

концентрация *f* растворенного вещества B amount-of-substance concentration of substance B

координирующая группа *f* coordinating group

коротковолновая граница *f* непрерывного спектра short-wavelength limit

короткоживущий изотоп *m* short-lived radioisotope

корпускулярное излучение *n* corpuscular radiation

корректор *m* фона background corrector

корреляционная матрица *f* correlation matrix

корреляция *f* correlation

косвенное титрование *n* indirect titration

косвенный метод *m* indirect method

коэффициент *m* активности вещества B activity coefficient of substance B

коэффициент *m* вариации coefficient of variation

коэффициент *m* диффузии diffusion coefficient

коэффициент *m* ёмкости mass distribution ratio

коэффициент *m* контрастности gamma

коэффициент *m* корреляции coefficient of correlation

коэффициент *m* обогащения enrichment factor

коэффициент *m* обратного рассеяния backscattering factor

коэффициент *m* ослабления attenuation coefficient

коэффициент *m* очистки decontamination factor

коэффициент *m* переноса заряда charge-transfer coefficient

коэффициент *m* побочной реакции side-reaction coefficient

коэффициент *m* поглощения 1. absorptance 2. specific decadic absorption coefficient

коэффициент *m* прилипания sticking probability

коэффициент *m* пропускания transmission factor

коэффициент *m* распределения 1. separation factor 2. concentration distribution ratio 3. partition coefficient 4. sputtering yield

коэффициент *m* распределения D_g distribution coefficient D_g

коэффициент *m* распределения D_s distribution coefficient D_s

коэффициент *m* распределения D_v volume distribution coefficient

коэффициент *m* распыления sputtering yield

коэффициент *m* регрессии regression coefficient

коэффициент *m* селективности selectivity coefficient

коэффициент *m* экстинкции specific decadic absorption coefficient

коэффициент *m* экстракции concentration distribution ratio (*in extraction*)

край *m* поглощения absorption edge

край *m* полосы band head

кратер *m* распыления crater

кривая *f* нагревания heating curve

кривая *f* проскакивания breakthrough curve

кривая *f* проявления elution curve

кривая *f* радиоактивного распада decay curve

кривая *f* титрования titration curve

кривая *f* титрования редокс redox titration curve

кривая *f* титрования с осаждением precipitation titration curve

кривая *f* ток-напряжение polarization curve

кривая *f* Хартера и Дриффильда emulsion characteristic curve

кривая *f* чувствительности весов sensitivity curve of balance

кривая *f* экстракции extraction curve

кривая *f* элюирования elution curve

криволинейная регрессия *f* curvilinear regression

криоскопическое определение *n* молярной массы cryoscopic determination of the molar mass

кристалл-анализатор *m* analysing crystal

кристаллизатор *m* crystallizing dish

кристаллизационная вода *f* water of crystallization

кристаллизационное перенапряжение *n* crystallization overpotential

кристаллизация *f* crystallization

кристаллическая мембрана *f* crystalline membrane

кристаллический мембранный электрод *m* crystalline membrane ion-selective electrode

кристаллический осадок *m* crystalline precipitate

кристаллический электрод *m* crystalline membrane ion-selective electrode

критерий *m* test

критерий *m* значимости test of significance

критерий *m* оптимальности плана эксперимента criterion of optimality of experimental design

критерий *m* Снедекора *F*-test

критерий *m* согласия test for goodness of fit

критерий *m* согласия χ^2 chi-squared test of goodness of fit

критерий *m* Стьюдента Student's test

критическая область *f* rejection region

критическое определение *n* critical determination

критическое отношение *n* пересыщения critical supersaturation ratio

круговая поляризация *f* circular polarization
круговая хроматография *f* circular chromatography
круговой дихроизм *m* circular dichroism
крутильные колебания *npl* twisting vibrations
кулометр *m* coulometer
кулонографиметрия *f* coulogravimetry
кулонометр *m* coulometer
кулонометрический анализ *m* coulometric analysis
кулонометрическое титрование *n* controlled-current coulometry
кулонометрия *f* coulometry
кулонометрия *f* при постоянном потенциале controlled-potential coulometry
кулонометрия *f* при постоянном токе controlled-current coulometry
кулонометрия *f* с капающим электродом dropping electrode coulometry
кулонометрия *f* с контролируемым потенциалом controlled-potential coulometry
кулонометрия *f* с контролируемым потенциалом и потенциометрической индикацией конечной точки controlled-current coulometry with potentiometric end-point detection
кулонометрия *f* с контролируемым током controlled-current coulometry
кулоностатический метод *m* coulostatic method
кумуляция *f* импульсов pile-up
купелирование *n* cupellation
купеляция *f* cupellation
кюри *m* curie

Л

лабораторные весы *pl* laboratory balance
лазер *m* laser
лазер *m* на красителе dye laser
лазерная рамановская спектроскопия *f* laser Raman spectroscopy
лазерная спектроскопия *f* комбинационного рассеяния laser Raman spectroscopy
лазерная флуоресцентная спектроскопия *f* laser fluorescence spectroscopy
ламинарное пламя *n* laminar flame
лампа *f* накаливания incandescent lamp
лампа *f* накаливания с вольфрамовой нитью tungsten incandescent lamp
лампа *f* Нернста Nernst glower
лампа *f* с вольфрамовой нитью tungsten incandescent lamp
лампа *f* с вольфрамовой спиралью tungsten incandescent lamp
лампа *f* с полым катодом hollow-cathode lamp
лампа *f* с разрядом в парах металлов metal-vapour lamp
латинский квадрат *m* latin square
левовращающее вещество *n* laevorotatory substance
левополяризованный по кругу свет *m* left-handed circularly polarized light
лигандообменная хроматография *f* ligand-exchange chromatography
линеаризация *f* linearization
линейная дисперсия *f* linear dispersion

линейная поляризация *f* plane polarization
линейная регрессия *f* linear regression
линейный десятичный показатель *m* поглощения linear decadic absorption coefficient
линейный коэффициент *m* ослабления linear attenuation coefficient
линейный показатель *m* поглощения linear decadic absorption coefficient
линейный спектр *m* line spectrum
линии *fpl* спектра комбинационного рассеяния света Raman lines
линии *fpl* спектра Рамана Raman lines
линия *f* рэлеевского рассеяния Rayleigh line
линия *f* элемента сравнения internal reference line
литий-германиевый детектор *m* lithium-drifted germanium detector
литий-кремнёвый детектор *m* lithium-drifted silicon detector
логарифмически нормальное распределение *n* logarithmic-normal distribution
лодочка *f* для сжигания combustion boat
лодочка *f* для сожжения combustion boat
ложка *f* для взятия навески weighing scoop
локальный анализ *m* local analysis
лучеиспускание *n* radiation
γ-лучи *mpl* gamma radiation
люминесценция *f* luminescence

М

магниевая жёсткость *f* magnesium hardness
магнитный резонанс *m* magnetic resonance
макроанализ *m* macroanalysis
макрокомпонент *m* macro-component
макропористый ионит *m* macroporous exchanger
макропористый ионообменник *m* macroporous ion exchanger
макроскопическое сечение *n* macroscopic cross section
макроскопическое эффективное сечение *n* macroscopic cross section
максимум *m* пика peak maximum
манометрические методы *mpl* manometric methods
манометрический газовый анализ *m* manometric gas analysis
марганцевомедный катализатор *m* Hopcalite
маскирование *n* ионов masking of ions
маскирующий агент *m* masking agent
масса *f* покоя rest mass
масс-анализатор *m* mass analyser
масс-анализатор *m* с магнитным полем magnetic-deflection mass spectrometer
масс-спектр *m* mass spectrum
масс-спектр *m* отрицательных ионов negative-ion mass spectrum
масс-спектр *m* положительных ионов positive-ion mass spectrum
масс-спектрограф *m* mass spectrograph
масс-спектрометр *m* mass spectrometer
масс-спектрометрия *f* mass spectrometry
масс-спектроскоп *m* mass spectroscope
масс-спектроскопия *f* mass spectrometry

масс-спектроскопия *f* вторичных ионов secondary ion mass spectroscopy

масс-спектроскопия *f* с искровым источником spark source mass spectroscopy

массовая концентрация *f* mass concentration of substance B

массовое отношение *n* распределения mass distribution ratio

массовый коэффициент *m* ослабления mass attenuation coefficient

массообмен *m* mass exchange

математическая статистика *f* statistics

математическое ожидание *n* expected value

материал *m* для набивки колонки column packing

материнский нуклид *m* parent nuclide

матрица *f* matrix

матрица *f* ионита ion exchanger matrix

матрица *f* плана design matrix

матрица *f* планирования design matrix

матричный эффект *m* matrix effect

маятниковые колебания *mpl* rocking vibrations

мгновенное γ-излучение *n* prompt gamma radiation

мгновенные нейтроны *mpl* prompt neutrons

медиана *f* sample median

медленные нейтроны *mpl* slow neutrons

медный кулонометр *m* copper coulometer

межзональная область *f* interzonal region

межлабораторные исследования *npl* interlaboratory research

межэлектродное расстояние *n* analytical gap

мембрана *f* из жидкого ионита liquid ion exchanger membrane

мембрана *f* на основе нейтральных переносчиков neutral-carrier membrane

мембрана *f* с электрическим нейтральным лигандом neutral-carrier membrane

мембранная разность *f* потенциалов membrane potential

мембранное равновесие *n* membrane equilibrium

мембранный потенциал *m* membrane potential

мембранный электрод *m* membrane ion-selective electrode

меркуриметрия *f* mercurimetry

меркурометрическое титрование *n* mercurometric titration

меркурометрия *f* mercurometry

мёртвое время *n* dead time

мёртвый объём *m* extra-column volume

мёртвый объём *m* датчика intensitive volume of a detector

металлиндикаторы *mpl* metal indicators

металлический проводник *m* electronic conductor

металлический электрод *m* metal-metal ion electrode

металлоокисный электрод *m* oxide electrode

металлооксидный электрод *m* oxide electrode

металлофлуоресцентный индикатор *m* metallofluorescent indicator

металлохромный индикатор *m* metallochromic indicator

метастабильное равновесие *n* электрода metastable equilibrium of an electrode

метастабильный ион *m* metastable ion

метастабильный пик *m* metastable ion peak

метастабильный распад *m* metastable decomposition

метка *f* marker

метод *m* абсорбции ядерного излучения nuclear radiation absorptiometry

метод *m* анализа по поглощению β-лучей beta-particle absorptiometry

метод *m* антисовпадений anticoincidence technique

метод *m* висячей капли hanging-drop method

метод *m* внутреннего стандарта internal-standard method

метод *m* Грана Gran's plot

метод *m* добавления стандарта standard addition method

метод *m* добавок standard addition method

метод *m* Дюма Dumas method

метод *m* Дюма-Прегля Dumas-Pregl method

метод *m* изотопного разведения isotope dilution analysis

метод *m* изотопных радиоактивных индикаторов radioactive tracer method

метод *m* Кариуса Carius method

метод *m* Кельвина Kelvin method

метод *m* Ковеля Covell's method

метод *m* колебаний method of swings

метод *m* концентрирования concentration method

метод *m* коротких колебаний method of short swings

метод *m* коэффициентов влияния empirical coefficient method

метод *m* Крейга countercurrent distribution

метод *m* Кьельдаля Kjeldahl method

метод *m* Либиха Liebig's method

метод *m* максимального правдоподобия maximum likelihood method

метод *m* Мора Mohr's method

метод *m* наименьших квадратов least-squares method

метод *m* обогащения concentration method

метод *m* освобождения радиоиндикатора radio-release method

метод *m* ослабления полного отражения attenuated total reflectance

метод *m* основной линии base-line method

метод *m* основных параметров fundamental parameter method

метод *m* отраженной волны reflected wave method

метод *m* переменного угла выхода variable take-off angle method

метод *m* плавления fusion method

метод *m* Поггендорфа Poggendorff's compensation method

метод *m* применения мониторов monostandard method

метод *m* проходящей волны transmitted wave method

метод *m* развёртки магнитного поля field-sweep method

метод *m* Раста Rast method

метод *m* совпадений coincidence technique

метод *m* сравнения reference method

метод *m* **Унтерзаухера определения кислорода** Unterzaucher method for oxygen determination

метод *m* **Фольгарда определения галогенидов** Volhard method for determination of halides

метод *m* **Фольгарда определения марганца** Volhard method for determination of manganese

метод *m* **частотной развёртки** frequency-sweep method

метод *m* **Шенигера** Schöniger method

методическая ошибка *f* error of method

методы *mpl* **анализа по поглощению** γ-**излучения** gamma-ray absorptiometry

методы *mpl* **исследования поверхности** surface sensitive techniques

PH-метр *m* PH-meter

PH-метрический электрод *m* PH electrode

PH-метрия *f* PH-metry

механический захват *m* mechanical entrapment

меченое соединение *n* labelled compound

мешающее вещество *n* interfering substance

мешающие ионы *mpl* interfering ions

миграционный ток *m* migration current

миграция *f* **ионов** migration of ions

микроазотометр *m* micronitrometer

микроанализ *m* 1. microanalysis 2. submicro-analysis

микробюретка *f* microburette

микровесы *pl* microchemical balance

микроволновая область *f* microwave region

микроволновая плазма *f* microwave plasma

микроволновая спектроскопия *f* microwave spectroscopy

микроволновый спектр *m* microwave spectrum

микроволны *fpl* microwaves

микрокомпонент *m* micro-component

микрокристаллоскопия *f* microcrystalloscopic analysis

микрокристаллоскоповый анализ *f* microcrystalloscopic analysis

микрокруговая хроматография *f* microcircular chromatography

микрокулонометрия *f* polarographic coulometry

микрометод *m* milligram method

микронитрометр *m* micronitrometer

микропипетка *f* micropipette

микроскопическое сечение *n* microscopic cross section

микроскопическое эффективное сечение *n* microscopic cross section

микроскопия *f* **фотоэлектронов** photoelectron microscopy

микроследы *mpl* microtraces

микрофотометр *m* microphotometer

миллеровские индексы *mpl* Miller indices

милликулонометрия *f* polarographic coulometry

миллионная доля *f* parts per million

минерализация *f* mineralization

минералогический эффект *m* mineralogical effect

минимальное детектируемое количество *n* detection limit

минимальное определяемое количество *n* detection limit

мишень *f* target

мишень *f* **из трития** tritium target

многокомпонентный анализ *m* multiple analysis

множественная регрессия *f* multiple regression

множественный коэффициент *m* **корреляции** coefficient of multiple correlation

мода *f* mode

модель *f* **с фиксированными эффектами** fixed effects model

модель *f* **со случайными эффектами** random effects model

модифицированное активное твёрдое вещество *n* modified active solid

модифицированное уравнение *n* **Нернста** Nikolsky equation

модифицированный слой *m* altered layer

модуляционная полярография *f* modulation polarography

мокрый путь *m* wet method

мокрый путь *m* **анализа** wet method

молекулярная диффузия *f* molecular diffusion term

молекулярная продольная диффузия *f* molecular diffusion term

молекулярная спектроскопия *f* molecular spectroscopy

молекулярное вращение *n* molar rotation

молекулярное излучение *n* molecular radiation

молекулярные колебания *npl* molecular vibrations

молекулярные сита *npl* molecular sieves

молекулярный ион *m* molecular ion

молекулярный пик *m* molecular peak

молекулярный пучок *m* molecular beam

молекулярный спектр *m* molecular spectrum

молекулярный спектрохимический анализ *m* molecular spectrochemical analysis

моль *m* 1. mole 2. gram molecule

мольная рефракция *f* molar refraction

мольная электропроводность *f* **электролита** molar conductivity of electrolyte

мольный десятичный коэффициент *m* **погашения** molar absorptivity

мольный коэффициент *m* **погашения** molar absorptivity

моляльность *f* molality

молярная абсорбируемость *f* molar absorptivity

молярная доля *f* mole fraction of substance B

молярная ионная электропроводность *f* molar conductivity of ionic species

молярная ионная электропроводность *f* **при бесконечном разбавлении** limiting molar conductivity of ionic species

молярная концентрация *f* amount-of-substance concentration of substance B

молярная масса *f* molar mass

молярная рефракция *f* molar refraction

молярная электрическая проводимость *f* **электролита** molar conductivity of electrolyte

молярная электропроводность *f* **иона** molar conductivity of ionic species

молярная электропроводность *f* **при бесконечном разбавлении** molar conductivity of electrolyte

молярная электропроводность *f* **электролита** molar conductivity of electrolyte

молярная электропроводность f электролита при бесконечном разбавлении limiting molar conductivity of electrolyte

молярность f amount-of-substance concentration of substance В

молярный десятичный показатель m поглощения molar absorptivity

молярный показатель m поглощения molar absorptivity

монослой m monolayer

монофункциональный ионообменник m monofunctional ion exchanger

монохроматическое излучение n monochromatic radiation

монохроматор m monochromator

монохроматор m возбуждения excitation monochromator

мостик m для измерения электропроводности conductometer

мостик m Уитстона Wheatstone bridge

мощность f излучения radiant flux

мощность f критерия power of a test

мультиплет m multiplet

мурексидная реакция f murexide reaction

муфельная печь f muffle furnace

Н

набивка f колонки column packing

набивная колонка f packed column

набухание n swelling

навеска f weighed amount

нагрузка f load

надкритическая флюидная хроматография f supercritical fluid chromatography

надтепловые нейтроны mpl epithermal neutrons

наиболее эффективная оценка f most efficient estimator

наложение n импульсов pile-up

наложение n линий line coincidence

напряжение n Вольта Volta potential difference

напряжение n Гальвани Galvani potential difference

напряжение n разложения decomposition voltage

насадочная колонка f packed column

насыщение n saturation

насыщение n камеры chamber saturation

насыщенная камера f saturated chamber

насыщенный каломельный электрод m saturated-calomel electrode

насыщенный план m saturated design

насыщенный раствор m saturated solution

неводное титрование n non-aqueous titration

недеструктивный анализ m non-destructive analysis

неизотопный носитель m non-isotopic carrier

неисправленное время n удерживания total retention time

неисправленный удерживаемый объём m retention volume

нейтрализация f neutralization

нейтральный фильтр m neutral-density filter

нейтронно-абсорбционный метод m анализа neutron absorptiometry

нейтронный активационный анализ m neutron activation analysis

нейтронный генератор m neutron generator

нейтроноактивационный анализ m neutron activation analysis

нейтроны mpl деления fission neutrons

некарбонатная жёсткость f non-carbonate hardness

некогерентное рассеяние n Compton effect

нелинейная регрессия f non-linear regression

немецкий градус m жёсткости воды German degree of water hardness

ненасыщенная камера f unsaturated chamber

ненасыщенный элемент m Вестона unsaturated Weston cell

необратимая коагуляция f irreversible coagulation

необратимая электродная реакция f irreversible electrode reaction

необратимый окислительно-восстановительный индикатор m irreversible redox indicator

необратимый редокс-индикатор m irreversible redox indicator

необратимый электрод m irreversible electrode

непараметрический критерий m non-parametric test

неплоские колебания npl out-of-plane vibrations

неподвижная жидкая фаза f stationary liquid phase

неподвижная фаза f stationary phase

неполяризуемый электрод m ideal non-polarisable electrode

неполяризующийся электрод m ideal non-polarisable electrode

непрерывная случайная величина f continuous random variable

непрерывная экстракция f continuous extraction

неразрешённый пик m unresolved peak

неразрушающий анализ m non-destructive analysis

нерегистрирующий спектрофотометр m non-recording spectrophotometer

нернстовская электродная функция f Nernstian electrode response

нернстовский наклон m Nernstian slope

несмещённая оценка f unbiased estimator

неспаренный электрон m unpaired electron

несущая пластинка f support plate

неупругое рассеяние n inelastic scattering

неупругое рассеяние n медленных электронов inelastic low energy electron diffraction

неупругое столкновение n inelastic collision

нефелометр m nephelometer

нефелометрическое титрование n nephelometric titration

нефелометрия f nephelometry

нивелирующий растворитель m levelling solvent

нивелирующий эффект m levelling effect

нийол m Nujol

нисходящая хроматография f descending development

нисходящее проявление n descending development

нисходящий метод m descending development

нитрометр m nitrometer

нихромная спираль f coil of Nichrome wire
ножничные колебания npl scissoring vibrations
номинальный линейный поток m nominal linear flow
нормаль f standard sample
нормальная дисперсия f normal dispersion
нормальная импульсная полярография f pulse polarography
нормальное распределение n normal distribution
нормальность f normality
нормальные колебания npl normal vibrations
нормальные уравнения npl normal equations
нормальные условия npl standard conditions
нормальный водородный электрод m standard hydrogen electrode
нормальный потенциал m standard electrode potential
нормальный электродный потенциал m standard electrode potential
нормальный элемент m standard cell
нормальный элемент m Вестона Weston standard cell
носитель m carrier
носитель m с контролируемой поверхностной пористостью superficially porous support
нуйол m Nujol
нулевая ветвь f Q branch
нулевая гипотеза f null hypothesis
нулевая линия f baseline
нуль m шкалы zero point of the scale

О

обертон m overtone band
обессоленная вода f deionized water
область f измерений experimental region
область f недодержек under-exposure region
область f отбрасывания rejection region
область f отпечатка пальцев fingerprint region
область f передержек over-exposure region
обменная адсорбция f exchange adsorption
обнаружение n detection
обнаружение n выделенных газов evolved gas detection
обнаружение n излучения detection of radiation
обозначенный объём m designated volume
У-образная трубка f U-tube
образование n зародышей nucleation
образование n производных derivatization
образование n сетчатых молекул cross-linking
образование n хвоста tailing
образцовый раствор m standard reference solution
обратимая гальваническая цепь f reversible galvanic cell
обратимая коагуляция f reversible coagulation
обратимая цепь f reversible galvanic cell
обратимая электродная реакция f reversible electrode reaction
обратимый гальванический элемент m reversible galvanic cell
обратимый окислительно-восстановительный индикатор m reversible redox indicator
обратимый редокс-индикатор m reversible redox indicator
обратимый электрод m reversible electrode

обратимый элемент m reversible galvanic cell
обратная продувка f backflushing
обратное рассеяние n backscattering
обратное титрование n back-titration
обращённо-фазная хроматография f reversed-phase chromatography
обугливание n фильтра charring of filter paper
обходный дозатор m by-pass injector
общая жёсткость f воды total water hardness
общая константа f образования cumulative stability constant
общая проба f gross sample
общее время n удерживания total retention time
общий порядок m реакции order of reaction
общий ток m net current
общий удерживаемый объём m retention volume
объём m выборки sample size
объём m жидкой неподвижной фазы liquid phase volume
объём m жидкой фазы liquid phase volume
объём m задержки подвижной фазы mobile phase hold-up volume
объём m колонки column volume
объём m неподвижной фазы stationary phase volume
объём m пика элюирования peak elution volume
объём m подвижной застойной фазы stationary liquid volume
объём m пор stationary liquid volume
объём m слоя bed volume
объём m твёрдого наполнителя solid volume
объём m твёрдого тела solid volume
объём m твёрдой фазы solid volume
объём m удерживания retention volume
объём m удерживания несорбирующегося вещества mobile phase hold-up volume
объёмная доля f volume fraction of substance B
объёмная ёмкость f volume capacity
объёмная ёмкость f слоя bed volume capacity
объёмная скорость f F_a подвижной фазы volumetric flow rate F_a
объёмная скорость f F_c подвижной фазы volumetric flow rate F_c
объёмная скорость f потока F_a volumetric flow rate F_a
объёмная скорость f потока F_c volumetric flow rate F_c
объёмный анализ m titrimetric analysis
объёмный газовый анализ m volumetric gas analysis
объёмный коэффициент m распределения volume distribution coefficient D_V
одиночная линия f singlet line volume
одновременное определение n simultaneous determination
одновременные методы mpl simultaneous techniques
однолучевой спектрофотометр m single-beam spectrophotometer
односторонний критерий m one-sided test
одноцветный индикатор m one-colour indicator
одориметрия f odorimetry
озоление n пробы ashing technique
окисление n oxidation
окислитель m oxidant

окислительно-восстановительное титрование *n* oxidation-reduction titration

окислительно-восстановительный индикатор *m* oxidation-reduction indicator

окислительно-восстановительный потенциал *m* oxidation-reduction potential

окислительно-восстановительный электрод *m* redox electrode

окиснортутный электрод *m* mercury-mercuric oxide electrode

окклюзия *f* occlusion

окраска *f* пламени flame coloration

окружность *f* изображения focusing circle

окружность *f* фокусировки focusing circle

оксидиметрическое титрование *n* oxidimetric titration

оксидиметрия *f* oxidimetry

олеофильный ионит *m* oleophilic ion exchanger

олеофильный ионообменник *m* oleophilic ion exchanger

омическая поляризация *f* pseudo-resistance overpotential

омическое перенапряжение *n* pseudo-resistance overpotential

оперативная погрешность *m* operative error

оперативная характеристика *f* operating characteristic function

операционный усилитель *m* operational amplifier

определение *n* determination

определение *n* жёсткости воды мыльным раствором Clark's method

определяемое вещество *n* determinand

определяемый минимум *m* limit of determination

оптимизация *f* optimization

оптическая активность *f* optical activity

оптическая ось *f* optical axis

оптическая плотность *f* 1. blackening 2. absorbance

оптическая скамья *f* optical bench

оптическая спектроскопия *f* optical spectroscopy

оптическая чистота *f* optical purity

оптически активное соединение *n* optically active substance

оптический квантовый генератор *m* laser

оптический интерферометр *m* optical interferometer

оптический спектр *m* optical spectrum

оптический фильтр *m* optical filter

оптическое вращение *n* optical rotation

опыт *m* experiment

органический ионит *m* ion-exchange resin

органолептическая оценка *f* organoleptic assessment

ортогональный план *m* orthogonal design

осадительное титрование *n* precipitation titration

осадительный индикатор *m* precipitation indicator

осадок *m* precipitate

осадочный мембранный электрод *m* precipitate-based ion-selective electrode

осаждающий реактив *m* precipitating agent

осаждение *n* precipitation

осаждение *n* из гомогенного раствора precipitation from homogeneous solution

освещённость *f* illuminance

осколки *mpl* деления fission fragments

ослабление *n* излучения attenuation of radiation

основа *f* matrix

основание *n* пика peak base

основная линия *f* baseline

основная полоса *f* fundamental vibration band

основная форма *f* анионообменника base form of an anion exchange

основное вещество *n* standard substance

основной компонент *m* major constituent

основной ток *m* residual current

остаточная дисперсия *f* residual variance

основный растворитель *m* protophilic solvent

остаточный ток *m* residual current

оствальдовское созревание *n* Ostwald ripening

остывание *n* radioactive cooling

осушитель *m* desicant

осциллографическая полярография *f* oscillopolarography

осциллографическая полярография *f* с однократной развёрткой потенциала single-sweep polarography

осциллографический полярограф *m* oscillopolarograph

осциллометрическое титрование *n* high-frequency conductometric titration

осциллометрия *f* high-frequency conductometry

осциллополярография *f* oscillopolarography

отдача *f* рентгеновской флуоресценции X-ray fluorescence yield

отделение *n* separation

отклик *m* потенциала электрода electrode response

отклонение *n* в магнитном поле magnetic deflection

открываемый минимум *m* limit of detection

открываемый предел *m* limit of detection

открытая колонка *f* open-tube column

открытая трубчатая колонка *f* open-tube column

отмывание *n* от пыли elutriation

относительная активность *f* вещества B relative activity of substance B

относительная атомная масса *f* relative atomic mass

относительная диэлектрическая проницаемость *f* relative permittivity

относительная молекулярная масса *f* relative molecular mass

относительная объёмная набухаемость *f* volume swelling ratio

относительная ошибка *f* relative error

относительное пересыщение *n* relative supersaturation

относительное удерживание *n* relative retention

относительный показатель *m* преломления relative refractive index

относительный электродный потенциал *m* electrode potential

отношение *n* масса/заряд mass-to-charge ratio

отношение *n* пересыщения supersaturation ratio

отношение *n* распределения concentration distribution ratio

отношение *n* сигнал-шум signal-to-noise ratio
отношение *n* фаз phase ratio
отражательная решётка *f* reflection rating
отрицательная ветвь *f* P-branch
отрицательный ион *f* negative ion
охлаждение *n* radioactive cooling
оценивание *n* с помощью доверительного интервала interval estimation
оценка *f* estimator
ошибка *f* воспроизведения reproducibility error
ошибка *f* второго рода error of second kind
ошибка *f* выборочного обследования sampling error
ошибка *f* метода error of method
ошибка *f* первого рода error of first kind
ошибка *f* титрования titration error

П

падение *n* напряжения electric potential difference
память *f* электрода hysteresis
пара *f* линий line pair
параметр *m* parameter
параметр *m* растворимости solubilityp arameter
партия *f* материала bulk of material
пассивация *f* электрода passivation of an electrode
пелликулярный ионообменник *m* pellicular ion exchanger
пелликулярный носитель *m* superficially porous support
пептизация *f* peptization
первичная проба *f* primary sample
первичная реакционная зона *f* primary combustion zone
первичное рентгеновское излучение *n* primary X-radiation
первичный элемент *m* primary cell
первичный эталон *m* primary standard
первый обертон *m* first overtone
перегруппировочный ион *m* rearrangement ion
перекристаллизация *f* recrystallization
перекрывание *n* линий line coincydence
перекрытие *n* линий line coincidence
переменно-токовая полярография *f*
  1. alternating-current polarography
  2. conventional alternating-current polarography
переменно-токовая полярография *f* второй гармоники second-harmonic ac polarography
переменно-токовая полярография *f* высших гармоник higher-harmonic ac polarography
переменно-токовая полярография *f* высших гармоник с фазовой селекцией higher-harmonic ac polarography with phase sensitive rectification
переменно-токовая хронопотенциометрия *f* alternating-current chronopotentiometry
перенапряжение *n* overpotential
перенапряжение *n* водорода hydrogen overpotential
перенапряжение *n* диффузии diffusion overpotential
перенапряжение *n* кристаллизации crystallization overpotential

перенапряжение *n* перехода activation overpotential
перенапряжение *n* реакции reaction overpotential
перенапряжение *n* сопротивления resistance overpotential
перенапряжение *n* стадии переноса заряда activation overpotential
переосаждение *n* reprecipitation
перестраиваемый лазер *m* tunable laser
пересыщенный раствор *m* supersaturated solution
пересыщение *n* supersaturation
переходное время *n* transition time
переходное равновесие *n* transient radioactive equilibrium
переходное радиоактивное равновесие *n* transient radioactive equilibrium
период *m* жизни капли drop time
период *m* полураспада half-life period
периодическая экстракция *f* periodic extraction
периодические весы *pl* undamped balance
перл *m* буры borax bead
перл *m* фосфорной соли microcosmic salt bead
перманганатометрия *f* manganometry
печь *f* сжигания combustion furnace
пик *m* peak
пик *m* воздуха air peak
пик *m* вылета pair escape peak
пик *m* вылета двух квантов double escape peak
пик *m* вылета одного кванта single escape peak
пик *m* двойного вылета double escape peak
пик *m* метастабильного иона metastable ion peak
пик *m* обратного рассеяния backscatter peak
пик *m* общей абсорбции full energy peak
пик *m* одиночного вылета single escape peak
пик *m* полного поглощения full energy peak
пик *m* суммарной энергии sum peak
пик *m* утечки пары pair escape peak
пик *m* утечки пары электронов pair escape peak
пикнометр *m* pycnometer
пиковая ёмкость *f* peak capacity
пиковый ток *m* peak current
пипетка *f* pipette
пипетка *f* Лунге-Рея Lunge-Rey weighing pipette
пипетка *f* с одной меткой transfer pipette
пиролиз *m* pyrolysis
пиролизёр *m* pyrolyzer
пирохимический анализ *m* pyrochemical analysis
плавень *m* flux
плавление *n* melting
плазма *f* plasma
плазменная горелка *f* 1. plasma jet 2. plasma torch
плазменная очистка *f* plasma cleaning
плазменная струя *f* plasma jet
плазменное анодирование *n* plasma anodization
плазменное озоление *n* plasma ashing
плазменное травление *n* plasma etching
плазменный факел *m* plasma torch
пламенная абсорбционная спектроскопия flame absorption spectroscopy
пламенная атомно-абсорбционная спектроскопия *f* flame atomic absorption spectroscopy

пламенная атомно-флуоресцентная спектроскопия *f* flame atomic fluorescence spectroscopy

пламенная атомно-эмиссионная спектроскопия *f* flame atomic emission spectroscopy

пламенная флуоресцентная спектроскопия *f* flame fluorescence spectroscopy

пламенная фотометрия *f* flame emission spectroscopy

пламенно-ионизационный детектор *m* flame ionization detector

пламенно-фотометрический детектор *m* flame photometric detector

пламенный спектрофотометр *m* flame spectrophotometer

пламенный фотометр *m* flame photometer

пламя *n* flame

план *m* второго порядка second order design

план *m* для смесей mixture design

план *m* первого порядка first order design

план *m* эксперимента exact design

планирование *n* эксперимента experimental design

пластинка-подложка *f* support plate

пластинка *f* с углублениями spotting plate

платинирование *n* электродов platinization of electrodes

платинированный асбест *m* platinized asbestos

платинированный уголь *m* platinized carbon

платиновый электрод *m* platinum electrode

плато *n* счётчика plateau of a counter

плёночный ионит *m* pellicular ion exchanger

плёночный ионообменник *m* pellicular ion exchanger

плёночный носитель *m* superficially porous support

плоёный фильтр *m* fluted filter

плоская решётка *f* plane grating

плоские колебания *npl* in-plane vibrations

плоскостнополяризованный свет *m* plane-polarized light

плоскостные колебания *npl* in-plane vibrations

плоскость *f* колебаний plane of vibration

плоскость *f* поляризации plane of polarization

плотность *f* вероятности probability density function

плотность *f* потока flux density

плотность *f* распределения probability density function

плотность *f* электрического тока current density

площадь *f* пика peak area

пневматический распылитель *m* pneumatic nebulizer

побочный компонент *m* minor constituent

поверхностная ионизация *f* surface ionization

поверхностная плотность *f* заряда surface charge density

поверхностная плотность *f* электрического заряда surface charge density

поверхностно-активное вещество *n* surfactant

поверхностно-барьерный полупроводниковый детектор *m* surface barrier detector

поверхностно-пористый носитель *m* superficially porous support

поверхностное покрытие *n* surface coverage

поверхностное рассеяние *n* молекулярного пучка molecular beam surface scattering

поверхностный электрический потенциал *m* фазы *β* surface electric potential of phase *β*

поверхность *f* surface

поверхность *f* отклика response surface

повторно дистиллированная вода *f* redistilled water

повторяемость *f* repeatability

погашение *n* absorbance

поглотитель *m* absorbent

поглотительная трубка *f* absorption tube

подвижная фаза *f* mobile phase

подвижность *f* иона B electric mobility of ion B

подтепловые нейтроны *mpl* cold neutrons

показатель *m* индикатора indicator exponent

показатель *m* чувствительности Санделля Sandell's sensitivity index

показатель *m* поглощения specific decadic absorption coefficient

показатель *m* экстинкции specific decadic absorption coefficient

полевой транзистор *m* field effect transistor

полевой транзистор *m* селективный к газам gas-sensitive field effect transistor

поликислота *f* polyacid

полифункциональный ионообменник *m* polyfunctional ion exchanger

полихроматическое излучение *n* polychromatic radiation

полиэлектрод *m* complex electrode

полиэлектролит *m* polyelectrolyte

полная проводимость *f* admittance

полное внутреннее отражение *n* total internal reflection

полное сопротивление *n* impedance

полностью твердофазный ионоселективный электрод *m* all-solid-state ion-selective electrode

полный анализ *m* complete analysis

полный объём *m* подвижной фазы total liquid volume

полный объём *m* растворителя в колонке total liquid volume

полный факторный план *m* complete factorial design

положительная ветвь *f* R-branch

положительный ион *m* positive ion

полоса *f* zone

полоса *f* B B band

полоса *f* K K band

полоса *f* R R band

полоса *f* основных колебаний fundamental vibration band

полоса *f* переноса заряда charge-transfer band

полоса *f* сравнения reference band

полоса *f* элюирования peak (*in chromatography*)

полосатый спектр *m* band spectrum

полуколичественный анализ *m* semiquantitative analysis

полумикроанализ *m* mesoanalysis

полумикроаналитические весы *pl* semimicrochemical balance

полумикрометод *m* centigram method

полупроводниковый детектор *m* semiconductor detector

полупроводниковый Ge-(Li) детектор *m*
lithium-drifted germanium detector

полупроводниковый Si-(Li) детектор *m*
lithium-drifted silicon detector

полуширина *f* линии half-intensity width

полуэлемент *m* electrode (*in electrochemistry*)

поляризационная кривая *f* polarization curve

поляризация *f* света polarization of light

поляризация *f* электрода polarization of
electrode

поляризованный свет *m* polarized light

поляриметр *m* polarimeter

поляриметрия *f* polarimetry

полярограмма *f* polarogram

полярограф *m* polarograph

полярографическая волна *f* polarographic wave

полярографическая кривая *f* polarogram

полярографическая кулонометрия *f*
polarographic coulometry

полярографическая хроноамперометрия *f*
polarographic chronoamperometry

полярографические максимумы *mpl*
polarographic maxima

полярографический анализ *m* polarographic
analysis

полярографический сдвиг *m* polarographic
wave

полярографическое титрование *n*
polarographic titration

полярография *f* polarography

полярография *f* Калоусека Kalousek
polarography

полярография *f* переменного тока
alternating-current polarography

полярография *f* постоянного тока direct current
polarography

полярография *f* с измерением переменного
напряжения alternating voltage polarography

полярография *f* с многократной развёрткой
потенциала multisweep polarography

полярография *f* с наложением периодически
меняющегося напряжения alternating-current
polarography

полярография *f* с нарастающим зарядом
incremental-charge polarography

полярография *f* с однократной развёрткой
потенциала single-sweep polarography

полярография *f* с прямоугольным напряжением
square-wave polarography

полярография *f* с развёрткой тока
current-scanning polarography

полярография *f* с треугольной развёрткой
потенциала triangular-wave polarography

полярография *f* со ступенчатой развёрткой
потенциала staircase polarography

полярография *f* со ступенчатым изменением
заряда incremental-charge polarography

полярометрическое титрование *n* polarometric
titration

поправка *f* на разбавление dilution-correction
factor

поправочный фактор *m* на градиент давления
pressure-gradient correction factor

популяция *f* population

порапак *m* Porapak

пористый носитель *m* totally porous support

пористый электрод *m* porous cup electrode

порог *m* коагуляции coagulation concentration

пороговая энергия *f* threshold energy

порошкограмма *f* X-ray powdered-crystal
pattern

порядок *m* спектра spectral order

порядок *m* реакции по данному веществу order
of reaction with respect to a given substance

порядок *m* реакции order of reaction

последние линии *fpl* raies ultimes

последовательный анализ *m* sequential analysis

последовательный критерий *m* sequential test

последующее осаждение *n* postprecipitation

постепенная регрессия *f* stepwise regression

постоянная жёсткость *f* permanent hardness

постоянная *f* кондуктометрической ячейки cell
constant of a conductivity cell

постоянная масса *f* constant mass

постоянная *f* Ридберга Rydberg constant

постоянная *f* Фарадея Faraday constant

постоянная *f* ячейки cell constant
of a conductivity cell

постоянно-токовая полярография *f* direct
current polarography

потенциал *m* выделения deposition potential

потенциал Гальвани inner electric potential of
phase

потенциал *m* Доннана membrane potential

потенциал *m* жидкостного соединения liquid-
junction potential

потенциал *m* коррозии corrosion potential

потенциал *m* нулевого заряда potential at the
point of zero charge

потенциал *m* оседания sedimentation potential

потенциал *m* покоя open-circuit electrode
potential

потенциал *m* полуволны half-wave potential

потенциал *m* разложения decomposition
potential

потенциал *m* седиментации sedimentation
potential

потенциал *m* течения streaming potential

потенциалопределяющая реакция *f*
potential-determining reaction

потенциометр *m* potentiometer

потенциометрический анализ *m* potentiometric
analysis

потенциометрический коэффициент *m*
селективности potentiometric selectivity
coefficient

потенциометрическое кулонометрическое
титрование *n* potentiometric coulometric
titration

потенциометрическое титрование *n*
potentiometric titration

потенциометрическое титрование *n*
с контролируемым током controlled-current
potentiometric titration

потенциометрическое титрование *n*
с контролируемым током и двумя
индикаторными электродами controlled-
current potentiometric titration with two
indicator electrodes

потенциометрическое титрование *n*
с регистрацией второй производной
second-derivative potentiometric titration

потенциометрическое титрование *n*
с регистрацией обратной производной inverse
derivative potentiometric titration
потенциометрия *f* potentiometry
потенциометрия *f* с контролируемым током
controlled-current potentiometry
потенциометрия *f* с контролируемым током
с двумя индикаторными электродами
controlled-current potentiometry with two
indicator electrodes
потенциостат *m* potentiostat
потенциостатическая кулонометрия *f*
controlled-potential coulometry
потенциостатический метод *m* potentiostatic
method
поток *m* излучения radiant flux
почернение *n* blackening
правила *npl* отбора selection rules
правило *n* трёх сигм three-sigma rule
правильность *f* accuracy
правовращающее вещество *n* dextrorotatory
substance
правополяризованный по кругу свет *m* right-
handed circularly polarized light
практическая удельная ёмкость *f* practical
specific capacity
предел *m* детектирования detection limit
предел *m* исключения exclusion limit
предел *m* обнаружения 1. detection limit
2. limit of detection
предел *m* определения limit of determination
пределы *mpl* измерения весов range of
applicability of a balance
предельная концентрация *f* concentration limit
предельная молярная электропроводность *f*
иона limiting molar conductivity of ionic
species
предельная молярная электропроводность *f*
электролита limiting molar conductivity of
electrolyte
предельная нагрузка *f* capacity
предельная эквивалентная электропроводность
f иона limiting equivalent conductivity of ionic
species B
предельная эквивалентная электропроводность
f электролита limiting equivalent conductivity
of electrolyte
предельное разбавление *n* dilution limit
предельное уравнение *n* Дебая-Хюккеля
Debye-Hückel limiting law
предельный закон *m* Дебая-Гюккеля
Debye-Hückel limiting law
предельный закон *m* Дебая-Хюккеля
Debye-Hückel limiting law
предельный закон *m* Онзагера
Debye-Hückel-Onsager's limiting law for
conductivity
предельный ток *m* limiting current
предколоночный объём *m* extra-column volume
предметный столик *m* Кофлера для
микроскопа Kofler micro melting point
apparatus
преимущественное распыление *n* preferential
sputtering
преломление *n* refraction
преломление *n* света light refraction

преобразование *n* Зейделя Seidel transformation
прецизионная потенциометрия *f* с нулевой
точкой differential potentiometry
прибор *m* для нанесения тонких слоёв spreader
прибор *m* для определения температуры
кипения boiling point apparatus
прибор *m* Парнаса и Вагнера Parnas-Wagner
apparatus
приведение *n* слоя в равновесие layer
equilibration
приведённая высота *f* эквивалентная
теоретической тарелке reduced plate height
приведённая скорость *f* потока подвижной фазы
reduced velocity of the mobile phase
приведённое время *n* удерживания adjusted
retention time
приведённые величины *fpl* reduced parameters
приведённые переменные *fpl* reduced parameters
приведённый объём *m* удерживания adjusted
retention volume
призма *f* optical prism
прикатодный слой *m* cathode layer
прилипание *n* adhesion
примесь *f* minor constituent
природная распространённость *f* изотопов
natural isotopic abundance
проба *f* test (*in chemical analysis*)
проба *f* паяльной трубкой blow-pipe test
пробирка *f* Несслера Nessler measuring cylinder
пробка *f* Шенигера Schöniger stopper
проверка *f* гирь calibration of weights
проводимость *f* conductivity
проводник *m* второго рода ionic conductor
проводник *m* первого рода electronic conductor
проводник *m* с ионной проводимостью ionic
conductor
проводник *m* с электронной проводимостью
electronic conductor
проволочный электрод *m* с чувствительным
покрытием coated wire ion-selective electrode
проволочный электрод *m*
с электродноактивным покрытием coated
wire ion-selective electrode
прозрачная решётка *f* transmission grating
произведение *n* растворимости solubility
product
производная вольтамперометрия *f* derivative
voltametry
производная дилатометрия *f* derivative
thermodilatometry
производная импульсная полярография *f*
derivative pulse polarography
производная полярография *f* derivative
polarography
производная термогравиметрия *f* derivative
thermogravimetry
производная хронопотенциометрия *f* derivative
chronopotentiometry
производное потенциометрическое титрование
n derivative potentiometric titration
производственный анализ *m* process analysis
пройденное расстояние *n* подвижной фазы
mobile phase distance
прокаливание *n* осадка ignition of precipitate
промах *m* gross error

промывка *f* scrubbing
промывка *f* осадка washing of precipitate
промывная жидкость *f* wash liquid
промывная склянка *f* со стеклянной пористой пластинкой gas washing bottle with sintered head
промывной раствор *m* washing solution
проникающая хроматография *f* permeation chromatography
пропорциональный счётчик *m* proportional counter tube
пропускание *n* transmission factor
проскок *m* breakthrough
простая реакция *f* elementary reaction
простая электродная реакция *f* simple electrode reaction
противоионы *mpl* counter-ions
противоточная экстракция *f* countercurrent extraction
противоточное распределение *n* countercurrent distribution
противоэлектрод *m* **1.** auxiliary electrode (*in electrochemistry*) **2.** counter electrode (*in arc and spark spectroscopy*)
протолитический растворитель *m* protolytic solvent
протонный магнитный резонанс *m* proton magnetic resonance
протоногенный растворитель *m* protogenic solvent
протофильный растворитель *m* protophilic solvent
проточный пропорциональный газовый счётчик *m* proportional gas-flow counter
профиль *m* распределения по глубине depth profile
процент *m* титрования per cent titrated
процент *m* экстракции extraction percentage
процентная ошибка *f* percentage error
проявительная хроматография *f* elution chromatography
проявление *n* development
проявление *n* хроматограммы development of a chromatogram
пружинные весы *pl* spring balance
прямое титрование *n* direct titration
прямой метод *m* direct method
прямоточная горелка *f* direct-injection burner
пятно *n* spot

Р

рабочее напряжение *n* burning voltage
рабочий электрод *m* working electrode
равновесный потенциал *m* equilibrium electrode potential
равновесный электродный потенциал *m* equilibrium electrode potential
радиальная хроматография *f* circular chromatography
радиатор *m* emitter
радиация *f* radiation
радиоактивационный реагент *m* radioactive reagent
радиоактивное превращение *n* radioactive decay

радиоактивность *f* radioactivity
радиоактивный индикатор *m* radioactive tracer
радиоактивный криптонат *m* radioactive kryptonate
радиоактивный распад *m* radioactive decay
радиоактивный ряд *m* decay chain
радиоизотопный рентгенофлуоресцентный анализ *m* radioisotope-excited X-ray fluorescence analysis
радиоиндикатор *m* radioactive tracer
радиолюминесценция *f* radioluminescence
радиометрический анализ *m* radiometric analysis
радиометрическое титрование *n* radiometric titration
радиометрия *f* radiometry
радиохимическая чистота *f* radiochemical purity
радиохимический анализ *m* radiochemical analysis
радиохимический выход *m* radiochemical yield
радиохимическое разделение *n* radiochemical separation
радиохроматография *f* radiochromatography
радиочастотная полярография *f* radio-frequency polarography
радиоэлектрофорез *m* radioelectrophoresis
разбавитель *m* diluent
разветвление *n* branching decay
разветвлённый распад *m* branching decay
разделение *n* пиков peak resolution
разделённое пламя *n* separated flame
размах *m* range
размывание *n* полосы band broadening
размывание *n* пятна band broadening, spot broadening
размытие *n* фронта fronting
разностная амперометрия *f* differential amperometry
разностная вольтамперометрия *f* differential voltammetry
разностная импульсная полярография *f* differential pulse polarography
разностная полярография *f* differential polarography
разностная потенциометрия *f* differential potentiometry
разностное потенциометрическое титрование *n* differential potentiometric titration
разность *f* вольта-потенциалов Volta potential difference
разность *f* гальвани-потенциалов Galvani potential difference
разность *f* электрических потенциалов electric potential difference
разность *f* электрических потенциалов для гальванического элемента electric potential difference of a galvanic cell
разрешающая сила *f* **1.** resolution (*of a spectrometer*) **2.** resolution (*of a photographic emulsion*)
разрешающая способность *f* **1.** resolution (*of a photographic emulsion*) **2.** resolution (*of a spectrometer*)
разрешающая способность *f* кристалл-анализатора crystal resolving power
разрешающее время *n* resolving time

разрешение *n* пиков peak resolution
разрешённый переход *m* allowed transition
разряд *m* в полом катоде hollow-cathode discharge
разрядная лампа *f* discharge lamp
разрядная полярография *f* incremental-charge polarography
разрядная трубка *f* с парами металлов metal-vapour lamp
рамановская спектроскопия *f* Raman spectroscopy
рамановское смещение *n* Raman frequency shift
рандомизация *f* randomization
распределение *n* χ^2 chi-squared distribution
распределение *n* Бернулли binomial distribution
распределение *n* Гаусса normal distribution
распределение *n* Пуассона Poisson distribution
распределение *n* Снедекора *F*-distribution
распределение *n* Стьюдента Student's distribution
распределительная хроматография *f* partition chromatography
распределительная хроматография *f* с обращёнными фазами extraction chromatography
распространение *n* дисперсии variance propagation
распространённость *f* изотопа abundance
распыление *n* 1. nebulization 2. sputtering
распылитель *m* nebulizer
рассеяние *n* scattering
рассеяние *n* света light scattering
расслаивание *n* растворителя solvent demixing
расслоение *n* растворителя solvent demixing
расстояние *n* миграции растворителя mobile phase distance
расстояние *n* пройденное фронтом растворителя mobile phase distance
раствор *m* solution
раствор *m* сравнения 1. comparison solution 2. reference solution
растворение *n* 1. digestion 2. dissolution
растворённое вещество *n* solute
растворимость *f* solubility
растворитель *f* solvent
растровый электронный микроскоп *m* scanning electron microscope
расширение *n* пика peak broadening
расширение *n* полосы band broadening
расширение *n* пятна spot broadening
расширенное уравнение *n* Дебая и Хюккеля Debye-Hückel equation
реагент *m* reagent
реагент *m* для микроанализа microanalytical reagent
реактанс *m* reactance
реактив *m* reagent
реактив *m* Несслера Nessler reagent
реактив *m* Фелинга Fehling's reagent
реактив *m* Фишера Karl Fischer reagent
реактив *m* чистый для анализа analytical reagent
реактив *m* Чугаева Chugaev reagent
реактивная бумага *f* indicator paper
реактивная проводимость *f* susceptance

реактивное распыление *n* reactive sputtering
реактивное рассеяние *n* молекулярного пучка molecular beam reactive scattering
реактивное сопротивление *n* reactance
реакционное перенапряжение *n* reaction overpotential
реакционный слой *m* reaction layer
реакция *f* reaction
реакция *f* кислота-основание acid-base reaction
реакция *f* окисления-восстановления redox reaction
реакция *f* переноса заряда charge-transfer step
реакция *f* Цейзеля Zeisel reaction
реальный потенциал *m* conditional electrode potential
ребристая воронка *f* ribbed funnel
регистрирующий спектрофотометр *m* recording spectrophotometer
регрессионный анализ *m* regression analysis
регрессия *f* regression function
редокс-индикатор *m* oxidation-reduction indicator
редокс-полимер *m* redox polymer
редокс-потенциал *m* oxidation-reduction potential
редокс-электрод *m* redox electrode
редоксибуфер *m* oxidation-reduction buffer
редоксиионообменник *m* redox ion exchanger
редуктометрия *f* reductometry
резко выделяющееся наблюдение *n* outlier
резонанс *m* Ферми Fermi resonance
резонансная полость *f* cavity resonator
резонансная флуоресценция *f* resonance fluorescence
резонансное комбинационное рассеяние *n* света resonance Raman effect
резонансные нейтроны *mpl* resonance neutrons
резонансный захват *m* resonance capture
резонансный максимум *m* resonance peak
резонансный пик *m* resonance peak
резонатор *m* cavity resonator
релаксационное торможение *n* relaxation-time effect
релаксационные методы *mpl* transient methods
релаксационный эффект *m* relaxation-time effect
релаксация *f* relaxation
рельс *m* optical bench
рентгеновская спектроскопия *f* индуцированная ионной бомбардировкой ion induced X-ray spectroscopy
рентгеновская трубка *f* X-ray tube
рентгеновская флуоресцентная спектроскопия *f* X-ray fluorescence spectroscopy
рентгеновская флуоресценция *f* X-ray fluorescence
рентгеновская фотоэлектронная спектроскопия *f* X-ray photoelectron spectroscopy
рентгеновские лучи *mpl* X-radiation
рентгеновский анализ *m* с возбуждением ускоренными ионами particle induced X-ray emission
рентгеновский гониометр *m* X-ray goniometer
рентгеновский дифрактометр *m* X-ray diffractometer
рентгеновский микроанализатор *m* electron microprobe X-ray analyser

рентгеновский пик *m* утечки X-ray escape peak

рентгеновский спектрометр *m* X-ray spectrometer

рентгеновский спектрохимический анализ *m* X-ray spectrochemical analysis

рентгеновский флуоресцентный спектр *m* X-ray fluorescence spectrum

рентгеновский характеристический спектр *m* X-ray characteristic spectrum

рентгеновское излучение *n* X-radiation

рентгенограмма *f* X-ray pattern

рентгенограмма *f* вращения кристалла X-ray rotating-crystal pattern

рентгенограмма *f* колебания X-ray oscillating-crystal pattern

рентгенодифракционный анализ *m* X-ray diffraction analysis

рентгеноспектральный флуоресцентный анализ *m* X-ray fluorescence analysis

рентгенофлуоресцентное излучение *n* X-ray fluorescence radiation

рентгенофлуоресцентный анализ *m* с волновой дисперсией wavelength dispersive X-ray fluorescence analysis

рентгенофлуоресцентный анализ *m* с энергетической дисперсией energy dispersive X-ray fluorescence analysis

рентгенофлуоресценция *f* X-ray fluorescence

рефрактометр *m* refractometer

рефрактометрия *f* refractometry

рефракция *f* refraction

реэкстрагент *m* stripping solution

реэкстракция *f* back-extraction

ридберг *m* rydberg

рифлёная воронка *f* ribbed funnel

ротатабельность *f* rotatability

ротатабельный план *m* rotatable design

ртутно-окисный электрод *m* mercury-mercuric oxide electrode

ртутно-сульфатный электрод *m* mercury-mercurous sulphate electrode

ртутный висячий электрод *m* hanging mercury drop electrode

ртутный капающий электрод *m* dropping-mercury electrode

ртутный капельный электрод *m* dropping-mercury electrode

ртутный элемент *m* mercury dry cell

рэлеевское рассеяние *n* Rayleigh scattering

ряд *m* напряжений electrochemical series

ряд *m* потенциалов electrochemical series

С

самообращение *n* линии self-reversal

самопоглощение *n* self-absorption

самопроизвольная электрогравиметрия *f* internal electrogravimetry

самоэкранирование *n* self-shielding

самоэмитирующий электрод *m* self-electrode

сборник фракций *m* fraction collector

сборные колонки *fpl* coupled columns

сверхвысокий вакуум *m* ultra-high vacuum

сверхкритическая флюидная хроматография *f* supercritical fluid chromatography

сверхтонкая структура *f* hyperfine structure

свет *m* light

свет *m* поляризованный по кругу влево left-handed circularly polarized light

свет *m* поляризованный по кругу вправо right-handed circularly polarized light

световая энергия *f* luminous energy

световой поток *m* luminous flux

светорассеяние *n* light scattering

светофильтр *m* light filter

светочувствительный полупроводник *m* photoconductor

свидетель *m* comparison solution

свободный объём *m* interstitial volume

сгорание *n* combustion

сегрегация *f* segregation

седиментационные весы *pl* sedimentation balance

седиментационный анализ *m* sedimentation analysis

седиментация *f* sedimentation

седиментометрический анализ *m* sedimentation analysis

селективная проницаемость *f* permselectivity

селективная реакция *f* selective reaction

селективное элюирование *n* selective elution

селективность *f* реагента selectivity of a reagent

селективный ионообменник *m* selective ion exchanger

селективный ионочувствительный электрод *m* ion-selective electrode

селективный реагент *m* selective reagent

сенсибилизированный ионоселективный электрод *m* sensitized ion-selective electrode

сенсорный анализ *m* sensory analysis

серебряный электрод *m* silver electrode

середина *f* размаха mid-range

серийный анализ *m* routine analysis

сефадекс *m* G Sephadex G

сечение *n* активации activation cross section

сечение *n* поглощения absorption cross section

сжигание *n* combustion

сжигание *n* фильтра incineration of filter paper

сжигательная трубка *f* combustion tube

сила *f* света luminous intensity

сила *f* тока electric current

силилирование *n* silylation

силилирующий реактив *m* silylation reagent

силитовый стержень *m* Globar

силовая константа *f* force constant

симметричные колебания *npl* symmetrical vibrations

симплекс *m* simplex

симплекс-план *m* simplex design

синглет *m* singlet line

синий декстран *m* Blue Dextran

синтетический органический ионит *m* ion-exchange resin

система *f* полос band group

система *f* предварительного смешения premix burner

система *f* редокс redox system

систематическая ошибка *f* systematic error

систематическая погрешность *f* systematic error

систематический анализ *m* systematic qualitative analysis

систематический качественный анализ *m* systematic qualitative analysis

ситовый анализ *m* sieve analysis

сканирующая спектроскопия *f* оже-электронов Auger-electron microscopy

скелетные колебания *npl* skeletal vibrations

склянка *f* Дрекселя Drechsel gas washing bottle

склянка *f* Фридрихса Friedrichs gas washing bottle

скорость *f* вытекания ртути из капилляра flow rate of mercury

скорость *f* образования центров кристаллизации rate of nucleation

скорость *f* света в вакууме velocity of light in vacuum

скорость *f* счёта counting rate

следовой компонент *m* trace constituent

слепой опыт *m* blank test

сложная электродная реакция *f* stepwise electrode reaction

слой *m* Гуи-Чэпмена diffuse double layer

слой *m* половичного поглощения half-thickness

случайная величина *f* random variable

случайная выборка *f* random sample

случайная модель *f* random effects model

случайная ошибка *f* random error

случайное событие *n* random event

смешанный индикатор *m* mixed indicator

смешанный кристалл *m* mixed crystal

смешанный потенциал *m* mixed polyelectrode potential

смешанный потенциал *m* полиэлектрода mixed polyelectrode potential

смешанный растворитель *m* mixed solvent

смешанный слой *m* mixed bed

смешанный электрод *m* complex electrode

смещение *n* bias of estimator

смещение *n* комбинационного рассеяния света Raman frequency shift

собачка *f* piggie

собирание *n* collection

совместное образование *n* зародышей simultaneous nucleation

содержание *n* content

соединённые колонки *fpl* coupled columns

сожжение *n* в водородно-кислородном пламени oxy-hydrogen flame method

сожжение *n* в пустой трубке empty tube combustion

сожжение *n* со вспышкой flash combustion

сокращённый анализ *m* short analysis

солевая ошибка *f* salt error

солевая форма *f* ионообменника salt form of an ion exchanger

солевой мостик *m* salt bridge

солевой эффект *m* salt effect

солеобразующая группа *f* acidic group

сольватохромия *f* solvatochromism

солюбилизационная хроматография *f* solubilization chromatography

соосаждение *n* coprecipitation

соотношение *n* пик-фон peak-to-Compton ratio

сопротивление *n* массообмену mass transfer term

сопротивление *n* массопереносу mass transfer term

сопряжённая реакция *f* conjugate reaction

сопряжённые методы *mpl* coupled simultaneous techniques

сопряжённые одновременные методы *mpl* coupled simultaneous techniques

сорбат *m* sorbate

сорбент *m* sorbent

сорбция *f* sorption

состоятельная оценка *f* consistent estimator

сосуд *m* для взвешивания weighing bottle

спектр *m* spectrum

спектр *m* возбуждения excitation spectrum

спектр *m* испускания emission spectrum

спектр *m* комбинационного рассеяния Raman spectrum

спектр *m* комбинационного рассеяния света Raman spectrum

спектр *m* нейтронов neutron spectrum

спектр *m* ослабления полного отражения attenuated total reflectance spectrum

спектр *m* поглощения absorption spectrum

спектр *m* с разрешением во времени time-resolved spectrum

спектр *m* УФ ultraviolet spectrum

спектр *m* ядерного излучения nuclear radiation spectrum

спектр *m* ядерного магнитного резонанса высокого разрешения high-resolution nuclear magnetic resonance spectrum

спектральная линия *f* spectral line

спектральная полоса *f* spectral band

спектральная серия *f* spectral series

спектрально чистый реактив *m* spectrally pure reagent

спектральный анализ *m* spectrochemical analysis

спектрограмма *f* spectrogram

спектрограф *m* spectrograph

спектрографический анализ *m* spectrographic analysis

спектрометр *m* optical spectrometer

спектрометр *m* γ-излучения gamma-ray spectrometer

спектрометр *m* ядерного магнитного резонанса nuclear magnetic resonance spectrometer

спектрометр *m* ЯМР nuclear magnetic resonance spectrometer

спектрометрия *f* spectrometry

спектрополяриметр *m* spectropolarimeter

спектрополяриметрия *f* spectropolarimetry

спектроскоп *m* spectroscope

спектроскопический анализ *m* spectrochemical analysis

спектроскопия *f* spectroscopy

спектроскопия *f* внешнего потенциала appearance-potential spectroscopy

спектроскопия *f* валентных электронов valence level electron spectroscopy

спектроскопия *f* внешнего потенциала электронов индуцированных рентгеновским цзлучением X-ray electron appearance potential spectroscopy

спектроскопия *f* затухающего потенциала disappearance potential spectroscopy

спектроскопия *f* комбинационного рассеяния Raman spectroscopy спектроскопия *f* Мёссбауэра Mössbauer spectroscopy

спектроскопия *f* нейтрализации ионов ion neutralization spectroscopy

спектроскопия *f* низкоэнергетических характеристических потерь electron-energy loss spectroscopy

спектроскопия *f* обратного рассеяния быстрых ионов high-energy ion scattering spectroscopy

спектроскопия *f* обратного рассеяния ионов low energy ion scattering spectroscopy

спектроскопия *f* обратного рассеяния медленных ионов low energy ion scattering spectroscopy

спектроскопия *f* оже-электронов Auger-electron microscopy

спектроскопия *f* оже-электронов индуцированных ионной бомбардировкой ion-induced Auger electron spectroscopy

спектроскопия *f* оже-электронов индуцированных рентгеновским излучением X-ray induced Auger electron spectroscopy

спектроскопия *f* потерь проходящей энергии transmission energy loss spectroscopy

спектроскопия *f* рентгеновских фотоэлектронов X-ray photoelectron spectroscopy

спектроскопия *f* характеристических потерь энергии transmission energy loss spectroscopy

спектроскопия *f* электронного парамагнитного резонанса electron paramagnetic resonance spectroscopy

спектроскопия *f* электронов electron spectroscopy

спектроскопия *f* электронов Оже Auger electron spectroscopy

спектроскопия *f* ядерного магнитного резонанса nuclear magnetic resonance spectroscopy

спектрофлуориметр *m* spectrofluorimeter

спектрофлуориметрия *f* spectrofluorimetry

спектрофлуорометр *m* spectrofluorimeter

спектрофлуорометрия *f* spectrofluorimetry

спектрофотометр *m* spectrophotometer

спектрофотометрическое титрование *n* spectrophotometric titration

спектрофотометрия *f* spectrophotometry

спектрохимический анализ *m* spectrochemical analysis

спектрохимический буфер *m* spectrochemical buffer

спектрохимический носитель *m* spectrochemical carrier

специфическая адсорбция *f* specific adsorption

специфическая поверхность *f* specific surface area

специфическая реакция *f* specific reaction

специфический реагент *m* specific reagent

специфическое элюирование *n* specific elution

специфичная реакция *f* specific reaction

специфичность *m* реагента specificity of a reagent

сплавление *n* в бомбе Парра Parr bomb fusion

сплошной спектр *m* continuous spectrum

среднее абсолютное отклонение *n* mean deviation

среднее *n* арифметическое arithmetic mean

среднее *n* геометрическое geometric mean

среднее инфракрасное излучение *n* middle-infrared

среднее стандартное отклонение *n* standart deviation of the mean

среднеионный коэффициент *m* активности mean activity coefficient of electrolyte

средний ионный диаметр *m* mean ionic diameter

средний ионный коэффициент *m* активности mean activity coefficient of electrolyte

средний коэффициент *m* активности электролита mean activity coefficient of electrolyte

средняя активность *f* электролита mean activity of electrolyte

средняя ионная активность *f* mean activity of electrolyte

средняя концентрация *f* электролита mean concentration of electrolyte

средняя лабораторная проба *f* average sample

стадия *f* процесса elementary step

стадия *f* процесса разряда charge-transfer step

стадия *f* разряда-ионизации charge-transfer step

стандартизация *f* standardization

стандартизированная случайная величина *f* standardized random variable

стандартизированное нормальное распределение *n* standardized normal distribution

стандартная константа *f* скорости ецектродизй реакйи standard rate constant of elektrode reaktion

стандартная проба *f* standard sample

стандартная э.д.с. standard electromotive force

стандартная электродвижущая сила *f* standard electromotive force

стандартное отклонение *n* standard deviation

стандартные вещества *npl* для элементного анализа standarts for elemental analysis

стандартные условия *npl* standard conditions

стандартный водородный электрод *m* standard hydrogen electrode

стандартный метод *m* standardized method

стандартный потенциал *m* standard electrode potential

стандартный раствор *m* standard solution

стандартный электродный потенциал *m* standard electrode potential

стандартный элемент *m* standard cell

стандартный элемент *m* Вестона Weston standard cell

старение *n* ageing

старение *n* коллоидных растворов ageing of sols

стартовая линия *f* starting line

стартовая точка *f* starting line

pH- стат *m* pH-stat

статистика *f* statistic

статистическая вероятность *f* statistical probability

статистическая гипотеза *f* statistical hypothesis

статистический контроль *m* качества statistical quality control

статистический критерий *m* test

стационарная фаза *f* stationary phase

стационарные методы *mpl* steady-state methods

стационарный ртутный капельный электрод m hanging mercury drop electrode

стекло n с контролируемой пористостью controlled-porosity glass

стекло-углеродный электрод m glassy carbon electrode

стеклянная воронка f glass funnel

стеклянный мембранный электрод m glass membrane electrode

стеклянный электрод m glass membrane electrode

степень f ассоциации ионов association degree of ions

степень f заполнения поверхности surface coverage

степень f электролитической диссоциации apparent degree of dissociation

стехиометрическая точка f equivalence-point

стимулирование n десорбции ионной бомбардировкой ion impact desorption

сток m drain

Стокгольмская конвенция f sign convention

Стокгольмское соглашение n sign convention

стоксовы линии fpl Stokes lines

струйчатый ртутный электрод m streaming mercury electrode

ступенчатая константа f образования stepwise stability constant

ступенчатое элюирование n stepwise elution

ступенчатый ослабитель m step filter

ступенчатый фильтр m step filter

ступень f step on an integral chromatogram

ступень f на интегральной хроматограмме step on an integral chromatogram

субстехиометрический метод m в изотопном разбавлении substoichiometric isotope dilution analysis

сульфатно-ртутный электрод m mercury-mercurous sulphate electrode

суммарный пик m sum peak

суммарный порядок m реакции order of reaction

сурьмяный электрод m antimony electrode

суспендирование n в таблетке бромистого калия pellet technique

суспендирование n вещества в таблетке бромистого калия pellet technique

суспендирование n твёрдого вещества в нуйоле mull technique

суспензионное заполнение n колонки slurry packing

сусцептометрия f susceptometry

сухой ионит m absolutely dry ion exchanger

сухой ионообменник m absolutely dry ion exchanger

сухой путь m dry method

сухой путь m анализа dry method

сухой ртутный элемент m mercury dry cell

сушильный агент m desiccant

сферическая аберрация f spherical aberration

схема f распада decay scheme

схема f фрагментации fragmentation pattern

сцинтилляционный β-спектрометр m beta scintillation spectrometer

сцинтилляционный детектор m scintillation detector

сцинтилляционный спектрометр m scintillation spectrometer

сцинтилляционный счётчик m scintillation counter

счётная трубка f counter tube

счётная характеристика f plateau of a counter

2π-счётчик m 2π-counter

4π-счётчик m 4π-counter

счётчик m Гейгера Geiger-Müller counter tube

счётчик m Гейгера-Мюллера Geiger-Müller counter tube

счётчик m излучений radiation counter

счётчик m пузырьков bubble gauge

счётчик m радиоактивного излучения radiation counter

счётчик m с геометрией 2π 2π-counter

счётчик m с геометрией 4π 4π-counter

счётчик m с каналом well-type scintillation counter

счётчик m с каналом для образцов well-type scintillation counter

счётчик m с окном window counter

сшивание n полимера cross-linking

сэндвич-камера f S-chamber

Т

таблетка f бромистого калия potassium bromide disk

таблицы fpl случайных чисел random number tables

таламидный электрод m thalamid electrode

таллиевоамальгамный-хлорталлиевый электрод m thalamid electrode

таст-полярография f Tast polarography

твёрдая мембрана f solid-state membrane

твердотельная мембрана f solid-state membrane

твердотельный электрод m solid electrode

твёрдый мембранный электрод m solid-state membrane ion-selective electrode

твёрдый носитель m solid support

твёрдый раствор m mixed crystal

твёрдый электрод m solid electrode

темновой ток m dark current

температура f колонки column temperature

температура f разделения separation temperature

температура f удерживания retention temperature

температурно-програмированная десорбция f temperature programmed desorption

температурный коэффициент m temperature coefficient

тензаметрия f measurement of nonfaradaic admittance

теоретическая конечная точка f equivalence-point

теоретическая удельная ёмкость f total specific capacity

теория f ассоциации ионов Бьеррума Bjerrum's theory of ionic association

теория f вероятностей probability theory

теория f Дебая и Гюккеля Debye-Hückel's theory of strong electrolytes

теория f Дебая и Хюккеля Debye-Hückel's theory of strong electrolytes

теория *f* ионной ассоциации Бьеррума Bjerrum's theory of ionic association

теория *f* сильных электролитов Дебая и Хюккеля Debye-Hückel's theory of strong electrolytes

теория *f* сольватации Борна Born's theory of solvation

теория *f* строения двойного электрического слоя theory of double layer

теория *f* электролитической диссоциации Аррениуса и Оствальда Arrhenius' and Ostwald's theory of electrolytic dissociation

термализация *f* нейтронов neutron thermalization

термическая десорбция *f* thermal desorption

термическая диссоциация *f* thermal dissociation

термическая ионизация *f* thermal ionization

термические нейтроны *mpl* thermal neutrons

термический анализ *m* thermal analysis

термоаналитическая кривая *f* thermal curve

термовесы *pl* thermobalance

термогазоволюметрический анализ *m* thermovolumetric analysis

термогальванический элемент *m* galvanic thermocell

термогравиметрия *f* thermogravimetry

термограмма *f* thermal curve

термодесорбция *f* temperature programmed desorption

термодетектор *m* thermal radiation detector

термодиффузионный потенциал *m* thermodiffusion potential

термоионизационный ионный источник *m* thermal emission ion source

термоионизация *f* thermal ionization

термокондуктометрическая ячейка *f* thermal conductivity detector

термокондуктометрический детектор *m* thermal conductivity detector

термоманометрический анализ *m* thermomanometric analysis

термопара *f* thermocouple

термоэлемент *m* thermocouple

технические весы *pl* chemical balance

тиндалиметрия *f* nephelometry

титанометрия *f* titanometry

титр *m* titre

титрант *m* titrant

титриметрический анализ *m* titrimetric analysis

титриметрический газовый анализ *m* gas titration analysis

титриметрический фактор *m* пересчёта titrimetric conversion factor

титрование *n* titration

титрование *n* до мёртвой точки amperometric titration with two indicator electrodes

титрование *n* с применением двух индикаторов double-indicator titration

тлеющий разряд *m* glow discharge

ток *m* заряжения capacity current

ток *m* насыщения limiting current

ток *m* обмена exchange current

толстая мишень *f* thick target

толщина *f* ионной атмосферы radius of ionic atmosphere

толщина *f* поглощающего слоя path length

тонкая мишень *f* thin target

тонкая структура *f* fine structure

тонкий образец *m* thin-film specimen

тонкослойная хроматография *f* thin-layer chromatography

торзионные весы *pl* torsion balance

торзионные колебания *npl* twisting vibrations

тормозное излучение *n* bremsstrahlung

точечная оценка *f* point estimation

точка *f* конца титрования end-point of titration

точка *f* равновесия rest point

точка *f* эквивалентности equivalence-point

точка *f* эксперимента experimental point

точки *fpl* перегиба inflection points

точки *fpl* перегиба пика inflection points

точность *f* precision

точность *f* весов precision of a balance

точность *f* взвешивания precision of a weighing

точный план *m* эксперимента exact design

транспортное число *n* иона B transport number of ionic species B

тренд *m* drift

третье приближение *n* теории Дебая-Гюккеля Debye-Hückel-Brönsted equation

триметилхлорсилан *m* trimethylchlorosilane

тритиевая мишень *f* tritium target

трубка *f* для сожжения combustion tube

трубка *f* для фильтрования filter tube

турбидиметр *m* turbidimeter

турбидиметрическое титрование *n* turbidimetric titration

турбидиметрия *f* turbidimetry

турбулентное пламя *n* turbulent flame

тушение *n* quenching

У

угловая дисперсия *f* angular dispersion

угловая дисперсия *f* кристалл-анализатора angular dispersion of analysing crystal

угол *m* блеска blaze angle

угол *m* выхода take-off angle

угол *m* скольжения glancing angle

удельная активность *f* specific activity

удельная поверхность *f* specific surface area

удельная рефракция *f* specific refraction

удельная электрическая проводимость *f* electrical conductivity

удельная электрическая проводимость *f* электролита electrolytic conductivity

удельная электропроводность *f* электролита electrolytic conductivity

удельное вращение *n* specific rotation

удельное электрическое сопротивление *n* resistivity

удельный десятичный показатель *m* поглощения specific decadic absorption coefficient

удельный объём *m* удерживания specific retention volume

удельный показатель *m* поглощения specific decadic absorption coefficient

удельный удерживаемый объём *m* specific retention volume

удерживающий агент *m* hold-back carrier

удерживающий носитель *m* hold-back carrier

ультразвуковой распылитель *m* ultrasonic nebulizer
ультрамикроанализ *m* ultramicro analysis
ультрамикровесы *pl* ultramicro balance
ультрафиолет *m* ultraviolet
ультрафиолетовая область *f* ultraviolet region
ультрафиолетовая фотоэлектронная спектроскопия *f* ultraviolet photoelectron spectroscopy
ультрафиолетовое излучение *n* ultraviolet
ультрафиолетовый спектр *m* ultraviolet spectrum
универсальный индикатор *m* universal indicator
унифицированная атомная единица *f* unified atomic mass unit
унифицированный метод *m* standardized method
упругое рассеяние *n* elastic scattering
упругое рассеяние *n* медленных электронов low energy electron diffraction
упругое столкновение *n* elastic collision
уравнение *n* Бутлера-Вольмера Butler-Volmer equation
уравнение *n* ван Деемтера van Deemter equation
уравнение *n* Гендерсона Henderson's equation
уравнение *n* Дебая-Гюккеля-Онзагера Debye-Hückel-Onsager's limiting law for conductivity
уравнение *n* Дебая и Хюккеля Debye-Hückel equation
уравнение *n* Ильковича Ilkovič equation
уравнение *n* Котрелла Cottrell equation
уравнение *n* Левича Levich equation
уравнение *n* Нернста Nernst equation
уравнение *n* Онзагера Debye-Hückel-Onsager's limiting law for conductivity
уравнение *n* Санда Sand equation
уравнение *n* Тафеля Tafel's equation
уравнение *n* Шевчика Randles-Ševčik equation
уравновешивание *n* слоя layer equilibration
уровень *m* значимости significanse level
уровень *m* шумов noise level
условная константа *f* скорости электродной реакции conditional rate constant of electrode reaction
условный потенциал *m* conditional electrode potential
условный электродный потенциал *m* conditional electrode potential
УФ-фотоэлектронная спектроскопия *f* ultraviolet photoelectron spectroscopy

Ф

фазовое отношение *n* phase ratio
фазовое титрование *n* phase titration
фазовый анализ *m* phase analysis
фактор *m* factor
фактор *m* асимметрии пика asymmetry factor of a peak
фактор *m* извлечения recovery factor
фактор *m* насыщения saturation factor
фактор *m* пересчёта analytical factor
фактор *m* разделения separation factor
фактор *m* селективности potentiometric selectivity coefficient

фактор *m* шероховатости roughness factor
фактор *m* эквивалентности для кислотно--основных реакций equivalence factor in acid-base reactions
фактор *m* эквивалентности для окислительно-восстановительных реакций equivalence factor in redox reactions
фактор *m* эквивалентности компонента B equivalence factor of component B
факторный анализ *m* factor analysis
фарадеевский ток *m* faradaic current
фарадеевское выпрямление *n* высокого уровня high-level faradaic rectification
ферментный электрод *m* enzyme-substrate electrode
физическая адсорбция *f* physisorption
фиксированные ионы *mpl* fixed ions
фикспара *f* линий fixation pair
фильтр *m* рентгеновского излучения X-ray filter
фильтр *m* со стеклянной пористой пластинкой filter funnel with sintered disk
фильтрат *m* filtrate
фильтровальный тигель *m* Гуча Gooch crucible
фильтрование *n* filtration
фильтровый флуориметр *m* filter fluorimeter
флеш-десорбция *f* flash desorption
флокуляция *f* coagulation
флуоресцентная молекулярная спектроскопия molecular fluorescence spectroscopy
флуоресцентная спектроскопия *f* fluorescence spectroscopy
флуоресцентная спектроскопия *f* пламени flame fluorescence spectroscopy
флуоресцентное рентгеновское излучение *n* X-ray fluorescence radiation
флуоресцентный анализ *m* fluorimetry
флуоресцентный индикатор *m* fluorescent indicator
флуоресцентный спектр *m* fluorescence spectrum
флуоресцентный спектр *m* возбуждения fluorescence excitation spectrum
флуоресцентный спектрохимический анализ *m* fluorescence spectrochemical analysis
флуоресцентный эмиссионный спектр *m* fluorescence emission spectrum
флуоресценция *f* fluorescence
флуориметр *m* fluorimeter
флуориметрическое титрование *n* fluorimetric titration
флуориметрия *f* fluorimetry
флуорометр *m* fluorimeter
флуорометрия *f* fluorimetry
флюс *m* flux
фокусирующий рентгеновский спектрометр *m* X-ray focusing spectrometer
фон *m* детектора background of a device
фон *m* радиации background radiation
фоновый электролит *m* supporting electrolyte
формальность *f* formality
формальный потенциал *m* conditional electrode potential
формальный электродный потенциал *m* conditional electrode potential
формула *f* Гендерсона Henderson's equation

формула *f* Льюиса-Саржента Lewis-Sargent's relation
формула *f* Тафеля Tafel's equation
фосфоресцентный спектр *m* возбуждения phosphorescence excitation spectrum
фосфоресцентный эмиссионный спектр *m* phosphorescence emission spectrum
фосфоресценция *f* phosphorescence
фотоактивационный анализ *m* photoactivation analysis
фотоактивация *f* photoactivation
фотографическая пластинка *f* photographic plate
фотографическая широта *f* latitude
фотодесорбция *f* photodesorption
фотодетектор *m* photodetector
фотодиод *m* photodiode
фотоионизационный ионный источник *m* photoionization ion source
фотоионизация *f* photoionization
фотоколориметр *m* photocolorimeter
фотолюминесценция *f* photoluminescence
фотометр *m* photometer
фотометрическое титрование *n* photometric titration
фотометрия *f* photometry
фотон *m* photon
фотонейтрон *m* photoneutron
фотопик *m* photoelectric peak
фотополярография *f* photopolarography
фотоприёмник *m* photodetector
фотопроводник *m* photoconductor
фототок *m* photocurrent
фототранзистор *m* phototransistor
фототриод *m* phototransistor
фотоумножитель *m* photomultiplier tube
фоточасть *f* photofraction
фотоэлектрический детектор *m* photodetector
фотоэлектрон *m* photoelectron
фотоэлектронная линия *f* photoelectron line
фотоэлектронная спектроскопия *f* photoelectron spectroscopy
фотоэлектронный умножитель *m*
1. photomultiplier tube 2. electron multiplier
фотоэлемент *m* photocell
фотоэлемент *m* с запирающим слоем photovoltaic cell
фотоэффект *m* photoelectric effect
фрагментация *f* fragmentation
фрагментный ион *m* fragment ion
фракционная дистилляция *f* с носителем carrier distillation
фракционное осаждение *n* fractional precipitation
фракционный анализ *m* fractional analysis
фракция *f* fraction
французский градус *m* жёсткости воды French degree of water hardness
фронт *m* подвижной фазы mobile phase front
фронт *m* растворителя mobile phase front
фронтальная хроматография *f* frontal chromatography
фронтальный анализ *m* frontal chromatography
фронтальный метод *m* frontal chromatography
функциональная группа *f* functional group

функциональный анализ *m* functional group analysis
функция *f* возбуждения excitation function
функция *f* распределения distribution function
функция *f* регрессии regression function

X

характеристическая кривая *f* эмульсии emulsion characteristic curve
характеристическая частота *f* characteristic frequency
характеристическое рентгеновское излучение *n* characteristic X-radiation
характерная реакция *f* identification reaction
хелат *m* chelate compound
хелатная ионообменная смола *f* chelating resin
хелатная смола *f* chelating resin
хелатное кольцо *n* chelate ring
хелатный ионообменник *m* chelating ion exchanger
хелатообразующий лиганд *m* chelating agent
хемиионизация *f* chemi-ionization
хемилюминесцентный индикатор *m* chemiluminescent indicator
хемилюминесценция *f* chemiluminescence
хемосорбция *f* chemisorption
химическая адсорбция *f* chemisorption
химическая ионизация *f* chemical ionization
химическая цепь *f* chemical cell
химически связанная фаза *f* chemically bonded stationary phase
химически чистый реактив *m* chemically pure reagent
химический анализ *m* chemical analysis
химический кулонометр *m* chemical coulometer
химический потенциал *m* chemical potential of species B in phase α
химический сдвиг *n* affinity
химический эквивалент *m* chemical equivalent
химический эффект *m* chemical effect
химическое перенапряжение *n* reaction overpotential
химическое сродство *n* affinity
хингидронный электрод *m* quinhydrone electrode
хлоросеребряный электрод *m* silver-silver chloride electrode
холодные нейтроны *mpl* cold neutrons
холостое титрование *n* blank titration
холостой опыт *m* blank test
холостой раствор *m* blank solution
хроматическая аберрация *f* chromatic aberration
хроматограмма *f* chromatogram
хроматограф *m* chromatograph
хроматографировать chromatograph
хроматографическая камера *f* chromatographic chamber
хроматографическая колонка *f* chromatographic column
хроматографический анализ *m* chromatographic analysis
хроматографический слой *m* chromatographic bed
хроматография *f* chromatography

хроматография *f* **в открытой трубке** open-tube chromatography

хроматография *f* **в тонком слое** thin-layer chromatography

хроматография *f* **на бумаге** paper chromatography

хроматография *f* **на колонке** column chromatography

хроматография *f* **на молекулярных ситах** molecular-sieve chromatography

хроматография *f* **обмена лигандов** ligand-exchange chromatography

хроматография *f* **с высаливанием** salting-out chromatography

хроматография *f* **с обращёнными фазами** reserved-phase chromatography

хроматография *f* **с программированием потока** flow-programmed chromatography

хроматография *f* **с программированием температуры** temperature-programmed chromatography

хроматометрия *f* chromatometry

хроматополярография *f* chromatopolarography

хроматотермография *f* chromathermography

хромосорб *m* Chromosorb

хромофор *m* chromophore

хромофорная группа *f* chromophore

хроноамперометрия *f* chronoamperometry

хроноамперометрия *f* **с двойным ступенчатым изменением потенциала** double-potential-step chronoamperometry

хроноамперометрия *f* **с капающим электродом с линейной развёрткой потенциала/ напряжения** dropping electrode chronoamperometry with linear potential/ voltage sweep

хроноамперометрия *f* **с линейной развёрткой потенциала** chronoamperometry with linear potential sweep

хронокулонометрия *f* chronocoulometry

хронокулонометрия *f* **с двойным ступенчатым изменением потенциала** double-potential step chronocoulometry

хронокулонометрия *f* **со ступенчатым изменением потенциала** chronocoulometry

хронопотенциометрия *f* chronopotentiometry

хронопотенциометрия *f* **с заданной развёрткой тока** programmed-current chronopotentiometry

хронопотенциометрия *f* **с линейной развёрткой тока** chronopotentiometry with linear current sweep

хронопотенциометрия *f* **с наложением переменного тока** chronopotentiometry with superimposed alternating current

хронопотенциометрия *f* **с переменным напряжением** alternating-voltage chronopotentiometry

хронопотенциометрия *f* **с прерыванием тока** current-cessation chronopotentiometry

хронопотенциометрия *f* **с реверсом тока** current-reversal chronopotentiometry

хронопотенциометрия *f* **со ступенчатым изменением тока** current-step chronopotentiometry

Ц

цвиттерион *m* ampholyte ion

целит *m* Celite

цена *f* **деления шкалы** value of a division

центрифугирование *n* centrifuging

цеолиты *mpl* zeolites

цепочка *f* **распадов** decay chain

цепь *f* **без переноса** cell without transference

цепь *f* **с переносом** cell with transference

цериметрическое титрование *n* cerimetric titration

церометрия *f* cerimetry

циклическая вольтамперометрия *f* **с треугольной развёрткой потенциала** cyclic triangular-wave voltametry

циклическая полярография *f* **с треугольной развёрткой потенциала** cyclic triangular-wave polarography

циклическая хронопотенциометрия *f* cyclic chronopotentiometry

циклическая хронопотенциометрия *f* **с реверсом тока** cyclic current-reversal chronopotentiometry

циклическая хронопотенциометрия *f* **со ступенчатым изменением тока** cyclic current-step chronopotentiometry

циклический комплекс *m* chelate compound

цилиндр *m* **Генера** Hehner measuring cylinder

циркулярно поляризованный свет *m* circularly polarized light

Ч

часовое время *n* clock time

частичная проба *f* primary sample

частичная электродная реакция *f* partial electrode reaction

частичный анодно-катодный ток *m* partial anodic/cathodic current

частота *f* frequency

частота *f* **возбуждения** exciting frequency

черенковское излучение *n* Cerenkov radiation

число *n* **коагуляции** coagulation concentration

число *n* **переноса иона B** transport number of ionic species B

число *n* **степеней свободы** number of degrees of freedom

число *n* **теоретических тарелок** number of theoretical plates

число *n* **Фарадея** Faraday constant

число *n* **эффективных теоретических тарелок** number of effective plates

чистое время *n* **удерживания** net retention time

чистота *f* **реагента** purity of a reagent

чистый реактив *m* pure reagent

чистый удерживаемый объём *m* net retention volume

чувствительность *f* sensitivity

чувствительность *f* **аналитической реакции** sensitivity of analytical reaction

чувствительность *f* **весов** balance sensitivity

чувствительность *f* **детектора** detector sensitivity

чувствительный объём *m* датчика sensitive volume of a detector
чувствительный элемент *m* electrochemical sensor

Ш, Щ

ширина *f* пика на половине высоты peak width at half-height
ширина *f* пика у основания peak width at base
ширина *f* спектральной линии line width
широта *f* эмульсии latitude
шкала *f* δ delta scale
шкала *f* активности по Брокману Brockmann scale of activity
шоттовский тигель *m* crucible with sintered disk
штифт *m* Глобара Globar
штифт *m* Нернста Nernst glower
шум *m* noise level
щелевая горелка *f* slot-burner
щель *f* slit
щель *f* Соллера Soller collimator

Э

эбулиоскопическое определение *n* молярной массы ebullioscopic determination of the molar mass
э.д.с. гальванического элемента electromotive force of a galvanic cell
эквивалентная ионная электропроводность *f* equivalent conductivity of ionic species B
эквивалентная ионная электропроводность *f* при бесконечном разбавлении limiting equivalent conductivity of ionic species B
эквивалентная электрическая проводимость *f* электролита equivalent conductivity of electrolyte
эквивалентная электропроводность *f* иона equivalent conductivity of ionic species B
эквивалентная электропроводность *f* электролита equivalent conductivity of electrolyte
эквивалентная электропроводность *f* электролита при бесконечном разбавлении limiting equivalent conductivity of electrolyte
эквиэлюотропный ряд *m* equieluotropic series
экзоэлектронная спектроскопия *f* exoelectron spectroscopy
эксикатор *m* desiccator
эксклюзионная хроматография *f* ионов ion-exclusion chromatography
эксклюзия *f* exclusion
эксперимент *m* experiment
экспериментально измеренный ток *m* observed current
экспоненциальное распределение *n* exponential distribution
экспозиция *f* radiant exposure
экстинкция *f* absorbance
экстрагент *m* extractant
экстрагирование *n* extraction
экстрагирующее средство *n* extractant
экстракт *m* extract
экстрактор *m* extractor

экстракционная хроматография *f* extraction chromatography
экстракционное титрование *n* extractive titration
экстракционный индикатор *m* extraction indicator
экстракционный реагент *m* extracting agent
экстракция *f* extraction
экстракция *f* из твёрдой фазы leaching
электрическая дуга *f* electrical arc
электрическая подвижность *f* иона B electric mobility of ion B
электрическая проводимость *f* 1. conductance 2. electrical conduction
электрическая проводимость электролита conductance of electrolyte
электрический проводник *m* electrical conductor
электрический ток *m* electric current
электрическое напряжение *n* electric potential difference
электрическое напряжение *n* электрохимической цепи electric potential difference of a galvanic cell
электрическое сопротивление *n* resistance
электроактивное вещество *n* electroactive substance
электроанализ *m* electroanalysis
электровесовой анализ *m* electrogravimetry
электрогенерирование *n* титранта electrochemical generation of a titrant
электрогравиметрия *f* electrogravimetry
электрогравиметрия *f* при постоянной плотности тока constant-current electrogravimetry
электрогравиметрия *f* при регулируемом потенциале controlled-potential electrogravimetry
электрогравиметрия *f* с контролируемым потенциалом controlled-potential electrogravimetry
электрография *f* electrography
электрод *m* electrode
pH электрод *m* pH electrode
электрод *m* второго рода electrode of the second kind
электрод *m* второго типа electrode of the second kind
электрод *m* из угольной пасты carbon-paste electrode
электрод *m* на основе субстрата энзима enzyme-substrate electrode
электрод-носитель *m* supporting electrode
электрод *m* первого рода electrode of the first kind
электрод *m* первого типа electrode of the first kind
электрод *m* с воздушным зазором air-gap electrode
электрод *m* с воздушным промежутком air-gap electrode
электрод *m* с газовым зазором air-gap electrode
электрод *m* с гетерогенной мембраной heterogeneous membrane ion-selective electrode
электрод *m* с гомогенной мембраной homogeneous membrane ion-selective electrode

электрод *m* с жёсткой матрицей rigid matrix electrode

электрод *m* с жидкой мембраной liquid membrane ion-selective electrode

электрод *m* с мембраной из жидкого ионита liquid ion-exchange membrane electrode

электрод *m* с мембраной на основе нейтральных переносчиков neutral-carrier membrane ion-selective electrode

электрод *m* с твёрдой мембраной solid-state membrane ion-selective electrode

электрод *m* сравнения reference electrode

электрод *m* третьего рода electrode of the third kind

электрод *m* третьего типа electrode of the third kind

электрод *m* чувствительный к газам gas sensing electrode

электродвижущая сила *f* гальванического элемента electromotive force of a galvanic cell

электродиализ *m* electrodialysis

электродная реакция *f* electrode reaction

электродная функция *f* electrode response

электродное напряжение *n* elektrode potential

электродный потенциал *m* electrode potential

электродный процесс *m* electrode process

электродный процесс *m* в условиях медленной химической реакции kinetic controlled process

электродный процесс *m* лимитируемый стадией диффузии diffusion controlled process

электродный процесс *m* лимитируемый стадией разряда-ионизации electroactivation control

электродный процесс *m* лимитируемый стадией химической реакции kinetic controlled process

электрокапиллярность *f* electrocapillary phenomena

электрокапиллярные явления *npl* electrocapillary phenomena

электрокинетические явления *npl* electrokinetic phenomena

электрокинетический потенциал *m* electrokinetic potential

электролиз *m* electrolysis

электролиз *m* при контролируемом потенциале controlled potential electrolysis

электролиз *m* при постоянной силе тока constant current electrolysis

электролизёр *m* electrolytic cell

электролизная ячейка *f* electrolytic cell

электролит *m* electrolyte

электролитическая диссоциация *f* electrolytic dissociation

электролитическая проводимость *f* ionic conduction

электролитическая ячейка *f* electrolytic cell

электролитический мостик *m* salt bridge

электролитический проводник *m* ionic conductor

электролитический сифон *m* salt bridge

электролитическое осаждение *n* electrodeposition

электролюминесценция *f* electroluminescence

электромагнетический спектр *m* electromagnetic spectrum

электромагнитное излучение *n* electromagnetic radiation

электромеханические весы *pl* electromechanical balance

электрон-вольт *m* electronvolt

электронная оже-спектроскопия *f* Auger-electron spectroscopy

электронная полоса *f* electronic band

электронная проводимость *f* electronic conduction

электронная спектроскопия *f* electronic spectroscopy

электронная спектроскопия *f* для химического анализа X-ray photoelectron spectroscopy

электронная эмиссия *f* electron emission

электронно-захватный детектор *m* electron capture detector

электронное стимулирование *n* десорбции ионов electron stimulated desorption of ions

электронное стимулирование *n* десорбции нейтральных частиц electron stimulated desorption of neutrals

электронный парамагнитный резонанс *m* electron paramagnetic resonance

электронный переход *m* $n - \pi^*$ $n - \pi^*$ electronic transition

электронный переход *m* $n - \sigma^*$ $n - \sigma^*$ electronic transition

электронный переход *m* $\pi - \pi^*$ $\pi - \pi^*$ electronic transition

электронный переход *m* $\sigma - \pi^*$ $\sigma - \pi^*$ electronic transition

электронный спектр *m* electronic spectrum

электронный спиновый резонанс *m* electron paramagnetic resonance

электронообменник *m* electron exchanger

электроосаждение *n* electrodeposition

электроосаждение *n* металлов electrodeposition of metals

электроосмос *m* electro-osmosis

электропроводимость *f* electrical conduction

электростатический эффект *m* effect of electrostatic charges

электротермический атомизатор *m* electrothermal atomizer

электрофорез *m* electrophoresis

электрофорезная диаграмма *f* electrophorogram

электрофоретический эффект *m* electrophoretic effect

электрофоретическое торможение *n* electrophoretic effect

электрофорограмма *f* electrophorogram

электрохимическая термоцепь *f* galvanic thermocell

электрохимическая цепь *f* electrochemical cell

электрохимическая цепь *f* с переносом cell with transference

электрохимический анализ *m* electroanalysis

электрохимический датчик *m* electrochemical sensor

электрохимический двойной слой *m* electrochemical double layer

электрохимический потенциал *m* компонента B в фазе α electrochemical potential of ionic component B in phase α

электрохимический ряд *m* активности
металлов electrochemical series
электрохимический ряд *m* напряжений
electrochemical series
электрохимический ряд *m* потенциалов
electrochemical series
электрохимический сенсор *m* electrochemical
sensor
электрохимический эквивалент *m*
electrochemical equivalent
электрохимия *f* electrochemistry
элемент *m* Голея Golay pneumatic detector
элемент *m* Лекланше Leclanché cell
элементарная единица *f* elementary entity
элементарная реакция *f* elementary reaction
элементарное событие *m* elementary event
элементарный анализ *m* elementary analysis
(*of organic compounds*)
элементный анализ *m* 1. elemental analysis
2. elemental analysis (*of organic compounds*)
элементный микроанализ *m* методом Прегля
Pregl procedure
эллипсометрия *f* ellipsometry
эллиптическая поляризация *f* elliptical
polarization
эллиптически поляризованный свет *m*
elliptically polarized light
элюат *m* eluate
элюент *m* eluent
элюентная хроматография *f* elution
chromatography
элюирование *n* elution
элюирущая сила *f* растворителя solvent
strength ε^0
элюирущая способность *f* растворителя solvent
strength ε^0
элюоотропная серия *f* eluotropic series
элюоотропный ряд *m* eluotropic series
элюционная кривая *f* elution curve
элюционная хроматография *f* elution
chromatography
элюция *f* elution
эманационный метод *m* emanometric method
эмиссионная молекулярная спектроскопия *f*
molecular emission spectroscopy
эмиссионная рентгеновская спектроскопия *f*
X-ray emission spectroscopy
эмиссионная спектроскопия *f* emission
spectroscopy
эмиссионная спектроскопия *f* пламени flame
emission spectroscopy
эмиссионный монохроматор *m* emission
monochromator
эмиссионный спектр *m* emission spectrum
эмиссионный спектрохимический анализ *m*
emission spectrochemical analysis
эмпирическое распределение *n* empirical
distribution
энергетическая освещённость *f* irradiance

энергетическое разрешение *n*
гамма-спектрометра energy resolution of
a gamma spectrometer
энергия *f* активации activation energy
энергия *f* излучения radiant energy
энзимный электрод *m* enzyme-substrate
electrode
энтальпиметрическое титрование *n*
thermometric titration
энтальпиметрия *f* enthalpimetry
эпитепловые нейтроны *mpl* epithermal neutrons
эталон *m* 1. standard sample 2. reference sample
эталонное вещество *n* reference material
эталонный элемент *m* standard cell
эталонный элемент *m* Вестона Weston
standard cell
эффект *m* Вина Wien's effect
эффект *m* Дебая-Фалькенгагена
Debye-Falkenhagen's effect
эффект *m* диссоциации в поле dissociation field
effect
эффект *m* Дорна sedimentation potential
эффект *m* заряжения charging effect
эффект *m* избирательного возбуждения
enhancement effect
эффект *m* Комптона Compton effect
эффект *m* Мёссбауэра Mössbauer effect
эффект *m* Оже Auger effect
эффект *m* разбавления dilution effect
эффект *m* размера зерна grain size effect
эффект *m* Рамана Raman effect
эффект *m* Тиндаля Tyndall effect
эффективная глубина *f* effective depth
эффективное сечение *n* cross section
эффективность *f* атомизации efficiency of
atomization
эффективность *f* детектора intrinsic detector
efficiency
эффективность *f* колонки column performance
эффективность *f* распыления efficiency of
nebulization
эффективность *f* счёта counting efficiency
эффективность *f* счётчика counting efficiency
эффективный радиус *m* ионной атмосферы
radius of ionic atmosphere
эффективный слой *m* effective layer
эффлюент *m* effluent

Я

явление *n* Зеемана Zeeman effect
ядерная чистота *f* nuclear purity
ядерное деление *n* fission
ядерный магнитный резонанс *m* nuclear
magnetic resonance
ЯМР-спектроскопия *f* nuclear magnetic
resonance spectroscopy
яркость *f* luminance
ячейка *f* Голея Golay pneumatic detector